Classics in Mathematics

Charles B. Morrey, Jr.

Multiple Integrals in the Calculus of Variations

Charles Bradfield Morrey Jr. (1907 – 1984) received a PhD degree in mathematics from Harvard University, Cambridge, MA, in 1931, and was awarded a National Research Council Fellowship afterwards that brought him to Princeton University, Princeton, NJ, from 1931 to 1932, and to the Rice Institute, Houston, TX, from 1932 to 1933. In 1933 he took a position at the Department of Mathematics at the University of California, Berkeley, where he was actively involved in teaching and research until his retirement in 1973.
Morrey was an outstanding mathematician, known also as a very effective and successful teacher. He solved Hilbert's Nineteenth Problem, and also contributed considerably to the solution of questions raised in Problem No. 20. Following in the footsteps of Leonida Tonelli, Morrey became the founder of the modern Calculus of Variations, and the present treatise is the mature fruit of his achievements. This and his other books have had and continue to have a wide influence on the teaching of mathematics.
Morrey's scientific accomplishments and his administrative abilities were complemented by his gift for friendship, his charming sense of humour, and his continuing concern for people, for mathematics, and for music.

[Sources: S. Hildebrandt, and "In memoriam" by J.L. Kelley, D.H. Lehmer, R.M. Robinson, University of California]

Charles B. Morrey, Jr.

Multiple Integrals in the Calculus of Variations

Reprint of the 1966 Edition

Springer

Originally published as Vol. 130 of the series *Grundlehren der mathematischen Wissenschaften*

ISBN 978-3-540-69915-6 e-ISBN 978-3-540-69952-1

DOI 10.1007/978-3-540-69952-1

Classics in Mathematics ISSN 1431-0821

Library of Congress Control Number: 2008932928

Mathematics Subject Classification (2000): 49-xx, 58Exx, 35N15, 46E35, 46E39

Cover design: WMXDesign GmbH, Heidelberg

Printed on acid-free paper

9 8 7 6 5 4 3 2 1

springer.com

Die Grundlehren der mathematischen Wissenschaften

in Einzeldarstellungen
mit besonderer Berücksichtigung
der Anwendungsgebiete

Band 130

Herausgegeben von

J. L. Doob · E. Heinz · F. Hirzebruch · E. Hopf
H. Hopf · W. Maak · S. MacLane
W. Magnus · D. Mumford · F. K. Schmidt · K. Stein

Geschäftsführende Herausgeber

B. Eckmann und B. L. van der Waerden

Multiple Integrals in the Calculus of Variations

Charles B. Morrey, Jr.
Professor of Mathematics
University of California, Berkeley

Springer-Verlag Berlin Heidelberg New York 1966

Library of Congress Catalog Card Number 66-24 365

Printed in Germany

Title No. 5113

Preface

The principal theme of this book is "the existence and differentiability of the solutions of variational problems involving multiple integrals." We shall discuss the corresponding questions for single integrals only very briefly since these have been discussed adequately in every other book on the calculus of variations. Moreover, applications to engineering, physics, etc., are not discussed at all; however, we do discuss *mathematical* applications to such subjects as the theory of harmonic integrals and the so-called "$\bar{\partial}$-Neumann" problem (see Chapters 7 and 8). Since the plan of the book is described in Section 1.2 below we shall merely make a few observations here.

In order to study the questions mentioned above it is necessary to use some very elementary theorems about convex functions and operators on Banach and Hilbert spaces and some special function spaces, now known as "SOBOLEV spaces". However, most of the facts which we use concerning these spaces were known before the war when a different terminology was used (see CALKIN and MORREY [5]); but we have included some powerful new results due to CALDERON in our exposition in Chapter 3. The definitions of these spaces and some of the proofs have been made simpler by using the most elementary ideas of distribution theory; however, almost no other use has been made of that theory and no knowledge of that theory is required in order to read this book. Of course we have found it necessary to develop the theory of linear elliptic systems at some length in order to present our desired differentiability results. We found it particularly essential to consider "weak solutions" of such systems in which we were often forced to allow discontinuous coefficients; in this connection, we include an exposition of the DE GIORGI—NASH—MOSER results. And we include in Chapter 6 a proof of the analyticity of the solutions (on the interior and at the boundary) of the most general non-linear analytic elliptic system with general regular (as in AGMON, DOUGLIS, and NIRENBERG) boundary conditions. But we confine ourselves to functions which are analytic, of class C^∞, of class C_μ^n or C^n (see § 1.2), or in some Sobolev space H_p^m with m an integer ≥ 0 (except in Chapter 9). These latter spaces have been defined for all real m in a domain (or manifold) or on its boundary and have been used by many authors in their studies of linear systems. We have not included a study of these spaces since (i) this book is already sufficiently long, (ii) we took no part in this development, and (iii) these spaces are adequately discussed in other *books* (see A. FRIEDMAN [2], HORMANDER [1], LIONS [2]) as well as in many papers (see § 1.8 and papers by LIONS and MAGENES).

The research of the author which is reported on in this book has been partially supported for several years by the Office of Naval Research under contract Nonr 222(62) and was partially supported during the year 1961—62, while the author was in France, by the National Science Foundation under the grant G—19782.

Berkeley, August 1966

CHARLES B. MORREY, JR.

Contents

Chapter 1

Introduction

1.1. Introductory remarks 1
1.2. The plan of the book: notation 2
1.3. Very brief historical remarks 5
1.4. The EULER equations 7
1.5. Other classical necessary conditions 10
1.6. Classical sufficient conditions 12
1.7. The direct methods 15
1.8. Lower semicontinuity 19
1.9. Existence 23
1.10. The differentiability theory. Introduction 26
1.11. Differentiability; reduction to linear equations 34

Chapter 2

Semi-classical results

2.1. Introduction 39
2.2. Elementary properties of harmonic functions 40
2.3. WEYL's lemma 41
2.4. POISSON's integral formula; elementary functions; GREEN's functions 43
2.5. Potentials 47
2.6. Generalized potential theory; singular integrals 48
2.7. The CALDERON-ZYGMUND inequalities 55
2.8. The maximum principle for a linear elliptic equation of the second order 61

Chapter 3

The spaces H_p^m and H_{p0}^m

3.1. Definitions and first theorems 62
3.2. General boundary values; the spaces $H_{p0}^m(G)$; weak convergence . . . 68
3.3. The DIRICHLET problem 70
3.4. Boundary values 72
3.5. Examples; continuity; some SOBOLEV lemmas 78
3.6. Miscellaneous additional results 81
3.7. Potentials and quasi-potentials; generalizations 86

Chapter 4

Existence theorems

4.1. The lower-semicontinuity theorems of SERRIN 90
4.2. Variational problems with $f = f(p)$; the equations (1.10.13) with $N = 1$, $B_i = 0$, $A^\alpha = A^\alpha(p)$ 98
4.3. The borderline cases $k = \nu$ 105
4.4. The general quasi-regular integral 112

Chapter 5

Differentiability of weak solutions

5.1. Introduction 126
5.2. General theory; $\nu > 2$ 128
5.3. Extensions of the DE GIORGI-NASH-MOSER results; $\nu > 2$ 134
5.4. The case $\nu = 2$ 143
5.5. L_p and SCHAUDER estimates 149
5.6. The equation $a \cdot \nabla^2 u + b \cdot \nabla u + c\,u - \lambda u = f$ 157
5.7. Analyticity of the solutions of analytic linear equations 164
5.8. Analyticity of the solutions of analytic, non-linear, elliptic equations 170
5.9. Properties of the extremals; regular cases 186
5.10. The extremals in the case $1 < k < 2$ 191
5.11. The theory of LADYZENSKAYA and URAL'TSEVA 194
5.12. A class of non-linear equations 203

Chapter 6

Regularity theorems for the solutions of general elliptic systems and boundary value problems

6.1. Introduction 209
6.2. Interior estimates for general elliptic systems 215
6.3. Estimates near the boundary; coerciveness 225
6.4. Weak solutions 242
6.5. The existence theory for the DIRICHLET problem for strongly elliptic systems 251
6.6. The analyticity of the solutions of analytic systems of linear elliptic equations 258
6.7. The analyticity of the solutions of analytic nonlinear elliptic systems 266
6.8. The differentiability of the solutions of non-linear elliptic systems; weak solutions; a perturbation theorem 277

Chapter 7

A variational method in the theory of harmonic integrals

7.1. Introduction 286
7.2. Fundamentals; the GAFFNEY-GÅRDING inequality 288
7.3. The variational method 293
7.4. The decomposition theorem. Final results for compact manifolds without boundary 295
7.5. Manifolds with boundary 300
7.6. Differentiability at the boundary 305
7.7. Potentials, the decomposition theorem 309
7.8. Boundary value problems 314

Chapter 8

The $\bar{\partial}$-Neumann problem on strongly pseudo-convex manifolds

8.1. Introduction 316
8.2. Results. Examples. The analytic embedding theorem 320
8.3. Some important formulas 328
8.4. The Hilbert space results 333
8.5. The local analysis 337
8.6. The smoothness results 341

Chapter 9

Introduction to parametric Integrals; two dimensional problems

9.1. Introduction. Parametric integrals 349
9.2. A lower semi-continuity theorem 354
9.3. Two dimensional problems; introduction; the conformal mapping of surfaces 362
9.4. The problem of Plateau 374
9.5. The general two-dimensional parametric problem 390

Chapter 10

The higher dimensional Plateau problems

10.1. Introduction 400
10.2. ν surfaces, their boundaries, and their Hausdorff measures 407
10.3. The topological results of Adams 414
10.4. The minimizing sequence; the minimizing set 421
10.5. The local topological disc property 439
10.6. The Reifenberg cone inequality 459
10.7. The local differentiability 474
10.8. Additional results of Federer concerning Lebesgue ν-area 480
Bibliography 494
Index 504

Multiple Integrals in the Calculus of Variations

Chapter 1

Introduction

1.1. Introductory remarks

The principal theme of these lectures is "the existence and differentiability of the solutions of variational problems involving multiple integrals." I shall discuss the corresponding questions for single integrals only very briefly since these have been adequately discussed in every book on the calculus of variations (see, for instance, AKHIEZER [1], BLISS [1], BOLZA [1], CARATHEODORY [2], FUNK [1], PARS [1]. Moreover, I shall not discuss applications to engineering, physics, etc., at all, although I shall mention some *mathematical* applications.

In general, I shall consider integrals of the form

$$I(z, G) = \int_G f[x, z(x), \nabla z(x)]\, dx \tag{1.1.1}$$

where G is a domain,

$$x = (x^1, \ldots, x^\nu), \quad z = (z^1, \ldots, z^N), \; dx = dx^1 \ldots dx^\nu, \tag{1.1.2}$$

$z(x)$ is a vector function, ∇z denotes its gradient which is the set of functions $\{z^i_{.\alpha}\}$, where $z^i_{.\alpha}$ denotes $\partial z^i/\partial x^\alpha$, and $f(x, z, p)$ $(p = \{p^i_\alpha\})$ is generally assumed continuous in all its arguments. The integrals

$$\int_a^b \sqrt{1 + (dz/dx)^2}\, dx \quad \text{and} \quad \iint_G \left[\left(\frac{\partial z}{\partial x^1}\right)^2 + \left(\frac{\partial z}{\partial x^2}\right)^2\right] dx^1\, dx^2$$

are familiar examples of integrals of the form (1.1.1.) in which $N = 1$ in both cases, $\nu = 1$ in the first case, $\nu = 2$ in the second case and the corresponding functions f are defined respectively by

$$f(x, z, p) = \sqrt{1 + p^2}, \qquad f(x, z, p) = p_1^2 + p_2^2$$

where we have omitted the superscripts on z and p since $N = 1$. The second integral is a special case of the *Dirichlet integral* which is defined in general by

$$D(z, G) = \int_G |\nabla z|^2\, dx, \; f(x, z, p) = |p|^2 = \sum_{i, \alpha} (p^i_\alpha)^2. \tag{1.1.3}$$

Another example is the *area integral*

$$A(z, G) = \iint_G \sqrt{\left[\frac{\partial(z^2, z^3)}{\partial(x^1, x^2)}\right]^2 + \left[\frac{\partial(z^3, z^1)}{\partial(x^1, x^2)}\right]^2 + \left[\frac{\partial(z^1, z^2)}{\partial(x^1, x^2)}\right]^2}\, dx^1\, dx^2 \tag{1 1.4}$$

which gives the area of the surface

$$(1.1.5) \qquad z^i = z^i(x^1, x^2), \quad (x^1, x^2) \in G, \quad i = 1, 2, 3.$$

It is to be noticed that the area integral has the special property that it is invariant under diffeomorphisms (1—1 differentiable mappings, etc.) of the domain G onto other domains. This is the first example of an *integral in parametric form*. I shall discuss such integrals later (in Chapters 9 and 10).

I shall also discuss briefly integrals like that in (1.1.1) but involving derivatives of higher order. And, of course, the variational *method* has been used in problems which involve a "functional" not at all like the integral in (1.1.1); as for example in proving the Riemann mapping theorem where one minimizes sup $|f(z)|$ among all schlicht functions $f(z)$ defined on the given simply connected region G for which $f(z_0) = 0$ and $f'(z_0) = 1$ at some given point z_0 in G.

We shall consider only problems in which the domain G is fixed; variations in G may be taken care of by transformations of coordinates. We shall usually consider problems involving fixed boundary values; we shall discuss other problems but will not derive the *transversality* conditions for such problems.

1.2. The plan of the book: notation

In this chapter we attempt to present an overall view of the principal theme of the book as stated at the beginning of the preceding section. However, we do not include a discussion of integrals in parametric form; these are discussed at some length in Chapters 9 and 10. The material in this book is not presented in its logical order. A possible logical order would be § 1.1—1.5, Chapter 2, Chapter 3, §§ 5.1—5.8, § 5.12, Chapter 6, §§ 1.6—1.9, §§ 4.1, 4.3, 4.4. Then the reader must skip back and forth as required among the material of § 1.10, 1.11, 4.2, 5.9, 5.10 and 5.11. Then the remainder of the book may be read substantially in order. Actually, Chapters 7 and 8 could be read immediately after § 5.8.

We begin by presenting background material including derivations, under restrictive hypotheses, of Euler's equations and the classical necessary conditions of Legendre and Weierstrass. Next, we include a brief and incomplete presentation of the classical so-called "sufficiency" conditions, including references to other works where a more complete presentation may be found.

The second half of this chapter presents a reasonably complete outline of the existence and differentiability theory for the solutions of variational problems. This begins with a brief discussion of the development of the direct methods and of the successively more general classes of "admissible" functions, culminating in the so-called "Sobolev spaces".

These are then defined and discussed briefly after which two theorems on lower-semicontinuity are presented. These are not the most general theorems possible but are selected for the simplicity of their proofs which, however, assume that the reader is willing to grant the truth of some well-known theorems on the Sobolev spaces. The relevant theorems about these spaces are proved in Chapter 3 and more general lower-semicontinuity and existence theorems are presented in Chapter 4.

In Section 1.10 the differentiability results are stated and some preliminary results are proved. In Section 1.11, an outline of the differentiability theory is presented. It is first shown that the solutions are "weak solutions" of the Euler equations. The theory of these non-linear equations is reduced to that of linear equations which, initially, may have discontinuous coefficients. The theory of these general linear equations is discussed in detail in Chapter 5. However, the higher order differentiability for the solutions of *systems* of Euler equations required the same methods as are used in studying systems of equations of higher order. Accordingly, we present in Chapter 6 many of the results in the two recent papers of AGMON, DOUGLIS, and NIRENBERG ([1], [2]) concerning the solutions and weak solutions of such systems. Both the L_p-estimates and the SCHAUDER-type estimates (concerning HÖLDER continuity) are presented. We have included sections in both Chapters 5 und 6 proving the analyticity, including analyticity at the boundary, of the solutions of both linear and non-linear analytic elliptic equations and systems; the most general "properly elliptic" systems with "complementing boundary conditions" (see § 6.1) are treated. The proof of analyticity in this generality is new. In Chapter 2 we present well-known facts about harmonic functions and generalized potentials and conclude with proofs of the CALDERON-ZYGMUND inequalities and of the maximum principle for the solutions of second order equations.

In Chapters 7 and 8, we present applications of the variational method to the HODGE theory of harmonic integrals and to the so-called $\bar{\partial}$-NEUMANN problem for exterior differential forms on strongly pseudo-convex complex analytic manifolds with boundary. In Chapter 9, we present a brief discussion of ν-dimensional parametric problems in general and then discuss the two dimensional Plateau problem in Euclidean space and on a Riemannian manifold. The chapter concludes with the author's simplified proof of the existence theorem of CESARI [4], DANSKIN, and SIGALOV [2] for the general two dimensional parametric problem and some incomplete results concerning the differentiability of the solutions of such problems. In Chapter 10, we present the author's simplification of the very important resent work of REIFENBERG [1], [2], and [3] concerning the higher dimensional PLATEAU problem and the author's extension of these results to varieties on a Riemannian manifold.

Notations. For the most part, we use standard notations. G and D will denote domains which are bounded unless otherwise specified. We denote the boundary of D by ∂D and its closure by $\bar{D}$. We shall often use the notation $D \subset\subset G$ to mean that $\bar{D}$ is compact and $\bar{D} \subset G$. $B(x_0, R)$ denotes the ball with center at x_0 and radius R. γ_ν and Γ_ν denote the ν-measure and $(\nu - 1)$-measure of $B(0,1)$ and $\partial B(0,1)$, respectively. We often denote $\partial B(0,1)$ by Σ. Most of the time (unless otherwise specified) we let R_q be q-dimensional number space with the usual metric and abbreviate $B(0,R)$ to B_R, denote by σ the $(\nu - 1)$-plane $x^\nu = 0$, and define

$$\begin{aligned} &R_\nu^+ = \{x \mid x^\nu > 0\}, && R_\nu^- = \{x \mid x^\nu < 0\} \\ &G_R = B_R \cap R_\nu^+, && \Sigma_R = \partial B_R \cap R_\nu^+, \quad \sigma_R = B_R \cap \sigma \qquad (1.2.1) \\ &G_R^- = B_R \cap R_\nu^-, && \Sigma_R^- = \partial B_R \cap R_\nu^-. \end{aligned}$$

If S is a set in R_q, $|S|$ denotes its Lebesgue q-measure; if x is a point, $d(x, S)$ denotes the distance of x from S. We define

$$[a, b] = \{x \mid a^\alpha \le x^\alpha \le b^\alpha, \quad \alpha = 1, \ldots, \nu, \ x \in R_\nu\}.$$

In the case of boundary integrals, we often use dx'_α to denote $n_\alpha \, dS$ where dS is the surface area and n_α is the α-th component of the *exterior normal*. We say that a function $u \in C^n(G)$ iff (if and only if) u and its partial derivatives of order $\le n$ are continuous on G and $u \in C^n(\bar{G})$ iff $u \in C^n(G)$ and each of its derivatives of order $\le n$ can be extended to be continuous on $\bar{G}$. If $0 < \mu \le 1$, $u \in C_\mu^n(G)$ (or $C_\mu^n(\bar{G})$) $\Leftrightarrow$ (i.e. iff) $u \in C^n(G)$ (or $C^n(\bar{G})$) and all the derivatives of order $\le n$ satisfy a Hölder (Lipschitz if $\mu = 1$) condition on each compact subset of G (or on the whole of $\bar{G}$ as extended). If $u \in C_\mu^0(\bar{G})$, then $h_\mu(u, \bar{G}) = \sup |x_2 - x_1|^{-\mu} \cdot |u(x_2) - u(x_1)|$ for x_1 and $x_2 \in \bar{G}$ and $x_1 \ne x_2$. A domain G is said to be of class C^n (or C_μ^n, $0 < \mu \le 1$) iff G is bounded and each point P_0 of ∂G is in a neighborhood $\mathfrak{n}$ on $\bar{G}$ which can be mapped by a $1-1$ mapping of class C^n (or C_μ^n), together with its inverse, onto $G_R \cup \sigma_R$ for some R in such a way that P_0 corresponds to the origin and $\mathfrak{n} \cup \partial G$ corresponds to σ_R. If $u \in C^1(G)$, we denote its derivatives $\partial u/\partial x^\alpha$ by $u_{,\alpha}$. If $u \in C_2(G)$, then $\nabla^2 u$ denotes the tensor $u_{,\alpha\beta}$ where α and β run independently from to ν. Likewise $\nabla^3 u = \{u_{,\alpha\beta\gamma}\}$, etc., and $|\nabla^2 u|^2 = \sum_{\alpha,\beta} |u_{,\alpha\beta}|^2$, etc. If G is also of class C^1, then Green's theorem becomes (in our notations)

$$\int_G u_{,\alpha}(x) \, dx = \int_{\partial G} u \, n_\alpha \, dS = \int_{\partial G} u \, dx'_\alpha.$$

Sometimes when we wish to consider u as a function of some single x^α, we write $x = (x^\alpha, x'_\alpha)$ and $u(x) = u(x^\alpha, x'_\alpha)$ where x'_α denotes the remaining x^β. One dimensional or $(\nu - 1)$-dimensional integrals are then indicated as might be expected. We often let α denote a "multi-index", i.e. a vector $(\alpha_1, \ldots, \alpha_\nu)$ in which each α_i is a non-negative integer. We then define

$$|\alpha| = \alpha_1 + \cdots + \alpha_\nu, \; D^\alpha u = \frac{\partial^{|\alpha|} u}{(\partial x^1)^{\alpha_1} \ldots (\partial x^\nu)^{\alpha_\nu}} \; (u \in C^{|\alpha|}(G))$$

$$\alpha! = (\alpha_1!) \ldots (\alpha_\nu!), \; C_\alpha = \frac{|\alpha|!}{\alpha!}, \xi^\alpha = (\xi^1)^{\alpha_1} \ldots (\xi^\nu)^{\alpha_\nu}.$$

Using this notation

$$|\nabla^m u|^2 = \sum_{|\alpha|=m} C_\alpha |D^\alpha u|^2.$$

We shall denote constants by C or Z with or without subscripts. These constants will, perhaps depend on other constants; in this case we may write $C = C(h, \mu)$ if C depends only on h and μ, for example. However, even though we may distinguish between different constants in some discussion by inserting subscripts, there is no guarantee that C_2, for example, will always denote the same constant. We sometimes denote the support of u by spt u. We denote by $C_c^\infty(G)$, $C_c^n(G)$, and $C_{\mu c}^n(G)$ the sets of functions in $C^\infty(G)$, $C^n(G)$, or $C_\mu^n(G)$, respectively, which have support in G (i.e. which vanish on and near ∂G). But it is handy to say that u has support in $G_R \cup \sigma_R \Leftrightarrow u$ vanishes on and near Σ_R (see 1.2.1); we allow $u(x)$ to be $\neq 0$ on σ_R.

1.3. Very brief historical remarks

Problems in the calculus of variations which involve only single integrals ($\nu = 1$) have been discussed at least since the time of the BERNOULLI's. Although there was some early consideration of double integrals, it was RIEMANN who aroused great interest in them by proving many interesting results in function theory by assuming DIRICHLET's *principle* which may be stated as follows: *There is a unique function which minimizes the* DIRICHLET *integral among all functions of class C^1 on a domain G and continuous on $\bar{G}$ which takes on given values on the boundary ∂G and, moreover, that function is harmonic on G.*

RIEMANN's work was criticized on the grounds that just because the integral was bounded below among the competing functions it didn't follow that the greatest lower bound was *taken on* in the class of competing functions. In fact an example was given of a (1-dimensional) integral of the type (1.1.1) for which there is no minimizing function and another was given of continuous boundary values on the unit circle such that $D(z, G) = +\infty$ for every z as above having those boundary values.

The first example is the integral (see COURANT [3])

$$I(z, G) = \int_0^1 \left[1 + \left(\frac{dz}{dx}\right)^2\right]^{1/4} dx, \; G = (0,1), \tag{1.3.1}$$

the admissible functions z being those $\in C^1$ on $[0,1]$ with

$$z(0) = 0 \text{ and } z(1) = 1.$$

Obviously $I(z, G) > 1$ for every such z, $I(z, G)$ has no upper bound and if we define

$$z_r(x) = \begin{cases} 0 & , 0 \leq x \leq r \qquad , 0 < r < 1, \\ -1 + [1 + 3(x-r)^2/(1-r)^2]^{1/2}, & r \leq x \leq 1 \end{cases}$$

we see that $I(z_r, G) \to 1$ as $r \to 1^-$.

The second example is the following (see COURANT [3]): It is now known that Dirichlet's principle holds for a circle and that each function harmonic on the unit circle has the form

$$(1.3.2) \qquad w(r, \Theta) = \frac{a_0}{2} + \sum_{n=1}^{\infty} r^n (a_n \cos n\Theta + b_n \sin n\Theta), \ (a_n, b_n \text{ const}),$$

in polar coordinates and that the Dirichlet integral is

$$(1.3.3) \qquad D(w, G) = \pi \sum_{n=1}^{\infty} n(a_n^2 + b_n^2)$$

provided this sum converges. But if we define

$$a_n = k^{-2} \text{ if } n = k! \quad, a_n = b_n = 0 \text{ otherwise,}$$

we see that the series in (1.3.2) converges uniformly but that in (1.3.3) reduces to

$$\pi \sum_{k=1}^{\infty} \frac{k!}{k^4}$$

which diverges.

DIRICHLET's principle was established rigorously in certain important cases by HILBERT, LEBESGUE [2] and others shorly after 1900. That was the beginning of the so-called "direct methods" of the calculus of variations of which we shall say more later.

There was renewed interest in one dimensional problems with the advent of the MORSE theory of the critical points of functionals in which M. MORSE generalized his theory of critical points of functions defined on finite-dimensional manifolds [1] to certain functionals defined on infinite-dimensional spaces [2], [3]. He was able to obtain the MORSE inequalities between the numbers of possibly "unstable" (i.e. critical but not minimizing) geodesics (and unstable minmal surfaces) having various indices (see also MORSE and TOMPKINS, [1]—[4]). Except for the latter (which could be reduced to the case of curves), MORSE's theory was applied mainly to one-dimensional problems. However, within the last two years, SMALE and PALAIS and SMALE have found a modification of MORSE's theory which is applicable to a wide class of multiple integral problems.

Variational methods are beginning to be used in differential geometry. For example, the author and Eells (see MORREY and EELS, MORREY,

[11] and Chapter 7) developed the HODGE theory ([1], [2]) by variational methods (HODGE's original idea [1]). HÖRMANDER [2], KOHN [1], SPENCER (KOHN and SPENCER), and the author (MORREY [19], [20]) have applied variational techniques to the study of the $\bar{\partial}$-Neumann problem for exterior differential forms on complex analytic manifolds (see Chapter 8; the author encountered this problem in his work on the analytic embedding of real-analytic manifolds (MORREY [13]). Very recently, EELLS and SAMPSON have proved the existence of "harmonic" mappings (i.e. mappings which minimize an intrinsic Dirichlet integral) from one compact manifold into a manifold having negative curvature. Since the inf. of this integral is zero if the dimension of the compact manifold >2, they found it necessary to use a gradient line method which led to a non-linear system of parabolic equations which they then solved; the curvature restriction was essential in their work.

1.4. The Euler equations

After a number of special problems had been solved, EULER deduced in 1744 the first general necessary condition, now known as EULER's equation, which must be satisfied by a minimizing or maximizing arc. His derivation, given for the case $N=\nu=1$, proceeds as follows: Suppose that the function z is of class C^1 on $[a,b]$ $(=G)$ minimizes (for example) the integral $I(z,G)$ among all similar functions having the same values at a and b. Then, if ζ is *any* function of class C^1 on $[a,b]$ which vanishes at a and b, the function $z+\lambda\zeta$ is, for every λ, of class C^1 on $[a,b]$ and has the same values as z at a and b. Thus, if we define

$$\varphi(\lambda)=I(z+\lambda\zeta,G)=\int_a^b f[x,z(x)+\lambda\zeta(x),z'(x)+\lambda\zeta'(x)]\,dx \tag{1.4.1}$$

φ must take on its minimum for $\lambda=0$. If we assume that f is of class C^1 in its arguments, we find by differentiating (1.4.1) and setting $\lambda=0$ that

$$\int_a^b \{\zeta'(x)\cdot f_p[x,z(x),z'(x)]+\zeta(x)f_z[x,z(x),z'(x)]\}\,dx=0 \qquad \left(f_p=\frac{\partial f}{\partial p},\ \text{etc.}\right). \tag{1.4.2}$$

The integral in (1.4.2) is called the *first variation* of the integral I; it is supposed to vanish for every ζ of class C^1 on $[a,b]$ which vanishes at a and b. *If we now assume that f and z are of class C^2 on $[a,b]$* (EULER had no compunctions about this) we can integrate (1.4.2) by parts to obtain

$$\int_a^b \zeta(x)\cdot\left\{f_z-\frac{d}{dx}f_p\right\}dx=0,\quad f_p=f_p[x,z(x),\nabla z(x)],\ \text{etc.} \tag{1.4.3}$$

Since (1.4.3) holds for all ζ as above, it follows that the equation

$$\frac{d}{dx} f_p = f_z \tag{1.4.4}$$

must hold. This is *Euler's equation* for the integral I in this simple case. If we write out (1.4.4) in full, we obtain

$$f_{pp} \cdot z'' + f_{pz} z' + f_{px} = f_z \tag{1.4.5}$$

which shows that Euler's equation is non-linear and of the second order. It is, however, linear in z''; equations which are linear in the derivatives of highest order are frequently called *quasi-linear*. The equation evidently becomes singular whenever $f_{pp} = 0$. Hence *regular* variational problems are those for which f_{pp} never vanishes; in that case, it is assumed that $f_{pp} > 0$ which turns out to make minimum problems more natural than maximum problems.

It is clear that this derivation generalizes to the most general integral (1.1.1) provided that f and the minimizing (or maximizing, etc.) function z is of class C^2 on the closed domain G which has a sufficiently smooth boundary. Then, if z minimizes I among all (vector) functions of class C^1 with the same boundary values and ζ is any such vector which vanishes on the boundary or G, it follows that $z + \lambda\zeta$ is a "competing" or "admissible" function for each λ so that if φ is defined by

$$\varphi(\lambda) = I(z + \lambda\zeta, G) \tag{1.4.6}$$

then $\varphi'(0) = 0$. This leads to the condition that

$$\int_G \sum_{i=1}^{N} \left\{ \sum_{\alpha=1}^{\nu} \zeta^i_{,\alpha} f_{p^i_\alpha} + \zeta^i f_{z^i} \right\} dx = 0 \tag{1.4.7}$$

for all ζ as indicated. The integral in (1.4.7) is the *first variation* of the general integral (1.1.1). Integrating (1.4.7) by parts leads to

$$\int_G \sum_{i=1}^{N} \zeta^i \cdot \left\{ f_{z^i} - \sum_{\alpha=1}^{\nu} \frac{\partial}{\partial x^\alpha} f_{p^i} \right\} dx = 0.$$

Since this is zero for all vectors ζ, it follows that

$$\sum_{\alpha=1}^{\nu} \frac{\partial}{\partial x^\alpha} f_{p^i_\alpha} = f_{z^i}, \quad i = 1, \ldots, N \tag{1.4.8}$$

which is a quasi-linear system of partial differential equations of the second order. In the case $N = 1$, it reduces to

$$\sum_{\alpha=1}^{\nu} \frac{\partial}{\partial x^\alpha} f_{p_\alpha} = f_z, \quad \text{or}$$
$$\sum_{\alpha,\beta=1}^{\nu} f_{p_\alpha p_\beta} z_{,\alpha\beta} + \sum_{\alpha=1}^{\nu} (f_{p_\alpha z} z_{,\alpha} + f_{p_\alpha x^\alpha}) = f_z. \tag{1.4.9}$$

The equation (1.4.9) is evidently singular whenever the quadratic form

(1.4.10) $$\sum_{\alpha,\beta} f_{p_\alpha p_\beta}(x, z, p)\,\lambda_\alpha \lambda_\beta\,*$$

in λ is degenerate.

We notice from (1.4.5) that if $N = \nu = 1$ and f depends only on p and the problem is regular, then Euler's equation reduces to

$$z'' = 0.$$

In general, if f depends only on $p(= p^i_\alpha)$, Euler's equation has the form

$$\sum_{j,\alpha,\beta} f_{p^i_\alpha p^j_\beta}\, z^j_{,\alpha\beta} = 0, \quad i = 1, \ldots, N$$

and every linear vector function is a solution. In particular, if $N = 1$ and $f = |p|^2$, Euler's equation is just Laplaces equation

$$\Delta z \equiv \sum_\alpha z_{,\alpha\alpha} = 0.$$

In case $f = (1 + |p|^2)^{1/4}$ as in the first example in § 1.3, we see that

$$4 f_{pp} = (2 - p^2)(1 + p^2)^{-7/4}$$

which is not always > 0. On the other hand $f_{pp} > 0$ if $|p| < \sqrt{2}$ so classical results which we shall discuss later (see § 1.6) show that the linear function $z(x) = x$ minimizes the integral among all arcs having $|z'(x)| \leq \sqrt{2}$.

We now revert to equation (1.4.9). If we take, for instance, $N = 1$, $\nu = 2$, $f = p_1^2 - p_2^2$, then (1.4.9) becomes

$$z_{,11} - z_{,22}\left(= \frac{\partial^2 z}{\partial (x^1)^2} - \frac{\partial^2 z}{\partial (x^2)^2}\right) = 0$$

which is of *hyperbolic type*. Moreover, the integral (1.1.1) with this f obviously has no minimum or maximum, whatever boundary values are given for z. Anyhow, it is well known that boundary value problems are not natural for equations of hyperbolic type. If $\nu > 2$ a greater variety of types may occur, depending on the signature of the quadratic form (1.4.10). A similar objection occurs in all cases except those in which the form (1.4.10) is *positive definite* or negative definite; we shall restrict ourselves to the case where it is positive definite. In this case Euler's equation is of *elliptic type*. The choice of this condition on f is re-enforced by analogy with the case $\nu = 1$; in that case $f_{pp} \geq 0$ *implies the convexity* (see § 1.8) *of f as a function of p* for each (x, z) and the non-negative definiteness of the form (1.4.10) is equivalent to the convexity of f as a function of $p_1, \ldots, p_\nu$ for each set $(x^1, \ldots, x^\nu, z)$. Our choice is re-enforced further by the classical derivation given in the next section.

---

* Greek indices are summed from 1 to ν and Latin indices are summed from 1 to N. Hereafter we shall usually employ the summation convention in which repeated indices are summed and summation signs omitted.

1.5. Other classical necessary conditions

Suppose that f is of class C^2 in its arguments, that $N = 1$, that z is of class C^1 on the closure of G, and that z minimizes $I(z, G)$ among all functions Z of the same class which coincide with z on the boundary and are sufficiently close to z in the C^1 norm, i.e.

$$|Z(x) - z(x)| \leq \delta, \quad |\nabla Z(x) - \nabla z(x)| \leq \delta, \quad x \text{ on } G.$$

In classical terminology, we say that z furnishes a *weak relative minimum* to $I(z, G)$. We shall show that this implies the non-negative definiteness of the form (1.4.10) when $z = z(x)$ and $p = \nabla z(x)$. We note that our hypotheses imply that for each ζ of the type above, vanishing on the boundary, the function $\varphi(\lambda)$, defined by (1.4.6) is of class C^2 for $|\lambda| \leq \lambda_0 (> 0)$ and has a relative minimum at $\lambda = 0$. This implies that

$$\begin{aligned} &\varphi''(0) = \int_G \Big[\sum_{\alpha,\beta} a^{\alpha\beta}(x)\,\zeta_{,\alpha}\zeta_{,\beta} + 2\sum_\alpha b^\alpha(x)\,\zeta\zeta_{,\alpha} + c(x)\,\zeta^2\Big]\,dx \geq 0 \\ &a^{\alpha\beta}(x) = f_{p_\alpha p_\beta}[x, z(x), \nabla z(x)], \quad b^\alpha = f_{p_\alpha z}, \quad c = f_{zz}, \end{aligned} \tag{1.5.1}$$

for all ζ as described. The integral in (1.5.1) is called the *second variation* of the integral (1.1.1). By approximations, it follows that (1.5.1) holds for all LIPSCHITZ functions ζ which vanish on the boundary. Now let us select a point $x_0 \in G$ and a unit vector λ, and let us choose new coordinates y related to x by a transformation

$$y^\gamma = \sum_\alpha d_\alpha^\gamma (x^\alpha - x_0^\alpha), \quad x^\alpha - x_0^\alpha = \sum_\gamma d_\alpha^\gamma y^\gamma, \quad \lambda_\alpha = d_\alpha^1 \tag{1.5.2}$$

where d is a constant orthogonal matrix so that λ is the unit vector in the y^1 direction, and define

$$\begin{aligned} \omega(y) &= \zeta[x(y)], \quad {}'a^{\gamma\delta}(y) = a^{\alpha\beta}[x(y)]\,d_\alpha^\gamma d_\beta^\delta, \\ {}'b^\gamma(y) &= b^\alpha[x(y)]\,d_\alpha^\gamma, \quad {}'c(y) = c[x(y)]. \end{aligned} \tag{1.5.3}$$

Then if G' denotes the image of G,

$$\varphi''(0) = \int_{G'} \Big[\sum_{\gamma,\delta} {}'a^{\gamma\delta}(y)\,\omega_{,\gamma}\,\omega_{,\delta} + 2'b^\gamma\,\omega\cdot\omega_{,\gamma} + {}'c\,\omega^2\Big]\,dy \geq 0.$$

Now, choose $0 < h < H$ so small that the support of $\omega \subset G'$, where

$$\omega(y^1, \ldots, y^\nu) = \begin{cases} (h - |y^1|)(1 - r/H), & \text{if } |y^1| \leq h, \ 0 \leq r \leq H \\ 0 & \text{, otherwise} \end{cases} \tag{1.5.4}$$

$$r^2 = (y^2)^2 + \cdots + (y^\nu)^2.$$

Then if we divide $\varphi''(0)$ by the measure of the support of ω and then let h and $H \to 0$ so that $h/H \to 0$, we conclude that

$$'a^{11}(0) = a^{\alpha\beta}(x_0)\,d_\alpha^1 d_\beta^1 = a^{\alpha\beta}(x_0)\,\lambda_\alpha\lambda_\beta \geq 0$$

which is the stated result. This is called the *Legendre condition*.

If we repeat this derivation for the case of the general integral (1.1.1), assuming $f \in C^2$ of course, we obtain

(1.5.1′)
$$\varphi''(0) = \int_G \Big\{ \sum_{i,j} \sum_{\alpha,\beta} a_{ij}^{\alpha\beta} \zeta^i_{,\alpha} \zeta^j_{\beta} + 2 \sum_\alpha b_{ij}^{\alpha} \zeta^i \zeta^j_{,\alpha} + c_{ij} \zeta^i \zeta^j \Big\} dx \geq 0$$
$$a_{ij}^{\alpha\beta}(x) = f_{p_\alpha^i p_\beta^j}[x, z(x), \nabla z(x)], \quad b_{ij}^{\alpha} = f_{z^i p_\alpha^j}, \quad c_{ij} = f_{z^i z^j}.$$

Making the change of variables (1.5.2) and (1.5.3) and setting

$$\omega^i(y^1, \ldots, y^\nu) = \xi^i \omega(y^1, \ldots, y^\nu), \quad (i = 1, \ldots, N)$$

where ξ is an arbitrary constant vector and ω is defined by (1.5.4), and letting h and $H \to 0$ as above, we obtain

(1.5.5)
$$\sum_{i,j,\alpha,\beta} f_{p_\alpha^i p_\beta^j}[x_0, z(x_0), \nabla z(x_0)] \lambda_\alpha \lambda_\beta \xi^i \xi^j \geq 0 \text{ for all } \lambda, \xi,$$

which is known as the *Legendre-Hadamard condition* (HADAMARD [1]). In this case, we say that the integral (1.1.1) or the integrand f is *regular* if the inequality holds in (1.5.5) for all $\lambda \neq 0$ and $\xi \neq 0$. It turns out that the system (1.4.8) of Euler's equations is *strongly elliptic* in the sense defined by NIRENBERG [2].

Let us suppose, now, that $f \in C^2$ everywhere and that $z \in C^1$ on $\bar{G}$ and minimizes $I(z, G)$, as given by (1.1.1), among all such functions with the same boundary values. A simple approximation argument shows that z minimizes I among all LIPSCHITZ functions with the same boundary values. Let us choose $x_0 \in G$ and a unit vector λ, and let us introduce the y coordinates as in (1.5.2) and let us define (using part of the notation of (1.5.4)

$$\zeta_h^i(x) = \xi^i \omega_h[y(x)]$$
$$\omega_h(x) = \begin{cases} (y^1 + h)\varphi(r h^{-1/2}) & , \quad -h \leq y^1 \leq 0 \ , \quad 0 \leq r \leq h^{1/2}, \\ h^{1/2}(h^{1/2} - y^1) \cdot \varphi(r h^{-1/2}), & \quad 0 \leq y^1 \leq h^{1/2}, \quad 0 \leq r \leq h^{1/2}, \\ 0 & , \quad \text{otherwise} \end{cases}$$

(1.5.6)

where $\varphi \in C^\infty$ on $[0,1]$ with $\varphi(0) = 1$ and $\varphi(\varrho) = 0$ for ϱ near 1. Since the first variation vanishes, we have

$$\int_G [f(x, z + \zeta_h, \nabla z + \nabla \zeta_h) - f(x, z, \nabla z) - \zeta_h^i f_{z^i} - \zeta^i_{h,\alpha} f_{p_\alpha^i}] dx \geq 0,$$

(1.5.7)
$$f_{z^i} = f_{z^i}(x, z, \nabla z), \quad f_{p_\alpha^i} = f_{p_\alpha^i}(x, z, \nabla z).$$

We notice first that the integrand is $O(h)$ (since ζ_h and $\nabla \zeta_h$ are both small) for $x \in R_h^2$ where $0 \leq y^1 \leq h^{1/2}$, $0 \leq r \leq h^{1/2}$. By setting $y^1 = h\eta^1$, $r = h^{1/2}\varrho$ in $R_h^1(-h \leq y^1 \leq 0, 0 \leq r \leq h^{1/2})$, dividing by $h^{(\nu+1)/2}$ and letting $h \to 0$, we obtain

$$\int_{\varrho \leq 1} \Big\{ [f(x_0, z_0, p_{0\alpha}^i + \lambda_\alpha \xi^i \varphi(\varrho)] - f(x_0, z_0, p_0) -$$
$$- \sum_{i,\alpha} \lambda_\alpha \xi^i \varphi(\varrho) f_{p_\alpha^i}(x_0, z_0, p_0) \Big\} d\eta_1' \geq 0.$$

We may now choose a sequence $\{\varphi_n\}$ so $\varphi_n(\varrho) \to 1$ boundedly. This leads to

$$f(x_0, z_0, p^i_{\alpha 0} + \lambda_\alpha \xi^i) - f(x_0, z_0, p_0) - \lambda_\alpha \xi^i f_{p^i_\alpha}(x_0, z_0, p_0) \geq 0 \tag{1.5.8}$$

which is the *Weierstrass condition* (see GRAVES). In case $N = 1$, (1.5.8) yields the following more familiar form of this condition:

$$E(x, z, \nabla z, P) = f(x, z, P) - f(x, z, \nabla z) - (P_\alpha - z_{,\alpha}) f_{p_\alpha}(x, z, \nabla z) \geq 0 \tag{1.5.9}$$

for all P and all x. The function $E(x, z, p, P)$ here defined is known as the *Weierstrass E-function.*

HESTENES and MACSHANE studied these general integrals in cases where $\nu = 2$. HESTENES and E. HÖLDER studied the second variation of these integrals. DEDECKER studied the first variation of very general problems on manifolds.

1.6. Classical sufficient conditions

A detailed account of classical and recent work in this field is given in the recent book by FUNK, pp. 410—433) where other references are given. I shall give only a brief introduction to this subject.

It is clear that the positiveness of the second variation along a function z guarantees that z furnishes a relative minimum to $I(Z, G)$ among all $Z(= z$ on $\partial G)$ in any finite dimensional space. However, if $N = 1$, a great deal more can be concluded, namely that z furnishes a *strong relative minimum* to I, i.e. minimizes $I(Z, G)$ among all $Z \in C^1(\bar{G})$ with $Z = z$ on ∂G for which $|Z(x) - z(x)| < \delta$ for some $\delta > 0$ regardless of the values of the derivatives. WEIERSTRASS was the first to prove such a theorem but his proof was greatly simplified by the use of HILBERT's invariant integral. Of course, the original proof was for the case $N = \nu = 1$; we present briefly an extension to the case $N = 1$, ν arbitrary.

Suppose G is of class C^2_μ, $z \in C^2_\mu(\bar{G})$, and f and f_p are of class C^3_μ in their arguments, $0 < \mu < 1$ (see § 1.2), and suppose that the second variation, as defined in (1.5.1), >0 for each $\zeta \in C^1_c(\bar{G})$ (compact support). By a straightforward approximation, it follows that the second variation is defined for all $\zeta \in H^1_{20}(G)$ (see § 1.8). If we call the integral (1.5.1) $I_2(z; \zeta; G)$ we see from the theorems of § 1.8 below that I_2 is lower-semicontinuous with respect to weak convergence in $H^1_{20}(G)$. Moreover, from the assumed positive definiteness of the form (1.4.10), it follows from the continuity of the $a^{\alpha\beta}(x)$ (they $\in C^1_\mu(\bar{G})$ in fact) that there exist $m_1 > 0$ and M_1 such that

$$a^{\alpha\beta}(x) \lambda_\alpha \lambda_\beta \geq m_1 |\lambda|^2, \quad \sum_{\alpha\beta} [a^{\alpha\beta}(x)]^2 \leq M_1^2. \tag{1.6.1}$$

Then, from the SCHWARZ and CAUCHY inequalities, we conclude that there is a K such that

$$(1.6.2) \qquad I_2(z;\zeta;G) \geq \frac{m_1}{2}\int_G |\nabla\zeta|^2\,dx - K\int_G \zeta^2\,dx.$$

Since weak convergence in $H^1_{20}(G)$ implies strong convergence in $L_2(G)$ (RELLICH's theorem, Theorem 3.4.4), it follows that there is a ζ_0 in $H^1_{20}(G)$ (actually $C^2_\mu(\bar{G})$) which minimizes I_2 among all $\zeta \in H^1_{20}(G)$ for which $\int_G \zeta^2\,dx = 1$. Since we have assumed $I_2 > 0$ for every $\zeta \neq 0$, it follows that

$$(1.6.3) \qquad I_2(z;\zeta;G) \geq \lambda_1 \int_G \zeta^2\,dx, \; \lambda_1 > 0.$$

From the theory of §§ 5.2—5.6, it follows that there is a unique solution ζ of *Jacobi's equation*

$$(1.6.4) \qquad L\zeta \equiv \frac{\partial}{\partial x^\alpha}(a^{\alpha\beta}\zeta_{,\alpha}) + (b^\alpha_{.\alpha} - c)\zeta = 0$$

with given smooth boundary values. It is to be noted that JACOBI's equation is just the Euler equation (z fixed) corresponding to I_2. It is also the equation of variation of the Euler equation for the original I, i.e.

$$(1.6.5) \qquad L\zeta = \frac{\partial}{\partial\varrho}\left\{\frac{\partial}{\partial x^\alpha} f_{p_\alpha}[x, z+\varrho\zeta, \nabla z + \varrho\nabla\zeta] - f_z[\text{same}]\right\}_{\varrho=0}.$$

It follows from Theorems 6.8.5 and 6.8.6 that there is a unique solution of the Euler equation for all sufficiently near (in $C^2_\mu(\partial G)$) boundary values, in particular for the boundary values $z + \varrho$, and that $z = z(\varrho)$ satisfies an ordinary differential equation

$$(1.6.6) \qquad \frac{dz}{d\varrho} = F(z)$$

in the Banach space $(C^2_\mu(\bar{G}))$, where $F(z)$ denotes the solution ζ of Jacobi's equation (1.6.4) with $z = z(\varrho)$ for which $\zeta = 1$ on ∂G. We shall show below that this solution ζ cannot vanish on $\bar{G}$ for ϱ sufficiently small; it is sufficient to do this for $\varrho = 0$, when $z(\varrho)$ = our solution z, on account of the continuity.

So, let ζ_1 be this solution. If $\zeta_1(x) < 0$ anywhere, then the set where this holds is an open set D and $\zeta_1 = 0$ on $\partial D(\subset G)$. Since ζ_1 is a solution on D, $I(\zeta_1, D) \leq 0$ since ζ_1 is minimizing on D. (D may not be smooth, but see Chapters 3—5). But if we set $\zeta = \zeta_1$ on D and $\zeta = 0$, otherwise, $\zeta \in H^1_{20}(G)$ so (1.6.3) holds and we must have $\zeta \equiv 0$. Hence $\zeta_1(x) \geq 0$ everywhere. Now, suppose $\zeta_1(x_0) = 0$. From Theorem 6.8.7, it follows that we may choose R so small that $B(x_0, R) \subset G$ and there is a non-vanishing solution ω of (1.6.4) on $B(x_0, R)$. Letting $\zeta_1 = \omega v$, we see that v satisfies the equation

$$a^{\alpha\beta} v_{,\alpha\beta} + (a^{\alpha\beta}_{,\beta} + 2\omega^{-1} a^{\alpha\beta} \omega_{,\beta}) v_{,\alpha} = 0$$

and $v(x_0) = 0$. But from the maximum principle as proved by *E.* HOPF [1] (see § 2.8), it follows that v cannot have a minimum interior to G. Accordingly $\zeta_1 \neq 0$ anywhere in $\bar{G}$.

Therefore it is possible to embed our solution z in a *field of extremals*. That is, there is a 1 parameter family $Z(x, \varrho)$ of solutions of Euler's equation where $Z(x, 0) = z(x)$, our given solution. $Z(x, \varrho) \in C^1(\bar{G} \times [-\varrho_0, \varrho_0])$ and $\in C^2_\mu(\bar{G})$ as a function of x for each ϱ with $z_\varrho > 0$. Consequently there are functions $P_\alpha(x, z)$ on the set Γ, where

$$\Gamma: x \in \bar{G},\ Z(x, -\varrho_0) \leq z \leq Z(x, \varrho_0), \tag{1.6.7}$$

which act as *slope-functions* for the field, i.e.

$$Z_{,\alpha}(x, \varrho) = P_\alpha[x, Z(x, \varrho)]. \tag{1.6.8}$$

By virtue of the facts that $Z(x, \varrho)$ satisfies Euler's equation for each ϱ, that (1.6.8) holds, and that if $(x, z) \in \Gamma$, then $z = Z(x, \varrho)$ for a unique ϱ on $[-\varrho_0, \varrho_0]$, we find that

$$\begin{gathered} f_z - f_{p_\alpha p_\beta}(P_{\beta x^\alpha} + P_{\beta z} P_\alpha) - f_{p_\alpha z} P_\alpha - f_{p_\alpha x^\alpha} = 0 \\ f_z = f_z[x, z, P(x, z)],\ \text{etc.},\ (x, z) \in \Gamma. \end{gathered} \tag{1.6.9}$$

Let us define

$$\begin{gathered} I^*(z, G) = \int_G f^*(x, z, \nabla z)\, dx, \\ f^*(x, z, p) = f_{p_\alpha}[x, z, P(x, z)] \cdot [p_\alpha - P_\alpha(x, z)] + f[x, z, P(x, z)]. \end{gathered} \tag{1.6.10}$$

We observe that

$$\begin{gathered} f^*_z = [p_\alpha - P_\alpha(x, z)] \cdot \{f_{p_\alpha z} + f_{p_\alpha p_\beta} P_{\beta z}\} - P_{\alpha z} f_{p_\alpha} \\ + f_z + f_{p_\alpha} P_{\alpha z}; \quad f^*_{p_\alpha} = f_{p_\alpha}[x, z, P(x, z)]. \end{gathered} \tag{1.6.11}$$

Thus, if $z \in C^1(\bar{G})$ and $(x, z(x)) \in \Gamma$ for $x \in \bar{G}$, we see that

$$f^*_z[x, z(x), \nabla z(x)] - \frac{\partial}{\partial x^\alpha} f^*_{p_\alpha}[x, z(x), \nabla z(x)] = 0. \tag{1.6.12}$$

Accordingly *the integral $I^*(z, G)$ has the same value for all such z which have the same boundary values.* Moreover, if $z(x) \equiv Z(x, \varrho)$ for some ϱ, then

$$\begin{gathered} \nabla z(x) = P[x, z(x)],\quad f^*[x, z(x), \nabla z(x)] = f[x, z(x), \nabla z(x)], \\ I^*(z, G) = I(z, G). \end{gathered} \tag{1.6.13}$$

This integral $I^*(z, G)$ is known as *Hilbert's invariant integral*. Therefore, if $z \in C^1(\bar{G})$ and $(x, z(x)) \in \Gamma$ for all $x \in \bar{G}$, and $z(x) = z_0(x)$ on ∂G, then

$$\begin{aligned} I(z, G) - I(z_0, G) &= I(z, G) - I^*(z_0, G) = I(z, G) - I^*(z, G) \\ &= \int_G E[x, z(x), P\{x, z(x)\}, \nabla z(x)]\, dx \end{aligned} \tag{1.6.14}$$

where $E(x, z, P, p)$ is the Weierstrass E-function defined in (1.5.9). *Thus z_0 minimizes $I(z, G)$ among all such z, and hence furnishes a strong relative minimum to I, provided that*

$$E[x, z, P(x, z), p] \geq 0, \quad (x, z) \in \Gamma, \ p \text{ arbitrary}. \tag{1.6.15}$$

This same proof shows that if (1.6.15) *holds for all (x, z, p) in some domain R, where all the (x, z) involved $\in \Gamma$, then $I(z_0, G) \leq I(z, G)$ for all z for which $[x, z(x), \nabla z(x)] \in R$ for all $x \in \bar{G}$.*

In the cases $\nu = 1$, the *Jacobi condition* is frequently stated in terms of "conjugate points". A corresponding condition for $\nu > 1$ is that the JACOBI equation has no non-zero solution which vanishes on the boundary ∂D of any sub-domain $D \subset G$; D may coincide with G or may not be smooth, in which case we say u vanishes on $\partial D \Leftrightarrow u \in H_{20}^1(D)$. The most interesting condition is that there exist a non-vanishing solution ω on $\bar{G}$; we have seen above that this is implied by the positivity of the second variation. If we then set $\zeta = \omega u$, where $u = 0$ on ∂G, then the reader may easily verify that

$$I_2(z; \zeta; G) = \int_G [\omega^2 a^{\alpha\beta} u_{,\alpha} u_{,\beta} - u^2 \omega L \omega] \, dx > 0$$

for all $u \in H_{20}^1(G)$, since $L\omega = 0$.

In cases where $\nu > 1$ and $N > 1$ it is still true (if we continue to assume the same differentiability for G, f, and z) that (1.6.2) holds with I_2 defined by (1.5.1′) even in the general regular case where (1.5.5) holds with the inequality for $\lambda \neq 0$ and $\xi \neq 0$. This is proved in § 5.2. So it is still true that if $I_2 > 0$ for all ζ, the Euler equations have a unique solution for sufficiently nearby boundary values. It is more difficult (but possible) to show that there is an N-parameter field of extremals and then it turns out that such a field does not lead so easily to an invariant integral. By allowing slope functions $P_\alpha^i(x, z)$ which are not "integrable" (i.e. there may not be z's such that $z_{,\alpha}^i = P_\alpha^i(x, z)$), WEYL [1] (see also DE DONDER) developed a comparatively simple field theory and showed the existence of his types of fields under certain conditions. His theory is succesful if f is convex in all the p_α^i. To treat more general cases, CARATHEODORY [1], BOERNER, and LE PAGE have introduced more general field theories. The latter two noticed that exterior differential forms were a natural tool to use in forming the analog of Hilbert's invariant integral. However, the sufficient conditions developed so far are rather far from the necessary conditions and many questions remain to be answered.

1.7. The direct methods

The necessary and sufficient conditions which we have just discussed have presupposed the existence and differentiability of an extremal. In the cases $\nu = 1$, this was often proved using the existence theorems for ordinary differential equations. However, until recently, corresponding theorems for partial differential equations were not available so the direct methods were developed to handle this problem and to obtain results in the large for one dimensional problems.

As has already been said, HILBERT [1] and LEBESGUE [2] had solved the Dirichlet problem by essentially direct methods. These methods were exploited and popularized by TONELLI in a series of papers and a book ([1], [2], [3], [4], [5], [7], [8]), and have been and still are being used by many others. The idea of the direct methods is to show (i) that the integral to be minimized is lower-semicontinuous with respect to some kind of convergence, (ii) that it is bounded below in the class of "admissible functions," and (iii) that there is a *minimizing sequence* (i.e. a sequence $\{z_n\}$ of admissible functions for which $I(z_n, G)$ tends to its infimum in the class) which converges in the sense required to some admissible function.

Tonelli applied these methods to many single integral problems and some double integral problems. In doing this he found it expedient to use uniform convergence (at least on interior domains) and to allow absolutely continuous functions (satisfying the given boundary conditions) as admissible for one dimensional problems; and he defined what he called absolutely continuous functions of two variables ([6]) to handle certain double integral problems (see the next section). In the double integral problems ($N = 1$, $\nu = 2$), he found it expedient to require that $f(x, z, p)$ satisfy conditions such as

(1.7.1) $$m|p|^k - K \leq f(x, z, p), \quad k \geq 2, \quad m > 0, \text{ where}$$

(1.7.2) $$f(x, z, p) \geq 0 \quad \text{and} \quad f(x, z, 0) = 0 \quad \text{if} \quad k = 2,$$

in order to obtain equicontinuous minimizing sequences. However, Tonelli was not able to get a general theorem to cover the case where f satisfies (1.7.1) with $1 < k < 2$. Moreover, if one considers integrals in which $\nu > 2$, one soon finds that one would have to require $k > \nu$ in order to ensure that the functions in any minimizing sequence would be equicontinuous, at least on interior domains (see Theorem 3.5.2). To see this, one needs only to notice that the functions

$$\log\log(1 + 1/|x|), \quad 1/|x|^h, \quad 0 < |x| < 1$$

are limits of C^1 functions z_n in which

$$\int |\nabla z_n|^\nu \, dx \quad \text{and} \quad \int |\nabla z_n|^k \, dx \quad \text{for} \quad k < \nu/(h + 1)$$

are uniformly bounded over the unit ball G. In the "borderline case" $k = \nu$, it is possible, in case f satisfies the supplementary condition (1.7.2), to replace an arbitrary minimizing sequence of continuous functions by a minimizing sequence each member of which is *monotone in the sense of Lebesgue* (i.e. takes on its max. and min. values on the boundary of each compact sub-domain); the new sequence is equicontinuous on interior domains (see ε 4.3).

However, even this Lebesgue smoothing process does not work in general for $1 < k < \nu$. In order to get a more complete existence theory, the writer and CALKIN (MORREY [5], [6], [7]) found it expedient to allow as admissible, functions which are still more general than Tonelli's functions and to allow correspondingly more general types of convergence. The new spaces of functions can now be identified with the Banach spaces $H_p^1(G)$ (see the next section) (or the Sobolev spaces $W_p^1(G)$) which have been and still are being used by many writers in many different connections (see § 1.8). In this way, the writer was able to obtain very general existence theorems. Unfortunately the solution shown to exist was known only to be in one of these general spaces and hence wasn't even known to be continuous, let alone of class C^2! So these existence theorems in themselves have only minor interest. However, at the same time,[1] the writer was able to show in the case $\nu = 2$ (N arbitrary) that these very general solutions were, in fact, of class C^2 after all provided that f satisfied the conditions (1.10.8) below with $k = \nu = 2$. A greatly simplified presentation of this old work is to be found in the author's paper [15]; recent developments have permitted further simplifications and extensions which we shall discuss later.

In general, it is still not known that the solutions are continuous if f satisfies (1.7.1) with $1 < k < \nu$. In the case $\nu = 2$, $N = 1$ TONELLI [8] showed that the solutions in this case are continuous if f is such that there is a unique minimizing function in the small. More recently, SIGALOV ([1], [3]) showed that the solution surfaces always posess conformal maps (possibly with vertical segments) in the case $\nu = 2$, $N = 1$. In the case $\nu = 2$, $N = 1$, $f = f(p)$ it was proved a long time ago by A. HAAR (see also RADO [2]) that there is a unique minimizing function z which is defined on a strictly convex domain G and which satisfies a Lipschitz condition with constant L, provided that any linear function which coincides with the given boundary values at three different points on the boundary has slope $\leq L$. This result has recently been generalized (not quite completely) by GILBARG and STAMPACCHIA [3]. The author has completed the extension of HAAR's results and has extended those

[1] This work was completed during the year 1937—38 and the author lectured on it in the seminar of Marston Morse during the spring of 1938 and also reported on this work in an invited lecture to the American Mathematical Society at its meeting in Pasadena, California, on December 2, 1939 [6]. The necessary theorems about the H_p^1 spaces (called $\mathfrak{P}_p$ at that time) (see § 1.8) were published in the papers referred to above by CALKIN and the author [5]. The remainder of the work was first published in a paper (MORREY (7]) which was released in December 1943; the manuscript had been approved for publication in 1939.

of STAMPACCHIA and GILBARG. These results are presented and proved in § 4.2. The results include GILBARG's existence theorem for equations of the form (1.10.13) with $N = 1$, $B_i = 0$, $A^\alpha = A^\alpha(p)$. The advantages of these theorems are that one can restrict one's self to LIPSCHITZ functions z and no assumption has to be made about how $f(x, z, p)$ behaves as $|p| \to \infty$. The convexity assumption for all (x, z, p) is suggested by the conditions in §§ 1.4—1.6.

In the existence theorems mentioned above, the author considered integrals of the form (1.1.1) in which ν and N are arbitrary but in which f is convex in all the p^i_α; with this convexity assumption, no difficulties were introduced in the proofs by allowing $N > 1$. The results have been extended and the old proofs greatly simplified by SERRIN ([1], [2]); we shall present (in § 1.8) a simple lower-semicontinuity proof based on some of his ideas and on some ideas of TONELLI. However, for $N > 1$, the proper condition would be to assume that (1.5.8) and/or (1.5.5) held for all (x, z, p, λ, ξ). The author has studied these general integrals ([9]) and found that if f satisfies the conditions

$$m\, V^k - K \le f(x, z, p) \le M\, V^k, \quad |f_p(x, z, p)| \le M\, V^{k-1}, \quad k > 1$$
$$|f_x|, |f_z| \le M\, V^k, \quad m > 0, \quad V = (1 + |p|^2)^{1/2}$$

then a necessary and sufficient condition that $I(z, G)$ be lower semicontinuous on the space $H^1_k(G)$ with respect to uniform convergence is that f be quasi-convex as a function of p. A function $f(p)$, $p = \{p^i_\alpha\}$ is quasi-convex if and only if it is continuous and

$$\int_G f[p_0 + \nabla \zeta(x)]\, dx \ge f(p_0) \cdot |G|, \quad \zeta \in C^1_c(G);$$

that is, linear vectors give the absolute minimum to $I(z, G)$ among all z with such boundary values (note that linear functions always satisfy EULER's equation if $f \in C^2$). A *necessary* condition for quasi-convexity is just (1.5.5). The author showed that (1.5.5) is *sufficient* for quasi-convexity if $f(p)$ is of one of the two following forms

$$(1)\ f(p) = a^{\alpha\beta}_{ij}\, p^i p^j \qquad (a^{\alpha\beta}_{ij}) \text{ const.}$$
$$(2)\ f(p) = F(D_1, \ldots, D_{\nu+1}), \quad N = \nu + 1$$

where F is homogeneous of degree 1 in the D_i and D_i is the determinant of the submatrix obtained by omitting the i-th column of the p^i_α matrix; or if for each p, there exist alternating constant tensors A^α_i, $A^{\alpha\beta}_{ij}$, ..., such that

$$f(p + \pi) \ge f(p) + A^\alpha_i \pi^i_\alpha + A^{\alpha\beta}_{ij} \pi^i_\alpha \pi^j_\beta + \cdots + A^{\alpha_1 \ldots \alpha_\mu}_{i_1 \ldots i_\mu} \pi^{i_1}_{\alpha_1} \ldots \pi^{i_\mu}_{\alpha_\mu}$$

for all π. Under some additional conditions the integral is lower-semicontinuous with respect to weak convergence in $H^1_k(G)$. We discuss these integrals in § 4.4. Recently Norman MEYERS has extended the author's results to higher order integrals and has improved them somewhat. The proofs are very similar to those of the author for the first order integrals and we shall not present them.

Finally, it should be pointed out that the ν-dimensional area integral in the non-parametric case with $N > 1$ is not convex in all the p_α^i but is a regular integrand in the general sense; for $\nu = N = 2$, the integral is

$$I(z, G) = \iint_G \sqrt{1 + (z_x^1)^2 + (z_y^1)^2 + (z_x^2)^2 + (z_y^2)^2 + \left(\frac{\partial(z^1, z^2)}{\partial(x, y)}\right)^2}\, dx\, dy.$$

$$f(x, z, p) = [1 + (p_1^1)^2 + (p_2^1)^2 + (p_1^2)^2 + (p_2^2)^2 + (p_1^1 p_2^2 - p_2^1 p_1^2)^2]^{1/2}.$$

The integrand of a parametric problem is never regular since it is necessarily degenerate but a large class of such integrands do satisfy (1.5.5) and (1.5.8) with the equality allowed (whenever $f \in C^2$). We shall speak about parametric problems in Chapter 9.

1.8. Lower semicontinuity

We begin with a brief discussion of the spaces of admissible functions which we shall use; a more extended discussion including complete proofs is given in Chapter 3. It is convenient to call these spaces SOBOLEV spaces; in addition to the brevity of this designation, it is appropriate since he proved some important results concerning these functions [1] and popularized them in his book [2]. However, he was by no means the first person to use these functions. Beppo LEVI was probably the first to use (in (1906) admissible functions which required the use of the Lebesgue integral to express $I(z, G)$; his functions (of two variables) were continuous, absolutely continuous in each variable for almost all values of the other, and their first partial derivatives were in L_2. In discussing the area of surfaces, TONELLI [6] introduced his notion of absolutely continuous functions (ACT) in 1926. His definition was identical with that of Levi except that the partial derivatives were required only to be in L_1; he used these functions to discuss the double integral problems mentioned above. But meanwhile in 1920, G. C. EVANS ([1], [2]) had encountered more general functions, essentially those we now use, in his study of potential theory. RELLICH proved the compactness in L_2 of bounded sets in H_2^1. SOBOLEV proved his well known results on the L_p-properties of these functions in 1938 [1]. In 1940, J. W. CALKIN and the author ([5]) proved many of the fundamental properties of these functions stated below. Since the war ARONSZAJN and SMITH have studied these functions in great detail [1]. No doubt many others have studied these functions. Recently these functions have been used by many people in many different connections (see, for instance, DENY, FRIEDRICHS [1], [2], [3], FUBINI, JOHN [2], LAX, MORREY [1], [10], MORREY and EELLS,

NIKODYM, SHIFFMAN, SIGALOV ([2]); their use is now standard in partial differential equations (see FRIEDMAN [2], HÖRMANDER [1], LIONS [1]).

Definition 1.8.1. We say that a function z *is of class* $H_r^1(G)$, $r \geq 1$, if and only if z is of class $L_r(G)$ and there exist functions $p_1, \ldots, p_\nu$, also of class $L_r(G)$, such that

$$(1.8.1) \quad \int_G g(x)\, p_\alpha(x)\, dx = -\int_G g_{,\alpha}(x)\, z(x)\, dx, \; g \in C_c^1(G), \alpha = 1, \ldots, \nu.$$

Remarks. It is clear that the functions p_α are uniquely determined up to null functions and that if z is of class $H_r^1(G)$ and $z^*(x) = z(x)$ almost everywhere, then z^* is of class $H_r^1(G)$ and the same functions p_α will do for z^*.

Definition 1.8.2. As in the case of the L_r-spaces, the elements of the *space* $H_r^1(G)$ are classes of equivalent functions of class $H_r^1(G)$. We denote the classes of equivalent functions p_α by $z_{,\alpha}$ and call them the *distribution derivatives* of the element z. An element $z \in$ *space* $H_r^m(G)$ if and only if z and its distribution derivatives up to order $m-1$ are successively seen to $\in H_r^1(G)$.

Remark. Naturally we may regard an element z in $H_r^1(G)$ as a distribution and then the distribution corresponding to $z_{,\alpha}$ would be the derivative of z in the distribution sense.

Definition 1.8.3. φ is *a Friedrichs mollifier* if and only if $\varphi \in C_c^\infty(R_\nu)$, $\varphi(x) \geq 0$, spt φ (i.e. the support of φ) $\subset B(0,1)$, and

$$\int_{B(0,1)} \varphi(x)\, dx = 1.$$

If u is locally summable on an open set G, we define its *φ-mollified function* u_ϱ by

$$u_\varrho(x) = \int_{B(x,\varrho)} u(\xi)\, \varphi_\varrho^*(\xi - x)\, d\xi, \quad x \in G_\varrho = \{x \in G \mid B(x,\varrho) \subset G\},$$

$$\varphi_\varrho^*(\eta) = \varrho^{-\nu} \varphi[\varrho^{-1}\eta].$$

The following theorems are almost evident and are proved in Chapter 3 (in fact, Theorem B is evident):

Theorem A. *The space* $H_p^1(G)$ *is a Banach space under the norm*

$$\|z\|_{p,G}^1 = \|z\|_{p,G}^0 + \sum_{\alpha=1}^{\nu} \|z_{,\alpha}\|_{p,G}^0$$

$$(\|\varphi\|_{p,G}^0 = \|\varphi\|_{L_p(G)}).$$

Theorem B. (a) *If* $u \in L_p(G)$ *and* u_ϱ *denotes its* φ*-mollified function, then* $u_\varrho \in C^\infty(G_\varrho)$ *and* $u_\varrho \to u$ *in* $L_p(D)$ *for each* $D \subset\subset G$.

(b) *If* $u \in H_p^1(G)$, *then* $u_{\varrho,\alpha}(x) = (u_{,\alpha})_\varrho(x)$ *for* $x \in G_\varrho$ *so that* $u_\varrho \to u$ *and* $u_{\varrho,\alpha} \to u_{,\alpha}$ *in* $L_p(D)$ *for each* $D \subset\subset G$.

(c) *The convergence in* (a) *and* (b) *holds for almost all* x.

The lower semicontinuity theorems in this section depend on some well-known theorems on convex functions which we now define:

Definition 1.8.4. A set S in a linear space is said to be *convex* $\Leftrightarrow$ the segment $P_1P_2 \subset S$ whenever P_1 and $P_2 \in S$. A function φ is said to be *convex* on the convex set $S \Leftrightarrow$

$$\varphi[(1-\lambda)\,\xi_1 + \lambda\xi_2] \leq (1-\lambda)\,\varphi(\xi_1) + \lambda\varphi(\xi_2),\ \xi_1, \xi_2 \in S, \quad 0 \leq \lambda \leq 1.$$

Remark. Evidently φ is convex on the convex set $S \Leftrightarrow$ the set of points (ξ, ζ) for which $\xi \in S$ and $\zeta \geq \varphi(\xi)$ is convex.

We now state the characteristic property of convex functions:

Lemma 1.8.1. *A necessary and sufficient condition that φ be convex on the open convex set $S \subset R_P$ is that for each ζ in S there exists a linear function $a_p\,\xi^p + b$ such that*

$$(1.8.2) \qquad \varphi(\zeta) = a_p\,\zeta^p + b, \quad \varphi(\xi) \geq a_p\,\xi^p + b, \quad \xi \in S.$$

If φ is of class C^1 on S, this condition is equivalent to

$$(1.8.3) \qquad E(\zeta, \xi) \equiv \varphi(\xi) - \varphi(\zeta) - (\xi^p - \zeta^p)\,\varphi_p(\zeta) \geq 0, \quad \xi, \zeta \in S.$$

If φ is of class C^2 on S, this condition is equivalent to

$$\varphi_{pq}(\zeta)\,\eta^p\,\eta^p \geq 0,$$

for all ζ and all η.

Remark. The function $E(\zeta, \xi)$ in (1.8.3) is seen to be the WEIERSTRASS E-function.

Our first lower-semicontinuity theorem depends on Jensen's inequality which we now state:

Lemma 1.8.2 (*Jensens's* inequality). *Suppose φ is convex on R_p, S is a set, μ is a non-negative bounded measure on S, and the functions $\xi^p \in L_1(S, \mu)$, $p = 1, \ldots, P$. Then*

$$(1.8.4) \qquad \begin{aligned} \varphi(\xi^1, \ldots, \xi^P) &\leq [\mu(S)]^{-1} \int_S \varphi[\xi^1(x), \ldots, \xi^P(x)]\,d\mu, \\ \zeta^p &= [\mu(S)]^{-1} \int_S \xi^p\,d\mu. \end{aligned}$$

Remark. I.e. φ (average) $\leq$ average of φ.

Proof. Choose a_p (Lemma 1.8.1) so that

$$\varphi(\zeta) + a_p(\xi^p - \zeta^p) \leq \varphi(\xi) \quad \text{for all } \xi$$

and then average over S.

We can now state and prove our first lower semicontinuity theorem; in this general form, it is due to SERRIN [2]:

Theorem 1.8.1. *Suppose that $f = f(p)$ is non-negative and convex for all $p = \{p_\alpha^i\}$ and suppose that z and each $z_n \in H_1^1(G)$ and that $z_n \rightharpoonup z$* (i.e. *tends weakly to*) *z in $L_1(D)$ for each $D \subset\subset G$. Then $I(z, G)$ and $I(z_n G)$ are each finite or $+\infty$ and*

$$I(z, G) \leq \liminf_{n \to \infty} I(z_n, G)$$

Proof. The first conclusion is obvious. Let $D \subset\subset G$; we may suppose that $D \subset G_a$ for some $a > 0$. Let φ be a mollifier and, for $0 < \varrho < a$, let z_ϱ and $z_{n\varrho}$ be the φ-mollified functions (of class C^∞ on $\bar{D}$). From Fatou's lemma, Theorem B, and equation (1.8.4), we conclude that (since $f[\nabla z_\varrho(x)] \to f[\nabla z(x)]$ a.e.)

$$I(z, D) \leq \liminf_{\varrho \to 0} I(z_\varrho, D). \tag{1.8.5}$$

Let us now define

$$F(x) = f[\nabla z(x)], \quad F_n(x) = f[\nabla z_n(x)].$$

Then, from Jensen's inequality with $d\mu = \varphi_\varrho^*(\xi - x)\, d\xi$ (see Equation (1.8.4)), we conclude that

$$f[\nabla z_\varrho(x)] \leq F_\varrho(x), \quad f[\nabla z_{n\varrho}(x)] \leq F_{n\varrho}(x), \quad x \in D, \quad 0 < \varrho < a.$$

Using this and Theorem B (setting $F_n(x) = 0$ outside G) we conclude that

$$I(z_{n\varrho}, D) \leq \int_D F_{n\varrho}(x)\, dx \leq I(z_n, G). \tag{1.8.6}$$

The weak convergence in $L_1(D')$ for each $D' \subset\subset G$ implies that $\nabla z_{n\varrho}$ converges uniformly on D to ∇z_ϱ for each ϱ, $0 < \varrho < a$. Thus

$$I(z_\varrho, D) = \lim_{n \to \infty} I(z_{n\varrho}, D) \leq \liminf_{n \to \infty} I(z_n, G). \tag{1.8.7}$$

Combining (1.8.5) and (1.8.7), we conclude that

$$I(z, D) \leq \liminf_{n \to \infty} I(z_n, G)$$

from which the result follows easily using the arbitrariness of D.

We now give a simple proof of lower semicontinuity for a wide class of integrals but using a more restricted type of convergence which is, however, sufficiently general for the existence theory. The hypothesis that the $f_{p_\alpha^i}$ be continuous can be removed rather easily. More general theorems have been proved by SERRIN (see Chapter 4).

Theorem 1.8.2. *Suppose $f = f(x, z, p)$ and the $f_{p_\alpha^i}$ are continuous with $f(x, z, p) \geq 0$ for all (x, z, p), suppose f is convex in p for each (x, z), and suppose $z_n \rightharpoonup z$ in $H_1^1(D)$ for each $D \subset\subset G$. Then the conclusions of Theorem 1.8.1 hold.*

Proof. Choose $D \subset\subset G$. The weak convergence in $H_1^1(D')$ for each $D' \subset\subset G$ implies the strong convergence of z_n to z in $L_1(D)$ (see Theorem 3.4.4). By choosing a subsequence, still called z_n, in which $I(z_n, G) \to$ its former lim inf, we may assume that $z_n(x) \to z(x)$ a.e. on D. We now suppose $I(z, D) < +\infty$. Then, for each $\varepsilon > 0$, there is a compact subset $S \subset D$ on which z and ∇z (i.e. representatives) are continuous, on which z_n converges uniformly to z, and which is such that

$$I(z, S) \geq I(z, D) - \varepsilon \tag{1.8.8}$$

(if $I(z, D) = +\infty$, we may take $I(z, S) > M$, arbitrary). Then, from Lemma 1.8.1, we conclude that

$$f[x, z_n(x), \nabla z_n(x)] \geq f[x, z_n(x), \nabla z(x)] + f_p[x, z(x), \nabla z(x)] \cdot [\nabla z_n(x) - \nabla z(x)] + \{f_p[x, z_n(x), \nabla z(x)] - f_p[x, z(x), \nabla z(x)]\} \cdot [\nabla z_n(x) - \nabla z(x)]. \tag{1.8.9}$$

The weak convergence implies (see Theorem 3.2.4 (a)) the weak convergence of ∇z_n to ∇z in $L_1(D)$ which implies, in turn, that

$$\lim_{n\to\infty} \int_S f_p[x, z(x), \nabla z(x)] \cdot [\nabla z_n(x) - \nabla z(x)]\, dx = 0. \tag{1.8.10}$$

The uniform convergence of z_n to z on S, together with the uniform boundedness of the L_1 norms of ∇z_n and ∇z, implies that

$$\lim_{n\to\infty} \int_S \{f_p[x, z_n, \nabla z] - f_p[x, z, \nabla z]\} \cdot [\nabla z_n - \nabla z]\, dx = 0. \tag{1.8.11}$$

Hence, from (1.8.8—1.8.11), we conclude that

$$I(z, D) - \varepsilon \leq \lim_{n\to\infty} \int_S f[x, z_n(x), \nabla z(x)]\, dx \leq \liminf_{n\to\infty} I(z_n, G).$$

The theorem follows easily.

1.9. Existence

If f satisfies the condition (1.7.1) with $k > 1$, an existence theorem can easily be deduced from the lower-semicontinuity theorems of § 1.8 and Theorems 3.4.4 and 3.4.5. The following simple lemma enables us to prove easily a more general existence theorem (Theorem 1.9.1 below)

Lemma 1.9.1. *Suppose that $f_0(p)$ is continuous and that*

$$\lim_{p\to\infty} |p|^{-1} f_0(p) = +\infty. \tag{1.9.1}$$

Then, for each M, there exists a function φ such that $\varphi(\varrho) > 0$ for $\varrho > 0$ and $\varphi(\varrho) \to 0$ as $\varrho \to 0$ such that

$$\int_e |p(x)|\, dx \leq \varphi(|e|)^* \tag{1.9.2}$$

for every measurable $e \subset G$ and every vector $p(x)$ satisfying

$$\int_G f_0[p(x)]\, dx \leq M. \tag{1.9.3}$$

Proof. We define $\varphi(\varrho)$ as the sup of the left member of (1.9.2) for all e such that $|e| \leq \varrho$ and p satisfying (1.9.3). If $\varphi(\varrho)$ does not $\to 0$ as $\varrho \to 0$, $\exists$ an $\varepsilon_0 > 0$ and sequences $\{e_n\}$ and $\{p_n\}$ such that $|e_n| \to 0$ and $\int_{e_n} |p_n(x)|\, dx \geq \varepsilon_0$. We define

$$\psi(\sigma) = \inf |p|^{-1} f_0(p): \text{ for } |p| \geq \sigma.$$

* We often use $|e|$ to denote the measure of the set e.

Clearly $\psi(\sigma) \to +\infty$ as $\sigma \to +\infty$. For each n, let g_n be the subset of e_n where $|p_n(x)| \geq \sigma_n = \varepsilon_0/2|e_n|$. Obviously

$$\int_{e_n - g_n} |p_n(x)\, dx \leq \sigma_n \cdot |e_n| = \varepsilon_0/2.$$

Consequently, since $f_0(p) \geq |p| \cdot \psi(\sigma)$ for $|p| \geq \sigma$, we obtain

$$(1.9.4) \qquad \int_{g_n} |p_n(x)|\, dx \geq \varepsilon_0/2, \quad \int_{g_n} f_0[p_n(x)]\, dx \geq \psi(\sigma_n) \cdot \varepsilon_0/2.$$

Since $\sigma_n \to +\infty$ and $\psi(\sigma_n) \to +\infty$, (1.9.4) contradicts (1.9.3).

We now state and prove our principal existence theorem.

Theorem 1.9.1. *We suppose that* (i) f *and the* $f_{p^i_\alpha}$ *are continuous in their arguments*; (ii) f *is convex in* p *for each* (x, z); (iii) *there is a function* f_0 *satisyfing the conditions of Lemma* 1.9.1 *such that* $f(x, z, p) \geq f_0(p)$ *for all* (x, z, p); (iv) F^* *is a* (non-empty) *family of vector functions which is compact with respect to weak convergence in* $H^1_1(G)$; (v) F *is a family, closed under weak convergence in* $H^1_1(G)$, *such that each* z *in* F *coincides on* ∂G *with a function* z^* *in* F^* (i.e. $z - z^* \in H^1_{1,0}(G)$, see § 3.2); (vi) $I(z_0, G) < +\infty$ *for some* $z_0 \in F$; (vii) G *is bounded. Then* $I(z, G)$ *takes on its minimum for some* z *in* F.

Remarks. Since we have not made any assumptions on G other than boundedness and since the admissible functions are not continuous, the most convenient way to specify the boundary values of a function is to state that $z - z^* \in H^1_{10}(G)$ for some given z^*. Thus the family F^* defines, so to speak, the class of boundary values being allowed. Of course, F^* could consist of a single function z^*.

Proof of the theorem. Let $\{z_n\}$ be a minimizing sequence; we may assume that $I(z_n, G) \leq I(z_0, G) = M$. Using (iii) and Lemma 1.9.1, we conclude that the set functions $\int_e |\nabla z_n(x)|\, dx$ are uniformly bounded and uniformly absolutely continuous. From (v), we conclude for each n that there is a $z^*_n \in F^*$ such that $w_n = z_n - z^*_n \in H^1_{10}(G)$. From (iv) we conclude that there is a subsequence, still called $\{n\}$, such that $z^*_n \rightharpoonup$ some z^* in F^*. From Theorem 3.2.4 (a) with $p = 1$, it follows that the set functions $\int_e |\nabla z^*_n|\, dx$ are uniformly bounded and uniformly absolutely continuous so that the same is true for $\{w_n\}$. From Theorem 3.2.4 and the corollary to Theorem 3.2.1, it follows that there is a further subsequence, still called $\{n\}$, such that $w_n \rightharpoonup w \in H^1_{10}(G)$. Then, of course $z_n \rightharpoonup z = z^* + w$ in $H^1_1(G)$ and the theorem follows from Theorem 1.8.2. Actually our hypotheses do not imply that $f(x, z, p) \geq 0$; however, it is clear that f_0 takes on its minimum, so that f is bounded below and there is no loss in generality (since G is bounded) in assuming that $f \geq 0$.

Remark 1. If f satisfies (1.7.1) with $k > 1$, it obviously satisfies (iii) above.

Remark 2. In the case $\nu = 1$, one concludes immediately the uniform absolute continuity of the (unique in this case) absolutely continuous representatives $\bar{z}_n$ and hence their equicontinuity. The weak convergence implies uniform convergence, the limit function being obviously absolutely continuous.

Remark 3. In Theorem 1.9.1, F could be, for example, the set of all z such that $z - z^* \in H^1_{10}(G)$, for some z^* in F^* (F^* assumed given), and z satisfies a system of equations of the form

$$(1.9.5) \qquad a^{\alpha}_{ij}[x, z(x)]\, z^j_{.\alpha}(x) + b_{ij}[x, z(x)]\, z^j(x) + c_i[x, z(x)] = 0, \quad i = 1, \ldots, P,$$

where the a^{α}_{ij}, b_{ij}, and c_i are everywhere bounded and continuous. For, suppose that $\{z_n\}$ is any sequence in $F \ni z_n \rightharpoonup z$ in $H^1_1(G)$ and let $\{z_r\}$ be an arbitrary sub-sequence of $\{z_n\}$. From Theorem 3.4.4 (applied to each smooth domain $D \subset\subset G$), it follows that $z_s(x) \to z(x)$ almost everywhere for some subsequence $\{z_s\}$ of $\{z_r\}$. If v^i, $i = 1, \ldots, P$, are arbitrary bounded functions, then

$$a^{\alpha}_{ij}[x, z_s(x)]\, v^i(x), \quad b_{ij}[x, z_s(x)]\, v^i(x), \quad c_i[x, z_s(x)]\, v^i(x)$$

converge almost everywhere and boundedly to their limits. By replacing z in (1.9.5) by z_s, multiplying by v^i and summing, integrating the result, and passing to the limit using the weak convergence, we find that $z \in F$.

Remark 4. If z is the vector function defined by

$$z^i_{\alpha}(x) = D^{\alpha} w^i(x), \quad 0 \le |\alpha| \le m_i - 1, \quad i = 1, \ldots, N,$$

the integral $I(z, G)$ is equal to an integral $J(w, G)$ of a variational problem involving derivatives of order $\le m_i$ of w^i. Then Theorem 1.9.1 for I implies a corresponding theorem for J. In fact the existence theorem of I. GEL'MAN can be deduced immediately in this way.

In order to obtain more meaningful boundary value problems, it is convenient to restrict ourselves to Lipschitz domains G. Using the general compactness theorems of § 3.4 we can prove the following boundedness lemma (by contradiction):

Lemma 1.9.2 (cf. Theorem 3.6.4). *Suppose G is Lipschitz, σ is an open subset of G, and τ is an open subset of ∂G. Then there are constants* $C_1(G, \sigma, r, \nu)$ *and* $C_2(G, \tau, r, \nu)$ (depending only on the quantities indicated) *such that*

$$\left.\begin{aligned} \|z\|^1_{r,G} &\le C_1 \cdot (\|\nabla z\|^0_{r,G} + \|z\|^0_{1,\sigma}) \\ \|z\|^1_{r,G} &\le C_2 \cdot (\|\nabla z\|^0_{r,G} + \|z\|^0_{1,\tau}) \end{aligned}\right\} \begin{cases} z \in H^1_r(G), \\ r \ge 1. \end{cases}$$

Remarks. This lemma implies that if F is any family of functions z in $H^1_r(G)$ for which the $L_r(G)$ norms $\|\nabla z\|^0_{r,G}$ of the first derivatives are bounded, then the L_r norms of the functions will be bounded if their L_1-norms either over a fixed open subset σ of G or a fixed open subset τ

of ∂G are bounded. The family F^* could then be replaced by a compact (in $L_r(\partial G)$) family of functions really defined only on ∂G. As an example of the use of our general existence theorem in connection with Lemma 1.9.2, we mention the problem of Plateau for a surface of least area which has a given arc Γ, with end points on a manifold M, as part of its boundary, the rest of its boundary being required to lie on M. On account of conformal mapping (see Chapter 9), it is sufficient to minimize the Dirichlet integral among all $z \in H_2^1[B(0,1)]$ whose boundary values along $\partial^+ B(0,1)$ (the closed upper semicircle) are continuous and give a $1-1$ continuous parametric representation of Γ, in which the point $(0,1)$ corresponds to a fixed point of Γ, and whose (possibly discontinuous) values on $\partial^- B(0,1)$ lie on M (almost everywhere). It turns out (see § 9.3) that the mappings from $\partial^+ B(0,1)$ to Γ are equicontinuous for any minimizing sequence. Since they are uniformly bounded $\| z_n \|_{2,B}^1$ are uniformly bounded and a subsequence, still called $\{z_n\}$ converges weakly in $H_2^1[B(0,1)]$ and uniformly along $\partial^+ B(0,1)$ to a minimizing function z. However, z (harmonic and conformal inside $B(0,1)$ may not be continuous along $\partial^- B(0,1)$ if M is allowed to have edges, as has been shown in an example of Courant ([2], [3]. However see, LEWY [2]).

Remarks. The theorems of this section and the preceding one can be carried over to integrals involving derivatives of higher order of the form

$$I(z, G) = \int_G f[x, L(z), M(z)]\, dx$$

where $M(z)$ denotes all the derivatives $D^\alpha z^i$ of highest order where $|\alpha| = m_i$ and $L(z)$ denotes all those derivatives where $0 \leq |\alpha| < m_i$. It is assumed that f is continuous in its arguments and convex and differentiable in the set of arguments $M(z)$. There are essentially no differences in the proofs. Of course each vector z in F (Theorem 1.9.1) would be such that $z^i - z^{i*} \in H_{10}^{m_i}(G)$ for some z^* in F^*, $f_0 = f_0[M(z)]$, etc. FICHERA has observed that *lower-semicontinuity theorems* for such integrals can be obtained when $M(z)$ consists only of *some combinations* of the derivatives of highest order and $L(z)$ consists only of *some* of those of lower order.

1.10. The differentiability theory. Introduction

Of perhaps greater interest than the existence theory is the theory of the differentiability of the solutions. In this chapter, we shall confine ourselves to the non-parametric case; we shall discuss the parametric case in Chapters 9 and 10.

The first result about differentiability was that of LICHTENSTEIN who proved in 1912 [1] that a solution z of class C'' of a regular double integral problem ($\nu = 2, N = 1$) in which f is analytic is of class C'''

and hence analytic by the famous theorem of S. BERNSTEIN ([1]). The same conclusion was shown to hold (a) if z is of class C^1 with Hölder continuous derivatives by E. Hopf in 1929 ([2]) and (b) if z is Lipschitz by thewriter in 1938 (MORREY [4]). Using the latter result, it follows that the Lipschitz solutions (mentioned above) obtained by A. Haar for the case $f = f(p)$, $\nu = 2$, $N = 1$, G strictly convex, are analytic if f is. But except for problems whose Euler equations are linear and certain integrals where f is of the form

$$f(x, z, p) = a(x, z) \, |p|^2 + 2\sum_{\alpha=1}^{2} b^\alpha(x, z) p_\alpha + c\,(x, z) \; (\nu = 2, N = 1)$$

treated by HIRSCHFELD there were no other results in which the solutions which were shown to exist were shown to be differentiable until the work of the writer ([6], [7] see footnote on p. 17) in which the entre existence and differentiability program was carried through for essentially the class of problems in which f satisfies the conditions (1.10.8) below with $\nu = 2$, N arbitrary, and $k = 2$; although N was allowed to be > 1, the convexity hypothesis on f was retained, that is

$$m_1 \sum_{i,\alpha} |\pi^i_\alpha|^2 \le \sum_{i,j,\alpha,\beta} f_{p^i_\alpha p^j_\beta} (x, z, p) \pi^i_\alpha \pi^j_\beta \le M_1 \sum_{i,\alpha} |\pi^i_\alpha|^2, \quad 0 < m_1 \le M_1.$$

Later ([8]), the writer applied these results to prove the differentiability of his solutions of the problem of Plateau on a Riemann manifold. A greatly simplified version of this old work is to be found in [15], recent developments have permitted still further simplifications and some extensions.

The methods used in this differentiability theory with $\nu = 2$ would not generalize to larger values of ν and it was not until the recent results of DE GIORGI [1] and NASH [3], as simplified still more by MOSER that the existence and differentiability program could be carried through for problems in which $\nu > 2$. The methods of NASH and DE GIORGI were entirely different; NASH obtained his results as a by-product of his work on parabolic equations and confined himself to bounded solutions whereas DE GIORGI dealt only with elliptic equations but allowed solutions in L_2. Using MOSER's simplification of DE GIORGI's work, a student E. R. BULEY was able in the spring of 1960 to obtain the results stated in Theorem 1.10.4 (ii) for the cases where f satisfies (1.10.7) or (1.10.7') and the author was able to obtain those in (ii) (the (1.10.8) cases), (iii), and (iv). At about the same time, LADYZENSKAYA and URAL'-TSEVA ([1], [2], [3]) obtained their results stated in Theorem 1.10.4 (v) which include the author's results. These results are discussed further below and most of the details are proved in Chapter 5. An unfortunate feature of the DE GIORGI-NASH results is that they have been proved only for $N = 1$.

We now show how to prove differentiability in the case $\nu = 1$, N arbitrary. We do not attempt to prove the most general theorem.

Theorem 1.10.1. *Suppose that $\nu = 1$, that $f \in C^2$ everywhere, that f satisfies (1.7.1) with $k > 1$, and that there is a positive function $M_1(R)$ such that*

$$|f_p(x, z, p)|, |f_z(x, z, p)| \leq M_1(R)(1 + |p|^k) \text{ for } x^2 + |z|^2 \leq R^2. \tag{1.10.1}$$

Suppose z minimizes $I(z, G)$ $(G = (a, b))$ among all A.C. functions with the same boundary values. Then $z \in C^2[a, b]$ and satisfies Euler's equations.

Proof. Let ζ be any Lipschitz vector which vanishes at a and b. Since f satisfies (1.7.1), it follows that z is A.C. and $\|z'\|_k^0 < \infty$. Since ζ is bounded and z and ζ are absolutely continuous, we see from (1.10.1) that

$$\zeta^{i\prime}(x) \cdot f_{p^i}[x, z(x) + \lambda\zeta(x), z'(x) + t\zeta'(x)] + \zeta^i(x) \cdot f_{z^i}[x, z + \lambda\zeta, z' + \lambda\zeta']$$

is dominated for $|\lambda| \leq 1$ by the fixed summable function

$$2Q \cdot M_1(R)\,[1 + 2^{k-1}(Q^k + |z'(x)|^k)]$$

where Q is a bound for $|\zeta'(x)|$ and $|\zeta(x)|$ and R is one for $x^2 + |z + \lambda\zeta|^2$ for $|\lambda| \leq 1$; here we used the inequality

$$(a + b)^k \leq 2^{k-1}(a^k + b^k), \quad k \geq 1.$$

Thus the function $\varphi(\lambda)$ defined in (1.4.6) is of class C' for $|\lambda| < 1$ and $\varphi'(0) = 0$ so that (1.4.7) holds.

Now (1.4.7) is of the form

$$\int_a^b (\zeta^{i\prime} A_i + \zeta^i B_i)\, dx = 0, \quad \zeta^i(a) = \zeta^i(b) = 0 \tag{1.10.2}$$

where A_i and B_i are summable. Let π^i be an arbitrary bounded measurable vector such that

$$\int_a^b \pi^i(x)\, dx = 0 \tag{1.10.3}$$

Any Lipschitz ζ vanishing at a and b is of the form

$$\zeta^i(y) = \int_a^y \pi^i(x)\, dx = -\int_y^b \pi^i(x)\, dx \tag{1.10.4}$$

Breaking the integral (1.10.2) into the integrals of its two terms, replacing x by y in the second, then substituting (1.10.4) for ζ, interchanging orders of integration, and recombining, we conclude from (1.10.2) that

$$\int_a^b \pi^i(x) \left[A_i(x) - \int_a^x B_i(y)\, dy \right] dx = 0. \tag{1.10.5}$$

Since this holds for all π satisfying (1.10.3), it follows from a well-known lemma that

$$f_{p^i}[x, z(x), z'(x)] = A_i(x) = \int_a^x B_i(y)\, dy + C_i \quad (C_i = \text{const.}) \text{ a.e.}, \tag{1.10.6}$$

in which the right member is A.C. Since I is regular, the equations (1.10.6) can be solved for the $'z^i$ in terms of x, $z(x)$, and the integral. Thus $'z^i$ is equivalent to an A.C. function. Thus $'z^i$ is easily seen to be A.C. Hence both sides of (1.10.6) are A.C and the right side $\in C^1$. Consequently $z \in C^1$ so $z \in C^2$ and EULER's equations follow as usual.

For $\nu > 1$, the theory is not so simple of course. Rather than trying to present the most general conditions under which each part of the demonstration can be carried through, we state now two sets of conditions on the integral function f under which differentiability results have been obtained.

Common Condition. *$f \in C^2_\mu$ in its arguments or f and $f_p \in C^{n-1}_\mu$ for some $n \geq 3$ and some μ with $0 < \mu < 1$.*

The use of Hölder conditions in connection with elliptic differential equations is very common. The desirability of their use arises in potential theory (see Chapter 2).

Besides the common condition above, we require f to satisfy one of the following sets of conditions for all (x, z, p):

$$(1.10.7) \begin{cases} m V^k - K \leq f(x, z, p) \leq M V^k, \\ |f_p|^2 + |f_{px}|^2 + |f_z|^2 + |f_{zx}|^2 \leq M_1^2 V^{2k-2} \\ |f_{pz}|^2 + |f_{zz}|^2 \leq M_1^2 V^{2k-4} \\ m_1 V^{k-2} |\pi|^2 \leq \sum f_{p^i_\alpha p^j_\beta}(x, z, p) \pi^i_\alpha \pi^j_\beta \leq M_1 V^{k-2} |\pi|^2, \\ 0 < m \leq M, \ k > 1, \ V = (1 + |z|^2 + |p|^2)^{1/2}, \ 0 < m_1 \leq M_1, \end{cases}$$

$$|f_{pp}|^2 = \sum (f_{p^i_\alpha p^j_\beta})^2, \quad |f_{pz}|^2 = \sum (f_{p^i_\alpha z^j})^2, \quad |\pi|^2 = \sum (\pi^i_\alpha)^2, \quad \text{ect.}$$

(1.10.7′) Same as (1.10.7) but $f = f(x, p)$, $V = (1 + |p|^2)^{1/2}$.

$$(1.10.8) \begin{cases} m V^k - K \leq f(x, z, p) \leq M V^k, \\ |f_z|^2 + |f_{zx}|^2 + |f_{zz}|^2 \leq M(R) \, V^{2k} \\ |f_p|^2 + |f_{pz}|^2 + |f_{px}|^2 \leq M(R) \, V^{2k-2} \\ m_1(R) \, V^{k-2} |\pi|^2 \leq \sum f_{p^i_\alpha p^j_\beta} \pi^i_\alpha \pi^j_\beta \leq M_1(R) V^{k-2} |\pi|^2, \end{cases}$$

$$k \geq \nu, \ 0 < m \leq M, \ 0 < m_1(R) \leq M_1(R), \ V = (1 + |p|^2)^{1/2}$$
$$|x|^2 + |z|^2 \leq R^2$$

Remarks. We notice that (1.10.8) reduces to (1.10.7′) in case f does not depend on z (except for the R-condition which is somewhat meaningless if z is not present since our domains G are bounded). To see the difference between (1.10.7) and (1.10.8), we notce that the function f defined by

$$f(x, z, p) = [1 + a^{\alpha\beta}_{ij}(x, z) p^i_\alpha p^j_\beta]^{k/2}$$

satisfies the conditions (1.10.8) but not (1.10.7) if the $a^{\alpha\beta}_{ij} \in C^2_\mu$ or C^{n-1}_μ if $n \geq 3$ and the quadratic form is positive definite and bounded above and below in the obvious way.

The first step in the differentiability theory is to note continuity properties of the minimizing functions as follows:

Theorem 1. 10. 2. *In all cases if $k > \nu$ the minimizing functions are Hölder continuous on interior domains and at any boundary points in the neighborhood of which ∂G is Lipschitz. If $k = \nu$, the minimizing functions are Hölder continuous on interior domains and*

$$\left\{\int_{B(x_0,r)} V^\nu \, dx\right\}^{2/\nu} \leq C \cdot (\|V\|_\nu^1)^2 (r/R)^\mu \text{ if } B(x_0, r) \subset B(x_0, R) \subset G. \tag{1.10.9}$$

If $k = \nu = 2$, and f satisfies

$$m\,|p|^2 \leq f(x, z, p) \leq M\,|p|^2 \tag{1.10.8*}$$

G is bounded by a finite number of disjoint Jordan curves, $z^* \in H_2^1(G)$, and z^* is continuous on $\bar{G}$, then any minimizing function with $z - z^* \in H_{20}^1(G)$ is continuous on $\bar{G}$. If *$k = \nu$, G is Lipschitz, the boundary values are continuous, and f satisfies the supplementary conditions*

$$f(x, z, p) \geq 0,\ f(x, z, p^1, \ldots, p^{r-1}, 0, p^{r+1}, \ldots, p^N) = 0, \quad r = 1, \ldots, N (p^i = \{p_\alpha^i\},\ i \text{ fixed}) \tag{1.10.10}$$

then any minimizing function with those boundary values is continuous on $\bar{G}$.

Proof. If $k > \nu$ and $I(z, G)$ is finite, it follows from the first assumptions in (1.10.7) and (1.10.8) that the $H_k^1(G)$ norm of the minimizing function is finite. That z is continuous on the interior follows from the corollary to Theorem 3.5.1. The continuity on the boundary is obtained by first "flattening out" a piece of the boundary by a bi-Lipschitz map (see the definition in § 1.2) and then reflecting the function.

In the case $\nu = 2 = k$ where f satisfies (1.10.8*), the minimizing vector z satisfies

$$m\,D[z, B(x_0 R)] \leq I[z, B(x_0, R)] \leq I[H, B(x_0, R)] \leq M\,D[H, B(x_0, R)] \tag{1.10.11}$$

where H is the harmonic function coinciding with z on $\partial B(x_0, R)$. If we let $\varphi(r) = D[z, B(x_0, r)]$, take polar coordinates about x_0, expand z in a Fourier series

$$z = \frac{a_0(r)}{z} + \sum_n [a_n(r) \cos n\theta + b_n(r) \sin n\theta], \tag{1.10.12}$$

and use the formula (1.3.3) for $D[H, B_R]$, we find that

$$\varphi(R) = \pi \int_0^R r \left\{\frac{a_0'^2}{2} + \sum_n (a_n'^2 + b_n'^2) + \frac{n^2}{r^2}(a_n^2 + b_n^2)\right\} dr$$
$$\leq K\pi \sum_n n[a_n^2(R) + b_n^2(R)] \leq K \cdot R \cdot \varphi'(R), \quad K = M/m$$

from which (1.10.9) follows with $\mu = 1/K$.

The condition (1.10.11) holds, obviously, with $B(x_0, R)$ replaced by any other sub-domain of G, the resulting condition being invariant under conformal mappings. The two-dimensional results on continuity at the boundary involve this idea and are carried out in § 4.3. The proof above can be generalized to the general case $k = \nu$; the harmonic function H is replaced by the function

$$H(r, \theta) = \bar{z} + (r/R)^\lambda [z(R, \theta) - \bar{z}], \quad 0 \leq r \leq R,$$

where (r, θ) are spherical coordinates, θ denotes a point on $\partial B(0,1)$, and $\bar{z}$ is the average of $z(r, \theta)$ over $\partial B(0,1)$. The remaining results are proved in § 4.3.

We must prove following interesting theorem:

Theorem 1.10.3. *If f is of class C^2 and satisfies (1.10.7) or (1.10.7') for some $k > 2$, then $I(z) = I(z, G)$ is of class C^2 over $H_k^1(G)$. If, instead, f satisfies (1.10.8) for some $k > 2$, then $I(z)$ is of class C^2 over the space ${}^*H_k^1(G) = H_k^1(G) \cap C^0(\bar{G})$, the norm in this space being*

$$\max_{x \in G} |z(x)| + \| \nabla z \|_{k, G}^0 .$$

If we merely have $1 < k \leq 2$, then $I(z)$ is of class C^1 over the corresponding space. In either caes, if z_0 minimizes $I(z)$ among all z with given boundary values, then the first variation, i.e. the first differential $I_1(z_0, \zeta)$, is zero, where $I_1(z, \zeta)$ is defined by (1.4.7).

Proof. We shall prove only the first statement; the others are proved similarly. For almost all x, we may write

$$f(x, z + \zeta, p + \pi) = f(x, z, p) + \zeta^i \cdot f_{z^i} + \pi_\alpha^i f_{p_\alpha^i} + \frac{1}{2}\Big[f_{z^i z^j} \zeta^i \zeta^j +$$

$$+ 2 f_{p_\alpha^i z^j} \pi_\alpha^i \zeta^j + f_{p_\alpha^i p_\beta^j} \pi_\alpha^i \pi_\beta^j\Big] + \varepsilon_{ij}^{\alpha\beta} \pi_\alpha^i \pi_\beta^j + 2 \varepsilon_{ij}^\alpha \pi_\alpha^i \zeta^j +$$

$$+ \varepsilon_{ij} \zeta^i \zeta^j, \; \varepsilon_{ij}^{\alpha\beta} = \int_0^1 (1 - t) \left\{ f_{p_\alpha^i p_\beta^j} [x, z + t\zeta, p + t\pi] - f_{p_\alpha^i p_\beta^j} (x, z, p) \right\} dt$$

and ε_{ij}^α and ε_{ij} are given by similar formulas. It is clear that if $z_n \to z$ in $H_k^1(G)$, then $f_p[z_n] \to f_p[z]$ and $f_z[z_n] \to f_z[z]$ strongly in L_r[1], $r = k/(k-1)$, $f_{pp}[z_n] \to f_{pp}[z]$, $f_{pz}[z_n] \to f_{pz}[z]$, and $f_{zz}[z_n] \to f_{zz}[z]$ strongly in L_s, $s = k/(k-2)$, and the $\varepsilon_{ij}^{\alpha\beta}(z_s)$, etc., $\to 0$ strongly in L_s. Thus we see that

$$I(z + \zeta) = I(z) + I_1(z, \zeta) + \frac{1}{2} I_2(z, \zeta) + R(z, \zeta)$$

---

[1] To see this, let $\{q\}$ be any subsequence of $\{n\}$. Then there is a subsequence $\{s\}$ of $\{q\}$ such that $z_s(x) \to z(x)$ and $\nabla z_s(x) \to \nabla z(x)$ almost everywhere so that $f_p[z_s] \to f_p[z]$ for almost all x. We have $|f_p[z_s]|^r$ and $|f_z[z_s]|^r$ dominated by $C \cdot M_1 V_s^k(x)$ and the strong convergence of z_s to z in $H_k^1(G)$ implies the uniform absolute continuity of the set functions $\int_e V_s^k(x)\, dx$. A similar argument holds for f_{pp}, ε_{ij}, etc.

where I, I_1, and I_2 are continuous in both variables and

$$|R(z, \zeta)| \leq \varepsilon(z, \zeta) \cdot ||\zeta||^2$$

where $\varepsilon(z', \zeta) \to 0$ uniformly as $||\zeta|| \to 0$ for $||z' - z|| \leq$ some $h > 0$.

Definition. Any vector $z \in H_k^1(G)$ for which $I_1(z, \zeta) = 0$ for all ζ as in Theorem 1.10.3 is called an *extremal.*

We can now state our principal results:

Theorem 1.10.4. *We assume that f is regular and satisfies the "common condition." Then:*

(i) *If $N = 1$ and z is Lipschitz and is an extremal, then $z \in C_\mu^n(D)$ for each $D \subset\subset G$.* (Recall the existence theorem with $f = f(p)$ and G strictly convex mentioned at the end of § 1.7).

(ii) *If f satisfies* (1.10.7) *or* (1.10.7') *with $N = 1$, ν arbitrary, and $k \geq 2$ or if f satisfies* (1.10.8) *with $N = 1$ and $k > \nu$, then every extremal $\in C_\mu^n(D)$ for all $D \subset\subset G$.*

(iii) *If f satisfies* (1.10.8) *with $N = 1$, ν arbitrary, and $k = \nu$ or with N arbitrary and $k = \nu = 2$, then any extremal which $\in C_\mu^0(D)$ for $D \subset\subset G$ and which satisfies the Dirichlet growth condition* (1.10.9) $\in C_\mu^n(D)$ *for $D \subset\subset G$.*

(iv) *If f satisfies* (1.10.7) *or* (1.10.7') *with $N = 1$, ν arbitrary, and $1 < k < 2$, and if $z^* \in H_k^1(G)$, then there exists an extremal $z \in C_\mu^n(D)$ for all $D \subset\subset G$ and which minimizes $I(Z, G)$ among all Z such that $Z - z^* \in H_{k0}(G)$.*

(v) *If f satisfies* (1.10.8) *with $N = 1$, ν arbitrary, and $k > 1$, then any bounded extremal $\in C_\mu^n(D)$ for all $D \subset\subset G$.*

(vi) *In all cases above, the extremal $\in C^\infty(D)$ for $D \subset\subset G$ if $f \in C^\infty$ and is analytic on G if f is analytic.*

Remark 1. The restriction $N = 1$ limits the applicability of the results very seriously. The removal of this restriction would not only increase the applicability of our results in such fields as Differential Geometry and Topology but is essential for the proof of the differentiability of the solutions of variational problems of higher order. The removal of this restriction involves the extension of the De Giorgi-Nash results to systems of equations.

Remark 2. If $f(x, z, p)$ is quadratic in z and p, the Euler equations are linear and more detailed results are available in this case.

The results in (iii) with N arbitrary and $k = \nu = 2$ are those obtained before the war (1937—38) by the author and referred to earlier. The results in (ii)—(iv) other than those just mentioned, were obtained early in 1960 in conjunction with a student (E. R. Buley). The remarkable results in (v) were obtained at about the same time by Ladyzenskaya and Ural'tseva ([1], [2], [3]). Their results include those of ours in (ii)—(iv) above which refer to the cases (1.10.8) with $N = 1$ and $k \geq \nu$ since we have seen in Theorem 1.10.2 above that the solutions are continuous in these cases. The theory of LADYZENSKAYA and URAL'TSEVA is discussed briefly in § 5.11; they have treated these and other details in their recent book [3]. However, the results in (ii) and (iv) in the cases (1.10.7) and (1.10.7′) are of interest because the continuity of the solutions doesn't follow from the existence theory in the cases where $1 < k < \nu$.

Remarks. The functions z which we are calling extremals are also called *weak solutions* of the Euler equations

$$\frac{\partial}{\partial x^\alpha} f_{p_\alpha^i} = f_{z^i}.$$

It was noticed by LADYZENSKAYA and URAL'TSEVA that identical proofs could be used to prove the differentiability of weak solutions of equations of the form

$$\frac{\partial}{\partial x^\alpha} A_i^\alpha = B_i, \quad A_i^\alpha = A_i^\alpha(x, z, p), \quad B_i = B_i(x, z, p); \tag{1.10.13}$$

that is, solutions z of the equations

$$\int_G (\zeta_\alpha^i A_i^\alpha + \zeta^i B_i)\, dx = 0 \tag{1.10.14}$$

for the ζ in Theorem 1.10.3. The equations (1.10.14) are obtained from (1.10.13) formally by multiplying (1.10.13) by ζ, integrating, and then integrating by parts; and, of course, the equations (1.10.13) are obtained from (1.10.14) formally by following the derivation of Euler's equations given above. The assumptions on the A_i^α and B_i corresponding to (1.10.7), (1.10.7′), and (1.10.8) are:

$$\left\{\begin{array}{l} |A|^2 + |A_x|^2 + |B|^2 + |B_x|^2 \leq M_1^2 V^{2k-2}, \\ |A_z|^2 + |A_p|^2 + |B_z|^2 + |B_p|^2 \leq M_1^2 V^{2k-4}, \\ m_1 V^{k-2} |\pi|^2 \leq A_{i p_\beta^j}^\alpha (x, z, p) \pi_\alpha^i \pi_\beta^j, \\ m_1 > 0, \; k > 1, \; V = (1 + |z|^2 + |p|^2)^{1/2} \end{array}\right. \tag{1.10.7''}$$

(1.10.7‴) The same as (1.10.7″) with $A = A(x, p)$, $B = 0$, $V = (1 + |p|^2)^{1/2}$, $k > 1$.

$$\left\{\begin{array}{l} |B|^2 + |B_z|^2 + |B_x|^2 \leq M_1^2(R)\, V^{2k} \\ |A|^2 + |A_z|^2 + |A_x|^2 + |B_p|^2 \leq M_1(R)\, V^{2k-2} \\ m_1(R)\, V^{k-2} |\pi|^2 \leq A_{i p_\beta^j}^\alpha \pi_\alpha^i \pi_\beta^j, \quad |A_p| \leq M_1(R)\, V^{k-2} \\ m_1(R) > 0, \; V = (1 + |p|^2)^{1/2}. \end{array}\right. \tag{1.10.8''}$$

In all cases, it is assumed that the A_i^α and $B_i \in C_\mu^1$ if $n = 2$ or the $A_i^\alpha \in C_\mu^{n-1}$ and $B_i \in C_\mu^{n-2}$ if $n \geq 3$. And, of course, it is not assumed that $A_{i p_\beta^j}^\alpha = A_{j p_\alpha^i}^\beta$.

Theorem 1.10.4. *With these assumptions, the results* (ii), (iii), (v), *and* (vi) *hold if the assumption that* z *be an extremal be replaced by that that it be a weak solution in* $H_k^1(G)$ *of* (1.10.13) *and satisfy the additional conditions mentioned in the cases* (iii) *and* (v).

LADYZENSKAYA and URAL'TSEVA ([1], [2], [3]) and GILBARG and others have obtained existence theorems for equations of the type (1.10.13). But in the last two or three years, VIŠIK, MINTY, BROWDER ([3]) and LERAY-LIONS ([68)] have developed an existence theory for non-linear equations in Banach spaces which covers a wide class of equations including many of higher order and *all equations of the type* (1.10.14) *in which the* A_i^α *and* B_i *satisfy* (1.10.7″) *or* (1.10.7‴). Their theory yields solutions in $H_k^1(G)$ (or corresponding spaces in the higher order cases) rather than classical solutions and therefore takes the place of the existence theory rather than the differentiability theory. However, combined with our regularity results we *now have an existence theorem for any equation of the form* (1.10.13) *which satisfies the* (1.10.7″) *or* (1.10.7‴) *conditions with* $N = 1$. The relevant abstract theorem of Leray and Lions is stated and proved in § 5.12 where this existence theorem is proved.

1.11. Differentiability; reduction to linear equations

In this section we wish to give some indication as to how one goes about proving the results on differentiability which we stated in Theorem 1.10.4. We begin by applying a difference quotient process to the equations (1.10.14) in which we regard the solution as known. We shall assume first that the A_i^α and B_i satisfy (1.10.7″) with $k \geq 2$; we shall indicate the modifications for the case (1.10.8″).

To do this we choose an integer γ, $1 \leq \gamma \leq \nu$, let e_γ be the unit vector in the x^γ direction and define

$$\zeta_h^i(x) = h^{-1}[\zeta^i(x - h\,e_\gamma) - \zeta^i(x)], \quad z_h^i(x) = h^{-1}[z^i(x + h\,e_\gamma) - z^i(x)], \quad 0 < |h| < a \tag{1.11.1}$$

where ζ has support in a domain $D' \subset\subset G_a$. If we replace ζ by ζ_h in (1.10.14), make a change of coordinates x in the terms containing $\zeta(x - h\,e_\gamma)$ or $\nabla\zeta(x - h\,e_\gamma)$, we obtain the equation

$$\int_{D'} h^{-1}[\zeta_{,\alpha}^i \Delta A_i^\alpha + \zeta^i \Delta B_i]\,dx = 0, \tag{1.11.2}$$

$$\Delta A_i^\alpha = A_i^\alpha[x + h\,e_\gamma,\ z(x + h\,e_\gamma),\ \nabla z(x + h\,e_\gamma)] - A_i^\alpha[x, z(x),\ \nabla z(x)]$$

and ΔB_i is given by a similar formula. Now for each fixed h, $0 < |h| < a$, the two terms in ΔA_i^α and ΔB_i are defined and measurable for almost all x and for such x can be expressed in terms of the difference quotients $\Delta z^i/h$ and their derivatives using the integral form of the Theorem of the

Mean. If this is done (1.11.2) becomes

$$\int_{D'} A_h(\zeta^i_{,\alpha}[a^{\alpha\beta}_{hij} z^j_{h,\beta} + b^\alpha_{hij} z^j_h + e^{\alpha\gamma}_{hi} P_h] + \zeta^i[c^\alpha_{hij} z^j_{h,\alpha} + d_{hij} z^j_h + f^\gamma_{hi} P_h])\,dx$$
$$(1.11.3) \qquad = 0$$

where

$$A_h(x) = \int_0^1 [1 + |z(x) + t\Delta z|^2 + |p(x) + t\Delta p|^2]^{-1+k/2}\,dt$$

$$A_h(x)\, a^{\alpha\beta}_{hij}(x) = \int_0^1 A^\alpha_{i\, p^j_\beta}[x + t h e_\gamma,\ z(x) + t\Delta z,\ p(x) + t\Delta p]\,dt$$

$$A_h b^\alpha_{hij} = \int_0^1 A^\alpha_{i z^j}\,dt,\quad A_h e^{\alpha\gamma}_{hi} P_h = \int_0^1 A^\alpha_{i x^\gamma}\,dt$$

$$(1.11.4) \quad A_h P_h = \int_0^1 [1 + |z(x) + t\Delta z|^2 + |p(x) + t\Delta p|^2]^{(k-1)/2}\,dt$$

$$A_h c^\alpha_{hij} = \int_0^1 B_{i p^j_\alpha}\,dt,\quad A_h d_{hij} = \int_0^1 B_{i z^j}\,dt$$

$$A_h f^\gamma_{hi} P_h = \int_0^1 B_{i x^\gamma}\,dt,\quad p(x) = \nabla z(x)$$

$$\Delta z = z(x + h e_\gamma) - z(x),\quad \Delta p = p(x + h e_\gamma) - p(x).$$

In this case ($k \geq 2$), we see from (1.10.7″) that the functions A_h etc., are all measurable, the coefficients a_h, b_h, c_h, d_h, e_h, and f_h are all uniformly bounded (independently of h) and

$$(1.11.5) \quad m_1|\pi|^2 \leq a^{\alpha\beta}_{hij}(x)\pi^i_\alpha \pi^j_\beta,\ \sum [a_{hij}(x)]^2 \leq M_1^2,\ A_h(x) \geq 1$$

for all h with $0 < |h| < a$.

Next, we let $\qquad \zeta^i = \eta w^i,\ w = \eta z^i_h$

where $\eta = 1$ on D which we assume $\subset D'_a$, $\eta = 1 - 2a^{-1} d(x, D)$ for $0 \leq d(x, D) \leq a/2$, and $\eta = 0$ otherwise. Then we have

$$\zeta^i_{,\alpha} = \eta(w^i_{,\alpha} + \eta_{,\alpha} z^i_h),\ \eta z^j_{h,\beta} = w^j_{,\beta} - \eta_{,\beta} z^j_h.$$

If these ζ^i are substituted into (1.11.3), one obtains

$$\int_{D'} A_h\{(w^i_{,\alpha} + \eta_{,\alpha} z^i_h)\,[a^{\alpha\beta}_{hij}(w^j_{,\beta} - \eta_{,\beta} z^j_h) + b^\alpha_{hij} w^j + \eta e^{\alpha\gamma}_{hi} P_h] +$$
$$(1.11.6) \qquad + w^i[c^\alpha_{hij}(w^j_{,\alpha} - \eta_{,\alpha} z^j_h) + d_{hij} w^j + \eta f^\gamma_{hi} P_h]\}\,dx = 0.$$

Using (1.11.5), the Schwarz inequality, and the Cauchy inequality ($2|a\,b| \leq \varepsilon a^2 + \varepsilon^{-1} b^2,\ \varepsilon > 0$), (1.11.6) yields

$$(1.11.7) \quad \int_{D'} A_h |\nabla w|^2\,dx \leq C \int_{D'} A_h[|\nabla\eta|^2 |z_h|^2 + |w|^2 + \eta^2 P_h^2]\,dx.$$

From the definition of η, we then conclude that

$$(1.11.8) \quad \int_D A_h |\nabla z_h|^2\,dx \leq C \int_{D'} A_h[(a^{-2} + 1)|z_h|^2 + P_h^2]\,dx.$$

But, from Lemma 3.4.2 and Theorem 3.6.8, it follows that $A_h \to V^{k-2}$ in $L_r(D')$, $z_h \to z_{,\gamma}$ in $L_k(D')$ and $P_h \to V$ in $L_k(D')$ so that the right side of (1.11.8) is bounded independently of h. A straightforward argument, given in § 5.9, shows that we may let $h \to 0$ (through a subsequence) to obtain the following theorem:

Theorem 1.11.1. *Suppose that $z \in H_k^1(G)$ and is a weak solution of* (1.10.14), *where the A_i^α and B_i satisfy* (1.10.7″) *or* (1.10.7‴) *with $k \geq 2$. Then z and $U = V^{k/2} \in H_2^1(D)$ for each $D \subset\subset G$ and the vectors p_γ satisfy*

$$\int_D V^{k-2}\{\zeta_\alpha^i(a_{ij}^{\alpha\beta} p_{\gamma,\beta}^j + b_{ij}^\alpha p_\gamma^j + e_i^{\alpha\gamma} V) + \zeta^i(c_{ij}^\alpha p_{\gamma,\alpha}^j + d_{ij} p_\gamma^j + f_i^\gamma V)\}dx = 0 \tag{1.11.9}$$

$$\int_D V^{k-2}|\nabla p|^2 dx \leq C a^{-2} \int_D V^k dx, \quad D \subset D_a \subset\subset G, \quad \zeta \in Lip_c(D) \tag{1.11.10}$$

where

$$\begin{aligned} &V^{k-2}(x)\, a_{ij}^{\alpha\beta}(x) = A_{i v_\beta^j}^\alpha[x, z(x), p(x)], \\ &V^{k-2} b_{ij}^\alpha = A_{i z^j}^\alpha, \quad V^{k-1} e_i^{\alpha\gamma} = A_{i x^\gamma}^\alpha, \quad V^{k-2} c_{ij}^\alpha = B_{i p_\alpha^i}, \\ &V^{k-2} d_{ij} = B_{i z^i}, \quad V^{k-1} f_i^\gamma = B_{i x^\gamma}. \end{aligned} \tag{1.11.11}$$

In the case (1.10.8″) with $k \geq \nu$, we can prove in essentially the same manner **Theorem 1.11.1′** *which is the same as Theorem* 1.11.1 *except that b_{ij}^α, c_{ij}^α and f_i^γ must be replaced respectively by $b_{ij}^\alpha \cdot V$, $c_{ij}^\alpha V$, and $f_i^\gamma V$ and d_{ij} must be replaced by $V^2 d_{ij}$ and we must assume that z satisfies* 1.10.9). *Moreover we conclude also that*

$$\int_D V^{k+2} dx \leq C a^{-2} \int_{D'} V^k dx. \tag{1.11.12}$$

The proof in this case is more difficult but is simplified by using a lemma of Ladyzenskaya and Ural'tseva [2] (see Lemma 5.9.1). There is no result corresponding to (iv) of Theorem 1.10.4 for the variational problem. In the case of the variational problem we have the following result:

Theorem 1.11.1″. *Suppose that f satisfies* (1.10.7) *or* (1.10.7′), *with $1 < k < 2$, and $z^* \in H_k^1(G)$. Then there is a z which minimizes $I(Z, G)$ among all Z such that $Z - z^* \in H_k^1(G)$ and which is such that its derivatives $\in H_k^1(D)$ and satisfy* (1.11.9) *and* (1.11.10) *on each $D \subset\subset G_a$ and the function $U = V^{k/2} \in H_2^1(D)$ for each such D.*

The difficulty in the proof of this arises from the fact that A_h always ≤ 1 and this causes trouble in the difference-quotient procedure. To get around this we minimize $I(z, G)$ with $D(z, G) \leq K$. For each K, we obtain certain bounds independent of h and so we may let $h \to 0$. Then we may allow $K \to \infty$. The special argument used in this proof is presented in § 5.10.

In all cases, the next theorem to be proved is the following:

Theorem 1.11.2. *Suppose that $N = 1$ and that f or the A^α and B and z satisfy the hypotheses of Theorems* 1.11.1, 1.11.1′, *or* 1.11.1″. *Then the* *function* $W = U^\lambda$ *satisfies an inequality of the form*

$$\int_G [\zeta_{,\alpha}(a^{\alpha\beta} W_{,\beta} + B^\alpha P W) + \zeta(C^\alpha P W_{,\alpha} + D P^2 W)]\,dx \leq 0 \tag{1.11.13}$$
$$\text{if } \zeta \in Lip_c(G),\ \zeta(x) \geq 0 \text{ on } G,$$

for some λ, $1 \leq \lambda < 2$, *where the* $a^{\alpha\beta}$ *satisfy* (1.11.5) $(N = 1)$, *the coefficients* B, C *and* D *are bounded and measurable, and* $P \in L_\nu(G)$ *and satisfies*

$$\left\{\int_{B(x_0,r)\cap\Delta} P^\nu\,dx\right\}^{2/\nu} \leq Z r^\varrho, \quad D \subset\subset G \tag{1.11.14}$$

where Z *depends on* Δ, G, *and* $\|P\|^0_{\nu,G}$.

This is proved in § 5.9. The idea is to set

$$\zeta = \psi V^{-\varepsilon} p_\gamma, \quad \psi \in Lip_c(G), \quad \psi(x) \geq 0 \text{ on } G,$$

in equations (1.11.9) (technical lemmas allow this). It turns out to be possible to choose ε small enough so that $W = V^{k-\varepsilon} = U^\lambda$ satisfies (1.11.13). In the cases (1.10.7), etc., $P = 1$ whereas $P = V$ in the cases (1.10.8). The inequality (1.11.14) follows in the cases (1.10.8), etc., from the Hölder inequality in case $k > \nu$ and from Theorem 1.10.2 in the case $k = \nu$. The idea of studying the solutions of inequalities like (1.11.13) is basic to MOSER's simplification [1] of the DE GIORGI-NASH theory. This is presented in §§ 5.3 and 5.4.

But by combining the results of Theorems 1.11.1, 1.11.1′, 1.11.1″, and 1.11.2, we conclude that the function U satisfies the hypotheses of Theorem 5.3.1 and hence U and therefore z and all the p_γ are bounded on each domain $D \subset\subset G$. Then the equations (1.11.9) take the form (5.3.21). It follows from the theorems in § 5.3 that the p_γ are Hölder-continuous on interior domains. Then the equations (1.11.9) take the form (5.2.2) with Hölder-continuous coefficients and it follows from Theorem 5.5.3 that the derivatives of the p_γ are Hölder-continuous on interior domains from which it follows that $z \in C^2_\mu(G)$.

The results under (v) of Ladyzenskaya and Ural'tseva for bounded weak solutions, when $1 < k < \nu$, are somewhat more difficult and are presented in § 5.11. But as soon as the first derivatives are seen to be bounded on interior domains, the proof that $z \in C^2_\mu$ in the case $n = 2$ is the same as that sketched above.

Once the solution $(N = 1)$ z is known to $\in C^2_\mu(D)$ for each $D \subset\subset G$, it then follows that z satisfies the equation (1.10.13) which becomes

$$\begin{gathered} a^{\alpha\beta}(x, z, \nabla z) z_{,\alpha\beta} = g(x, z, \nabla z) \\ a^{\alpha\beta}(x, z, p) = \tfrac{1}{2}(A^\alpha_{p_\beta} + A^\beta_{p_\alpha}), \quad g = B - A^\alpha_z p_\alpha - A^\alpha_{x^\alpha} \end{gathered} \tag{1.11.15}$$

where in the case of a variational problem, this is Euler's equation and

$$A^\alpha = f_{p_\alpha}, \quad B = f_z.$$

If f satisfies the "common condition" with $n \geq 3$, i.e. if f and $f_p \in C_\mu^{n-1}$, in the variational case, or if $A \in C_\mu^{n-1}$ and $B \in C_\mu^{n-2}$ in general, it follows that the $a^{\alpha\beta}$ and $g \in C_\mu^{n-2}$ and the differentiability of z stated in Theorem 1.10.4 follows from a repeated application of Theorem 5.6.3 as follows: Since $z \in C_\mu^2(D)$ for each $D \subset\subset G$, it follows that $a^{\alpha\beta}$ and $g \in C_\mu^1(D)$ for such D so that $z \in C_\mu^3(D)$ for such D by Theorem 5.6.3. Then $a^{\alpha\beta}$ and $g \in C_\mu^2(D)$ for such D and hence $z \in C_\mu^4(D)$ for such D. The result follows by induction. The C^∞ result follows but the analyticity requires a separate proof. The analyticity proof for this case ($N = 1$) is presented in § 5.8 where references are given.

It is to be noticed that the first theorem above which does not hold for systems is Theorem 1.11.2 (see the remark after the proof of Theorem 1.11.2 in § 5.9). This enables us to show that solutions are Lipschitz on interior domains. But even if this could be proved, the De Giorgi-Nash-Moser theory (presented in §§ 5.3 and 5.4) has not been extended to systems so we would still be unable to prove the Hölder continuity of the p_γ, or even their continuity.

In case a part of ∂G is smooth and the boundary values are smooth along that part, there is a gap in the differentiability results. By making a correspondingly smooth (C_μ^n, $n \geq 1$) change of variables and subtracting off a smooth function having the given boundary values, one reduces the equation (1.10.14) (or (1.10.13)) to one of the same type in which one works in a hemisphere and the solution is supposed to vanish (in some H_{k0}^1 sense) on the flat part σ_R ($x^\nu = 0$) of the hemisphere G_R (see § 1.2). The transformed equation or system has the same properties as the original except possibly for the bounds. One then carries out the difference quotient procedure of Theorem 1.11.1 in the *tangential directions* and shows that the $p_\gamma \in H_2^1(G_r)$ for $r < R$ and satisfy the equations (1.11.9), $\gamma = 1, \ldots, \nu - 1$. Since it is true that $p_{\nu,\gamma} = p_{\gamma,\nu}$ for $\gamma < \nu$, all the derivatives $p_{\nu,\gamma}$ with $\gamma < \nu \in L_2(G_r)$. Since we already have interior differentiability we may solve the equations (1.10.13) for the $z^i_{,\nu\nu} = p^i_{\nu,\nu}$ in terms of the others. Thus all the $p_{\gamma,\nu}$, $\gamma = 1, \ldots, \nu$, satisfy (1.11.9) and (1.11.10) follows. However, Theorem 1.11.2 has not been shown, in all cases, to hold in such a boundary neighborhood. But the following theorems cover what is known to the author about the boundary behavior of solutions:

Theorem 1.11.3. *Suppose a portion γ of ∂G, containing a point P_0, is of class C_μ^n, $n \geq 1$, suppose that the A_i^α and B_i satisfy the common condition and any one of the sets* (1.10.7″), (1.10.7‴), *or* (1.10.8″) *for some $\nu \geq 2$ and $k > 1$, and suppose that $z \in H_k^1(G)$ and is a solution of* (1.10.14). *If, in addition, $N = 1$, z satisfies a Lipschitz condition in a neighborhood $\Gamma \cup \gamma$ ($\Gamma \subset G$) of P_0 on $\bar{G}$ and vanishes along γ, then z is of class C_μ^n in any smaller neighborhood $\mathfrak{N}$ of P_0 on $\bar{G}$. If $N > 1$, the same conclusion holds if* *we know that* $z \in C^1(\Gamma \cup \gamma)$. *If* γ, A, *and* $B \in C^\infty$ *or analytic, respectively, so is* z *on* $\mathfrak{N}$.

Theorem 1.11.4. *Suppose* G *is bounded,* $\varphi \in H_2^1(G)$, $N = 1$, $B = 0$, *and the* $A^\alpha = A^\alpha(p) \in C^1$ *and satisfy the inequalities* (1.10.7‴) *with* $k = 2$. *Then there exists a unique solution* z *of Equation* (1.10.14) *such that* $z - \varphi \in H_{20}^1(G)$. *If* G *is of class* C^2 *and* $\varphi \in C^2(\bar{G})$ *then* $z \in H_r^2(G)$ *for any* r. *If the* $A^\alpha \in C_\mu^1$, G *is of class* C_μ^2 *and* $\varphi \in C_\mu^2(\bar{G})$, *then* $z \in C_\mu^2(\bar{G})$. *If we merely know that the* $A^\alpha(p) \in C_\mu^1$ *and satisfy the condition*

$$A^\alpha_{p_\beta}(p)\,\lambda_\alpha\lambda_\beta > 0 \quad \text{if} \quad \lambda \neq 0, \quad p \in R_\nu$$

and if G *is of class* C_μ^2 *and* $\varphi \in C_\mu^2(\bar{G})$ *then there exists a unique solution* $z \in C_\mu^2(\bar{G})$ *of Equation* (1.10.13) *such that* $z = \varphi$ *on* ∂G.

This is just a combination of Lemma 4.2.5 and Theorem 4.2.3. Additional boundary value results are found in Theorems 4.2.1 and 4.2.2.

For $N = 1$ and the p_γ all bounded, the equations (1.11.9) take the form (5.3.21) with G replaced by G_R, assuming that the above-mentioned transformations have been made. Since each p_γ with $\gamma < \nu$ vanishes along σ_R, it follows from Lemma 5.3.5 that these $p_\gamma \in C_\mu^0(G_r)$ for each $r < R$ and, in fact these p_γ each satisfy the "Dirichlet growth" condition (5.3.27). Since the derivatives of p_ν are all determined as above from those of the other p_γ, it follows that p_ν also satisfies such a condition and hence $\in C_\mu^0(\bar{G}_r)$ for each $r < R$ (Theorem 3.5.2). The C^∞ results are evident and the analyticity is proved in § 5.8. The results for $N > 1$ follow from those in §§ 6.8, 6.4, and 6.7.

Chapter 2

Semi-classical results

2.1. Introduction

In this chapter, we begin by proving some of the elementary properties of harmonic functions. A proof of Weyl's lemma (Weyl [2]) is inserted in § 2.3 for later reference; the proof is included at that point since it is closely related to the mean value property. Then in § 2.4 the classical notions of Green's functions and elementary functions are introduced and these notions and Poisson's integral formula for the circle and halfplane are carried over to the ν-dimensional case. In § 2.5, the study of potential functions is begun and the formulas for and continuity properties of their first derivatives are derived. In § 2.6, the formulas for and continuity properties of the first derivatives of certain "generalized potential functions" are studied and the Hölder continuity of the second derivatives of ordinary potentials of Hölder continuous density functions follow from the general results; an example of a continuous density function whose potential is not everywhere of class C^2 is given. In § 2.7 we present a proof of the now famous inequalities of CALDERON and ZYGMUND ([1]) and ([2]) for singular integrals; we confine ourselves to reasonably smooth functions in order to retain the essential simplicity of their proofs. The chapter concludes with the proof of the maximum principle for certain second order elliptic equations which was given by E. HOPF ([1]).

2.2. Elementary properties of harmonic functions

Definition 2.2.1. u is *harmonic* on $G \longleftrightarrow u \in C^2(G)$ and

$$\Delta u(x) \equiv \sum_{\alpha=1}^{\nu} u_{,\alpha\alpha}(x) = 0, \quad x \in G.$$

Theorem 2.2.1. *If u is harmonic on G, $D \subset\subset G$, D is class C^1, and $\overline{B(x_0, R)} \subset G$, then*

$$\int_{\partial D} u_{,\alpha}\, dx'_\alpha = 0 \tag{2.2.1}$$

$$u(x_0) = \Gamma_\nu^{-1} \int_{\partial B(x_0, R)} u(x)\, d\Sigma \tag{2.2.2}$$

$$u(x_0) = |B(x_0, R)|^{-1} \int_{B(x_0, R)} u(x)\, dx. \tag{2.2.3}$$

Proof. The first follows from GREEN's theorem. If $D = B(x_0, r)$, $0 < r < R$, (2.2.1) takes the form

$$\int_{\partial B(x_0, R)} u_r\, dS = 0 \tag{2.2.4}$$

where u_r denotes the radial derivative. Let (r, p) be polar coordinates with pole at x_0, $r = |x - x_0|$ and p on $\Sigma = \partial B(0,1)$. Then (2.2.4) is equivalent to

$$\int_{\Sigma} v_r(r, p)\, d\Sigma = 0, \quad v(r, p) = u(x_0 + r\zeta_p), \quad \zeta_p = \overrightarrow{0p}. \tag{2.2.5}$$

Integrating (2.2.5) from 0 to R yields (2.2.2). Using (2.2.2) with R replaced by r, we obtain (2.2.3) by multiplying both sides of (2.2.2) by $\Gamma_\nu r^{\nu-1}$ and integrating with respect to r from 0 to R.

Theorem 2.2.2. *Suppose u is harmonic on the domain G, $x_0 \in G$, u takes on its maximum value at x_0. Then $u(x) \equiv u(x_0)$.*

The proof follows easily from the mean value Theorem 2.2.1 and the connectedness of G. The following important uniqueness theorem follows immediately from the maximum principle:

Theorem 2.2.3. *There is at most one function u which is continuous on $\bar{G}$ and harmonic in G which takes on given continuous boundary values on ∂G* (G being a bounded domain).

Theorem 2.2.4. *If u is continuous and* (2.2.3) *holds for every $B(x_0, R) \subset\subset G$, then u is harmonic. A harmonic function has derivatives of all orders which are harmonic.*

Proof. Since (2.2.3) holds, we see that $u \in C^1(G^0)$ with

$$u_{,\alpha}(x) = |B(x_0, R)|^{-1} \int_{\partial B(x_0, R)} u(x)\, dx'_\alpha \text{ if } B(x_0, R) \subset\subset G, \quad \alpha = 1, \ldots, \nu.$$

But then by GREEN's theorem, we see that $u_{,\alpha}$ also satisfies (2.2.3). By induction, we see that all derivatives are continuous and satisfy (2.2.3).

So, suppose $x_0 \in G$. Expanding in TAYLOR's series, we obtain

$$\begin{aligned} u(x) - u(x_0) &= (x^\alpha - x_0^\alpha)\, u_{,\alpha}(x_0) + \\ &+ \frac{1}{2!}(x^\alpha - x_0^\alpha)(x^\beta - x_0^\beta)\, u_{,\alpha\beta}(x_0) + R(x_0, x) \end{aligned} \tag{2.2.6}$$

$$|R(x, x_0)| \leq M \cdot |x - x_0|^3 \text{ if } |x - x_0| \leq R, \quad B(x_0, R) \subset\subset G.$$

It follows, by integrating (2.2.6) over $B(x_0, r)$, dividing by $r^2\, |B(x_0, r)|$, using (2.2.3), and letting $r \to 0$, that $\Delta u(x_0) = 0$.

From this one easily obtains the following corollary:

Corollary. *If $u \in C^1(G)$ and satisfies* (2.2.4) (i.e. (2.2.1)) *for each sphere $B(x_0, r) \subset G$, then u is harmonic on G.*

For, by repeating part of the proof of Theorem 2.2.1, one concludes that (2.2.3) holds for each such $B(x_0, r)$.

Theorem 2.2.5. *Suppose each $u_n \in C^0(\bar{G})$ and is harmonic in G, and suppose that the sequence $\{u_n\}$ converges uniformly on $\bar{G}$ to a function u. Then $u \in C^0(\bar{G})$ and is harmonic on G.*

For each u_n satisfies (2.2.3) so the limit function does also.

The reader can prove the following inequalities:

Theorem 2.2.6. *If u is harmonic and in $L_p(G)$, then*

$$|u(x)|^p \leq |B(x, r)|^{-1} \int_{B(x, r)} |u(y)|^p\, dy \text{ if } B(x, r) \subset G.$$

Theorem 2.2.7. *If u is harmonic and $u \in L_2(G)$, then $|\nabla u(x)| \leq C(\nu)\, \|u\|_2^0\, \delta_x^{-1-\nu/2}$ where δ_x denotes the distance from x to ∂G.*

Theorem 2.2.8. *If u is harmonic on G, $|u(x)| \leq M$ there, δ_x denotes the distance of x from ∂G for $x \in G$, there exists a constant C, depending only on ν, such that*

$$|\nabla^k u(x)| \leq k!\, e^{k-1}\, C^k\, M\, \delta_x^{-k}, \quad k \geq 1.$$

2.3. Weyl's lemma

We begin with a generalization of the Lemma of du BOIS-RAYMOND:

Lemma 2.3.1. *Suppose $f \in L_1$ on the interval $[a, b]$ of R_1 and suppose that*

$$\int_a^b f(x)\, g(x)\, dx = 0 \tag{2.3.1}$$

for all $g \in C_c^\infty[a, b]$. *Then* $f(x) = 0$ *almost everywhere. If* (2.3.1) *holds only for all* $g \in C_c^\infty[a, b]$ *for which*

$$\int_a^b g(x)\,dx = 0, \tag{2.3.2}$$

then $f(x) = a$ *const. almost everywhere.*

Proof. In the first case, it follows by approximations that (2.3.1) holds for all g which are bounded and measurable on $[a, b]$ from which the first result follows immediately by setting $g(x) = sgn\, f(x)$. In the second case, let $g_1 \in C_c^\infty[a, b]$ with

$$\int_a^b g_1(x)\,dx = 1.$$

Then if g is any function $\in C_c^\infty[a, b]$, we have

$$g(x) = g^*(x) + g_1(x)\int_a^b g(y)\,dy, \quad \int_a^b g^*(x)\,dx = 0$$

and, of course, $g^* \in C_c^\infty[a, b]$. The second result follows.

Theorem 2.3.1 (WEYL's lemma, WEYL [2]). *Suppose that* $u \in L_1(D)$ *for each* $D \subset\subset G$ *and satisfies*

$$\int_G u(x)\,\Delta v(x)\,dx = 0 \tag{2.3.3}$$

for all $v \in C_c^\infty(G)$. *Then u is equivalent to a harmonic function.*

Proof. Suppose $\overline{B(x_0, a)} < G$, $0 < \varepsilon < a$, $\varphi(r) \in C^\infty$ for $r \geq 0$, $\varphi(r) = 0$ for $r \geq a$, $\varphi(r) = \varphi(\varepsilon)$ if $0 \leq r \leq \varepsilon$, and

$$v(x) = \varphi(|x - x_0|), \quad X(r) = \int_{\partial B(x_0, r)} u(x)\,d\Sigma. \tag{2.3.4}$$

Then $v \in C_c^\infty(G)$ and $X \in L_1[\varepsilon, a]$. Then (2.3.3) becomes

$$\int_\varepsilon^a X(r)\left[\frac{d}{dr}(r^{\nu-1}\varphi')\right]dr = \int_\varepsilon^\alpha X(r)\,\psi(r)\,dr = 0,$$

where ψ may be any function $\in C_c^\infty[\varepsilon, a]$ for which (2.3.2) holds (with $[a, b] = [\varepsilon, a]$). It follows that $X(r) = a$ constant a.e. on $[\varepsilon, a]$. Since ε and a are arbitrary this is true for $0 < r < \delta(x_0, \partial G)$; let us call that constant $\Gamma_\nu \cdot \bar{u}(x_0)$. Using the definition of $X(r)$ in (2.3.4), multiplying by $r^{\nu-1}$ and integrating, we obtain

$$\bar{u}(x_0) = |B(x_0, r)|^{-1}\int_{B(x_0, r)} u(x)\,dx, \quad \overline{B(x_0, r)} \subset G. \tag{2.3.5}$$

By holding r fixed, we see that $\bar{u}$ is continuous on G and by letting $r \to 0$ we see that $\bar{u}(x) = u(x)$ a.e. so that (2.3.5) holds with u replaced by $\bar{u}$ on the right. Thus $\bar{u}$ is harmonic on G.

2.4. Poisson's integral formula; elementary functions; Green's functions

Suppose G is a bounded domain of class C^1 and u and v are of class C^2 on $\bar{G} = G \cup \partial G$. Then, from GREEN's theorem, we obtain the formula

$$\int_G (u\Delta v - v\Delta u)\,dx = \int_{\partial G} \left(u\frac{\partial v}{\partial n} - v\frac{\partial u}{\partial n}\right) dS = \int_{\partial G} (uv_{,\alpha} - vu_{,\alpha})\,dx'_\alpha \tag{2.4.1}$$

n being the exterior normal.

Next, it is a well-known and easily verified fact that

$$f(y) = \begin{cases} |y|^{2-\nu}, & \text{if } \nu > 2 \\ \log|y| & \text{if } \nu = 2 \end{cases}$$

is harmonic if $y \neq 0$, Moreover

$$\int_{\partial B(0,\varrho)} \frac{\partial f}{\partial n}\,dS = \begin{cases} -(\nu-2)\int_{\partial B(0,\varrho)} \varrho^{1-\nu}\cdot dS = -(\nu-2)\,\Gamma_\nu, \text{ if } \nu > 2 \\ 2\pi, \text{ if } \nu = 2. \end{cases}$$

So, let us define

$$K_0(y) = \begin{cases} -(\nu-2)^{-1}\Gamma_\nu^{-1}|y|^{2-\nu}, & \text{if } \nu > 2 \\ (2\pi)^{-1}\log|y| & \text{if } \nu = 2 \end{cases} \tag{2.4.2}$$

Now, suppose G is bounded and of class C^1, $u \in C^2(\bar{G})$, $\Delta u(x) = f(x)$ on G, and $x_0 \in G$. Suppose $\overline{B(x_0, \varrho)} \subset G$ and we apply (2.4.1) to the domain $G - \overline{B(x_0, \varrho)}$ with $v(x) = K_0(x - x_0)$. Then we obtain

$$\int_{\partial B(x_0,\varrho)} \left(u\frac{\partial v}{\partial n} - v\frac{\partial u}{\partial n}\right) dS = \int_{\partial G} \left(u\frac{\partial v}{\partial n} - v\frac{\partial u}{\partial n}\right) dS + \int_{G-B(x_0,\varrho)} v(x)\,f(x)\,dx. \tag{2.4.3}$$

Letting $\varrho \to 0$ in (2.4.3), we obtain

$$u(x_0) = \int_{\partial G} \left(u\frac{\partial v}{\partial n} - v\frac{\partial u}{\partial n}\right) dS + \int_G K_0(x - x_0)\,f(x)\,dx \tag{2.4.4}$$
$$(v(x) = K_0(x - x_0)).$$

Definition 2.4.1. The function K_0, defined in (2.4.2) is called the *elementary function for Laplace's equation* $\Delta u = 0$.

If $f(x) = 0$, then (2.4.4) expresses $u(x_0)$ in terms of its boundary values and those of its normal derivative. However, from the maximum principle, a harmonic function is completely determined by its boundary values alone. If, in (2.4.4), we have $f(x) = 0$ and could take

$$v(x) = K_0(x - x_0) + H(x, x_0) = G_0(x_0, x) \tag{2.4.5}$$

where H is harmonic in x for each x_0 and is so chosen that $v = 0$ on ∂G, then (2.4.4) would express $u(x_0)$ in terms of its boundary values. Such a function v, if it exists, is called a GREEN's function for G with pole at x_0. By the maximum principle, the GREEN's function is unique if it exists at all.

From the discussion so far given, it follows that (*a*) if a GREEN's function v exists for a given domain G and point x_0 and (*b*) if u is of class C^2 on $\bar{G}$ and harmonic on G, then (2.4.4) expresses $u(x_0)$ in terms of its boundary values. If a GREEN's function could be found and if it could be shown to be harmonic in x_0 for each x in G, then the function u defined by (2.4.4) with $f = 0$ and u in the boundary integral replaced by a function u^* would be harmonic; it would then remain to show that $u(x) \to u^*(x_0)$ as $x \to x_0$ for each x_0 on ∂G. And, of course, proving the existence of the GREEN's function requires proving the existence of harmonic functions having given boundary values. This problem is called the *Dirichlet problem.*

Because of all the problems mentioned in the discussion above, the DIRICHLET problem is not usually solved by proving the existence of the GREEN's function. However, there are two cases where this is possible, namely the case when G is a sphere which, obviously, may be assumed to have center at the origin and the other is a half-space which we may assume to be that where $x^\nu > 0$. We now derive the GREEN's function for such a sphere.

Let $x \in B_R = B(0, R)$ and let x' be the point inverse to x with respect to B_R, that is, the point where

$$'x^\alpha = R^2 x^\alpha / |x|^2.$$

Using the spherical symmetry, it is easy to verify that the ratio $|\xi - x| / |\xi - x'|$ is the same for all ξ on ∂G_R so that

$$\text{(2.4.6)} \qquad \frac{|\xi - x|}{|\xi - x'|} = \frac{R - |x|}{\left(\frac{R^2}{|x|} - R\right)} = \frac{|x|}{R}.$$

Thus we note that if we define

$$\text{(2.4.7)} \qquad G(x, \xi) = \begin{cases} \frac{1}{2\pi} [\log|\xi - x| - \log|\xi - x'| - \log(|x|/R)], & \nu = 2 \\ -(\nu - 2)^{-1} \Gamma_\nu^{-1} [|\xi - x|^{2-\nu} - |\xi - x'|^{2-\nu} (|x|/R)^{2-\nu}] & \\ & \nu > 2 \end{cases}$$

then $G(x, \xi)$ is of the form (2.4.5). Moreover, by using the formulas for x' and ξ', we see that

$$\text{(2.4.8)} \qquad G(\xi, x) = G(x, \xi)$$

so that G is harmonic in x for each ξ and vanishes for ξ interior to B_R and $x \in \partial B_R$. Thus the function u defined by (2.4.4) with $f = 0$ and $v(\xi)$

$= G(x, \xi)$ is harmonic on B_R. Finally, since the function $u \equiv 1$ and G satisfy the hypotheses of the argument in the paragraph containing equations (2.4.3) and (2.4.4), we conclude that

$$\int_{\partial B_R} \frac{\partial G(x,\xi)}{\partial n(\xi)}\, dS(\xi) = 1, \quad x \in B_R. \tag{2.4.9}$$

By computation from (2.4.7), we see that

$$\begin{aligned}\frac{\partial G(x,\xi)}{\partial n(\xi)} &= R^{-1}\xi^\alpha G_{\xi^\alpha} = R^{-1}\Gamma_\nu^{-1}\xi^\alpha[|\xi - x|^{-\nu}(\xi^\alpha - x^\alpha)\\ &\quad - (|x|/R)^{2-\nu}|\xi - x'|^{-\nu}(\xi^\alpha - x'^\alpha)\\ &= R^{-1}\Gamma_\nu^{-1}[|\xi - x|^{-\nu}(R^2 - \xi\cdot x) - (|x|/R)^{-\nu}|\xi - x'|^{-\nu}(|x|^2 - \xi\cdot x)].\end{aligned}$$

For ξ on ∂B_R, we may use (2.4.6) to obtain

$$\frac{\partial G(x,\xi)}{\partial n(\xi)} = (\Gamma_\nu R)^{-1}|\xi - x|^{-\nu}(R^2 - |x|^2) > 0 \tag{2.4.10}$$

and thus obtain *Poisson's integral formula*

$$u(x) = (\Gamma_\nu R)^{-1}\int_{\partial B_R} |\xi - x|^{-\nu}(R^2 - |x|^2)\, u^*(\xi)\, d\xi \tag{2.4.11}$$

for the harmonic function u which takes on given values u^* on ∂B_R.

To see that $u(x) \to u^*(\xi_0)$ as $x \to \xi_0$ if u^* is continuous, $x \in B_R$, $\xi_0 \in \partial B_R$, we note from (2.4.9), (2.4.10), and (2.4.11) that

$$u(x) - u^*(\xi_0) = (\Gamma_\nu R)^{-1}\int_{\partial B_R} |\xi - x|^{-\nu}(R^2 - |x|^2)[u^*(\xi) - u^*(\xi_0)]\, d\xi. \tag{2.4.12}$$

To show that this difference $\to 0$, we break the integral on the right in (2.4.12) into integrals I_1 over $\partial B_R \cap B(\xi_0, \varrho)$ and I_2 over $\partial B_R - B(\xi_0, \varrho)$ where we may choose ϱ so that $|u^*(\xi) - u^*(\xi_0)| < \varepsilon/2$ for $\xi \in \partial B_R \cap B(\xi_0, \varrho)$, ε being given. The reader may complete the proof. Thus we have the following theorem:

Theorem 2.4.1. *There is a unique function u which is continuous on $\overline{B(0, R)}$, coincides with a given continuous function u^* on $\partial B(0, R)$, and is harmonic in $B(0, R)$. It is given in $B(0, R)$ by Poisson's integral formula* (2.4.11).

We can now prove the following important reflection principle for harmonic functions:

Theorem 2.4.2 (Reflection principle). *Suppose u is continuous on $\bar G$ and harmonic on the (possibly unbounded) domain G which lies in a half-space bounded by the hyperplane Π, suppose $\bar G \cap \Pi = S$, and suppose $u(x) = 0$ for $x \in S$. Let $\Gamma = G \cup G' \cup S^{(0)}$, where G' is the domain obtained by reflecting G in Π and $S^{(0)}$ is the non-empty set of interior (with respect to Π) points of S. Define $U(x) = u(x)$ for $x \in \bar G$ and define $U(x) = -u(x')$ for $x \in \bar\Gamma - \bar G$, where x' is the point obtained by reflecting x in Π. Then U is continuous on $\bar\Gamma$ and harmonic in Γ.*

Proof. That U is continuous on $\bar{\Gamma}$ and harmonic near each x_0 in $\Gamma - S^{(0)}$ is obvious. So suppose $x_0 \in S^{(0)}$ and choose R so small that $B(x_0, R) \subset \Gamma$ and let $G(x_0, R) = B(x_0, R) \cap G$. Let H be that harmonic function in $B(x_0, R)$ which coincides with U on $\partial B(x_0, R)$. From the symmetry in Poisson's integral formula, it follows that $H(x) = 0$ along $S^{(0)} \cap B(x_0, R)$ and hence $H(x) = u(x)$ on $\partial G(x_0, R)$. Thus $H(x) = u(x)$ on $G(x_0, R)$ and hence, using the symmetry in Poisson's integral formula, we conclude that $H(x) = U(x)$ on $B(x_0, R)$.

The following Liouville-type theorem is another immediate consequence of Poisson's integral formula:

Theorem 2.4.3. *A function which is harmonic and bounded on R_ν is a constant.*

Proof. For, by differentiating (2.4.11) with $u^* = u$ and letting $R \to \infty$, one finds easily that each $u_{,\alpha}(x) \equiv 0$.

Suppose we now consider the half-space $R_\nu^+ : x^\nu > 0$ and define

$$G(x, \xi) = K_0(x - \xi) - K_0('x - \xi) \quad \text{where} \quad 'x^\nu = -x^\nu, \quad 'x'_\nu = x'_\nu,$$

i.e. $'x$ is the reflection of x in $\Pi : x^\nu = 0$. We notice that

$$G(\xi, x) = G(x, \xi), \quad G(x, \xi) = 0 \text{ for } \xi \in \Pi.$$

Thus G is harmonic in ξ for each x and in x for each ξ and thus would appear to be a candidate for a GREEN's function for R_ν^+. Clearly $\partial G(x, \xi)/\partial n(\xi)$ is just $-\partial G(x, \xi)/\partial \xi^\nu$. Along $\xi^\nu = 0$, we have

$$\begin{aligned} -\partial G(x, \xi)/\partial \xi^\nu &= K_{0,\nu}(x - \xi) - K_{0,\nu}(x' - \xi) \\ &= 2\Gamma_\nu^{-1} x^\nu \cdot |x - \xi|^{-\nu} = L(x, \xi'_\nu), \quad \xi^\nu = 0. \end{aligned} \tag{2.4.13}$$

A straightforward computation shows that L is harmonic in x for each ξ'_ν, is positive, and

$$\int_{R_{\nu-1}} L(x, \xi'_\nu)\, d\xi'_\nu = 1. \tag{2.4.14}$$

Thus if $u^*(\xi'_\nu)$ is continuous and bounded and we define

$$u(x) = \int_{R_{\nu-1}} L(x, \xi'_\nu)\, u^*(\xi'_\nu)\, d\xi'_\nu, \tag{2.4.15}$$

we see, as in the case of Poisson's integral formula for $B(0, R)$, that $u(x)$ is harmonic in R_ν^+ *and bounded there by the sup* of $u^*(\xi'_\nu)$, is continuous on $\overline{R_\nu^+}$, and coincides with u^* along Π. We now prove:

Theorem 2.4.4. *There is a unique function u which is harmonic in R_ν^+ and continuous and bounded in $\overline{R_\nu^+}$ and which coincides on $x^\nu = 0$ with a given bounded continuous function u^*; u is given by* (2.4.15) *and* (2.4.13).

Proof. It remains only to prove the uniqueness. If u_1 and u_2 both satisfy all the conditions of the theorem, then $u = u_1 - u_2$ is bounded and continuous on $\overline{R_\nu^+}$ and vanishes on Π. If we then extend u by reflection as in Theorem 2.4.2, then U is harmonic and bounded everywhere and is therefore a constant which must be zero.

2.5. Potentials

In formula (2.4.4) with $v(x) = K_0(x - x_0) = K_0(x_0 - x)$, we see that if u is of class C^2 on G with

$$\Delta u(x) = f(x) \tag{2.5.1}$$

where G is of class C^1, then the boundary integrals are harmonic so that the function U defined by

$$U(x) = \int_G K_0(x - \xi) f(\xi)\, d\xi \tag{2.5.2}$$

would differ from u by a harmonic function and hence would also be a solution of (2.5.1).

Definitions 2.5.1. The equation (2.5.1) is known as *Poisson's equation* and the function U in (2.5.2) is called the *potential of* f.

Unfortunately, if f is merely continuous, it does not follow that U is necessarily of class C^2; we shall give an example of this in § 2.6.

Since there is no difference in the proofs, we shall consider the more general integrals

$$U(x) = \int_G K(x - \xi) f(\xi)\, d\xi . \tag{2.5.3}$$

The following definition is useful in this section and in the study of higher order differential equations:

Definition 2.5.2. A function $\varphi(y)$ is *essentially homogeneous of degree* p in y if φ is positively homogeneous in case $p < 0$ or, if p is an integer ≥ 0, φ has the form

$$\varphi(y) = \varphi_0(y) \log|y| = \varphi_1(y) \tag{2.5.4}$$

where φ_0 is a homogeneous *polynomial* of degree p and φ_1 is positively homogeneous of degree p.

Theorem 2.5.1. *Suppose f is bounded and measurable with $|f(x)| \leq M$ and has compact support $\subset G$, $K \in C^2(R_\nu - \{0\})$ and K is essentially homogeneous of degree $2 - \nu$, and U is defined by* (2.5.3).

Then $U \in C^1$ everywhere with

$$U_{,\alpha}(x) = \int_G K_{,\alpha}(x - \xi) f(\xi)\, d\xi, \quad |\nabla U(x)| \leq M \| \nabla K \|^0_{1,\Sigma} \cdot \Delta \tag{2.5.5}$$

$$|\nabla U(x_2) - \nabla U(x_1)| \leq \begin{cases} M\varrho [3 \| \nabla K \|^0_{1,\Sigma} + \\ + \| \nabla^2 K \|^0_{1,\Sigma} \log(1 + \Delta/\varrho)], & \text{if } \varrho \geq \delta \\ M\varrho \| \nabla^2 K \|^0_{1,\Sigma} \log(1 + \Delta/\delta), & \text{if } \varrho \leq \delta, \end{cases} \tag{2.5.6}$$

where $\varrho = |x_1 - x_2|$, δ is the distance of G from the segment $x_1 x_2$ and Δ is the diameter of G.

Remark. This theorem holds if U and f are vector functions and K is a matrix function. The proof is the same with indices inserted.

Proof. We first assume that $f \in C_c^0(G)$, the proof in the general case follows by approximations. We define

$$U_\varrho(x) = \int_{G-B(x,\varrho)} K(x-\xi) f(\xi)\, d\xi. \tag{2.5.7}$$

It is easy to see that U_ϱ converges uniformly to U on any bounded set and that

$$U_{\varrho,\alpha}(x) = -\int_{\partial B(x,\varrho)} K(x-\xi) f(\xi)\, d\xi'_\alpha + \int_{G-B(x,\varrho)} K_{,\alpha}(x-\xi) f(\xi)\, d\xi; \tag{2.5.8}$$

this holds even if $B(x, \varrho) \not\subset G$, since we may first replace G by a large region $G' \supset G \cup B(x, \varrho)$. Since ∇K is homogeneous of degree $1-\nu$, it is easy to see that the vector ∇U_ϱ tends uniformly on any bounded set to the vector V defined by

$$V_\alpha(x) = U_{,\alpha}(x) = \int_G K_{,\alpha}(x-\xi) f(\xi)\, d\xi \tag{2.5.9}$$

from which (2.5.5) follows easily.

Next, define

$$V_{\alpha\varrho}(x) = \int_{G-B(x,\varrho)} K_{,\alpha}(x-\xi) f(\xi)\, d\xi. \tag{2.5.10}$$

Then

$$|V_\varrho(x) - V(x)| \le \begin{cases} M \cdot \| \nabla K \|^0_{1,\Sigma}\, \varrho & \text{if} \quad B(x, \varrho) \cap G \neq \Phi \\ 0 & \text{if} \quad B(x, \varrho) \cap G = \Phi. \end{cases} \tag{2.5.11}$$

Differentiating as in (2.5.8), we obtain

$$V_{\alpha\varrho,\beta}(x) = -\int_{\partial B(x,\varrho)} K_{,\alpha}(x-\xi) f(\xi)\, d\xi'_\beta + \int_{G-B(x,\varrho)} K_{,\alpha\beta}(x-\xi) f(\xi)\, d\xi.$$

Thus, if we denote the distance from x to G by δ_1, we see that

$$|\nabla V_\varrho(x)| \le \begin{cases} M[\| \nabla K \|^0_{1,\Sigma} + \| \nabla^2 K \|^0_{1,\Sigma} \log(1 + \Delta/\varrho), \\ \qquad\qquad\qquad \text{if} \quad B(x, \varrho) \cap G \neq \Phi \\ M \| \nabla^2 K \|^0_{1,\Sigma} \log(1 + \Delta/\delta_1), \quad \text{if} \quad B(x, \varrho) \cap G = \Phi. \end{cases} \tag{2.5.12}$$

The result (2.5.6) follows by setting $|x_2 - x_1| = \varrho$, observing that

$|V(x_2) - V(x_1)| \le |V(x_2) - V_\varrho(x_2)| + |V(x_1) - V_\varrho(x_1)| + |V_\varrho(x_2) - V_\varrho(x_1)|$

and using (2.5.12) to estimate $|V_\varrho(x_2) - V_\varrho(x_1)|$.

2.6. Generalized potential theory; singular integrals

From Theorem 2.5.1, it follows that the first derivatives of potential functions are given by formulas of the form

$$V(x) = \int_G \Gamma(x-\xi) f(\xi)\, d\xi \tag{2.6.1}$$

where Γ satisfies the conditions

(2.6.2) (a) Γ is positively homogeneous of degree $1-\nu$, and
(b) $\Gamma \in C^1(R_\nu - \{0\})$.

We shall begin with a study of such integrals. All the results hold if V and f are vector functions and Γ is a matrix function.

By repeating the second paragraph of the proof of Theorem 2.5.1 with $K_{,\alpha}$ replaced by Γ, we conclude the following theorem:

Theorem 2.6.1. *If Γ satisfies (2.6.2), f satisfies the conditions of Theorem 2.5.1, and V is given by (2.6.1), then V is continuous.*

Theorem 2.6.2. *Suppose Γ satisfies (2.6.2) and $\Gamma \in C^2(R_\nu - \{0\})$, f is bounded and measurable on the bounded domain G and $f \in C^\mu(\bar{D})$ for every $D \subset\subset G$, and V is defined by (2.6.1). Then $V \in C^{1+\mu}$ on each $D \subset\subset G$ and*

$$\begin{aligned} V_{,\alpha}(x) &= C_\alpha f(x) + W_\alpha(x), \\ W_\alpha(x) &= \lim_{\varrho \to 0} \int_{G - B(x,\varrho)} \Gamma_{,\alpha}(x-\xi) f(\xi)\, d\xi, \quad x \in G \\ C_\alpha &= -\int_{\partial B(0,1)} \Gamma(-y)\, dy'_\alpha, \end{aligned} \tag{2.6.3}$$

the convergence being uniform on any $D \subset\subset G$.

Proof. We define $V_\varrho(x)$ by (2.5.10), with ∇K replaced by Γ, and obtain the inequality corresponding to (2.5.11) and the formula

$$V_{\varrho,\alpha}(x) = -\int_{\partial B(x,\varrho)} \Gamma(x-\xi) f(\xi)\, d\xi'_\alpha + W_{\alpha\varrho}(x) \tag{2.6.4}$$

where $W_{\alpha\varrho}$ is defined by the integral over $G - B(x,\varrho)$ in (2.6.3). It is clear that the first term on the right in (2.6.4) tends to $C_\alpha f(x)$, uniformly on any $D \subset\subset G$. Next, choose a domain D' of class C^1 such that $D \subset\subset D' \subset\subset G$ and break up the integral for $W_{\alpha\varrho}$ in (2.6.4) into an integral over $G - D'$ plus one over $D' - B(x,\varrho)$. In the latter integral and in the first term on the right in (2.6.4), write $f(\xi) = f(x) + [f(\xi) - f(x)]$. Using the fact that $\Gamma_{,\alpha}(x-\xi) = -\partial\Gamma(x-\xi)/\partial\xi^\alpha$ and integrating over $D' - B(x,\varrho)$, we obtain

$$\begin{aligned} V_{\varrho,\alpha}(x) &= C_\alpha(x, D') \cdot f(x) + \int_{G-D'} \Gamma_{,\alpha}(x-\xi) f(\xi)\, d\xi - \\ &- \int_{\partial B(x,\varrho)} \Gamma(x-\xi)[f(\xi) - f(x)]\, d\xi'_\alpha + \int_{D'-B(x,\varrho)} \Gamma_{,\alpha}(x-\xi)[f(\xi) - f(x)]\, d\xi \\ C_\alpha(x, D') &= -\int_{\partial D'} \Gamma(x-\xi)\, d\xi'_\alpha, \quad x \in D. \end{aligned} \tag{2.6.5}$$

$C_\alpha(x, D')$ and the integral over $G - D' \in C^1(D')$ and the last two terms in (2.6.5) tend uniformly to their limits. Thus the convergence of $W_{\alpha\varrho}$ to W_α in (2.6.3) is uniform on D.

Differentiating $W_{\alpha\varrho}$ and using the ideas employed to obtain (2.6.5), we obtain

$$W_{\alpha\varrho,\beta}(x) = C_{\beta,\alpha}(x, D')f(x) + \int_{G-D'} \Gamma_{,\alpha\beta}(x-\xi)f(\xi)\,d\xi -$$

$$(2.6.6)\qquad -\int_{\partial B(x,\varrho)} \Gamma_{,\alpha}(x-\xi)[f(\xi)-f(x)]\,d\xi'_\beta + \int_{D'-B(x,\varrho)} \Gamma_{,\alpha\beta}(x-\xi)[f(\xi)-f(x)]\,d\xi.$$

The first two terms are bounded and are independent of ϱ; the last two are $0(\varrho^{\mu-1})$. Thus, if $\varrho = |x_1 - x_2|$ and $B(x,\varrho) \subset D'$ for every x on the segment $x_1 x_2$, we see that

$$|W_\varrho(x_2) - W_\varrho(x_1)| \leq C\varrho^\mu.$$

Using (2.6.5) and the fact that

$$V_{\varrho,\alpha}(x) - V_{,\alpha}(x) = -\int_{\partial B(x,\varrho)} \Gamma(x-\xi)[f(\xi)-f(x)]\,d\xi'_\alpha + W_{\varrho\alpha}(x) - W_\alpha(x)$$

we obtain

$$|W_\varrho(x) - W(x)| \leq C\varrho^\mu \quad \text{if} \quad B(x,\varrho) \subset D'.$$

The HÖLDER continuity follows.

Theorem 2.6.3. *Suppose Γ satisfies the conditions* (2.6.2). *Then*

$$(2.6.7)\qquad \int_{\partial B(0,1)} \Gamma_{,\alpha}(y)\,d\Sigma(y) = 0, \quad \alpha = 1, \ldots, \nu.$$

Proof. Let A_α denote the left side of (2.6.7). Then, from the homogeneity, we conclude that

$$A_\alpha \log 2 = \int_1^2 \left[\int_{\partial B(0,R)} \Gamma_{,\alpha}(y)\,dS(y)\right] dR = \int_{B(0,2)-B(0,1)} \Gamma_{,\alpha}(y)\,dy$$
$$= \int_{\partial B(0,2)} \Gamma(y)\,dy'_\alpha - \int_{\partial B(0,1)} \Gamma(y)\,dy'_\alpha = 0.$$

From Theorems 2.6.2 and 2.6.3, we see that the first derivatives of functions V defined by (2.6.1), and hence the second derivatives of the potentials (2.5.2) (or (2.5.3)) are given by the formulas (2.6.3), where the W_α are given by the *singular* (or CAUCHY *principal value*) *integrals* of the form

$$(2.6.8)\qquad W(x) = \lim_{\varrho\to 0} \int_{G-B(x,\varrho)} \Delta(x-\xi)f(\xi)\,d\xi = \lim_{\varrho\to 0} W_\varrho(x)$$

where Δ satisfies the conditions

(2.6.9)
(a) Δ is positively homogeneous of degree $-\nu$
(b) $\Delta \in C^1(R_\nu - \{0\})$
(c) $\int_\Sigma \Delta(y)\,dS(y) = 0$.

A proof like that of Theorem 2.6.2 yields the following theorem:

Theorem 2.6.4. *If G and f satisfy the conditions of Theorem* 2.6.2, *Δ satisfies* (2.6.9), *and W is given by* (2.6.8), *then $W \in C^\mu(\bar{D})$ for every $D \subset\subset G$. If $f \in C^\mu(R_\nu)$ and has compact support, then $W \in C^\mu(R_\nu)$.*

Proof. The last follows since G can be arbitrarily large. Using (2.6.9) (c), we see that if $0 < \sigma < \varrho$, then

$$W_\sigma(x) - W_\varrho(x) = \int\limits_{B(x,\varrho)-B(x,\sigma)} \Delta(x-\xi)\,[f(\xi) - f(x)]\,d\xi.$$

Clearly we may allow $\sigma \to 0$ to obtain

$$|W(x) - W_\varrho(x)| \leq C_1\,\varrho^\mu, \quad B(x,\varrho) \subset D' \subset\subset G.$$

Proceeding as in the proof of Theorem 2.6.2, we obtain

$$\begin{aligned} W_{\varrho,\alpha}(x) = &\int\limits_{G-D'} \Delta_{,\alpha}(x-\xi) f(\xi)\,d\xi - f(x) \int\limits_{\partial D'} \Delta(x-\xi)\,d\xi'_\alpha - \\ &- \int\limits_{\partial B(x,\varrho)} \Delta(x-\xi)\,[f(\xi) - f(x)]\,d\xi'_\alpha + \\ &+ \int\limits_{D'-B(x,\varrho)} \Delta_{,\alpha}(x-\xi)\,[f(\xi) - f(x)]\,d\xi, \quad x \in D, \\ &\quad (D \subset\subset D' \subset\subset G). \end{aligned}$$

Thus if $|x_1 - x_2| = \varrho$ and $\overline{B(x,\varrho)} \subset D'$ for every x on the segment $x_1\,x_2$, we find, as before, that

$$|W_\varrho(x_1) - W_\varrho(x_2)| \leq C_2\,\varrho^\mu.$$

The result now follows.

We now prove some theorems which will be useful later and which yield additional information in the case that Δ also satisfies

$$\Delta(-y) = \Delta(y). \tag{2.6.10}$$

Theorem 2.6.5. *Suppose Δ satisfies* (2.6.9) *and* (2.6.10). *Then*

$$\int\limits_{B(x_0,a)-B(x_1,\varrho)} \Delta(x-\xi)\,d\xi = 0, \quad x \in B(x_1,\varrho) \subset\subset B(x_0,a) \tag{2.6.11}$$

$$\int\limits_{\partial B(x_0,a)} \Delta(x-\xi)\,d\xi'_\alpha = 0, \quad x \in B(x_0,a). \tag{2.6.12}$$

Proof. For each $\eta \in \partial B(0,1)$, let the ray through x in the direction η intersect $\partial B(x_1,\varrho)$ in the point ξ_1 and $\partial B(x_0,a)$ in the point ξ_0 and define

$$\psi_0(\eta) = |\xi_0 - x|, \quad \psi_1(\eta) = |\xi_1 - x|. \tag{2.6.13}$$

From plane geometry, we conclude that

$$\begin{aligned} \psi_0(\eta)\cdot\psi_0(-\eta) &= a^2 - |x - x_0|^2, \\ \psi_1(\eta)\cdot\psi_1(-\eta) &= \varrho^2 - |x - x_1|^2. \end{aligned} \tag{2.6.14}$$

Taking polar coordinates with pole at x, we conclude that

$$\int\limits_{B(x_0,a)-B(x_1,\varrho)} \Delta(x-\xi)\,d\xi = \int\limits_\Sigma \Delta(\eta)\,[\log\psi_0(\eta) - \log\psi_1(\eta)]\,d\Sigma(\eta)$$

$$= \frac{1}{2}\int\limits_\Sigma \Delta(\eta)\,[\log\psi_0(\eta) + \log\psi_0(-\eta) - \log\psi_1(\eta) - \log\psi_1(-\eta)]\,d\Sigma(\eta)$$

$$= 0$$

on account of (2.6.10), (2.6.9) (c) and (2.6.14).

From (2.6.11), we conclude that

$$\int_{B(x_0,a)-B(x_1,\varrho)} \Delta_{,\alpha}(x-\xi)\,d\xi = 0$$

(2.6.15)
$$= -\int_{\partial B(x_0,a)} \Delta(x-\xi)\,d\xi'_\alpha + \int_{\partial B(x_1,\varrho)} \Delta(x-\xi)\,d\xi'_\alpha, \quad x \in B(x_1,\varrho).$$

Using (2.6.15) with $x_1 = x$, we conclude that

$$\int_{\partial B(x_0,a)} \Delta(x-\xi)\,d\xi'_\alpha = \int_{\partial B(x,\varrho)} \Delta(x-\xi)\,d\xi'_\alpha = \varrho^{-1}\int_{\partial B(0,1)} \eta^\alpha \Delta(\eta)\,dS(\eta) = 0$$

on account of (2.6.10).

Theorem 2.6.6. (a) *Any two points x_1 and x_2 in B_R can be joined by a path $x = x(s)$, $0 \le s \le l$, in B_R such that*

$$\int_0^l (R - |x(s)|)^{\mu-1}\,ds \le (2\mu^{-1}+1)\cdot|x_1-x_2|^\mu, \quad 0 < \mu \le 1.$$

(b) *There is a constant $C(\mu)$ such that any two points of G_R can be joined by a path $x = x(s)$, $0 \le s \le l$, in G_R such that*

$$\int_0^l \{\delta[x(s)]\}^{\mu-1}\,ds \le C(\mu)\cdot|x_1-x_2|^\mu, \quad 0 < \mu \le 1,$$

where $\delta(x)$ denotes the distance of x from $\partial G_R = \Sigma_R \cup \sigma_R$.

Proof. (a) Is easily verified as follows: Let $\varrho = |x_1 - x_2|$. If $R \le \varrho < 2R$, choose the polygonal path $x_1\,0\,x_2$. If $0 < \varrho < R$ and $|x_1| \le R-\varrho$, $|x_2| \le R-\varrho$, choose the segment $x_1 x_2$. If $|x_1| > R-\varrho$, $|x_2| \le R-\varrho$, choose the polygonal path $x_1 x_3 x_2$ where x_3 is on $0\,x_1$ with $|x_3| = R-\varrho$. If $|x_1| > R-\varrho$, $|x_2| > R-\varrho$, choose the polygonal path $x_1 x_3 x_4 x_2$ where x_3 is on $0\,x_1$, x_4 is on $0\,x_2$, and $|x_3| = |x_4| = R-\varrho$.

(b) It is also sufficient to prove this for $0 < |x_1 - x_2| = \varrho = kR$ where $0 < k < 1/3$, say. For such ϱ, the set S_ϱ of x such that $\delta(x) \ge \varrho$ is the part of the ball $|x| \le R-\varrho$ where $x^\nu(=y) \ge \varrho$. For such ϱ, we choose the paths $x_1 x_2$ if x_1 and $x_2 \in S_\varrho$, $x_1 x_3 x_2$ if $x_1 \notin S_\varrho$ and $x_2 \in S_\varrho$, and $x_1 x_3 x_4 x_2$ if $x_1 \notin S_\varrho$ and $x_2 \notin S_\varrho$; here x_3 is the nearest point of S_ϱ to x_1 and x_4 is that nearest x_2. A straightforward analysis verifies the result in all cases.

Theorem 2.6.7. *Suppose that Δ satisfies* (2.6.9) *and* (2.6.10). *Suppose that $f \in C^\mu(\bar{B}_R)$, and suppose W is given by* (2.6.8). *Then $W \in C^\mu(\bar{B}_R)$ and*

(2.6.16)
$$|W(x)| \le [2^\mu\cdot\mu^{-1}\|\Delta\|^0_{1,\Sigma}]\,h_\mu(f)\,R^\mu$$
$$h_\mu(W) \le C(\nu,\mu)[\|\Delta\|^0_{1,\Sigma} + \|\nabla\Delta\|^0_{1,\Sigma}]\,h_\mu(f).$$

Proof. From Theorem 2.6.5, it follows that

$$W(x) = \int_{B_R} \Delta(x-\xi)[f(\xi)-f(x)]\,d\xi$$

from which the first inequality in (2.6.16) follows immediately.

To prove the second inequality, we extend f to $\bar{B}_{2R}$ so that $\max f$ and $h_\mu(f)$ are unaltered and write

$$W = W_1 - W_2, \quad W_1(x) = \int_{B_{2R}} \Delta(x-\xi)\, f(\xi)\, d\xi \tag{2.6.17}$$
$$W_2(x) = \int_{B_{2R}-B_R} \Delta(x-\xi)\, f(\xi)\, d\xi .$$

Differentiating $W_{1\varrho}$ as in (2.6.6), we obtain

$$W_{1\varrho,\alpha}(x) = -f(x)\int_{\partial B_{2R}} \Delta(x-\xi)\, d\xi'_\alpha - \int_{\partial B(x,\varrho)} \Delta(x-\xi)[f(\xi)-f(x)]\, d\xi'_\alpha +$$
$$+ \int_{B_{2R}-B(x,\varrho)} \Delta_{,\alpha}(x-\xi)[f(\xi)-f(x)]\, d\xi, \quad x\in B_R, \quad 0<\varrho<R, \tag{2.6.18}$$

in which the first term vanishes. From (2.6.18), we obtain

$$|W_{1\varrho}(x_2) - W_{1\varrho}(x_1)| \le 2h_\mu(f)[\|\Delta\|^0_{1,\Sigma} + (1-\mu)^{-1}\|\nabla\Delta\|^0_{1,\Sigma}]\,|x_2-x_1|^\mu$$
$$(|x_2-x_1| = 2\varrho). \tag{2.6.19}$$

From Theorem 2.6.5 and equation (2.6.17), it follows that

$$|W_{1\varrho}(x) - W_1(x)| = \left|\int_{B(x,\varrho)} \Delta(x-\xi)[f(\xi)-f(x)]\, d\xi\right| \le h_\mu(f)\|\Delta\|^0_{1,\Sigma}\cdot\mu^{-1}\varrho^\mu. \tag{2.6.20}$$

The result for W_1 follows from (2.6.19) and (2.6.20).

To prove the second inequality for W_2, we first note that

$$\nabla W_2(x) = \int_{B_{2R}-B_R} \nabla\Delta(x-\xi)\, f(\xi)\, d\xi \tag{2.6.21}$$
$$= \int_{B_{2R}-B_R} \nabla\Delta(x-\xi)[f(\xi)-f(x)]\, d\xi, \quad x\in B_R$$

using Theorem 2.6.5 (or (2.6.15)). From (2.6.21), we conclude that

$$|\nabla W_2(x)| \le \int_{R_\nu - B(x,\delta)} |\nabla\Delta(x-\xi)|\cdot|f(\xi)-f(x)|\, d\xi \tag{2.6.22}$$
$$\le (1-\mu)^{-1} h_\mu(f)\cdot\|\nabla\Delta\|^0_{1,\Sigma}\cdot(R-|x|)^{\mu-1}$$
$$(\delta = R - |x|).$$

From (2.6.22), Theorem 2.6.6 (a), and the fact that $|W_2(x_2) - W_2(x_1)|$ is dominated by the integral of $|\nabla W_2|$ over any path from x_1 to x_2, the result for W_2 follows.

Corollary 1. *Suppose that $f\in C^\mu(\bar{B}_R)$ and U is its potential. Then $U\in C^{2+\mu}(\bar{B}_R)$ and there is a $C(\mu,\nu)$ such that*

$$\Delta U(x) = f(x), \quad x\in B_R, \quad h_\mu(\nabla^2 U) \le C\, h_\mu(f). \tag{2.6.23}$$

Proof. From Theorem 2.5.1, it follows that $U\in C^1$ with

$$U_{,\alpha}(x) = \int_{B_R} K_{0,\alpha}(x-\xi)\, f(\xi)\, d\xi .$$

Then, from Theorem 2.6.2 with $\Gamma = K_{0,\alpha}$, we see that $\nabla U \in C^1_\mu(B_r)$ for each $r < R$, with

$$(2.6.24) \qquad \begin{aligned} U_{,\alpha\beta}(x) &= C_{\alpha\beta} f(x) + \lim_{\varrho\to 0} \int_{B_R - B(x,\varrho)} K_{0,\alpha\beta}(x-\xi) f(\xi)\, d\xi \\ C_{\alpha\beta} &= \int_{\partial B(0,1)} K_{0,\alpha}(y)\, y^\beta\, d\sum = \nu^{-1} \Delta_{\alpha\beta} \end{aligned}$$

as one finds, using the formulas (2.4.2) for K_0. Since each $K_{0,\alpha\beta}$ satisfies (2.6.9) and (2.6.10), the second result in (2.6.23) follows. The first result follows from (2.6.24) and the fact that $\Delta K_0(y) = 0$ if $y \neq 0$.

By combining this result with POISSON's integral formula, we obtain the following corollary:

Corollary 2. *Suppose $f \in C^0_\mu(\bar{B}_R)$ and u^* is continuous on ∂B_R. Then there is a unique function u which is continuous on $\bar{B}_R$, coincides with u^* on ∂B_R, $\in C^2_\mu(B_r)$ for each $r < R$, and satisfies* POISSON's *equation* (2.5.1) *on B_R.*

We shall now give an example of a function $f \in C^0(\bar{B}_a)$ the potential of which is not in $C^2(B_a)$. To do this, we first note that if f is bounded and measurable on B_a and $f \in C^\mu[\overline{B(x_0, R)}]$ ($\bar{B}(x_0, R) \subset B_a$), then its potential $V \in C^2_\mu[B(x_0, R)]$ since one can write

$$V(x) = \int_{B(x_0,R)} K_0(x-\xi) f(\xi)\, d\xi + \int_{B_a - B(x_0,R)} K_0(x-\xi) f(\xi)\, d\xi$$

in which the first term $\in C^2_\mu[\overline{B(x_0, R)}]$ and the second is harmonic. Thus we have the formulas

$$V_{,\alpha\beta}(x) = C_{0\alpha\beta} f(x) + \int_{B_a} K_{0,\alpha\beta}(x-\xi)[f(\xi) - f(x)]\, d\xi \quad (x \in B(x_0, R)).$$

So now we define $f(0) = 0$ and

$$f(x) = [\log|x|]^{-1} |x|^{-2} x^1 x^2, \quad x \neq 0, \quad 0 < a < 1.$$

Then its potential $\in C^2(B_a - \{0\})$ and the formulas (2.6.22) hold if $x \neq 0$; but $V_{,12}(x) \to +\infty$ as $x \to 0$.

Remarks. However, if f satisfies a DINI *condition* on a domain $\bar{G}$, then $V \in C^2(G)$. Indeed if, in Theorem 2.6.2, f satisfies a DINI condition on each $\bar{D} \subset G$, then ∇V is continuous and given by (2.6.3), the convergence being uniform on any $D \subset\subset G$. Similarly if, in Theorem 2.6.4, f satisfies a DINI condition on each $\bar{D} \subset G$, then $W \in C^0(G)$. However W does not necessarily satisfy a DINI condition. f is said to satisfy a DINI condition on a compact set S if there is a positive, non-decreasing, continuous function $\varphi(\varrho)$ such that

$$|f(x_1) - f(x_2)| \leq \varphi(|x_1 - x_2|), \quad x_1, x_2 \in S, \quad \text{and}$$

$$\int_0^a \varrho^{-1} \varphi(\varrho)\, d\varrho < \infty, \quad a > 0.$$

2.7. The Calderon-Zygmund inequalities

These very useful and now well-known inequalities are based on the following well-known (see ZYGMUND) inequality for the HILBERT transform which is actually the one dimensional case of the general inequalities. As was stated in § 2.1, we shall restrict ourselves to HÖLDER-continuous functions. The proofs are greatly simplified; moreover these results are sufficient for our later purposes.

Theorem 2.7.1. *Suppose* $f \in C^0_{\mu c}(R^1)$ $(0 < \mu < 1)$. *Then the* HILBERT *transform* g, *defined by*

$$g(x) = \lim_{\varrho \to 0^+} \frac{1}{\pi} \int\limits_{|x-\xi|>\varrho} \frac{f(\xi)\,d\xi}{x-\xi} \tag{2.7.1}$$

$\in C^0_\mu(R^1)$ *and, for each* $p > 1$, $g \in L_p(R_1)$ *and*

$$\|g\|^0_p \leq C(p) \cdot \|f\|^0_p. \tag{2.7.2}$$

Proof. The integral (2.7.1) is the one dimensional case of the integral (2.6.8). Accordingly $g \in C^0_\mu(R_1)$ by Theorem 2.6.4.

It is sufficient to prove the result (2.7.2) for $f(x) \geq 0$ since we may always write $f = f_1 - f_2$ where each $f_i \in C^0_{\mu c}(R_1)$ and is non-negative. So we define

$$u(x, y) = \frac{1}{\pi} \int\limits_{-\infty}^{\infty} \frac{y f(\xi)\,d\xi}{(x-\xi)^2 + y^2} \tag{2.7.3}$$

$$v(x, y) = \frac{1}{\pi} \int\limits_{-\infty}^{\infty} \frac{(x-\xi) f(\xi)\,d\xi}{(x-\xi)^2 + y^2}. \tag{2.7.4}$$

Since (2.7.3) is just POISSON's integral formula for the upper half-plane (see § 2.4) we see that u is harmonic for $y > 0$ and is continuous for $y \geq 0$ with

$$u(x, 0) = f(x).$$

Since for some $A > 0$, $f \in C^0_\mu$ on $|x| \leq A + 1$ and $f(x) = 0$ for $|x| \geq A$, we see, by writing $f(\xi) = f(x) + [f(\xi) - f(x)]$ in (2.7.4) that $v(x, y)$ tends uniformly in x as $y \to 0^+$ to $g(x)$ for $|x| \leq A + 1/2$, the convergence for larger x being obvious. Moreover v is a conjugate harmonic function to u. If we let $F(z) = u + i\,v$, then

$$F(z) = \frac{1}{\pi i} \int\limits_{-\infty}^{\infty} \frac{f(\xi)\,d\xi}{\xi - z}. \tag{2.7.5}$$

Now, consider the function $[F(z)]^p$ for $y \geq 0$, where w^p denotes that branch which is real and positive when w is (remember $u \geq 0$ since $f \geq 0$). From (2.7.5), we see that

$$|F(z)|^p \leq M \cdot |z|^{-p} \text{ for } |z| > A + 1,$$

say. A simple limit argument using CAUCHY's theorem on hemispheres G_R shows that

$$(2.7.6)\qquad \int_{-\infty}^{\infty} [F(x)]^p\,dx = \int_{-\infty}^{\infty} [f(x) + i\,g(x)]^p\,dx = 0.$$

By integrating along the segment of length $u = f(x)$ in the w plane from $i\,v$ to $(u + i\,v)$, we obtain for each x

$$(2.7.7)\qquad (f + i\,g)^p - (i\,g)^p = p \int_{\text{segment}} w^{p-1}\,dw$$

$$(2.7.8)\qquad \begin{aligned} \left| p \int w^{p-1}\,dw \right| &\leq p\,|f(x)| \cdot [f^2 + g^2]^{(p-1)/2} \\ &\leq p \cdot Z(p)\,(|f|^p + |f| \cdot |g|^{p-1}). \end{aligned}$$

Integrating (2.7.7) from $-\infty$ to ∞ and using (2.7.6) and (2.7.8), we obtain

$$(2.7.9)\qquad \left| \int_{-\infty}^{\infty} [i\,g(x)]^p\,dx \right| \leq C_1(p)\,[(\|f\|_p^0)^p + \|f\|_p^0 \cdot (\|g\|_p^0)^{p-1}].$$

Since $[i\,g(x)]^p = |g(x)|^p \exp\left[\pm \frac{i\,p\,\pi}{2}\right]$ and $|R(w)| \leq |w|$, we see that

$$\left| \cos \frac{p\,\pi}{2} \right| \cdot (\|g\|_p^0)^p \leq C_1(p)\,[(\|f\|_p^0)^p + \|f\|_p^0 \cdot (\|g\|_p^0)^{p-1}]$$

from which the result follows for any p for which $\cos(p\,\pi/2) \neq 0$.

We take care of those cases as follows: We note that $\cos(p\,\pi/2) \neq 0$ if $1 < p \leq 2$. So suppose $p > 2$ and let p' be the conjugate exponent $(p^{-1} + p'^{-1} = 1)$; then $1 < p' < 2$. Let $h \in C_c^\mu(R_1)$ and let k be its HILBERT transform. Then

$$(2.7.10)\qquad \|k\|_{p'}^0 \leq C(p')\,\|h\|_{p'}^0.$$

Moreover, it is easy to see that

$$(2.7.11)\qquad \int_{-\infty}^{\infty} h(x)\,g(x)\,dx = \int_{-\infty}^{\infty} -\,k(x)\,f(x)\,dx \leq C(p')\,\|f\|_p^0 \cdot \|h\|_{p'}^0$$

by applying the HÖLDER inequality to the right side of (2.7.11) and using (2.7.10). Since the functions h are everywhere dense in $L_{p'}$, the theorem follows in this case also with $C(p') = C(p)$.

Theorem 2.7.2. *Suppose $f \in C_{\mu c}^0(R_\nu)$, Δ satisfies (2.6.9) and*

$$(2.7.12)\qquad \Delta(-y) = -\Delta(y),$$

and suppose

$$(2.7.13)\qquad g(x) = \lim_{\varrho \to 0^+} \int_{R_\nu - B(x,\varrho)} \Delta(x - \xi)\,f(\xi)\,d\xi.$$

Then $g \in C_\mu^0(R_\nu) \cap L_p(R_\nu)$ for each $p > 1$ and

$$(2.7.14)\qquad \|g\|_p^0 \leq C(p)\,\|\Delta\|_{1,\Sigma}^0 \cdot \|f\|_p^0$$

The hypothesis that $\Delta \in C^1(R_\nu - \{0\})$ is not necessary; it is sufficient for $\Delta \in C^0(R_\nu - \{0\})$.

Proof. That $g \in C^0_\mu(R_\nu)$ follows from Theorem 2.6.4 and that $g \in$ each $L_p(R_\nu)$ follows since $g(x) = 0(|x|^{-\nu})$ at ∞.

Now, if we write $\xi = x + r\eta$ where $r > 0$ and $\eta \in \Sigma$ and use (2.7.12), we conclude that

$$g(x) = \frac{1}{2}\int_\Sigma \Delta(\eta)\, h(x, \eta)\, d\Sigma(\eta), \qquad (2.7.15)$$
$$h(x, \eta) = \lim_{\varrho \to 0^+} \int_{|r|>\varrho} - r^{-1} f(x + r\eta)\, dr.$$

For each η, let us write $x = x_0 + s\eta$ where $x_0 \cdot \eta = 0$. Then

$$h(x_0 + s\eta, \eta) = \lim_{\varrho \to 0^+} \int_{|s-t|>\varrho} \frac{\varphi(t; x_0, \eta)\, dt}{s - t}, \qquad (2.7.16)$$
$$\varphi(t; x_0, \eta) = f(x_0 + t\eta)$$

From Theorem 2.7.1, we conclude that $h(x_0 + s\eta, \eta) \in L_p$ in s for each (x_0, η) with

$$\|\psi\|^0_p \le C_p \cdot \|\varphi\|^0_p, \quad \psi(s; x_0, \eta) = h(x_0 + s\eta; \eta). \qquad (2.7.17)$$

By raising (2.7.17) to the $p-th$ power and integrating, we obtain

$$\int_{R_\nu} |h(x, \eta)|^p\, dx \le C^p_p (\|f\|^0_p)^p \quad (dx = dx_0\, ds). \qquad (2.7.18)$$

The result (2.7.14) follows from (2.7.15) and (2.7.18) by first applying the Hölder inequality with measure $\frac{1}{2}|\Delta(y)|\, d\Sigma(y)$ to (2.7.15) to estimate $|g(x)|^p$ and then integrating with respect to x.

Before we can prove the theorem corresponding to Theorem 2.7.2 when Δ satisfies (2.6.10) instead of (2.7.12), we need the following useful lemmas:

Theorem 2.7.3. *If $u \in C^1_{\mu c}(G)$, then*

$$u(x) = \int_G K_{0,\alpha}(x - \xi)\, u_{,\alpha}(\xi)\, d\xi \qquad (2.7.19)$$
$$= \Gamma_\nu^{-1} \int_G |x - \xi|^{-\nu} (x^\alpha - \xi^\alpha)\, u_{,\alpha}(\xi)\, d\xi.$$

Proof. We may assume $G \subset B(x, A)$, $u = 0$ in $B(x, A) - G$. Taking polar coordinates at x, the integral is just

$$\int_0^A \left[\int_\Sigma - u_r(r, \eta)\, d\Sigma(\eta)\right] dr.$$

Lemma 2.7.1. *If x_1 and $x_2 \in B_A$,*

$$\int_{B_R} |\xi - x_1|^{1-\nu} \cdot |\xi - x_2|^{1-\nu}\, d\xi = \begin{cases} -C_0^2 K_0(x_2 - x_1) + \varepsilon(x_1, x_2, R), & \text{if } \nu > 2 \\ -(2\pi)^2 K_0(x_2 - x_1) + C_1 + 2\pi \log R + \varepsilon(x_1, x_2, R), & \nu = 2 \end{cases} \qquad (2.7.20)$$

where C_0 *and* C_1 *are constants and* $\varepsilon(x_1, x_2, R)$ *converges uniformly on* $B_A \times B_A$ *to* 0 *as* $R \to \infty$.

Proof. If $\nu > 2$, it follows that as $R \to \infty$, the integral tends to a function of $|x_2 - x_1|$ only which is homogeneous of degree $2 - \nu$ and so must be a negative multiple of K_0. Clearly also

$$\varepsilon(x_1, x_2, R) = -\int_{R_\nu - B_R} |\xi - x_1|^{1-\nu} \cdot |\xi - x_2|^{1-\nu} d\xi$$

converges as stated.

In the case $\nu = 2$, let $\bar{x} = (x_1 + x_2)/2$ and $\varrho = |x_2 - x_1|$ and suppose $R > A + \varrho$. Then

$$\text{(2.7.21)} \quad \begin{aligned} \int_{B(\bar{x}, R-|\bar{x}|)} |\xi - x_1|^{-1} \cdot |\xi - x_2|^{-1} d\xi &\leq \int_{B_R} |\xi - x_1|^{-1} \cdot |\xi - x_2|^{-1} d\xi \\ &\leq \int_{B(\bar{x}, R+|\bar{x}|)} |\xi - x_1|^{-1} \cdot |\xi - x_2|^{-1} d\xi. \end{aligned}$$

Evaluating the integral on the left, we obtain

$$C_2 + 2\pi \log[(R - |\bar{x}|)/\varrho] + \varphi[(R - |\bar{x}|)/\varrho]$$

where

$$C_2 = \int_{B(\bar{x}, \varrho)} |\xi - x_1|^{-1} \cdot |\xi - x_2|^{-1} d\xi = \int_{B(0,1)} \left|\xi - \frac{1}{2} e_1\right|^{-1} \cdot \left|\xi + \frac{1}{2} e_1\right|^{-1} d\xi \qquad (e_1 = (1,0))$$

$$\begin{aligned} \varphi(S) &= \int_{B(\bar{x}, R-|\bar{x}|) - B(\bar{x}, \varrho)} (|\xi - x_1|^{-1} \cdot |\xi - x_2|^{-1} - |\xi - \bar{x}|^{-2}) d\xi \\ &= \int_{B(0,S) - B(0,1)} (|\xi - e_1/2|^{-1} \cdot |\xi + e_1/2|^{-1} - |\xi|^{-2}) d\xi, \\ &\qquad S = (R - |\bar{x}|)/\varrho. \end{aligned}$$

In like manner, the right integral in (2.7.21) is given by

$$C_2 + 2\pi \log[(R + |\bar{x}|)/\varrho] + \varphi[(R + |\bar{x}|)/\varrho].$$

Clearly $\varphi(S)$ tends to a limit φ_0 as $S \to \infty$. If we take $C_1 = C_2 + \varphi_0$, then

$$\begin{aligned} 2\pi \log(1 - |\bar{x}|/R) + \varphi[(R - |\bar{x}|)/\varrho] - \varphi_0 &\leq \varepsilon \\ &\leq 2\pi \log(1 + |\bar{x}|/R) + \varphi[(R + |\bar{x}|)/\varrho] - \varphi_0. \end{aligned}$$

Theorem 2.7.4. *Theorem 2.7.2 holds, with* $\|\Delta\|_{1\Sigma}^{0}$ *replaced by* $M_0 + M_1$ *in* (2.7.14), *if* Δ *satisfies* (2.6.9) *and* (2.6.10) *instead of* (2.6.9) *and* (2.7.12). *Here* M_0 *and* M_1 *are the respective maxima of* $|\Delta(x)|$ *and* $|\nabla \Delta(x)|$ *for* x *on* $\partial B(0,1)$.

Proof. We suppose first that $f \in C_c^\infty(R_\nu)$, the general case follows by approximations. Let us define

$$\text{(2.7.22)} \qquad h(x) = \int_{R_\nu} R(x - \xi) f(\xi) d\xi, \quad R(y) = C_0^{-1} |y|^{1-\nu},$$

$C_0 = 2\pi$ *if* $\nu = 2$ and otherwise is the constant of Lemma 2.7.1. Then, by Theorem 2.6.4, $h \in C^1_\mu(R_\nu)$ with

$$h_{,\alpha}(x) = \lim_{\varrho \to 0^+} \int_{R_\nu} R_{,\alpha}(x-\xi) f(\xi)\, d\xi = \int_{R_\nu} R(x-\xi) f_{,\alpha}(\xi)\, d\xi \tag{2.7.23}$$

from which it follows that $h \in C^\infty(R_\nu)$. Since f has compact support, it follows that

$$\nabla h(x) = 0(|x|^{-\nu}),\ \Delta h(x) = 0(|x|^{-\nu-1}) \text{ at } \infty. \tag{2.7.24}$$

Next, we define

$$k(x) = -\int_{R_\nu} R(x-\xi)\Delta h(\xi)\, d\xi = -\int_{B_R} R(x-\xi)\Delta h(\xi)\, d\xi + \varepsilon_R(x) \tag{2.7.25}$$

where ε_R converges uniformly to 0 on any compact set on account of (2.7.24). Since the support of f is in some B_A, we may write (since $\int_{B_A} \Delta f(\eta)\, d\eta = 0$)

$$\begin{aligned} k(x) &= \varepsilon_R(x) - \int_{B_A}\left[\int_{B_R} R(x-\xi) R(\xi-\eta)\, d\xi\right] \Delta f(\eta)\, d\eta \\ &= \varepsilon_{1R}(x) + \int_{B_A} K_0(x-\eta)\Delta f(\eta)\, d\eta \\ &= \varepsilon_{1R} + \int_{B_A} K_{0,\alpha}(x-\eta) f_{,\alpha}(\eta)\, d\eta = \varepsilon_{1R}(x) + f(x), \\ \varepsilon_{1R}(x) &= \varepsilon_R(x) - \int_{B_A} \varepsilon_1(x, \eta, R)\Delta f(\eta)\, d\eta \end{aligned}$$

for all $\nu \geq 2$. Since k and f are independent of R, $\varepsilon_{1R} \equiv 0$. Hence, substituting the right side of (2.7.25) for $f(x)$ in (2.7.13), we obtain

$$\begin{aligned} g(x) &= \lim_{\varrho \to 0}\left\{\lim_{\sigma \to 0} \int_{R_\nu - B(x,\varrho)} \Delta(x-\xi)\left[\int_{R_\nu - B(x,\sigma)} - R(\xi-\eta)\Delta h(\eta)\, d\eta\right] d\xi\right. \\ &= \lim_{\varrho \to 0}\left\{\lim_{\sigma \to 0} \int_{R_\nu - B(x,\sigma)} \Delta h(\eta)\left[\int_{R_\nu - B(x,\varrho)} - \Delta(x-\xi) R(\xi-\eta)\, d\xi\right] d\eta.\right. \end{aligned} \tag{2.7.26}$$

If we now set $\xi = x - \omega$ on the right in (2.7.26), we obtain

$$\begin{aligned} g(x) &= -\lim_{\varrho \to 0}\left\{\lim_{\sigma \to 0} \int_{R_\nu - B(x,\sigma)} S(x-\eta;\ \varrho)\Delta h(\eta)\, d\eta\right\} \\ S(y;\ \varrho) &= \int_{R_\nu - B_\varrho} \Delta(\omega) R(y-\omega)\, d\omega. \end{aligned} \tag{2.7.27}$$

From the homogeneity, we conclude that

$$\begin{aligned} S(y;\varrho) &= |y|^{1-\nu} S(\eta;\varrho/|y|),\ \eta = |y|^{-1} y,\ |\eta| = 1 \\ S(-y;\varrho) &= S(y;\varrho). \end{aligned} \tag{2.7.28}$$

Now, if $|\eta| = 1$ and $|\omega| \geq 3/2$, then $|\eta - \omega| \geq |\omega|/3$. If $|\eta| = 1$ and $1/2 \leq |\omega| \leq 3/2$, then $|\Delta(\omega)| \leq 2^\nu M_0$ if $|\eta - \omega| \leq 1/2$ and $R(\eta - \omega)$

$\leq 2^{\nu-1} \cdot C_0$ if $|\eta - \omega| \geq 1/2$, M_0 being the max. of $|\Delta(\omega)|$ for $|\omega| = 1$. Finally, if $0 < \varrho < 1/2$,

$$S(\eta;\varrho) = \int_{R_\nu - B(0,1/2)} \Delta(\omega) R(\eta - \omega)\,d\omega + \int_{B(0,1/2)-B(0,\varrho)} \Delta(\omega)\,[R(\eta-\omega) - R(\eta)]\,d\omega \tag{2.7.29}$$

and as $\varrho \to 0$, $S(\eta;\varrho)$ converges uniformly for η on Σ to $S(\eta)$. From the analysis above, we see that

$$|S(y;\varrho)| \leq C(\nu)\, M_0\, |y|^{1-\nu}, \quad \lim_{\varrho\to 0} S(y;\varrho) = S(y) \tag{2.7.30}$$

so that we may let $\sigma \to 0$ and then $\varrho \to 0$ in (2.7.27) to yield

$$g(x) = \int_{R_\nu} S(x-\eta)\, \Delta h(\eta)\, d\eta. \tag{2.7.31}$$

Now, in (2.7.29), we assume that $|\eta|$ is near 1 and $0 < \varrho < 1/2$. By defining

$$S_1(\eta;\varrho,\tau) = \int_{R_\nu - B(0,1/2) - B(\eta;\tau)} \Delta(\omega)\, R(\eta-\omega)\, d\omega,$$

differentiating with respect to η^α and letting $\tau \to 0$, we find that $S(\eta;\varrho)$ is of class C^1 in η and that

$$\begin{aligned} S_{\eta^\alpha}(\eta;\varrho) &= \int_{R_\nu - B(0,1/2) - B(\eta,1/3)} \Delta(\omega)\, R_{,\alpha}(\eta-\omega)\, d\omega + \int_{B(\eta,1/3)} [\Delta(\omega) - \Delta(\eta)]\, R_{,\alpha}(\eta-\omega)\, d\omega \, + \\ &+ \int_{B(0,1/2)-B(0,\varrho)} \Delta(\omega)\,[R_{,\alpha}(\eta-\omega) - R_{,\alpha}(\eta)]\,d\omega; \quad (0<\varrho<1/2) \end{aligned} \tag{2.7.32}$$

in this development, we used (2.6.9) (c) for $R_{,\alpha}$. In (2.7.32), we may let $\varrho \to 0$ and conclude from the uniform convergence in η for $|\eta|$ near 1 that $S \in C^1(R_\nu - \{0\})$ and that

$$|\nabla S(y)| \leq [C_1(\nu)\, M_0 + C_2(\nu)\, M_1]\, |y|^{-\nu} \tag{2.7.33}$$

where M_1 is a bound for $|\nabla \Delta(\omega)|$ for $|\omega| = 1$. Thus, from (2.7.31), we deduce that

$$-g(x) = \lim_{\varrho \to 0} \int_{R_\nu - B(x,\varrho)} S_{,\alpha}(x-\eta)\, h_{,\alpha}(\eta)\, d\eta.$$

From (2.7.28), etc., we see that ∇S and ∇R satisfy (2.7.12). From (2.7.23) and Theorem 2.7.2 we conclude that $\nabla h \in L_p(R_\nu)$ with

$$\|\nabla h\|_p^0 \leq C(p,\nu)\, \|f\|_p^0. \tag{2.7.34}$$

From (2.7.34), we see that ∇h can be approximated strongly in L_p by functions $k_n \in C_c^\infty(R_\nu)$ satisfying (2.7.34) uniformly so that, by Theorem 2.7.2 again, g is the strong limit (uniform on any bounded set) in L_p of functions g_n so that

$$\|g\|_p^0 \leq C^*(p,\nu)\, \|\nabla S\|_{1,\Sigma}^0 \cdot \|\nabla h\|_p^0$$

from which the result follows, since $\|\nabla S\|_{1,\Sigma}^0 \leq C(\nu) \cdot (M_0 + M_1)$ from (2.7.33).

From Theorems 2.7.2 and 2.7.4, we obtain the corollary.

Theorem 2.7.5. *If Δ satisfies (2.6.9), $f \in C^0_{\mu c}(R_\nu)$, and g is defined by (2.7.13), then $g \in C^0_\mu(R_\nu) \cap L_p(R_\nu)$ for every $p > 1$ and*

$$\| g \|^0_p \leq C(\nu, p)\,(M_0 + M_1)\,\| f \|^0_p$$

where M_0 and M_1 are defined in the statement of Theorem 2.7.4.

2.8. The maximum principle for a linear elliptic equation of the second order

We conclude this chapter with the well known proof of this principle due to E. Hopf ([1]).

Theorem 2.8.1. *Suppose that $u \in C^2(G)$, that $a^{\alpha\beta}$, b^α, c and $f \in C^0(G)$, c and f satisfy*

$$c(x) \equiv 0, \quad f(x) \geq 0, \quad x \in G \tag{2.8.1}$$

and that u is a solution of the equation

$$a^{\alpha\beta}(x)\, u_{,\alpha\beta}(x) + b^\alpha(x)\, u_{,\alpha}(x) + c(x)\, u(x) = f(x), \quad x \in G, \tag{2.8.2}$$

assumed to be elliptic on G. Suppose that $x_0 \in G$ and that u takes on its maximum value at x_0. Then $u(x) = u(x_0)$ for all x on the domain G.

Proof. Suppose $u(x) \not\equiv u(x_0) = M$. Then the set where $u(x) < M$ is open. There is a ball $B(x_2, R) \subset G$ such that $u(x) < M$ for $x \in \overline{B(x_2, R)} - \{x_1\}$, where $x_1 \in \partial B(x_2, R)$ and $u(x_1) = M$. Finally, there is a ball $\overline{B(x_1, R_1)} \subset G$ with $R_1 < R$. Let $S_i = \overline{B(x_2, R)} \cap \partial B(x_1, R_1)$ and $S_e = \partial B(x_1, R_1) - \overline{B(x_2, R)}$, so that $S_i \cup S_e = \partial B(x_1, R_1)$. Then

$$u(x) \leq M - \varepsilon \quad \text{on } S_i \quad \text{and } u(x) \leq M \text{ on } S_e$$

for some $\varepsilon > 0$.

Now, let

$$h(x) = e^{-\gamma r^2} - e^{-\gamma R^2}, \quad r = |x - x_2|.$$

Letting $L\varphi$ stand for $a^{\alpha\beta}\varphi_{,\alpha\beta} + b^\alpha \varphi_{,\alpha}$, we see that

$$e^{\gamma r^2} L h = 4\gamma^2 a^{\alpha\beta}(x^\alpha - x_2^\alpha)(x^\beta - x_2^\beta) - 2\gamma[a^{\alpha\beta}\delta_{\alpha\beta} + b^\alpha(x^\alpha - x_2^\alpha)].$$

We can choose γ so large that $Lh(x) > 0$ in $\overline{B(x_1, R_1)}$. Finally

$$h(x) < 0 \text{ on } S_e, \quad h(x_1) = 0. \tag{2.8.3}$$

Let

$$v(x) = u(x) + \delta h(x), \quad \delta > 0, \tag{2.8.4}$$

where δ is small enough so that $v(x) < M$ on S_i. From (2.8.3), we see that $v(x) < M$ on $\partial B(x_1, R_1)$, $v(x_1) = M$ so that v has a maximum at a point x_3 in $B(x_1, R_1)$ while $L(v) > 0$ there.

But this would imply that (since all $v_{,\alpha}(x_3) = 0$)

$$a^{\alpha\beta}(x_3)\, v_{,\alpha\beta}(x_3) > 0 \text{ but } v_{,\alpha\beta}(x_3)\,\eta^\alpha\eta^\beta \leq 0 \text{ for all } \eta. \tag{2.8.5}$$

Now, we may define new variables y and w by the rotation

$$y^\gamma = c^\gamma_\alpha(x^\alpha - x_3^\alpha), \quad \zeta^\gamma = c^\gamma_\alpha \eta^\alpha, \quad w(y) = v(x)$$

where the matrix c is chosen so that $c^{-1}\, a(x_3)\, c = \acute{a}(0)$ is diagonal. Then (2.8.5) is equivalent to

$$(2.8.6)\qquad \sum_{\gamma=1}^{\nu} \acute{a}^{\gamma\gamma}(0)\, w_{,\gamma\gamma}(0) > 0 \quad \text{but} \quad w_{,\gamma\delta}(0)\, \zeta^{\gamma}\, \zeta^{\delta} \leq 0 \quad \text{for all } \zeta$$

where all the $\acute{a}^{\gamma\gamma} > 0$. But the first inequality in (2.8.6) implies that some one $w_{,\gamma\gamma}(0) > 0$ which contradicts the second.

Corollary. *If u and the coefficients satisfy the conditions of Theorem* 2.8.1 *except that we require $c(x) \leq 0$, then u cannot have a positive maximum. If, also, $f(x) \equiv 0$, then u has neither a positive maximum nor a negative minimum.*

Chapter 3

The spaces H_p^m and H_{p0}^m

3.1. Definitions and first theorems

In this chapter we collect statements and proofs of the theorems which we need concerning these functions. As we stated in Chapter 1, these and similar spaces have been discussed at great length by many authors and have certainly proved their worth in connection with the study of differential equations.

Definition 3.1.1. A function u is said to be *of class H_p^m* on G iff u is of class L_p on G and there exist functions r_α of class L_p on G, $\alpha = \alpha_1, \ldots, \alpha_\nu$, $0 \leq |\alpha| \leq m$, such that

$$(3.1.1)\qquad \int_G g(x)\, r_\alpha(x)\, dx = (-1)^{|\alpha|} \int_G D^\alpha g(x)\, u(x)\, dx, \quad g \in C_c^\infty(G).$$

The functions r_α (or rather the classes of equivalent functions determined by them) are called the *distribution derivatives* of u and r_α is hereafter denoted by $D^\alpha u$ or $u_{,\alpha}$; we make the convention that $D^\alpha u = u$ if $|\alpha| = 0$.

Remarks. It is clear that if u is of class H_p^m on G, its distribution derivatives are determined only up to additive null functions; moreover if u^* differs from u by a null function, then u^* is also of class H_p^m on G and has the same distribution derivatives. The definitions above extend to vector functions and to complex-valued functions. The LEBESGUE derivatives (SAKS p. 106) of the set functions $\int_e r_\alpha(x)\, dx$ are called the *generalized derivatives* of u at points where they exist.

Theorem 3.1.1. *The space $H_p^m(G)$ of classes of equivalent vector functions of class H_p^m on G with norm defined by*

$$(3.1.2)\qquad \|u\|_p^m = \left\{ \int_G \left[\sum_{i=1}^{N} \sum_{0 \leq |\alpha| \leq m} C_\alpha \left| D^\alpha u^i \right|^2 \right]^{p/2} dx \right\}^{1/p} \quad (u = u^1, \ldots, u^N)$$

is a BANACH *space. If $p = 2$, the space is a* HILBERT *space if we define*

$$(u, v)_2^m = \int_G \sum_{i=1}^{N} \sum_{0 \le |\alpha| \le m} C_\alpha D^\alpha u^i \overline{D^\alpha v^i}\, dx. \tag{3.1.3}$$

In (3.1.2) *and* (3.1.3), *C_α denotes the multinomial coefficient $|\alpha|!/\alpha_1! \ldots \alpha_\nu!$*

Proof. The only property requiring proof is the completeness. So suppose $\{u_n\}$ is a CAUCHY sequence in H_p^m. Then each of the sequences $D^\alpha u_n$, with $0 \le |\alpha| \le m$, is a CAUCHY sequence in $L_p(G)$ and so converges in $L_p(G)$ to some vector function $r_\alpha (= u$ if $|\alpha| = 0)$. But, for each α with $0 \le |\alpha| \le m$, it follows that (3.1.1) holds in the limit so that the r_α are the corresponding distribution derivatives of u.

Remark 1. It will be convenient to call these elements of H_p^m functions and to say that u is continuous, harmonic, etc., iff some representative of the class forming the element has these properties. Naturally, also, there are many different topologically equivalent norms which could be used.

Remark 2. The spaces $H_p^m(G)$ have been defined for all real m (see, for example LIONS [1]) and interesting results in the theory of differential equations have been obtained using these and other spaces such as those introduced by ARONSZAJN and SMITH [1], [2], CALDERON and others. A discussion of these matters, though interesting, would lead us rather far afield into functional analysis and is not really relevant to our discussion. Accordingly, we shall not define them here. However, a special class is used in Chapter 8.

Theorem 3.1.2. (a) *If $u \in H_p^m(G)$ and $D \subset G$, then $u|_D \in H_p^m(D)$.*

(b) *Suppose u is defined on the whole of G and each point of G is in a domain D such that $u|_D \in H_p^m(D)$. Then there are functions r_α defined on G such that $(D^\alpha u|_D)(x) = r_\alpha(x)$ for almost all x on D, and any $D \subset\subset G$.*

(c) *If, in* (b), *the $r_\alpha \in L_p(G)$, then $u \in H_p^m(G)$.*

(d) *If $u \in H_p^m(G)$, then $\nabla^k u \in H_p^{m-k}(G)$, $1 \le k \le m$.*

(e) *If $u \in H_p^m(G)$ and $\nabla^m u \in H_p^r(G)$, then $u \in H_p^{m+r}(G)$.*

(f) *If $u \in H_p^m(G)$ and $g \in C_{1,c}^{m-1}(G)$* (i. e. *$g \in C_c^{m-1}(G)$*

and all its derivatives of order $\le m-1$ are LIPSCHITZ, *then* (3.1.1) *holds with $r_\alpha = D^\alpha u \equiv u_{,\alpha}$.*

(g) *If $u \in L_p(G)$, u is absolutely continuous in each variable (on segments in G) for almost all values of the other variables, and if its first partial derivatives (which consequently exist a.e. and are measurable) $\in L_p(G)$, then $u \in H_p^1(G)$ and its partial and generalized derivatives coincide almost everywhere.*

Proof. Parts (a) — (e) are obvious and part (f) follows by a straightforward approximation of the function g. To prove (g), we notice that if $g \in C_c^\infty(G)$, then $g(x) \cdot u(x)$ has the absolute continuity properties of u

and has compact support in G. From FUBINI's theorem, it follows that we may take the r_α as the partial derivatives $\partial u/\partial x^\alpha$ in (3.1.1).

We recall the definitions of a mollifier φ and the φ-mollified functions u_ϱ which were given in Chapter 1 (Definition 1.8.3).

Theorem 3.1.3. *Suppose $u \in H_p^m(G)$, φ is a mollifier, and u_ϱ is the φ-mollified function of u. Then $u_\varrho \to u$ in $H_p^m(D)$ for each $D \subset\subset G$ and*

$$(3.1.4) \qquad \psi_{\alpha\varrho}(x) = D^\alpha u_\varrho(x) \text{ if } \psi_\alpha = u_{,\alpha}, \quad 1 \leq |\alpha| \leq m.$$

If $m = 1$ and $u_{,\alpha}(x) = 0$ a.e. on G, $\alpha = 1, \ldots, \nu$, and G is a domain, then $u(x) = const.$ a.e. on G.

Proof. If $\overline{B(x,\varrho)} \subset G$, the function $g_{x\varrho}$ defined on G by

$$g_{x\varrho}(\xi) = \varphi_\varrho^*(\xi - x) \qquad (\varphi_\varrho^*(y) = \varrho^{-\nu}\varphi(\varrho^{-1}y))$$

$\in C_c^\infty(G)$. Hence, using Theorem B, § 1.8 and the definitions, we see that

$$D^\alpha u_\varrho(x) = \int_G (-1)^{|\alpha|} g_{x\varrho,\alpha}(\xi)\, u(\xi)\, d\xi = \int_G g_{x\varrho}(\xi)\, D^\alpha u(\xi)\, d\xi = \psi_{\alpha\varrho}(x).$$

The remaining statements follow from Theorem B, § 1.8 and the connectedness of G.

Remark. It is not known (and the writer believes it is not true) that a function $u \in H_p^m$ on a domain G with sufficiently wild boundary can be approximated over the whole of G by functions of class C^m on a domain containing $G \cup \partial G$. We shall prove this, however, for a surprisingly wide class of domains in § 3.4.

The following theorem is an immediate consequence of Theorem 3.1.1, 3.1.2 b and c, and 3.1.3 and the details of its proof are left to the reader.

Theorem 3.1.4. *If $u \in H_p^m(G)$ and $\zeta \in C_1^{m-1}(G)$ and ζ and its derivatives are bounded on G, then $\zeta u \in H_p^m(G)$ and its derivatives are obtained from those of ζ and u by their usual formulas.*

Proof. For if φ is a mollifier and ζ_ϱ and u_ϱ are the φ-mollified functions we conclude that $D^\alpha \zeta_\varrho$ converges almost everywhere and boundedly to $D^\alpha \zeta$ and $D^\alpha u_\varrho$ converges strongly in L_p to $D^\alpha u$ on each $D \subset\subset G$, $0 \leq |\alpha| \leq m$.

Definition 3.1.2. A mapping $T\colon x = x(y)$ of class C^m or C_1^{m-1} of a domain H onto a domain G is said to be *regular* iff it is $1 - 1$ and all the derivatives of order $\leq m$ of the mapping functions for T and T^{-1} are uniformly bounded.

Theorem 3.1.5. *Suppose that $x = x(y)$ is a regular mapping of class C^m of a domain H onto G, suppose $u \in H_p^m(G)$, and suppose $v(y) = u[x(y)]$. Then $v \in H_p^m(H)$ and, if all of the generalized derivatives $D^\alpha u(x_0)$ exist at $x_0 = x(y_0)$, then all of the generalized derivatives $D^\alpha v(y_0)$ exist and are connected with those of u at x_0 by the usual rules of the calculus.*

Proof. From Theorem 3.1.3, it follows that we may approximate strongly to u in $H_p^m(\Delta)$ for each $\Delta \subset \subset G$. If D is the counter-image of such a Δ, the transformed functions converge strongly in $H_p^m(D)$ to v. The last statement follows easily since a regular family of sets (see SAKS, p. 106) about y_0 corresponds to one about x_0.

We now digress to give a simple proof of the famous theorem of RADEMACHER concerning LIPSCHITZ functions.

Theorem 3.1.6. *Suppose u satisfies a* LIPSCHITZ *condition on G. Then u possesses a total differential almost everywhere on G.*

Proof. Let Z_0 be the set of points where one of the generalized derivatives of u fails to exist. Let E be an everywhere dense denumerable set of points ζ on Σ. With each such ζ, let $x = x_\zeta(y)$ be a rotation which carries the vector e_1 in the y-space into ζ and let $v_\zeta(y) = u[x_\zeta(y)]$. The generalized (first) derivatives of v_ζ exist at *all points* y for which $x_\zeta(y) \notin Z_0$. Also, for each ζ, there is a set Z_ζ of measure 0 such that the *partial* derivative $\partial v_\zeta/\partial y^1 = D_{y^1} v$ if $x_\zeta(y) \notin Z_\zeta$ (Theorem 3.1.2 g). If $x_0 \notin Z_0 \cup (\cup Z_\zeta)$, then all the *directional derivatives* in the directions ζ exist and are connected with the generalized derivatives $D^\alpha u(x_0)$ by the usual formulas. At any such point, it is easy to see, using the LIPSCHITZ condition that u has a total differential.

From this, we conclude the following generalization of Theorem 3.1.5:

Theorem 3.1.7. *Suppose $x = x(y)$ is a regular mapping of class $C_1^{m-1}(m \geq 1)$ of H onto G, $u \in H_p^m(G)$, and $v(y) = u[x(y)]$. Then $v \in H_p^m(H)$ and if all the generalized derivatives $D^\alpha u(x_0)$ exist and all those of the x^ν at y_0 exist, and if $x_0 = x(y_0)$, then all the generalized derivatives $D^\alpha v(y_0)$ exist and are given by their usual formulas.*

Proof. Since $x = x(y)$ is of class C^{m-1}, $v \in H_p^{m-1}(D)$ on each $D \subset \subset G$. If $m > 1$, we conclude from that theorem that any derivative $D^\alpha v$ of order $|\alpha| \leq m - 1$ is given by a formula of the form

$$D^\alpha v(y) = \sum_{\substack{\beta \leq \alpha \\ |\beta| \geq 1}} A_\beta^\alpha(y)\, u_{,\beta}[x(y)] \tag{3.1.5}$$

where $1 \leq \beta \leq \mathrm{m} - 1$ and the A_β^α are polynomials in the derivatives of the x^ν of order $|\alpha| + 1 - |\beta| \leq m - 1$ and hence are LIPSCHITZ. Since the $u_{,\beta}$ all $\in H_p^1(G)$ at least, the theorem will follow from Theorem 3.1.2(g) and the special case $m = 1$.

If u_ϱ is a mollified function of u, defined on G_ϱ, if $D \subset \subset$ the counter image of G_ϱ, and $v_\varrho^*(y) = u_\varrho[x(y)]$, then v_ϱ^* is LIPSCHITZ and

$$v_{\varrho,\alpha}^*(y) = u_{\varrho,\beta}[x(y)] \cdot x_{,\alpha}^\beta(y) \quad \text{(a.e.)} \tag{3.1.6}$$

It is clear that we may let $\varrho \to 0$ and conclude $v \in H_p^1(D)$ with $v_{,\alpha}$ given by the limit of the right side of (3.1.6). If, now the generalized derivatives $u_{,\beta}(x_0)$ and $\bar{x}_{,\alpha}^\beta(y_0)$ all exist and e runs through a regular family of sets at y_0 and E is the image of e, these forming a regular family at x_0, we conclude:

$$
\begin{aligned}
&\left| |e|^{-1} \int_e [v_{,\alpha}(y) - \bar{u}_{,\beta}(x_0)\, \bar{x}^{\beta}_{.\alpha}(y_0)]\, dy \right| \\
(3.1.7) \qquad &\le |e|^{-1} \int_e |u_{,\beta}[x(y)] - \bar{u}_{,\beta}(x_0)| \cdot |x^{\beta}_{.\alpha}(y)|\, dy + \\
&+ \left| |e|^{-1} u_{,\beta}(x_0) \cdot \int_e |x^{\beta}_{.\alpha}(y) - \bar{x}^{\beta}_{.\alpha}(y_0)| \right| dy.
\end{aligned}
$$

The second term on the right clearly tends to zero and the first does also since it is

$$\le C |E|^{-1} \int_E |u_{,\beta}(x) - \bar{u}_{,\beta}(x_0)|\, dx$$

where C depends on the LIPSCHITZ constants.

We now prove two theorems concerned with the absolute continuity properties of functions $\in H_p^1(G)$.

Lemma 3.1.1. *Each class of functions u in $H_p^1(G)$ contains a representative u_0 which for each α is absolutely continuous in x^α on each segment in G with endpoints in $G \cup \partial G$ for almost all values of x'_α and u_0 tends to limits as (x^α, x'_α) tends to the end points of any such segment.*

Proof. If $R = [a, b]$ is any rational cell in G, there is a sequence of values of $\varrho \to 0$ such that

$$\lim_{\varrho \to 0} \int_{a^\alpha}^{b^\alpha} [|u_\varrho(x^\alpha, x'_\alpha) - u(x^\alpha, x'_\alpha)|^p + |u_{\varrho,\alpha}(x^\alpha, x'_\alpha) - u_{,\alpha}(x^\alpha, x'_\alpha)|^p]\, dx = 0 \tag{3.1.8}$$

for almost all x'_α, $\alpha = 1, \ldots, \nu$. For any x'_α for which (3.1.8) holds, the u_ϱ are uniformly absolutely continuous and hence tend uniformly to an absolutely continuous function. The result follows by ordering the rational cells in G and choosing successive subsequences.

Theorem 3.1.8. *Each class of functions in $H_p^1(G)$ ($\subset H_1^1(G)$) contains a representative $\bar{u}$ which has the absolute continuity properties of u_0 in the lemma and which retains these properties under a regular* LIPSCHITZ *mapping. In fact we may define $\bar{u}(x_0)$ as the* LEBESGUE *derivative of $\int_e u(x)\, dx$ for each x_0 for which that derivative exists.*

Proof. Since regular families of sets correspond under regular LIPSCHITZ maps, we see that $\bar{u}[x(y)] = \overline{u[x(y)]}$ and so it suffices to show that $\bar{u}$ has these properties in some one coordinate system. Also, it is clear that $\bar{u}(x_0)$ is defined if there is an A such that

$$\lim_{h \to 0} (2h)^{-\nu} \int_{x_0 - h}^{x_0 + h} |u(x) - A|\, dx = 0 \tag{3.1.9}$$

in which case $\bar{u}(x_0) = A$.

Now, let u_0 be a representative as in Lemma 3.1.1 and let $[a, b] \subset (A, B) \subset [A, B] \subset G$. Using FUBINI's theorem and LEBESGUE's theorem, it follows that for each α, there is a set Z_α of x'_α of measure 0 such that if $x'_{\alpha 0}$ is not in Z_α then $u_0(x^\alpha, x'_{\alpha 0})$ is A.C. in $[A^\alpha, B^\alpha]$ and

$$\lim_{h\to 0} (2h)^{1-\nu} \int_{x'_{\alpha 0}-h}^{x'_{\alpha 0}+h} \int_{A^\alpha}^{x^\alpha} |u_{0,\alpha}(\xi)|\, d\xi = \int_{A^\alpha}^{x^\alpha} |u_{0,\alpha}(\xi^\alpha, x'_{\alpha 0})|\, d\xi^\alpha \tag{3.1.10}$$

for a given dense denumerable set of x^α on $[A^\alpha, B^\alpha]$, including A^α and B^α and

$$\lim_{h\to 0} (2h)^{-\nu} \int_{x_0-h}^{x_0+h} |u_0(\xi) - u_0(x_0)|\, d\xi = 0 \quad (x_0 = (x_0^\alpha, x'_{\alpha 0})) \tag{3.1.11}$$

for *almost every* x_0^α on $[A^\alpha, B^\alpha]$. Since the functions of x^α in (3.1.10) are all monotone and continuous, the convergence in (3.1.10) *is uniform*.

Now, let $x_0 = (x_0^\alpha, x'_{\alpha 0})$ be *any* point for which $x'_{\alpha 0}$ is not in Z_α and $a^\alpha \le x_0^\alpha \le b^\alpha$ and let $\varepsilon > 0$. Using the uniform convergence in (3.1.10) and the absolute continuity of $u_0(x^\alpha, x'_{\alpha 0})$, we see that we can choose an $x_1^\alpha > x_0^\alpha$ on (A^α, B^α) such that (3.1.11) holds at $(x_1^\alpha, x'_{\alpha 0})$ and such that

$$|u_0(x_1^\alpha, x'_{\alpha 0}) - u_0(x_0^\alpha, x'_{\alpha 0})| < \varepsilon/3,$$

$$\int_{x^\alpha_0}^{x^\alpha_1} |u_{0,\alpha}(\xi^\alpha, x'_{\alpha 0})|\, d\xi^\alpha < \xi/3$$

$$(2h)^{-\nu} \int_{x^\alpha_0}^{x^\alpha_1} dx^\alpha \int_{x^\alpha-h}^{x^\alpha+h} d\xi^\alpha \int_{x'_{\alpha 0}-h}^{x'_{\alpha 0}+h} |u_{,\alpha}(\xi^\alpha, x'_\alpha)|\, dx'_\alpha < \varepsilon/3, \quad 0 < h < h_0.$$

Then (setting $\delta = x_1^\alpha - x_0^\alpha$)

$$\left| (2h)^{-\nu} \int_{x_0-h}^{x_0+h} |u_0(\xi) - u_0(x_0)|\, d\xi - (2h)^{-\nu} \int_{x^\alpha_1-h}^{x^\alpha_1+h} \int_{x'_{\alpha 0}-h}^{x'_{\alpha 0}+h} |u_0(\xi) - u_0(x_1^\alpha, x'_{\alpha 0})|\, d\xi \right|$$

$$\le |u_0(x_1^\alpha, x'_{\alpha 0}) - u_0(x_0)| + (2h)^{-\nu} \int_{x_0-h}^{x_0+h} |u_0(\xi+\delta) - u_0(\xi)|\, d\xi < 2\varepsilon/3,$$

$$0 < h < h_0,$$

so that (3.1.11) holds and $\bar{u}(x^\alpha, x'_{\alpha 0})$ is defined and $= u_0(x^\alpha, x'_{\alpha 0})$ for *all* x^α on $[a^\alpha, b^\alpha]$.

Definition 3.1.3. A function u is *absolutely continuous in the sense of* Tonelli (A.C.T.) *on* G if and only if $u \in H_1^1(G)$ and u is continuous on G.

Corollary. *If u is* A.C.T. *on G, it has the absolute continuity properties described in Lemma* 3.1.1 *and if $x = x(y)$ is a regular* Lipschitz *map of a domain H onto G and $v(y) = u[x(y)]$, then v is* A.C.T. *on G.*

Theorem 3.1.9. *Suppose $F \in C^1(R_P)$, suppose each $u^p \in H^1_{\lambda_p}(G)$ for some $\lambda_p \ge 1$, $p = 1, \ldots, P$, suppose*

$$U(x) = F[u^1(x), \ldots, u^P(x)], \qquad V_\alpha(x) = \sum_{p=1}^{P} F_{,p}[\boldsymbol{u}(x)]\, u^p_{,\alpha}(x) \tag{3.1.12}$$

and suppose U and the $V_\alpha \in L_\lambda(G)$ for some $\lambda \le 1$. Then $U \in H_\lambda^1(G)$ and

$$U_{,\alpha}(x) = V_\alpha(x) \quad \text{a.e.}$$

If the u^p are A.C.T., *so is U.*

Proof. For each p, let $\bar{u}^p$ be a representative of the class u^p which has the absolute continuity properties described in Theorem 3.1.8 and suppose $\bar{U}(x) = F[\bar{u}(x)]$. Then $\bar{U}$ has these absolute continuity properties, is equivalent to U, and its partial derivatives $\partial \bar{U}/\partial x^\alpha = V_\alpha$ almost everywhere. The result follows from the hypotheses and Theorem 3.1.2(g).

We conclude with two inequalities which will prove useful later (the inequality $\left|\int v(x)\,dx\right| \le \int |v(x)|\,dx$ for vector functions v is useful):

Theorem 3.1.10. *Suppose $z \in H_p^m(G)$, $D \subset\subset G_\varrho$, and φ is any non-negative mollifier. Then*

$$\text{(a)} \quad \int_D \left\{ \int_{B(x,\varrho)} \left[\sum_{j=0}^{m-1} |\nabla^j z(\xi) - \nabla^j z(x)|^2 \right]^{p/2} \varphi_\varrho^*(\xi - x)\, d\xi \right\} dx \le (\varrho \|z\|_{p,G}^m)^p$$

$$\text{(b)} \quad \|z_\varrho|_D - z|_D\|_{p,D}^{m-1} \le \varrho \|z\|_{p,G}^m \quad (p \ge 1),$$

z_ϱ denoting the mollified functions.

Proof. It is easy to see that (b) follows from (a) and (3.1.2). By regarding the derivatives $D^\alpha z^i$ with $|\alpha| \le m-1$ as components of a vector, (a) is reduced to the special case where $m = 1$. There is a domain $D' \subset\subset G$ such that $D \subset D'_\varrho$. Using the mollified functions, we may approximate to $z|_{D'}$ by functions of class C^1, so we may assume $z \in C^1(D')$. Then

$$|z(\xi) - z(x)|^p \le \varrho^p \int_0^1 |\nabla z[x + t(\xi - x)]|^p\, dt, \quad \xi \in B(x,\varrho), \quad x \in D.$$

Thus, we obtain

$$\begin{aligned} &\int_D \left\{ \int_{B(x,\varrho)} |z(\xi) - z(x)|^p \varphi_\varrho^*(\xi - x)\, d\xi \right\} dx \\ &\le \varrho^p \int_0^1 dt \int_{B_\varrho} \varphi_\varrho^*(\zeta) \left[\int_D |\nabla z(x + t\zeta)|^p\, dx \right] d\zeta \\ &= \varrho^p \int_0^1 dt \int_{B_\varrho} \varphi_\varrho^*(\zeta) \left[\int_{D_{t\zeta}} |\nabla z(y)|^p\, dy \right] d\zeta \le (\varrho \|z\|_{p,G}^1)^p \end{aligned}$$

if D_ξ denotes, for each fixed vector ξ, the set of all $y = x + \xi$ where $x \in D$; clearly $D_{t\zeta} \subset D' \subset G$ for each (t, ζ) considered.

3.2. General boundary values; the spaces $H_{p0}^m(G)$; weak convergence

We begin by defining the spaces $H_{p0}^m(G)$:

Definition 3.2.1. The space $H_{p0}^m(G)$ is the closure in $H_p^m(G)$ of the set of functions $C_c^\infty(G)$.

Theorems 3.2.1 (POINCARÉ's inequality). *Suppose* $G \subset B(x_0, R)$ *and* $u \in H_{p0}^m(G)$. *Then*

$$\int_G |\nabla^k u(x)|^p \, dx \leq p^{k-m} R^{(m-k)p} \int_G |\nabla^m u(x)|^p \, dx, \quad 0 \leq k \leq m. \tag{3.2.1}$$

Proof. We shall prove this for $m = 1$, $k = 0$; the general result follows easily by induction, since if $u \in H_{p0}^m(G)$, $\nabla^k u \in H_{p0}^{m-k}(G)$ ($= L_p(G)$ if $k = m$). From Definition 2.3.1, it suffices to prove the theorem in the case that $u \in C_c^1[B(x_0, R)]$. Taking polar coordinates (r, ζ) (ζ on $\partial B(0,1)$) with pole at x_0 and setting $v(r, \zeta) = u(x_0 + r\zeta)$, we obtain

$$\begin{aligned} \int_\Sigma |v(r,\zeta)|^p \, d\Sigma(\zeta) &= \int_\Sigma |v(r,\zeta) - v(R,\zeta)|^p \, d\Sigma(\zeta) \\ &\leq (R-r)^{p-1} \int_r^R \int_\Sigma |v_{,r}(s,\zeta)|^p \, ds \, d\Sigma(\zeta) \end{aligned} \tag{3.2.2}$$

using the HÖLDER inequality. The result (3.2.1) follows for $m = 1$, $k = 0$ from (3.2.2) and the fact that

$$\int_{B(x_0,R)} |u(x)|^p \, dx = \int_0^R r^{\nu-1} \int_\Sigma |v(r,\zeta)|^p \, dr \, d\Sigma.$$

Corollary. *The space* $H_{p0}^m(G)$ *is a closed linear subspace of* $H_p^m(G)$ *and, if* G *is bounded, the norm* $\|u\|_{p0}^m$ *defined by*

$$\|u\|_{p0}^m = \left\{\int_G |\nabla^m u(x)|^p \, dx\right\}^{1/p} \tag{3.2.3}$$

is topologically equivalent to the norm $\|u\|_p^m$ *for* u *on* $H_{p0}^m(G)$. *If* $p = 2$, *the inner product may be taken on* H_{20}^m *to be*

$$(u, v)_{20}^m = \int_G \sum_{i=1}^N \sum_{|\alpha|=m} C_\alpha u_{,\alpha}(x) \overline{v_{,\alpha}(x)} \, dx \quad (u = u^1, \ldots, u^N, \text{ etc.}). \tag{3.2.4}$$

For functions in H_p^m on arbitrary regions, we have the following theorems:

Theorem 3.2.2. (a) *Suppose* $u \in H_{p0}^m(G)$ *and* $V(x) = u(x)$ *for* $x \in G$ *and* $V(x) = 0$ *elsewhere. Then* $V \in H_p^m(R_\nu)$ *and* $V \in H_{p0}^m(\Delta)$ *for any open set* $\Delta \supset G$. *Moreover* $D^\alpha V(x) = D^\alpha u(x)$ *on* G *and* $D^\alpha V(x) = 0$ *for* $x \in R_\nu - G$ *if* $0 \leq |\alpha| \leq m$ *(almost everywhere).*

(b) *Suppose* $u \in H_p^m(G)$, $D \subset G$, $v \in H_p^m(D)$, $v - u|_D \in H_{p0}^m(D)$, $U(x) = v(x)$ *on* D, *and* $U(x) = u(x)$ *on* $G - D$. *Then* $U \in H_p^m(G)$, $U - u \in H_{p0}^m(G)$, *and* $D^\alpha U(x) = D^\alpha v(x)$ *on* D *and* $D^\alpha U(x) = D^\alpha u(x)$ *on* $G - D$ *if* $0 \leq |\alpha| \leq m$ (a.e.).

(c) *If* $u \in H_p^1(G)$, E *is measurable,* $E \subset G$, *and* $u(x) = const.$ *a.e. on* E, *then* $\nabla u(x) = 0$ *a.e. on* E (see MORREY [16], p. 254).

Proof. (a) follows immediately from Definition 3.2.1 and the theorems, of Section 3.1.1. If we define $V(x) = v(x) - u(x)$ on D and $V(x) = 0$

elsewhere, then the conclusions of part (a) hold with G replaced by D. But then $U(x) = u(x) + V(x)$ on G so the results in (*b*) follow. Part (c) follows by choosing an absolutely continuous representative $\bar{u}$ of u.

Theorem 3.2.3. If $p > 1$, *the most general linear functional in* $H_p^m(G)$ *has the form*

$$f(u) = \int_G \sum_{i=1}^{N} \sum_{0 \leq |\alpha| \leq m}' A_i^\alpha(x)\, u_{,\alpha}^i(x)\, dx \tag{3.2.5}$$

where the $A_i^\alpha \in L_q(G)$ *where* $p^{-1} + q^{-1} = 1$. *If* $p = 1$, *each linear functional has the form* (3.2.5) *in which the* A_i^α *are bounded and measurable.*

Proof. It is clear that any expression (3.2.5) defines a linear functional *on* H_p^m. Conversely, let B_p be the space of all tensors $\{\varphi_\alpha^i\}$ where each $\varphi_\alpha^i \in L_p(G)$ with norm

$$\|\varphi\| = \left\{\int_G \left[\sum_{i=1}^{N} \sum_{0 \leq |\alpha| \leq m} C_\alpha |\varphi_\alpha^i(x)|^2\right]^{p/2} dx\right\}^{1/p}.$$

Then, from Theorem 3.1.1, it follows that the subspace M of all tensors φ where $\varphi_\alpha^i = u_{,\alpha}^i$ and $u \in H_p^m(G)$ is a closed linear submanifold of B_p. If we define $F_1(\varphi) = f(u)$ for such tensors φ, we have $\|F_1\| = \|f\|$ and F_1 can be extended (Hahn-Banach Theorem) to a linear functional F over B_p with the same norm. But any linear functional F on B_p has the form

$$F(\varphi) = \int_G \sum_{i=1}^{N} \sum_{0 \leq |\alpha| \leq m} A_i^\alpha(x)\, \varphi_\alpha^i(x)\, dx$$

where the A_i^α have the stated properties.

From Theorem 3.2.3, we immediately obtain:

Theorem 3.2.4. (a) *A necessary and sufficient condition that* u_n *converges weakly to* $u (u_n \rightharpoonup u)$ *in* $H_p^m(G)$ *is that each component* $u_{n,\alpha}^i$ *of* u_n *converges weakly in* $L_p(G)$ *to* $u_{,\alpha}^i$.

(b) *If* $u_n \rightharpoonup u$ *in* $H_p^m(G)$ *then* $u_n \rightharpoonup u$ *in* $H_p^m(D)$ *for* $D \subset G$ (i.e. $u_n|_D \rightharpoonup u|_D$).

(c) *If* $u_n \rightharpoonup u$ *in* $H_p^m(G)$, $x = x(y)$ *is a regular transformation* $\in C_1^{m-1}$ *of* H *onto* G, $v_n(y) = u_n[x(y)]$, *and* $v(y) = u[x(y)]$, *then* $v_n \rightharpoonup v$ *in* $H_p^m(H)$.

(d) *If* $u_n \rightharpoonup u$ *in* $H_p^m(G)$, $\zeta \in C_1^{m-1}(G)$, *and all the* $D^\alpha \zeta$ *with* $0 \leq |\alpha| \leq m$ *are uniformly bounded on* G, *then* $\zeta u_n \rightharpoonup \zeta u$ *in* $H_p^m(G)$.

(e) *If* $p > 1$, *bounded sets in* $H_p^m(G)$ *are conditionally compact with respect to weak convergence in* $H_p^m(G)$.

3.3. The Dirichlet problem

In this section, we interrupt our study of the spaces H_p^m and H_{p0}^m in order to illustrate the variational method for proving the existence of the solutions of certain differential equations. We also establish DIRICHLET's principle.

Definition 3.3.1. If u and $u^* \in H_p^1(G)$ and $u - u^* \in H_{p0}^1(G)$, we say that *u and u^* coincide* on ∂G or u and u^* *have the same boundary values on* ∂G. If $u \in H_{p0}^1(G)$, we say that u vanishes on ∂G.

Remark. The content of the definition depends on p (but see Theorem 3.6.2).

Theorem 3.3.1. (a) *Suppose that $u \in H_1^1(D)$ for each $D \subset\subset G$ and satisfies*

$$\int_G \zeta_{,\alpha} u_{,\alpha} \, dx = 0, \quad \zeta \in C_c^\infty(G). \tag{3.3.1}$$

Then u is harmonic.

(b) *If $u \in H_2^1(G)$ and satisfies* (3.3.1), *then u minimizes the* Dirichlet *integral*

$$D(u, G) = \int_G |\nabla u|^2 \, dx \tag{3.3.2}$$

among all functions with the same boundary values.

(c) *If G is bounded and $u^* \in H_2^1(G)$, there exists a unique function $u = u^*$ on ∂G which minimizes $D(u, G)$ among all such functions. That function is harmonic.*

Proof. (a) The hypotheses imply that u is locally summable on G. Since the $\zeta_{,\alpha} \in C_c^\infty(G)$ if ζ does, we conclude from (3.3.1) and (3.1.1) that

$$\int_G u \Delta \zeta \, dx = 0, \quad \zeta \in C_c^\infty(G).$$

The result follows from Weyl's lemma (§ 2.3).

(b) If $u \in H_2^1(G)$ it follows from Definition 3.2.1 that (3.3.1) holds for all $\zeta \in H_{20}^1(G)$. So if $u^* \in H_2^1(G)$, $u^* = u$ on ∂G, and we set $\zeta = u^* - u$ so $u^* = u + \zeta$ and $\zeta \in H_{20}^1(G)$, we conclude that

$$D(u^*, G) = D(u, G) + 2\int_G \zeta_{,\alpha} u_{,\alpha} \, dx + D(\zeta, G) > D(u, G)$$

unless $\zeta = 0$.

(c) Let $\{u_n\}$ be a minimizing sequence and let $\zeta_n = u^* - u_n$. Each $\zeta_n \in H_{20}^1(G)$ and $D(\zeta_n, G)$ is uniformly bounded. By Poincare's inequality $\|\zeta_n\|_{2,G}^1$ and hence $\|u_n\|_{2,G}^1$ is uniformly bounded. Thus a subsequence, still called $u_n \rightharpoonup u$. Since the norm of an element in a Banach space ($L_2(G)$ in this case) is lower semicontinuous with respect to weak convergence, it follows that the Dirichlet integral $(=\sum (\|u_{,\alpha}\|_{2,G}^0)^2)$ has this property so that $D(u, G) \le \liminf_{n\to\infty} D(u_n, G)$. But $u \in H_2^1(G)$ and $\zeta_n \rightharpoonup \zeta = u^* - u$ so that $\zeta \in H_{20}^1(G)$ and $u = u^*$ on ∂G. Thus u minimizes the Dirichlet integral. So if $\zeta \in H_{20}^1(G)$, $u + \lambda\zeta = u^*$ on ∂G for all λ and

$$D(u + \lambda\zeta, G) = D(u, G) + 2\lambda \int_G \zeta_{,\alpha} u_{,\alpha} \, dx + \lambda^2 D(\zeta, G).$$

Since $\lambda = 0$ gives the minimum, (3.3.1) holds and the results follow.

3.4. Boundary values

In this section we introduce the class of strongly LIPSCHITZ domains and prove that if G is strongly LIPSCHITZ and $u \in H_p^m(G)$ *for some arbitrary* $m \geq 1$, then u can be approximated in $H_p^m(G)$ by functions each of class $C^\infty(\Gamma)$ for some $\Gamma \supset G$. We then prove a variant CALDERON's extension theorem for such domains and discuss bundary values on the boundaries of smoother regions. For the case $m = 1$, some of the results are extended to ordinary LIPSCHITZ domains i.e. domains of class C_1^0. An example is given of a LIPSCHITZ domain which is not strongly LIPSCHITZ.

Definition 3.4.1. A domain G is said to be *strongly* LIPSCHITZ iff G is bounded and each point x_0 of ∂G is in a neighborhood $\mathfrak{R}$ which is the image *under a rotation and translation* of axes of a domain $|y_\nu'| < R$, $|y^\nu| < 2LR$ in which x_0 corresponds to the origin, $\mathfrak{R} \cap \partial G$ corresponds to the locus of $y^\nu = f(y_\nu')$ where f satisfies a LIPSCHITZ condition with constant L, and $\mathfrak{R} \cap G$ corresponds to the set of y where $|y_\nu'| < R$ and $f(y_\nu') < y^\nu < 2LR$.

Remark. A bounded domain is strongly LIPSCHITZ iff it has a regular boundary in the sense of CALDERON (p. 45).

Lemma 3.4.1. *Any bounded open convex domain is strongly* LIPSCHITZ.

Sketch of proof. Let $B(x_0, R)$ be the largest sphere $\subset G$, the given domain. Let $x_1 \in \partial G$. From the convexity, it follows that any point interior to any segment $x_1 x$ with $x \in B(x_0, R)$ lies in G. If we choose axes so the y^ν axis runs along the ray $x_1 x_0$, it is easy to see that a part of ∂G near x_0 can be represented in the desired form.

We need the following well-known lemma:

Lemma 3.4.2. *Suppose* $f \in L_p(R_\nu)$, *e is a unit vector, and* f_h *is defined by* $f_h(x) = f(x + h\,e)$. *Then* $f_h \to f$ *in* $L_p(R_\nu)$.

Proof. Let $\varepsilon > 0$. There is a $g \in C_c^0(R_\nu)$ such that $\|g_h - f_h\| = \|g - f\| < \varepsilon/3$. Since g is also uniformly continuous, there is a $\delta > 0$ such that $\|g_h - g\| < \varepsilon/3$ for $|h| < \delta$.

Theorem 3.4.1. *Suppose G is strongly* LIPSCHITZ *and* $u \in H_p^m(G)$. *Then there is a sequence* $\{u_n\}$, *each* $\in C^\infty(G_n)$ *where* $G \subset\subset G_n$, *such that* $u_n \to u$ *in* $H_p^m(G)$.

Proof. $G \cup \partial G$ can be covered by a finite number of open sets $\mathfrak{R}_i$, each of which is either a cell with $\overline{\mathfrak{R}}_i \subset G$ or a boundary neighborhood of the type described in the definition. There is a finite partition of unity $\{\zeta_j\}$, each of which has support in some one $\mathfrak{R}_i$ and is of class C^∞ everywhere; it can be arranged that $\zeta_1(x) + \cdots + \zeta_J(x) = 1$ on a domain $\Gamma \supset G \cup \partial G$. For each j, let $u_j(x) = \zeta_j(x)\,u(x)$. For each j for which ζ_j has support in G, u_j can be approximated by mollified functions which are 0 near ∂G and can be extended.

So, let us consider a j where ζ_j has support in a boundary neighborhood $\mathfrak{R}$, and let $v_j(y) = u_j(x)$ and for $|y_\nu'| < R$, let us extend $v_j(y) = 0$

for $y^\nu \geq 2LR$. Clearly, the function w_{jn} defined by $w_{jn}(y) = v_j(y + n^{-1} e_\nu) \in H_p^m(\mathfrak{R}_n^+)$ where $\mathfrak{R}_n^+$ is the part of $\mathfrak{R}$ where $y^\nu > f(y'_\nu) - n^{-1}$ and also $w_{jn} \to v_j$ in $H_p^m(\mathfrak{R}^+)$ where $\mathfrak{R}^+$ is where $y^\nu > f(y'_\nu)$. There is a sequence $\varrho_n \to 0$ such that $\varrho_n < (2L\, n)^{-1}$ and $<$ distance of the support of ζ_j from $\partial \mathfrak{R}$, and also so small that if we define $v_{jn} = w_{jn\varrho_n}$, then $v_{jn} \to v_j$ on $\mathfrak{R}^+$ and each $v_{jn} \in C^\infty$ on $\mathfrak{R}_{2n}^+$. If we define $u_{jn}(x) = 0$ for x not in $\mathfrak{R}$ and equal to the transform of $v_{jn}(y)$ for x in $\mathfrak{R}$, we see that $u_{jn} \in C^\infty$ in a domain $\supset G \cup \partial G$. We define $u_n = \sum u_{nj}$ on their common domain.

Lemma 3.4.3. *Suppose $0 < s < \nu$ and S is a measurable set of finite measure. Then*

$$\int_S |x - y|^{-s} dy \leq \Gamma_\nu (\nu - s)^{-1} c^{\nu - s}, \quad \gamma_\nu c^\nu = |S| = |B(x, \sigma)|.$$

Proof. Since $|x - y|^{-s} \leq \sigma^{-s}$ if $y \notin B(x, \sigma)$ and $|x - y|^{-s} > \sigma^{-s}$ if $y \in B(x, \sigma)$ and since $|B(x, \sigma)| = |S|$, it follows that

$$\int_S |x - y|^{-s} dy \leq \int_{B(x,\sigma)} |x - y|^{-s} dy = \Gamma_\nu (\nu - s)^{-1} \sigma^{\nu - s}.$$

We next prove the following important theorem:

Theorem 3.4.2. *Suppose K is essentially homogeneous* (see Def. 2.5.2) *of degree $m - \nu$ $(m > 0)$ and $\in C^{m+1}[R_\nu - \{0\}]$ and suppose f is defined on R_ν and u is defined by*

$$u(x) = \int_{R_\nu} K(x - y) f(y)\, dy. \tag{3.4.1}$$

Then

(a) *If $f \in C_c^\mu(R_\nu)$, $u \in C_\mu^m(R_\nu)$, and*

$$D^\alpha u(x) = \int_{R_\nu} K_{,\alpha}(x - y) f(y)\, dy, \quad 0 \leq |\alpha| \leq m - 1 \tag{3.4.2}$$

$$D^\alpha u(x) = C_\alpha f(x) + \lim_{\varrho \to 0} \int_{E_\nu - B(x,\varrho)} K_{,\alpha}(x - y) f(y)\, dy, \quad |\alpha| = m, \tag{3.4.3}$$

$$C_\alpha = -\int_\Sigma K_{,\beta}(-\eta)\eta_\gamma\, d\Sigma, \quad \alpha = \beta + \gamma, \quad |\beta| = m - 1, \quad |\gamma| = 1.$$

(b) *If $f \in L_p(R_\nu)$ and has compact support $\subset G \subset\subset R_\nu$, then $u \in H_p^m(D)$ for any bounded D, the formulas* (3.4.2) *hold almost everywhere, and*

$$\| u \|_{p,D}^m \leq C(\nu, N, m, p, D, K) \cdot \| f \|_{p,G}^0.$$

Remark. Formula (3.4.3) holds in (b) also, almost everywhere, but we shall not prove this. Further properties of the functions in (3.4.1) in case (b) will follow most easily from certain Sobolev type theorems to be given in §§ 3.5 and 3.7.

Proof. Part (a) follows from the methods and results of §§ 2.5 and 2.6.

To prove (b), let us assume that $f \in L_p$. Clearly $K_{,\alpha}$ is essentially homogeneous of degree $m - |\alpha| - \nu$. If this is > 0, then $K_{,\alpha}$ is continuous and it follows easily by approximating to f that $D^\alpha u$ is continuous and given by (3.4.2). If $m - |\alpha| - \nu = 0$, then $K_{,\alpha}(x, y) = C \log|x - y| + K_{1\alpha}(x - y)$ where C is a constant and $K_{1\alpha}$ is positively homogeneous of degree 0. In that case, if $p > 1$, the right side of (3.4.2) is bounded by a constant times $\| f \|_p^0$; if $p = 1$, it follows that the right side $\in L_r$ for every r with r norm bounded by a constant times $\| f \|_p^0$.

So, let us assume that $m - |\alpha| - \nu = -s$, $0 < s < \nu$. Let U_α denote the right side of (3.4.2). Then, for almost all x,

$$| U_\alpha(x) | \leq M_{|\alpha|} \int_G | x - y |^{-s} | f(y) | \, dx$$

$$(3.4.4) \quad | U_\alpha(x) |^p \leq M_{|\alpha|}^p \left[\int_G | x - y |^{-s} dy \right]^{p-1} \int_G | x - y |^{-s} | f(y) |^p dy$$
$$\leq M_{|\alpha|}^p \cdot C^{p-1}(\nu, s) \cdot | G |^{(p-1)(1-s/\nu)} \int_G | x - y |^{-s} | f(y) |^p dy$$

by Lemma 3.4.3. From (3.4.4) and the preceding discussion, it follows that $\| U_\alpha \|_{p,D}^0 \leq C \| f \|_p^0$ for $0 \leq |\alpha| \leq m - 1$. So, suppose D is a fixed bounded domain and we approximate f strongly in L_p by functions $f_n \in C_c^\mu(G)$. Then all the $U_{n\alpha} \to U_\alpha$ in $L_p(D)$. Moreover, from the CALDERON-ZYGMUND inequalities, it follows that the $D^\alpha u_n$ converge strongly in $L_p(R)$ to some limits U_α when $|\alpha| = m$. The results follow.

Theorem 3.4.3. (Variant of CALDERON's Extension Theorem) *Suppose G is strongly* LIPSCHITZ *and $G \subset\subset D$. There is a linear bounded extension operator $\mathfrak{E}$ which carries $H_p^m(G)$ into $H_{p0}^m(D)$ and which has the property that if $v = \mathfrak{E} u$, then $v(x) = u(x)$ for $x \in G$.*

Proof. We begin as in the proof of Theorem 3.4.1 assuming, as we may, that each $\mathfrak{N}_i \subset D$. For each j for which ζ_j has support $\subset G$, we define $\mathfrak{E}_j u$ as the extension of $u_j(x)$ to D obtained by setting $u_j(x) = 0$ for $x \in D - G$.

For a j for which ζ_j has support in a boundary neighborhood $\mathfrak{N}$, we define $\mathfrak{E}_j u$ first for $u \in C^\infty(G)$ and show that it is bounded. Then $\mathfrak{E} u = \sum \mathfrak{E}_j u$. To define $\mathfrak{E}_j u$ for such j and u, we define $v_j(y) = u_j[x(y)]$ for $y \in \mathfrak{N}^+$ (notation of the proof of Theorem 3.4.1), extend the function f to the whole space to have LIPSCHITZ constant L and extend $v_j(y) = 0$ for $y^\nu > f(y'_\nu)$ where it is not already defined; clearly $v_j(y) \in C^\infty$ in that domain U. Now, let $y \in U$ and let ζ be a unit vector with $| \zeta'_\nu | < L^{-1} \zeta^\nu$ and let $w(r, \zeta) = v_j(y + r\zeta)$. Then

$$(3.4.5) \qquad w_j(0) = \int_{4LR}^{0} \frac{(-r)^{m-1}}{(m-1)!} D_r^m w_j(r, \zeta) \, dr = v_j(y).$$

Let us choose a function $\omega_j(z)$ which is homogeneous of degree 0, of class $C^\infty[R_\nu - \{0\}]$, which has its support in the cone $z^\nu > L | z'_\nu |$ and is such that its integral over $\sum$ is 1. Multiplying (3.4.5) by $\omega(\zeta)$ and integrating over $\sum$ we obtain

$$v_j(y) = \int_{R_\nu} \sum_{|\alpha|=m} K_\alpha(y-z)\, v_j^\alpha(z)\, dz$$

(3.4.6) $$v_j^\alpha(z) = \begin{cases} D^\alpha v(z), & z \in U \\ 0, & z \notin U \end{cases}$$

$$K_\alpha(z) = \frac{|z|^{-\nu}}{(m-1)!} \sum_{|\alpha|=m} C_\alpha z^\alpha \omega_j(z) \quad (C_\alpha = |\alpha|!/\alpha_1! \ldots \alpha_\nu!)$$

where $K_\alpha \in C^\infty[R_\nu - \{0\}]$ and is positively homogeneous of degree $m - \nu$.

To define $\mathfrak{E}_j u$, we begin by extending v_j to the whole space by (3.4.6) and letting $u_j^*(x) = v_j(y)$. Then $u_j^*(x) = u_j(x)$ for $x \in G$ and $u_j^* \in H_p^m(\Delta)$ for any bounded domain Δ, by Theorem 3.4.2 with

$$\| u_j^* \|_{p,\Delta}^m \leq C(\nu, N, m, p, \Delta) \cdot \| u \|_{p,G}^m .$$

Finally we define $u_j^{**} = \mathfrak{E}_j u$ by

$$u_j^{**}(x) = \zeta(x)\, u_j^*(x)$$

where $\zeta \in C_c^\infty(D)$ and $\zeta(x) = 1$ on G. The result follows.

A special case of the following Theorem was proved by Rellich.

Theorem 3.4.4. *Suppose G is strongly* Lipschitz *and $m \geq 1$. Then bounded subsets of $H_p^m(G)$ are conditionally (sequentially) compact as subsets of $H_p^{m-1}(G)$. If $u_n \rightharpoonup u$ in $H_p^m(G)$, then $u_n \to u$ in $H_p^{m-1}(G)$. The theorem is true for any bounded domain G if we replace the spaces $H_p^m(G)$ and $H_p^{m-1}(G)$ by $H_{p0}^m(G)$ and $H_{p0}^{m-1}(G)$, respectively.*

Proof. Suppose $G \subset\subset D \subset\subset R_\nu$, $G \subset D_{\varrho 0}$, $\mathfrak{E}$ is the extension operator of Theorem 3.4.3, and $v_n = \mathfrak{E} u_n$. Then, from Theorem 3.1.10 we conclude that

(3.4.7) $$\| v_{n\varrho} - v_n \|_{p,G}^{m-1} \leq \varrho \| v_n \|_{p,D}^m \leq M \varrho .$$

First, let us suppose that $p > 1$. Then, according to Theorem 3.2.4(e), $\exists$ a subsequence, still called $\{v_n\}$, such that $v_n \rightharpoonup v$ in $H_p^m(D)$. It follows from the formula in Definition 1.8.3 for $v_{n\varrho}$ and v_ϱ and the weak convergence that $D^\alpha v_{n\varrho}$ converges uniformly to $D^\alpha v_\varrho$ on $\bar{G}$, for $0 \leq |\alpha| \leq$ $\leq m - 1$. Then, taking norms on $H_p^{m-1}(G)$, we obtain

$$\begin{aligned} \| u_n - u \| &\leq \| u_n - u_{n\varrho} \| + \| u_{n\varrho} - u_\varrho \| + \| u_\varrho - u \| \\ &\leq 2M\varrho + \| u_{n\varrho} - u_\varrho \| \; (u = v|_G, \; u_n = v_n|_G), \\ &(u_{n\varrho} = v_{n\varrho}|_G, \; u_\varrho = v_\varrho|_G) \end{aligned}$$

using (3.4.7). The result follows easily. If $p = 1$, all the $D^\alpha v_{n\varrho}$ are equicontinuous and uniformly bounded so that a subsequence, still called $\{u_n\}$ exists so that the $D^\alpha u_{n\varrho}$ converge uniformly to some functions $\varphi_{\alpha\varrho}$ for each of a sequence of $\varrho \to 0$. Since (3.4.7) still holds the sequences

$\varphi_{\alpha\varrho}(|\alpha| \leq m-1)$ form a CAUCHY sequence in $L_p(G)$ and so tend to limit functions φ_α in $L_p(G)$. If we set $u = \varphi_\alpha$ when $|\alpha| = 0$ then we see that $u \in H_p^{m-1}(G)$, $\varphi_\alpha = u_{,\alpha}$, $\varphi_{\alpha\varrho} = u_{\varrho,\alpha}$ and (3.4.7) holds for u_ϱ and u. The result again follows, but now, we do not know that $u \in H_p^m$. However, if $u_n \rightharpoonup u$ in H_1^m, then this u must be the same as the preceding one. The last statement is now evident since we may assume $\bar{G} \subset D$ where D is strongly LIPSCHITZ.

Remarks. It is clear from Theorem 3.1.5 and 3.1.7 (change of variable theorems) how to define the elements of the spaces $H_p^m(\mathfrak{M})$ for a compact manifold $\mathfrak{M}$, with or without boundary, of class C_1^{m-1}. For a given finite covering $\mathfrak{U}$ of $\mathfrak{M}$ by coordinate patches τ_i with domains G_i in R_ν and ranges $\mathfrak{N}_i$ in $\mathfrak{M}$ one can define a norm by

$$(3.4.8) \qquad (\| u \|_{p\mathfrak{U}}^m)^p = \sum_i (\| u_i \|_{p,G_i}^m)^p, \quad u_i(x) = u[\tau_i(x)]$$

for example and any two such norms are topologically equivalent. In the case $p = 2$, the corresponding inner product would be

$$(3.4.9) \qquad (u, v)_{\mathfrak{U}}^m = \sum_i \int_{G_i} \sum_{0 \leq |\alpha| \leq m} C_\alpha D^\alpha u_i D^\alpha v_i \, dx.$$

The definitions can be carried over for tensors on $\mathfrak{M}$, but then, usually, $\mathfrak{M}$ must be of a higher class than C_1^{m-1} since the relations between the components in different coordinate systems usually involve at least the first derivatives of the coordinate transformations (see Chapters 7 and 8). Also, an extension can be made to non-compact manifolds which possess coverings in which no point of $\mathfrak{M}$ is in more than K of the $\mathfrak{N}_i$, K being independent of the point; but then the topological equivalence of two norms (3.4.8) is not usually true. If G is a bounded domain of class C_1^{m-1}, then we may consider ∂G as a manifold of class C_1^{m-1}, the structure being defined by the mappings allowed in § 1.2, Notations.

Theorem 3.4.5. *If G is bounded and of class C_1^{m-1} the functions $u \in C_1^{m-1}(\bar{G})$ are dense in any space $H_p^m(G)$ with $p \geq 1$ and there is a bounded operator B from $H_p^m(G)$ into $H_p^{m-1}(\partial G)$ such that $B u = u|_{\partial G}$ whenever $u \in C_1^{m-1}(\bar{G})$. If $u_n \rightharpoonup u$ in $H_p^m(G)$, then $u_n \to u$ in $H_p^{m-1}(\partial G)$. If $p > 1$, the mapping B is compact.*

Proof. To prove the first statement, we select a finite covering of $\bar{G}$ by neighborhoods $\mathfrak{N}_i$, each of which is a cell with closure interior to G or is a boundary neighborhood in the sense of the definition in § 1.2, Notations; in the latter case we suppose that $\mathfrak{N}_i$ is mapped onto $\Gamma_1 \cup \sigma_1$ by a regular map $x = x_i(y)$ of class C_1^{m-1}, Γ_1 being the set $0 < y^\nu < 1$, $|y'_\nu| < 1$. We select a partition of unity $\{\zeta_s\}$, $s = 1, \ldots, S$, of class $C_1^{m-1}(\bar{G})$, each ζ_s having support in some $\mathfrak{N}_i$. Now, suppose $u \in H_p^m(G)$, let $\zeta_s u = u_s$, and let $v_s(y) = u_s[x_i(y)]$ for $y \in \Gamma_1$ in case $\mathfrak{N}_i$ is a boundary neighborhood. For those s for which $\bar{\mathfrak{N}}_i \subset G$, we can approximate to u_s, by $\{u_{ns}\}$, in which each $u_{ns} \in C_c^\infty(G)$. For the other s, we can approximate to v_s by similar v_{ns} on $\Gamma_1 \cup \sigma_1$, using Theorem 3.4.1; clearly each v_{ns} may be chosen to vanish near $G \cap \partial\Gamma_1$ since v does. The first statement follows by setting $u_n = \sum u_{ns}$ where, of course $u_{ns}[x_i(y)] = v_s(y)$ when $y \in \Gamma_1$ and $u_{ns}(x) = 0$ elsewhere.

Next, suppose $u \in C_1^{m-1}(\bar{G})$ and let u_s and v_s be defined as above. We define $B_s u = u_s|_{\partial G}$. Then, clearly $Bu = u|_{\partial G}$. For each s for which $\mathfrak{N}_i \subset G$, $B_s u = 0$. So let s be one the remaining indices; evidently $v_s = U_s u$, U_s being bounded. Since $v_s \in C_1^{m-1}(\bar{\Gamma}_1)$, we have

$$\int_{\sigma_1} \left\{\sum_{j=0}^{m-1} |\nabla^j v_s(y^\nu, y'_\nu) - \nabla^j v_s(0, y'_\nu)|^2\right\}^{p/2} dy'_\nu$$

$$\leq (y^\nu)^{p-1} \int_0^{y^\nu} \int_{\sigma_1} \left\{\sum_{j=0}^{m} |\nabla^j v_s(\eta^\nu, y'_\nu)|^2\right\}^{p/2} dy'_\nu \, d\eta^\nu \leq \varepsilon(y^\nu) \tag{3.4.10}$$

$$\lim_{\varrho \to 0+} \varepsilon(\varrho) = 0, \quad 0 \leq \varepsilon(y^\nu) \leq \varepsilon(1) = \|v_s\|_{p,\Gamma_1}^m$$

so that $\|v_s(0, y'_\nu)\|_{p,\sigma_1}^{m-1}$ is evidently bounded, since y^ν is arbitrary in (3.4.10). For $u \in H_p^m(G)$, it is easy to see that $B_s u = w_s$ where $w_s = 0$ on $\partial G - \mathfrak{N}_i$ and w_s is the transform of $\bar{v}_s(0, y'_\nu)$ on $\partial G \cap \mathfrak{N}_i$ where $\bar{v}_s$ is an A.C. representative of v_s (Theorem 3.1.8).

Now, suppose $u_n \rightharpoonup u$ in $H_p^m(G)$. Then each $v_{ns} \rightharpoonup v_s$ and (3.4.10) holds with a function $\varepsilon(\varrho)$ which is independent of n, since in case $p = 1$, the set functions $\int_e |\nabla^j v_{ns}|^p dy$ are uniformly absolutely continuous. From the fact (Theorem 3.4.4) that $v_{ns} \to v_s$ in $H_p^{m-1}(\Gamma_1)$ (since Γ_1 is strongly LIPSCHITZ) it follows easily from (3.4.10) that the limiting values $\bar{v}_{ns}(0, y'_\nu) \to \bar{v}_s(0, y'_\nu)$ in $H_p^{m-1}(\sigma_1)$. The last statement now follows from Theorem 3.2.4(e).

The case $m = 1$ in Theorem 3.4.4 can be generalized to LIPSCHITZ domains, i.e. those of class C_1^0.

Theorem 3.4.6. *If $m = 1$, Theorem 3.4.4 holds for* LIPSCHITZ *domains, i.e. those of class C_1^0.*

Proof. Using the notation of the proof of the preceding theorem, we conclude, since the interior $\mathfrak{N}_i$ are strongly LIPSCHITZ as is Γ_1, that if $\{u_n\}$ is any bounded sequence in $H_p^m(G)$, then a subsequence $\{q\}$ of $\{n\}$ exists such that all the $u_{qs} \to u_s$ in $H_p^{m-1}(G)$ if $\bar{\mathfrak{N}}_i \subset G$ and $v_{qs} \to v_s$ in $H_p^{m-1}(\Gamma_1)$ and hence $u_{qs} \to u_s$ in $H_p^{m-1}(G)$. If $u_n \rightharpoonup u$ in $H_p^m(G)$, then $u_{ns} \to u_s$ and $v_{ns} \to v_s$. The theorem follows.

An example. That not every LIPSCHITZ domain is strongly LIPSCHITZ is seen by the following example of a bi-LIPSCHITZ map in the case $\nu = 2$. Let (r, θ) and (R, Θ) be polar coordinates in R_2. We define the mapping

$$T: R = r, \quad \Theta = \Theta - \log r, \quad 0 \leq \Theta \leq \pi, \quad 0 < r \leq 1,$$
$$T\,(\text{origin}) = \text{origin}$$

By computing the derivatives X_x, etc., where

$$X = R\cos\Theta,\ Y = R\sin\Theta,\ x = r\cos\theta,\ y = r\ \sin\theta$$

one easily verifies that the map is bi-LIPSCHITZ.

Remarks. We have not proved an extension theorem like Theorem 3.4.3 for $m = 1$ with LIPSCHITZ domains.

3.5. Examples; continuity; some Sobolev lemmas

As was pointed out in the introduction (§ 1.8), the functions in $H_p^m(G)$ have been studied from many different points of view by many writers. We shall not go deeply into real variable properties of these functions here but will merely give a few examples to indicate the generality of the functions allowed and then shall prove some SOBOLEV-type lemmas which will indicate when such functions are continuous or have several continuous derivatives.

It is easy to verify that the function f defined by

$$f(x) = |x|^{-h},\ x \in B(0,1),\ (h+m)\cdot p < \nu, \tag{3.5.1}$$

$\in H_p^m$. From this, it can be easily shown that any function f defined by

$$f(x) = \int_G |x-\xi|^{-h}\,d\mu(e_\xi),\quad (h+m)\cdot p < \nu$$

where G is bounded and μ is a finite measure on $G, \in H_p^m$ on any bounded domain D. (3.5.1) shows also that, for a given m and p, the wildness of functions of class H_p^m increases with the dimension ν. We now prove a SOBOLEV-type lemma guaranteeing continuity.

Theorem 3.5.1. *Suppose $u \in H_p^m(G)$ where G is bounded and strongly* LIPSCHITZ *and $m > \nu/p$. Then u is continuous on G and there is a constant $C(\nu, m, p, G)$ such that*

$$|u(x)| \le C\cdot|G|^{-1/p}\left\{\sum_{j=0}^{m-1}\frac{\Delta^j}{j!}\|\nabla^j u\|_p^0 + (m-\nu/p)^{-1}\frac{\Delta^m}{(m-1)!}\|\nabla^m u\|_p^0\right\} \tag{3.5.2}$$
$$(\Delta = \operatorname{diam} G).$$

If G is also (see Lemma 3.4.1) *convex, we may take $C = 1$; u may be a vector function.*

Proof. If G is strongly LIPSCHITZ, u may be extended to $\in H_{p0}^m(D)$ where D is a given open set $\supset G$ with $\|u\|_{pD}^m \le C_1(\nu, m, p, G, D)\cdot\|u\|_{pG}^m$ and then may be extended to be 0 outside D. So, by allowing a factor C, we may replace G by its convex cover. Since u may be approximated in $H_p^m(G_\varrho)$ by functions $\in C^m$, for each $\varrho > 0$, we may assume $G = G_\varrho$ and $u \in C^m(G)$ and then let $\varrho \to 0$ in (3.5.2).

By expanding in a TAYLOR's series with remainder about each y in G and then integrating with respect to y over G, we obtain (for notation, see § 5.7 below)

$$(3.5.3)\qquad |G|\cdot u(x) = \sum_{j=0}^{m-1}(j!)^{-1}(-1)^j\int_G \nabla^j u(y)\cdot(y-x)^j\,dy +$$

$$+\int_0^1 (-1)^m[(m-1)!]^{-1}t^{m-1}\left[\int_G \nabla^m u[x+t(y-x)]\cdot(y-x)^m\,dy\right]dt.$$

The inequality follows by applying the HÖLDER inequality to each term, first setting $z = x + t(y-x)$ to handle the last integral; when this is done the last integral is dominated by

$$[\Delta^m/(m-1)!]\cdot\int_0^1 t^{m-1-\nu}\left[\int_{G(x,t)}|\nabla^m u(z)|\,dz\right]dt$$

where $G(x, t)$ consists of all $z = x + t(y-x)$ for $y \in G$ and $|G(x,t)| = t^\nu|G|$.

Corollary. *If $u \in H_p^1(G)$ with $p > \nu$, then u is continuous.*

We now present a "DIRICHLET growth" theorem guaranteeing continuity (MORREY [4] and [7]):

Theorem 3.5.2. *Suppose $u \in H_p^1[B(x_0, R)]$, $1 \le p \le \nu$, and suppose*

$$(3.5.4)\qquad \int_{B(x,r)}|\nabla u|^p\,dx \le L^p(r/\delta)^{\nu-p+p\mu},\quad 0 \le r \le \delta = R - |x - x_0|,$$
$$0 < \mu < 1$$

for every $x \in B(x_0, R)$. Then $u \in C_\mu^0[B(x_0, r)]$ for each $r < R$ and

$$(3.5.5)\qquad |u(\xi) - u(x)| \le C\,L\,\delta^{1-\nu/p}(|\xi - x|/\delta)^\mu$$
$$\text{for}\quad |\xi - x| \le \delta/2,\quad C = C(\nu, p, \mu).$$

Proof. By approximations, we may assume $u \in C^1(B_R)$. Let x and ξ be given, let $\varrho = |\xi - x|/2$, and $\bar{x} = (\xi + x)/2$. For each $\eta \in B(\bar{x}, \varrho)$, we observe that

$$u(\eta) - u(\xi) = (\eta^\alpha - \xi^\alpha)\int_0^1 u_{,\alpha}[\xi + t(\eta - \xi)]\,dt$$

$$|u(\eta) - u(\xi)| \le 2\varrho\int_0^1 |\nabla u[\xi + t(\eta - \xi)]|\,dt.$$

Averaging over $B(\bar{x}, \varrho)$ we obtain

$$(3.5.6)\qquad |B(\bar{x}, \varrho)|^{-1}\int_{B(\bar{x},\varrho)}|u(\eta) - u(\xi)|\,d\eta$$

$$\le 2\varrho|B(\bar{x}, \varrho)|^{-1}\int_{B(\bar{x},\varrho)}\left[\int_0^1 |\nabla u[\xi + t(\eta - \xi)]|\,dt\right]d\eta.$$

Interchanging the order of integration, setting $y = \xi + t(\eta - \xi)$ and noting that y ranges over $B(\bar{x}_t, t_\varrho)$ where $\bar{x}_t = (1-t)\,\xi + t\,\bar{x}$ which is at a distance $\delta_t \ge \delta - 2\varrho + t\varrho \ge \delta/2$ from $\partial B(x_0, R)$ and then using

(3.5.4) and the HÖLDER inequality we obtain

$$(3.5.7)\quad \int\limits_{B(\bar{x}_t, t\varrho)} |\nabla u(y)|\, dy \le C_1 L (\delta/2)^{1-\mu-\nu/p} (t\varrho)^{\nu-1+\mu} \quad C_1 = C(\nu, p).$$

Thus we see that the right side of (3.5.6)

$$(3.5.8)\quad \le 2 C_2 \varrho^{1-\nu} L (\delta/2)^{1-\mu-\nu/p} \varrho^{\nu-1+\mu} \int\limits_0^1 t^{\mu-1} d\mu, \quad C_2 = C_2(\nu, p).$$

Using the same result for the average of $|u(\eta) - u(x)|$, we obtain the result.

We now prove another SOBOLEV Lemma (SOBOLEV [1], NIRENBERG [3]):

Theorem 3.5.3. *Suppose $u \in H_p^1(R_\nu)$ with $1 \le p < \nu$, and has compact support. Then $u \in L_r(R_\nu)$, where $r = \nu p/(\nu - p)$ and*

$$(3.5.9)\quad \| u \|_r^0 \le t \prod_{\alpha=1}^{\nu} (\| u_{,\alpha} \|_p^0)^{1/\nu} \le \nu^{-1/2} t \|\nabla u\|_p^0,$$
$$1 \le p < \nu, \quad t = p(\nu - 1)/(\nu - p);$$

u may be a vector function.

Proof. It is sufficient to prove this for $u \in C_c^1(R_\nu)$. The case where $p > 1$ can be proved by applying the inequality (3.5.9) for $p = 1$ to the function v defined by

$$v(x) = |u(x)|^t.$$

For, if this is done, we obtain

$$(3.5.10)\quad \int |u(x)|^r dx = \int |v(x)|^s dx \le \prod_{\alpha=1}^{\nu} (\| v_{,\alpha} \|_1^0)^{s/\nu}$$
$$\le t^s (\| u \|_r^0)^{r s (p-1)/p} \prod_{\alpha=1}^{\nu} (\| u_{,\alpha} \|_p^0)^{s/\nu}, \quad (s = \nu/(\nu - 1))$$

since $r = p(t-1)/(p-1)$. The first result in (3.5.9) follows easily since

$$r - [r s (p-1)/p] = s$$

and the second result follows from an elementary inequality.

We shall prove the inequality for $p = 1$ and $\nu = 3$, the proof for $\nu = 2$ or $\nu > 3$ is similar. Clearly

$$|u(x, y, z)| \le \int\limits_{-\infty}^{\infty} |u_{,1}(\xi, y, z)|\, d\xi, \quad \int\limits_{-\infty}^{\infty} |u_{,2}(x, \eta, z)|\, d\eta,$$
$$\int\limits_{-\infty}^{\infty} |u_{,3}(x, y, \zeta)|\, d\zeta.$$

Thus

$$|u(x, y, z)|^{3/2} \le \left(\int\limits_{-\infty}^{\infty} |u_{,1}(\xi, y, z)|\, d\xi \right)^{1/2}$$
$$\left(\int\limits_{-\infty}^{\infty} |u_{,2}(x, \eta, z)|\, d\eta \right)^{1/2} \left(\int\limits_{-\infty}^{\infty} |u_{,3}(x, y, \zeta)|\, d\zeta \right)^{1/2}.$$

Integrating first with respect to x then y, and using the SCHWARZ inequality, we obtain

$$\int_{-\infty}^{\infty}|u(x,y,z)|^{3/2}dx \leq \left(\int_{-\infty}^{\infty}|u_{,1}(\xi,y,z)|d\xi\right)^{1/2}\left(\iint_{-\infty}^{\infty}|u_{,2}(x,\eta,z)|dx\,d\eta\right)^{1/2}\left(\iint_{-\infty}^{\infty}|u_{,3}(x,y,\zeta)|dx\,d\zeta\right)^{1/2}$$

$$\iint_{-\infty}^{\infty}|u(x,y,z)|^{3/2}dx\,dy \leq \left(\iint_{-\infty}^{\infty}|u_{,1}(\xi,y,z)|d\xi\,dy\right)^{1/2}\left(\iint_{-\infty}^{\infty}|u_{,2}(x,\eta,z)|dx\,d\eta\right)^{1/2}\left(\iiint_{-\infty}^{\infty}|u_{,3}(x,y,\zeta)|dx\,dy\,d\zeta\right)^{1/2}$$

from with the result follows by integrating with respect to z.

Theorem 3.5.4. *There is a constant $C(\nu)$ such that if $u \in H_p^1[B(x_0,R)]$, there exists a function $U \in H_{p0}^1[B(x_0, 2R)]$ such that $U(x) = u(x)$ for $x \in B(x_0, R)$ and*

$$\|\nabla U\|_{p,2R}^0 \leq C \cdot {}'\|u\|_{p,R}^1, \quad \textit{where} \quad ('\|u\|_{p,R}^1)^p = \int_{B_R}(|\nabla u|^2 + R^{-2}u^2)^{p/2}dx.$$

Proof. From considerations of homogeneity, it follows that it is sufficient to prove this for the unit ball $B(0,1)$. But then the result follows form Theorem 3.4.3 (the extension theorem).

Theorem 3.5.5. *If G is LIPSCHITZ and u (a vector) $\in H_p^1(G)$, $1 \leq p < \nu$, then $u \in L_r(G)$, where $r = \nu p/(\nu - p)$ and*

$$\|u\|_r^0 \leq C(\nu, p, G) \cdot \|u\|_p^1. \tag{3.5.11}$$

If $u \in H_\nu^1(G)$, then (3.5.11) *holds for each $p < \nu$. If $G = B(x_0, R)$, we may replace* (3.5.11) *by*

$$\|u\|_r^0 \leq C(\nu, p) \cdot {}'\|u\|_p^1. \tag{3.5.12}$$

Proof. The last statement follows from Theorem 3.4.6 (the Extension Theorem for $m = 1$). If $u \in H_\nu^1(G)$ then, of course $u \in H_p^1(G)$ for each $p < \nu$. In order to prove the first statement, we cover $\bar{G}$ by neighborhoods $\mathfrak{N}_i$ and choose a partition of unity $\{\zeta_s\}$ as in the proof of Theorem 3.4.5. For each s for which the support of $\zeta_s \subset G$, (3.5.11) holds for u_s. For the remaining s, we may extend $v_s(y)$ by positive reflection $(v_s(y^\nu, y'_\nu) = v_s(-y^\nu, y'_\nu))$ and then v_s has compact support in $|y^\nu| < 1$, $|y'_\nu| < 1$. So that (3.5.11) hold for such v_s and hence also for the corresponding u_s.

3.6. Miscellaneous additional results

In this section, we include a selection of theorems which are useful in discussing boundary value problems in the calculus of variations and a few results useful in the case of differential equations of higher order.

Theorem 3.6.1. *Suppose $u \in H_p^1(G)$ and $\bar{u}$ is the absolutely continuous representative of Theorem 3.1.8. Then*

(a) *if $R = [a, b]$ is any cell in G, $\bar{u}|_{\partial R}$ is a representative of $B_R u$, the boundary value operator of Theorem 3.4.5 for R;*

(b) *if $x = x(y)$ is a bi-*LIPSCHITZ *map of a part $\overline{\mathfrak{N}}$ of $\bar{G}$ onto $\Gamma_R \cup \sigma_R$ in which $\bar{\sigma}_R$ corresponds to $\mathfrak{N} \cap \partial G$, $\bar{v}(y) = \bar{u}[x(y)]$, and $\varphi(y'_\nu) = \psi[x(0, y'_\nu)]$ for $y'_\nu \in \bar{\sigma}_R$, where $\psi = B\,u$, then $\bar{v}(y^\nu, y'_\nu) \in L_p(\varrho_R)$ for each y^ν, varies continuously with y^ν as an element of $L_p(\sigma_R)$ and, for almost every y'_ν, $\bar{v}(y^\nu, y'_\nu) \to \bar{v}(0, y'_\nu)$ as $y^\nu \to 0^+$ and $\bar{v}(0, y'_\nu)$ is a representative of $\varphi(y'_\nu)$.*

(c) *If $\overline{B(x_0, R)} \subset G$, then $\bar{u}|_{\partial B(x_0, R)}$ is a representative of $B\,u$, B being the boundary operator for u; $\bar{u}|_{\partial B(x_0, R)} \in H_p^1[\partial B(x_0, R)]$ for almost all R for which $\overline{B(x_0, R)} \subset G$.*

Proof. This follows from Theorem 3.1.8 combined with the proof of Theorem 3.4.5.

Theorem 3.6.2. *Suppose G is* LIPSCHITZ. *Then $u \in H_{p0}^1(G)$ if $u \in H_p^1(G)$ and $B\,u = 0$ a.e. on ∂G. Consequently, if $u \in H_{p0}^1(G) \cap H_q^1(G)$, then $u \in H_{q0}^1(G)$.*

Proof. The second statement follows from the first. To prove the first, cover $\bar{G}$ by neighborhoods $\mathfrak{N}_i$, choose a partition of unity ζ_s, and define u_s and v_s as in the proof of Theorem 3.4.5. By using the mollified functions $u_{s\varrho}$, we see that those u_s for which ζ_s has support interior to G can be approximated strongly in $H_p^1(G)$ by functions $\in C_c^\infty(G)$. So let s be one of the other indices. From Theorem 3.6.1, it follows that the function $\bar{v}_s(y)$ extended to be 0 for $y^\nu \leq 0 \in H_p^1(\Gamma_1 \cup \Gamma_1^-)$. From Lemma 3.4.2, it follows that $w_n \to v_s$ in $H_p^1(\Gamma_1)$ if $w_n(y^\nu, y'_\nu) = v_s(y^\nu - n^{-1} h, y'_\nu)$ and the support of each $w_n \in \Gamma_1$ if h is small enough. By mollifying the w_n, one obtains a sequence v_{ns}, each $\in C_c^\infty(\Gamma_1)$ such that $v_{ns} \to v_s$ in $H_p^1(\Gamma_1)$, and hence one obtains a sequence $u_{ns} \in Lip_c(G)$, etc.

Theorem 3.6.3. (a) *Suppose G is* LIPSCHITZ, *$u \in H_p^1(G)$, and u is continuous on $\bar{G}$. Then $u|_{\partial G}$ is a representative of $B\,u$.*

(b) *Suppose G is* LIPSCHITZ, *$u_n \rightharpoonup u$ in $H_p^1(G)$, $B\,u_n$ has a continuous representative φ_n for each n, and $\varphi_n \to \varphi$ uniformly on ∂G. Then φ is a representative of $B\,u$.*

Proof. (a) follows from Theorem 3.6.1. (b) follows from Theorem 3.4.5.

Theorem 3.6.4. *Suppose G is* LIPSCHITZ. *Then there are constants $C_1(\nu, p, \sigma, G)$ and $C_2(\nu, p, \tau, G)$ such that*

$$\int_G |u|^p\,dx \leq C_1 \left[\int_G |\nabla u|^p\,dx + \left\{\int_\sigma |B\,u|\,dS\right\}^p\right]$$

$$\int_G |u|^p\,dx \leq C_2 \left[\int_G |\nabla u|^p\,dx + \left\{\int_\tau |u|\,dx\right\}^p\right]$$

for every $u \in H_p^1(G)$; here σ and τ are subsets of positive measure of ∂G and G, respectively.

Proof. We prove the first, the proof of the second is similar. If the statement were not true, there would exist a sequence $\{u_n\}$, each $u_n \in H_p^1(G)$ such that

$$(3.6.1) \qquad \int_G |u_n|^p dx > n \left[\int_G |\nabla u_n|^p dx + \left\{ \int_\sigma |B u_n| dS \right\}^p \right].$$

On account of the homogeneity, we may assume that $\|u_n\|_{p,G}^1 = 1$ for each n. Thus there is a subsequence, still called $\{u_n\}$, such that $u_n \to u_0$ in $L_p(G)$ and $B u_n \to 0$ in $L_p(\sigma)$ (Theorem 3.4.6). Thus (see (3.6.1)) $\nabla u_n \to 0$ in $L_p(G)$, so that $u_n \to u_0$ in $H_p^1(G)$ so that $B u_n \to B u_0$ in $L_1(\sigma)$, on account of Theorem 3.4.5. But, from Theorem 3.1.3, it follows that $\bar{u}_0 =$ const. which must be 0 since $B u_0 = 0$ on σ. But this contradicts the fact that $\|u_n\|_p^1 = 1$ for each n.

Theorem 3.6.5. *Suppose G is* Lipschitz. *Then there are constants $C_1(\nu, p, G)$ and $C_2(\nu, p, G, c)$ such that*

$$\int_G |u|^p dx \le C_1 \int_G |\nabla u|^p dx \quad \text{if} \quad \bar{u} = |G|^{-1} \int_G u\, dx = 0$$

$$\int_G |u|^p dx \le C_2 \int_G |\nabla u|^p dx \quad \text{if} \quad |S| \ge c|G| \quad (c > 0)$$

for every $u \in H_p^1(G)$; here S is the set where $u(x) = 0$. In case $G = B(x_0, R)$, we may write

$$C_1(\nu, p, G) = C_1(\nu, p)\, R^p, \quad C_2(\nu, p, G, c) = C_2(\nu, p, c)\, R^p.$$

Proof. Again we prove only the second, the proof of the first being similar. The last statement follows by homogeneity. If the statement were not true, there would exist a sequence $\{u_n\}$ such that $\|u_n\|_p^1 = 1$ and $u_n \to u_0$ in $L_p(G)$ and

$$\int_G |u_n|^p dx > n \int_G |\nabla u_n|^p dx, \quad |S_n| \ge c|G| > 0$$

for each n. Thus $\nabla u_n \to 0$ in $L_p(G)$ so $u_n \to u_0$ in $H_p^1(G)$ and $u_0 =$ const. on account of Theorem 3.1.3 and $\bar{u}_0 \ne 0$ since $\nabla u_0 = 0$ but $\|u_0\|_p^1 = 1$. But since $u_n = 0$ on S_n, we conclude from the convergence that

$$0 = \lim_{n \to \infty} \int_G |u_n - u_0|^p dx = \lim_{n \to \infty} \int_{S_n} |u_n - u_0|^p dx = \lim_{n \to \infty} |\bar{u}_0|^p \cdot |S_n|$$

which contradicts the fact that $|S_n| \ge c|G| > 0$ for all n.

Theorem 3.6.6. *Suppose G is a* Lipschitz *domain and $u \in H_p^1(G)$ where $p > \nu$. Then $\bar{u} \in C_\mu^0(\bar{G})$ and*

$$(3.6.2) \qquad h_\mu(\bar{u}, \bar{G}) \le C(\nu, p, G) \cdot \|u\|_p^1, \quad \mu = 1 - \nu/p.$$

The corresponding result holds if $\bar{G}$ is replaced by a compact Lipschitz *manifold (i.e. one of class C_1^0) with or without boundary, the norm being any one of the norms discussed in the remarks after Theorem 3.4.4.*

Proof. We prove the first, that of the second being similar. The theorem follows for any interior domain D from Theorems 3.5.1 and 3.5.2 and the HÖLDER inequality. The theorem follows for $\bar{G}$ by covering $\bar{G}$ with neighborhoods $\mathfrak{N}_i$, choosing a partition of unity $\{\zeta_s\}$, and considering the functions u_s and v_s of Theorem 3.4.5. Since Γ_1 is convex and each interior $\mathfrak{N}_i$ is also, the theorem follows for each u_s on $\bar{G}$.

Theorem 3.6.7. *Suppose, for each n, that $u_n \in H_p^1(G)$ with $p > 1$ and is continuous on $\bar{G}$. Suppose also that $\|\nabla u_n\|_{p,G}^0$ is uniformly bounded and u_n converges uniformly to u on $\bar{G}$. Then $u_n \rightharpoonup u$ in $H_p^1(G)$. If G is* LIPSCHITZ, *the same result follows if each u_n is continuous on G and u_n converges uniformly to u on each $D \subset\subset G$.*

Proof. We prove the first statement; the proof of the second is similar if one makes use of Theorem 3.6.4. The hypotheses imply that $\|u_n\|_{p,G}^1$ are uniformly bounded. Let $\{q\}$ be any subsequence of $\{n\}$. There is a further subsequence $\{r\}$ of $\{q\}$ such that $u_r \rightharpoonup$ some u_0 in $H_p^1(G)$ (Theorem 3.2.4(e)). But then we conclude from Theorem 3.4.6 that $u_r \to u_0$ in $L_p(D)$ for each LIPSCHITZ $D \subset\subset G$, so that $u = u_0$. Hence the whole sequence $u_n \rightharpoonup u$ in $H_p^1(G)$.

Theorem 3.6.8. (a) *Suppose $\varphi \in L_p(G)$, $D \subset G_a$, e is a unit vector, and ψ_h is defined on D by*

$$\psi_h(x) = \int_0^1 \varphi(x + t\,h\,e)\,dt, \quad |h| < a.$$

Then $\psi_h \in L_p(D)$ for each such h and $\psi_h \to \varphi$ in $L_p(D)$.

(b) *Suppose $u \in H_p^1(G)$, $D \subset G_a$, e_γ is the unit vector in the x^γ direction and v_h is defined on D by*

$$v_h(x) = h^{-1}[u(x + h\,e_\gamma) - u(x)].$$

Then $v_h \to u_{,\gamma}$ in $L_p(D)$.

Proof. (a) Let $F(x, u) = \varphi(x + u\,e)$, $|u| < a$. Then $F \in L_p(D)$ for each u and is measurable in (x, u) over $D \times (-a, a)$. Also

$$\int_{-a}^{a}\left[\int_D |F(x, u)|^p\,dx\right] du = \int_{-a}^{a}\left[\int_{D(u)} |\varphi(y)|^p\,dy\right] du \le 2a \int_G |\varphi(y)|^p\,dy$$

since $D(u) \subset G$ for each u on $(-a, a)$, $D(u)$ being the set of $y = x + u$ for $u \in (-a, a)$. Hence $F \in L_p(\Gamma)$, $\Gamma = (-a, a) \times D$. Now

$$\psi_h(x) = \int_0^1 F(x, t\,h)\,dt = h^{-1}\int_0^h F(x, u)\,du$$

so $\psi_h(x)$ is defined a.e. on D and $\in L_1(D)$ by FUBINI's theorem. Also

$$|\psi_h(x)|^p \le h^{-1}\left|\int_0^h |F(x, u)|^p\,du\right|.$$

Consequently $\psi_h \in L_p$ for each h; if $h = 0$, $\psi_h(x) = \varphi(x)$. Now, from Lemma 3.4.2, it follows that

$$\lim_{u \to 0} \int_D |F(x, u) - \varphi(x)|^p \, dx = 0.$$

Accordingly, we see that

$$\psi_h(x) - \varphi(x) = h^{-1} \int_0^h [F(x, u) - \varphi(x)] \, du$$

$$\int_D |\psi_h(x) - \varphi(x)|^p \, dx \leq h^{-1} \int_0^h \left\{ \int_D |F(x, u) - \varphi(x)|^p \, dx \right\} du$$

from which the result follows.

(b) Let $\bar{u}$ be the representative of Theorem 3.1.8. Then v_h is given almost everywhere by

$$v_h(x) = \int_0^1 \bar{u}_{,\gamma}(x + t h e_\gamma) \, dt.$$

The result follows from part (a).

We conclude this section with three theorems which are useful in the study of differential equations of higher order. There are a great variety of additional theorems like those above and those below which can be proved by the methods of this section.

Theorem 3.6.9. *Suppose G is bounded and strongly* Lipschitz, $m > 1$ *and* $p \geq 1$. *Then, for each j, $1 \leq j \leq m - 1$, and each $\varepsilon > 0$, there is a constant $C(\nu, m, p, G, j, \varepsilon)$ such that*

$$(3.6.3) \qquad \|\nabla^j u\|_p^0 \leq \varepsilon \|\nabla^m u\|_p^0 + C \|u\|_p^0, \quad u \in H_p^m(G).$$

Proof. For otherwise there is a sequence $\{u_n\}$ with $u_n \in H_p^m(G)$ and $\|u_n\|_p^m = 1$ such that

$$(3.6.4) \qquad \|\nabla^j u_n\|_p^0 > \varepsilon \|\nabla^m u_n\|_p^0 + n \|u_n\|_p^0.$$

From Theorem 3.4.4, it follows that we may assume that $u_n \to u$ in $H_p^{m-1}(G)$. Since $\|\nabla^j u_n\|_p^0 \leq \|u_n\|_p^{m-1} \leq \|u_n\|_p^m = 1$, it follows that $u_n \to u$ in $L_p(G)$, so $u = 0$. But then $\|\nabla^j u_n\|_p^0 \to 0$ so that $\|\nabla^m u_n\|_p^0$ and hence $\|u_n\|_p^m \to 0$.

Theorem 3.6.10. *If $u \in H_p^m(G)$, there is a unique polynomial P of degree* $\leq m - 1$ (or $= 0$) *such that the average over G of each $D^\alpha(u - P)$ is 0 if* $0 \leq |\alpha| \leq m - 1$.

This is easily proved by induction on m.

Theorem 3.6.11. *Suppose G is a strongly* Lipschitz *domain and $G \subset B(x_0, R)$. Then there is a constant $C(\nu, m, p, G)$ such that*

$$\|\nabla^j u\|_p^0 \leq C R^{m-j} \|\nabla^m u\|_p^0, \quad 0 \leq j \leq m - 1$$

for every $u \in H_p^m(G)$ such that the average over G of each $D^\alpha u$ is 0 for $0 \leq |\alpha| \leq m - 1$.

Proof. We first assume that $G \subset B(0,1)$. Suppose the theorem is not true. Then there is a j, $0 \leq j \leq m-1$, and a sequence $\{u_n\}$ of functions of the type described such that

$$\|\nabla^j u_n\|_p^0 > n \|\nabla^m u_n\|_p^0, \quad \|u_n\|_p^m = 1. \tag{3.6.5}$$

A subsequence, still called $\{u_n\}$ converges in $H_p^{m-1}(G)$ to some u. From (3.6.5) it follows that $\nabla^m u_n \to 0$ in $L_p(G)$ so that $u_n \to u$ in $H_p^m(G)$ and $\nabla^m u = 0$. By induction on m and the use of mollifiers, it can be shown that u is a polynomial of degree $\leq m-1$. Clearly the averages over G of each $D^\alpha u$ is 0 if $0 \leq |\alpha| \leq m-1$. From Theorem 3.6.10, it follows that $u = 0$. Since $u_n \to u$ in $H_p^m(G)$, this contradicts (3.6.5).

3.7. Potentials and quasi-potentials; generalizations

In this section, we introduce the notion of "quasipotentials" and deduce some of their differentiability properties as well as some additional such properties of potentials.

Definition 3.7.1. Suppose e is a vector $\in L_p(G)$, $p > 1$. We define its *quasi-potential* u by

$$u(x) = -\int_G K_{0,\alpha}(x-\xi)\, e^\alpha(\xi)\, d\xi = \Gamma_\nu^{-1} \int_G |\xi - x|^{-\nu} (\xi^\alpha - x^\alpha)\, e^\alpha(\xi)\, d\xi. \tag{3.7.1}$$

Theorem 3.7.1. *Suppose $e \in L_p(G)$, $p > 1$, u is its quasi-potential, and V^α is the potential of e^α. Then*

(a) *Each $V^\alpha \in H_p^2(D)$ and $u \in H_p^1(D)$ with*

$$\|\nabla^2 V\|_{p,D}^0 \leq C(\nu,p)\, \|e\|_{p,G}^0, \qquad \|\nabla u\|_{p,D}^0 \leq C(\nu,p)\, \|e\|_{p,G}^0$$
$$\|\nabla V\|_{p,D}^0 \leq C(\nu,p) \cdot \Delta \cdot \|e\|_{p,G}^0, \qquad \|u\|_{p,D}^0 \leq C(\nu,p) \cdot \Delta \cdot \|e\|_{p,G}^0$$

for any bounded domain D, Δ being the diameter of D;

(b) $u(x) = -V^\alpha_{,\alpha}(x)$, $x \in G$ (a.e.),

(c) $\int_G v_{,\alpha}(u_{,\alpha} + e^\alpha)\, dx = 0$, $v \in Lip_c(G)$

(d) *if $G = B_R$ and $e \in C_\mu^0(\bar{B}_R)$, then $u \in C_\mu^1(\bar{B}_R)$ and*

$$h_\mu(\nabla u, \bar{B}_R) \leq C(\nu,\mu)\, h_\mu(e, \bar{B}_R) \quad (0 < \mu < 1).$$

Proof. (a) Follows immediately from Theorem 3.4.2 and its proof. (b) follows from (3.7.1), Theorem 3.4.2, and the definitions. To prove (c), let $\{e_n\} \in C_c^\infty(G)$ and suppose $e_n \to e$ in $L_p(G)$. For each n, we see that

$$u_n(x) = \lim_{\varrho \to 0} \int_{G - B(x,\varrho)} e_n^\alpha(\xi) \frac{\partial}{\partial \xi^\alpha} K_0(x-\xi)\, d\xi = -\int_G K_0(x-\xi)\, e^\alpha_{n,\alpha}(\xi)\, d\xi$$

so that

$$\Delta u_n = -e^\alpha_{n,\alpha}$$

and (c) holds for each n. The result follows by passing to the limit. (d) follows from Theorems 2.6.2 and 2.6.7.

Theorem 3.7.2. *If $u \in H_{10}^1(G)$, then $u(x)$ is given a. e. on G by*

$$u(x) = -\Gamma_\nu^{-1} \int_G |\xi - x|^{-\nu} (\xi^\alpha - x^\alpha)\, u_{,\alpha}(\xi)\, d\xi = \int_G K_{0,\alpha}(x - \xi)\, u_{,\alpha}(\xi)\, d\xi.$$

(2.7.19)

Proof. We have seen in Theorem 2.7.3 that this holds if $u \in C_{\mu c}^1(G)$. So, let $U(x)$ denote the right side of (2.7.19) and let $u_n \in C_{\mu c}^1(G)$ for each n and $u_n \to u$ in $H_{10}^1(G)$. We note that

$$|U(x)| \le \Gamma_\nu^{-1} \int_G |\xi - x|^{1-\nu} \cdot |\nabla u(\xi)|\, d\xi$$

$$\int_G |U(x)|\, dx \le \int_G |\nabla u(\xi)| \cdot \left[\Gamma_\nu^{-1} \int_G |\xi - x|^{1-\nu}\, dx\right] d\xi \le g \cdot \|\nabla u\|_{1,G}^0$$

$$\gamma_\nu g^\nu = |G|. \tag{3.7.2}$$

Using (3.7.2) for the difference $u_n - U$, we see that $u_n \to U$ in $L_1(G)$.

Theorem 3.7.3. (a) *Suppose that $f \in L_1(G)$ and satisfies*

$$\int_{G \cap B(x_0, r)} |f(\xi)|\, d\xi \le L \cdot r^{\nu-2+\lambda}, \quad 0 < \lambda < 1 \tag{3.7.3}$$

for each ball $B(x_0, r)$, and suppose V is its potential. Then $V \in H_2^1(D)$ for any bounded domain D and satisfies

$$\int_{B(x_0, r)} |\nabla V|^2\, dx \le C^2(\nu, \mu) \cdot L^2 r^{\nu-2+2\lambda}, \quad x_0 \in G, \quad 0 \le r \le R \tag{3.7.4}$$

where R is the diameter of G. Moreover

$$\int_G v(x) f(x)\, dx = -\int_G v_{,\alpha}(x)\, V_{,\alpha}(x)\, dx, \quad v \in H_{20}^1(G). \tag{3.7.5}$$

(b) *If $\nu > 2$ and $f \in L_{2s'}(G)$, $s' = \nu/(\nu + 2)$, then $V \in H_2^1(D)$ for any bounded D and*

$$\int_D |\nabla V|^2\, dx \le C^2(\nu) \cdot (\|f\|_{2s'}^0)^2. \tag{3.7.6}$$

If, also, f satisfies

$$\int_{B(x_0, r)} |f(x)|\, dx \le L_1 R^{\tau-1} (r/R)^{\nu-2+\mu}, \quad 0 \le r \le R, \quad B(x_0, R) \subset G \tag{3.7.7}$$

$$0 < \mu < 1, \quad L_1 \ge \|f\|_{2s'}^0, \quad \tau = \nu/2$$

for all x_0 in G, then

$$\int_{B(x_0, r)} |\nabla V|^2\, dx \le C(\nu, \mu)\, L_1^2 (r/R)^{\nu-2+2\mu}. \tag{3.7.8}$$

(c) *If $f \in L_2(B_R)$ and satisfies a condition*

$$\int_{B(x_0, r) \cap B_R} |f(x)|^2\, dx \le L^2 r^{\nu-2+2\mu}$$

for every $B(x_0, r)$, then $V \in C_\mu^1(\bar{B}_R)$ and

$$h_\mu(f, B_R) \le C(\nu, \mu)\, L.$$

Proof. (a) Let $W_\alpha(x)$ be the right side of the equation

$$V_{,\alpha}(x) = -\Gamma_\nu^{-1} \int_G |\xi - x|^{-\nu} (\xi^\alpha - x^\alpha) f(\xi)\, d\xi. \tag{3.7.9}$$

By proceeding as in the proof of the preceding theorem, we easily conclude that V and $W_\alpha \in L_1(D)$ for any bounded D. By approximating to f in $L_1(G)$ by f_n in $C^0_{\mu c}(G)$, we see as in the preceding theorem that $V \in H^1_1(D)$ for any bounded D and that (3.7.9) holds a.e.

Now, we select $x_0 \in G$ and write

$$f(\xi) = f_1(\xi) + f_2(\xi) \text{ where } f_1(\xi) = f(\xi) \text{ in } B(x_0, 2r)$$

and $f_1(\xi) = 0$ elsewhere and let V_k be the potential of f_k. Let

$$\varphi_2(\varrho; x) = \int_{B(x,\varrho)\cap[G - B(x_0, 2r)]} |f(\xi)|\, d\xi. \tag{3.7.10}$$

Then, from (3.7.9) for V_2 we have

$$\begin{aligned} |\nabla V_2(x)| &\le \Gamma_\nu^{-1} \int_{G - B(x_0, 2r)} |\xi - x|^{1-\nu} |f(\xi)|\, d\xi = \Gamma_\nu^{-1} \int_r^R \varrho^{1-\nu} \varphi_2'(\varrho; x)\, d\varrho \\ &\le L\Gamma_\nu^{-1} \left[R^{\lambda-1} + (\nu - 1) \int_r^R \varrho^{\lambda-2}\, d\varrho \right] \\ &\le (\nu - 1)\Gamma_\nu^{-1} (1-\lambda)^{-1} L r^{\lambda-1}, \quad x \in B(x_0; r) \end{aligned}$$

since obviously

$$\varphi_2(\varrho; x) \le \begin{cases} 0, & 0 \le \varrho \le r \\ L\varrho^{\nu-2+\lambda} & r \le \varrho \le R \end{cases} \qquad (x \in B(x_0; r)).$$

Accordingly

$$\int_{B(x_0, r)} |\nabla V_2(x)|^2\, dx \le C(\nu, \lambda)\, L^2 r^{\nu-2+2\lambda}.$$

From the SCHWARZ inequality, we obtain

$$\begin{aligned} &|\nabla V_1(x)|^2 \le \Gamma_\nu^{-1} I_1 I_2, \quad \text{where } 0 \le \sigma < \lambda \text{ and} \\ &I_1 = \Gamma_\nu^{-1} \int_{B(x_0, 2r)} |\xi - x|^{\sigma-\nu} |f(\xi)|\, d\xi, \\ &I_2 = \int_{B(x_0, 2r)} |\xi - x|^{2-\nu-\sigma} |f(\xi)|\, d\xi \end{aligned} \tag{3.7.11}$$

since $f_1(\xi) = 0$ outside $B(x_0, 2r)$. In order to evaluate I_2 define

$$\varphi_1(\varrho; x) = \int_{B(x,\varrho)\cap B(x_0, 2r)\cap G} |f(\xi)|\, d\xi.$$

Then we see that

$$\varphi_1(\varrho; x) \le \begin{cases} L\varrho^{\nu-2+\lambda} & 0 \le \varrho \le r, \\ L(2r)^{\nu-2+\lambda} & \varrho \ge r, \end{cases} \qquad x \in B(x_0; r).$$

Proceeding as with φ_2, we see that

$$I_2 \leq C(\nu,\lambda)(\lambda-\sigma)^{-1} L r^{\lambda-\sigma}, \quad x \in B(x_0, r). \tag{3.7.12}$$

Integrating (3.7.11) over $B(x_0, r)$ and using (3.7.12), we see that

$$\int_{B(x_0,r)} |\nabla V_1(x)|^2 \, dx \leq \sigma^{-1}(2r)^\sigma L (2r)^{\nu-2+\lambda} C(\nu,\lambda,\sigma) L r^{\lambda-\sigma}.$$

Since this holds for any σ with $0 < \sigma < \lambda$, we see that

$$\int_{B(x_0,r)} |\nabla V_1(x)|^2 \, dx \leq C(\nu,\lambda) L^2 r^{\nu-2+2\lambda}.$$

Using (3.7.9), we see that

$$-\int_G v_{,\alpha}(x) V_{,\alpha}(x)\, dx = \Gamma_\nu^{-1} \int_G \int_G |\xi - x|^{-\nu} (\xi^\alpha - x^\alpha) v_{,\alpha}(x) f(\xi)\, dx\, d\xi$$

$$= \int_G v(\xi) f(\xi)\, d\xi$$

using Theorem 3.7.2 (with x and ξ interchanged).

To prove (b) we note first that it follows from Theorem 3.4.2 that $V \in H^1_{2s'}(D)$ for any bounded D with $\|\nabla^2 V\|^0_{2s', R_\nu} \leq C(\nu) \|f\|^0_{2s'}$. If we take any A, we find, as in the proof of Theorem 3.4.2 that

$$|\nabla V(x)|^{2s'} \leq g^{2s'-1} \cdot \Gamma_\nu^{-1} \int_G |\xi - x|^{1-\nu} |f(\xi)|^{2s'}\, d\xi$$

$$(\gamma_\nu g^\nu = |G|)$$

$$\int_{B_{2A}} |\nabla V(x)|^{2s'}\, dx \leq g^{2s'-1}(2A) \int_G |f(\xi)|^{2s'}\, d\xi.$$

So let $W_\alpha = \eta V_{,\alpha}$ where $\eta = 1$ on B_A on $\eta = 0$ on and near ∂B_{2A}. Then $W \in H^1_{2s'}(R_\nu)$ and has compact support, and

$$W_{\alpha,\beta}(x) = \eta V_{,\alpha\beta} + \eta_{,\beta} V_{,\alpha}$$

$$\|\nabla W\|^0_{2s',2A} \leq \|\nabla^2 V\|^0_{2s',R_\nu} + C \cdot (g/A)^{1-1/2s'} \|f\|^0_{2s'}.$$

Hence, for each A, we find that $\nabla V \in L_2(B_A)$, using Theorem 3.5.3. By letting $A \to +\infty$, we find that

$$\int_{R_\nu} |\nabla V|^2\, dx \leq C^2(\nu) (\|f\|^0_{2s'})^2.$$

To prove the second statement, choose a point $x_0 \in G$ and let R be the distance of x_0 from ∂G. Write

$$f(x) = f_1(x) + f_2(x)$$

where $f_2(x) = 0$ in $B(x_0, R/2)$ and $f_1(x) = 0$ in $G - B(x_0, R/2)$, and let V_k be the potential of f_k. Clearly $f_2 \in L_{2s'}(G)$ with $\|f_2\|^0_{2s',G} \leq \|f\|^0_{2s',G}$, so $V_2 \in H^1_2(D)$ on any bounded D. But since V_2 is harmonic on $B(x_0, R/2)$, it satisfies a condition (3.7.8) with $\mu = 1$ and $L_1 = \|f\|^0_{2s'}$. Finally, it is easy to see that f_1 satisfies the condition of part (a) with L replaced by $L_1 R^{-\mu-\tau+1}$. The result follows.

(c) Choose $x_0 \in B_R$ and $r > 0$. Write $f = f_1 + f_2$ where $f_1(x) = f(x)$ on $B(x_0, 2r)$ and $f_1(x) = 0$ elsewhere; let V_k be the potential of f_k. Then, from Theorem 3.7.1 or Theorem 3.4.2, we conclude that

$$\int_{B(x_0,r)} |\nabla^2 V_1|^2 dx \leq \int_{R_\nu} |\nabla^2 V_1|^2 dx \leq Z_1 L^2 r^{\nu-2+2\mu}.$$

Moreover, we conclude that V_2 is harmonic on $B(x_0, 2r)$ and

$$|\nabla^2 V_2(x)| \leq Z_2 \int_{B_R - B(x_0,2r)} |\xi - x|^{-\nu} |f(\xi)| \, d\xi, \quad x \in B(x_0, r).$$

If we define

$$\varphi_2(\varrho; x) = \int_{B(x,\varrho)\cap[B_R - B(x_0,2r)]} |f(\xi)| \, d\xi$$

we find that

$$|\nabla^2 V_2(x)| \leq Z_2 \int_r^\infty \varrho^{-\nu} \varphi_2'(\varrho; x) \, d\varrho \leq \nu Z_2 \int_r^\infty \varrho^{-\nu-1} \varphi_2(\varrho; x) \, d\varrho \leq Z_3 L r^{\mu-1}$$

since from the SCHWARZ inequality, it follows that

$$\varphi_2(\varrho; x) \leq \begin{cases} \gamma_\nu^{1/2} L \varrho^{\nu-1+\mu}, & \varrho \leq R \\ \gamma_\nu^{1/2} L R^{\nu-1+\mu}, & \varrho \geq R. \end{cases}$$

Then ∇V satisfies the growth condition for ∇V_1 so the result follows from Theorem 3.5.2.

Chapter 4

Existence theorems

4.1. The lower-semicontinuity theorems of Serrin

In Chapter 1. we proved two theorems concerning the lower-semicontinuity of integrals $I(z, G)$ (Theorems 1.8.1 and 1.8.2). Although each of the functions considered was required to be in the space $H_1^1(D)$ for each $D \subset\subset G$, it was necessary in the first theorem only to require that the *functions* z_n converged weakly in $L_1(D)$ to z for each $D \subset\subset G$, nothing being required concerning their derivatives. In the second, however, we required $z_n \rightharpoonup z$ in $H_1^1(D)$ for each $D \subset\subset G$ which implies strong convergence in L_1 and the uniform absolute continuity of the set functions $\int_e |\nabla z_n| \, dx$ on such domains D. The writer, in his work referred to in Chapter 1, required essentially this latter type of convergence in his lower-semicontinuity theorems. In this section we present some of SERRIN's recent generalizations ([1], [2]) of these theorems and some of those of TONELLI in which SERRIN merely requires that $z_n \to z$ (strongly) in $L_1(D)$ for each $D \subset\subset G$. Our general hypotheses in this section concerning f are as follows:

$$
(4.1.1)\qquad \begin{aligned}
&\text{(i)}\ \ f(x, z, p) \textit{ is continuous for all } (x, z, p)\ (N \text{ arbitrary}),\\
&\text{(ii)}\ \ f(x, z, p) \geq 0 \textit{ for all } (x, z, p),\\
&\text{(iii)}\ \ f(x, z, p) \textit{ is convex in } p \textit{ for each } (x, z).
\end{aligned}
$$

For much of this section, we shall abbreviate (x, z) to t.

We need the following well known lemma which is an immediate consequence of the definition given in § 1.

Lemma 4.1.1. *Suppose $\{f\}$ is a family of convex functions defined on a convex set with $f(p) \leq f_0(p) < +\infty$ for each p in S. Then $\sup f(p)$ is a convex function $\leq f_0(p)$ on S.*

Suppose that $f(t, p)$ satisfies (4.1.1) $(t = (x, z))$ and that its derivatives f_p are also continuous. For each $L > 0$, we define

$$
(4.1.2)\qquad f_L(t, p) = \max\left\{0, \sup_{|q| \leq L}[F(t, p, q)]\right\}
$$

where

$$
F(t, p, q) = f(t, q) + (p - q) \cdot f_p(t, q).
$$

Lemma 4.1.2. *The function $f_L(t, p)$ defined above has the following properties:*

(i) *f_L satisfies the conditions* (4.1.1);

(ii) *$f_L(t, p) \leq f(t, p)$ for all (t, p), the equality holding if $|p| \leq L$;*

(iii) *for each compact subset $\sum$ of the t-space, there is a constant A and a function $\lambda(\sigma)$, > 0 for $\sigma > 0$ with $\lambda(\sigma) \to 0$ as $\sigma \to 0$, such that*

$$
\begin{gathered}
f_L(t, p) \leq A(1 + |p|)\\
|f_L(s, p) - f_L(t, p)| \leq \lambda(|s - t|) \cdot (1 + |p|), \quad s, t \in \sum\\
|f_L(t, p_2) - f_L(t, p_1)| \leq A\,|p_2 - p_1|.
\end{gathered}
$$

Proof. Obviously $f_L \geq 0$. Using Lemma 1.8.1, we see that 0 and $F(t, p, q) \leq f(t, p)$ for each t and q, so that f_L is convex and $\leq f$. Since $F(t, p, q)$ is uniformly continuous on any compact set, it follows that f_L is continuous. Obviously $f_L(t, p) = f(t, \mathrm{p})$ for $|p| \leq L$. Now, clearly uniform inequalities of the form (iii) hold for each function $F(t, p, q)$ when $|q| \leq L$ and s and t are in $\sum$; also the sup in (4.1.2) is taken on for some q, depending on t and p. Clearly the first inequality in (iii) follows. To prove the second define

$$
f_{0L}(t, p) = \max_{|q| \leq L} F(t, p, q).
$$

Then

$$
\begin{gathered}
f_L(t, p) = \max\{0, f_{0L}(t, p)\},\\
|f_L(s, p) - f_L(t, p)| \leq |f_{0L}(s, p) - f_{0L}(t, p)|.
\end{gathered}
$$

Also

$$
\begin{aligned}
&f_{0L}(s, p) = F[s, p, q(s, p)],\\
&f_{0L}(t, p) = F[t, p, q(t, p)] \geq F[t, p, q(s, p)]\\
&f_{0L}(s, p) - f_{0L}(t, p) \leq F[s, p, q(s, p)] - F[t, p, q(s, p)]\\
\leq &\max_{|q| \leq L}\{|f(s, q) - f(t, q)| + |p - q| \cdot |f_p(s, q) - f_p(t, q)|\}.
\end{aligned}
$$

One obtains the same result with s and t interchanged. The third result is proved similarly.

Definition. An integrand function f is *normal* if and only if

$$(4.1.3) \qquad f(x, z, p) \to +\infty \quad \text{as} \quad |p| \to +\infty \quad \text{for each } (x, z).$$

Lemma 4.1.3. *If f is normal and $\sum$ is any compact subset of the (x, z) space, there is a constant a such that*

$$f(x, z, p) \geq a\,|p| - 1, \quad a > 0, \quad (x, z) \in \textstyle\sum.$$

Proof. Suppose this is not true. Denote (x, z) by t. Then I sequences $\{t_n\}$ and $\{p_n\}$ such that

$$f(t_n, p_n) < \frac{1}{n}\,|p_n| - 1.$$

Since $f(t, \text{p}) \geq 0$, we see that $R_n = |p_n| \to \infty$. Let $\pi_n = R_n^{-1} p_n$; we may, by taking a subsequence, suppose that $t_n \to t_0$ and $\pi_n \to \pi_0$. Since f is convex in p for each t, we have

$$(4.1.4) \quad f(t_n, R\,\pi_n) < \left(1 - \frac{R}{R_n}\right) f(t_n, 0) + \frac{R}{R_n}\left(\frac{1}{n} R_n - 1\right), \quad 0 \leq R \leq R_n.$$

Holding R fixed and letting $n \to \infty$ in (4.1.4) we conclude that

$$f(t_0, R\,\pi_0) \leq f(t_0, 0), \quad 0 \leq R < \infty$$

which contradicts (4.1.3).

Lemma 4.1.4. *Suppose $g(x, z, p)$ satisfies (4.1.1), vanishes outside a compact set $\sum = S_0 \times T_0$ in the (x, z) space, and has continuous partial derivatives g_p which are Lipschitz continuous in x, z, and p and satisfy $|g_p| \leq M$ for some M. Suppose z_n and $z \in H_1^1(D)$ and $z_n \to z$ in $L_1(D)$ for each $D \subset\subset G$. Then*

$$I_g(z, G) \leq \liminf_{n \to \infty} I_g(z_n, G).$$

Proof. Let $\varepsilon > 0$ and choose ϱ so small that $S_0 < G_\varrho$ and

$$(4.1.5) \qquad \int_{S_0} |\nabla z - \nabla z_\varrho|\, dx \leq \varepsilon$$

z_ϱ being the φ mollified function for some mollifier φ. Now the WEIERSTRASS E-function associated with g is defined by

$$E(x, z, p, q) = g(x, z, q) - g(x, z, p) - (q - p) \cdot g_p(x, z, p).$$

Since g is convex and $|g_p| \leq M$, we conclude that

$$0 \leq E(x, z, p, q) \leq 2\,M\,|p - q|.$$

Making the abbreviations $\nabla z_n = p_n$, $\nabla z_\varrho = p_\varrho$, $\nabla z = p$, we obtain

$$\begin{aligned} & g(x, z_n, p_n) - g(x, z, p) \\ (4.1.6) \quad & \geq g(x, z_n, p_\varrho) + (p_n - p_\varrho) \cdot g_p(x, z_n, p_\varrho) - g(x, z, p_\varrho) \\ & \quad - (p - p_\varrho) \cdot g_p(x, z, p_\varrho) - 2M\,|p - p_\varrho|. \end{aligned}$$

where $\quad 2M\,|p - p_\varrho| \geq g(x, z, p) - g(x, z, p_\varrho) - (p - p_\varrho) \cdot g_p(x, z, p_\varrho).$

Now, from the dominated convergence theorem and FATOU's lemma it follows that

$$(4.1.7)\qquad \begin{aligned} &\liminf_{n\to\infty} \int_{S_0} [g(x, z_n, p_\varrho) - g(x, z, p_\varrho)]\, dx \geq 0, \\ &\lim_{n\to\infty} \int_{S_0} p_\varrho \cdot [g_p(x, z_n, p_\varrho) - g_p(x, z, p_\varrho)]\, dx = 0. \end{aligned}$$

Now, let us write

$$Q(x, z) = g_p[x, z, p_\varrho(x)].$$

From our hypotheses, it follows that Q is continuous and has support in $S_0 \times T_0$ and

$$|Q(x_1, z) - Q(x_2, z)| \leq M' |x_1 - x_2|.$$

We may approximate uniformly to Q by similar functions of class C^1, all with the same M'; so we may assume $Q \in C^1$. Define

$$P(x) = \int_{z(x)}^{z_n(x)} Q(x, \zeta)\, d\zeta.$$

Then, from Theorem 3.1.9, it follows that $P \in H_1^1(G)$ and has compact support in G, and so

$$\int_G \nabla P(x)\, dx = \int_G \{Q[x, z_n(x)] \nabla z_n - Q(x, z) \nabla z\}\, dx + \\ + \int_G \left\{\int_z^{z_n} Q_x(x, \zeta)\, d\zeta\right\} dx = 0.$$

From this it follows that

$$(4.1.8)\qquad \left|\int_{S_0} [g_p(x, z_n, p_\varrho)\, p_n - g_p(x, z, p_\varrho)\, p]\, dx\right| \leq M' \int_{S_0} |z_n - z|\, dx.$$

Integrating (4.1.6) over S_0 and using (4.1.5), (4.1.7) and (4.1.8), we obtain

$$I(z_n, G) - I(z, G) \geq -\varepsilon_n - 2M\,\varepsilon - M' \int_{S_0} |z_n - z|\, dx,$$

$$\lim_{n\to\infty} \varepsilon_n = \lim_{n\to\infty} M' \int_{S_0} |z_n - z|\, dx = 0.$$

The lemma follows easily from this and the arbitrariness of ε.

Lemma 4.1.5. *If φ is convex on an open convex set S, it satisfies a uniform* LIPSCHITZ *condition on any compact subset of S.*

This is well known.

Lemma 4.1.6. *Suppose $f(t, p)$ is strictly convex in p for each t* (and satisfies (4.1.1)), *suppose $\sum$ is a compact set in the t-space, and suppose K, L, and R are positive numbers. Then there exists a positive constant $\varkappa$ such that*

$$(4.1.9)\qquad f(t, p) - f(t, q) - A \cdot (p - q) \geq \varkappa$$

whenever $t \in \sum$, $|p| \leq K$, $|q| \leq L$, $|p - q| \geq R$, and A is a vector such that $f(t, r) - f(t, q) - A \cdot (r - q) \geq 0$ for all r (Lemma 1.2.2).

Proof. Let $\{t_n\}$, $\{p_n\}$, $\{q_n\}$, $\{A_n\}$ be chosen to satisfy the conditions and so that the left side of (4.1.9) tends to its inf. Since $\sum$ is compact and f satisfies a uniform Lipschitz condition in p for p on any bounded part of space, it follows that the A_n are uniformly bounded.

We now prove our first lower-semicontinuity theorem (SERRIN [2], Theorem 12):

Theorem 4.1.1. *Suppose that $f(x, z, p)$ satisfies one of the following conditions* (in addition to (4.1.1)):

A. *f is normal.*

B. *f is strictly convex.*

C. *The derivatives f_x, f_p, and f_{px} are continuous.*

Then if z_n and $z \in H_1^1(D)$ and $z_n \to z$ in $L_1(D)$ for each $D \subset\subset G$, it follows that

$$I(z, G) \leq \liminf I(z_n, G).$$

Proof. We shall prove the theorem in case A first. Let S and T be compact subsets of the x and z spaces, respectively, let $L > 0$, let S_0 and T_0 be bounded open sets containing S and T, respectively, and let $\varepsilon > 0$. From Lemma 4.1.3, it follows that there is a constant $a > 0$ such that

$$f(x, z, p) \geq a|p| - 1, \quad x \in \bar{S}_0, \ z \in \bar{T}_0, \ |p| \leq L + 1. \tag{4.1.10}$$

Let φ be a mollifier in (x, z, p) space and let f_ϱ be the φ-mollified function of f where ϱ is chosen so small that

$$|f(x, z, p) - f_\varrho(x, z, p)| \leq \varepsilon \quad \text{for} \quad (x, z) \in \bar{S}_0 \times \bar{T}_0, \ |p| \leq L + 1. \tag{4.1.11}$$

Next, we let $\alpha(x, z)$ be of class C^1 everywhere, have support in $S_0 \times T_0$ and equal 1 on $S \times T$; then define

$$g'(x, z, p) = \alpha(x, z) f_{\varrho L}(x, z, p). \tag{4.1.12}$$

It is clear that g' satisfies the conclusions (i) and (iii) of Lemma 4.1.2 and vanishes for (x, z) outside $S_0 \times T_0$.

Now, letting $(x, z) = t$, $\sum_0 = S_0 \times T_0$, $\sum = S \times T$

$$g'(t, p) \leq f_{\varrho L}(t, p) \leq f_\varrho(t, p) \leq f(t, p) + \varepsilon \quad \text{if } t \in \textstyle\sum_0, \ |p| \leq L + 1. \tag{4.1.13}$$

So now, suppose $|p| > L + 1$. Then

$$\begin{aligned} f_{\varrho L}(t, p) &= f_\varrho(t, q_p) + (p - q_p) \cdot f_{\varrho p}(t, q_p) \quad \text{for some } |q_p| \leq L \\ &= f_\varrho(t, q_p) + S \cdot (r_p - q_p) \cdot f_{\varrho p}(t, q_p) \quad \left(S = \frac{|p - q_p|}{|r_p - q_p|} > 1\right) \\ &= S[f_\varrho(t, q_p) + (r_p - q_p) \cdot f_{\varrho p}(t, q_p)] + (1 - S) f_\varrho(t, q_p) \\ &\leq S f_\varrho(t, r_p) + (1 - S) f_\varrho(t, q_p) \\ &\leq S[f(t, r_p) + \varepsilon] + (1 - S)[f(t, q_p) - \varepsilon] \\ &\leq (2S - 1)\varepsilon + f(t, p) \end{aligned} \tag{4.1.14}$$

using the convexity and the fact that $S > 1$. Here, r_p is the intersection of the segment $q_p p$ with $\partial B(0, L+1)$. But now let t_p be the intersection of the ray $\underline{0}\,p$ with $\partial B(0, L+1)$ and let s_p be the point on the *line* $\underline{0}\,p$ such that $q_p s_p \| r_p t_p$. Then a simple geometric argument shows that $s_p \in \overline{B(0, L)}$ so that

$$S = \frac{|p - s_p|}{|t_p - s_p|} \leq |p| - L. \tag{4.1.15}$$

If, finally, we let $g(t, p) = g'_\sigma(t, p)$ (mollified) with σ sufficiently small we see, using the bound $|g'(t, p)| \leq A(1 + |p|)$, that

$$\begin{aligned} g(t, p) &\leq f_{\varrho L}(t, p) + \varepsilon(1 + |p|) \leq f(t, p) + \varepsilon(3 + |p|) \\ &\leq f(t, p) + C\varepsilon[1 + f(t, p)] \\ |g(t, p) - f(t, p)| &\leq C\varepsilon \quad \text{for} \quad t \in \Sigma, \ |p| \leq L \end{aligned} \tag{4.1.16}$$

using (4.1.10) for $t \in \Sigma_0$ and the fact that $g(t, p) = 0$ otherwise. Using the conclusions (iii) of Lemma 4.1.2 for g', we see that g satisfies the hypotheses of Lemma 4.1.4.

Now, suppose $I(z, G) < +\infty$ and $\eta > 0$. There are compact sets $S \subset G$ in the x space, T in the z space, and a number L so large that if S^* is the subset of S where $z \in T$ and $|p| \leq L$, then

$$I(z, S^*) > I(z, G) - \eta \tag{4.1.17}$$

(if $I(z, G) = +\infty$, let M be arbitrarily large and S, T, and L can be found so that $I(z, S^*) > M$). Then, using (4.1.16)

$$\begin{aligned} I(z, G) &< I(z, S^*) + \eta \leq I_g(z, S^*) + \eta + C\varepsilon|S^*| \\ &\leq I_g(z, G) + \eta + C\varepsilon|S^*| \leq \liminf_{n\to\infty} I_g(z_n, G) + \eta + C\varepsilon|S^*| \\ &\leq (1 + C\varepsilon) \liminf_{n\to\infty} I(z_n, G) + \eta + C\varepsilon(1 + |G|). \end{aligned}$$

Since η and ε are arbitrary, the result follows in this case.

In case B, we proceed in the same way. But, in this case we find, using Lemma 4.1.6 for f_ϱ that (4.1.14) can be replaced by

$$f_{\varrho L}(t, p) \leq f(t, p) + (2S - 1)\varepsilon - \varkappa S \tag{4.1.14'}$$

where $\varkappa$ is for f_ϱ where $t \in \Sigma_0$, $|q| \leq L$, $|r| \leq L + 1$, and $|q - r| \geq 1$ since, (4.1.14)

$$f_\varrho(t, q_p) + (r_p - q_p) \cdot f_{\varrho p}(t, q_p) \leq f_\varrho(t, r_p) - \varkappa.$$

Since a convex function is Lipschitz, it follows that the supporting vectors A in Lemma 4.1.6 are just $f_p(t, q)$ for almost all q; consequently one can mollify both sides of (4.1.9) to see that the $\varkappa$ above is that for f with L replaced by $L + \varrho$ and $R = 1$ replaced by $1 - 2\varrho$. Thus, if we choose $\varepsilon < \varkappa/2$, we see that (4.1.16) can be replaced by

$$\begin{aligned} g(t, p) &\leq f(t, p) + C\varepsilon \\ |g(t, p) - f(t, p)| &\leq C\varepsilon \quad \text{for} \quad t \in \Sigma, \ |p| \leq L. \end{aligned}$$

The rest of the proof is the same.

In case C, we proceed by defining

$$g'(x, z, p) = \alpha(x, z) f_L(x, z, p)$$

where $\alpha(x, z)$ was discussed above. By inspecting the proof of Lemma 4.1.2, we see that g' satisfies the conclusions of that lemma with $\lambda(\sigma) \leq B \cdot \sigma$ if z is not varied and

$$g'(x, z, p) \leq f(x, z, p), \text{ equality holding if } (x, z) \in \sum \text{ and } |p| \leq L.$$

If we now *mollify with respect to p only* we obtain (since $1 + |p|$ satisfies a LIPSCHITZ condition for all p) that $g(x, z, p) = g'_\sigma(x, z, p)$ satisfies all the conditions of Lemma 4.1.4 and

$$|g(x, z, p) - g'(x, z, p)| \leq A\,\sigma \quad \text{for all} \quad (x, z, p).$$

The rest of the proof is the same.

We now remove the restriction in Theorem 1.8.2 that f_p be continuous and generalize that result somewhat (see SERRIN [2], Theorem 13 and MORREY [7], Chapter III, Theorem 4.1):

Theorem 4.1.2. *If f satisfies (4.1.1) only and z_n and $z \in H_1^1(D)$ with $\|z_n\|_{1,D}^1$ and $\|z\|_{1,D}^1$ uniformly bounded and $z_n \to z$ in $L_1(D)$, all for each $D \subset\subset G$, then*

$$I(z, G) \leq \liminf_{n\to\infty} I(z_n, G).$$

Proof. Choose $D \subset\subset G$. If $\liminf I(z_n, G) = +\infty$, there is nothing to prove. Otherwise, let $\varepsilon > 0$ and M be an upper bound for $\|z_n\|_{1,D}^1$ and $\|z\|_{1,D}^1$. Define

$$g(x, z, p) = f(x, z, p) + \varepsilon\,|p|.$$

Then g is normal. Using Theorem 4.1.1, we obtain

$$\begin{aligned} I(z, D) &\leq I_g(z, D) \leq \liminf_{n\to\infty} I_g(z_n, D) \\ &\leq \varepsilon M + \liminf_{n\to\infty} I(z_n, D) \leq \varepsilon M + \liminf_{n\to\infty} I(z_n, G). \end{aligned}$$

The theorem follows.

If, in Theorem 4.1.2, we require only that $z_n \to z$ in $L_1(D)$ (without the uniform boundedness of $\|z_n\|_{1,D}^1$), then the conclusion of the theorem doesn't necessarily hold. An example, due to ARONSZAJN, is to be found in the book by PAUC.

A third and perhaps less interesting theorem, stated below, requires the following definitions:

Definitions. An integrand f *is of type I* if and only if there are two positive functions $\lambda(\sigma)$ and $\mu(\sigma)$ defined for $\sigma > 0$ with

$$\lim_{\sigma\to 0}\lambda(\sigma) = \lim_{\sigma\to 0}\mu(\sigma) = 0, \; \mu(\sigma) \leq B\,\sigma \text{ for } \sigma \text{ large,}$$

such that

$$|f(x, z, p) - f(y, w, p)| \leq \lambda(|x - y|) \cdot [1 + f(x, z, p)] + \mu(|z - w|)$$

for all x, y, z, w, p. An integrand is *of type II* if and only if there is a function λ with the properties above such that

$$|f(x, z, p) - f(y, w, p)| \leq \lambda(|x - y| + |u - v|)\,[1 + f(x, z, p)].$$

Remarks. Important examples of integrands of type I and II are

$$f = f(x, p) = A(x)\,F(p),\ f = f(x, z, p) = A(x, z)\,F(p),$$

respectively, where $A(x)$ and $A(x, z)$ are continuous functions with some positive lower bound.

Theorem 4.1.3. (a) *Suppose f is of type I. Then the conclusion of Theorem* 4.1.1 *holds.*

(b) *Suppose f is of type II, z is continuous on G, z_n and $z \in H_1^1(D)$ for each $D \subset\subset G$, and z_n converges in measure to z on G. Then* $I(z, G) \leq \liminf I(z_n, G)$.

We refer to reader to SERRIN's paper (SERRIN [2]) for the proof. In that paper SERRIN defines the functional $I(z, G)$ in much the same way as LEBESGUE defined the area of a surface:

$$\mathfrak{J}(z, G) = \inf \left\{\liminf_{n\to\infty} I(z_n, G_n)\right\}$$

for all sequences $\{z_n\}$ in which each $z_n \in C^1(G_n)$, $G_n \subset G$, $U\, G_n = G$, and $z_n \to z$ in $L_1(D)$ for each $D \subset\subset G$. The scope of the discussion is enlarged by allowing "weakly differentiable" functions z in the discussion. A vector function z is *weakly differentiable* on D if and only if it $\in L_1(D)$ and there exist *measures* μ_α^i such that

$$\int_D z^i\,\omega_{i,\alpha}\,dx = -\int_D \omega_i\,d\mu_\alpha^i \quad \text{for all} \quad \omega \in C_c^\infty(D);$$

in case z is continuous, it is of bounded variation in the sense of TONELLI (bounded variation in the ordinary sense if $\nu = 1$). SERRIN then investigates in some detail the relation between the integrals $I(z, G)$ and $I(z, G)$. In many cases it is shown that $I(z, G) \leq I(z, G)$, the equality holding when z is "strongly differentiable", i.e. $z \in H_1^1\,(D)$ for each $D \subset\subset G$. As an example, if we allow functions of bounded variation and define

$$I(z, G) = \int_a^b \sqrt{1 + z'^2}\,dx \quad (G = (a, b)),$$

then $I(z, G)$ is just the length of the arc $z = z(x)$, if z is continuous. In this case, one can say that $I(z, G) \leq \mathfrak{J}(z, G)$ and the equality holds if and only if z is absolutely continuous. For details of the results, the reader is referred to the paper of SERRIN already referred to.

4.2. Variational problems with $f = f(p)$; the equations (1.10.13) with $N = 1$, $B_i = 0$, $A^\alpha = A^\alpha(p)$

In this section, we prove the theorems stated below. The first represents a complete extension of the results of A. HAAR discussed in Chapter 1. STAMPACCHIA [3][1] required some additional smoothness of the domain G and of the boundary values as did GILBARG.

Definition 4.2.1. Let G be a bounded strictly convex domain. A function φ defined on ∂G is said to satisfy *the bounded slope condition with bound M (BSM condition)* iff with each $x_0 \in \partial G$, there exist two linear functions

$$l_k(x) \equiv \varphi(x_0) + a_{k\alpha}(x^\alpha - x_0^\alpha), \quad |a_k| \leq M, \quad k = 1,2$$

such that $l_1(x) \leq \varphi(x) \leq l_2(x)$ for all x on ∂G.

Theorem 4.2.1. *Suppose G is bounded and strictly convex, φ satisfies the BSM condition on ∂G, $f = f(p)$ is convex and $I(z, G)$ is given by* (1.1.1). *Then there is a function z_0 which coincides with φ on ∂G, which satisfies a Lipschitz condition with constant M on $\bar{G}$, and which minimizes $I(z, G)$ among all Lipschitz functions $z = \varphi$ on ∂G. If f is of class C^2 with*

$$f_{p_\alpha p_\alpha}(p)\, \lambda_\alpha \lambda_\beta > 0 \quad \text{if} \quad \lambda \neq 0, \tag{4.2.1}$$

then z_0 is unique and minimizes $I(z, G)$ among all $z \in H_1^1(G)$ such that $z - z_0 \in H_{10}^1(G)$. If $\varphi \in H_p^2(G)$ with $p \geq 2$, and G is of class C_1^1, then $z_0 \in H_p^2(G)$. If $f \in C_\mu^n$, C^∞, or analytic, respectively, then so is z_0 on G. If, also, G and $\varphi \in C_\mu^n$, C^∞, or analytic, then so is z_0 on $\bar{G}$.

Theorem 4.2.2. *Suppose G and φ satisfy the hypotheses of Theorem* 4.2.1 *and suppose that the $A^\alpha(p) \in C_\mu^1$, $0 < \mu < 1$, and satisfy*

$$A_{p_\beta}^\alpha(p)\, \lambda_\alpha \lambda_\beta > 0 \quad \text{if} \quad \lambda \neq 0. \tag{4.2.2}$$

Then there is a unique Lipschitz solution z_0 of the equation

$$\int_G \zeta_{,\alpha} A^\alpha(\nabla z)\, dx = 0, \quad \zeta \in Lip_c G \tag{4.2.3}$$

which coincides with φ on ∂G; z_0 satisfies a Lipschitz with coefficient M on $\bar{G}$. If the $A^\alpha \in C_\mu^n$ with $n \geq 1$, then $z \in C_\mu^{n+1}(G)$. If G and the boundary data also $\in C_\mu^{n+1}$, $z \in C_\mu^{n+1}(\bar{G})$. The C^∞ and analytic cases also hold.

Theorem 4.2.3. *Suppose G is of class C_μ^2 and $\varphi \in C_\mu^2(\bar{G})$ and suppose the $A^\alpha \in C_\mu^1$ and satisfy*

$$m_1(p)|\lambda|^2 \leq A_{p_\beta}^\alpha(p)\, \lambda_\alpha \lambda_\beta \leq M_1(p)|\lambda|^2, \quad m_1(p) > 0, \quad M_1(p) \leq \gamma m_1(p). \tag{4.2.4}$$

Then there exists a unique solution z_0 of (4.2.3) $\in C_\mu^2(\bar{G})$. *The higher differentiability results of Theorem* 4.2.2 *hold.*

Our method is first to solve the equations (4.2.3) in the cases where G, φ, and the A^α are reasonably smooth (these include the Euler equations for the variational problem) and satisfy (4.2.4) with $m(p)$ and

[1] He has recently improved his results

$M(p)$ constants. We then obtain bounds for the first derivatives which enable us to handle the general cases by approximations.

The proofs of the first statements in Theorem 4.2.1 depend on the following two lemmas:

Lemma 4.2.1. *Suppose $f(p)$ is convex for all p. Then there exists a sequence $\{f_n\}$ of convex functions such that each $\in C^\infty(R_\nu)$, $f_n(p) \leq f(p)$ for all n and p, and $f_n(p)$ converges uniformly to $f(p)$ on any bounded set in R_ν.*

Proof. Since f has a supporting plane at 0 (Lemma 1.8.1 (a)) we may subtract that off and assume that $f(p) \geq 0$ and $f(0) = 0$. Define $g_n(p)$ as f_{L0} was defined in the proof of Lemma 4.1.2 with $L = n$. It follows (since $f(0) = 0$, $f(p) \geq 0$) that $f_{L0} = f_L$ and so satisfies the conclusions of that lemma. From the Lipschitz condition in (iii), it follows that, if $g_{n\sigma}$ denotes the $\varphi - \sigma$-mollified function for some mollifier φ, then $g_{n\sigma} \in C^\infty$ and

$$g_{n\sigma}(p) \leq g_n(p) + A\sigma \leq f(p) + A\sigma, \quad |p| \leq n.$$

We may clearly take

$$f_n(p) = g_{n\sigma_n}(p) - A\sigma_n, \quad \sigma_n = 1/nA.$$

Lemma 4.2.2. *Suppose f and f_n are convex, $f_n(p) \leq f(p)$ for each n and p, and $f_n \to f$ uniformly on any bounded set. Suppose that z_n and z all satisfy a Lipschitz condition with coefficient M on the bounded region $\bar{G}$ and that $z_n \to z$ uniformly on $\bar{G}$. Suppose $z_n = z$ on ∂G and suppose, for each n, that z_n minimizes $I_n(Z, G)$ among all Lipschitz $Z = z$ on ∂G. Then z minimizes $I(Z, G)$ among all such Z.*

Proof. Let $d = \inf I(Z, G)$ among all $Z = z$ on ∂G. Let $\varepsilon > 0$ and choose a $Z = z$ on ∂G such that $I(Z, G) < d + \varepsilon$. Then

$$I_n(z_n, G) \leq I_n(Z, G) \leq I(Z, G) < d + \varepsilon.$$

Then
$$\limsup_{n\to\infty} I_n(z_n, G) \leq d.$$

On the other hand (using the uniform Lipschitz condition and the uniform convergence of f_n to f),

$$I(z, G) \leq \liminf_{n\to\infty} I(z_n, G) \leq \liminf_{n\to\infty} [I_n(z_n, G) + \varepsilon_n] \leq d$$
$$(\lim \varepsilon_n = 0).$$

Lemma 4.2.3. *Suppose G is bounded and strictly convex and suppose φ satisfies the BSM condition on ∂G. Let $\Gamma = \{(x, z) \mid x \in \partial G, z = \varphi(x)\}$ and let Σ be the convex cover of Γ. Then there are convex functions φ^- and $-\varphi^+$ on $\bar{G}$ such that*

$$\Sigma = \{(x, z) \mid x \in \bar{G}, \varphi^-(x) \leq z \leq \varphi^+(x)\}. \tag{4.2.5}$$

φ^- and φ^+ satisfy Lipschitz conditions with coefficient M on $\bar{G}$ and

$$\varphi^-(y) = \varphi(y) = \varphi^+(y) \quad \text{for } y \in \partial G. \tag{4.2.6}$$

If $\varphi^-(x_0) - \varphi^+(x_0) < 0$ for some $x_0 \in G$, then $\varphi^-(x) - \varphi^+(x) < 0$ for all $x \in G$.

Proof. The hypotheses on G and φ imply that Σ has the form (4.2.5) where φ^- and $-\varphi^+$ are convex and (4.2.6) holds. Suppose x_1 and $x_2 \in \bar{G}$ and suppose x_3 and x_4 are the intersections of the line $x_1 x_2$ with ∂G. At x_3 we have linear functions l_3^- and l_3^+ of slopes $\leq M$ such that

$$l_3^-(x) \leq \varphi^-(x) \leq \varphi^+(x) \leq l_3^+(x), \quad x \in \bar{G}, \quad l_3^-(x_3) = \varphi(x_3) = l_3^+(x_3)$$

with the same at x_4. It follows easily from the elementary properties of convex functions of one variable (along x_1, x_2) that

$$|\varphi^-(x_1) - \varphi^-(x_2)| \leq M \cdot |x_1 - x_2|, \quad |\varphi^+(x_1) - \varphi^+(x_2)| \leq M\,|x_1 - x_2|.$$

The last statement is evident.

Lemma 4.2.4. *Suppose the $A^\alpha \in C_\mu^n$ (or C^∞), $n \geq 1$, and satisfy* (4.2.2). *Then, for each R, there exists a vector $A_R^\alpha \in C_\mu^n$ (or C^∞) which satisfies* (4.2.4) *with m_1 and M_1 constants and also satisfies*

$$|A_{Rp}(p)| \leq M_1(R) \quad \text{for all} \quad p,\ A_R^\alpha(p) = A^\alpha(p) \quad \text{if} \quad |p| \leq R.$$

Proof. From (4.2.2) and the continuity we have

$$A^\alpha_{p_\beta}(p)\,\lambda_\alpha \lambda_\beta \geq m_1(r), \quad |A_p(p)| \leq M_1(r) \quad \text{for} \quad |p| \leq r. \tag{4.2.7}$$

Let $\omega \in C^\infty(R_1)$ with $\omega(s) = 0$ for $s \leq 1$, $\omega(s) = 1$ for $s \geq 2$, and $\omega'(s) \geq 0$ for all s. Clearly $m_1(r)$ is non-increasing and $M_1(r)$ is non-decreasing. Let us define

$$A_R^\alpha(p) = [1 - \omega(r/R)]\,A^\alpha(p) + c(R)\,\omega(r/R)\,p_\alpha, \quad |p| = r, \tag{4.2.8}$$

where $c = c(R)$ is to be chosen. By computation we obtain

$$A^\alpha_{R\,p_\beta}(p)\,\lambda_\alpha\lambda_\beta = (1-\omega)\,A^\alpha_{p_\beta}\lambda_\alpha\lambda_\beta + c\,\omega\,|\lambda|^2 + s\,\omega'(c\,H^2 - r^{-1}\,H\,A^\alpha\lambda_\alpha)$$
$$s = r/R, \quad H = |p|^{-1}(p\cdot\lambda).$$

Since for large r, $|A(p)| \leq 2\,M_1(r)\cdot r$, we find using (4.2.7) that

$$A^\alpha_{R\,p_\beta}\lambda_\alpha\lambda_\beta \geq [2k(1-\omega) + c\,\omega]\,|\lambda|^2 + s\,\omega'(c\,H^2 - 2M_1|H|\,|\lambda|),$$
$$2k = m_1(2R). \tag{4.2.9}$$

If we take $c \geq 2\,k$, we see that the right side of (4.2.9) $\geq k\,|\lambda|^2$ for all r on $[R, 2\,R]$ if

$$4k\,c(s\,\omega') \geq 4M_1^2(s\,\omega')^2 \quad (M_1 = M_1(2R))$$

which holds if $k\,c \geq M_1^2 \cdot \max(s\,\omega'(s))$.

Lemma 4.2.5. *Suppose G is bounded, $\varphi \in H_2^1(G)$, the $A^\alpha(p) \in C^1$ and satisfy* (4.2.4) *with m_1 and M_1 constant and the condition*

$$|A_p(p)| \leq M_1. \tag{4.2.10}$$

Then there exists a unique solution of Equation (4.2.3) *such that $z - \varphi \in H_{20}^1(G)$. If G is of class C^2 and $\varphi \in C^2(\bar{G})$, then $z \in H_r^2(G)$ for every r. If G is of class C_μ^2, the A^α are of class C_μ^1, and $\varphi \in C_\mu^2(\bar{G})$, then $z \in C_\mu^2(\bar{G})$. Additional theorems concerning differentiability on the interior and at the boundary hold as stated in Theorem* 1.10.4 *and* 1.11.3.

Proof. We use Theorem 5.12.2 to prove the existence. We set

$$z = \varphi + u, \quad A^\alpha(x, p) = A^\alpha[\nabla\varphi(x) + p], \tag{4.2.11}$$

and define $A(u)$ and $\mathfrak{A}(u, v)$ by

$$\mathfrak{A}(u, v) = A(v), \quad (A(v), w) = \int_G w_{,\alpha} A^\alpha(\nabla\varphi + \nabla v)\, dx. \tag{4.2.12}$$

We conclude, from (4.2.4) and (4.2.10) that

$$A^\alpha(q) = A^\alpha(0) + q_{,\beta}\int_0^1 A^\alpha_{p_\beta}(tq)\, dt, \quad |A(q)| \le M_1(1 + |q|),$$

$$p_\alpha A^\alpha[\nabla\varphi(x) + p] = p_\alpha A^\alpha[\nabla\varphi(x)] + p_\alpha p_\beta \int_0^1 A_{p_\beta}[\nabla\varphi(x) + tp]\, dt$$
$$\ge m_1|p|^2 - M_1|p|\cdot(1 + |\nabla\varphi(x)|).$$

$$\begin{aligned}&\{A^\alpha[\nabla\varphi(x) + p_2] - A^\alpha[\nabla\varphi(x) + p_1]\}(p_{2\alpha} - p_{1\alpha}) \\ &= (p_{2\alpha} - p_{1\alpha})(p_{2\beta} - p_{1\beta})\int_0^1 A^\alpha_{p_\beta}[\nabla\varphi(x) + \\ &+ (1 - t)p_1 + tp_2]\, dt \ge m_1|p_2 - p_1|^2.\end{aligned} \tag{4.2.13}$$

Thus the $A^\alpha(x, p)$ satisfy the conditions (5.12.13) and (5.12.14). Also

$$(A(v), v) = \int_G v_{,\alpha} A^\alpha(x, \nabla v)\, dx \ge m_1(\|v\|^1_{20})^2 - M_1\|v\|^1_{20}(|G| + \|\nabla\varphi\|^0_2)$$

so that A satisfies the coerciveness condition (5.12.3). Thus A satisfies all the conditions of Theorem 5.12.2 so there exists a solution $u \in H^1_{20}(G)$ of Equation (4.2.3). The solution is easily seen, using (4.2.13), to be unique. The interior differentiability results follow from the results of § 1.11 and Chapter 5.

We must now prove the differentiability at the boundary assuming G of class C^2 and $\varphi \in C^2(\bar{G})$. To that end, let $x_0 \in \partial G$ and map a neighborhood $\mathfrak{N}$ of x_0 on $\bar{G}$ onto $G_R \cup \sigma_R$ (see §§ 1.2, 5.1, etc.) using a diffeomorphism of class C^2 so that x_0 corresponds to 0 and $\mathfrak{N} \cap \partial G$ corresponds to σ_R. Then Equation (4.2.3) assumes the form

$$\int_{G_R} v_{,\alpha} A^\alpha(x, \nabla u)\, dx = 0, \quad v \in C^1_c(G_R) \tag{4.2.14}$$

where the new $A^\alpha \in C^1$ and still satisfy the stated condition, possibly with different bounds. We may rewrite (4.2.14) in the form

$$\int_{G_R} v_{,\alpha}[a_1^{\alpha\beta}(x)\, u_{,\beta} + A^\alpha(x, 0)]\, dx = 0, \quad a_1^{\alpha\beta}(x) = \int_0^1 A^\alpha_{p_\beta}[x, t\nabla u(x)]\, dt$$

and the $a_1^{\alpha\beta}$ are bounded and measurable and satisfy

$$a_1^{\alpha\beta}(x)\lambda_\alpha\lambda_\beta \ge m_1|\lambda|^2, \quad |a(x)| \le M_1 \tag{4.2.15}$$

and the $A^\alpha(x, 0) \in C^1$. From Lemma 5.3.5, it follows that $u \in C^0_\mu(\bar{G}_r)$ for each $r < R$ and satisfies a Dirichlet growth condition as in that lemma

Since this is true for each x_0, we conclude that

$$(4.2.16)\qquad \int\limits_{B(x_0,r)\cap G} |\nabla u|^2\,dx \le L^2 r^{\nu-2+2\mu}, \quad 0<\mu<1$$

for all $B(x_0, r)$.

We now revert to Equations (4.2.14). Applying the difference quotient procedure of § 1.11 in a tangential direction, we see that $u_h = h^{-1}[u(x + h\,e_\gamma) - u(x)]$ satisfies

$$\int\limits_{G_R} v_{,\alpha}(a_h^{\alpha\beta}\,u_{h,\beta} + e_h^\alpha)\,dx = 0,$$

where $a_h^{\alpha\beta}$ and e_h^α are given by formulas like (1.11.4), the $a_h^{\alpha\beta}$ satisfy (4.2.15) uniformly for small h, $\|u_h\|_{2,G(r)}^1$ is uniformly bounded for each $r<R$, and e_h^α satisfies a condition

$$(4.2.17)\qquad \int\limits_{B(y_0,r)\cap G_R} |e_h|^2\,dx \le L^2 r^{\nu-2+2\mu}, \quad 0\le r\le R.$$

It follows, again from Lemma 5.3.5 that the u_h are uniformly Hölder continuous on each $\bar{G}_r$ and ∇u_h satisfies (4.2.17) with R replaced by r with an L and μ independent of h. Thus we may let $h\to 0$ and find that the ∇p_γ with $\gamma<\nu$ satisfy these conditions. Since $p_{\nu,\gamma} = p_{\gamma,\nu}$ and we can solve for $p_{\nu,\nu}$ in terms of the others, we see that ∇p_ν also satisfies (4.2.17) as do the p_γ, and therefore $p_\gamma \in C_\mu^0(\bar{G}_r)$ for each $r<R$. The higher differentiability results follow from Theorem 1.10.3.

Lemma 4.2.6. *Suppose G is of class C_μ^2, $z\in C_\mu^1(\bar{G})$, and z satisfies an equation of the form*

$$\int\limits_G v_{,\alpha}\,a^{\alpha\beta}(x)\,z_{,\beta}\,dx = 0, \quad v\in H_{20}^1(G), \quad a^{\alpha\beta}\in C_\mu^0(\bar{G}).$$

Then z takes on its maximum and minimum values on ∂G.

Proof. Approximate to the $a^{\alpha\beta}$ by $a_n^{\alpha\beta}\in C_\mu^1(\bar{G})$ and let z_n be the solution coinciding on ∂G with z of the equations

$$\frac{\partial}{\partial x}\,a_n^{\alpha\beta}(x)\,z_{n,\beta} = 0.$$

Then $z_n\in C_\mu^1(\bar{G})$, $z_n\in C_\mu^2(D)$ for each $D\subset\subset G$, each z_n takes on its maximum and minimum values on ∂G and $z_n\to z$ in $C^1(G)$ (see §§ 5.2, 5.5, and 5.6).

Proof of Theorems 4.2.1 and 4.2.2. In case φ coincides on ∂G with a linear function $l(x)$, then $l(x)$ is the desired solution in either case. So we assume that this is not the case.

We first prove Theorem 4.2.2. We first assume that the $A^\alpha\in C^\infty$ and satisfy the conditions in Lemma 4.2.5. Let ψ be a mollifier. For each n, we define

$$'\varphi_n^\pm(x) = \varphi^\pm(x) \mp M\,n^{-1} \mp n^{-1}|x-x_0|^2, \quad x\in\bar{G}, \quad x_0\in G,$$
$$\varphi_n^\pm(x) = {}'\varphi_{\varrho_n}^\pm(x), \quad x\in\bar{G}_{\varrho_n}, \quad \varrho_n = 1/n.$$

Then φ_n^- and $-\varphi_n^+ \in C^\infty(G_{\varrho_n})$ and are convex. Let

$$G_n = \{x \mid \varphi_n^+ - \varphi_n^- > 0\}, \quad \varphi_n(x) = \varphi_n^+(x) = \varphi_n^-(x), \quad x \in \partial G_n.$$

Then G_n is convex and of class C^∞, $G_n \subset G_{n+1}$, $G_n \to G$, and $\varphi_n^\pm$ converge nicely to $\varphi^\pm$ on any $D \subset\subset G$. For each n, let z_n be the solution of Lemma 4.2.5 with φ replaced by φ_n^-, say. Each $z_n \in C^\infty(\bar{G}_n)$. By a difference quotient device like that in § 1.11, starting from (4.2.3) we find that each $z_{n,\gamma} = p_{n\gamma}$ satisfies

$$\int_{G_n} v_{,\alpha}[a^{\alpha\beta}(\nabla z_n)\, p_{n\gamma,\beta}]\, dx = 0, \quad v \in H^1_{20}(G_n) \quad a^{\alpha\beta}(q) = A^\alpha_{p_\beta}(q)$$

and so each linear combination of the $p_{n\gamma}$ takes on its maximum on ∂G_n so that $|\nabla z_n|$ takes on its maximum on ∂G_n.

Now, since φ_n^- is convex, we conclude that

$$\frac{\partial}{\partial x^\alpha} A^\alpha(\nabla \varphi_n^-) \equiv a_n^{\alpha\beta}(x)\, \varphi^-_{n,\alpha\beta} \geq 0, \quad a_n^{\alpha\beta}(x) = \frac{1}{2}\left[A^\alpha_{p_\beta}(\nabla \varphi_n^-) + A^\beta_{p_\alpha}(\nabla \varphi_n^-)\right],$$

as is seen by rotating axes. Accordingly

$$\frac{\partial}{\partial x^\alpha}[A^\alpha(\nabla z_n) - A^\alpha(\nabla \varphi_n^-)] = \frac{\partial}{\partial x^\alpha}\, {}'a_n^{\alpha\beta}(x)\,(z_{n,\beta} - \varphi^-_{n,\beta}) \leq 0,$$

$${}'a_n^{\alpha\beta}(x) = \int_0^1 A^\alpha_{p_\beta}[(1-t)\,\nabla \varphi_n^- + t\,\nabla z_n]\, dt,$$

so that $z_n - \varphi_n^- \geq 0$ by the maximum principle. In like manner $z_n \leq \varphi_n^+$ on $\bar{G}_n$. Consequently $|\nabla z_n| \leq M_n$ on ∂G_n and hence on $\bar{G}_n$ and $M_n \to M$.

Then a subsequence, still called $\{z_n\}$, is such that z_n converges uniformly on each $D \subset\subset G$ to a Lipschitz function $z = \varphi$ on ∂G and z satisfies a Lipschitz condition with coefficient M on G. Moreover, on any $D \subset\subset G$, it follows from the theorems of §§ 5.2, 5.5, and 5.6 that the derivatives of all orders converge uniformly on D to those of z. Thus z is a solution. If the $A^\alpha \in C^\infty$ but do not satisfy the conditions (4.2.4) with m_1 and M_1 constants, we may find functions $A^\alpha_R(p)$ which $= A^\alpha(p)$ for $|p| \leq R$, $R > M$, which do satisfy (4.2.4). For $R > M$, the solution z obtained above for A_R is just that for A. Finally we can approximate any $A^\alpha \in C^{n-1}_\mu$ by those $\in C^\infty$. This proves Theorem 4.2.2. The higher differentiability results along ∂G follow from Theorem 1.11.3.

In case f is regular and of class C^2, we conclude from Theorem 4.2.2 just proved that there is a solution z_0 of Euler's equation on G where z satisfies a Lipschitz condition with coefficient M on $\bar{G}$ and $z_0 \in H^2_r(D)$ for each $D \subset\subset G$ and each $r \geq 2$. It is immediately evident that z_0 is the unique function minimizing $I(Z, G)$ among all Lipschitz functions $Z = z_0 = \varphi$ on ∂G. The higher differentiability results follow as above. If f is not smooth, we may approximate it from below as in Lemma 4.2.1 and complete the proof of Theorem 4.2.1 by using Lemma 4.2.2 and the uniform Lipschitz bound M.

Proof of Theorem 4.2.3. Let us first assume that the $A^\alpha \in C^1_\mu$ and satisfy the conditions of Lemma 4.2.5. Then there is a unique solution $z \in C^2_\mu(\bar{G})$ with $z = \sigma\varphi$ on G of Equation 4.2.3 for each σ, $0 \leq \sigma \leq 1$. As above we see that ∇z takes on its maximum on ∂G (for each σ). We note that $z(x, 0) \equiv 0$ and, by difference quotienting with respect to σ, we see that z and ∇z vary continuously with σ. In fact $z_\sigma \in C^1_\mu(\bar{G})$ and satisfies

$$(4.2.18)\qquad \int_G v_{,\alpha} A^\alpha_{p_\beta}(\sigma\nabla\varphi + \nabla u)\, z_{\sigma,\beta}\, dx = 0, \quad v \in C^1_c(G), \quad z = \sigma\varphi + u.$$

Now, let us suppose that the A^α satisfy the conditions (4.2.4) with $m_1(p)$ and $M_1(p)$ variable. Let us choose $R > \max|\nabla\varphi(x)|$ and define the A^α_R as in (4.2.8) so Lemma 4.2.4 holds. Then, as long as $|\sigma\nabla\varphi + \nabla u| = |\nabla z(x,\sigma)| \leq R$, u is a solution of

$$a^{\alpha\beta}(x,\sigma)\, u_{,\alpha\beta}(x,\sigma) = -\sigma a^{\alpha\beta}\varphi_{,\alpha\beta}, \quad z(x,\sigma) = \sigma\varphi(x) + u(x,\sigma)$$
$$a^{\alpha\beta}(x,\sigma) = \{A^\alpha_{p_\beta}[\nabla z(x,\sigma)] + A^\beta_{p_\alpha}[\nabla z(x,\sigma)]\}\{m_1(\nabla z) + M_1(\nabla z)\}.$$
(4.2.19)

Then we notice that

$$(4.2.20)\qquad m_2(\gamma)|\lambda|^2 \leq a^{\alpha\beta}(x,\sigma)\lambda_\alpha\lambda_\beta \leq \gamma m_2(\gamma), \quad m_2(\gamma) = 2/(1+\gamma)$$

for such (x, σ). From (4.2.19) and (4.2.20), it follows that

$$|L u(x,\sigma)| \leq \sigma \cdot C(\nu,\gamma)\, \|\nabla^2\varphi\|^0_0, \quad L w \equiv a^{\alpha\beta} w_{,\alpha\beta}$$

where $\|\psi\|^0_0 = \max|\psi(x)|$.

Now, let $x_0 \in \partial G$. Since G is of class C^2, there is a $\varrho > 0$, independent of x_0, such that there is a unique ball $\overline{B(x_1, \varrho)}$ such that $B(x_1,\varrho) \cap \bar{G} = \{x_0\}$. If we define

$$w(x) = K \cdot P(\varrho^{-t} - r^{-t}), \quad r = |x - x_1|, \quad P = \|\varphi\|^0_0,$$

we see that

$$L w = K P t r^{-t-2}\Big[\sum_\alpha a^{\alpha\alpha} - (t+2) r^{-2} a^{\alpha\beta}(x^\alpha - x^\alpha_1)(x^\beta - x^\beta_1)\Big]$$
$$\leq K P t m_2(\gamma) r^{-t-2}[\nu\gamma - (t+2)]$$
$$|\nabla w(x)| = K P t \varrho^{-t-1}, \quad w(x_0) = 0, \quad w(x) \geq 0 \text{ for } x \in \bar{G}.$$

Clearly we may choose $K(\nu, \gamma, D)$ and $t(\nu, \gamma, D)$ so that

$$L w(x) \leq -C(\nu,\gamma) P, \quad x \in \bar{G}.$$

Then we see that

$$L(\sigma w \pm u) \leq 0 \text{ in } G, \quad \sigma w \pm u \geq 0 \text{ on } \partial G.$$

The maximum principle implies that $\sigma w \pm u \geq 0$ in $\bar{G}$ and so, since $\sigma w \pm u = 0$ at x_0, it follows that

$$(4.2.21)\qquad |\nabla u(x_0)| = \left|\frac{\partial u}{\partial n}(x_0)\right| \leq \sigma|\nabla w(x_0)| \leq \sigma C_1(\nu,\gamma,D)\cdot P,$$

provided that $\max(\sigma |\nabla \varphi| + |\nabla u|) \leq R$. But if we take

$$R \geq \|\nabla \varphi\|_0^0 + C_1 \|\nabla^2 \varphi\|_0^0$$

the bound (4.2.21) will hold for all x on $\bar{G}$ and all σ, $0 \leq \sigma \leq 1$. This proves the theorem.

4.3. The borderline cases $k = \nu$

In this section, we complete the proof of the continuity properties of the solutions of minimum problems which were stated in Theorem 1.10.2. We assume that

(i) $f(x, z, p)$ satisfies (4.1.1); and

(ii) there exist numbers $m > 0$ and K such that

$$f(x, z, p) \geq m V^\nu - K \quad \text{for all} \quad (x, z, p),\ V = (1 + |p|^2)^{1/2}. \tag{4.3.1}$$

Lemma 4.3.1. *Suppose $\zeta \in H_k^1(\Sigma)$, $k > 1$, and $\bar{\zeta}$ is its average over Σ. Then there is a constant $C = C(\nu, k)$ such that*

$$\int_\Sigma \{|\zeta(\sigma) - \bar{\zeta}|^2 + |\nabla_\sigma \zeta(\sigma)|^2\}^{k/2} d\Sigma(\sigma) \leq C \int_\Sigma |\nabla_\sigma \zeta(\sigma)|^k d\Sigma(\sigma).$$

Proof (k and ν fixed). If this is not so, $\exists \{\zeta_n\} \in H_k^1(\Sigma)$ such that

$$\int_\Sigma \{|\zeta_n(\sigma) - \bar{\zeta}_n|^2 + |\nabla_\sigma \zeta_n|^2\}^{k/2} d\Sigma(\sigma) > n \int_\Sigma |\nabla_\sigma \zeta_n|^k d\Sigma.$$

On account of the homogeneity, we may assume the left side $= 1$ for every n. Since the H_k^1 norms of the functions $\zeta_n - \bar{\zeta}_n$ are all 1, a subsequence $\rightharpoonup$ some ζ_0 in $H_k^1(\Sigma)$. But then $\zeta_n - \bar{\zeta}_n$ tends strongly in $H_k^1(\Sigma)$ to ζ_0 since $\nabla_\sigma z_n \to 0$ in $L_k(\Sigma)$. Thus ζ_0 is a constant, which must be 0 since its integral over Σ is 0. But this contradicts the fact that $\|\zeta_n - \bar{\zeta}_n\|_{k,\Sigma}^1 = 1$.

Theorem 4.3.1. *If f satisfies* (i) *and* (ii) *above and also*

$$f(x, z, p) \leq M V^\nu \tag{4.3.2}$$

and if z minimizes $I(z, G)$ among all vectors in $H_\nu^1(G)$ having the same boundary values, then z is Hölder continuous on interior domains and satisfies

$$\left\{\int_{B(x_0, r)} V^\nu dx\right\}^{2/\nu} \leq C \cdot (\|V\|^0)^2 \cdot (r/R)^\mu \quad \text{if} \quad r \leq R,\ B(x_0, R) \subset G, \tag{1.10.9}$$

where C and μ depend only on m, M, K, and ν.

Proof. The hypotheses (4.3.1) and (4.3.2) imply that

$$m |p|^\nu - K \leq f(x, z, p) \leq M_1 |p|^\nu + M_1,\ M_1 = 2^{\nu-1} M. \tag{4.3.3}$$

Let us define

$$\varphi(r) = \int_{B_r} |p|^\nu dx, \quad B_r = B(x_0, r). \tag{4.3.4}$$

Using (4.3.3) and the fact that z is minimizing we find that

$$(4.3.5)\qquad m\,\varphi(r) \leq I(z, B_r) + K\,|B_r| \leq I(Z, B_r) + K\,|B_r| \leq (K + M_1)\,|B_r| + M_1 \int_{B_r} |\nabla Z|^\nu\,dx$$

where Z is any function $= z$ on ∂B_r.

Now, for almost every r, $z \in H_\nu^1(\partial B_r)$ and is therefore essentially continuous. For such r, let

$$Z(s, \sigma) = \bar{z} + (s/r)\,[z(r, \sigma) - \bar{z}]$$

(s, σ) being "spherical coordinates", σ being on $\partial B(0,1)$ and $\bar{z}$ being the average of $z(r, \sigma)$ over $\partial B(0,1)$. Since

$$|\nabla Z|^2 = Z_s^2 + s^{-2}|\nabla_\sigma Z|^2 = r^{-2}\{[z(r, \sigma) - \bar{z}]^2 + |\nabla_\sigma z(r, \sigma)|^2\},$$

we find that

$$\int_{B_r} |\nabla Z|^\nu\,dx = \nu^{-1} \int_{\partial B(r,\sigma)} \{[z(r, \sigma) - \bar{z}]^2 + |\nabla_\sigma z(r, \sigma)|^2\}^{\nu/2}\,d\Sigma$$

$$(4.3.6)\qquad \leq C(\nu) \int_{\partial B(r,\sigma)} |\nabla_\sigma z(r, \sigma)|^\nu\,d\Sigma \leq C\,r\,\varphi'(r)$$

using Lemma 4.3.1. From (4.3.5) we conclude that

$$(4.3.7)\qquad \varphi(r) \leq A\,r\,\varphi'(r) + B\,r^\nu, \quad A = M_1/m, \quad B = \gamma_\nu(K + M_1)/m.$$

From (4.3.7) and the absolute continuity of φ it follows that

$$\varphi(r) \leq (r/R)^{1/A}\,\varphi(R) + \frac{B}{A\nu - 1}\,(r^{1/A}\,R^{\nu - 1/A} - r^\nu)$$

$$(4.3.8)\qquad \int_{B_r} V^\nu\,dx \leq 2^{\nu-1}\,[\gamma_\nu r^\nu + \varphi(r)]$$

from which the result (1.10.9) easily follows. The HÖLDER continuity follows from this and Theorem 3.5.2.

Theorem 4.3.2. *Suppose* $k = \nu = 2$ *and* f *satisfies*

$$(1.10.8^*)\qquad m\,|p|^2 \leq f(x, z, p) \leq M\,|p|^2, \quad 0 < m \leq M,$$

instead of (4.3.1) *and* (4.3.2). *Then if* G *is bounded by* $k \geq 1$ *disjoint* JORDAN *curves and* $z^* \in H_2^1(G)$ *and is continuous on* $\bar{G}$, *then any minimizing function* z *with* $z - z^* \in H_{20}^1(G)$ *is continuous on* $\bar{G}$.

Proof. Since (1.10.8*) implies (4.3.1) and (4.3.2), the continuity on the interior follows from Theorem 4.3.1. If, in the argument in (4.3.5) we replace B_r by an arbitrary domain $g \subset G$, we conclude, using (1.10.8*) that

$$(4.3.9)\qquad D(z, g) \leq (M/m)\,D(H_g, g)$$

H_g denoting the harmonic function coinciding with z on ∂g.

Now, let $P_0 \in \partial G$; suppose it is on the bounding curve Γ. If Γ is the outer boundary map the interior of Γ conformally (using the RIEMANN mapping theorem, see for example Chapter 9) onto a domain G' in the upper half-plane so that P_0 corresponds to the origin, a part of ∂G corresponds to σ_a and a part of G corresponds to G_a. If Γ is an inner boundary, the same result may be obtained by first performing an inversion with respect to a point interior to Γ. The transformed function $'z^*$ is continuous on $'\bar{G}$ and $\in H_2^1('G)$ and the transformed function $'z$ is continuous in $'G$, $\in H_2^1('G)$, satisfies (4.3.9) for subdomains $'g \subset 'G$, and $'z - 'z^* \in H_{20}^1('G)$; all these things follow from Theorem 3.1.5, the invariance of the DIRICHLET integral under conformal mappings, and an approximation.

Let us extend $'z^*$, $'z$ and $'w = 'z - 'z^*$ by the formulas

$$'z^*(x^1, x^2) = 'z^*(x^1, -x^2), \quad 'w(x^1, x^2) = -'w(x^1, -x^2),$$
$$'z(x^1, x^2) = 'z^*(x^1, x^2) + 'w(x^1, x^2) \tag{4.3.10}$$
$$(x^1, x^2) \in 'G^-$$

where $'G^-$ is the reflected region. From Theorems 3.1.2g, 3.1.8 and its corollary, it follows that $'z^*$, $'z$, and $'w \in H_2^1('G_0)$ and $'z^*$ is continuous on $'\bar{G}_0 = '\bar{G} \cup '\bar{G}^-$ ($'G_0 = 'G \cup 'G^- \cup$ part of x^1 axis on $\partial 'G$). Let φ be a mollifier and let $'z_\varrho$, etc., be the mollified functions. Now, for each R, $0 < R < a$, there is a set of $\bar{r}$ of positive measure between R/e and R such that ($'z$ is already A.C. in σ for almost all r, see the corollary to Theorem 3.1.8) $'z(\bar{r}, \sigma)$ is A.C. in σ with $'z_\sigma(\bar{r}, \sigma) \in L_2$ and

$$\int_0^{2\pi} |'z_\sigma(\bar{r}, \sigma)|^2 d\sigma \leq 2 \int_{R/e}^{R} \int_0^{2\pi} r(|'z_r|^2 + r^{-2}|'z_\sigma|^2)\, dr\, d\sigma \tag{4.3.11}$$
$$\leq 2D('z, B_R) \leq \varepsilon^2(R)/2\pi.$$

Using the method of proof of Theorem 3.1.8, we conclude that we may choose an $\bar{r}$ satisfying (4.3.11) and also the condition that $'z_\varrho(\bar{r}, \sigma)$ converges uniformly to $'z(\bar{r}, \sigma)$ and $'z_{\varrho,\sigma}(\bar{r}, \sigma) \to 'z_{,\sigma}(\bar{r}, \sigma)$ in L_2 for a sequence of $\varrho \to 0$. And, on $\sigma_{\bar{r}}$, $'z_\varrho(x^1, 0) = 'z_\varrho^*(x^1, 0)$ and so converges uniformly in x^1 to $'z^*(x^1, 0)$.

Now, let us choose an arbitrary point P_1 in $G_{\bar{r}}$ and map $G_{\bar{r}}$ conformally onto $B(0,1)$ so that P_1 corresponds to the origin. Then the transformed function $''z_\varrho$ is continuous on $\overline{B(0,1)}$, $\in H_2^1(B_1)$, $''z_\varrho \to ''z$ in $H_2^1(B_1)$ and $''z_\varrho(1, \varphi)$ converges uniformly on $\partial B(0,1)$ to a function $''\zeta(\varphi)$ which is the transform of $'z(\bar{r}, \sigma)$ on the part of $\partial B(0,1)$ corresponding to $\sum_{\bar{r}}$ and $'z^*(x^1, 0)$ on the part of $\partial B(0,1)$ corresponding to $\sigma_{\bar{r}}$. It follows from Theorem 3.4.5 that $''z$ has the continuous boundary values $''\zeta$. Moreover

$$\operatorname{Osc} {''\zeta} \leq \omega(R) + \int_0^{2\pi} |'z_\sigma(\bar{r}, \sigma)|\, d\sigma \leq \omega(R) + \varepsilon(R) \tag{4.3.12}$$

using the SCHWARZ inequality; here $\omega(R)$ is the oscillation of z^* along σ_R. Finally, $''z$ still satisfies (4.3.9) for subregions $''g \subset B(0,1)$ which we shall now take as circles $B_r = B(0, r)$. Replacing H_g by Z as in (4.3.5) we obtain

$$\varphi(r) = D(''z, B_r) \le C \cdot (M/m)\, r\, \varphi'(r) \quad (C = C(2))$$

from which we conclude that

$$\varphi(r) \le \varphi(1)\, r^{1/L}, \quad \varphi(1) \le \varepsilon^2(R)/2\pi.$$

Setting

$$\psi(r) = \int_0^r \int_0^{2\pi} s^{1/2} |''z_s(s, \varphi)|\, ds\, d\varphi,$$

$$\chi(r) = \int_0^r \int_0^{2\pi} |''z_s(s, \sigma)|\, ds\, d\sigma \quad (L = M/m)$$

we find, successively

$$\psi^2(r) \le 2\pi r\, \varphi(r) \le \varepsilon^2(R) r^{1+1/L}, \quad \psi(r) \le \varepsilon(R) r^{(L+1)/2L}$$

$$\chi(r) = \int_0^r s^{-1/2}\, \psi'(s)\, ds = r^{-1/2}\, \psi(r) + \frac{1}{2} \int_0^r s^{-3/2}\, \psi(s)\, ds$$

$$\le \varepsilon(R) \cdot (L+1)\, r^{1/2L}.$$

Thus we see that (since $''z$ is continuous inside)

$$\int_0^{2\pi} |''z(1, \varphi) - ''z(0)|\, d\varphi = \int_0^{2\pi} |''\zeta(\varphi) - ''z(0)|\, d\varphi \tag{4.3.13}$$

$$\le \chi(1) \le (L+1)\, \varepsilon(R).$$

Using (4.3.12) and (4.3.13) we see that $|''\zeta(\varphi) - ''z(0)| < \varepsilon$ if R is small enough. Since P_1 was arbitrary, the continuity at P_0 follows.

Before developing the notion of a continuous function monotone in the sense of LEBESGUE, which was defined in the introduction, we introduce the following notations and definitions:

Notations. For scalar functions u continuous on $\bar{G}$, we define

$$M(u, S) = \max_{x \in S} u(x), \quad m(u, S) = \min_{x \in S} u(x),$$

$$S \text{ compact} \subset \bar{G}$$

$$\omega(u, S) = M(u, S) - m(u, S)$$

$$\left.\begin{aligned} \mu^+(u, D) &= M(u, \bar{D}) - M(u, \partial D) \\ \mu^-(u, D) &= m(u, \partial D) - m(u, \bar{D}) \end{aligned}\right\} D \text{ a domain} \subset G.$$

Definition. For a scalar function u continuous on $\bar{G}$ and a real number w, we define $T^+(u, w)$ as that function U such that $U(x) = w$ whenever $x \in$ a domain D for which $u(y) > w$ for all y in D and $u(y) = w$ on ∂D and $U(x) = u(x)$ otherwise. We define $T^-(u, w)$ as that function U such that $U(x) = w$ whenever $x \in D$ on which $u(y) < w$, etc.

The reader can easily verify the facts stated in the following lemma:

Lemma 4.3.2. *Suppose u is continuous on $\bar{G}$, w is a real number, and $U = T^+(u, w)$. Then*

(i) $U(x) \leq u(x)$ *for* $x \in \bar{G}$ *and* $U(x) = u(x)$ *for* $x \in \partial G$;

(ii) $\omega(U, S) \leq \omega(u, S)$ *for every continuum* $S \subset \bar{G}$;

(iii) $\mu^+(U, D) \leq \mu^+(u, D)$ *and* $\mu^-(U, D) \leq \mu^-(u, D)$

for every $D \subset G$;

(iv) *if* $M(U, \partial D) \leq w$, *then* $M(U, \bar{D}) \leq w$.

Corresponding results hold if $U = T^-(u, w)$.

For (i) is obvious and the others follow easily from the fact that the set of x where $U(x) < u(x)$ is an open set $\Omega \subset G$ which is the union of domains D on each of which $u(x) > w$ with $u(x) = w$ on ∂D; $U(x) = w$ on Ω.

Theorem 4.3.3. *Suppose f satisfies* (4.1.1), (4.3.1), *and the supplementary conditions*

$$f(x, z, p) \geq 0, \quad f(x, z, p^1, \ldots, p^{r-1}, 0, p^{r+1}, \ldots, p^N) = 0,$$

$$r = 1, \ldots, N, \quad p^i = \{p^i_\alpha\}, \; i \text{ fixed.} \tag{1.3.10}$$

Suppose z is continuous on $\bar{G}$ and $\in H^1_\nu(G)$. Then there is a function $z_0 \in H^1_\nu(G)$ which coincides with z on ∂G, for which $I(z_0, G) \leq I(z, G)$, and of which each component is monotone in the sense of LEBESGUE.

Proof. We shall show that we may replace each component z^i in turn by a function z^i_0 which is monotone in the sense of LEBESGUE, reducing the integral at each step. We begin with z^1.

For each n and each i, $1 \leq i \leq 2^{n-1}$, we define w^+_{ni}, ${}^+z^1_{ni}$, and ${}^+z^1_n$ by induction as follows:

$$w^+_{ni} = M - 2^{-n} \cdot i \cdot (M - m), \quad M = M(z^1, \bar{G}), \quad m = m(z^1, \bar{G})$$

$${}^+z^1_1 = T^+(z^1, w^+_{11}), \quad {}^+z^1_n = {}^+z^1_{n, 2^n - 1},$$

$${}^+z^1_{n+1,1} = T^+({}^+z^1_n, w^+_{n+1,1}), \quad {}^+z^1_{n+1,i+1} = T^+({}^+z^1_{n+1,i}, w^+_{n+1,i+1}).$$

Clearly each ${}^+z^1_n$ is continuous on $\bar{G}$. We note that if D is a domain in which $M({}^+z^1_n, \partial D) > w^+_{n+1,1}$, then ${}^+z^1_{n+1,1}(x) \leq {}^+z^1_n(x) \leq M$ for $x \in \bar{D}$. Also, if D is a domain in which $w^+_{n+1,2} < M({}^+z^1_{n+1,1}, \partial D) \leq w^+_{n+1,1}$, then $M({}^+z^1_{n+1,1}, \bar{D}) \leq w^+_{n+1,1}$ (otherwise $\exists$ an x_0 in D where ${}^+z^1_{n+1,1}(x_0) > w^+_{n+1,1}$ and hence a sub-domain Δ in which ${}^+z^1_{n+1,1}(x) > w^+_{n+1,1}$ with the equality on the boundary). Thus, if D is a domain in which $M({}^+z^1_{n+1,1}, \partial D) > w^+_{n+1,2}$, then $\mu^+({}^+z^1_{n+1,1}, D) \leq 2^{-n-1}(M - m)$.

Now, suppose we have proved that

$$\mu^+({}^+z^1_{n+1,i}, D) \leq 2^{-n-1} \cdot (M - m) \quad \text{whenever}$$

$$M({}^+z^1_{n+1,i}, \partial D) > w^+_{n+1,i+1} \; (i < 2^{n+1} - 1). \tag{4.3.14}$$

Then suppose $M({}^+z^1_{n+1,i+1}, \partial D) > w^+_{n+1,i+1}$. Then, from (4.3.14), we conclude that $M({}^+z^1_{n+1,i+1}, \bar{D}) \leq M({}^+z^1_{n+1,i}, \bar{D}) \leq M({}^+z^1_{n+1,i}, \partial D) +$

$+ 2^{-n-1}(M - m)$; and also if $w^+_{n+1,i+2} < M(^+z^1_{n+1,i+1}, \partial D) \leq w^+_{n+1,i+1}$, then $M(^+z^1_{n+1,i+1}, \bar{D}) \leq w^+_{n+1,i+1}$. Thus we see that (4.3.14) holds with i replaced by $i + 1$.

Thus it follows that, for each n, $^+z^1_{n+1}(x) \leq {^+z^1_n}(x)$ for $x \in \bar{G}$, the equality holding if $x \in \partial G$. It is also true that the $^+z^1_n$ are equicontinuous on $\bar{G}$ (property (ii) of the lemma) and, from (4.3.14) for $i = 2^{n+1} - 1$, it follows that

$$\mu^+(^+z^1_n, D) \leq 2^{-n}(M - m) \quad \text{for every } D \subset G.$$

Hence the $^+z^1_n$ converge uniformly to a function $^+z^1$ for which

$$\mu^+(^+z^1, D) = 0 \quad \text{for every } D \subset G. \tag{4.3.15}$$

Obviously the integral is reduced at each step and from the lower-semicontinuity we conclude that it is reduced if z^1 is replaced by $^+z^1$. Then, starting with $^+z^1$, we may form the functions $^-z^1_{ni}$ and $^-z^1_n$ in the analogous way using T^- and setting

$$w^-_{ni} = m + 2^{-n} \cdot i \cdot (M - m).$$

From the lemma it follows that $\mu^+(^-z^1_n, D) = 0$ for each D and n so that the $^-z^1_n$ converge uniformly to the desired function z^1_0 which is monotone in the sense of LEBESGUE, since

$$\mu^-(z^1_0, D) = \mu^+(z^1_0, D) = 0 \quad \text{for every } D \subset G.$$

Theorem 4.3.4. *Any family of functions $z \in H^1_\nu(G)$, each of which is continuous on G and monotone in the sense of* LEBESGUE, *for which $\|\nabla z\|^0_{\nu,G}$ is uniformly bounded, is equicontinuous on each $D \subset\subset G$. If G is Lipschitz, each z is continuous on $\bar{G}$, and the functions are equicontinuous along ∂G, then the functions are equicontinuous on $\bar{G}$.*

Proof. Since $\omega[z, B(x_0, r)] = \omega[z, \partial B(x_0, r)]$, we see that the latter function is non-decreasing in r. From Theorem 3.5.2 for $\partial B(0,1)$ with ν replaced by $\nu - 1$, it follows that

$$\begin{aligned} &\{\omega[z, B(x_0, \delta)]\}^\nu \leq \{\omega[z, \partial B(x_0, r)]\}^\nu \\ &\leq C^\nu_\nu \int_{\partial B(x_0, r)} |\nabla_\sigma w(r, \sigma)|^\nu d\Sigma, \\ &\delta \leq r \leq a, \ B(x_0, a) \subset G, \ (w(r, \sigma) = z(x)) \end{aligned} \tag{4.3.16}$$

for almost r. Multiplying (4.3.16) by r^{-1} and integrating from δ to a yields

$$\begin{aligned} \{\omega[z, B(x_0, \delta)]\}^\nu \log(a/\delta) &\leq C^\nu_\nu \int_\delta^a r^{-1} \int_{\partial B(x_0, r)} |\nabla_\sigma w(r, \sigma)|^\nu d\Sigma\, dr \\ &\leq C^\nu_\nu \int_0^a \int_{\partial B(x_0, r)} r^{\nu-1}[|w_r(r, \sigma)|^2 + r^{-2}|\nabla_\sigma w(r, \sigma)|^2]^{\nu/2} dr\, d\Sigma \\ &\leq (C_\nu \|\nabla z\|^0_{\nu,G})^\nu \end{aligned} \tag{4.3.17}$$

which proves the first result.

Suppose, now, that G is LIPSCHITZ, z is continuous on $\bar{G}$, and z is monotone. Let $x_0 \in \partial G$ and map a neighborhood $\mathfrak{N}$ of x_0 onto G_a by a bi-LIPSCHITZ map $x = x(y)$ so that x_0 is carried into the origin. Let y_1 be any point on $\sigma_{a/2}$, let $R = a/2$, let $\varepsilon > 0$, let $\varrho (< R)$ be so small that the oscillation of $\zeta(\zeta(y) = z[x(y)])$ along the part of σ_a with $|y - y_1| \leq \varrho$ is $\leq \varepsilon/2$, let $\delta < \varrho$, and suppose that $\omega[\zeta, G(y_1, \delta)] \geq \varepsilon$. Then

$$\omega[\zeta, \partial G(y_1, r)] \geq \varepsilon, \quad \delta \leq r \leq R,$$

so that

$$\omega[\zeta, \textstyle\sum(y_1, r)] \geq \varepsilon/2, \ \delta \leq r \leq \varrho, \ (\textstyle\sum(y_1, r) = G_a \cap \partial B(y_1, r). \tag{4.3.18}$$

Then, proceeding as in (4.3.16) and (4.3.17), we find that

$$(\varepsilon/2)^\nu \log(\varrho/\delta) \leq (C_\nu \|\nabla \zeta\|^0_{\nu, G_a})^\nu \leq (K_\nu \|\nabla z\|^0_{\nu, G})^\nu \tag{4.3.19}$$

where K_ν depends only on G and ν. Thus if the z are equicontinuous along ∂G, we may first choose ϱ small enough, independently of z, and then, using (4.3.19), choose δ small enough so that $\omega[\zeta, G(y_1, \delta)] < \varepsilon$.

Remark. It is easy to see that Theorem 4.3.4. holds for vector functions z if we interpret

$$\omega(z, S) = \sup_{x_1, x_2 \in S} |z(x_2) - z(x_1)|$$

and define a continuous vector function to be monotone in the sense of LEBESGUE if and only if

$$\omega(z, \bar{D}) = \omega(z, \partial D) \quad \text{for every } D \subset G.$$

From Theorems 4.3.3 and 4.3.4, we easily deduce the following Theorem:

Theorem 4.3.5. *Suppose G is Lipschitz, $z^* \in H^1_\nu(G)$, z^* is continuous on $\bar{G}$, and f satisfies* (4.1.1), (4.3.1), *and* (1.10.10). *Then there exists a function $z \in H^1_\nu(G)$ which is continuous on $\bar{G}$ and coincides with z^* on ∂G and minimizes $I(Z, G)$ among all such Z. If $z^* \in H^1_\nu(G)$ and is continuous on G, there is a similar function z which minimizes $I(Z, G)$ among all similar Z such that $Z - z^* \in H^1_{\nu 0}(G)$.*

Proof. In each case, let $\{z_n\}$ be a minimizing sequence. In the first case, each z_n may be replaced by a z_{0n}, in which each component is monotone, and the $\{z_{0n}\}$ form an equicontinuous minimizing sequence. In the second case, we choose an expanding sequence $\{G_n\}$ $(\bar{G}_n \subset G_{n+1})$ of LIPSCHITZ domains such that $U\, G_n = G$; and replace z_n by a function z_{0n} such that $z_{0n}(x) = z_n(x)$ for $x \in G - G_n$ and $z_0(x)$ is monotone on G_n. The resulting minimizing sequence is equicontinuous on each $D \subset\subset G$ by Theorem 4.3.4.

Remarks. If is clear that theorems like the main existence Theorem 1.9.1, involving variable boundary values can be proved. We leave their formulation and proof to the reader.

4.4. The general quasi-regular integral

In this section, we study integrals in which f is of class C^2 and satisfies the general LEGENDRE-HADAMARD condition

$$(1.5.5) \qquad f_{p^i_\alpha p^j_\beta}(x, z, p)\,\lambda_\alpha \lambda_\beta \xi^i \xi^j \geq 0 \quad (\text{all } x, z, p, \lambda, \xi).$$

We shall be interested in the way such functions vary with p, so we shall suppress the arguments x and z for the present. We begin by generalizing such functions slightly as follows:

Definition 4.4.1. We say that a function $\varphi(p)$ is *quasi-convex*[1] if and only if $\varphi(p^i_\alpha + \lambda_\alpha \xi^i)$ is convex in λ for each p and ξ and convex in ξ for each p and λ.

Lemma 4.4.1. *Suppose φ is convex on an open convex set S. Then φ is continuous on S. If $|\varphi(\xi)| \leq M$ on S, then φ satisfies a uniform* LIPSCHITZ *condition with constant $2\,M/d$ on S_d. If φ_n is convex on S for each n and $\varphi_n(\xi) \to \varphi(\xi)$ for each ξ, the convergence is uniform on each compact subset of S. If φ is convex on the interval $[a, b]$ in R_1 and $a < c < d < b$, then φ is bounded on $[a, b]$ by a number which depends only on a, b, c, d and its values at these points.*

The proof is left to the reader.

Theorem 4.4.1. *If f is quasi-convex everywhere, it satisfies a uniform* LIPSCHITZ *condition on any bounded part of space. If p is given, there exis constants A^α_j such that*

$$(4.4.1) \qquad f(p^j_\alpha + \lambda_\alpha \xi^j) \geq f(p^j_\alpha) + A^\alpha_j \lambda_\alpha \xi^j \quad \text{for all } \lambda, \xi.$$

If f is also of class C^1, then $A^\alpha_j = f_{p^j_\alpha}(p)$. If f is of class C^2, then f is quasi-convex if and only if (1.5.5) *holds.*

Proof. If is clear that if f is quasi-convex, it is convex in each single p^i_α for fixed values of the others. Thus it is sufficient to prove that a function $\varphi(\xi^1, \ldots, \xi^P)$ which is convex in each ξ^p satisfies a uniform LIPSCHITZ condition on each hypercube $R_0 : |\xi^p| \leq M$. Let R denote the hypercube $|\xi^p| \leq M + 1$. In the case $P = 2$, φ, being one dimensionally convex, is continuous on ∂R and ∂R_0 and is bounded there by a number Q. Hence φ satisfies a uniform LIPSCHITZ condition in ξ^1 and ξ^2 with constant $2\,Q$ on R_0. In the case $P = 3$, we similarly get φ continuous on each face of ∂R_0 and ∂R. and bounded by a number Q on the part where the varying ξ^p satisfy $|\xi^p| \leq M$. Then again we obtain the result that φ satisfies a LIPSCHITZ condition in each ξ^p of bound $2\,Q$ on R_0. The result may be proved by induction.

The last statement is obvious (using Lemma (1.8.1)). It is clear (from the definition of convexity) that any mollified function φ_ϱ is also quasi-

[1] The terminology here is slightly different here than in the writer's paper (MORREY [9]) where functions φ satisfying this condition are called weakly quasi-convex; the functions φ which we call strongly quasi-convex below were called quasi-convex in the papers cited.

convex and $\in C^\infty$ and so satisfies (1.5.5). From Lemma 1.8.1, it follows that

$$\varphi_\varrho(p^j_\alpha + \lambda_\alpha \xi^j) \geq \varphi_\varrho(p^j_\alpha) + \varphi_{\varrho\, p^j_\alpha} \lambda_\alpha \xi^j.$$

Since φ satisfies a uniform LIPSCHITZ condition near p, we may choose a sequence of $\varrho \to 0$ so that the numbers $\varphi_{\varrho\, p^j_\alpha}(p)$ tend to limits A^α_j, thus verifying (4.4.1).

We now obtain a *necessary* condition for lower-semicontinuity:

Definition 4.4.2. We say that z_n *converges Lipschitz to* z *on* G if and only if z_n converges uniformly to z and the z_n and z satisfy a uniform LIPSCHITZ condition (which may depend on the sequence).

Remarks. It is clear that the LIPSCHITZ convergence of z_n to z on G implies that $z_n \rightharpoonup z$ in $H^1_s(G)$ for any $s \geq 1$ but does not imply the strong convergence of z_n to z in any $H^1_s(G)$.

Theorem 4.4.2. *Suppose* $I(z, G)$ *is lower-semicontinuous with respect to this type of convergence at any* z *on any* G *and* f *is continuous. Then*

$$\int_G f[x_0, z_0, p_0 + \nabla \zeta(x)]\, dx \geq f(x_0, z_0, p_0) \cdot m(G) \tag{4.4.2}$$

for any constant (x_0, z_0, p_0), *any bounded domain* G, *and any* LIPSCHITZ *vector* ζ *which vanishes on* ∂G.

Proof. Let x_0 be any point, R be the cell $x^\alpha_0 \leq x^\alpha \leq x^\alpha_0 + h$, z_0 be any vector of class C^1 on $R \cup \partial R$, Q be the cell $0 \leq x^\alpha \leq 1$, and ζ be any vector which satisfies a uniform LIPSCHITZ condition over the whole space and is periodic of period 1 in each x^α.

For each n, define $\zeta_n(x)$ on R by

$$\zeta^j_n(x) = n^{-1} h\, \zeta^j[n\, h^{-1}(x - x_0)].$$

Then the ζ^j_n tend to zero in our sense. Then, for each n, $I(z_0 + \zeta_n, R)$ can be written as a sum of integrals over the sub-hypercubes of R of side $n^{-1} h$. If r is one of these the integral over it is

$$n^{-\nu} h^\nu \int_Q f[x_1 + n^{-1} h\, \xi,\ z_n(x_1 + n^{-1} h\, \xi),$$
$$p_0(x_1 + n^{-1} h\, \xi) + \nabla \zeta(\xi)]\, d\xi,$$

where

$$r: x^\alpha_1 \leq x^\alpha \leq x^\alpha_1 + n^{-1} h,\ x^\alpha_1 = x^\alpha_0 + k^\alpha n^{-1} h,\ 0 \leq k^\alpha \leq n - 1$$
$$z_n(x) = z_0(x) + \zeta_n(x),\ x^\alpha = x^\alpha_1 + n^{-1} h\, \xi^\alpha,\ 0 \leq \alpha \leq 1.$$

Thus we see that

$$\lim_{n\to\infty} I(z_0 + \zeta_n, R) = \int_R \left\{ \int_Q f[x, z_0(x), p_0(x) + \nabla \zeta(\xi)]\, d\xi \right\} dx \geq I(z_0, R).$$

By letting z_0 and p_0 be arbitrary constant vectors, setting $z_0(x) = z_0 + p_{0\alpha} \cdot (x_\alpha - x^\alpha_0)$, dividing by $m(R) = h^\nu$ and letting $h \to 0$, we obtain (4.4.2) for $G = Q$ and ζ periodic of period 1 in each x^α. But if G is any bounded domain and ζ vanishes on ∂G, we may choose a hypercube Q' of side t containing G and extend $\zeta(x)$ to be zero in $Q' - G$ and of period t in each x^α. Then a simple change of variable obtains the result in general.

Theorem 4.4.3. *If f satisfies the conclusion of Theorem 4.4.2, then f is quasi-convex (in p).*

Proof. Clearly we may suppress (x_0, z_0). We next note that any mollified function f_ϱ again satisfies the same conditions and $\in C^\infty$. Then it follows from the derivation in the introduction that (1.5.5) holds for f_ϱ. It then follows from Lemma 4.4.1 that f is quasi-convex.

Definition 4.4.3. If f satisfies the conclusion of Theorem 4.4.2, we say that f *is strongly quasi-convex in p*; if f depends only on p, we say that f is strongly quasi-convex.

The following theorem is interesting. However, we shall not prove it here. The proof is much like but simpler than that of the more general lower-semicontinuity Theorem 4.4.5 below. A proof is to be found in the writer's paper referred to above (MORREY [9]).

Theorem 4.4.4. *If f is continuous and strongly quasi convex in p, then $I(z, G)$ is lower-semicontinuous with respect to* LIPSCHITZ *convergence.*

We wish now to prove a lower semi-continuity theorem which involves weak convergence. The theorem is proved after four lemmas.

Definition 4.4.4. Suppose $\zeta \in H_s^1(G)$ and suppose R is a cell with $\bar{R} \subset G$. Then ζ is said to be *strongly in $H_s^1(\partial R)$* if and only if the boundary values $B\zeta$, as defined in § 3.4, $\in H_s^1(\partial R)$ and also there is a sequence $\{\zeta_n\}$ of class C^1 on $\bar{R}$ such that $\zeta_n \to \zeta$ in $H_s^1(R)$ and $\zeta_n \to B\zeta$ in $H_s^1(\partial R)$.

Remark. In this instance a representative of the boundary values $B\zeta$ is obtained by just taking the restriction to ∂R of an absolutely continuous representative ζ of ζ.

Lemma 4.4.2. *Suppose $\zeta \in H_s^1(G)$. For each α, $1 \le \alpha \le \nu$, let (a^α, b^α) be the open projection of G on the x^α axis. Then there exists sets Z^α of measure zero such that if $R: c^\alpha \le x^\alpha \le d^\alpha$ $(\alpha = 1, \ldots, \nu)$ is any closed cell in G with*

$$c^\alpha \in (a^\alpha, b^\alpha) - Z^\alpha \quad \text{and} \quad d^\alpha \in (a^\alpha, b^\alpha) - Z^\alpha, \ \alpha = 1, \ldots, \nu$$

then ζ is strongly in $H_s^1(\partial R)$.

Proof. This is proved using the method of proof of Theorem 3.1.8.

Lemma 4.4.3. *Suppose R is a cell with edges $2h^1, \ldots, 2h^\nu$ and center x_0. Let*

$$h = \min_{1 \le \alpha \le \nu} h^\alpha, \ L = h^{-1}(h^\alpha \cdot h^\alpha)^{1/2}$$

Suppose also that $0 < k < h$, that $\zeta^* \in H_s^1(D)$ for some $D \supset \bar{R}$ and ζ^* is strongly in $H_s^1(\partial R)$ with

$$\|\zeta^*\|_{s,\partial R}^0 \le k, \quad \|\nabla \zeta^*\|_{s,\partial R}^0 \le M \quad (s \ge 1).$$

Then there is a function $\zeta \in H_s^1(R)$ which coincides with ζ^* on ∂R, is zero except on a set of measure

$$m(R) \cdot [1 - (1 - h^{-1} k)^\nu],$$

and satisfies

$$\int_R |\nabla \zeta|^s \, dx \leq C \cdot k \cdot (1 + M^s)$$

where C depends only on s and L.

Proof. For each $x \in \bar{R}$, $x \neq x_0$, let $x^*(x)$ be the intersection of the ray $\overrightarrow{x_0 x}$ with ∂R, and for each $x \in \bar{R}$ define

$$r(x) = \begin{cases} 0 & (x = x_0) \\ |x^*(x) - x_0|^{-1} \cdot |x - x_0| & (x \neq x_0) \end{cases}$$

Let $\Pi_\alpha^\pm$ be the pyramid in $\bar{R}$ with vertex x_0 and base the face $F_\alpha^\pm$ where

$$x^\alpha = x_0^\alpha \pm h^\alpha.$$

On the pyramid Π_ν^+, introduce coordinates $\xi^1, \ldots, \xi^{\nu-1}, r$ by

$$x^\nu = x_0^\nu + r\, h^\nu, \; x^\gamma = x_0^\gamma + r\, \xi^\gamma \quad (0 \leq r \leq 1, \; \gamma = 1, \ldots, \nu - 1).$$

Then, if r and ξ^γ are considered as functions of x, we have

$$r(x) = r, \; x^*(x) = [\xi^1(x) + x_0^1, \ldots, \xi^{\nu-1}(x) + x_0^{\nu-1}, h^\nu + x_0^\nu].$$

Similar coordinate systems may be set up on each of the other $\Pi_\alpha^\pm$.

Define

$$\varphi(r) = \begin{cases} 0 & (0 \leq r \leq 1 - k h^{-1}), \\ h k^{-1}(r - 1 + k h^{-1}) & (1 - k h^{-1} \leq r \leq 1). \end{cases}$$

Choose a sequence ζ_n^* satisfying the conditions of Definition 4.4.4; and for each n, define

$$\zeta_n(x) = \varphi[r(x)] \cdot \zeta_n^*[x^*(x)].$$

The each $\zeta_n(x)$ is LIPSCHITZ on $\bar{R}$.

By computing the derivatives of ζ_n in Π_ν^+, using the SCHWARZ inequality and the fact that each $|\xi^\gamma| \leq h^\gamma$, we find that

$$|\nabla \zeta_n(x)|^2 \leq (1 + 2L^2) r^{-2} \varphi^2(r) |\nabla \zeta_n^*(\xi)|^2 + 2(h^\nu)^{-2} \varphi'^2 |\zeta_n^*(\xi)|^2$$
$$(\xi \in F_\nu^+).$$

Using the facts that $\varphi(r) = \varphi'(r) = 0$ for $r \leq 1 - h^{-1} k$ and that

$$\frac{\partial(\xi^1, \ldots, \xi^{\nu-1}, r)}{\partial(x^1, \ldots, x^\nu)} = (h^\nu)^{-1} r^{1-\nu} \quad \text{on} \quad \Pi_\nu^+$$

we obtain

$$r^{-1} \varphi(r) \leq 1, \; \varphi'(r) = k^{-1} h, \; 1 - h^{-1} k \leq r \leq 1, \; h^\nu \geq h.$$

$$\int_{\Pi_\nu^+} |\nabla \zeta_n|^s dx \leq C_1(s, L) k \int_{F_\nu^+} [|\nabla \zeta_n^*|^s + k^{-s} |\zeta_n^*|^s] \, dS.$$

Also

$$\int_{\Pi_\nu^+} |\zeta_n|^s dx = \int_{1-kh^{-1}} h^\nu r^{\nu-1} \varphi^s(r)\, dr \int_{F_\nu^+} |\zeta_n^*|^s dS \leq L k \int_{F_\nu^+} |\zeta_n^*|^s dS.$$

Adding these results for all the $\Pi_\alpha^\pm$, we obtain the result for each n; and also $\|\zeta_n\|_{s,R}^1$ is uniformly bounded. Thus, we may extract a subsequence which tends weakly in $H_s^1(R)$ to some function $\zeta \in H_s^1(R)$. Since each $\zeta_n = \zeta_n^*$ on ∂R, ζ_n^* tends strongly in L_s to ζ^* on ∂R and $\zeta = \zeta^*$ on ∂R. From the lower semicontinuity, the result follows.

Lemma 4.4.4. *Suppose f is strongly quasi-convex and suppose for all p_1 and p_2 that*

$$|f(p_2) - f(p_1)| \leq K(|p_1|^{s-1} + |p_2|^{s-1} + 1)\, |p_2 - p_1|.$$

Then if p_0 is any constant vector, D is any bounded domain, and $\zeta \in H_{s0}^1(D)$, it follows that $f[p_0 + \pi(x)]$ is summable over D and

$$\int_D f[p_0 + \pi(x)]\, dx \geq m(D) \cdot f(p_0).$$

Proof. There exists a sequence of functions ζ_n, each of class C^1 on D and vanishing on and near ∂D, such that $\zeta_n \to \zeta$ in $H_s^1(D)$. For each n and almost all x on D, we have

$$|f[p_0 + \pi_n(x)] - f[p_0 + \pi(x)]|$$

$$\leq K[\pi_n(x) - \pi(x)] \cdot [1 + |p_0 + \pi_n(x)|^{s-1} + |p_0 + \pi(x)|^{s-1}]. \quad (4.4.3)$$

Using the HÖLDER inequality, and so on, and the strong convergence in $H_s^1(D)$, we see that

$$\lim_{n\to\infty} \int_D f[p_0 + \pi_n(x)]\, dx = \int_D f[p_0 + \pi(x)]\, dx.$$

Since f is quasi-convex, the result follows.

Lemma 4.4.5. *Suppose that f satisfies the hypotheses of Lemma 4.4.4. Suppose also that each $\zeta_n \in H_s^1(D)$ and is strongly in $H_s^1(\partial R)$ where R is a cell with $\bar{R} \subset D$, and that*

$$\zeta_n \to 0 \text{ in } L_s(\partial R), \ \|\zeta_n\|_{s,\partial R}^1 \leq M, \ \|\zeta_n\|_{s,R}^1 \leq M.$$

Then for each p_0, $f[p_0 + \pi_n(x)]$ is summable for all sufficiently large n, and

$$\liminf_{n\to\infty} \int_R f[p_0 + \pi_n(x)]\, dx \geq m(R) \cdot f(p_0), \quad \pi_{n\alpha}^i(x) = \zeta_{n x^\alpha}^i(x).$$

Proof. Let K be the number in Lemma 4.4.4. For each n, let

$$k_n = \|\zeta_n\|_{s,\partial R}^0$$

and let h be the quantity of Lemma 4.4.3 for R. Since $k_n \to 0$, we have $k_n < h$ for all $n >$ some n_1. For each such n, let η_n be the function of Lemma 4.4.3 which coincides on ∂R with ζ_n, and let

$$\chi_n = \zeta_n - \eta_n, \quad \varkappa_{n\alpha}^i = \eta_{n x^\alpha}^i, \quad \omega_{n\alpha}^i = \chi_{n x^\alpha}^i, \quad (\pi_n = \varkappa_n + \omega_n).$$

Then, since $\chi_n = 0$ on ∂R, we have

$$\int_R f[\pi_0 + \omega_n(x)]\, dx \geq m(R)\, f(p_0).$$

We also have (using Lemma 4.4.3)

(4.4.4) $\quad \| \varkappa_n \|_{s,R}^0 \to 0, \quad \| \pi_n \|_{s,R}^0 \leq M, \quad \| \omega_n \|_{s,R}^0 \leq M + \| \varkappa_n \|_{s,R}^0.$

As in (4.4.3), we see that, for each n, and almost all x on D,

$$|f[p_0 + \omega_n(x) + \varkappa_n(x)] - f[p_0 + \omega_n(x)]|$$
$$\leq K \cdot |\varkappa_n(x)| \cdot (|p_0 + \omega_n(x) + \varkappa_n(x)|^{s-1} + |p_0 + \omega_n(x)|^{s-1} + 1).$$

Using the Hölder inequality (4.4.4) and so on, we see that

$$\lim_{n\to\infty} \int_R |f[p_0 + \pi_n(x)] - f[p_0 + \omega_n(x)]|\, dx = 0,$$

from which the result follows.

Theorem 4.4.5. *Suppose f is continuous in (x, z, p) and strongly quasi-convex in p. Suppose also that there are numbers m and K, $K > 0$, such that*

(i) $f(x, z, p) \geq m$
(ii) $|f(x, z, p_2) - f(x, z, p_1)| \leq K[1 + [p_1|^{s-1} + |p_2|^{s-1}] \cdot |p_2 - p_1|$
(iii) $|f(x_2, z_2, p) - f(x_1, z_1, p)| \leq K(1 + |p|^s)\, [|x_2 - x_1| + |z_2 - z_1|]$

for all (x, z, p) with various subscripts.

Suppose also that $z_n \rightharpoonup z_0$ (weakly) in $H_s^1(G)$ and that either

(a) *each z_n and z_0 are continuous on G and z_n converges uniformly to z_0 on each closed set interior to G, or*

(b) *the set functions $D_s(z_n, e)$ are uniformly absolutely continuous on each closed set interior to G, where*

$$D_s(z_n, e) = \int_e |\nabla z_n|^s\, dx.$$

Then

$$I(z_0, G) \leq \liminf_{n\to\infty} I(z_n, G).$$

Remark. If $s = 1$, weak convergence in G implies the hypothesis (b). The conditions (ii) and (iii) are closely related to the corresponding ones in (1.10.8).

Proof. We note first that hypothesis (ii) implies

(4.4.5) $\quad |f(x, z, p) - f(x, z, 0) \leq K|p|\, (1 + |p|^{s-1}).$

Also, hypothesis (iii) similarly implies

(4.4.6) $\quad |f(x, z, 0) - f(0, 0, 0)| \leq K(|x| + |z|).$

Thus, for all (x, z, p), we have

(4.4.7) $\quad |f(x, z, p)| \leq |f(0, 0, 0)| + K(|x| + |z| + |p|^s + |p|).$

Therefore $I(z_0, G)$ and the $I(z_n, G)$ are uniformly bounded.

We shall select subsequences $\{z_{rt}\}$ and form gratings $\mathfrak{G}_r$ as defined below[1]. We begin by selecting a subsequence, still called $\{z_n\}$ such that $I(z_n, G) \to$ its former liminf. For each α, $1 \leq \alpha \leq \nu$, let (a^α, b^α) be the open interval projection of G on the x^α axis, we let Z^α be the union of all the sets (for the z_n) in Lemma 4.4.2, and for each α, P, and (r, t) (or n), we let E^α_{rtP} be the set of all $x^\alpha \in (a^\alpha, b^\alpha) - Z^\alpha$ such that $\|\bar{z}_{rt}\|^1_{s, G(x^\alpha)} \leq P$, $G(x^\alpha)$ being the set of all x'_α such that $(x^\alpha, x'_\alpha) \in G$. Suppose that M is a uniform bound for $\|z_n\|^1_{s,G}$ there being one on account of the weak convergence. Evidently

$$|E^\alpha_{rtP}| \geq (b^\alpha - a^\alpha) - MP^{-1}.$$

Now we choose P_1 so that $MP_1^{-1} < (b^1 - a^1)/3$ and define

$$E_1^1 = \bigcap_{S=1}^{\infty} \bigcup_{n=S}^{\infty} E^1_{nP_1}.$$

Evidently $|E_1^1| > 2(b^1 - a^1)/3$ so that there is a point $x^1_{11} \in E_1^1$ which is in the middle third of the interval (a^1, b^1). By definition, x^1_{11} is in $E^1_{nP_1}$ for infinitely many n; so we let $z_{1t} = z_{n_t}$ where $\{n_t\}$ are just these n arranged in order. Next, we choose P_2 so that $MP_2^{-1} < (b^2 - a^2)/3$ and define

$$E_2^2 = \bigcap_{S=1}^{\infty} \bigcup_{n=S}^{\infty} E^2_{1nP_2}.$$

As before, there is an $x^2_{21} \in E_2^2$ and in the middle third of (a^2, b^2) and we define $z_{2t} = z_{1n_t}$ where $\{n_t\}$ are those integers for which $x^2_{21} \in E^2_{1n_tP_2}$. This process is continued until $\alpha = \nu$. We define $x^\alpha_{\alpha 0} = a^\alpha$, $x^\alpha_{\alpha 2} = b^\alpha$, $\alpha = 1, \ldots, \nu$. Next, we define $x^1_{\nu+1, 2i} = x^1_{1,i}$ for $i = 0, 1, 2$, and choose $P_{\nu+1}$ so large that $MP^{-1}_{\nu+1} < (1/5)$ of the length of the shorter interval (x^1_{10}, x^1_{11}) and (x^1_{11}, x^1_{12}). We define $E^1_{\nu+1} = \bigcap_{S=1}^{\infty} \bigcup_{t=S}^{\infty} E^1_{\nu t P_2}$. Then there is a point $x^1_{\nu+1,1} \in E^1_{\nu+1}$ and in the middle fifth of the interval (x^1_{10}, x^1_{11}). We choose $z'_{\nu+1,t} = z_{\nu n'_t}$ where n'_t are those integers for which $x^1_{11} \in E^1_{\nu n'_t P_2}$, define $'E^1_{\nu+1} = \bigcap_{S=1}^{\infty} \bigcup_{t=S}^{\infty} E^1_{\nu, n'_t, P_2}$, choose $x^1_{\nu+1,3}$ in the middle fifth of the interval (x^1_{11}, x^1_{12}) and in $'E^1_{\nu+1}$ and define $z_{\nu+1,t} = z'_{\nu+1, r_t}$ where r_t are those t such that $x^1_{\nu+1,3} \in E^1_{\nu, n'_{r_t}, P_2}$. This process is repeated for $\alpha = 2, \ldots, \nu$ using the middle fifth of each interval $(x^\alpha_{\alpha 0}, x^\alpha_{\alpha 1})$ and $(x^\alpha_{\alpha 1}, x^\alpha_{\alpha 2})$. The we define $x^1_{2\nu+1, 2i} = x^1_{\nu+1, i}$, $i = 0, \ldots, 4$, choose $P_{2\nu+1}$ so large that $MP^{-1}_{2\nu+1} < (1/9)$ the length of any of the intervals $(x^1_{\nu+1, i-1}, x^1_{\nu+1, i})$. This process is continued indefinitely taking the middle 9th of each interval for $\alpha = 2, \ldots, \nu$, then $1/17$ for $\alpha = 1, \ldots, \nu$, and in general the middle $1/(2^n + 1)$ of each interval. We let $\mathfrak{G}_k$ denote the $2^{\nu k}$ cells whose faces lie along hyperplanes $x^\alpha = x^\alpha_{ri}$, $i = 0, \ldots, 2^k$, $r = (k-1)\nu + \alpha$. We now let $\{z_n\}$ denote the diagonal sequence. Then if R is any cell of any

[1] This procedure is essentially due to TONELLI (see TONELLI [8]). It is also similar to the $\varepsilon - \delta$ grating process used by L. C. YOUNG ([1]).

$\mathfrak{G}_k$ for which $\bar{R} \subset G$, we have each z_n strongly in $H_s^1(\partial R)$ with $\| \bar{z}_n \|_{s,\partial R}^1$ uniformly bounded and, of course (since that was true of the original sequence) $\bar{z}_n \to \bar{z}$ in $L_s(\partial R)$. Moreover, the quantity L of Lemma 4.4.3 is uniformly bounded for all cells in any $\mathfrak{G}_k$.

Now, we first consider the alternative (a). Let ε be any positive number. For each k, let D_k be the union of all the cells of $\mathfrak{G}_k$ which are interior to G. Since f is bounded below and $I(z_0, G)$ is finite, we first choose k_1 so large that

$$\begin{aligned} &I(z_n, G - D_{k1}) > -\varepsilon/5 \quad (n = 1, 2, \ldots) \\ &I(z_0, D_{k1}) > I,(z_0\, G) - \varepsilon/5. \end{aligned} \tag{4.4.8}$$

For this k_1, let $R_1, \ldots, R_q$ be the cells of D_{k1} and for each $k \geq k_1$, let

$$R_{ki}(i = 1, \ldots, q \cdot 2^{\nu(k-k_1)})$$

be the cells of $\mathfrak{G}_k$ in D_{k_1}. For each k, define $x_k^*(x)$, $z_k^*(x)$, $p_k^*(x)$ on D_{k1} by defining them on each cell R of $\mathfrak{G}_k$ to be equal, respectively, to the average over R of x, $z_0(x)$, and $p_0(x)$. Then, from hypotheses (ii) and (iii), we conclude that

$$\begin{aligned} &|f[(x, z_0(x), p_0(x)] - f[x_k^*(x), z_k^*(x), p_k^*(x)]| \\ &\leq K(|p_0(x)|^s + 1) \cdot (|x - x_k^*(x)| + |z_0(x) - z_k^*(x)|) + \\ &+ K[|p_0(x)|^{s-1} + |p_k^*(x)|^{s-1} + 1] \cdot |p_0(x) - p_k^*(x)|; \end{aligned} \tag{4.4.9}$$

the method of proof is similar to that of (4.4.7). If we let

$$\zeta_n = z_n - z_0, \quad \pi_n = p_n - p_0,$$

we see similarly that

$$\begin{aligned} &|f[x, z_0(x), p_0(x) + \pi_n(x)] - f[x_k^*(x), z_k^*(x), p_k^*(x) + \pi_n(x)]| \\ &\leq K(|p_n(x)|^s + 1)\,(|x - x_k^*(x)| + |z_0(x) - z_k^*(x)|) + \\ &+ K(|p_n(x)|^{s-1} + |p_k^*(x) + \pi_n(x)|^{s-1} + 1) \cdot |p_0(x) - p_k^*(x)|; \end{aligned} \tag{4.4.10}$$

$$\begin{aligned} &|f[x, z_n(x), p_n(x)] - f[x, z_0(x), p_n(x)]| \\ &\leq K(|p_n(x)|^s + 1) \cdot |z_n(x) - z_0(x)|. \end{aligned} \tag{4.4.11}$$

Now, by the HÖLDER inequality on each R_{ki}, we see that

$$\int_{D_{k_1}} |p_k^*(x)|^s\, dx \leq \int_{D_{k_1}} |p_0(x)|^s\, dx. \tag{4.4.12}$$

By applying the MINKOWSKI inequality, we see that the integrals

$$\int_{D_{k_1}} |\pi_n(x)|^s\, dx, \quad \int_{D_{k_1}} |p_k^*(x) + \pi_n(x)|^s\, dx \tag{4.4.13}$$

are uniformly bounded. Finally,

$$\lim_{k \to \infty} \int_{D_{k_1}} |p_0(x) - p_k^*(x)|^s\, dx = 0 \tag{4.4.14}$$

Hence, using (4.4.9)—(4.4.14), we may choose a k so large that

$$\int_{D_{k_1}} |f[x, z_0(x), p_0(x)] - f[x_k^*(x), z_k^*(x), p_k^*(x)]|\,dx < \varepsilon/5, \tag{4.4.15}$$

$$\int_{D_{k_1}} |f[x, z_0(x), p_n(x)] - f[x_k^*(x), z_k^*(x), p_k^*(x) + \pi_n(x)]|\,dx < \varepsilon/5 \quad (n = 1, 2, \ldots), \tag{4.4.16}$$

and then choose n_1 so large that

$$\int_{D_{k_1}} |f[x, z_n(x), p_n(x)] - f[x, z_0(x), p_n(x)]|\,dx < \varepsilon/5, \quad n > n_1. \tag{4.4.17}$$

Since $x_k^*(x)$, $z_k^*(x)$, $p_k^*(x)$ are constant on each R_{ki}, it follows from Lemma 4.4.5 that

$$\liminf_{n\to\infty} \int_{D_{k_1}} f[x_k^*(x), z_k^*(x), p_k^*(x) + \pi_n(x)]\,dx \geq \int_{D_{k_1}} f[x_k^*(x), z_k^*(x), p_k^*(x)]\,dx. \tag{4.4.18}$$

Using (4.4.8) and (4.4.15)—(4.4.18), we see that

$$\liminf_{n\to\infty} I(z_n, G) \geq I(z_0, G) - \varepsilon.$$

The result follows in this case.

We now consider the alternative (b). For each natural number q, we define

$$f_q(x, z, p) = [1 - a_q(x, z)]\,f(x, z, p) + m \cdot a_q(x, z),$$

$$a_q(x, z) = \begin{cases} 0 & (0 \leq R \leq q), \\ 3(R-q)^2 - 2(R-q)^3 & (q \leq R \leq q+1), \\ 1 & (R \geq q+1), \end{cases} \quad R = (|x|^2 + |z|^2)^{1/2}.$$

It is easy to see that each f_q satisfies hypotheses (i)—(iii) with the same m and some K_q. Moreover f_q is independent of (x, z) for $R \geq q + 1$, and also

$$f_q(x, z, p) \leq f_{q+1}(x, z, p), \quad \lim_{q\to\infty} f_q(x, z, p) = f(x, z, p).$$

Thus it is sufficient to prove the lower semicontinuity for each q.

For a fixed q, we note that we may replace $|z_0(x) - z_k^*(x)|$ by $\varphi_k(x)$ in (4.4.9) and (4.4.10) and $|z_n(x) - z_0(x)|$ by $\psi_n(x)$ in (4.4.11), where

$$\varphi_k(x) = \min(|z_0(x) - z_k^*(x)|, 2q + 2),$$

$$\psi_n(x) = \min(|z_n(x) - z_0(x)|, 2q + 2).$$

From the uniform boundedness of the φ_k and ψ_n (q fixed), the uniform absolute continuity of the set function $D_s(z_n, e)$, and the facts that

$$\lim_{k\to\infty} \varphi_k(x) = 0, \quad \lim_{n\to\infty} \psi_n(x) = 0$$

almost everywhere, it follows that the argument can be carried through as before for each fixed q.

Using substantially the same proof, we can prove the following theorem, the hypotheses of which are closely related to the corresponding conditions in (1.10.7).

Theorem 4.4.6. *Suppose that f satisfies the hypotheses of Theorem 4.4.5 with conditions* (ii) *and* (iii) *replaced by*

$$\text{(ii')}\qquad |f(x, z_1, p_1) - f(x, z_2, p_2)| \le K(1 + |z_1|^{s-1} + |p_1|^{s-1} + |z_2|^{s-1} + |p_2|^{s-1}) (|z_2 - z_1| + |p_2 - p_1|)$$

$$\text{(iii')}\qquad |f(x_2, z, p) - f(x_1, z, p)| \le K(1 + |z|^s + |p|^s) \cdot |x_2 - x_1|.$$

Then $I(z, G)$ is lower-semicontinuous with respect to weak convergence in $H_s^1(G)$. A corresponding theorem holds if $f = f(x, p)$ and z_1, z_2, and z are omitted in (ii') *and* (iii').

Proof. The right sides of (4.4.9), (4.4.10), and (4.4.11) can be replaced respectively by

$$\text{(4.4.9')}\qquad K(|p_0(x)|^s + |z_0(x)|^s + 1) \cdot |x - x_k^*(x)| + + K(1 + |z_0|^{s-1} + |p_0|^{s-1} + |z_k^*|^{s-1} + |p_k^*|^{s-1}) \cdot (|z_0 - z_k^*| + |p_0 - p_k^*|),$$

$$\text{(4.4.10')}\qquad K(|p_n(x)|^s + |z_0(x)|^s + 1) \cdot |x - x_k^*(x)| + + K(1 + |z_0|^{s-1} + |p_n|^{s-1} + |z_k^*|^{s-1} + |p_k^* + \pi_n|^{s-1}) (|z - z_k^*| + |p_0 - p_k^*|),$$

$$\text{(4.4.11')}\qquad K(1 + |z_0|^{s-1} + |z_n|^{s-1} + 2|p_n|^{s-1}) |z_n - z_0|.$$

Inequalities (4.4.12), (4.4.13), and (4.4.14) hold with similar inequalities involving z_0, z_n, and z_k^*. Also $z_n \to z_0$ in $L_s(G)$. Consequently the remainder of the proof may be carried over.

We can now prove an existence theorem similar to Theorem 1.9.1 for quasi-convex f.

Theorem 4.4.7. *Suppose that f satisfies the hypotheses of Theorem 4.4.6 for some $s > 1$ or satisfies those of Theorem 4.4.5 with $s > \nu$, condition* (i) *being replaced in both cases by*

$$\text{(4.4.19)}\qquad f(x, z, p) \ge m|p|^s - m_1, \quad m > 0.$$

Suppose also that F^* is a family of vector functions which is compact with respect to weak convergence in $H_s^1(G)$ and that F is a non-empty family, closed under weak convergence in $H_s^1(G)$, such that each z in F coincides on ∂G with a z^* in F^*. Then $I(z, G)$ takes on its minimum in F.

Proof. Let $\{z_n\}$ be a minimizing sequence, suppose $z_n - z_n^* = w_n \in H_{s0}^1(G)$, z_n^* being in F^*. From (4.4.19) and our hypothesis on F^*, it follows that $\|\nabla z_n\|_s^0$, $\|\nabla z_n^*\|_s^0$, and $\|\nabla w_n\|_s^0$ are uniformly bounded. From Poincaré's inequality, it follows that $\|w_n\|_s^0$ is uniformly bounded.

Consequently, for a subsequence of n, $z_n^* \rightharpoonup z^*$ in F^* and $w_n \rightharpoonup w$ in $H^1_{s0}(G)$, so that $z_n \rightharpoonup z = z^* + w$ in $H^1_s(G)$ and $z = z^*$ on ∂G. If f satisfies the hypotheses of Theorem 4.4.6, I is lower-semicontinuous. If f satisfies those of Theorem 4.4.5, then the z_n are equicontinuous on interior domains and we may assume z_n converges uniformly to z on such domains. Again, I is lower-semicontinuous and z is minimizing.

Remarks. We have seen in Theorems 4.4.2 through 4.4.7 that it is the *strong* quasi-convexity which plays the important role in the lower-semicontinuity and existence theory. *It is an unsolved problem to prove or disprove the theorem that every quasi-convex function of p is strongly quasi-convex.* We now prove a general sufficient condition for strong quasi-convexity and then prove two theorems giving examples of special forms of f for which quasi-convexity implies strong quasi-convexity.

Lemma 4.4.6. *Suppose $\nu \geq 2$ and $\zeta^1, \ldots, \zeta^{\nu-1} \in C^2(G)$. Then*

$$\sum_{\alpha=1}^{\nu} (-1)^\alpha \frac{\partial}{\partial x^\alpha}\left[\frac{\partial(\zeta^1, \ldots, \zeta^{\alpha-1}, \zeta^\alpha, \ldots, \zeta^{\nu-1})}{\partial(x^1, \ldots, x^{\alpha-1}, x^{\alpha+1}, \ldots, x^\nu)}\right] = 0. \tag{4.4.20}$$

Proof. This is proved by induction on ν. If $\nu = 2$, (4.4.20) is just

$$-\frac{\partial}{\partial x^1}\left(\frac{\partial \zeta^1}{\partial x^2}\right) + \frac{\partial}{\partial x^2}\left(\frac{\partial \zeta^1}{\partial x^1}\right) = 0$$

which is true. So, suppose the theorem to be true for $2 \leq \nu \leq K$ and then let $\zeta^1, \ldots, \zeta^K$ be functions of $(x^1, \ldots, x^{K+1}) \in C^2(G)$. Then, if we call S the resulting sum on the left in (4.4.20), we obtain

$$\begin{aligned} S = &\sum_{\alpha=1}^{K} (-1)^\alpha \frac{\partial}{\partial x^\alpha}\left[\sum_{\beta=1}^{K} (-1)^{\beta+K} \frac{\partial \zeta^\beta}{\partial x^{K+1}} \cdot \frac{\partial(\zeta^1, \ldots, \zeta^{\beta-1}, \zeta^{\beta+1}, \ldots, \zeta^K)}{\partial(x^1, \ldots, x^{\alpha-1}, x^{\alpha+1}, \ldots, x^K)}\right] + \\ &+ (-1)^{K+1} \frac{\partial}{\partial x^{K+1}} \frac{\partial(\zeta^1, \ldots, \zeta^K)}{\partial(x^1, \ldots, x^K)} = (-1)^{K+1} \frac{\partial}{\partial x^{K+1}} \frac{\partial(\zeta^1, \ldots, \zeta^K)}{\partial(x^1, \ldots, x^K)} + \\ &+ \sum_{\alpha,\beta=1}^{K} (-1)^{\alpha+\beta+K} \frac{\partial^2 \zeta^\beta}{\partial x^\alpha \partial x^{K+1}} \cdot \frac{\partial(\zeta^1, \ldots, \zeta^{\beta-1}, \zeta^{\beta+1}, \ldots, \zeta^K)}{\partial(x^1, \ldots, x^{\alpha-1}, x^{\alpha+1}, \ldots, x^K)} + \\ &+ \sum_{\beta=1}^{K} (-1)^{\beta+K} \frac{\partial \zeta^\beta}{\partial x^{K+1}}\left[\sum_{\alpha=1}^{K} (-1)^\alpha \frac{\partial}{\partial x^\alpha} \frac{\partial(\zeta^1, \ldots, \zeta^{\beta-1}, \zeta^{\beta+1}, \ldots, \zeta^K)}{\partial(x^1, \ldots, x^{\alpha-1}, x^{\alpha+1}, \ldots, x^K)}\right], \end{aligned}$$

the last term of which vanishes by our induction hypothesis. But the first two terms also cancel as one sees by using the well-known method of differentiating a determinant.

Lemma 4.4.7. *Suppose that the functions $\zeta^1, \ldots, \zeta^\mu$, $\mu \leq \nu$, satisfy a uniform Lipschitz condition on $\bar{G}$ and that one of them vanishes on ∂G. Then*

$$\int_G \frac{\partial(\zeta^1, \ldots, \zeta^\mu)}{\partial(x^1, \ldots, x^\mu)}\, dx = 0. \tag{4.4.21}$$

Proof. Suppose $\zeta^\mu = 0$ on ∂G. Choose a large cell R containing $\bar{G}$ in its interior, extend ζ to satisfy the same LIPSCHITZ condition over the whole space with $\zeta^\mu = 0$ outside G, let φ be a mollifier, and ζ_ϱ be the mollified functions. Clearly the integral in (4.4.21) formed with ζ_ϱ converges to that for ζ as $\varrho \to 0$, $\zeta_\varrho^\mu = 0$ on and near ∂R, and $\zeta_\varrho \in C^\infty$. So we may as well assume $\zeta \in C_c^\infty(R)$ and $G = R$. *Then*

$$\int_R \frac{\partial(x^1, \ldots, x^\mu)}{\partial(\zeta^1, \ldots, \zeta^\mu)} dx = \int_R \sum_{\alpha=1}^{\mu} (-1)^{\mu+\alpha} \zeta^\mu_{x^\alpha} Q^\alpha \, dx$$

$$= \int_{\partial R} \zeta^\mu \sum_{\alpha=1}^{\mu} (-1)^{\mu+\alpha} Q^\alpha \, dx'_\alpha - \int_R (-1)^\mu \zeta^\mu \sum_{\alpha=1}^{\mu} (-1)^\alpha \frac{\partial}{\partial x^\alpha} Q^\alpha \, dx,$$

where

$$Q^\alpha = \frac{\partial(\zeta^1, \ldots, \zeta^{\alpha-1}, \zeta^\alpha, \ldots, \zeta^{u-1})}{\partial(x^1, \ldots, x^{\alpha-1}, x^{\alpha+1}, \ldots, x^\mu)},$$

the last equality holding by GREEN's theorem. But the boundary integral vanishes since $\zeta^\mu = 0$ on ∂R, and the integrand in the second integral vanishes on R by Lemma 4.4.6.

Definition 4.4.5. A form

$$A^{\alpha_1 \ldots \alpha_\mu}_{i_1 \ldots i_\mu} \pi^{i_1}_{\alpha_1} \ldots \pi^{i_\mu}_{\alpha_\mu}$$

is said to be *alternating* if and only if the coefficient is zero unless all the α 's and all the i's are distinct and the interchange of two α's or two i's change its sign.

Theorem 4.4.8. *A sufficient condition that f be strongly quasi-convex is that for each p there exist alternating forms*

$$A^\alpha_i \pi^i_\alpha, \; A^{\alpha\beta}_{ij} \pi^i_\alpha \pi^j_\beta, \; \ldots, \; A^{\alpha_1, \ldots, \alpha_\nu}_{i_1, \ldots, i_\nu} \pi^{i_\nu}_{\alpha_1} \ldots \pi^{i_\nu}_{\alpha_\nu}$$

with constant coefficients such that for all π we have

$$f(p + \pi) \geq f(p) + A^\alpha_i \pi^i_\alpha + \cdots + A^{\alpha_1, \ldots, \alpha_\nu}_{i_1, \ldots, i_\nu} \pi^{i_1}_{\alpha_1} \ldots \pi^{i_\nu}_{\alpha_\nu}.$$

Proof. This is an immediate consequence of the preceding lemma.

Theorem 4.4.9. *If the a_{jk} are constants and*

$$f(p) = a^{\alpha\beta}_{jk} p^j_\alpha p^k_\beta, \tag{4.4.22}$$

a necessary and sufficient condition that f be strongly quasi-convex is that

$$a^{\alpha\beta}_{jk} \lambda_\alpha \lambda_\beta \xi^j \xi^k \geq 0 \tag{4.4.23}$$

for all λ and ξ.

Proof. If $\zeta = 0$ on D^*, we see from Lemma 4.4.7 with $\mu = 1$ that

$$\int_G f[p + \pi(x)] \, dx = f(p) \, m(G) + \int_G a^{\alpha\beta}_{jk} \pi^j_\alpha(x) \, \pi^k_\beta(x) \, dx. \tag{4.4.24}$$

The condition (4.4.23) is just (1.5.5) and so is a necessary condition. But if we introduce the FOURIER transforms (see VAN HOVE)

$$\hat{\zeta}^j(y) = (2\pi)^{-\nu/2} \int_G e^{-i x \cdot y} \zeta^j(x) \, dx,$$

the integral on the right in (4.4.24) becomes

$$Re \int_{-\infty}^{\infty} a_{jk}^{\alpha\beta}\, y^\alpha\, y^\beta\, \hat{\zeta}^j\, \bar{\hat{\zeta}}^k\, dy \geq 0$$

if the condition (4.4.23) holds. Thus (4.4.23) is sufficient in this case.

Lemma 4.4.8. *Suppose*

$$\sum_{i=1}^{m} \sum_{j=1}^{n} a_{ij}\, x^i\, y^j = 0$$

for all x and y for which

$$\sum_{i=1}^{m} \sum_{j=1}^{n} b_{ij}\, x^i\, y^j = 0.$$

Then there is a constant K such that

$$a_{ij} = K\, b_{ij} \quad (i = 1, \ldots, m;\ j = 1, \ldots, n).$$

Proof. We may introduce new variables ξ and η by

$$x = c\,\xi, \quad y = d\eta,$$

c and d being nonsingular matrices. Let a and b be the matrices of the original forms and A and B those of the transformed forms. Then

$$A = c'\, a\, d, \quad B = c'\, b\, d \quad (c'_{ij} = c_{ji}).$$

We shall show that there is a scalar K such that $A = KB$. We may assume that

$$B_{ii} = 1 \ \ (i = 1, \ldots, r);\ B_{ij} = 0 \text{ otherwise},\ r \leq m, n,$$

unless $B = 0$ in which case $A = 0$ also and the theorem holds. By taking $\eta^s = 1$, $\eta^j = 0$ $(j \neq s,\ s = 1, \ldots, n)$ in turn we see that

$$A_{is} = 0\,(i = 1, \ldots, m,\ s > r);\, A_{is} = 0\,(i \neq s,\ s = 1, \ldots, r,\ i = 1, \ldots, m).$$

Then, by choosing $1 \leq s < t \leq r$ and setting $\eta^s = \eta^t = 1$, $n^j = 0$, $j \neq s$, $j \neq t$, we have

$$(A_{is} + A_{it})\,\xi^i = 0 \quad \text{for all } \xi \text{ with } \xi^s + \xi^t = 0.$$

Thus there exists a constant $K(s, t)$ such that

$$A_{ss} + A_{st} = K(s, t), \quad A_{ts} + A_{tt} = K(s, t).$$

Hence

$$A_{11} = A_{22} = \cdots = A_{rr} = K,$$

so that $A = KB$.

Theorem 4.4.10. *Suppose that* $N = \nu + 1$ *and*

$$f(p) = F(X_1, \ldots, X_{\nu+1}), \tag{4.4.25}$$

where F is continuous in the X_i *and*

$$X_i = (-1)^{\nu+1+i} \frac{\partial(z^1, \ldots, z^{i-1}, z^{i+1}, \ldots, z^{\nu+1})}{\partial(x^1, \ldots, x^{i-1}, x^i, \ldots, x^\nu)} \quad (i = 1, \ldots, \nu + 1).$$

Then f is strongly quasi-convex in p if and only if F is convex in the X_i.

Proof. If F is convex in the X_i, it follows from Theorem 4.4.8 that f is strongly quasi-convex in p.

Hence suppose f is given by (4.4.25) and is strongly quasi-convex in p. If

$$\Delta X_k = X_k(p^i_\alpha + \lambda_\alpha \xi^i) - X_k(p^i_\alpha),$$

then it is easily seen that

$$\Delta X_k = X_{k p^i_\alpha} \lambda_\alpha \xi^i. \tag{4.4.26}$$

Also, since

$$p^k_\beta X_k = 0 \quad (\beta = 1, \ldots, \nu),$$

we have

$$p^k_\beta X_{k p^i_\alpha} = -\delta^\alpha_\beta X_i. \tag{4.4.27}$$

Now, choose a set of X_i not all zero and choose any p such that

$$X_i(p) = X_i.$$

Since f is strongly quasi-convex and hence quasi-convex, there are constants A^i_α such that

$$f(p^i_\alpha + \lambda_\alpha \xi^i) \geq f(p) + A^\alpha_i \lambda_\alpha \xi^i.$$

Since f depends only on the X_i, we must have

$$A^\alpha_i \lambda_\alpha \xi^i \leq 0 \quad \text{for all } \lambda, \xi \text{ with} \quad X_{k p^i_\alpha} \lambda_\alpha \xi^i = 0 \quad (k = 1, \ldots, \nu + 1). \tag{4.4.28}$$

Obviously, then, the equality must hold in (4.4.28). Using (4.4.26) and (4.4.27), we see that

$$\begin{aligned} p^k_\beta \Delta X_k &= -\lambda_\beta (X_i \xi^i), \quad \beta = 1, \ldots, \nu). \\ X_k \Delta X_k &= D^\alpha_i \lambda_\alpha \xi^i, \quad D^\alpha_i = X_k X_{k p^i_\alpha}. \end{aligned} \tag{4.4.29}$$

Hence since the determinant of the coefficients of the ΔX_k is not zero, we must have

$$A^\alpha_i \lambda_\alpha \xi^i = 0 \tag{4.4.30}$$

for all λ, ξ for which

$$X_i \xi^i = 0 \quad \text{and} \quad D^\alpha_i \lambda_\alpha \xi^i = 0. \tag{4.4.31}$$

Now, since not all the X_i are zero, assume $X_k \neq 0$. Then (solving for ξ^k in terms of the ξ^i with $i \neq k$) we obtain

$$\sum_{i \neq k} (A^\alpha_i X_k - A^\alpha_k X_i) \lambda_\alpha \xi^i = 0 \tag{4.4.32}$$

for all λ, ξ for which

$$\sum_{i \neq k} (D^\alpha_i X_k - D^\alpha_k X_i) \lambda_\alpha \xi^i = 0. \tag{4.4.33}$$

From the preceding lemma, it follows that there is a constant K such that

$$A^\alpha_i X_k - A^\alpha_k X_i = K(D^\alpha_i X_k - D^\alpha_k X_i). \tag{4.4.34}$$

Hence

$$A_i^\alpha = K D_i^\alpha + L^\alpha X_i, \quad L^\alpha = X_k^{-1}(A_k^\alpha - K D_k^\alpha). \tag{4.4.35}$$

From (4.4.29) and (4.4.35) it follows that

$$A_i^\alpha \lambda^\alpha \xi^i = K D_i^\alpha \lambda^\alpha \xi^i + L^\alpha \lambda_\alpha X_i \xi^i = C^k \Delta X_k, \tag{4.4.36}$$
$$C^k = (K X_k - L^\alpha p_\alpha^k).$$

Finally, if we are given any values of the ΔX_k, the quantities

$$h_i = p_i^k \Delta X_k \ (i = 1, \ldots, \nu) \quad \text{and} \quad h_{\nu+1} = X_k \Delta X_k$$

are determined and the ΔX_k are also uniquely determined by the h_i. Using (4.4.29), we may determine the λ_α in terms of the $h_i = (i = 1, \ldots, \nu)$, and substitute them into

$$h_{\nu+1} = X_k \Delta X_k = D_i^\alpha \lambda_\alpha \xi^i,$$

and we merely have to choose the ξ^i to satisfy the equation

$$(D_i^\alpha h_\alpha + h_{\nu+1} X_i) \xi^i = 0 \quad \text{with} \quad X_i \xi^i \neq 0;$$

this is always possible unless all the $D_i^\alpha h_\alpha = 0$. Thus, unless these linear relations in ΔX_i hold, we have

$$F(X + \Delta X) = f(p_\alpha^i + \lambda_\alpha \xi^i) \geq f(p) + A_i^\alpha \lambda_\alpha \xi^i = F(X) + C^k \Delta X_k.$$

The result follows in general by continuity.

Chapter 5

Differentiability of weak solutions

5.1. Introduction

In this chapter we supply the details which were omitted in the sketch of differentiability theory which was presented in Chapter 1. It was seen there that this theory involved a study of the solutions of generalized linear equations of the form

$$\int_G [\zeta_{,\alpha}(a^{\alpha\beta} u_{,\beta} + b^\alpha u + e^\alpha) + \zeta(c^\alpha u_{,\alpha} + du + f)]\, dx = 0, \quad \zeta \in Lip_c(G), \tag{5.1.1}$$

in which the $a^{\alpha\beta}$ are bounded and measurable and the coefficients b and $c \in L_\nu(G)$ and $d \in L_{\nu/2}(G)$, and satisfy

$$m|\lambda|^2 \leq a^{\alpha\beta}(x)\lambda_\alpha \lambda_\beta, \quad |a(x)| \leq M \text{ for a.e. } x \in G \text{ and all } \lambda; \tag{5.1.2}$$

$$\int_{B(x_0,r)\cap G} (|b|^2 + |c|^2 + |d|)^{\nu/2}\, dx \leq (C_0 r^{\mu_1})^{\nu/2}. \tag{5.1.3}$$

We do not assume that $a^{\beta\alpha} = a^{\alpha\beta}$ or that $b^\alpha = c^\alpha$. The proofs of the theorems in this generality are new.

Although it is not strictly necessary for the applications, we present in § 5.2 a general existence theory for such equations and prove an interior boundedness theorem and an approximation theorem; using these theorems a theorem on further L_2-type differentiability is proved. If the coefficients $\in C^\infty$, the solutions $\in C^\infty$ on interior domains. If a portion of $\partial G \in C^\infty$ and the boundary values $\in C^\infty$ along that portion, then the solution $\in C^\infty$ along that portion. These results follow by repeated application of the general theorems. This entire analysis carries over to systems, even those where the $a_{ij}^{\alpha\beta}$ merely satisfy the general Legendre-Hadamard type condition

$$(5.1.4) \qquad a_{ij}^{\alpha\beta}(x)\,\lambda_\alpha\,\lambda_\beta\,\xi^i\,\xi^j \geq m|\lambda|^2\,|\xi|^2, \quad |a(x)| \leq M, \quad m > 0,$$

corresponding to (1.5.5), provided in this case that the $a_{ij}^{\alpha\beta}$ *are continuous* at least.

In §§ 5.3 and 5.4, we present a much simplified generalization of the DE GIORGI-NASH-MOSER results from which the interior boundedness and Hölder continuity of the solutions of (5.1.1) follow. In § 5.5 we present L_p and Schauder estimates of the solutions of (5.1.1) under various hypotheses concerning the coefficients. In § 5.6, we treat the equation

$$(5.1.5) \qquad L u = a^{\alpha\beta} u_{,\alpha\beta} + b^\alpha u_{,\alpha} + c\,u = f;$$

in this case the coefficients $a^{\alpha\beta}$ must be at least continuous. Higher differentiability, both of L_p-type and Hölder continuous (Schauder) type are obtained in terms of the differentiability of f and the coefficients. In all these sections, regularity at the boundary is proved whenever possible.

In § 5.7, we prove that the solutions of linear analytic elliptic equations are analytic on the interior and along an analytic portion of the boundary in case the boundary values are analytic. The method consists merely in finding bounds for the derivatives. This method has been used by A. FRIEDMAN [1] for non-linear equations, using the device of GEVREY (see also for systems § 6.7). But the writer presents his proof for non-linear equations in § 5.8 (see also § 6.7 for systems) since it uses a different method and the author considers it more interesting.

In § 5.9, we supply the details of the proofs of Theorems 1.11.1, 1.11.1′, and 1.11.2 which were omitted in § 1.11. The special argument needed to prove Theorem 1.11.1″ is presented in § 5.10. The results of LADYZENSKAYA and URAL'TSEVA ([1], [2], [3]) mentioned under Theorem 1.10.4(v), with k restricted to be ≥ 2, are proved in § 5.11. In § 5.12, we present the version of LERAY-LIONS of an abstract existence theorem, involving "monotonicity", for the solutions of non-linear functional equations and apply it to show the existence of solutions of certain non-linear elliptic equations of higher order. BROWDER, MINTY, and VIŠIK have each written a great number of papers on this subject; we refer in that section to a recent paper of each.

5.2. General theory; $\nu > 2$

We begin by defining[1]

$$B(u,v) = B_1(u,v) + B_2(u,v), \quad L(v) = -\int_G (e^\alpha v_{,\alpha} + f v)\,dx$$

$$(5.2.1) \quad B_1(u,v) = \int_G v_{,\alpha} a^{\alpha\beta} u_{,\beta}\,dx, \quad C(u,v) = \int_G u v\,dx,$$

$$B_2(u,v) = \int_G (v_{,\alpha} b^\alpha u + v c^\alpha u_{,\alpha} + d u v)\,dx$$

and we consider the equation

$$(5.2.2) \qquad B(u,v) + \lambda C(u,v) = L(v)$$

which is the same as (5.1.1) except for the additional term $\lambda C(u,v)$. The equation (5.2.2) corresponds formally to the differential equation

$$(5.2.3) \qquad \frac{\partial}{\partial x^\alpha}(a^{\alpha\beta} u_{,\beta} + b^\alpha u) - c^\alpha u_{,\alpha} - du - \lambda u = f - e^\alpha_{,\alpha}.$$

Lemma 5.2.1. *Suppose $\nu > 2$, $q \in L_{\nu/2}(G)$ and satisfies*

$$(5.2.4) \qquad \int_{G\cap B(x_0,r)} |q|^{\nu/2}\,dx \le (C_0 r^{\mu_1})^{\nu/2} \quad \text{for all } B(x_0,r).$$

Then, for each ε, there is a C_1 depending only an ε, ν, μ_1, C_0, and R, such that

$$\int_G |q|\,|u|^2\,dx \le \varepsilon \int_G |\nabla u|^2\,dx + C_1 \int_G |u|^2\,dx,$$

$$u \in H^1_{20}(G), \quad \text{if} \quad G \subset B(x_1,R).$$

Moreover, there is a constant C_2, depending only on ν, such that

$$\int_G |q|\,|u|^2\,dx \le C_2 C_0 R^{\mu_1}\cdot(\|\nabla u\|_2^0)^2, \quad u \in H^1_{20}(G), \quad G \subset B(x_1,R).$$

Proof. For each r, there exists a finite sequence $\{\zeta_p\}$, $p = 1, \ldots, P$, each $C_c^\infty(G)$ and having support in some $B(x_p, r)$ such that

$$\sum_{p=1}^{P} \zeta_p^2(x) = 1, \quad x \in \overline{B(x_1,R)}.$$

Then, using the SOBOLEV lemma, Theorem 3.5.3, we obtain

$$\int_G |q|\cdot|u|^2\,dx = \int_G |q|\cdot|u|^2 \sum_{p=1}^{P} \zeta_p^2\,dx = \sum_p \int_{B(x_p,r)\,G} |q|\cdot|u_p|^2\,dx$$

$$\le (C_0 r^{\mu_1}) \sum_p \left\{\int_G |u_p|^{2s}\,dx\right\}^{1/s} \le C_0 Z_1 r^{\mu_1} \sum_p \int_G |\nabla u_p|^2\,dx$$

$$\le 2Z_1 C_0 r^{\mu_1} \int_G \sum_p (\zeta_p^2 |\nabla u|^2 + |\nabla\zeta_p|^2 |u|^2)\,dx \quad (u_p = \zeta_p u)$$

$$\le 2Z_1 C_0 r^{\mu_1} \int_G \left(|\nabla u|^2 + \left(\sum_p |\nabla\zeta_p|^2\right)|u|^2\right)dx$$

[1] If complex functions are allowed, replace v by $\bar{v}$.

from which the first result follows. The second result follows from the Hölder inequality, Theorem 3.5.3, and (5.2.4).

Theorem 5.2.1. *There exist numbers M_1 and λ_0, which depend only on $\nu(> 2)$, m, M, C_0, μ_1, and R such that*

$$(5.2.5)\qquad \begin{aligned} |B(u, v)| &\leq M_1 \|u\| \cdot \|v\|, \quad u, v \in H_{20}^1(G), \\ |B(u, u)| &\geq (m/2) \|u\|^2 - \lambda_0 C(u, u), \quad G \subset B(x_1, R). \end{aligned}$$

Moreover, there are constants C_3 and C_4, depending only on ν, such that

$$|B(u, u)| \geq [m - \varepsilon_1(R)] (\|\nabla u\|_2^0)^2 \geq [m - \varepsilon_1(R)] (1 + R^2/2)^{-1} (\|u\|)^2, \qquad (\|u\| = \|u\|_2^1)$$

$$\varepsilon_1(R) = C_3 (C_0 R^{\mu_1})^{1/2} + C_4 C_0 R^{\mu_1}.$$

Proof. This follows immediately from (5.1.2), (5.1.3), Lemma 5.2.1, and the Schwarz and Poincare (Theorem 3.2.1) inequalities.

Theorem 5.2.2 (Lemma of LAX and MILGRAM). *Suppose in a real (or complex) Hilbert space $\mathfrak{H}$, $B_0(u, v)$ is linear in u for each v and (conjugate) linear in v for each u, and suppose*

$$(5.2.6)\qquad \begin{aligned} &\text{(i)} \quad |B_0(u, v)| \leq M_1 \|u\| \cdot \|v\| \\ &\text{(ii)} \quad |B_0(u, u)| \geq m_1 \|u\|^2, \quad m_1 > 0. \end{aligned}$$

Suppose T_0 is defined by the condition

$$(5.2.7)\qquad B_0(u, v) = (T_0 u, v).$$

Then T_0 and T_0^{-1} are operators with bounds M_1 and m_1^{-1}, respectively.

Proof. It is clear that T_0 is a linear operator with bound M_1. From (5.2.6) (ii) and (5.2.7), we see that

$$m_1 \|u\|^2 \leq |B_0(u, u)| = |(T_0 u, u)| \leq \|u\| \cdot \|T_0 u\|$$

so that

$$\|T_0 u\| \geq m_1 \|u\|.$$

It follows easily that the range of T_0 is closed. If the range were not the whole space, there would be a v such that $B_0(u, v) = (T_0 u, v) = 0$ for every u. But, by setting $u = v$, if follows from (ii) that $v = 0$. Thus T_0^{-1} is a bounded operator with norm $\leq m_1^{-1}$.

Theorem 5.2.3. *Suppose the transformation U is defined on $H_{20}^1(G)$ by the condition that*

$$(5.2.8)\qquad C(u, v) = (U u, v)_{20}^1, \quad v \in H_{20}^1(G).$$

Then U is a completely continuous operator.

Proof. That U is an operator follows from Poincaré's inequality (Theorem 3.2.1), since

$$\|U u\| = \sup (U u, v) = \sup \int_G u \bar{v} \, dx \leq 2^{-1} R^2 \|u\| \quad \text{if} \quad \|v\| = 1.$$

Next, suppose $u_n \rightharpoonup u$ in $H^1_{20}(G)$. Then $u_n \to u$ in L_2 (Theorem 3.4.4) and

$$\| U(u_n - u)\| = \sup_v \int_G (u_n - u)\, \bar v\, dx \le 2^{-1/2} R \,\|v\| \cdot \|u_n - u\|_0 \to 0$$

so that U is compact.

Theorem 5.2.4. *If λ is not in a set $\mathfrak{b}$, which has no limit points (in the plane), the equation (5.2.2) has a unique solution u in $H^1_{20}(G)$ for each given e in $L_2(G)$ and f in $L_{2s'}(G)$, where $s' = \nu/(\nu+2)$. If $\lambda \in \mathfrak{b}$, there are solutions of (5.2.2) in which $u \neq 0$ and $e = f = 0$, but the manifold of these is finite dimensional. If λ_0 is defined as in Theorem 5.2.1, then no real number $\lambda_1 > \lambda_0$ is in $\mathfrak{b}$.*

Proof. Let us define λ_0 as in Theorem 5.2.1 and B_0 by

$$B_0(u, v) = B(u, v) + \lambda_0\, C(u, v)$$

and define T_0 by (5.2.7). Then, equation (5.2.2) is equivalent to

$$T_0\, u + (\lambda - \lambda_0)\, U\, u = w, \quad \text{where} \quad (w, v) = L(v), \tag{5.2.9}$$

L being a linear functional since $v \in L_{2s}(G)$ and $(2s)^{-1} + (2s')^{-1} = 1$. Moreover, from Theorem 5.2.1, it follows that B_0 satisfies the conditions of the Lemma of Lax and Milgram with $m_1 = m/2$. Accordingly T_0 has a bounded inverse so (5.2.9) is equivalent to

$$u + (\lambda - \lambda_0)\, T_0^{-1}\, U\, u = T_0^{-1}\, w.$$

Since $T_0^{-1}\, U$ is compact, the theorem follows from the Riesz theory of linear operators.

Theorem 5.2.5. *Suppose the coefficients, e, and f satisfy the conditions of Theorem 5.2.4. There is a constant C, depending only on ν, m, M, C_0, μ_1, and R, such that*

$$\|\nabla u\|^0_{2,D} \le C\,[a^{-1}\,\|u\|^0_{2,G} + \|e\|^0_{2,G} + \|f\|^0_{2s',G}],$$
$$D \subset G_a, \quad G \subset B(x_1, R),$$

for any solution u of (5.2.2) with $\lambda = 0$ in which $u \in L_2(G)$ and $u \in H^1_2(D)$ for each $D \subset\subset G$.

Proof. We define η as in the proof of Theorem 1.11.1 and define

$$v = \eta\, U, \quad U = \eta\, u.$$

Making these substitutions in (5.2.2), we obtain

$$\begin{aligned} B(U, U) + \int_G [a^{\alpha\beta}\,\eta_{,\alpha}\,\bar u\, U_{,\beta} - a^{\beta\alpha}\,\eta_{,\alpha}\, u\, \bar U_{,\beta} - a^{\alpha\beta}\,\eta_{,\alpha}\,\eta_{,\beta}\,|u|^2 + \\ + \eta\, e^{\alpha}\,(\bar U_{,\alpha} + \bar u\,\eta_{,\alpha}) + \eta_{,\alpha}\, u\,(b^{\alpha} - c^{\alpha})\,\bar U + \eta\, f\, \bar U]\, dx. \end{aligned} \tag{5.2.10}$$

Using (5.2.10), (5.2.5) for U, the Cauchy inequality, Lemma 5.2.1 for U, and the inequalities

$$\int_G |f\,U|\,dx \le \|f\|^0_{2s'} \cdot \|U\|^0_{2s} \le C\,\|f\|^0_{2s'} \cdot \|U\|^1_{20} \le \varepsilon\,(\|U\|^1_{20})^2 + C\,(\|f\|^0_{2s'})^2$$

we obtain the result.

Theorem 5.2.6. *Suppose* $G = G_R$, $R \leq 1$, *suppose the coefficients and* e *and* f *satisfy the conditions of Theorem* 5.2.5 *on* G, *suppose* $u \in L_2(G)$, $u \in H_2^1(G_r)$ *for each* $r < R$, *suppose* $u = 0$ *along* σ_R, *and suppose* u *is a solution of* (5.2.2) *with* $\lambda = 0$. *There is a constant* C, *depending only on* ν, m, M, C_0, *and* μ_1, *such that*

$$\|\nabla u\|_{2,r}^0 \leq C[(R-r)^{-1}\|u\|_{2,R}^0 + \|e\|_{2R}^0 + \|f\|_{2s',R}^0].$$

Proof. The proof is identical with the preceding proof except that we define $\eta(x) = 1$ if $|x| \leq r$, $\eta(x) = 1 - 2(R-r)^{-1}(|x|) - r)$ if $r \leq |x| \leq (r+R)/2$, $\eta(x) = 0$ if $|x| \geq (R+r)/2$. Then u, v, and U all vanish on ∂G_R.

Theorem 5.2.7. *Suppose that the coefficients* $a_n^{\alpha\beta}\, b_n^\alpha$, c_n^α, *and* d_n *all satisfy the conditions of Theorem* 5.2.5 *uniformly on* G *and converge almost everywhere to* $a^{\alpha\beta}$, b^α, c^α, *and* d *respectively, and suppose that* $e_n^\alpha \rightharpoonup e^\alpha$ *in* $L_2(G)$ *and* $f_n \rightharpoonup f$ *in* $L_{2s'}(G)$. *Suppose that* $u_n \rightharpoonup u$ *in* $H_2^1(G)$ *and that* u_n *is a solution of* $(5.2.2)_n$ *for each* n. *Then* u *is a solution of* (5.2.2) *with the limiting functions.*

Proof. For each fixed $v \in H_{20}^1(G)$, we see that

$$a_n^{\alpha\beta} v_{,\alpha} \to a^{\alpha\beta} v_{,\alpha}, \quad b_n^\alpha v_{,\alpha} \to b^\alpha v_{,\alpha}, \quad \text{etc.}$$

in $L_2(G)$, so that

$$B_n(u_n, v) \to B(u, v), \quad C(u_n, v) \to C(u, v), \quad L_n(v) \to L(v).$$

Theorem 5.2.8. *Suppose the coefficients and* e *and* f *satisfy the conditions of Theorem* 5.2.4 *and suppose that* $u \in H_2^1(G)$ *and satisfies* (5.2.2) *on* G. *Suppose also that* $a^{\alpha\beta} \in C_1^0(D)$, $b^\alpha \in H_\nu^1(D)$, $e^\alpha \in H_2^1(D)$, c^α *is bounded,* $d \in L_\nu(D)$, $f \in L_2(D)$, *and* ∇b *and* d *satisfy*

$$\int_{B(x_0,r)\cap D} (|\nabla b|^2 + |d|^2)^{\nu/2}\, dx \leq (C_0 r^{\mu_1})^{\nu/2}, \quad \mu_1 > 0 \tag{5.2.11}$$

for each $D \subset\subset G$. *Then* $u \in H_2^2(D)$ *for each such* D *where it satisfies* (almost everywhere) *the differential equation*

$$-\frac{\partial}{\partial x^\alpha}(a^{\alpha\beta} u_{,\beta} + b^\alpha u + e^\alpha) + c^\alpha u_{,\alpha} + du + f = 0. \tag{5.2.12}$$

If G *is of class* C_1^1, *the coefficients and* e *and* f *satisfy the conditions above on* G, *and* $u \in H_{20}^1(G)$, *then* $u \in H_2^2(G)$.

Proof. This is proved by the difference quotient procedure of Theorem 1.11.1. Let $D \subset\subset G$, choose D' and D'' so $D \subset\subset D' \subset\subset D'' \subset\subset G$ and $D' \subset D''_H$, let $v \in C_c^\infty(G)$ with support in D', let e_γ be the unit vector in the x^γ direction and define

$$\begin{aligned} v_h(x) &= h^{-1}[v(x - h e_\gamma) - v(x)] = \int_0^1 - v_{,\gamma}(x - t h e_\gamma)\, dt, \\ u_h(x) &= h^{-1}[u(x + h e_\gamma) - u(x)], \quad \varphi(x) = c^\alpha u_{,\alpha} + du + f, \\ &\qquad 0 < |h| < H. \end{aligned} \tag{5.2.13}$$

Replacing v by v_h in (5.2.2) and making the indicated changes of variable, we see that u_h satisfies the equation

$$(5.2.14)\qquad \int_{D'} v_{,\alpha}(a^{\alpha\beta} u_{h,\beta} + E^{\alpha}_{\gamma h})\, dx = 0, \quad v \in Lip_c D'$$

where

$$E^{\alpha}_{\gamma h}(x) = a^{\alpha\beta}_{h}(x)\, u_{,\beta}(x + h e_\gamma) + b^{\alpha}_{h}(x)\, u(x + h e_\gamma) + b^{\alpha}(x)\, u_h(x) +$$

$$(5.2.15)\qquad + e^{\alpha}_{h}(x) - \Delta^{\alpha}_{\gamma} \int_0^1 \varphi(x + t h e_\gamma)\, dt.$$

From our hypotheses and Theorem 3.5.2, it follows that b is continuous on D''. From our hypotheses and SOBOLEV's lemma (Theorems 3.5.3 and 3.5.5), it follows that $\varphi \in L_2(D'')$. Thus (see § 3.6) for a subsequence of $h \to 0$, $a^{\alpha\beta}_{h}(x)$ converges a.e. and boundedly to $a^{\alpha\beta}_{.\gamma}(x)$, $u_{,\beta}(x + h e_\gamma)$ converges in $L_2(D')$ to $u_{,\beta}(x)$, $b^{\alpha}_{h} \to b^{\alpha}_{.\gamma}$ in $L_\nu(D')$, $u(x + h e_\gamma) \to u(x)$ in $L_{2s}(D')$, $u_h(x) \to u_{,\gamma}$ in $L_2(D')$, $e^{\alpha}_{h} \to e^{\alpha}_{.\gamma}$ in $L_2(D')$, and $\int_0^1 \varphi(x + t h e_\gamma)\, dt \to \varphi$ in $L_2(D')$, so that

$$E^{\alpha}_{\gamma h} \to a^{\alpha\beta}_{.\gamma} u_{,\beta} + b^{\alpha}_{.\gamma} u + b^{\alpha} u_{,\gamma} + e^{\alpha}_{.\gamma} - \Delta^{\alpha}_{\gamma} \varphi \quad \text{in} \quad L_2(D').$$

It follows from Theorem 5.2.5 that $\|\nabla u_h\|^0_{2,D}$ is uniformly bounded, so that $u_h \rightharpoonup u_{,\gamma}$ in $H^1_2(D)$ for a further subsequence of $h \to 0$, since $u_h \to u_{,\gamma}$ in $L_2(D)$. From Theorem 5.2.7 it follows that $u_{,\gamma}$ satisfies the limiting equations on D. But then one may integrate by parts in (5.2.2) to arrive at (5.2.12).

In order to prove the last statement, we pick any point x_0 on ∂G and map a boundary neighborhood N of x_0 in the proper way onto G_R by a regular map of class C^1_1. Then (5.2.2) goes over into an equation of the same form on G_R. We may repeat the argument of the preceding paragraphs for each $\gamma \leq \nu - 1$, since each u_h vanishes along σ_r. Using Theorem 5.2.6 this time, we conclude that each $u_{,\gamma}$ with $\gamma \leq \nu - 1 \in H^1_2(G_r)$ for each $r < R$. This implies that all $u_{,\gamma\delta}$ with $\gamma \leq \nu - 1$ and $1 \leq \delta \leq \nu \in L_2(G_r)$. But now if $x^\nu > \varepsilon > 0$, $u_{,\nu}$ is also in H^1_2 and u satisfies (5.2.12). Since $a^{\nu\nu} > 0$, we can solve that equation for $u_{,\nu\nu}$ and thus conclude that it $\in L_2(G_r)$ also so that $u_{,\nu}$ also $H^1_2(G_r)$ so that $u \in H^2_2(G_r)$. Thus, since each point x_0 of ∂G is in a neighborhood on which $u \in H^2_2$, we conclude that $u \in H^2_2(G)$.

Corollary 1. *If, for some boundary neighborhood N of x_0 on $G \cup \partial G$, the part $N \cap \partial G$ is of class C^1_1 and if the coefficients and e and f satisfy the conditions of the Theorem on N, and if u is the solution of (5.2.2) in $H^1_{20}(G)$ then $u \in H^2_2(N')$ for each boundary neighborhood N' of x_0 with $N' \subset N$.*

In case the coefficients and e and $f \in C^\infty$ a repetition of the argument leads to the following result:

Corollary 2. *If the coefficients, e, and $f \in C^\infty$ on the interior of G and u is a solution of* (5.2.2), *then $u \in C^\infty$ interior to G. If G is of class C^∞, the coefficients, e, and $f \in C^\infty(\bar{G})$, and $u \in H^1_{20}(\bar{G})$, then $u \in C^\infty(\bar{G})$. Local results corresponding to that in Corollary* 1 *also hold.*

Further results are proved in § 5.5.

The entire discussion of this section carries over to systems of equations like (5.2.2) where (if complex functions are allowed)

$$B(u, v) = B_1(u, v) + B_2(u, v), \quad L(v) = -\int_G (e_i^\alpha \bar{v}^i_{,\alpha} + f_i \bar{v}^i)\, dx$$

$$(5.2.16) \qquad \begin{aligned} B_1(u, v) &= \int_G \bar{v}^i_{,\alpha} a^{\alpha\beta}_{ij} u^j_{,\beta}\, dx, \quad C(u, v) = \int_G u^i \bar{v}^i\, dx \\ B_2(u, v) &= \int_G [\bar{v}^i_{,\alpha} b^\alpha_{ij} u^j + \bar{v}^i(c^\alpha_{ij} u^j_{,\alpha} + d_{ij} u^j)]\, dx \end{aligned}$$

$$u = (u^1, \ldots, u^N), \quad v = (v^1, \ldots, v^N).$$

The assumptions on the coefficients analogous to (5.1.2) and (5.1.3) are that they are all measurable or in some L_p and satisfy

$$(5.2.17) \qquad m|\pi|^2 \le a^{\alpha\beta}_{ij}(x)\, \pi^i_\alpha \pi^j_\beta, \quad \text{all } \pi, \quad |a(x)| \le M, \quad m > 0$$

$$(5.2.18) \qquad \int_{B(x_0, r)\cap G} (|b|^2 + |c|^2 + |d|)^{\nu/2}\, dx \le (c_0 r^{\mu_1})^{\nu/2}.$$

In case the $a^{\alpha\beta}_{ij}$ are continuous on $\bar{G}$, (5.2.17) may be replaced by the more general Hadamard type condition (see (1.5.5)):

$$(5.2.19) \qquad m|\lambda|^2 |\xi|^2 \le a^{\alpha\beta}_{ij}(x) \lambda_\alpha \lambda_\beta \xi^i \xi^j, \quad |a(x)| \le M, \quad m > 0.$$

When this is done, we confine ourselves to $u \in H^1_{20}(G)$.

The reason why that can be done is because we see by taking the Fourier transform $\hat{u}$ of u that

$$\int_G a_{ij} \bar{u}^i_{,\alpha} u^j_{,\beta}\, dx = Re \int_R a^{\alpha\beta}_{ij} y^\alpha y^\beta \bar{\hat{u}}^i(y)\, u^j(y)\, dy$$

$$\ge m \int_{R_\nu} |y|^2 \cdot |\hat{u}(y)|^2\, dy = m \int_G |\nabla u|^2\, dx, \quad u \in H^1_{20}(G)$$

in case the $a^{\alpha\beta}_{ij}$ are *constants* satisfying (5.2.19). In case the $a^{\alpha\beta}_{ij}$ are continuous on some $\Gamma \supset \bar{G}$ we may, for each sufficiently small $r > 0$, cover $\bar{G}$ with a finite number of spheres $B(x_s, r)$, and can then choose ζ_s, $s = 1, \ldots, S, \in C^1_c[B(x_s, r)]$ so that $\zeta_1^2 + \cdots + \zeta_S^2 = 1$ on some domain

$\Delta \supset \bar{G}$. Then, if $\varepsilon(r)$ is a modulus of continuity of a, we find that

$$B_1(u,u) = \int_G \sum_s a_{ij}^{\alpha\beta}(\bar{u}^i_{s,\alpha} - \zeta_{s,\alpha}\bar{u}^i)(u^j_{s,\beta} - \zeta_{s,\beta}u^j)\,dx$$
$$\geq \frac{2}{3}\sum_s \int_G a_{ij}^{\alpha\beta}\bar{u}^i_{s,\alpha}u^j_{s,\beta}\,dx - C_1 M\int_G |u|^2\,dx$$
$$\geq \frac{2}{3}\sum_s \int_G a_{ij}^{\alpha\beta}(x_s)\bar{u}^i_{s,\alpha}u^j_{s,\beta}\,dx - C_2\varepsilon(r)\sum_s\int_G |\nabla u_s|^2\,dx$$
$$- C_1 M\int_G |u|^2\,dx \geq \frac{m}{2}\sum_s\int_G |\nabla u_s|^2\,dx - C_1 M\int_G |u|^2\,dx$$
$$\geq \frac{m}{4}\int_G |\nabla u|^2\,dx - C_3(r,m,M)\int_G |u|^2\,dx, \quad u_s = \zeta_s u.$$

5.3. Extensions of the de Giorgi-Nash-Moser results; $\nu > 2$

Lemma 5.3.1. *Suppose that* (i) $\omega \in H^1_{20}(G)$, (ii) ψ *and* $\psi\omega \in H^1_2(G)$, (iii) $\psi(x) \geq 1$ *on* G, *and* (iv) $\psi\omega_{,\alpha}$ *and* $\psi_{,\alpha}\omega \in L_2(G)$. *Then there exists a sequence* $\{\zeta_n\} \to \omega$ *in* $H^1_{20}(G)$, *in which each* $\zeta_n \in Lip_c(G)$, *such that* $\psi\zeta_{n,\alpha} \to \psi\omega_{,\alpha}$ *in* $L_2(G)$ *and* $\psi\zeta_n \to \omega\psi$ *in* $L_{2s}(G)$ *where* $s = \nu/(\nu-2)$. *If* $\omega(x) \geq 0$, *the* ζ_n *may be chosen* ≥ 0.

Proof. We prove the lemma in the case where $\omega(x) \geq 0$, the other case is easily proved by writing $\omega(x) = \omega^+(x) - \omega^-(x)$ where $\omega^\pm(x) \geq 0$. By using the fact that $\omega \in H^1_{20}(G)$ one sees by choosing absolutely continuous representatives of ψ and $\psi\omega$ that the latter function $\in H^1_{20}(\Gamma)$ on any $\Gamma \supset \bar{G}$ if we define it $= 0$ outside G. Thus $\psi\omega \in L_{2s}(G)$.

For each L, we define the truncated function ω_L by

$$\omega_L(x) = \begin{cases} \omega(x), & \text{if } x \in G - E_L, \\ L, & \text{if } x \in E_L \end{cases} \qquad E_L = \{x \mid \omega(x) \geq L\}. \tag{5.3.1}$$

Evidently

$$0 \leq \psi\cdot(\omega - \omega_L) \leq \psi\omega, \quad 0 \leq |\psi_{,\alpha}(\omega - \omega_L)| \leq |\nabla\psi|\cdot\omega,$$
$$\psi\cdot(\omega_{,\alpha} - \omega_{L,\alpha}) = \psi\omega_{,\alpha}\cdot\chi(E_L)$$

$\chi(E_L)$ being the characteristic function of E_L. Consequently

$$\begin{aligned} &\psi\cdot(\omega - \omega_L) \to 0 \quad \text{in} \quad L_{2s}(G) \\ &\psi_{,\alpha}\cdot(\omega - \omega_L) \to 0 \quad \text{and} \quad \psi(\omega_{,\alpha} - \omega_{L,\alpha}) \to 0 \quad \text{in} \quad L_2(G) \end{aligned} \tag{5.3.2}$$

as $L \to \infty$. Moreover, there exists a sequence $\{\zeta_n\}$, in which each $\zeta_n \in Lip_c G$ with $0 \leq \zeta_n(x) \leq L$, such that $\zeta_n \to \omega_L$ in $H^1_{20}(G)$ and $\zeta_n(x) \to \omega_L(x)$ and $\nabla\zeta_n(x) \to \nabla\omega_L(x)$ a.e. in G. Then it is easy to see that (i) $\psi\cdot(\omega_L - \zeta_n) \to 0$ in $L_{2s}(G)$, (ii) $(\psi - \psi_K)\cdot\nabla\omega_L(x) \to 0$ in $L_2(G)$ as $K \to \infty$, L being fixed and ψ_K being the truncated function of ψ, and (iii) that $\psi_K\cdot(\nabla\omega_L - \nabla\zeta_n) \to 0$ as $n \to \infty$, K and L being fixed. Consequently

$$\left.\begin{array}{l}\psi(\omega_L - \zeta_n) \to 0 \quad \text{in} \quad L_{2s}(G) \quad \text{and} \\ \psi(\nabla\omega_L - \nabla\zeta_n) \to 0 \quad \text{in} \quad L_2(G)\end{array}\right\} \quad \text{as } n \to \infty. \tag{5.3.3}$$

The lemma follows easily from (5.3.2) and (5.3.3).

Lemma 5.3.2. *Suppose that $U \in H_2^1(D)$ for each $D \subset\subset G$, that $U(x) \geq 1$ and $W = U^\lambda$ for some λ, $1 \leq \lambda < 2$, and that $P \in L_\nu(G)$. Then for each $D \subset\subset G$,*

$$U \in L_{2s}(D), \quad PU \in L_2(D), \quad P^2 U \in L_{2s'}(D), \quad s = \nu/(\nu - 2),$$
$$s' = \nu/(\nu + 2), \quad \nabla W, \quad PW, \quad P\nabla W \quad \text{and} \quad P^2 W \in L_1(D).$$

Proof. Suppose that $D \subset G_a$ and η is defined as usual (i.e. as in the proof of Theorem 1.11.1, etc.). Then $\eta U \in H_{20}^1(G)$ and it follows from the SOBOLEV lemma (Theorem 3.5.3) that $U \in L_{2s}(D)$. The remaining inequalities follows from the HÖLDER inequality.

Lemma 5.3.3. *Suppose* (i) *that $U \in H_2^1(D)$ for each $D \subset\subset G$,* (ii) *that $U(x) \geq 1$ in G,* (iii) *that $w = U^\tau \in L_2(G)$ for some $\tau \geq 1$, and* (iv) *that $W = U^\lambda$ satisfies (see* Lemma 5.3.2)

$$\int_G \zeta_{,\alpha}(a^{\alpha\beta} W_{,\beta} + b^\alpha W) + \zeta(c^\alpha W_{,\alpha} + dW)]\, dx \leq 0 \tag{5.3.4}$$
$$\text{for each } \zeta \in Lip_c(G) \text{ with } \zeta(x) \geq 0,$$

for some λ with $1 \leq \lambda < 2$; we suppose also that a, b, c, and d are measurable, a satisfies (5.1.2), *and b, c, and d satisfy* (5.1.3) *for $B(x_0, r) \subset G$. Then $w \in H_2^1(D)$ for each $D \subset\subset G$ and*

$$\int_{B(x_0,R)} |\nabla w|^2\, dx \leq C_2 \tau^\varrho a^{-2} \int_{B(x_0,R+a)} w^2\, dx, \quad B(x_0, R + a) \subset G, \tag{5.3.5}$$
$$\varrho = 1 + 4\mu_1^{-1}, \quad 0 < a \leq R,$$

where C_2 depends only on m, M, ν, λ, μ_1, C_1, and upper bounds for R and the coefficients: C_2 does not depend on τ.

Proof. It follows from the hypotheses, Lemma 5.3.2, and Lemma 5.3.1 with $\psi = U^{\lambda-1}$ and $\omega = \eta^2 U^{2-\lambda} U_L^{2\tau-2}$ that we may set

$$\zeta = \eta^2 U^{2-\lambda} U_L^{2\tau-2} \quad (\eta \in Lip_c G) \tag{5.3.6}$$

in (5.3.4), U_L being the truncated function (see (5.3.1)) of U. Since $U_{L,\alpha}(x) = 0$ a.e. on E_L, we conclude that

$$\zeta_{,\alpha} = \eta^2 U^{1-\lambda} U_L^{2\tau-2}[(2-\lambda) U_{,\alpha} + (2\tau - 2) U_{L,\alpha}] + 2\eta\eta_{,\alpha} U^{2-\lambda} U_L^{2\tau-2}. \tag{5.3.7}$$

The inequality (5.3.4) becomes (again using $\nabla U_L = 0$ on E_L)

$$\begin{aligned}\int_G \{\eta^2 U_L^{2\tau-2}[\lambda(2-\lambda)\nabla U\cdot a\cdot\nabla U + (2-\lambda) U b\cdot\nabla U + \lambda U c\cdot\nabla U + \\ + dU^2 + (2\tau - 2)(\lambda\nabla U_L\cdot a\cdot\nabla U_L + U_L b\cdot\nabla U_L)] + \\ + 2\eta U U_L^{2\tau-2}\nabla\eta\cdot(\lambda a\cdot\nabla U + bU)\}\, dx = 0.\end{aligned} \tag{5.3.8}$$

Using (5.1.2), the bounds for the coefficients, and the inequalities of SCHWARZ and CAUCHY, we conclude that

(5.3.9)
$$\int_G \eta^2 U_L^{2\tau-2}[|\nabla U|^2 + (\tau - 1)|\nabla U_L|^2]\,dx$$
$$\leq Z_1 \int_G (\eta^2 \tau P^2 + |\nabla \eta|^2) U_L^{2\tau-2} U^2\,dx. \quad (P^2 = |b|^2 + |c|^2 + |d|).$$

Now, let us define

(5.3.10)
$$w_L = \eta\, U U_L^{\tau-1}.$$

Then, as was done in (5.3.7), we find that

(5.3.11)
$$\nabla w_L = U U_L^{\tau-1} \nabla \eta + \eta U_L^{\tau-1}[\nabla U + (\tau - 1)\nabla U_L].$$

It follows from (5.3.9), (5.3.10), and (5.3.11) that

(5.3.12)
$$\int_G |\nabla w_L|^2\,dx \leq Z_2 \tau^2 \int_G P^2 w_L^2\,dx + Z_3 \tau \int_G |\nabla \eta|^2 U^2 \cdot U_L^{2\tau-2}\,dx.$$

Suppose that η is defined as usual with a replaced by $a/2$, G by $B(x_0, a) \subset G$, and D by $B(x_0, a/2)$. Then the inequalities

$$\int_{B(x_0,a)} w_L^2\,dx \leq (a^2/2) \int_{B(x_0,a)} |\nabla w_L|^2\,dx \quad (w_L \in H_{20}^1[B(x_0, a)])$$

(5.3.13)
$$\left\{\int_{B(x_0,a)} |w_L|^{2s}\,dx\right\}^{1/s} \leq C_3 \int_{B(x_0,a)} [|\nabla w_L|^2 + a^{-2} w_L^2]\,dx$$
$$\leq (C_3 + 1/2) \int_{B(x_0,a)} |\nabla w_L|^2\,dx, \quad s = \nu/(\nu - 2)$$

of POINCARÉ and SOBOLEV (Theorem 3.5.5) hold. Consequently

$$\int_{B(x_0,a)} P^2 w_L^2\,dx \leq \left\{\int_{B_a} P^\nu\,dx\right\}^{2/\nu} \left\{\int_{B_a} |w_L|^{2s}\,dx\right\}^{1/s} \leq Z_4 a^{\mu_1} \int_{B(x_0,a)} |\nabla w_L|^2\,dx.$$

(5.3.14)

From this and (5.3.12) we deduce that

(5.3.15)
$$\int_{B(x_0,a)} |\nabla w_L|^2\,dx \leq Z_5 \tau a^{-2} \int_{B(x_0,a)} U^2 U_L^{2\tau-2}\,dx, \quad 0 < a \leq \alpha,$$
$$\text{where} \quad 2 Z_2 Z_4 \tau^2 \alpha^{\mu_1} = 1.$$

We can now let $L \to \infty$ to conclude that

(5.3.16)
$$\int_{B(x_0,a/2)} |\nabla w|^2\,dx \leq Z_5 \tau a^{-2} \int_{B(x_0,a)} w^2\,dx, \quad 0 < a \leq \alpha.$$

Next, if $0 < a \leq \alpha$ and $B(x_0, R + a) \subset G$, we can cover $B(x_0, R)$ by a finite number of balls $B(x_i, a/2)$ such that each ball $B(x_i, a) \subset B(x_0, R + a)$ and there are no points in $B(x_0, R + a)$ which belong to more than $K(\nu)$ of the $B(x_i, a)$. Then from (5.3.16), we obtain

$$\int_{B(x_0,R)} |\nabla w|^2\,dx \leq \sum_i \int_{B(x_i,a/2)} |\nabla w|^2\,dx \leq Z_5 \tau a^{-2} \sum \int_{B(x_i,a)} w^2\,dx \leq K Z_5 \tau a^{-2} \int_{B(x_0,R+a)} w^2\,dx.$$

(5.3.17)

For larger values of a, (5.3.17) holds with a replaced by α. Thus we conclude that

$$\int_{B(x_0,R)} |\nabla w|^2\, dx \leq K Z_5 \tau \bar{a}^2 \alpha^{-2} a^{-2} \int_{B(x_0,R+a)} w^2\, dx,$$

$\bar{a}$ being an upper bound for $a(\leq R)$. The lemma follows by using the definition of α used in (5.3.15).

Theorem 5.3.1. *Suppose that the coefficients P, and U satisfy the hypotheses of Lemma 5.3.3 with $\tau = 1$. Then U is bounded on each domain $D \subset\subset G$ and*

$$|U(x)|^2 \leq C a^{-\nu} \int_{B(x_0,R+a)} |U(y)|^2\, dy, \quad x \in B(x_0, R) \tag{5.3.18}$$
$$0 < a \leq R, \quad B(x_0, R+a) \subset G$$

where C depends only on ν, m, M, C_0, and μ_1.

Proof. Let us define

$$s = \nu/(\nu-2),\ w_0 = U,\ w_n = U^{s^n},\ B_n = B(x_0, R + 2^{-n} a),\ W_n = \int_{B_n} w_n^2\, dx.$$

Using Lemma 5.3.3 we conclude in turn that $w_0 \in L_{2s}(B_1)$, $w_1 \in H_2^1(B_1)$, $w_2 = |w_1|^s \in L_2(B_2)$, $w_2 \in H_2^1(B_3)$, etc. Then, using the inequalities (5.3.5) and (5.3.13), we obtain

$$W_n^{1/s} = \left\{\int_{B_n} |w_{n-1}|^{2s}\, dx\right\}^{1/s} \leq C_3 \int_{B_n} (|\nabla w_{n-1}|^2 + R^{-2} w_{n-1}^2)\, dx$$
$$\leq 2 C_3 C_2 s^{\varrho n - \varrho} 4^n a^{-2} \int_{B_{n-1}} w_{n-1}^2\, dx = K_0 K_1^n W_{n-1}, \tag{5.3.19}$$
$$\text{where } K_0 = 2 C_3 C_2 s^{-\varrho} a^{-2}, \quad K_1 = 4 s^{\varrho}.$$

From this recurrence relation, we conclude that

$$W_n^{1/s^n} \leq K_0^{(1-s^{-1})^{-1}} \cdot K_1^{(1-s^{-1})^{-2}} \cdot W_0 = C a^{-\nu} W_0.$$

The theorem follows by letting $n \to \infty$.

Corollary. *Suppose the coefficients satisfy the conditions of Lemma 5.3.3, that $u \in L_2(G)$, and that $u \in H_2^1(D)$ for each $D \subset\subset G$ and satisfies*

$$\int_G [v_{,\alpha}(a^{\alpha\beta} u_{,\beta} + b^\alpha u) + v(c^\alpha u_{,\alpha} + du)]\, dx = 0, \quad v \in C_c^1(G).$$

Then there is a constant C, depending only on ν, m, M, C_0 and μ_1 such that

$$1 + u^2(x) \leq C a^{-\nu} \int_{B(x_0,R+a)} [1 + u^2(y)]\, dy, \quad x \in B(x_0, R),$$
$$0 < a \leq R, \quad B(x_0, R+a) \subset G.$$

Proof. Define $V^2(x) = 1 + u^2(x)$ and set

$$v = \psi V^{-1} u, \quad \psi \in Lip_c(G), \quad \psi(x) \geq 0 \text{ on } G$$

in the equation above. An easy approximation shows that this is legitimate. Since $V V_{,\alpha} = u u_{,\alpha}$ so

$$a^{\alpha\beta} V_{,\alpha} V_{,\beta} = V^{-2} u^2 a^{\alpha\beta} u_{,\alpha} u_{,\beta} \leq a^{\alpha\beta} u_{,\alpha} u_{,\beta},$$

one sees that V satisfies

$$\int_G \psi_{,\alpha}(a^{\alpha\beta} V_{,\alpha} + {}'b^\alpha V) + \psi({}'c^\alpha V_{,\alpha} + {}'d V)\, dx \leq 0$$

$${}'b^\alpha = (1 - V^{-2})\, b^\alpha, \quad {}'c^\alpha = c^\alpha + V^{-2}\, b^\alpha, \quad {}'d = (1 - V^{-2})\, d.$$

For the purpose of proving the differentiability of the solutions of the variational problems and the weak solutions of the other differential equations (1.10.13), it is sufficient to study equations of the forms

$$\left.\begin{aligned} &(5.3.20) \quad \int_G (\nabla\zeta \cdot a \cdot \nabla u + \zeta \cdot c \cdot \nabla u)\, dx = 0 \\ &(5.3.21) \quad \int_G [\nabla\zeta(a \cdot \nabla u + e) + \zeta(c \cdot \nabla u + f)]\, dx = 0 \end{aligned}\right\} \quad \zeta \in Lip_c G,$$

in which the $a^{\alpha\beta}$ satisfy (5.1.2), c satisfies (5.1.3) ($b = \bar{d} = 0$), $e \in L_2(G)$, and $f \in L_{2s'}(G)$, $s' = \nu/(\nu + 2)$, and e and f satisfy certain additional conditions; we suppose, of course, that $u \in H_2^1(D)$ for each $D \subset\subset G$. The more general equations (5.1.1) are interesting in themselves and are treated in a paper by the author (MORREY [14]).

Definition 5.3.1. A function $v \in H_2^1(D)$ for each $D \subset\subset G$ is a *sub-solution* of (5.3.20) if and only if

$$\int_G (\nabla\zeta \cdot a \cdot \nabla v + \zeta c \cdot \nabla v)\, dx \leq 0 \quad \text{for each } \zeta \in Lip_c(G), \quad \zeta(x) \geq 0. \tag{5.3.22}$$

Remarks. This condition is formally equivalent to the condition

$$\frac{\partial}{\partial x^\alpha} a^{\alpha\beta} v_{,\beta} - c^\alpha v_{,\alpha} \geq 0.$$

Lemma 5.3.4. *Suppose that* (i) *F is non-negative and convex on the interval* $[0, \infty)$, (ii) *$H = -e^{-F}$ is convex on that interval,* (iii) *u is a non-negative solution of* (5.3.20) *on G* (iv) *$V(x) = F(u(x))$, and* (v) *$v \in L_2(G)$. Then v is a sub-solution of* (5.3.20) *on G and*

$$\int_D |\nabla v|^2\, dx \leq C\, a^{-2}\, |G| \quad \text{if } D \subset G_a, \quad G \subset B(x_1, R)$$

where C depends only on ν, m, M, μ_1, C_0, and R.

Proof. First, we assume that

$$H \in C^2[0, \infty), \quad -1 \leq H(u) \leq -\varepsilon \quad (\varepsilon > 0)$$

and that H'' is bounded on $[0, \infty)$. Then $F \in C^2[0, \infty)$, and F, F', and F'' are bounded there with $F''(u) \geq [F'(u)]^2$. Let us set $\zeta = \eta^2 F'(u)$ in equation (5.3.20), where η is defined as usual. It follows that

$$\begin{aligned} 0 &= \int_G [2\eta\, \nabla\eta \cdot a \cdot \nabla v + \eta^2 F''(u)\, \nabla u \cdot a \cdot \nabla u + \eta^2\, c \cdot \nabla v]\, dx \\ &\geq \int_G [\eta^2(\nabla v \cdot a \cdot \nabla v + c \cdot \nabla v) + 2\eta\, \nabla\eta \cdot a \cdot \nabla v]\, dx, \end{aligned}$$

since $F'' \geq (F')^2$. Finally

$$\int_G \eta^2 |\nabla v|^2 \, dx \leq C \int_G (\eta^2 c^2 + |\nabla \eta|^2) \, dx \tag{5.3.23}$$

from which the inequality follows easily, using Lemma 5.2.1.

In the general case, H is convex with $-1 \leq H(u) < 0$ on $[0, \infty)$. It is easy to see that H can be approximated from below by functions H having the properties in the preceding paragraph. It follows that the functions $v_n(x) \to v(x)$ from below and hence strongly in $L_2(G)$. Clearly, also, $v_n \rightharpoonup v$ in $H_2^1(D)$ for each $D \subset\subset G$, on account of the inequality (5.3.23) which holds for each n. The inequality holds in the limit by lower-semicontinuity.

Theorem 5.3.2 (HARNACK type). *Suppose that* (i) *u is a non-negative solution of* (5.3.20) *on $B_{2R} \equiv B(x_0, 2R)$ and* (ii) *the set S where $u(x) \geq 1$ has measure $\geq c_1 |B_{2R}|$, $c_1 > 0$. Then*

$$u(x) \geq c_2 > 0 \quad \text{for} \quad x \in B_R$$

where c_2 depends only on ν, m, M, C_0, μ_1, and c_1.

Proof. There is a k, $1 < k < 2$, such that $|B_{2R} - B_{kR}| = (1/2) c_1 |B_{2R}|$. Then $|S \cap B_{kR}| \geq (1/2) c_1 |B_{kR}|$. Let us define $F(u) = \max[-\log(u + \varepsilon), 0]$, where $0 < \varepsilon < 1$. It is easy to see that F satisfies the hypotheses of Lemma 5.3.4. Consequently

$$\int_{B_{kR}} |\nabla v|^2 \, dx \leq C_1 R^{\nu-2} \quad \text{where} \quad v(x) = F[u(x)].$$

Since $v(x) = 0$ on S and $|S \cap B_{kR}| \geq (c_1/2) |B_{kR}|$, if follows from Theorem 3.6.5 that

$$\int_{B_{kR}} v^2 \, dx \leq C_2 R^{\nu}.$$

The theorem follows from this and Theorem 5.3.1.

Theorem 5.3.3. *Suppose u is a solution of* (5.3.20) *on G. Then $u \in C^0_{\mu_0}(G)$ where $0 < \mu_0 < 1$ and μ_0 depends only on ν, m, M, C_0, and μ_1. More precisely*

$$|u(x) - u(x_0)| \leq C L \delta^{-\tau} (|x - x_0|/R)^{\mu_0}, \quad x \in B(x_0, R), \quad \textit{where}$$
$$L = \|u\|^0_{2, R+\delta}, \quad B(x_0, R + \delta) \subset G, \quad \tau = \nu/2, \quad \delta \leq R,$$

and C depends only on ν, m, M, C_0, and μ_1.

Proof. It is sufficient to prove the inequality. It follows from the corollary to Theorem 5.3.1 that

$$|u(x)| \leq C_1 L \delta^{-\tau}, \quad x \in B_R \equiv B(x_0, R).$$

Let us define m^* and M^* as the essential inf. and sup. of $u(x)$ on B_R and let us choose $\overline{m}$ (unique) so that $|S^\pm| \leq |B_R|/2$, S^+ and S^- being the sets of points $x \in B_R$ for which $u(x) > \overline{m}$ and $u(x) < \overline{m}$, respectively. If $m^* < \overline{m} < M^*$, the functions $[M^* - u(x)]/(M^* - \overline{m})$ and

$[u(x) - m^*]/(\bar{m} - m^*)$ satisfy the hypotheses of Theorem 5.3.2 on B_R with $c_1 = 1/2$. It follows that $m_1 \leq u(x) \leq M_1$ for $x \in B_{R/2}$, where

$$m_1 = \bar{m} - h(\bar{m} - m^*), \quad M_1 = \bar{m} + h(M^* - \bar{m}), \quad h = 1 - c_2 < 1,$$

c_2 being the constant of Theorem 5.3.2 with $c_1 = 1/2$. The same results hold if $\bar{m} = m^*$ or $\bar{m} = M^*$ or both.

Now, let us define

$$\varphi(r) = [\text{ess sup}\, u(x)] - [\text{ess inf}\, u(x)] \quad \text{for} \quad x \in B_r, \quad r \leq R.$$

We conclude from the preceding paragraph that

$$\varphi(2^{-n} R) \leq h^n S, \quad S = 2\, C_1 L\, \delta^{-\tau}, \quad n = 1, 2, \ldots$$

Thus

$$\begin{aligned} \log \varphi(r) &\leq \log S - \log h + (n+1) \log h \\ &< \log(S/h) + [(\log h)/(\log 2)] \log(R/r), \text{ if} \\ &\quad n \log 2 \leq \log(R/r) < (n+1) \log 2. \end{aligned}$$

From this it follows that

$$\varphi(r) \leq h^{-1} S (r/R)^{\mu_0}, \quad \mu_0 = -\log h/\log 2.$$

Theorem 5.3.4. *There are constants $R_1 > 0$ and C which depend only on ν, m, M, C_0, and μ_1, such thst*

$$\|\nabla u\|_{2,r}^0 \leq C\, L (r/R)^{\tau - 1 + \mu_0}, \quad 0 \leq r \leq R, \quad L = \|\nabla u\|_{2,R}^0, \quad \tau = \nu/2,$$

for each R, $0 < R \leq R_1$, and each solution of (5.3.20) *with $\|\nabla u\|_{2,R}^0 < +\infty$.*

Proof. Evidently we may suppose that the average value of $u = 0$. From Theorem 3.6.5, we conclude that

$$\|u\|_{2,R}^0 \leq C\, L\, R.$$

From Theorem 5.3.3 we then obtain

$$\begin{aligned} |u(x) - u(x_0)| &\leq Z \cdot \|u\|_{2,R}^0 \cdot (R/2)^{-\tau} (|x - x_0|/R)^{\mu} \\ &\leq Z_2 L \cdot R^{1-\tau-\mu_0} \cdot |x - x_0|^{\mu_0}, \quad |x - x_0| \leq R/2. \end{aligned} \tag{5.3.24}$$

We define η as usual with a, G, D replaced by r, $B(x_0, 2r)$, and $B(x_0, r)$ respectively, and put

$$\zeta(x) = \eta^2 \cdot [u(x) - u(x_0)], \quad x \in B(x_0, 2r), \quad 0 < r \leq R/4 \tag{5.3.25}$$

in (5.3.20). We obtain

$$0 = \int_{B_{2r}} \{\eta^2 [\nabla u \cdot a \cdot \nabla u + c(u - u_0) \cdot \nabla u] + 2\eta (u - u_0)\, \nabla \eta \cdot a \cdot \nabla u\}\, dx.$$

The theorem follows easily by using (5.3.24) and the inequalities of Cauchy and Schwarz.

Theorem 5.3.5 (Existence and uniqueness in the small). *There is a constant $R_2 > 0$ which depends only on ν, m, M, C_0, and μ_1 such that there exists a unique solution U of* (5.1.1) *which* $\in H_{20}^1[B(x_0, R)]$ *for each $R \leq R_2$, $e \in L_2\, B(x_0, R)$, $f \in L_{2,s'}[B(x_0, R)]$, such that $B(x_0, R) \subset G$; this solution* *satisfies*

$$\|\nabla U\|^0_{2,R} \leq C(\|e\|^0_{2,R} + \|f\|^0_{2s',R})$$

where C depends only on the quantities above. If $u^* \in H^1_2[B(x_0, R)]$, *there is a unique solution H of the homogeneous equations* (5.1.1) *such that* $H - u^* \in H^1_{20}[B(x, R)]$. *This solution satisfies*

$$\|\nabla H\|^0_{2,R} \leq C \cdot {}'\|u^*\|^1_{2,R};$$

if $b = d = 0$, ${}'\|u^*\|^1_{2,R}$ *may be replaced by* $\|\nabla u^*\|^0_{2,R}$.

Proof. Using (5.1.3) and the notations of § 5.2, we see that

$$B_1(u, u) \geq m L^2, \quad L = \|\nabla u\|^0_{2,R}$$

$$|B_2(u, u)| \leq L \cdot \left[\int_{B(x_0,R)} (|b|^2 + |c|^2) u^2 \, dx\right]^{1/2} + \int_{B(x_0,R)} |d| \cdot u^2 \, dx$$

$$\leq (C R^{\mu_1/2} + C R^{\mu_1}) L^2$$

using also the SOBOLEV inequality (5.3.14). Thus

$$B(u, u) \geq (m/2) L^2$$

if $R \leq R_2$. Then B satisfies the hypotheses of the Lemma of LAX and MILGRAM. We note also that

$$|L(v)| \leq (\|e\|^0_{2,R} + C \|f\|^0_{2s',R})$$

where C is the $C_0^{1/2}$ in (5.3.14). The last statement follows from POINCARE's inequality (Theorem 3.2.1), Theorem 3.5.4 and the fact that $U = H - u^*$ satisfies (5.1.1) with

$$e^\alpha = a^{\alpha\beta} u^*_{,\alpha} + b^\alpha u^*, \quad f = c^\alpha u^*_{,\alpha} + d u^*.$$

Theorem 5.3.6. *Suppose that u is a solution of* (5.3.21), $u \in H^1_2(G)$, $e \in L_2(G)$, $f \in L_{2s'}(G)$ $(s' = \nu/(\nu + 2))$, *and e and f satisfy*

$$\int_{B(x_0,r)} |e|^2 \, dx \leq L_1^2 (r/R)^{\nu-2+2\mu}, \quad 0 < \mu < \mu_0, \quad 0 \leq r \leq R \leq R_2,$$

$$\int_{B(x_0,r)} |f| \, dx \leq L_2 R^{\tau-1} \cdot (r/R)^{\nu-2+\mu}, \quad B(x_0, R) \subset G. \tag{5.3.26}$$

Then $u \in C^0_\mu(G)$ *and, in fact, u satisfies a condition of the form*

$$\|\nabla u\|_r = \|\nabla u\|^0_{2,B(x_0,r)} \leq K(r/R)^{\tau-1+\mu}, \quad 0 < r \leq R \leq R_2. \tag{5.3.27}$$

Proof. Let V be the potential of f over G. Then it follows from Theorem 3.7.3 that V satisfies a condition like (5.3.27) and also that

$$\int_G (\nabla \zeta \cdot \nabla V + \zeta \cdot f) \, dx = 0, \quad \zeta \in H^1_{20}(G).$$

Thus equation (5.3.21) reduces to

$$\int_G [\nabla \zeta (a \cdot \nabla u + e - \nabla V) + \zeta c \cdot \nabla u] \, dx = 0$$

in which $e - \nabla V$ satisfies (5.3.26) with L_1 replaced by another L. Thus we may assume that $f \equiv 0$.

By virtue of Theorem 3.5.2, it is sufficient to prove that (5.3.27) holds for some K. Since $R \leq R_2$, we conclude, using Theorem 5.3.5 that $u = U + H$ on B_R where U is the solution of (5.3.21) on B_R which $\in H_{20}^1(B_R)$ and H is the solution of (5.3.20) such that $H - u \in H_{20}^1(B_R)$, and that

(5.3.28) $\|\nabla U\|_R \leq Z_1 \|e\|_R \leq Z_1 \|e\|_G, \quad \|\nabla H\|_R \leq Z_2 \|\nabla u\|_R \leq C_2 \|\nabla u\|_G.$

Then it follows from Theorem 5.3.4 that

$$\|\nabla H\|_r \leq C_3 \|\nabla u\|_G \cdot (r/R)^{\tau-1+\mu_0}.$$

Now, let us define $\varphi(s) = L^{-1} \sup \|\nabla U\|_{S s}$ for all e which satisfy (5.3.26) with L_1 replaced by L, R replaced by $S \leq R$, U being the solution of (5.3.21) $\in H_{20}^1(B_S)$. Next, choose an arbirtary e which satisfies (5.3.26) (L_1 replaced by L). We may write $U = U_S + H_S$ on B_S where U_S is the solution of (5.3.21) $\in H_{20}^1(B_S)$. Obviously e satisfies

$$\int_{B_r} |e|^2 \, dx \leq [L^2 (S/R)^{\nu-2+2\mu}] \cdot (r/S)^{\nu-2+2\mu}, \quad 0 \leq r \leq S.$$

Thus, using the ideas of (5.3.28) and the definition of φ, we conclude that

$$\|\nabla U_S\|_S \leq Z_1 L \cdot (S/R)^{\tau-1+\mu}, \quad \|\nabla H_S\|_S \leq Z_2 \|\nabla U\|_S \leq Z_2 L \cdot \varphi(S/R).$$

Now, suppose that $0 < r < S < R$. Then

$$\begin{aligned} \|\nabla U\|_r &\leq \|\nabla U_S\|_r + \|\nabla H_S\|_r \\ &\leq L (S/R)^{\tau-1+\mu} \varphi(r/S) + Z_3 L \, \varphi(S/R) \cdot (r/S)^{\tau-1+\mu_0}. \end{aligned}$$

Since e is arbitrary, we conclude (setting $s = r/R$, $t = S/R$) that

(5.3.29) $$\varphi(s) \leq t^{\tau-1+\mu} \varphi(s/t) + Z_3 \varphi(t) \cdot (s/t)^{\tau-1+\mu_0}.$$

Obviously φ is monotone and $\varphi(1) \leq Z_1$. So let us choose σ, $0 < \sigma < 1$. Then, obviously

$$\varphi(s) \leq S_0 s^{\tau-1+\mu}, \quad \sigma \leq s \leq 1, \quad S_0 \leq Z_1 \sigma^{-\tau+1-\mu}.$$

Using (5.3.29) with $\sigma^2 \leq s \leq \sigma$ and $t = \sigma^{-1} s$, we obtain

(5.3.30) $\varphi(s) \leq S_1 s^{\tau-1+\mu}, \quad \text{where} \quad S_1 = S_0(1 + Z_3 \omega), \quad \omega = \sigma^{\mu_0-\mu}.$

Since $S_1 \geq S_0$, (5.3.30) holds for $\sigma^2 \leq s \leq 1$. Using (5.3.29) with $\sigma^4 \leq s \leq \sigma^2$ and $t = \sigma^{-2} s$, we conclude that

$$\varphi(s) \leq S_2 s^{\tau-1+\mu}, \quad \sigma^4 \leq s \leq 1, \quad S_2 = S_0(1 + Z_3 \omega)(1 + Z_3 \omega^2).$$

By repeating the argument, we obtain

$$\varphi(s) \leq S s^{\tau-1+\mu}, \quad 0 \leq s \leq 1,$$
$$S = S_0(1 + Z_3 \omega)(1 + Z_3 \omega^2)(1 + Z_3 \omega^4) \ldots$$

from which the theorem follows immediately.

The following theorem concerning the boundary behavior of solutions of equations of the form (5.3.21) is very easily obtained. More general theorems concerning equations of the form (5.1.1) are presented in the author's paper referred to above (MORREY [14]).

Lemma 5.3.5. *Suppose u is a solution of (5.3.21) which $\in H_2^1(G_a)$ and which vanishes along σ_a. We suppose that e and f satisfy the hypotheses of Theorem 5.3.6 with $B(x_0, r)$ replaced by $B(x_0, r) \cap G_a$ for all x_0 such that $B(x_0, r) \subset B(0, a)$. Then, if we extend u to $B(0, a)$ by*

$$u(x^\nu, x'_\nu) = -u(-x^\nu, x'_\nu) \tag{5.3.31}$$

it follows that $u \in C_\mu^0(B_a)$ and satisfies (5.3.27) with $G = B_a$.

Proof. For, suppose $\zeta \in C_c(B_a)$, and we extend u, $a^{\alpha\nu}$, $a^{\nu\alpha}$, e^α, c^ν, and f, for $\alpha < \nu$, by (5.3.31) and extend $a^{\alpha\beta}$, $a^{\nu\nu}$, e^ν, and c^β, for α and $\beta < \nu$, by the formula

$$\varphi(x^\nu, x'_\nu) = \varphi(-x^\nu, x'_\nu). \tag{5.3.32}$$

Then

$$\begin{aligned} &\int_{B_a} [\nabla \zeta \cdot (a \cdot \nabla u + e) + \zeta(c \cdot \nabla u + f)]\, dx \\ &= \int_{G_a} [\nabla \zeta^* \cdot (a \cdot \nabla u + e) + \zeta^*(c \cdot \nabla u + f)]\, dx \end{aligned} \tag{5.3.33}$$

where $\zeta^*(x^\nu, x'_\nu) = \zeta(x^\nu, x'_\nu) - \zeta(-x^\nu, x'_\nu)$,

$\zeta^* = 0$ on ∂G_a.

Accordingly the extended function u satisfies (5.3.21) as extended to B_a and e and f satisfy conditions like (5.3.26) there. The lemma follows.

Theorem 5.3.7. *Suppose G is Lipschitz, u is the solution of (5.3.21) in $H_{20}^1(G)$, and $e \in L_2(G)$, $f \in L_{2s'}(G)$, and e and f satisfy*

$$\int_{B(x_0, r) \cap G} |e|^2\, dx \le L_1^2 r^{\nu-2+2\mu} \tag{5.3.34}$$

$$\int_{B(x_0, r) \cap G} |f|\, dx \le L_2 r^{\nu-2+\mu} \tag{5.3.35}$$

$$, \quad 0 \le r \le R_2, \quad 0 < \mu < \mu_0.$$

Then $u \in C_\mu^0(\bar{G})$ and satisfies a condition

$$\int_{B(x_0, r) \cap G} |\nabla u|^2\, dx \le K^2 r^{\nu-2+2\mu}, \quad K^2 \le C_1^2 L_1^2 + C_2^2 L_2^2 \tag{5.3.36}$$

where C_1 and C_2 depend only on ν, m, M, μ_1, C_0, μ, and G.

Proof. The condition for interior spheres is just Theorem 5.3.6. Each point of ∂G is in a set $N \subset \bar{G}$ which can be mapped on G_a by a bi-LIPSCHITZ map as in § 1.2, Notations. The form (5.3.21) is preserved by the mapping and the transformed e and f satisfy the conditions of Lemma 5.3.5. The result follows.

5.4. The case $\nu = 2$

The theory of §§ 5.2 and 5.3 carries over to case $\nu = 2$ with only rather minor modifications. In fact the theory can be greatly simplified in this case. However, there is a slight complication caused by the fact that the exponent $s' = \nu/(\nu + 2) = 1/2$, so the exponent $2s' = 1$

and we cannot conclude from the CALDERON-ZYGMUND inequalities that the second gradient $|\nabla^2 V|$ of the potential V of a function f in L_1 is again in L_1 and so cannot immediately conclude that $\nabla V \in L_2$ as in Theorems 3.7.1 and 3.7.3.

Lemma 5.4.1. *Suppose* $u \in H^1_{20}(G)$ *and* $q \in L_1(G)$ *and satisfies*

$$\int_{B(x_0,r)\cap G} |q(x)|\,dx \leq C_0 r^{\mu}, \quad \text{for all} \quad B(x_0, r). \tag{5.2.4}$$

Then $q \cdot u$ *and* $|q| \cdot |u|^2 \in L_1(G)$ *and*

$$\int_{B(x_0,r)\cup G} |q(x)\cdot u(x)|\,dx \leq C_1 \cdot C_0 \cdot \| \nabla u \|^0_{2,G} \cdot g^{\lambda/2} \cdot r^{\mu_1-\lambda/2}, \quad 0 < \lambda < \mu_1,\ \pi g^2 = |G| \tag{5.4.1}$$

$$\int_{B(x_0,r)\cap G} |q(x)| \cdot |u(x)|^2\,dx \leq C_2 \cdot C_0 (\| \nabla u \|^0_{2,G})^2 g^{\lambda/2} r^{\mu_1-\lambda/2}, \quad 0 < \lambda < \mu, \tag{5.4.2}$$

q and u may be tensors; C_1 and C_2 depend only on λ and μ_1.

Proof. It is sufficient to prove this for vectors $u \in C^1_c(G)$. Then

$$u^i(x) = -(2\pi)^{-1} \int_G |\xi - x|^{-2} (\xi^\alpha - x^\alpha)\, u^i_{,\alpha}(\xi)\, d\xi \tag{5.4.3}$$

using Theorem 2.7.3. Hence

$$\int_{B(x_0,r)\cap G} |q(x)\cdot u(x)|\,dx \leq (2\pi)^{-1} \int_{B(x_0,r)\cap G} \int_G |q(x)| \cdot |\xi - x|^{-1} \cdot |\nabla u(\xi)|\,d\xi\,dx. \tag{5.4.4}$$

Applying the SCHWARZ inequality to (5.4.4), we obtain

$$\int_{B(x_0,r)\cap G} |q(x)\cdot u(x)|\,dx \leq (2\pi)^{-1} \left[\int_G d\xi \int_{G\cap B(x_0,r)} |\xi - x|^{\lambda-2} \cdot |q(x)|\,dx \right]^{1/2} \times$$

$$\times \left[\int_G d\xi \int_{G\cap B(x_0,r)} |q(x)| \cdot |\xi - x|^{-\lambda} \cdot |\nabla u(\xi)|^2\,dx \right]^{1/2}, \quad 0 < \lambda < \mu_1. \tag{5.4.5}$$

Using Lemma 3.4.3, we see that

$$\int_G d\xi \int_{G\cap B(x_0,r)} |\xi - x|^{\lambda-2} \cdot |q(x)|\,dx \leq 2\pi \lambda^{-1} g^{\lambda} \cdot C_0 r^{\mu_1}. \tag{5.4.6}$$

Next, we define, for each ξ,

$$\varphi_\xi(\varrho) = \int_{B(\xi,\varrho)\cap B(x_0,r)\cap G} |q(x)|\,dx.$$

From our assumption on q, we see that

$$\varphi_\xi(\varrho) \leq C_0 \varrho^{\mu_1} \quad \text{and} \quad C_0 r^{\mu_1}.$$

Accordingly

$$\int_{B(x_0,r)\cap G} |\xi - x|^{-\lambda} \cdot |q(x)|\,dx \leq \int_0^\infty \varrho^{-\lambda} \varphi'_\xi(\varrho)\,d\varrho$$

$$= \int_0^\infty \lambda \varrho^{-\lambda-1} \varphi_\xi(\varrho)\,d\xi \leq \mu_1(\mu_1 - \lambda)^{-1} C_0 r^{\mu_1-\lambda}. \tag{5.4.7}$$

The first inequality follows easily from (5.4.5), (5.4.6), and (5.4.7). The second follows from two applications of the first with numbers λ' and λ'' with $\lambda' + \lambda'' = \lambda$.

Lemma 5.4.2. *Suppose q satisfies the condition of Lemma 5.4.1. Then, for each $\varepsilon > 0$, there are numbers C_3 and C_4 depending only on ε, ν, C_0, μ_1, and R such that*

$$\int_G |q| \cdot |u|^2 dx \leq C_4 C_0 R^{\mu_1} \int_G |\nabla u|^2 dx \qquad u \in H^1_{20}(G),$$

$$\int_G |q| \cdot |u|^2 dx \leq \varepsilon \int_G |\nabla u|^2 dx + C_3 \int_G |u|^2 dx \quad G \subset B(x_0, R).$$

Proof. The first follows by setting $g = r = R$ in (5.4.2). To prove the second, let $r > 0$ be given. There exists a set of ζ_p as in the proof of Lemma 5.2.1. Then, as in the proof of that lemma,

$$\begin{aligned}\int_G |q| \cdot |u|^2 dx &\leq \sum_p \int_G |q| \cdot |u_p|^2 dx \leq C C_0 r^{\mu_1} \sum_p \int_G |\nabla u_p|^2 dx \\ &\leq 2C C_0 r^{\mu_1} \int_G |\nabla u|^2 dx + 2C C_0 r^{\mu_1} \int_G \sum_p |\nabla \zeta_p|^2 |u|^2 dx\end{aligned}$$

using the first result and the fact that each ζ_p has support in some $B(x_p, r)$.

Using these lemmas instead of Lemma 5.2.1, we easily prove Theorem 5.2.1. In Theorems 5.2.4 and 5.2.5 it is necessary to require that f satisfy a condition of the form (5.2.4) in which case we find (using (5.4.1)) that

$$\int_G |f v| dx \leq C_1 C_0 \cdot \| \nabla v \|^0_{2,G} \cdot R^{\mu_1}$$

$$\int_G |f U| dx \leq C_1 C_0 \cdot \| \nabla U \|^0_{2,G} \cdot R^{\mu_1} \leq \varepsilon (\| U \|^1_{2,0})^2 + C \cdot \varepsilon^{-1} \cdot R^{2\mu_1} \cdot C_0^2$$

in the proofs of those theorems. In Theorems 5.2.5 and 5.2.6, the results are, respectively,

$$\| \nabla u \|^0_{2,D} \leq C [a^{-1} \| u \|^0_{2,G} + \| e \|^0_{2,G} + C_0 R^{\mu_1}],$$

$$\| \nabla u \|^0_{2,r} \leq C [(R - r)^{-1} \| u \|^0_{2,R} + \| e \|^0_{2,R} + C_0 R^{\mu_1}],$$

provided that f always satisfies (5.2.4). In Theorem 5.2.7, it must be assumed in addition that the f_n and f satisfy (5.2.4) uniformly. The statement and proof of Theorem 5.2.8 carry over, using Lemmas 5.4.1 and 5.4.2 to show that $d\,u$ and hence $\varphi \in L_2$.

The theorems in § 5.3 can be proved in much the same way as they were but replacing $\nu/(\nu - 2)$ by an arbitrarily large but fixed finite number $s > 2$; the Sobolev inequality (5.3.14) holds for any s in the case $\nu = 2$. However, the theory can be greatly simplified in the case $\nu = 2$ by using the original proof of the writer as modified to make use of recent simplifications and to take care of the lack of self-adjointness in the equations (we have not assumed $a^{\beta\alpha} = a^{\alpha\beta}$ nor $b^\alpha = c^\alpha$). We now present this simplified theory. *All the theorems and their proofs generalize immediately to vector functions except, of course, the last theorem which concerns sub-solutions and corresponds to Theorem 5.3.1.*

Theorem 5.4.1. *Suppose u is a solution $\in H_2^1(G)$ of (5.1.1) in which the coefficients satisfy (5.1.2) and (5.1.3) with $\nu = 2$. There are numbers C and μ_0, $0 < \mu_0 < \mu_1$, which depend only on ν, m, M, C_0, and μ_1 such that if e and f satisfy*

$$\int_{B(x_0,r)} |e|^2 dx \leq L_1^2 (r/a)^{2\mu}, \quad \int_{B(x_0,r)} |f|\, dx \leq L_2 (r/a)^{\mu} \tag{5.4.8}$$

$$0 < \mu < \mu_0, \ 0 \leq r \leq a, \ B(x_0, a) \subset G,$$

then $u \in C_\mu^0(G)$ and satisfies

$$\int_{B(x_0,r)} |\nabla u|^2 dx \leq C(L_1^2 + L_2^2 + L^2 a^{\mu_1})(r/a)^{2\mu}, \tag{5.4.9}$$

$$L^2 = ('\| u \|_{2,a}^1)^2 \equiv \int_{B(x_0,a)} (|\nabla u|^2 + a^{-2} |u|^2)\, dx,$$

$$B(x_0, r) \subset B(x_0, a) \subset G.$$

Proof. Let us consider a ball $B(x_0, a) \subset G$ and let us redefine u in $B(x_0, 2a) - B(x_0, a)$ so that the new $u \in H_{20}^1(B_{2a})$ and

$$\| u \|_{2,0,B(x_0,2a)}^1 \leq Z_1(\nu) L.$$

(Theorem 3.5.4). Let us define

$$E^\alpha = b^\alpha u + e^\alpha, \quad F = c^\alpha u_{,\alpha} + du + f, \quad x \in B(x_0, a).$$

E and F being 0 elsewhere. Then u, E, and F satisfy

$$\int_{B(x_0,a)} [v_{,\alpha}(a^{\alpha\beta} u_{,\beta} + E^\alpha) + vF]\, dx = 0, \quad v \in H_{20}^1(B_a), \tag{5.4.10}$$

$$\begin{aligned} \| E \|_{2,r}^0 &\leq L_1 (r/a)^\mu + (2^{\lambda/2} C_2 C_0)^{1/2} a^{\lambda/4} r^{(\mu_1 - \lambda/2)/2} L, \quad 0 < \lambda < \mu_1, \\ \| F \|_{1,r}^0 &\leq L_2 (r/a)^\mu + L [C_0^{1/2} r^{\mu_1/2} + C_1 C_0 (2a)^{\lambda/2} r^{\mu_1 - \lambda/2}], \end{aligned} \tag{5.4.11}$$

as one sees from (5.1.3), (5.4.8), and Lemma 5.4.1 with $G = B(x_0, 2a)$.

Now, let us choose (i) polar coordinates with pole at x_0, (ii) a representative $\bar{u}$ which is A.C. in θ for almost all r and (iii) an R, $0 < R < a$, such that $\bar{u}(R, \theta)$ is A.C. in θ with $\bar{u}_\theta \in L_2$. Write

$$v = u - H, \ H(r, \theta) = \overline{H} + (r/R)\, [\bar{u}(R, \theta) - \overline{H}], \quad \text{on } B_R$$

where $\overline{H}$ is the average of $\bar{u}(R, \theta)$ with respect to θ. We note that

$$\begin{aligned} \int_{B(x_0,R)} F v\, dx &\geq -Z_1 (L_2 + Z_2 L a^{\mu_1/2}) (R/a)^\mu \cdot \| \nabla v \|_{2,R}^0 \\ &\geq -\varepsilon (\| \nabla u \|_{2,R}^0)^2 - \varepsilon (\| \nabla H \|_{2,R}^0)^2 - (Z_3 L_2^2 + Z_4 L^2 a^{\mu_1}) \varepsilon^{-1} (R/a)^{2\mu}. \end{aligned} \tag{5.4.12}$$

Using (5.4.11) and (5.4.12), we conclude from (5.4.10) that

$$\int_{B(x_0,R)} |\nabla u|^2 dx \leq Z_5 \int_{B(x_0,R)} |\nabla H|^2 dx + Z_6 (L_1^2 + L_2^2 + L^2 a^{\mu_1}) (R/a)^{2\mu} \tag{5.4.13}$$

where we have assumed that $\mu < \mu_1/2$ and $\mu = (\mu_1/2) - (\lambda/4)$. If we now choose μ_0 so that

$$\mu_0 < \mu_1/2 \quad \text{and} \quad \mu_0 < 1/2\, Z_5$$

and if we set

$$\varphi(R) = \int_{B(x_0, R)} |\nabla u|^2 \, dx,$$

we see that

$$\int_{B(x_0, R)} |\nabla H|^2 \, dx = \int_0^{2\pi} \int_0^R r (H_r^2 + r^{-2} H_\theta^2) \, dr \, d\theta$$

$$= \frac{1}{2} \int_0^{2\pi} \{u_\theta^2(R, \theta) + [u(R, \theta) - \bar{u}]^2\} \, d\theta$$

$$\leq \int_0^{2\pi} u_\theta^2(R, \theta) \, d\theta \leq R\, \varphi'(R).$$

Then from (5.4.13), we conclude that

$$\varphi(R) \leq (2\mu_0)^{-1} R\, \varphi'(R) + Z_7 (R/a)^{2\mu}, \; Z_7 = Z_6 (L_1^2 + L_2^2 + L^2 a^{\mu_1})$$

from which we easily conclude (if $0 < \mu < \mu_0$), that

$$\varphi(R) \leq [\varphi(a) + Z_8 (L_1^2 + L_2^2 + L^2 a^{\mu_1}) (R/a)^{2\mu}.$$

On order to handle variational problems of degree other than 2 (i.e. $k \neq 2$ in (1.10.7) or (1.10.8)), it is necessary now to prove the lemmas corresponding to Lemmas 5.3.1—5.3.3 and then prove the theorem corresponding to Theorem 5.3.1.

Lemma 5.4.3. *Suppose* (i) *that* $\omega \in H_{20}^1(G)$, (ii) *that* ψ *and* $\psi\, \omega \in H_2^1(G)$, (iii) $\psi(x) \geq 1$ *on* G, (iv) $\psi\, \omega_{,\alpha}$ *and* $\psi_{,\alpha}\, \omega \in L_2(G)$, (v) $P \in L_2(G)$ *and satisfies*

(5.4.14) $$\int_{B(x_0, r) G} P^2(x) \, dx \leq C_0 r^{\mu_1}, \quad P(x) \geq 0.$$

Then there exists a sequence $\{\zeta_n\} \to \omega$ *in* $H_{20}^1(G)$, *in which each* $\zeta_n \in L i p_c(G)$, *such that* $\psi\, \zeta_{n,\alpha} \to \psi\, \omega_{,\alpha}$ *and* $P \psi \zeta_n \to P \psi \omega$ *in* $L_2(G)$. *If* $\omega(x) \geq 0$, *the* ζ_n *may be chosen* ≥ 0.

Proof. As we noted in the proof of Lemma 5.3.1, $\psi\, \omega \in H_{20}^1(\Gamma)$ on any $\Gamma \supset \bar{G}$ if we merely define $\psi\, \omega = 0$ on $\Gamma - G$. Thus it follows from Lemmas 5.4.1 and 5.4.2 that $P \psi\, \omega \in L_2(G)$. The remainder of the proof is essentially the same as that of Lemma 5.3.1.

Lemma 5.4.4. *Suppose that* $U \in H_2^1(D)$ *for each* $D \subset\subset G$, *that* $U(x) \geq 1$ *on* G, *that* $W(x) = [U(x)]^\lambda$ *for some* λ, $1 \leq \lambda < 2$, *and that* P *satisfies hypothesis* (v) *of Lemma* 5.4.3 *on each* $D \subset\subset G$ *with* $C_0 = C_0(D)$. *Then, for each such* D,

$$\nabla W, \; P W, \; P \nabla W, \quad \text{and} \quad P^2 w \in L_1(D).$$

Proof. For PU, $PU^{\lambda - 1}$, and $\nabla U \in L_2(D)$.

Theorem 5.4.2 (Existence and uniqueness in the small). *There is a constant $R_2 > 0$ which depends only on ν, m, M, C_0, and μ_1 such that there is a unique solution U of* (5.1.1) *which* $\in H^1_{20}[B(x_0, R)]$ *for each $R \leq R_2$, each $e \in L_2(B_R)$, and each $f \in L_1(B_R)$ satisfying* (5.2.4); *this solution satisfies*

$$\|\nabla U\|^0_{2,G} \leq C_1 \|e\|^0_{2,R} + C_2\, C_0\, R^{\mu_1},$$

where C_1 and C_2 depend only on the quantities above. If $u^* \in H^1_2[B(x_0, R)]$, and $R \leq R_2$, there exists a unique solution H of the homogeneous equation such that $H - u^* \in H^1_{20}[B(x_0, R)]$.

The proof is like that of Theorem 5.3.5.

Theorem 5.4.3. *Suppose* (i) *that $U \in H^1_2(D)$ for each $D \subset\subset G$,* (ii) *that $U(x) \geq 1$ in G,* (iii) *that $U \in L_2(G)$,* (iv) *that $W = U^\lambda$ satisfies* (5.3.4) *for some λ with $1 \leq \lambda < 2$,* (v) *that a satisfies* (5.1.2) *and that b, c, and d satisfy* (5.1.3) *on each $D \subset\subset G$ with $C_0 = C_0(D)$. Then $W \in H^1_2(D)$ and is bounded on each $D \subset\subset G$.*

Proof. Since $U \in H^1_2(D)$ for each $D \subset\subset G$, it follows that $U^\tau \in L_2(D)$ for every $\tau \geq 1$ and every $D \subset\subset G$. We may then repeat the first part of the proof of Lemma 5.3.3, setting

$$\zeta = \eta^2\, U^{2-\lambda} \cdot U_L^{2\lambda-2}, \quad w_L = \eta\, U\, U_L^{\lambda-1}$$

where η has support in D' and $D \subset\subset D'_a$. The result is equation (5.3.12) with G replaced by D' and τ by λ:

$$\int_{D'} |\nabla w_L|^2\, dx \leq Z_2\, \lambda^2 \int_{D'} P^2\, w_L^2\, dx + Z_3\, \lambda \int_{D'} |\nabla \eta|^2\, U^2\, U_L^{2\lambda-2}\, dx.$$

Using Lemma 5.4.2 with $\varepsilon Z_2 \lambda^2 \leq 1/2$, we find that

$$\int_{D'} |\nabla w_L|^2\, dx \leq \int_{D'} (Z_4\, \eta^2 + Z_5 |\nabla \eta|^2)\, U^2\, U_L^{2\lambda-2}\, dx. \tag{5.4.15}$$

It is now possible to let $L \to \infty$ and this obtain $W \in H^1_2(D)$.

But now, pick a ball $B(x_0, R) \subset D'$ where $R \leq R_2$ (which now depends on D') and let H be the solution of the homogeneous equation (5.1.1) which coincides with W on $\partial B(x_0, R)$. Let $w = W - H$. Then $w = 0$ on ∂B_R and satisfies (5.3.4) there. If we then set $\zeta = w^+$, we see that

$$\zeta_{,\alpha} = w_{,\alpha} \text{ and } \zeta = w \text{ if } w > 0$$
$$\zeta_{,\alpha} = \zeta = 0 \text{ if } w = 0 \qquad \text{(a.e.)}$$

Accordingly (5.3.4) with w and this ζ becomes

$$\int_{B(x_0, R)} [\nabla \zeta \cdot a \cdot \nabla \zeta + (b + c)\, \zeta \cdot \nabla \zeta + d \zeta^2]\, dx \leq 0$$

which implies that $\zeta = 0$ since $R \leq R_2$. Thus $w \leq 0$ or $W \leq H$ on $B(x_0, R)$. But from Theorem 5.4.1, it follows that H is Hölder continuous interior to $B(x_0, R)$.

5.5. L_p and Schauder estimates

We obtain further estimates concerning the solutions of equations (5.1.1) when the $a^{\alpha\beta}$ are continuous and the coefficients and e and f satisfy various supplementary conditions. In case the $a^{\alpha\beta}$, b^α, and $e^\alpha \in C^0_\mu$ and the c^α, d, and f satisfy certain supplementary conditions, we shall show that $\nabla u \in C^0_\mu$.

We shall obtain corresponding results for weak solutions of systems in § 6.4. These results were obtained in the two dimensional case in MORREY [7], Chapter VI, § 6, where a different method was developed to treat the case where the coefficients $a^{\alpha\beta}_{ij}$ were constants satisfying (5.2.17) and $b = c = d = 0$. Let us set

$$a^{11}_{ij} = a_{ij}, \quad a^{12}_{ij} = a^{21}_{ji} = b_{ij}, \quad a^{22}_{ij} = c_{ij}.$$

It turns out that if the $u^j \in H^1_2(G)$ and satisfy (5.1.1), i.e.

$$\int_G [\omega^i_x (a_{ij} u^j_x + b_{ij} u^j_y + d_i) + \omega^i_y (b_{ji} u^j_x + c_{ij} u^j_y + e_i)]\, dx\, dy = 0,$$

then there exist conjugate functions v_i such that

$$v_{ix} = -(b_{ji} u^j_x + c_{ij} u^j_y + e_i), \quad v_{iy} = a_{ij} u^j_x + b_{ij} u^j_y + d_i.$$

If we let w be the $2N$-vector $(u^1, \ldots, u^N, v^1, \ldots, v^N)$ and g be the $2N$-vector $(d_1, \ldots, d_N, e_1, \ldots, e_N)$, these equations become

$$A\, w_x + B\, w_y + g = 0$$

where A and B are $2N \times 2N$ square matrices having a particular form. By using the ellipticity and certain theorems on λ-matrices (see BÔCHER, §§ 91–96, it is seen that we may find affine transformations

$$w = D\, w' \text{ and } g' = C\, g$$

so that the transformed equations have the same form as the conjugate equations above but with

$$\sum_{i \neq j} ('a^2_{ij} + 2'b^2_{ij} + 'c^2_{ij}) \leq h^2$$

where h is arbitrarily small. Moreover, the maximum and minimum magnifications of the transformations and their inverses are bounded by a constant depending only on m, M, and h. An interesting generalization of this method was employed by C. MIRANDA ([1]), who introduced exterior differential forms (see Chapter 7) instead of the conjugate functions.

We begin by considering equations of the form (5.1.1) on spheres B_R or hemispheres G_R where the symmetric part of the $a^{\alpha\beta}_0$ matrix is just $\delta^{\alpha\beta}$ and prove local differentiability. A neighborhood of each interior point can be mapped on such a sphere by an affine map in which the maximum and minimum magnification and those of its inverse are uniformly bounded by a constant depending only on m and M. In the case of a boundary point, one can first perform the affine transformation as above, follow that by a rotation to make $x^\nu = 0$ the tangent plane, and follow that with a map of the form

$$(5.5.1) \quad 'x^\alpha = x^\alpha, \ \alpha = 1, \ldots, \nu - 1, \ 'x^\nu = x^\nu - f(x^1, \ldots, x^{\nu-1})$$

which carries a part of ∂G into the plane $\sigma : x^\nu = 0$. The antisymmetric part of the constant matrix $a_0^{\alpha\beta}$ may as well be omitted since

$$(5.5.2) \qquad \int_G v_{,\alpha} a_0^{\alpha\beta} u_{,\beta} \, dx = \int_G v_{,\alpha} \delta^{\alpha\beta} u_{,\beta} \, dx \quad \text{if} \quad v \in H_{20}^1(G)$$

as is easily seen by approximations, using Lemma 4.5.7.

Having mapped a neighborhood of x_0 onto B_{2R} or G_{2R}, we then alter the coefficients by defining

$$(5.5.3) \qquad \begin{aligned} a_R^{\alpha\beta}(x) &= a_0^{\alpha\beta} + \varphi(R^{-1}|x|)\,[a^{\alpha\beta}(x) - a_0^{\alpha\beta}(x)], \\ b_R^\alpha(x) &= \varphi \cdot b^\alpha, \quad c_R^\alpha = \varphi c^\alpha, \quad d_R = \varphi d, \end{aligned}$$

where $\varphi \in C^\infty(R_1)$, $\varphi(s) = 1$ for $s \leq 5/4$, $\varphi(s) = 0$ for $s \geq 7/4$ and φ is non-increasing. If u is a solution of (5.1.1) with support on B_R or $G_R \cup \sigma_R$, it is also a solution of

$$(5.5.4) \quad \int_G v_{,\alpha}(a_R^{\alpha\beta} u_{,\beta} + b_R^\alpha u + e^\alpha) + v(c_R^\alpha u_{,\alpha} + d_R u + f)\, dx = 0, \qquad v \in H_{20}^1(G), \quad G = B_{2R} \quad \text{or} \quad G_{2R}.$$

We then write

$$(5.5.5) \quad u = u_R + H_R, \ u_R = Q_{2R}[(a_R - a_0) \cdot \nabla u + b_R u + e] - P_{2R}[c_R \cdot \nabla u + d_R u + f]$$

where, in the case B_{2R}, Q_{2R} and P_{2R} are the respective quasipotentials, as defined in § 3.7, and potentials, except that in the case $\nu = 2$, we define $P_R(f)$ as the ordinary potential of f over B_R minus its average value. The reason for this definition of P_R in the case $\nu = 2$ is that if $V = P_R(f)$, then, in addition to the results in Theorem 3.7.1, we also have

$$\|V\|_q^0 \leq C(\nu, q)\, R^2 \|f\|_q^0$$

in all cases. This follows from the theorems of § 3.6.

In all cases, we see that $Q_{2R} \to 0$ as $x \to \infty$. This is true of P_{2R} also if $\nu > 2$; but if $\nu = 2$, P_{2R} might become logarithmically infinite. So if u has support in B_R, it follows that $H_R = 0$ if $\nu > 2$, or $H_R = \text{const.}$ if $\nu = 2$. In that case

$$(5.5.6) \quad u_R - T_R u_R = v_R = Q_{2R}(e + b_R H_R) - P_{2R}(d_R H_R + f)$$

$$T_R u_R = Q_{2R}[(a_R - a_0) \cdot \nabla u_R + b_R u_R] - P_{2R}(c_R \cdot \nabla u_R + d_R u_R).$$

It is then shown that if R is small enough, $\|T_R\| \leq 1/2$ in $H_2^1(B_{2R})$, where u and u_R are known to be, and also in a space $H_q^1(B_{2R})$ for some $q > 2$ or a space $C_\mu^1(\bar{B}_{2R})$. We then conclude that $u \in H_q^1(B_{2R})$ or $C_\mu^1(\bar{B}_{2R})$

as the case may be. A similar program is carried out for hemispheres G_{2R}. The general bounded domain G is handled by a partition of unity.

Definition 5.5.1. In the space $C^0_{\mu 0}(\bar{B}_{2R})$, we define $|||e|||^0_{\mu,2R} = h_\mu(e, \bar{B}_{2R}) + R^{-\mu}|||e|||^0_{2R}$, $|||e|||^0_{2R}$ being the norm in $C^0(\bar{B}_{2R})$. In the space $C^1_\mu(\bar{B}_{2R})$, we define

$$|||u|||^1_{\mu,2R} = |||\nabla u|||^0_{\mu,2R} + R^{-1}|||u|||^0_{\mu,2R}.$$

In the space $H^1_q(B_{2R})$, we define the norm

$$'\|u\|^1_{q,2R} = \|\nabla u\|^0_{q,2R} + (2R)^{-1}\|u\|^0_{q,2R}.$$

We define the space $L_{p\mu}(B_{2R})$ to consist of all $f \in L_p(B_{2R})$ such that

$$\int_{B(x_0,r)\cap B_{2R}} |f(x)|^p\,dx \le L^p r^{\nu-p+p\mu} \le L^p(2R)^{\nu-p+p\mu}$$

for every $B(x_0, r)$ and we define $\|f\|^0_{p,\mu,2R} = \inf. L$.

Then, from SOBOLEV's lemma (Theorem 3.5.3) and Theorems 3.7.1–3.7.3, we obtain the following results:

Theorem 5.5.1. *Q_R is a bounded operator from $L_q(B_R)$ to $H^1_q(B_R)$ for each $q > 1$, and from $C^0_{\mu 0}(\bar{B}_R)$ to $C^1_\mu(\bar{B}_R)$ with bounds depending on q and μ $(0 < \mu < 1)$, respectively, but not on R. P_R is a bounded operator from $L_p(B_R)$ to $H^2_p(B_R)$ and to $H^1_q(B_R)$ with bound independent of R if $1 < p < \nu$ and $q = p\nu/(\nu - p)$; P_R is also a bounded operator from $L_{p\mu}(B_R)$, $1 \le p \le \nu$, or $L_p(B_R)$ with $p = \nu/(1-\mu)$ into $C^1_\mu(\bar{B}_R)$, with bound independent of R. In fact if $u = Q_R(e)$ and $e \in L_q(B_R)$ then $u \in H^1_q(B_A)$ for any A and $\|\nabla u\|^0_{q,R_\nu} \le C\|e\|^0_{qR}$; if $e \in C^0_{\mu 0}(\bar{B}_R)$, then $u \in C^1_\mu(\bar{B}_A)$ with $h_\mu(\nabla u, \bar{B}_A) \le C\,h_\mu(e, \bar{B}_R)$ for every A; if $v = P_R(f)$ and $f \in L_p(B_R)$, $1 < p < \nu$, then $v \in H^1_q(B_A)$ for any A and $\|\nabla v\|^0_{q,R_\nu} \le C\|f\|^0_{pR}$ if $q = p\nu/(\nu - p)$; if $f \in L_{p,\mu}(B_R)$, $1 \le p \le \nu$, or to L_p with $p = \nu/(1 - \mu)$, then $v \in C^1_\mu(R_\nu)$ with $h_\mu(\nabla v, R_\nu) \le C\|f\|$, the norm being in the proper space. Finally, if $\nu = 2$, and f satisfies the condition*

$$\int_{B(x_0,r)\cap B_R} |f(x)|\,dx \le L\,r^{\mu_1}, \quad \text{all } B(x_0,r),\ \mu_1 > 0, \tag{5.5.7}$$

then $P_R(f) \in H^1_2(B_A)$ for any A and $\nabla P_R(f) \in L_{2,\mu}(B_A)$ for any A and

$$\|P_R(f)\|^1_{2,2R} \le CLR^{2\mu_1}, \quad \|\nabla P_R(f)\|^0_{2,\mu_1,2R} \le CL.$$

Definition 5.5.2. We say that the coefficients a, b, c, and d satisfy the H^1_q-conditions on a domain Γ if and only if the $a^{\alpha\beta}$ are continuous on Γ and the b^α, c^α, and d are measurable there and

(i) b^α and $c^\alpha \in L_\nu(\Gamma)$ and $d \in L_{\nu/2}(\Gamma)$ if $\nu/(\nu - 1) < q < \nu$ and $\nu > 2$

(ii) $$\int_{\Gamma\cap B(x_0,r)} (|b|^\nu + |c|^\nu + |d|^{\nu/2})\,dx \le L^\nu r^{\nu\mu_1},\ \mu_1 > 0,$$
for every $B(x_0, r)$, if $q = \nu \ge 2$,

(iii) $b^\alpha \in L_q(\Gamma)$, $c^\alpha \in L_\nu(\Gamma)$, and $d \in L_p(\Gamma)$, $p = \nu q/(\nu + q)$, if $q > \nu$.

We say that the coefficients *satisfy the C^1_μ-conditions on Γ* if and only if the $a^{\alpha\beta}$ and $b^\alpha \in C^0_u(\Gamma)$ and c^α and $d^\alpha \in L_1(\Gamma)$ and satisfy

$$\int_{B(x_0,r)\cap\Gamma} (|c(x)| + |d(x)|)\,dx \le L\,r^{\nu-1+\mu}, \quad x_0 \in R_\nu, \; r > 0. \tag{5.5.8}$$

It is desirable to prove the following lemmas:

Lemma 5.5.1. (a) *Suppose Γ is a bounded strongly Lipschitz domain and $\varphi \in L_1(\Gamma)$ and satisfies (5.5.7) with B_R replaced by Γ. Then, for each $\varepsilon > 0$, there is a $C(\varepsilon, \mu_1, \nu, \Gamma)$ such that*

$$\int_{\Gamma\cap B(x_0,r)} |\varphi(x)|\cdot|u(x)|\,dx \le C R^\varepsilon L\,'\|u\|_q^1 \cdot r^{\mu_1-\varepsilon}, \tag{5.5.9}$$
$$u \in H^1_\nu(\Gamma), \quad |\Gamma| = \gamma_\nu R^\nu$$

(b) *Suppose $u \in H^1_q(B_{2R})$ with $q > \nu$. Then*

$$|u(x)| \le C(\nu, q)\, R^{1-\nu/q}\,'\|u\|_q^1, \quad x \in \bar{B}_{2R}. \tag{5.5.10}$$

Proof. (a) On account of the extension theorem (Theorem 3.4.3), we may assume that $u \in H^1_{\nu 0}(\Gamma)$. Then, from Theorem 3.7.2, we conclude that

$$\int_{B(x_0,r)\cap\Gamma} |\varphi(x)|\,|u(x)|\,dx \le \Gamma_\nu^{-1} \int_{B(x_0,r)\cap\Gamma} dx \int_\Gamma |\varphi(x)|\cdot|\xi - x|^{1-\nu}\,|\nabla u(\xi)|\,d\xi. \tag{5.5.11}$$

Applying the Hölder inequality to the double integral on the right in (5.5.11), we see that our integral I on the left satisfies

$$I \le \Gamma_\nu^{-1} I_1^{(\nu-1)/\nu} \cdot I_2^{1/\nu}, \quad I_1 = \int_{B(x_0,r)\cap\Gamma} dx \int_\Gamma |\xi - x|^{-\nu+\sigma}\,|\varphi(x)|\,d\xi$$
$$I_2 = \int_{B(x_0,r)\cap\Gamma} dx \int_\Gamma |\xi - x|^{-\sigma(\nu-1)}\,|\varphi(x)|\cdot|\nabla u(\xi)|^\nu\,d\xi. \tag{5.5.12}$$

From Lemma 3.4.3, it follows that

$$I_1 \le \sigma^{-1}\Gamma_\nu(|\Gamma|/\gamma_\nu)^{\sigma/\nu} \cdot L\,r^{\mu_1} = \sigma^{-1}\Gamma_\nu R^\sigma L\,r^{\mu_1}. \tag{5.5.13}$$

In order to estimate I_2, we first note that

$$\int_{B(x_0,r)\cap\Gamma} |\xi - x|^{-\sigma(\nu-1)}\,|\varphi(x)|\,dx \le \int_0^\infty s^{-\sigma(\nu-1)}\,\varphi'(s;\xi)\,ds$$
$$= \sigma(\nu-1)\int_0^\infty s^{-1-\sigma(\nu-1)}\,\varphi(s;\xi)\,ds \tag{5.5.14}$$

where

$$\varphi(s;\xi) = \int_{B(\xi,s)\cap B(x_0,r)\cap\Gamma} |\varphi(x)|\,dx \le \begin{cases} L\,s^{\mu_1}, & 0 \le s \le r \\ L\,r^{\mu_1}, & s \ge r. \end{cases} \tag{5.5.15}$$

From (5.5.14) and (5.5.15), we find that

$$I_2 \le Z_1(\nu, \mu_1, \sigma)\, L\,r^{\mu_1-\sigma(\nu-1)}\,('\|u\|_\nu^1)^\nu, \quad 0 < \sigma < \mu_1/(\nu-1). \tag{5.5.16}$$

The result (5.5.9) follows by taking $\varepsilon = \sigma(\nu-1)/\nu < \mu_1/\nu$. From this, it follows that (5.5.9) holds for values of ε such that $\mu_1/\nu \le \varepsilon < \mu_1$.

(b) follows from Theorem 3.5.1 with $p = q > \nu$ and $m = 1$ and from the definition of the $'\|u\|^1_{q,2R}$.

Theorem 5.5.2. (a) *Suppose the coefficients a, b, c, and d satisfy the H_q^1-conditions on B_A where $q \geq \nu/(\nu-1)$ and the equality holds only if $\nu = 2$. Then for each $R \leq A/2$, the operator T_R is bounded on $H_q^1(B_{2R})$ and has a bound of the form $\varepsilon(R)$ where $\varepsilon(R) \to 0$ as $R \to 0$; $\varepsilon(R)$ depends on the moduli of continuity of the $a^{\alpha\beta}$ and on ν, m, and M and on*

$$\sup_{B(x_0,R)\cap B_A} \int (|b|^\nu + |c|^\nu + |d|^{\nu/2})\, dx \quad \textit{in case } (\text{i}) \text{ (of def. 5.5.2)},$$

L and μ_1 in case (ii), *and*

$$\sup_{B(x_0,R)\cap B_A} \int (|b|^q + |c|^\nu + |d|^p)\, dx \quad \textit{in case } (\text{iii}).$$

(b) *If the coefficients satisfy the C_μ^1-conditions on $\bar{B}_A$, then, for each $R \leq A/2$, T_R is bounded on $C_\mu^1(\bar{B}_{2R})$ with a bound of the form $C R^\mu$ where C depends on ν, m, M, L, μ, and the C_μ^0-norms of the coefficients a and b.*

Proof. We set

$$e_R = (a_R - a_0)\cdot\nabla u_R + b_R u_R, \quad f_R = c_R \cdot \nabla u_R + d_R u_R,$$
$$\eta(R) = \max |a_R(x) - a_0| \,.$$

In the H_q^1 case we see, using Lemma 5.5.1 (with $\varphi(x) = |b_R|^\nu + |c_R|^\nu + |d_R|^{\nu/2}$) and Theorem 3.5.5, that

(i) $\| e_R \|_q^0 \leq [\eta(R) + C \| b_R \|_\nu^0]\, '\| u_R \|_q^1, \ \| f_R \|_p^0 \leq C(\| c_R \|_\nu^0 + \| d_R \|_{\nu/2}^0)\, '\| u_R \|_q^1,$

(ii) $\| e_R \|_q^0 \leq [\eta(R) + C(L R^{\mu_1})^{1/\nu}]\, '\| u_R \|_\nu^1, \qquad \| f_R \|_{\nu/2}^0 \leq C(L R^{\mu_1})\, '\| u_R \|_\nu^1,$
$\nu > 2$

(iii) $\| e_R \|_q^0 \leq [\eta(R) + C \| b_R \|_q^0 R^{1-\nu/q}]\, '\| u_R \|_q^1,$
$\| f_R \|_p^0 \leq [\| c_R \|_\nu^0 + C \| d_R \|_p^0 R^{1-\nu/q}]\, '\| u_R \|_q^1$

in the respective cases; here we set $p = \nu q/(\nu + q)$ and omitted the subscript $2R$ on all the norms. If $\nu = 2$ in (ii), the e inequality still holds and f_R satisfies (5.5.7) with L replaced by $C L\, '\| u_R \|_2^1 (1 + L R^{\mu_1})$. In the C_μ^1 case, we note first that

$$|b_R(x_2) - b_R(x_1)| \leq |\varphi_R(x_2) - \varphi_R(x_1)| \cdot |b_R(x_2)| + \varphi_R(x_1) \cdot |b_R(x_2) - b_R(x_1)|$$
$$\leq (C B_0 \cdot R^{-\mu} + B_1)\, |x_2 - x_1|^\mu, \quad x_1, x_2 \in \bar{B}_{2R}, \tag{5.5.17}$$
$$\varphi_R(x) = \varphi(R^{-1}|x|), \quad |b_R(x)| \leq B_0, \quad x \in \bar{B}_{2R}, \quad B_0 = \max_{|x|\leq A} |b(x)|,$$
$$B_1 = h_\mu(b, \bar{B}_A).$$

Using this idea again, we obtain

$$|a_R(x_2) - a_R(x_1)| \leq 2A_1 |x_2 - x_1|^\mu, \quad A_1 = h_\mu(a, \bar{B}_A),$$
$$h_\mu(e_R, \bar{B}_{2R}) \leq {}'|\!|\!| u |\!|\!|_{\mu,2R}^1 [4 A_1 R^\mu + R B_0 (1 + C) + B_1 R^{1+\mu}],$$
$$\int_{B(x_0,r)\cap B_{2R}} |f_R(x)|\, dx \leq L R^\mu (1 + R) \cdot {}'|\!|\!| u |\!|\!|_{\mu,2R}^1 \cdot r^{\nu-1+\mu}.$$

The theorem follows from Theorem 5.5.1.

We can now prove the following interior boundedness theorem:

Theorem 5.5.3. (a) *Suppose that* (i) *the coefficients satisfy the* H_q^1 *and* $H_{q'}^1$ *conditions on the domain* G, *where* $\frac{\nu}{\nu-1} \leq q < q'$, (ii) *suppose* $e \in L_{q'}(D)$ *and* $f \in L_{p'}(D)$ *for each* $D \subset\subset G$, *where* $p' = \nu q'/(\nu + q') > 1$ *and* (iii) *suppose* $u \in H_q^1(D)$ *and satisfies* (5.2.2) (*with* $\lambda = 0$) *on such* D. *Then* $u \in H_{q'}^1(D)$ *for such* D.

(b) *Suppose that* (i) *the coefficients satisfy the* H_q^1 *and* C_μ^1 *conditions on* G, $q \geq \nu/(\nu - 1)$ *and* $0 < \mu < 1$; (ii) *suppose* $e \in C_\mu^0(D)$ *and* $f \in L_{1\mu}(D)$ *for each* $D \subset\subset G$, *and* (iii) *suppose* $u \in H_q^1(D)$ *and satisfies* (5.2.2) ($\lambda = 0$) *on such* D. *Then* $u \in C_\mu^1(D)$ *for each such* D.

Proof. (a) Each point x_0 in G is the center of an ellipse which can be mapped onto B_{2R} by an affine map as described above, where R is so small that the bound T_R is $\leq 1/2$ on $H_q^1(B_{2R})$ and on $H_{q'}^1(B_{2R})$ where we assume first that $q' \leq \nu q/(\nu - q)$, $q < \nu$. Let $\zeta \in C_c^1(G)$ and have support on the image of B_R and let U be the transform of ζu. Then, by replacing v by ζv in (5.2.2), we see that U has support on B_R and satisfies the transformed equations (5.2.2) ($\lambda = 0$) with e and f replaced respectively by

$$E_R^\alpha = \zeta e^\alpha - a_R^{\alpha\beta} \zeta_{,\beta} u, \quad F_R = \zeta f - c_R^\alpha \zeta_{,\alpha} u + \zeta_{,\alpha}(a_R^{\alpha\beta} u_{,\beta} + b_R^\alpha u + e^\alpha). \tag{5.5.18}$$

Since U has compact support, we see that

$$U = U_R + H_R, \quad U_R - T_R U_R = V_R = Q_{2R}(E_R + b_R H_R) - P_{2R}(F_R + d_R H_R) \quad (H_R = \text{const.}).$$

Since $q' \leq \nu q/(\nu - q)$, it follows that $p' \leq q$ so that $E \in L_{q'}(B_{2R})$ and $F \in L_{p'}(B_{2R})$ so that $U \in H_{q'}^1(B_{2R})$. Since it is clear that the coefficients satisfy the $H_{q''}^1$-conditions if they satisfy the $H_{q'}^1$-conditions and $q'' < q'$ or if they satisfy the C_μ^1 conditions we may use the result to prove the result for any q'. If $D \subset\subset G$, we may find a partition of unity on $\bar{D}$ each member of which has small support as above.

In the C_μ^1-case, we first show that $u \in H_{q'}^1(D)$ for some $q' > \nu$, for each $D \subset\subset G$. Then we see that $E \in C_\mu^0(D)$ so that $U \in C_\mu^1(\bar{B}_{2R})$. Thus, as above, $u \in C_\mu^1(D)$ for any $D \subset\subset G$.

We need the following lemma for later reference:

Lemma 5.5.2. *Suppose the coefficients satisfy the* H_q^1*-conditions on* B_A. *Then there is an* R_0, $0 < R_0 \leq A/2$, *such that if* u *has support on* B_R *and satisfies* (5.1.1) *there with* $e \in L_q$ *and* $f \in L_p$ *where* $\nu \geq 2$ *and* $p = q\nu/(q + \nu) > 1$, *then*

$$'\|u\|_{qR}^1 \leq C_1(\|e\|_{qR}^0 + \|f\|_{pR}^0), \quad 0 < R \leq R_0.$$

If $\nu = 2 = q$, *so* $p = 1$, *the same result holds with* $\|f\|_{p,R}^0$ *replaced by* $C L R^{\mu_1}$ *if* f *satisfies* (5.5.7). *If the coefficients satisfy the* C_μ^1*-conditions* *on $\bar{B}_A$, there is an R_1, $0 < R_1 \leq A/2$, such that if $u \in C^1_{\mu 0}(\bar{B}_R)$ and satisfies* (5.1.1) *there with $e \in C^0_\mu(\bar{B}_R)$ and $f \in L_{1,\mu}(B_R)$, then*

$$'\|u\|^1_{\mu R} \leq C_2('\|e\|^0_{qR} + \|f\|^0_{1,\mu,R}).$$

The constants R_0, R_1, C_1, and C_2 depend on the quantities specified in Theorem 5.5.2.

Proof. Since u has support in B_R, we have seen that

$$u = u_R + H_R \quad \text{so} \quad u_R - T_R u_R = v_R + w_R, \quad v_R = Q_{2R}(e) - P_{2R}(f),$$
$$w_R = Q_{2R}(b_R H_R) - P_{2R}(d_R H_R)$$

where $H_R = 0$ if $\nu > 2$ and is a constant if $\nu = 2$ and T_R has bound $\leq 1/2$ if $R \leq R_0$ or R_1. If $\nu = 2$, we see, since $\nabla u = \nabla u_R$ and u has compact support, that $'\|u\|^1_{qR} \leq C'\|u_R\|^1_{qR}$ so $'\|H_R\|^1_{qR} \leq C'\|u_R\|^1_{qR}$. Setting $p = q\nu/(q+\nu)$ we find as in the proof of Theorem 5.5.2 that if $\nu/(\nu-1) < q < \nu$, then

$$\|b_R H_R\|^0_{pR} \leq C'\|u_R\|^1_{qR}\|b_R\|^0_R, \quad \|d_R H_R\|^0_{pR} \leq C'\|u_R\|^1_{qR}\|d_R\|^0_{\nu/2,R}.$$

Since $\|b_R\|^0_{\nu R}$ and $\|d_R\|^0_{\nu/2,R} \to 0$ as $R \to 0$, the result follows in this case. The other cases are proved similarly.

Definition 5.5.1′. We define the norms and spaces of functions on G_{2R} as they were defined on B_{2R} but with B_{2R} replaced by G_{2R} and *the functions u* (in $C^1_\mu(\bar{G}_{2R})$ *or* $H^1_q(G_{2R})$) *being required to vanish along* σ_{2R}. The functions e^α are supposed to vanish along $\sum_{2R}$ but not necessarily along σ_{2R}. We define $P_{2R}(f)$ as the restriction to G_{2R} of the former $P_{2R}(\tilde{f})$ where $\tilde{f}$ is the extension to B_{2R} by "negative reflection"

$$\tilde{f}(x'_\nu, x) = \begin{cases} f(x'_\nu, x^\nu) & \text{if } x^\nu > 0 \\ -f(x'_\nu, -x^\nu) & \text{if } x^\nu < 0. \end{cases} \tag{5.5.19}$$

We define $Q_{2R}(e) = U$ by the formula

$$U(x) = -\int_{B_{2R}} K_{0,\alpha}(x-\xi)\,\tilde{e}^\alpha(\xi)\,d\xi + 2\sum_{\alpha=1}^{\nu-1} \int_{G_{2R}} K_{0,\alpha}(x-\xi)\,\tilde{e}^\alpha(\xi)\,d\xi \tag{5.5.20}$$

where we assume that e has been extended by "positive reflection"

$$\tilde{e}(x^\nu, x'_\nu) = \begin{cases} e(x^\nu, x'_\nu) & \text{if } x^\nu \geq 0, \\ e(-x^\nu, x'_\nu) & \text{if } x^\nu \leq 0. \end{cases} \tag{5.5.21}$$

K_0 is the elementary function for Laplace's equation.

Using the theorems above, we obtain the following result:

Theorem 5.5.1′. *Theorems* 5.5.1 *and* 5.5.2. *and Lemma* 5.5.2 *hold with B_R replaced by G_R; the H^1_q- and C^1_μ-conditions on the coefficients being defined as above.*

Proof. It remains only to prove that the second term in (5.5.20) $\in C^1_\mu(\bar{G}_A)$ for any A. Let this be denoted by V. For $x^\nu > 0$, we have

$$\nabla^2 V(x) = 2\sum_{\alpha=1}^{\nu-1} \int_{G_{2R}^-} \nabla^2 K_{0,\alpha}(x-\xi)\,\bar{e}^\alpha(\xi)\,d\xi$$
$$= 2\sum_{\alpha=1}^{\nu-1} \int_{R_\nu^-} \nabla^2 K_{0,\alpha}(x-\xi)\cdot[\bar{e}^\alpha(\xi) - \bar{e}^\alpha(x)]\,d\xi$$

since e vanishes along $\sum_R^-$ and the integral of $\nabla^2 K_{0,\alpha}$ over R_ν^- converges absolutely to 0. From this we conclude that

$$|\nabla^2 V(x)| \leq C(\nu,\mu)\, h_\mu(e, \bar{G}_{2R})\,(x^\nu)^{\mu-1}$$

from which the result follows, using Theorem 2.6.6.

By the method of proof of Theorem 5.5.3, using the result of Theorem 5.5.2 for G_{2R}, we can prove the following theorem:

Theorem 5.5.4′. *Suppose that $G \in C^1$, the coefficients satisfy the H^1_q- and $H^1_{q'}$-condition on a domain $\Gamma \supset \bar{G}$, where $\nu/(\nu-1) \leq q < q'$ and $p' = \nu q'/(\nu + q')$ (> 1), suppose $e \in L_{q'}(G)$ and $f \in L_{p'}(G)$, and suppose $u \in H^1_q(G)$ and is a solution of (5.1.1) on G. Then $u \in H^1_{q'}(G)$. If, also, $G \in C^1_\mu$, the coefficients satisfy the C^1_μ-conditions on Γ, $e \in C^0_\mu(\bar{G})$, and $f \in L_{1,\mu}(G)$, then $u \in C^1_\mu(G)$.*

We now prove the following a priori bound:

Theorem 5.5.5′. (a) *Suppose that $G \in C^1$ and that the coefficients satisfy the H^1_q-conditions on $\Gamma \supset G$ with $q \geq \nu/(\nu-1)$, the equality holding only if $\nu = 2$. Suppose also that $e \in L_q(G)$ and that $f \in L_p(G)$, where $p = q\nu/(\nu+q)$ (≥ 1) and where f satisfies (5.5.7) if $p = 1$. Suppose that u satisfies (5.1.1) on G. Then*

$$\|u\|^1_{q,G} \leq C(\|e\|^0_q + \|f\|^0_p + \|u\|^0_1) \tag{5.5.22}$$

where C depends only on ν, m, M, q, G, the moduli of continuity of a, and bounds for the coefficients; if $\nu = 2 = q$, so $p = 1$, the term $\|f\|^0_p$ must be replaced by L.

(b) *If $G \in C^1_\mu$, the coefficients satisfy the C^1_μ-conditions on $\Gamma \supset \bar{G}$, $e \in C^0_\mu(\bar{G})$ and $f \in L_{1,\mu}(G)$, then*

$$|||u|||^1_{\mu,G} \leq C(|||e|||^0_{\mu,G} + \|f\|^0_{1,\mu,G} + \|u\|^0_1). \tag{5.5.23}$$

Proof. We prove (a) with $p > 1$, the proof of (b) and the last case of (a) are similar. Each point x_0 of $\bar{G}$ is in a neighborhood $\mathfrak{N} \subset$ a neighborhood $\mathfrak{U}$ which can be mapped as described at the beginning of the section onto B_{2R} or $G_{2R} \cup \sigma_{2R}$ so that $\mathfrak{N}$ is mapped onto B_R or $G_R \cup \sigma_R$, where R is so small that Lemma 5.5.2 holds, and we may choose a partition of unity $\{\zeta_s\}$ of the proper class, each ζ_s having support in some one neighborhood $\mathfrak{N}$. Then, if u_s and $\tilde{\zeta}_s$ are the transforms of $\zeta_s u$ and ζ_s, respectively (on B_R or G_R), we see that u_s satisfies (5.1.1) with e and f replaced by E_s and F_s as given in (5.5.18) with the obvious changes of notation.

Now, suppose there is no such constant. Then there are sequences $\{u_n\}$, $\{e_n\}$, and $\{f_n\}$ such that

$$(5.5.24) \qquad \| u_n \|^1_{q,G} = 1, \quad \| e_n \|^0_{q,G} + \| f_n \|^0_{p,G} + \| u_n \|^0_{1,G} \to 0,$$

and $u_n \rightharpoonup u$ in $H^1_q(G)$. Since $u_n \to u$ in $L^1_q(G)$ and hence in $L^1_1(G)$ it follows that $u = 0$. Thus we conclude that $E_{ns} \to 0$ in $L_q(G)$, ($G = B_{2R}$ or G_{2R}) $F_{1ns} \to 0$ in $L_p(G)$ and $F_{2ns} \rightharpoonup 0$ in $L_q(G)$, $F_{2ns} = \zeta_{s,\alpha} a^{\alpha\beta}_R u_{n,\beta}$, $F_{1ns} = F_{ns} - F_{2ns}$. Thus $\nabla^2 P_{2R} F_{1ns} \to 0$ in $L_p(G)$ and $\nabla^2 P_{2R} F_{2ns} \rightharpoonup 0$ in $L_q(G)$, so that $P_{2R}(F_{ns}) \to 0$ in $H^1_q(G)$. But then $u_{ns} \to 0$ in $H^1_q(G)$ for each s (Lemma 5.5.2) so that $u_n \to 0$ in $H^1_q(G)$. But this contradicts (5.5.24).

Remarks. From Theorems 5.5.4 and Theorem 5.5.5 and the proof of Theorem 5.2.4 we conclude the following: *Suppose G, the coefficients, and e and f satisfy the H^1_2-conditions and either the conditions in Theorem 5.5.5 (a) or* (b) *and suppose $u \in H^1_2(G)$ and is a solution of* (5.2.2) *with $\lambda = \lambda_0$, where λ_0 is defined in Theorem* 5.2.1. *Then $u \in H^1_q(G)$ if $q \geq 2$ or $C^1_\mu(\bar{G})$, respectively, and the bounds* (5.5.22) *or* (5.5.23) *hold without the term $\| u \|^0_{1,G}$ on the right.* For if not, sequences $\{u_n\}$, $\{e_n\}$, and $\{f_n\}$ would exist satisfying (5.5.24) without the term $\| u_n \|^0_{1,G}$ on the right. But, from the proof of Theorem 5.2.4, it follows that $u_n \to 0$ in $H^1_{20}(G)$ and so in $L_1(G)$ and the remainder of the proof proceeds as before.

5.6. The equation $a \cdot \nabla^2 u + b \cdot \nabla u + c u - \lambda u = f$

Using the general theory of § 5.2 together with Theorem 5.2.8, we easily deduce the following preliminary existence theorem.

Theorem 5.6.1. *Suppose G is of class C^1_1, $a \in C^0_1$ on $\bar{G}$, b is bounded on G, $c \in L_\nu(G)$ and satisfies* (5.2.11) *there, and a satisfies* (5.1.2) *there. Then, if λ is not in a set $\mathfrak{b}$ without (finite) limit points, there is a unique solution of*

$$(5.6.1) \qquad L u - \lambda u \equiv (a \cdot \nabla^2 u + b \cdot \nabla u + c u) - \lambda u = f$$

which $\in H^2_2(G) \cap H^1_{20}(G)$ for each $f \in L_2(G)$. If $\lambda \in \mathfrak{b}$, there is a non-empty but finite-dimensional manifold of solutions in $H^2_2(G) \cap H^1_{20}(G)$ with $f = 0$.

Proof. Let us define $C(u, v)$ as in § 5.2, and define

$$B(u, v) = \int_G \{\bar{v}_{,\alpha} a^{\alpha\beta} u_{,\beta} + \bar{v}[(a^{\beta\alpha}_{,\beta} - b^\alpha) u_{,\alpha} - c u]\} dx,$$

$$L(v) = -\int_G f \bar{v}\, dx.$$

Then, from Theorem 5.2.8 it follows that (a) if λ is not in a set $\mathfrak{b}$, there is a unique solution u of (5.2.2) in $H^1_{20}(G)$, and (b) that solution (if it exists) $\in H^2_2(G)$ and satisfies the differential equation (5.2.12) (a.e.). But (5.2.12) reduces to (5.6.1) in our case.

In the theorem above, we were forced to assume that $a \in C^0_1(G)$ in order to be able to use the HILBERT space proof given above and in § 5.2.

It has long been known in the case $\nu = 2$ (see the article by L. LICHTENSTEIN in the Enzyklopädie der mathematischen Wissenschaften [3]) and was proved in 1934 by J. SCHAUDER for any ν that if the a, b, c, and f were merely HÖLDER continuous, then the solutions $\in C^2_\mu(G)$ and to $C^2_\mu(\bar{G})$ if $u = 0$ on ∂G and G is of class C^2_μ. We shall first obtain an existence theorem for the case where G is of class C^1_1, a is merely continuous and b and c satisfy the conditions above. We shall then obtain SCHAUDER'S results. There are some a priori bounds which have been obtained by CORDES ([1], [2], [3]) for the case that the $a^{\alpha\beta}$ are merely bounded and measurable with $b = c = 0$. Although these are important for the study of quasi-linear equations, the results are not completely general and require certain special techniques for their proofs. Accordingly we shall not consider these results.

Our method of proof is to reduce the problem to one on spheres or hemispheres B_{R0} or G_{R0} where $a_0^{\alpha\beta} = \Delta^{\alpha\beta}$ as in the preceding section; here we may take $a^{\beta\alpha} = a^{\alpha\beta}$. As in that section, we shall consider first functions with support in B_R or G_R. Without altering the fact that u satisfies (5.6.1) on B_R (or G_R), we may alter the coefficients as in (5.5.3). We then write, assuming that $\lambda = 0$ in (5.6.1),

$$(5.6.2)\quad u = u_R + H_R,\ u_R = P_R\,[f - (a_R - a_0)\cdot\nabla^2 u - b_R\cdot\nabla u - c_R u]$$

where in the B_R case, P_R is the potential as defined before, and in the G_R case, $P_R(f)$ denotes the potential of f^* where f^* is the extension of f to the whole of B_R by negative reflection:

$$(5.6.3)\qquad f^*(x^\nu, x'_\nu) = -f(-x^\nu, x'_\nu),\ x^\nu < 0;$$

f^* may not be continuous across $x^\nu = 0$. If $u \in H^2_{p0}(B_R)$ and spt. $u \subset B_R$, then we easily see that

$$(5.6.4)\quad u(x) = \int_{B_R} K_0(x-\xi)\,\Delta u(\xi)\,d\xi \quad \text{so} \quad H_R(x) = \begin{cases} 0 & \text{if } \nu > 2 \\ \text{const.} & \text{if } \nu = 2. \end{cases}$$

If $u \in H^2_p(G_R) \cap H^1_{p0}(G_R)$ and has spt. in $G_R \cup \sigma_R$, then this time

$$(5.6.5)\quad u(x) = u_R(x) = \int_{B_R} K_0(x-\xi)\,f^*(\xi)\,d\xi,\ \ f(x) = \Delta u(x),$$

and f^* is defined by (5.6.3); this follows from our definition since the average of $u_R = 0$ even in the two dimensional case. Thus if $\nu > 2$

$$(5.6.6)\qquad u_R - T_R u_R = v_R = P_R(f),$$
$$T_R u_R = P_R\,[(a_R - a_0)\cdot\nabla^2 u_R + b_R\cdot\nabla u_R + c_R u_R].$$

We shall assume $\nu > 2$; we have seen how to take care of H_R.

Definition 5.6.1. On the space $H^k_{p0}(B_{2R})$ or $H^k_p(G_{2R})$ we define $'\| u\|^k_{p0R}$ by

$$'\| u\|^k_{p0R} = \sum_{j=0}^{k} R^{-j}\,\|\nabla^{k-j} u\|^0_{pR};$$

on the space $C_\mu^k(\bar{B}_{2R})$ or $C_\mu^k(\bar{G}_{2R})$, we define

$$'|||u|||_{\mu R}^k = \sum_{j=0}^k [R^{-j} h_\mu(\nabla^{k-j} u, \Gamma_{2R}) + R^{-j-\mu} |||\nabla^{k-j} u|||_R],$$
$$\Gamma_{2R} = B_{2R} \quad \text{or} \quad G_{2R},$$

$|||\ |||$ being the norm in $C^0(\Gamma_{2R})$.

Definition 5.6.2. We say that the coefficients $a^{\alpha\beta}$, b^α, and c *satisfy the h-p-conditions on a region* Γ (h an integer ≥ 0) $\Leftrightarrow$ a is continuous and satisfies (5.1.2), b and c are measurable there, and

(i) if $h = 0$, and $p < \nu/2$, then $b \in L_\nu(\Gamma)$ and $c \in L_{\nu/2}(\Gamma)$ and

$$\int_{B(x_0,R)\cap\Gamma} (|b|^\nu + |c|^{\nu/2})\, dx \leq K R^{\mu_1}, \quad \mu_1 > 0; \tag{5.6.7}$$

(ii) if $h = 0$, and $\nu/2 \leq p < \nu$, then $b \in L_\nu(\Gamma)$ and $c \in L_p(\Gamma)$ and $|b|^\nu$ and $|c|^p$ satisfy a condition like (5.6.7);

(iii) if $h > 0$ and k is chosen so that $(k-1)p \leq \nu < kp$, then $\nabla^j a$, $\nabla^{j-1} b$, $\nabla^{j-2} c \in L_{(\nu/j)}$ for $j = 1, \ldots, k-1$ with

$$\int_{B(x_0,R)\cap\Gamma} (|\nabla^j a| + |\nabla^{j-1} b| + |\nabla^{j-2} c|)^{\nu/j}\, dx \leq K R^{\mu_1}$$

where meaningless terms are omitted ($\nabla^{-1} b$, etc.); if $j \geq k$, then $\nabla^j a$, $\nabla^{j-1} b$, and $\nabla^{j-2} c \in L_p$ with

$$\int_{B(x_0,R)\cap\Gamma} (|\nabla^j a| + |\nabla^{j-1} b| + |\nabla^{j-2} c|)^p\, dx \leq K R^{\mu_1}, \quad j = k, \ldots, h+2,$$

where $\nabla^{h+1} a$, $\nabla^{h+2} a$, $\nabla^{h+1} b$ are to be omitted.

The coefficients satisfy the h-μ-conditions on $\bar{\Gamma}$ $\Leftrightarrow$ they also $\in C_\mu^h(\bar{\Gamma})$.

Definition 5.6.3. We let E denote the set of bounds and moduli of continuity of the coefficients and their relevant derivatives; it will usually be clear from the context what quantities are included.

Lemma 5.6.1. (a) *Suppose* $G = B_{R_0}$ *or* G_{R_0}, $1 < p \leq q < \infty$, $h \geq 0$, *the coefficients satisfy the* 0-p- *and the* h-q-*conditions on* G, *and* $a^{\alpha\beta}(0) = \delta^{\alpha\beta}$. *Then there exists an* $R_1 > 0$, *depending only on* ν, h, p, q, *and* E *and a constant* C_1, *depending only on* ν, h, p, *and* q *such that if* $u \in H_{p0}^2(B_R)$ *then* $u \in H_{q0}^{h+2}(B_R)$ *and if* $u \in H_p^2(G_R) \cap H_{p0}^1(G_R)$, *then* $u \in H_q^{h+2}(G_R) \cap H_{q0}^1(G_R)$, *and*

$$'\|u\|_{qR}^{h+2} \leq C_1' \|Lu\|_{qR}^h, \quad \text{if} \quad 0 < R \leq R_1$$

and spt $u \subset B_R$ or $G_R \cup \sigma_R$, *respectively.*

(b) *If, also, the coefficients satisfy the* h-μ-*conditions on* $\bar{G}$, *there exists a constant* C_2, *depending only on* ν, h, p, and μ, *and an* $R_2 > 0$, *depending only on* ν, h, p, μ, *and* E *such that if* $u \in H_{p0}^2(B_R)$, *then* $u \in C_\mu^{h+2}(\bar{B}_R)$ *and if* $u \in H_p^2(G_R) \cap H_{p0}^1(G_R)$, *then* $u \in C_\mu^{h+2}(\bar{G}_R)$, *and*

$$'|||u|||_{\mu R}^{h+2} \leq C_2' |||Lu|||_{\mu R}^h, \quad \text{if} \quad 0 < R \leq R_2$$

and spt $u \subset B_R$ or $G_R \cup \sigma_R$, *respectively.*

Proof. We first show that P_R is a bounded transformation from $H^h_{p0}(B_{2R})$ to $H^{h+2}_p(B_{2R})$ and from $C^h_{\mu 0}(\bar{B}_{2R})$ to $C^{h+2}_\mu(\bar{B}_{2R})$. If $f \in C^\infty_c(B_{2R})$ and $u = P_R(f)$, then

$$\nabla^j u(x) = \int_{B_{2R}} K_0(x-\xi)\, \nabla^j f(\xi)\, d\xi, \quad 0 \leq j \leq h. \tag{5.6.8}$$

The result then follows from Theorem 3.7.1 for the H^{h+2}_p case and from Theorem 2.6.7 for the Hölder norm case.

If $f \in C^\infty(\bar{G}_{2R})$ and $\operatorname{spt} f \subset G_{2R} \cup \sigma_{2R}$, and if $u = P_R(f)$, then u is given by (5.6.5) so that

$$'\nabla^j u(x) = \int_{B_{2R}} K_0(x-\xi)\, '\nabla^j f^*(\xi)\, d\xi, \quad 0 \leq j \leq h,$$

where $'\nabla$ denotes the gradient with respect to the $(x^1, \ldots, x^{\nu-1})$ variables. This may be written in the form

$$'\nabla^j u = U_j - V_j, \quad U_j(x) = \int_{B_{2R}} K_0(x-\xi)\, \varphi_j(\xi)\, d\xi, \quad \varphi_j = {}'\nabla^j f^+,$$

$$V_j(x) = 2 \int_{G_R^-} K_0(x-\xi)\, \varphi_j(\xi)\, d\xi, \tag{5.6.9}$$

where f^+ is the extension of f by *positive* reflection:

$$f^+(x^\nu, x'_{\nu-1}) = f(-x^\nu, x'_{\nu-1}), \quad x^\nu < 0. \tag{5.6.10}$$

For each j, $\varphi_j \in L_p(B_{2R}) \cap C^0_\mu(\bar{B}_{2R})$ with

$$\|\varphi_j\|^0_{pR} \leq (2R)^{h-j}\, '\|f\|^h_{pR}, \quad '\|\!|\varphi_j|\!\|^0_{\mu R} \leq (2R)^{h-j}\, '\|\!|f|\!\|^h_{\mu R}. \tag{5.6.11}$$

The desired inequalities for U_j follow as before (with $h = 0$). The H^2_p results for V_j follow from Theorem 3.7.1. For $x^\nu > 0$, we have

$$\begin{aligned} \nabla' \nabla^2 V_j(x) &= 2 \int_{G_R^-} \nabla' \nabla^2 K_0(x-\xi)\, \varphi_j(\xi)\, d\xi \\ &= 2 \int_{R^-} \nabla' \nabla^2 K_0(x-\xi)\, [\varphi_j(\xi) - \varphi_j(x)]\, d\xi \end{aligned}$$

since the integral of $\nabla' \nabla^2 K_0(x-\xi)$ over R^-_ν converges absolutely to zero. Thus, using the fact that V_j is harmonic, we find that

$$|\nabla^3 V_j(x)| \leq C \cdot h_\mu(\varphi_j, \bar{B}_{2R}) \cdot (x^\nu)^{\mu-1}.$$

It follows from Theorem 2.6.6 that $V_j \in C^2_\mu(\bar{G}_{2R})$ with the desired bound. It follows easily that P_R is bounded in the G_{2R} cases also.

It is now easy to see, using (5.5.3) and (5.6.6), that

$$'\|T_R\|^{h+2}_{qR} \leq \varepsilon_1(R; \nu, h, q, E), \quad '\|\!|T_R|\!\|^{h+2}_{\mu R} \leq \varepsilon_2(R; \nu, h, \mu, E)$$

$$\lim_{R \to 0} \varepsilon_j(R; \ldots) = 0, \quad j = 1, 2.$$

Thus, if $R \leq R_1$, the norm of T_R in $H^2_p(\Gamma_{2R})$ and in $H^{h+2}_q(\Gamma_{2R})$ is $\leq 1/2$ ($\Gamma_{2R} = B_{2R}$ or G_{2R}), so there is a unique solution u_R of (5.6.6) in both of these spaces. Since the second space $\subset$ the first, these solutions must coincide. Since $u = u_R$, the result (a) follows. The proof of (b) is similar.

We can now prove the following a priori estimate:

Theorem 5.6.2. *Suppose that G is of class C_1^{h+1} and that the coefficients satisfy the h-p-conditions on $\bar{G}$ for some $p > 1$. Then $\exists$ a constant C_1, depending only on ν, h, p, E, and G such that*

$$(5.6.12) \quad \| u \|_{pG}^{h+2} \leq C_1 (\| L u \|_{pG}^h + \| u \|_{1G}^0), \quad u \in H_p^{h+2}(G) \cap H_{10}^1(G).$$

If G is of class C_μ^{h+2} and the coefficients satisfy the h-μ-conditions on $\bar{G}$, then there is a constant C_2, depending only on ν, μ, p, E, and G, such that

$$(5.6.13) \quad \||| u \|||_{\mu G}^{h+2} \leq C_2 (\||| L u \|||_{\mu G}^h) + \| u \|_{1G}^0), \quad u \in C_\mu^{h+2}(\bar{G}) \cap H_{10}^1(G).$$

Proof. We shall prove the second statement; the proof of the first is similar. Each point P of $\bar{G}$ is in a neighborhood on $\bar{G}$ which can be mapped as in § 5.5, by a mapping of class C_μ^{h+2}, onto either B_R or G_R in such a way that the coefficients of the transformed operator, still called L, satisfy the h-μ-conditions with $a^{\alpha\beta}(0) = \delta^{\alpha\beta}$, and R is small enough for the conclusions of Lemma 5.6.1 (b) to hold. Let $\{\zeta_s\}$, $s = 1, \ldots, S$, be a partition of unity in which each ζ_s has support in some such neighborhood. We let u_s be the transform of $\zeta_s u$ and L_s the transformed operator.

Now, suppose there were no constant C_2 satisfying (5.6.13). Then, there would be a sequence $\{u_n\} \in C_\mu^{h+2}(\bar{G}) \cap H_{10}^1(G)$ such that

$$(5.6.14) \qquad \||| u_n \|||_{\mu G}^{h+2} = 1, \quad \||| L u_n \|||_{\mu G}^h \to 0, \quad \| u_n \|_{1G}^0 \to 0$$

and we may assume that u_n, ∇u_n, and $\nabla^2 u_n$ converge uniformly to u, ∇u, and $\nabla^2 u$, respectively, where u must be zero on account of (5.6.14). Let u_{ns} be the transform of $\zeta_s u_n$. Then u_{ns} satisfies

$$(5.6.15) \qquad L_s u_{ns} = \zeta_s^* L_s u_n^* + M_s u_n^*,$$

where u_n^* is the transform of u_n and M_s involves only u_n^* and its first derivatives. From (5.6.14) and (5.6.15), it follows easily that $'\||| \zeta_s^* L_s u_n^* + M_s u_n^* \|||_{\mu R}^h \to 0$. But then $'\||| u_{ns} \|||_{\mu R}^{h+2} \to 0$ for each s. But then $\||| u_n \|||_{\mu G}^{h+2} \to 0$ contradicting (5.6.14).

We now prove the following differentiability theorem:

Theorem 5.6.3. (a) *Suppose that G is of class C_1^{h+1} and that the coefficients satisfy the $0 - p$ and $h - q$ conditions on G, where $h \geq 0$, $1 < p \leq q$, and either $h > 0$, $q > p$, or both. Then if $u \in H_p^2(G) \cap H_{p0}^1(G)$ and $L u \in H_q^h(G)$, it follows that $u \in H_q^{h+2}(G)$. If we know only that $u \in H_p^2(D)$ for each $D \subset\subset G$, it follows that $u \in H_q^{h+2}(D)$ for any such D.*

(b) *If G is of class C_μ^{h+2}, the coefficients satisfy the $0 - p$ and $h - \mu$ conditions on G, $u \in H_p^2(G) \cap H_{p0}^1(G)$, and $L u \in C_\mu^h(\bar{G})$, then $u \in C_\mu^{h+2}(\bar{G})$. If we know only that $u \in H_p^2(D)$ for each $D \subset\subset G$, we conclude that $u \in C_\mu^{h+2}(D)$ for such D. If G, the coefficients, and $L u \in C^\infty(\bar{G})$ then $u \in C^\infty(\bar{G})$ if $u \in H_p^2(G) \cap H_{p0}^1(G)$; if $u \in H_p^2(D)$ for $D \subset\subset G$, then $u \in C^\infty(G^{(0)})$.*

Proof. The C^∞ results in (b) follow from the C_1^{h+2} results. We prove (a) first. We first assume $q = p$, $h = 1$. We may assume the results and notations of the first paragraph of the proof of Theorem 5.6.2 where, however, we assume that the neighborhoods are small enough for the conclusions of Lemma 5.6.1 (a) to hold for $h = 0$ and for $h = 1$. Then u_s satisfies

$$L_s u_s = \zeta_s^* L_s u^* + M_s u^* \tag{5.6.16}$$

where M_s is of lower order as in (5.6.15). Thus $L_s u_s \in H_p^1(\Gamma_R)$ so that each $u_s \in H_p^3(\Gamma_R)$ so that $u \in H_p^3(G)$. A repetition of the argument proves the result for any h and the same p.

The argument for raising the exponent from p to $p' = \nu p/(\nu - p)$ (the Sobolev exponent) if $p < \nu$ is similar; since $M_s u^*$ is of lower order, $M_s u^* \in H_{p'}^h(\Gamma_R)$ if $u \in H_p^{h+2}(G)$. If $p \geq \nu$, then $M_s u^* \in H_q^h(\Gamma_R)$ for any q if $u^* \in H_p^{h+2}(\Gamma_R)$. A finite number of repetitions raises the exponent from p to any desired q. In part (b), we first raise the exponent to some $q > \nu$, at which time we can conclude that $M_s u^* \in C_\mu^h(\Gamma_R)$ if $u \in H_q^{h+2}(G)$.

The following special theorem and its method of proof are essentially due to NIRENBERG (see AGMON, DOUGLIS, and NIRENBERG [1] p. 693 and also §. 6.5 for a more general theorem).

Theorem 5.6.4. *Suppose that $G \in C_1^1$ and the coefficients satisfy the $0 - 2$-conditions $\bar{G}$. Then there exist numbers $\lambda_0 > 0$ and C, which depend only on ν, G, and E such that*

$$\|u\|_{2G}^2 \leq C \|L u - \lambda u\|_{2G}^0, \quad \lambda \text{ real}, \quad \lambda \geq \lambda_0, \quad u \in H_2^2(G) \cap H_{20}^1(G).$$

Proof. For any given $\eta_1 > 0$, each point x_0 is in a neighborhood or boundary neighborhood in which $|a(x) - a(x_0)| < \eta_1$. We choose a sequence ζ_s, $s = 1, \ldots, S$, such that each $\zeta_s \in C_1^1(\bar{G})$ with spt ζ_s in some one such neighborhood and $\zeta_1^2 + \cdots + \zeta_S^2 = 1$. We let $u_s = \zeta_s u$ and note that

$$\zeta_s L u = L u_s + M_s u$$

where each M_s is an operator of the first order. Then

$$-(L u, u)_2^0 = -\sum_{s=1}^{S} (\zeta_s L u, \zeta_s u)_2^0 = -\sum_s (L u_s, u_s)_2^0 + (M' u, u)_2^0 \tag{5.6.17}$$

where M' is of the first order. But

$$-(L u_s, u_s)_2^0 = -\int_G a_{0s}^{\alpha\beta} u_{s,\alpha\beta} u_s \, dx - \int_G (a^{\alpha\beta} - a_{0s}^{\alpha\beta}) u_{s,\alpha\beta} u_s \, dx -$$

$$-\int_G (b^\alpha u_{s,\alpha} + c u_s) u_s \, dx, \quad a_{0s}^{\alpha\beta} = a^{\alpha\beta}(x_{0s}) \tag{5.6.18}$$

where x_{0s} is in the small support of ζ_s. Since, for each s,

$$-\int_G a_{0s}^{\alpha\beta} u_{s,\alpha\beta} u_s \, dx = \int_G a_{0s}^{\alpha\beta} u_{s,\alpha} u_{s,\beta} \, dx \geq 0,$$

we see, by summing (5.6.18) with respect to s and using the facts that $u_s = \zeta_s$ u. etc., that

$$-(L u, u)_2^0 \geq -\int_G \left[\sum_s \zeta_s^2 (a^{\alpha\beta} - a_{0s}^{\alpha\beta}) u_{,\alpha\beta} u \, dx + (M u, u)_2^0\right.$$

where M is an operator of the first order. Thus

$$\begin{aligned} -(L u, u)_2^0 &\geq -\eta_1 \| \nabla^2 u \| \cdot \| u \| - C_1(\eta_1) (\| \nabla u \| + \| u \|) \cdot \| u \| \\ &\geq -2\eta_1 \| \nabla^2 u \| \cdot \| u \| - C(\eta_1) \cdot \| u \|^2. \quad (\| \varphi \| = \| \varphi \|_2^0). \end{aligned}$$

But, now for real λ,

$$\begin{aligned} \| L u - \lambda u \|^2 &= \| L u \|^2 + \lambda^2 \| u \|^2 - 2\lambda (L u, u) \\ &\geq \| L u \|^2 + \lambda^2 \| u \|^2 - 2\eta_1 (\| \nabla^2 u \|^2 + \lambda^2 \| u \|^2) - \\ &\quad - 2\lambda C(\eta_1) \| u \|^2 \geq \| L u \|^2 (1 - 2\eta_1 C_2) + \\ &\quad + [\lambda^2 (1 - 2\eta_1) - 2\lambda C(\eta_1) - 2\eta_1 C_2] \| u \|^2 \end{aligned}$$

where C_2 is related to the constant of Theorem 5.6.2. If we first choose η_1 so small that $(1 - 2\eta_1) \geq 1/2$ and $(1 - 2\eta_1 C_2) \geq 1/2$, we may then choose λ_0 so large that $\lambda^2/2 - 2\lambda C(\eta_1) - 2\eta_1 C_2 \geq 1/2$ for $\lambda \geq \lambda_0$. Then

$$\| L u - \lambda u \|^2 \geq \frac{1}{2} [\| L u \|^2 + \| u \|^2]$$

from which the theorem follows.

Theorem 5.6.5. *The conclusions of Theorem* 5.6.1 *hold under the hypotheses of Theorem* 5.6.4. *In fact no real $\lambda \geq \lambda_0$ is an eigenvalue. If G is of class C_μ^2 and the coefficients $\in C_\mu^0(\bar{G})$ with $c(x) \leq 0$, we may take $\lambda_0 = 0$.*

Proof. First, suppose $\lambda_1 \geq \lambda_0$. We may approximate to L by operators L_n whose coefficients satisfy our conditions uniformly with $a_n^{\alpha\beta}$ converging uniformly to $a^{\alpha\beta}$ on $\bar{G}$, each $a_n^{\alpha\beta} \in C_1^0(\bar{G})$. Define $L_{1n} = L_n - \lambda_1 I$ as an operator on $L_2(G)$ with domain all $u \in H_2^2(G)$ which vanish on ∂G. If, for some n, λ_1 were an eigenvalue for L_n, then L_{1n} would carry some non-zero element into 0 which would contradict Theorem 5.6.4. Thus λ_1 is not an eigenvalue for any n. Hence, if $f \in L_2(G)$, $L_{1n}^{-1}(f) = u_n$ is defined for each n and $\| u_n \|_2^2 \leq C \| f \|_2^0$ for all n. Hence a subsequence, still called $\{u_n\}$, $\rightharpoonup u$ in $H_2^2(G)$ and $u = 0$ on ∂G. Then $L_{1n} u_n \rightharpoonup L_1 u$ in L_2 (cf. the proof of Theorem 5.2.7), so that $L_1 u = f$.

Now, the equation $L u - \lambda u = f$ is equivalent to the equation

$$-L_0 u + (\lambda - \lambda_0) u = f \quad (L_0 = L - \lambda_0 I)$$

which is, in turn, equivalent to

$$u - (\lambda - \lambda_0) L_0^{-1} u = -L_0^{-1} f.$$

As an operator on $H_2^2 \cap H_{20}^1$, $L_0^{-1} u$ is compact, since weak convergence in H_2^2 implies strong convergence in L_2. The first part of the theorem follows from the Riesz theory of compact linear operators.

To prove the last statement, suppose $\lambda \geq 0$. If λ were an eigenvalue, then $L u - \lambda u = 0$ for some $u \in C^2_u(\bar{G})$ which vanishes on ∂G but is not $\equiv 0$. But since the coefficient of u in $L u - \lambda u$ is $c(x) - \lambda \leq 0$, this would contradict the maximum principle (Theorem 2.8.1).

5.7. Analyticity of the solutions of analytic linear equations

In this section, we present a simple proof of the analyticity of the solutions of (5.6.1) on the interior when the coefficients and f are analytic and show that solutions of such equations which vanish along an analytic part of the boundary can be continued analytically across the boundary. By using the mappings described in the proof of Theorem 5.6.2, which are analytic in these cases, the problem of local differentiability is reduced to that for solutions on spheres or hemispheres with $a_0^{\alpha\beta} = \Delta^{\alpha\beta}$. *The local differentiability theorems imply a global theorem in case the entire boundary ∂G is analytic.* The technique of proof is that given in a recent paper by MORREY and NIRENBERG; since the solutions are known already to be of class C^∞, it is sufficient to obtain bounds for the derivatives. In view of Lemma 5.7.2 below, it is sufficient to obtain bounds for their L_2-norms.

Lemma 5.7.1. *Suppose $u \in H^1_2(B_R)$, $u \in H^2_2(B_r)$ for $r < R$, and*

$$\Delta u = f \quad \text{on} \quad B_R. \tag{5.7.1}$$

Then there is a constant C_1 depending only on ν, such that

$$\int_{B_r} |\nabla^2 u|^2 \, dx \leq C_1 \left\{ \int_{B_{r+\delta}} f^2 \, dx + \int_{B_{r+\delta}} [\delta^{-2} |\nabla u|^2 + \delta^{-4} u^2] \, dx \right\} \tag{5.7.2}$$
$$0 < r < r + \delta < R, \quad 0 < \delta \leq r.$$

The lemma holds with B_R, B_r, and $B_{r+\delta}$ replaced by G_R, G_r, and $G_{r+\delta}$, respectively, and the spaces $H^1_2(B_R)$ and $H^2_2(B_r)$ replaced, respectively, by $H^{10}_2(G_R)$ and $H^{20}_2(G_R)$ ($u = 0$ along σ_R).

Proof. The second statement follows from the first by extending u and f to B_R by negative reflection. To prove the first, define η as usual with a replaced by δ, G by $B_{r+\delta}$, and D by B_r, and let

$$U = \eta \, u.$$

Then U has support in B_R, $U \in H^2_2(B_R)$, and U satisfies

$$\Delta U = \eta f + 2 \nabla \eta \cdot \nabla u + \Delta \eta \cdot u = F.$$

The result follows from Theorem 3.7.1.

Lemma 5.7.2. *Suppose $u \in C^\infty(B_R)$ [or $C^\infty(G_R \cup \sigma_R)$] and suppose that there are constants M and K such that*

$$\|\nabla^p u\|^0_{2,R} \leq M \cdot p! \, K^p R^{-p}, \quad p = 0, 1, 2, \ldots \tag{5.7.3}$$

Then u is analytic on $\bar{B}_R$ [or $\bar{G}_R$] and, in fact

$$|\nabla^p u(x)| \leq C_1(m, \nu) \cdot M \, p! \, (1.1 K)^p R^{-\tau-p}, \quad (\tau = \nu/2). \tag{5.7.4}$$

Proof. Applying the SOBOLEV lemma, Theorem 3.5.1, with $m = 1 + [\tau]$ ($\tau = \nu/2$) and $p = 2$, to $\nabla^q u$, we obtain

$$|\nabla^q u(x)| \le C R^{-\tau} \left\{ \sum_{j=0}^{m-1} \frac{(2R)^j}{j!} \|\nabla^{q+j} u\|_{2,R}^0 + (m-\tau)^{-1} \frac{(2R)^m}{(m-1)!} \|\nabla^{m+q} u\|_2^0 \right\} \tag{5.7.5}$$

where C is $\gamma_\gamma^{-1/2}$ or $2^{1/2} \gamma_\gamma^{-1/2}$, depending on whether we are considering B_R or G_R. Inserting the bounds (5.7.3) into (5.7.5), we obtain

$$|\nabla^p u(x)| \le C R^{-\tau} \cdot M \cdot R^{-p} K^p \left\{ \sum_{j=0}^{m-1} \frac{(2K)^j}{j!} (p+j)! + \right.$$

$$\left. + (m-\tau)^{-1} \frac{(2K)^m}{m-1!} (p+m)! \right\} \le C(m, \nu) \cdot M \cdot R^{-\tau-p} K^{m+p} \cdot p! \cdot \frac{(p+m)!}{p!\, m!}$$

from which (5.7.4) follows easily.

It is convenient to introduce the following notations:

Notations. For f and $u \in C^\infty(B_R)$, we define

$$\begin{aligned} M_{R,p}(f) &= (p!)^{-1} \sup_{R/2 \le r < R} (R-r)^{2+p} \|\nabla^p f\|_{2,r}^0, \quad p \ge 0 \\ N_{R,p}(u) &= [p!]^{-1} \sup_{R/2 \le r < R} (R-r)^{2+p} \|\nabla^{2+p} u\|_{2,r}^0, \quad p \ge -2 \\ [p!] &= p! \ \text{ if } \ p \ge 0, \quad [p!] = 1 \ \text{ if } \ p < 0. \end{aligned} \tag{5.7.6}$$

Remarks. From Lemma 5.7.2, it follows that the analyticity of u on B_R will follow from an inequality of the form

$$N_{R,p}(u) \le M \cdot K^p, \quad p \ge -2 \tag{5.7.7}$$

for some constants M and K. Our method of proof will consist in demonstrating (5.7.7).

Lemma 5.7.3. *Suppose u and $f \in C^\infty(B_R)$ and u satisfies (5.7.1) there. We suppose also that $M_{R,p}(f)$ and $N_{R,p}(u) < \infty$ for each p. Then there is a constant C_2, depending only on ν, such that*

$$N_{R,p}(u) \le C_2 [M_{R,p}(f) + N_{R,p-1}(u) + N_{R,p-2}(u)], \quad p \ge 0. \tag{5.7.8}$$

Proof. We apply Lemma 5.7.1 (for B_R) to each component of $\nabla^p u$, square, and add to obtain

$$\begin{aligned} [N_{R,p}(u)]^2 \le C_1 (p!)^{-2} \sup (R-r)^{4+2p} \Big[&\int_{B_{r+\delta}} |\nabla^p f|^2 \, dx + \\ &+ \int_{B_{r+\delta}} (\delta^{-2} |\nabla^{p+1} u|^2 + \delta^{-4} |\nabla^p u|^2) \, dx \Big]. \end{aligned} \tag{5.7.9}$$

We now obtain the following results using the notations

$$\int_{B_{r+\delta}} |\nabla^p f|^2 \, dx \le (p!)^2 M_{R,p}^2(f) \cdot (R-r-\delta)^{-(4+2p)} \tag{5.7.10}$$

$$\int_{B_{r+\delta}} |\nabla^{p+1} u|^2 \, dx \le \{[(p-1)!]\}^2 N_{R,p-1}^2(u) \cdot (R-r-\delta)^{-(2+2p)}, \quad p \ge 1$$

$$\int_{B_{r+\delta}} |\nabla^p u|^2 \, dx \le \{[(p-2)!]\}^2 N_{R,p-2}^2(u) \cdot (R-r-\delta)^{-2p}, \quad p \ge 2.$$

Setting

(5.7.11) $$\delta = (R - r)/(p + 1), \quad p \geq 1,$$

in (5.7.9) and (5.7.10), we obtain

$$[N_{R,p}(u)]^2 \leq C_1[(1 + p^{-1})^{4+2p} M^2_{R,p}(f) + (1 + p^{-1})^{4+2p} N^2_{R,p-1}(u) + \\ + (1 + p^{-1})^{2p}\{(p+1)^4/p^2(p-1)^2\} N^2_{R,p-2}(u)], \quad p \geq 2,$$

with analogous results in the cases $p = 1$ and $p = 0$ (taking $\delta = (R - r)/2$ in the latter case). The result follows.

Now, in order to prove the local differentiability, we write (5.6.1) (we absorb the term λu into $c u$) in the form

(5.7.12) $$\Delta u = F \equiv f - (a - a_0) \cdot \nabla^2 u - b \cdot \nabla u - c u.$$

In order to use Lemma 5.7.3, we shall obtain bounds for $M_{R,p}(F)$ in terms of $M_{R,p}(f)$ and the $N_{R,q}(u)$. As an aid in this work we state the following lemma, the proof of which is left to the reader:

Lemma 5.7.4. *Suppose V and W are tensors in $C^\infty(D)$. Then*

(5.7.13) $$|\nabla^p(V \cdot W)(x)| \leq \sum_{q=0}^{p} \binom{p}{q} |\nabla^{p-q} V(x)| \cdot |\nabla^q W(x)|, \quad x \in D.$$

Suggestion for the proof. By arranging the sets of common indices and the sets of remaining indices into single sequences, one may assume that

$$V = \{V_{ik}\}, \quad W = \{W_{jk}\}, \quad V \cdot W = \{\omega_{ij}\}, \quad \omega_{ij} = \sum_{k=1}^{q} V_{ik} \cdot W_{jk}, \\ i = 1, \ldots, n, \quad j = 1, \ldots, p.$$

Then, if α denotes the multi-index $(\alpha_1, \ldots, \alpha_\nu)$,

$$D^\alpha \omega_{ij} = \sum_{k=1}^{q} \sum_{\beta+\gamma=\alpha} D^\beta V_{ik} D^\gamma W_{jk}.$$

The result follows without much difficulty if one notes that

$$|\nabla^p V \cdot W|^2 = \sum_{\beta_1,\ldots,\beta_p=1}^{\nu} \sum_{i,j} |\omega_{ij,\beta_1,\ldots,\beta_p}|^2 = \sum_{i,j} \sum_{|\alpha|=p} \frac{|\alpha|!}{\alpha_1! \ldots \alpha_\nu!} |D^\alpha \omega_{ij}|^2.$$

Now, since a, b, and c are analytic, there are numbers $A \geq 2$, L, and $R_0 \leq 1$ such that

(5.8.14) $$\begin{aligned} &|\nabla^p f(x)|, \ |\nabla^p a(x)|, \ |\nabla^p b(x)|, \ |\nabla^p c(x)| \leq L\, p!\, A^p, \\ &|a(x) - a_0| \leq L\, A\, |x|, \end{aligned} \quad x \in B_{R_0}.$$

Theorem 5.7.1. *Suppose a, b, c, and f are analytic on $\bar{B}_{R_0}$ and satisfy* (5.7.14) *there, and suppose that $u \in H^2_2(B_{R_0})$ and is a solution of* (5.6.1) *there where $\lambda = 0$. Then there is an $R_7 < R_0$, which depends only on ν, A, L, and R_0, such that u is analytic in B_R for each $R \leq R_7$*

Proof. Suppose $0 < R < R_0$. From the results of §§ 5.5 and 5.6, it follows that $u \in C^\infty(\bar{B}_R)$ so that $N_{R,p}(u)$ is defined for every p. Obviously,

$M_{R,p}(f)$ is also defined for every p and, in fact,

$$(5.7.15)\qquad \begin{aligned} M_{R,p}(f) &= \sup_{R/2 \le r < R} (p!)^{-1} (R-r)^{2+p} \left\{ \int_{B_r} L^2 (p!)^2 A^{2p}\, dx \right\}^{\frac{1}{2}} \\ &\le Z_1(\nu) \cdot L\, R^{2+\tau} \cdot (A\, R)^p . \end{aligned}$$

Using (5.7.12), (5.7.14), and Lemma 5.7.4, we obtain

$$(5.7.16)\qquad \begin{aligned} |\nabla^p (F-f)| &\le L A R \cdot |\nabla^{p+2} u| + L\,|\nabla^{p+1} u| + L\,|\nabla^p u| + \\ &\quad + \sum_{q=1}^{p} \binom{p}{q} L A^q q! \left(|\nabla^{p-q+2} u| + |\nabla^{p-q+1} u| + |\nabla^{p-q} u|\right) \\ &= L A R\, |\nabla^{p+2} u| + L(1+pA)\, |\nabla^{p+1} u| + \\ &\quad + L \sum_{q=1}^{p+1} \{(p+1-q)!\}^{-1} L A^{q-1} \times \\ &\quad \times [1 + (p+1-q) A + (p+1-q)(p-q) A^2] \cdot |\nabla^{p+1-q} u| . \end{aligned}$$

Using the definition of $N_{R,p}(u)$, (5.7.15), (5.7.16), the fact that $A \ge 2$, and the fact that $M_{R,p}$ acts like a norm, we conclude that

$$(5.7.17)\qquad M_{R,p}(F) \le M_{R,p}(f) + L A R\, N_{R,p}(u) + 2L \sum_{q=1}^{p+2} (A\, R)^q N_{R,p-q}(u) .$$

Now if we use Lemma 5.7.3 with f replaced by F and choose $0 < R \le R_7$, where R_7 is chosen so that

$$C_2\, L\, A\, R_7 \le 1/2,$$

we conclude that

$$(5.7.18)\qquad \begin{aligned} N_{R,p}(u) &\le 2 C_2 Z_1 L\, R^{2+\tau} (A\, R)^p + 4 C_2 L \sum_{q=1}^{p+2} (A\, R)^q N_{R,p-q}(u) + \\ &\quad + 2 C_2 N_{R,p-1}(u) + 2 C_2 N_{R,p-2}(u) . \end{aligned}$$

We shall choose M and K so that (5.7.7) holds. Evidently, we can choose them so (5.7.7) holds for $p = -2$ and $p = -1$. Suppose (5.7.7) holds up to $p - 1$, $p \ge 0$. Then, from (5.7.18), we conclude that

$$\begin{aligned} N_{R,p}(u) &\le M K^p \cdot 2 C_2 \Big[M^{-1} Z_1 L\, R_7^{2+\tau} (A\, R_7/K)^p + K^{-1} + K^{-2} + \\ &\quad + 2L \sum_{q=1}^{p+2} (A\, R_7/K)^q \Big] \le M K^p \end{aligned}$$

for any $p \ge 0$ if we first choose R_7 so small that

$$2 C_2 M^{-1} Z_1 L\, R_7^{2+\tau} \le 1/2$$

(to take care of the case $p = 0$) and then choose K so large that

$$2 C_2 \left[K^{-1} + K^{-2} + 2L \sum_{q=1}^{\infty} (A\, R_7/K)^q \right] \le 1/2 .$$

In order to prove analyticity of u along σ_R when u vanishes there and f, a, b, and c are analytic near 0, we proceed in much the same way. But now, it is convenient to consider first only derivatives with respect to the x^α with $\alpha < \nu$ *since these all vanish along* σ_R so that Lemma 5.7.1 applies on G_R. We therefore denote x^ν by y, derivatives with respect to x^ν by D_y, and denote by ∇_x the gradient involving only derivatives with respect to the x^α with $\alpha < \nu$. As an aid in later estimates we state the following generalization of Lemma 5.7.4:

Lemma 5.7.4'. *Suppose V and W are tensors in $C^\infty(D)$. Then*

$$|\nabla_x^p(V\cdot W)| \leq \sum_{q=0}^{p} \binom{p}{q} |\nabla_x^q V| \cdot |\nabla_x^{p-q} W|,$$

$$|D_y^p \nabla_x^q (V\cdot W)| \leq \sum_{m=0}^{p}\sum_{n=0}^{q} \binom{p}{m}\binom{q}{n} |D_y^m \nabla_x^n V| \cdot |D_y^{p-m}\nabla_x^{q-n} W|,$$

and corresponding inequalities hold for

$$|D_y^p \nabla^q (V\cdot W)| \quad \text{and} \quad |\nabla_x^p \nabla^q (V\cdot W)|, \quad \text{etc.}$$

We now introduce the following notations:

Notations. For u and $f \in C^\infty(\bar{G}_R)$ with $u = 0$ along σ_R, we define

$$M'_{R,p}(f) = (p!)^{-1} \sup_{R/2 \leq r < R} (R-r)^{2+p} \|\nabla_x^p f\|_{2,r}^0$$

$$N'_{R,p}(u) = \begin{cases} (p!)^{-1} \sup\limits_{R/2\leq r<R} (R-r)^{2+p} \|\nabla^2 \nabla_x^p u\|_{2,r}^0, & p \geq 0 \\ \sup\limits_{R/2\leq r<R} (R-r)^{2+p} \|\nabla^{2+p} u\|_{2,r}^0, & p = -1, -2. \end{cases}$$

Lemma 5.7.3'. *Suppose u and $f \in C^\infty(\bar{G}_R)$, u satisfies (5.7.1) on G_R, and $u = 0$ along σ_R. There is a constant C_2', depending only on ν, such that*

$$N'_{R,p}(u) \leq C_2 [M'_{R,p}(f) + N'_{R,p-1}(u) + N'_{R,p-2}(u)].$$

Proof. We apply Lemma 5.7.1 for G_R to each component of $\nabla_x^p u$ and add, to obtain

$$\int_{G_r} |\nabla^2 \nabla_x^p u|^2\, dx \leq C_1 \Big\{ \int_{G_{r+\delta}} |\nabla_x^p f|^2\, dx + \int_{G_{r+\delta}} [\delta^{-2} |\nabla \nabla_x^p u|^2 + $$
$$+ \delta^{-4} |\nabla_x^p u|^2]\, dx \Big\}, \quad p \geq 0.$$

We note that ∇, ∇_x, and D_y all commute with one another and

$$|\nabla\nabla_x^p u|^2 \leq |\nabla^2 \nabla_x^{p-1} u|^2, \quad p\geq 1; \quad |\nabla\nabla_x^p u|^2 = |\nabla u|^2, \quad p = 0$$

$$|\nabla_x^p u|^2 \leq |\nabla^2\nabla_x^{p-2} u|^2, \quad p \geq 2; \quad |\nabla_x^p u| \leq |\nabla^p u|, \quad p = 0,1. \tag{5.7.19}$$

After taking (5.7.19) into account, the lemma is proved in exactly the same way as was Lemma 5.7.3.

In estimating derivatives of the form $D_y^{2+q} D_x^r u$, it is convenient to introduce the following notation:

Notation. $N'_{R,p,q}(u) = \sup\limits_{R/2\leq r<R} [(p+q)!]^{-1} (R-r)^{2+p+q} \|D_y^{q+2} \nabla_x^p u\|_{2,r}^0,$

$$u \in C^\infty(\bar{G}_R), \quad p \geq 0, \quad q \geq -2.$$

Theorem 5.7.1′. *Suppose a, b, c, and f are analytic on $\bar{G}_{R_0}$ and satisfy (5.7.14) there, and suppose $u \in H_2^{20}(G_{R_0})$ and is a solution of (5.6.1) there, where $\lambda = 0$. Then there is an $R_8 < R_0$, which depends only on ν, A, L, R_0, and corresponding quantities for certain ratios of the coefficients and f, such that u can be extended to be analytic in B_R for each $R \leq R_8$.*

Proof. As before, we know that $u \in C^\infty(\bar{G}_R)$ for each $R < R_0$. The proof proceeds as in that of Theorem 5.7.1, but estimating only the $\|\nabla^2 \nabla_x^p u\|_{2,r}^0$. The estimate (5.7.15) carries over without change. Instead of (5.7.16) we obtain

$$(5.7.20)\qquad \begin{aligned} |\nabla_x^p(F-f)| &\leq L A R |\nabla^2 \nabla_x^p u| + L|\nabla \nabla_x^p u| + L|\nabla_x^p u| + \\ &+ \sum_{q=1}^{p} \binom{p}{q} L A^q q! (|\nabla^2 \nabla_x^{p-q} u| + |\nabla \nabla_x^{p-q} u| + |\nabla_x^{p-q} u|). \end{aligned}$$

By using (5.7.19) and proceeding as before, we obtain first the estimate (5.7.17) with $M_{R,p}$ and $N_{R,p}$ replaced by $M'_{R,p}$ and $N'_{R,p}$, respectively. As before we then obtain

$$(5.7.21)\qquad N'_{R,p}(u) \leq M_1 K_1^p$$

which gives estimates for all derivatives $D_y^q \nabla_x^p u$ where $q \leq 2$.

In order to estimate the remaining derivatives[1], we note that we can solve 5.6.1 for $D_y^2 u$ in terms of the other derivatives.

$$(5.7.22)\qquad D_y^2 u = g + \sum_{\alpha=1}^{\nu-1} B^\alpha D_y D_\alpha u + \sum_{\alpha,\beta=1}^{\nu-1} C^{\alpha\beta} D_\alpha D_\beta u$$

in which g, the B^α, and $C^{\alpha\beta}$ are analytic on $\bar{G}_{R_0}$. We shall assume that g and the B^α and $C^{\alpha\beta}$ satisfy estimates of the form

$$(5.7.23)\qquad |D_y^q \nabla_x^p g|,\ |D_y^q \nabla_x^p B|,\ |D_y^q \nabla_x^p C| \leq p!\, q!\, L A^{p+q} \text{ in } \bar{G}_{R_0}.$$

We shall show that if R_8 is small enough, then

$$(5.7.24)\qquad N'_{R,p,q}(u) \leq \bar{M} \bar{K}^{p+q} \theta^p,\quad p \geq 0,\quad q \geq -2,\quad 0 < R \leq R_8,$$

where $\bar{M}$, $\bar{K}$, θ are fixed constants with $\bar{K} \geq 1$ and $\theta \leq 1/2$. Differentiating (5.7.22) and using the bounds (5.7.23) and Lemma 5.7.4′, we find that

$$(5.7.25)\qquad \begin{aligned} &|D_y^{2+} \nabla_x^p u| \leq p!\, q!\, L A^{p+q} + \\ &+ L \sum_{m=0}^{p} \sum_{n=0}^{q} \binom{p}{m}\binom{q}{n} m!\, n!\, A^{m+n} [|D_y^{q+1-n} \nabla_x^{p-m+1} u| + |D_y^{q-n} \nabla_x^{p-m+2} u|]. \end{aligned}$$

Noticing that $N'_{R,p,q}(u)$ acts like a norm, we next conclude that

$$(5.7.26)\qquad \begin{aligned} &N'_{R,p,q}(u) \leq L Z_2(\nu) R^{\tau+2} (A R)^{p+q} + \\ &+ L \sum_{m=0}^{p} \sum_{n=0}^{q} \binom{p}{m}\binom{q}{n} m!\, n!\, (p+q-m-n)!\, (A R)^{m+n} [(p+q)!]^{-1} \times \\ &\times [N'_{R,p-m+1,q-n-1}(u) + N'_{R,p-m+2,q-n-2}(u)]. \end{aligned}$$

[1] The remainder of this proof could be replaced by a dominating function argument like that used to prove the Cauchy-Kovalevsky Theorem. Or a theorem of Holmgren [1] may be employed.

Now, (5.7.24) will hold for $q = -2, -1$, and 0 and all $p \geq 0$ if R is small enough for (5.7.21) to hold and

$$\bar{M}\bar{K}^{p+q}\theta^p \geq M_1 K_1^{p+q}, \quad p \geq 0, \quad q = 0, -1, -2, \tag{5.7.27}$$

since it follows from the definitions that

$$N'_{R,p,q}(u) \leq N'_{R,p+q}(u) \quad \text{if} \quad p \geq 0, \quad q \leq 0. \tag{5.7.28}$$

Let us suppose, then that (5.7.24) holds for all q less than some positive integer, which we again denote by q, and all $p \geq 0$. Then, using (5.7.26) and (5.7.24) and the fact that, for $p \geq 0$ and $q \geq 0$, the factorial coefficient of each term in the double sum is ≤ 1, we conclude that

$$\begin{aligned} \bar{M}^{-1}\bar{K}^{-p-q}\theta^{-p}N'_{R,p,q}(u) &\leq \bar{M}^{-1}L\,R^{\tau+2}\left(\frac{A\,R}{\bar{K}\,\theta}\right)^p\left(\frac{A\,R}{\bar{K}}\right)^q + \\ &+ L(\theta+\theta^2)\sum_{m=0}^{p}\sum_{n=0}^{q}\left(\frac{A\,R}{\bar{K}\,\theta}\right)^m\left(\frac{A\,R}{\bar{K}}\right)^n. \end{aligned} \tag{5.7.29}$$

The right side of (5.7.29) will be ≤ 1 always ($R \leq R_8$) and (5.7.27) will hold if we choose R_8, θ, $\bar{K}$, and $\bar{M}$ in that order so that

$$L\,R_8^{\tau+2} \leq 1/2, \quad \theta = \min(1/2, 1/16L),$$
$$\bar{K}\theta = \max(K_1, 2A\,R_8), \quad \bar{M} = \max[1, M_1(\bar{K}/K_1)^2].$$

5.8. Analyticity of the solutions of analytic, non-linear, elliptic equations

In order to conclude the analyticity of the solutions of variational problems and of the quasi-linear equations mentioned in the introduction and treated in §§ 5.1—5.6, it is necessary to give a separate proof for the non-linear case since the methods of the preceeding section do not generalize immediately. The analyticity of the solutions of a single *linear* analytic elliptic equation in two variables was proved by HADAMARD in 1890. The analyticity of any solution $u(x, y)$ of class C^3 of a single *non-linear* analytic elliptic equation with $\nu = 2$ was first proved in a famous memoir of S. BERNSTEIN [1] in 1904. Other proofs of Bernstein's theorem were given by M. GEVREY, T. RADO [1], H. LEWY [1], and by BERNSTEIN himself [2]. In 1932, E. HOPF [3] proved the corresponding theorem for a single equation but with ν arbitrary. In 1939, PETROWSKY proved the theorem for systems which are almost as general as the most general elliptic systems yet considered (see §§ 6.2, 6.3). In 1958, MORREY [12] and A. FRIEDMAN [1] proved the theorem almost simultaneously but by different methods for the most general elliptic systems; analyticity at the boundary was proved for the solutions of strongly elliptic (see § 6.5) systems.

The methods of BERNSTEIN, GEVREY, RADÒ, and FRIEDMAN all involve estimating the magnitudes of the successive derivatives; by that method, FRIEDMAN was able to obtain corresponding results for quasi-analytic systems involving the quasianalytic classes of MANDELBROJT. Those of LEWY, HOPF, PETROWSKY, and MORREY all involved extension to the complex domain. LEWY used his theory of hyperbolic equations to carry out the extension and HOPF extended certain generalized potentials to the complex domain, a procedure which had been used by E. E. LEVI. The proof of the writer combines this idea with a simple type of implicit function theorem in a Banach space and the idea of § 5.5. We present this proof in this section. The proofs for more general systems will be sketched in § 6.7.

We consider a solution u of an equation of the form

$$\varphi(x, D^\alpha u) = 0, \quad 0 \leq |\alpha| \leq 2, \quad \alpha = (\alpha_1, \ldots, \alpha_\nu) \tag{5.8.1}$$

in which φ is analytic in its arguments (x, r_α) for x near x_0 and r_α near $D^\alpha u\,(x_0)$ and is real for real values, and the equation is elliptic at this point; that is, the linear equation

$$\sum_\alpha \varphi_{r_\alpha}(x_0, r_0) \cdot D^\alpha v = 0 \tag{5.8.2}$$

is elliptic. By a translation of axes, we may assume that $x_0 = 0$. Next, let $Q(x)$ be that quadratic polynomial such that

$$\nabla^k Q(0) = \nabla^k u(0), \quad k = 0, 1, 2. \tag{5.8.3}$$

Making the substitution

$$u = Q + v \tag{5.8.4}$$

in (5.8.1), we obtain a new equation of the same type in which $\nabla^k v(0) = 0$, $k = 0, 1, 2$. If we expand the new φ about the origin, we obtain a new equation which we can write in the form

$$a^{\alpha\beta} v_{,\alpha\beta} = b_1^\alpha v_{,\alpha} + c_1 v + \psi(x, D v) \quad (a^{\alpha\beta}, b_1^\alpha, c_1 \text{ const.}).$$

Finally, by making an affine transformation of the x space, we obtain (writing u again)

$$\Delta u = M u + \psi(x, D u), \quad M u = b^\alpha u_{,\alpha} + c u. \tag{5.8.5}$$

The Taylor expansion of ψ starts with a homogeneous linear function of x which is followed by terms in the x^α and r_α of second and higher orders.

We first suppose u defined and $\in C_\mu^2(\bar{B}_{R_0})$ and introduce the idea of § 5.5 in which we write

$$u = u_R + H_R, \quad u_R = P_R[M u + \psi(x, D u)], \quad H_R = u - u_R \tag{5.8.6}$$

where we make the following definitions:

Definition 5.8.1. The *space* $*C_0^\mu(B_R)$ consists of all $f \in C_\mu^0(\bar{B}_R)$ for which $f(0) = 0$, the norm $\|f\|_0$ being $h_\mu(f, B_R)$. The *space* $*C_\mu^2(B_R)$ consists of all $u \in C_\mu^2(B_R)$ for which $u(0) = \nabla u(0) = \nabla^2 u(0) = 0$, the norm $\|u\|_2$ being $h_\mu(\nabla^2 u, B_R)$. For $f \in *C_\mu^0(B_R)$, we define $P_R(f)$ as the potential of f minus the quadratic polynomial satisfying (5.8.3).

From the results in § 2.6, we obtain the following theorem:

Theorem 5.8.1. *P_R is a bounded operator from $*C^0_\mu(B_R)$ to $*C^2_\mu(B_R)$ with bound independent of R. If $F = P_R(f)$, then*

$$\Delta F = f. \tag{5.8.7}$$

Thus, in (5.8.6), H_R is harmonic.

Proof. The only thing not yet proved is (5.8.7). But, from its definition, $\Delta F(x) = f(x) + C$, C a constant. Substituting $x = 0$ gives $C = 0$.

Now, as in § 5.5, we regard H_R as known and shall try to find an equation for u_R. To that end, we define

$$\begin{aligned} v_R &= P_R[M\,H_R + \psi(x, D\,H_R)] \\ T_R(u_R; H_R) &= P_R[M\,u_R + \psi(x, D\,u_R + D\,H_R) - \psi(x, D\,H_R)]. \end{aligned} \tag{5.8.8}$$

We easily obtain the following theorem:

Theorem 5.8.2. *Suppose u is a solution $\in *C^2_u(B_{R_0})$ of (5.8.5) and suppose u_R, v_R, H_R, and T_R are defined as above. Then $\|u_R\|_2$, $\|v_R\|_2$, and $\|H_R\|_2$ are uniformly bounded by some number M_2 for $0 < R \le R_0$,*

$$T_R(0; H_R) = 0, \quad \lim_{R\to 0} \|v_R\|_2 = \lim_{R\to 0} \|u_R\|_2 = 0$$

$$\begin{aligned} &\|T_R(U_{1R}, H_R) - T_R(U_{2R}, H_R)\|_2 \le \varepsilon(R)\,\|U_{1R} - U_{2R}\|_2, \\ &\lim_{R\to 0} \varepsilon(R) = 0 \quad \text{if} \quad \|H_R\|, \|U_{1R}\|, \|U_{2R}\| \le M_2. \end{aligned} \tag{5.8.9}$$

Thus, if $0 < R \le R_1$, u_R is the only solution with $\|u_R\|_2 \le M_2$ of the equation

$$u_R - T_R(u_R; H_R) = v_R, \quad \|H_R\|_2 \le M_2, \quad H_R \text{ given.} \tag{5.8.10}$$

Proof. Obviously $\|u\|_2 \le L$ for $0 < R \le R_0$. Let $f = M\,u + \psi(x, D\,u)$. Then, for x, x_1, and x_2 on $\bar{B}_R$, we have

$$|\nabla^2 u(x)| \le L R^\mu, \ |\nabla u(x)| \le L R^{1+\mu}, \ |u(x)| \le L R^{2+\mu} \ \left(\mu \le \tfrac{1}{2}\right) \tag{5.8.11}$$

$$\begin{aligned} &|f(x_2) - f(x_1)| \le |b| \cdot |\nabla u(x_2) - \nabla u(x_1)| + |c| \cdot |u(x_2) - u(x_1)| + \\ &+ |\nabla_x \bar{\psi}| \cdot |x_2 - x_1| + |\nabla_r \bar{\psi}| \cdot \left[\sum_{k=0}^{2} |\nabla^k u(x_2) - \nabla^k u(x_1)|\right] \\ &\le Z_1 R^\mu \cdot |x_2 - x_1|^\mu \end{aligned} \tag{5.8.12}$$

where $|\nabla_x \bar{\psi}|$ and $|\nabla_r \bar{\psi}|$ are bounds for the indicated gradients for $x \in \bar{B}_R$ and the r_α in the range defined by (5.8.11); we know that $|\nabla_x \bar{\psi}|$ is bounded and $|\nabla_r \bar{\psi}| \le C R^\mu$ in that range. Thus $\|f\|_1 \to 0$ like R^μ and it follows that $\|u_R\|_2 \to 0$. Thus $\|H_R\|_2 \le L + \|u_R\|_2$ so $\|v_R\|_2 \to 0$. Obviously $T_R(0; H_R) = 0$. To get the other bound on T_R, we note that

$$\begin{aligned} &T_R(U_{1R}, H_R) - T_R(U_{2R}, H_R) \\ &= P_R[M(U_{1R} - U_{2R}) + \psi(x, D\,U_{1R} + D\,H_R) - \psi(x, D\,U_{2R} + D\,H_R)] \end{aligned}$$

and estimate the bracket as in (5.8.11) and (5.8.12) above. If we assume that R_1 is chosen so small that $\varepsilon(R) \le 1/2$ and $\|v_R\|_2 < M_2/2$, then it follows that the equation (5.8.10) has only one solution u_R with $\|u_R\|_2 \leq M_2$.

Now, after the manner of § 5.5, we wish to introduce certain spaces of *analytic* functions, which we shall denote by $^*C^0_\mu(B_{hR})$ and $^*C^2_\mu(B_{hR})$ and which are subspaces of $^*C^0_\mu(B_R)$ and $^*C^2_\mu(B_R)$, respectively, such that every harmonic function in $^*C^2_\mu(B_R)$ is also in $^*C^2_\mu(B_{hR})$ and P_R is a bounded operator from $^*C^0_\mu(B_{hR})$ to $^*C^2_\mu(B_{hR})$ as above. The functions in these new spaces can be extended to be analytic on the domains B_{hR} in the complex x-space which are defined below. Given f defined in some domain, we have to show how to extend its potential as given by (2.5.2) to that domain and verify the other properties.

Now $K_0(y)$ is a function of $|y|$ only and, for real y,

$$|y| = \left[\sum_{\alpha=1}^{\nu} (y^\alpha)^2\right]^{\frac{1}{2}}.$$

If, now, we consider complex vectors $y = y_1 + i\, y_2$ (y_1 and y_2 real vectors in R_ν), we see that

$$\sum_{\alpha=1}^{\nu} (y^\alpha)^2 = |y_1|^2 - |y_2|^2 + 2i\, y_1 \cdot y_2. \tag{5.8.13}$$

If we choose h, $0 < h < 1$, and assume that

$$|y_2| \leq h\,|y_1| \quad (y_1 \neq 0) \tag{5.8.14}$$

then

$$\left|\sum_{\alpha=1}^{\nu} (y^\alpha)^2\right| \geq Re \sum_{\alpha=1}^{\nu} (y^\alpha)^2 = |y_1|^2 - |y_2|^2 \geq (1-h^2)\,|y_1|^2. \tag{5.8.15}$$

Consequently, it is clear that we can extend $K_0(y)$ uniquely to be analytic in any region of the type (5.8.14); one encounters branch point trouble for $h \geq 1$.

In extending the formula (2.5.2) to complex values of x, it evidently will not do simply to let x be complex, keeping ξ real, because if $x = x_1 + i\, x_2$ and $x_2 \neq 0$, we note that

$$x - \xi = (x_1 - \xi) + i\, x_2$$

and $|x_2|$ is not $\leq h \cdot |x_1 - \xi|$ for ξ near x_1. Thus we must allow ξ to range over some ν-dimensional "surface" which evidently must pass through x, therefore depending on x, and must have ∂B_R as its boundary. Consequently, in order for $(x - \xi)$ to be in a domain (5.8.14) for ξ on such a surface, x must be restricted to a set of $x_1 + i\, x_2$ such that

$$|x_2| < h\,|x_1 - \xi| \quad \text{for all } \xi \text{ on } \partial B_R. \tag{5.8.16}$$

Since the minimum of the right side of (5.8.16) is just $R - |x_1|$, we see that $x = x_1 + i\, x_2$ must satisfy

$$|x_2| < h(R - |x_1|). \tag{5.8.17}$$

Next, any surface $S(x)$ over which we integrate must have the property that if $\xi = \xi_1 + i\,\xi_2 \in S(x)$, then $|x_2 - \xi_2| \leq h\,|x_1 - \xi_1|$. We may describe certain such surfaces by

$$S(x) : \xi = \xi(s; x), \quad |\xi_2(s, x) - x_2| \leq h\,|\xi_1(s, x) - x_1|, \quad s \in \bar{B}_R.$$

Finally, the integral over such a surface is to be interpreted property. We begin by defining the domains B_{hR} and the function spaces we shall use:

Definition 5.8.2. The domain B_{hR} in complex ν-space consists of all complex $x = x_1 + i\,x_2$ which satisfy (5.8.17). The *space* $^*C^0_\mu(B_{hR})$ consists of all $f \in {}^*C^0_\mu(B_R)$ which can be extended to be (single-valued) $\in C^0_\mu(\overline{B_{hR}})$ and analytic in B_{hR}, the norm being $h_\mu(f, B_{hR})$ (i.e. the Hölder constant on $\bar{B}_{hR}$). The *space* $^*C^2_\mu(B_{hR})$ consists of all $u \in {}^*C^2_\mu(B_R)$ which have similar extensions to $\bar{B}_{hR}$, the norm being $h_\mu(\nabla^2 u, B_{hR})$.

We now define our surface integrals.

Definition 5.8.3. Suppose $\mathfrak{F}$ is defined, and continuous, and complex-valued on a set D in complex ν-space, suppose $S : \xi = \xi(s)$, where $\xi(s)$ is a Lipschitz vector defined for $s \in \bar{G}$ where G is a Lipschitz domain in ν-space, and suppose that $\xi(s) \in D$ for $s \in \bar{G}$ except possibly at $s = s_0 \in G$ and $\mathfrak{F}[\xi(s)]$ is continuous except possibly at s_0. Then we say that S is *admissible* with respect to $\mathfrak{F}$ and define the integral

$$\int_S \mathfrak{F}(\xi)\, d\xi = \int_G \mathfrak{F}[\xi(s)]\, J(s)\, ds, \quad J = \frac{\partial(\xi^1, \ldots, \xi^\nu)}{\partial(s^1, \ldots, s^\nu)} \tag{5.8.18}$$

in case it is absolutely convergent.

Theorem 5.8.3. *Suppose* $f \in {}^*C^0_\mu(B_{hR})$, $x_0 \in B_{hR}$, *and* $S : \xi = \xi(s)$, $s \in \bar{B}_R$ *is admissible with respect to the function*

$$\mathfrak{F}(\xi; x_0) = K_0(x_0 - \xi)\, f(\xi) \tag{5.8.19}$$

and ξ *satisfies the additional conditions*

$$\xi_1(s) = s, \quad s \in \bar{B}_R, \quad \xi_2(s) = 0, \quad s \in \partial B_R. \tag{5.8.20}$$

Then the integral (5.8.18), *with* $\mathfrak{F}$ *defined by* (5.8.19), *exists. If* ξ *and* ξ^* *are both admissible and satisfy* (5.8.20), *the integrals have the same value. Finally, if for each* $x \in B_{hR}$, $S(x) : \xi = \xi(s; x)$ *is admissible with respect to* $\mathfrak{F}(\xi; x)$ *as defined in* (5.8.19) *and* $\xi(s; x)$ *satisfies* (5.8.20) *for each* x, *then the function* F *defined by*

$$F(x) = \int_{S(x)} K_0(x - \xi)\, f(\xi)\, d\xi \tag{5.8.21}$$

is analytic on B_{hR} *and*

$$F_{,\alpha}(x) = \int_{S(x)} K_{0,\alpha}(x - \xi)\, f(\xi)\, d\xi. \tag{5.8.22}$$

Remarks. We notice that in order for $S: \xi = \xi(s)$ to be admissible with respect to $\mathfrak{F}(\xi; x_0)$, ξ must satisfy also

$$\begin{aligned} &\xi_2(x_{10}) = x_{20}, \quad |\xi_2(s) - x_{20}| \leq h\,|s - x_{10}| \\ &|\xi_2(s)| < h(R - |s|). \end{aligned} \qquad s \in B_R \tag{5.8.23}$$

Proof. From the definition of K_0 and from (5.8.15), (5.8.20), and (5.8.23), and the fact that ξ is Lipschitz, it follows that $|f[\xi(s)]|$ and $|J(s)|$ are bounded and K_0 satisfies

$$\begin{aligned} &|K_0[x_0 - \xi(s)]| \leq Z_1 \cdot (1 - h^2)^{1-\nu/2}\,|s - x_{10}|^{2-\nu} \\ &|\nabla K_0[x_0 - \xi(s)]| \leq Z_2 \cdot (1 - h^2)^{(1-\nu)/2} \cdot |s - x_{10}|^{1-\nu} \end{aligned} \qquad (\nu > 2) \tag{5.8.24}$$

so that the integrals (5.8.21) and (5.8.22) converge absolutely. There is an obvious modificiation if $\nu = 2$.

If ξ and ξ^* are two admissible vectors $\in C^\infty(\bar{B}_R - \{x_{10}\})$, let $\xi(s,t) = (1-t)\,\xi(s) + t\,\xi^*(s)$ and define

$$\varphi(t, \varrho) = \int\limits_{B_R - B(x_{10}, \varrho)} K_0[x_0 - \xi(s,t)]\, f[\xi(s,t)] \cdot J(s,t)\, ds. \tag{5.8.25}$$

Differentiation with respect to t yields

$$\begin{aligned} \varphi_t(t,\varrho) &= \int\limits_{B_R - B(x_{10},\varrho)} \left\{ \sum_{\gamma=1}^{\nu} \mathfrak{F}_{\xi^\gamma}(\xi, x_0)\, \xi_t^\gamma \cdot J + \mathfrak{F}\, J_t \right\} ds \\ &= \int\limits_{B_R - B(x_{10},\varrho)} \sum_{\gamma=1}^{\nu} \frac{\partial(\xi^1, \ldots, \xi^{\gamma-1}, \mathfrak{F}\,\xi_t^\gamma\, \xi^{\gamma+1}, \ldots, \xi^\nu)}{\partial(s^1, \ldots, s^\nu)}\, ds, \\ \mathfrak{F}(\xi, x) &= K_0(x - \xi)\, f(\xi), \end{aligned} \tag{5.8.26}$$

as is seen by expanding the determinant with respect to the γ-th column and using the relations

$$\partial(\mathfrak{F}\,\xi_t^\gamma)/\partial s^\delta = \xi_t^\gamma\, \mathfrak{F}_{\xi^\varepsilon}\, \xi^\varepsilon_{s^\delta} + \mathfrak{F}\, \partial^2 \xi^\gamma / \partial s^\delta\, \partial t.$$

One sees in turn by using Lemma 4.5.6 and the proof of Lemma 4.5.7 (for complex-valued functions) that

$$\varphi_t(t,\varrho) = -\sum_{\gamma,\delta=1}^{\nu} (-1)^{\gamma+\delta} \int\limits_{\partial B(x_{10},\varrho)} \mathfrak{F} \cdot \xi_t^\gamma\, \frac{\partial \xi'_\gamma}{\partial s'_\delta}\, ds'_\delta. \tag{5.8.27}$$

From (5.8.24), (5.8.25) and (5.8.27), we see that $\varphi(t,\varrho) \to \varphi(t,0)$ and $\varphi_t(t,\varrho) \to 0 = \varphi_t(t,0)$ uniformly, so the integrals have the same value for ξ and ξ^*. If ξ and ξ^* are merely Lipschitz it is easy to see that one can approximate to each by C^∞ admissible functions.

To see the analyticity, pick $x_0 = x_{10} + i\,x_{20} \in B_{hR}$. It is easy to construct a family $S(x)$ of admissible surfaces $\xi = \xi(s,x)$ which $\in C^\infty$ as above in both s and x for x near x_0. Let us define

$$F(x,\varrho) = \int\limits_{B_R - B(x_1,\varrho)} \mathfrak{F}[\xi(s,x); x]\, J(s,x)\, ds,$$

where $\mathfrak{F}$ is defined in (5.8.26). Clearly for each sufficiently small $\varrho > 0$, $F(x, \varrho) \in C^1$ in x near x_0 and we obtain

$$(5.8.28)\quad \begin{aligned} F_{x_1\gamma}(x, \varrho) &= \int_{B_R - B(x_1,\varrho)} K_{0,\gamma}[x - \xi(s,x)] f[\xi(s,x)]\, ds - \\ &- \int_{\partial B(x_1,\varrho)} \mathfrak{F} \cdot J\, ds'_\gamma - \int_{\partial B(x_1,\varrho)} \sum_{\delta,\varepsilon=1}^{\nu} (-1)^{\delta+\varepsilon} \mathfrak{F}\, \xi^\delta_{x_1\gamma} (\partial \xi'_\delta / \partial s'_\varepsilon)\, ds'_\varepsilon \\ F_{x_2\gamma}(x, \varrho) &= i \int_{B_R - B(x_1,\varrho)} K_{0,\gamma}[x - \xi(s,x)] f[\xi(s,x)]\, ds - \\ &- \int_{\partial B(x_1,\varrho)} \sum_{\delta,\varepsilon=1}^{\nu} (-1)^{\delta+\varepsilon} \mathfrak{F}\, \xi^\delta_{x_2\gamma} (\partial \xi'_\delta / \partial s'_\varepsilon)\, ds'_\varepsilon \end{aligned}$$

by making use of the devices used in (5.8.26) and (5.8.27). Using the bounds (5.8.24), we see that $F(x, \varrho)$, $F_{x_1\gamma}(x, \varrho)$, and $F_{x_2\gamma}(x, \varrho)$ converge uniformly to F, $F_{x_1\gamma}$, and $F_{x_2\gamma}$ and that $F_{x_2\gamma} = i F_{x_1\gamma}$ so that F is analytic with derivative given by (5.8.22).

Definition 5.8.4. For $f \in {}^*C^0_\mu(B_{hR})$, we define $U = P_R(f)$ by the formula

$$(5.8.29)\qquad U(x) = -Q(x) + \int_{S(x)} K_0(x - \xi) f(\xi)\, d\xi, \quad x \in B_{hR}$$

where $S(x)$ is admissible for each x and Q is the quadratic polynomial satisfying (5.8.3).

We must now estimate $h^*_\mu(\nabla^2 U, B_{hR})$ when $U = P_R(f)$ and $f \in {}^*C^0_\mu(B_{hR})$; clearly we may assume, for this purpose, that U is given by the integral in (5.8.29). As in § 2.6, we estimate $\nabla^3 U(x)$ and then use a lemma like Theorem 2.6.6 which is included in the lemma we state and prove after introducing the following convenient notations:

Definition 5.8.5. For $x = x_1 + i x_2 \in B_{hR}$, we let $\delta(x)$ be the distance of x from ∂B_{hR} and $r(x)$ be the distance in the plane $x_2 = \text{const.}$ of x from the intersection of that plane with ∂B_{hR}. We define $S_1(x)$ to be the surface defined by

$$S_1(x):\quad \begin{aligned} &\xi_1 = \xi_{11}(s, x) = s, \quad s \in B_R - B(x_1, r), \quad r = r(x)/2 \\ &\xi_2 = \xi_{12}(s, x) = \begin{cases} (R - |x_1| - r - |s - x_1|)\, x_2/(R - |x_1| - 2r) \\ \qquad r \le |s - x_1| \le R - |x_1| - r \\ 0, \qquad s \in B_R - B(x_1, R - |x_1| - r). \end{cases} \end{aligned}$$

Lemma 5.8.1. (a) *If* $x = x_1 + i x_2 \in B_{hR}$, *then*

$$r(x) = R - |x_2|/h - |x_1|, \quad \delta(x) = h\, r(x)/\sqrt{1 + h^2}$$

(b) B_{hR} *is convex*

(c) *Any two points* z_1 *and* z_2 *of* B_{hR} *can be joined by a path* $x = x(s)$, $0 \le s \le l$, *in* B_{hR} *such that*

$$\int_0^l \{\delta[x(s)]\}^{\mu-1}\, ds \le C(\nu, \mu, h) \cdot |z_1 - z_2|^\mu.$$

Proof. (b) is evident. To prove the formula for $r(x)$, we note that $(\xi_1, x_2) \in B_{hR} \Leftrightarrow |x_2| < h(R - |\xi_1|)$ and $|\xi_1| < R \Leftrightarrow |\xi_1| < R - |x_2|/h$. To prove that for $\delta(x)$ we note that $(\xi_1, \xi_2) \in \partial B_{hR} \Leftrightarrow$

$$|\xi_1| \leq R, \quad |\xi_2| = h(R - |\xi_1|) \tag{5.8.30}$$

and the minimum of

$$|\xi_1 - x_1|^2 + |\xi_2 - x_2|^2$$

for all (ξ_1, ξ_2) satisfying (5.8.30) is just that of

$$(r - |x_1|)^2 + [h(R - r) - |x_2|]^2 \quad \text{for } 0 \leq r \leq R$$

which is just the square of the distance in R_2 from the point $(|x_1|, |x_2|)$ to the line $\zeta^2 = h(R - \zeta^1)$, (ζ^1, ζ^2) being the coordinates in R_2.

Using (a), it follows that the distance of $\bar{B}_{h,R-r}$ from ∂B_{hR} is $\geq hr/\sqrt{1 + h^2}$. Also if $(x_1, x_2) \in B_{hR}$, then $(t x_1, t x_2) \in B_{h,R-r}$ for all $t \leq 1 - r/R(r \leq R)$. Accordingly the desired path could be the polygon $z_1 z_3 z_4 z_2$ where z_3 and z_4 are the nearest points of $\bar{B}_{h,R-r}$ to z_1 and z_2 respectively, $r = |z_1 - z_2|$.

We need also the following two lemmas:

Lemma 5.8.2. *Suppose $x_0 \in B_{hR}$, $S_1(x_0)$ is the surface defined above and V is defined near x_0 by the formula*

$$V(x) = \int_{S_1(x_0)} K_0(x - \xi)\, d\xi.$$

Then for all x near x_0,

$$V(x) = -\nu^{-1} x_0^\alpha x^\alpha + \text{const.}$$

Proof. It is easy to see that for each x near x_0, a surface $S_2(x)$ can be defined by $\xi = \xi_2(s; x)$ with $s \in B(x_{10}, r)$, so that $\xi_2(s, x) = \xi_1(s, x_0)$ for $s \in \partial B(x_{10}, r)$, and the whole surface $S(x) = S_1(x_0) + S_2(x)$ is a admissible for x near x_0. Thus

$$V(x) + V_2(x) = V_0(x) = \int_{S(x)} K_0(x - \xi)\, d\xi, \quad V_2(x) = \int_{S_2(x)} K_0(x - \xi)\, d\xi.$$

But for x real, $V_0(x)$ just reduces to the potential of 1. Thus

$$V_0(x) = (2\nu)^{-1} \sum (x^\alpha)^2 + \text{const., and}$$
$$V_2(x) = (2\nu)^{-1} \sum_\alpha (x^\alpha - x_0^\alpha)^2 + \text{const.}$$

for the same reason, using a translation of axes. The lemma follows.

Lemma 5.8.3. *If $f \in {}^*C_\mu^0(B_{hR})$ and $L = \|f\|_0$, then*

$$|\nabla f(x_0)| \leq C_1(\nu, \mu, h)\, L \cdot [\delta(x_0)]^{\mu-1},$$
$$|\nabla^2 f(x_0)| \leq C_2(\nu, \mu, h) \cdot L \cdot [\delta(x_0)]^{\mu-2}, \quad x_0 \in B_{hR}.$$

Proof. There is a constant $\varkappa$, which depends only on ν and h, such that the polycylinder $|\xi^\alpha - x^\alpha| < \varkappa\, \delta(x)$, $\alpha = 1, \ldots, \nu$, lies in B_{hR}.

Since $f \in C^0_\mu(B_{hR})$, we have (using Cauchy's integral formula with $\gamma: |\xi - x| = \varkappa\,\delta(x)$)

$$f(x) = (2\pi i)^{-\nu} \int_\gamma \cdots \int_\gamma f(\xi)\, d\xi^1 \ldots d\xi^\nu / (\xi^1 - x^1) \ldots (\xi^\nu - x^\nu)$$

$$\bar{\zeta}^\alpha f_{,\alpha}(x) = (2\pi i)^{-\nu} \int_\gamma \cdots \int_\gamma \bar{\zeta}^\alpha [f(\xi) - f(x)] \times$$
$$\times\, d\xi^1 \ldots d\xi^\nu / (\xi^1 - x^1) \ldots (\xi^\nu - x^\nu)(\xi^\alpha - x^\alpha)$$

for any complex ζ^α. The first result follows from the fact that

$$|(\xi^\alpha - x^\alpha)^{-1} \bar{\zeta}^\alpha| \leq \|\zeta\| \cdot [\varkappa\,\delta(x)]^{-1}.$$

The second result is proved similarly.

We now prove the analog of Theorem 5.8.1 for B_{hR}:

Theorem 5.8.4. *$P_R(f)$ is a bounded operator from ${}^*C^0_\mu(B_{hR})$ into ${}^*C^2_\mu(B_{hR})$ with bound independent of R.*

Proof. It is sufficient to show that $\nabla^2 F \in C^0_\mu(\bar{B}_{hR})$, which we do by estimating $\nabla^3 F(x)$ in terms of $\delta(x)$, F being the potential as defined in Theorem 5.8.2. So, let $x_0 = x_{10} + i\,x_{20} \in B_{hR}$ and let $S_1(x_0)$ be the surface defined above. As was seen in the proof of Lemma 5.8.2, this surface may be completed to a surface $S(x)$ by adjoining a surface $S_2(x)$ with the same rim, if x is near enough to x_0. But the derivatives may be computed by differentiating only in real directions. So, we may choose $S_2(x)$ as the set of $\xi = s + i\,x_{20}$, where $s \in B(x_{10}, r)$, $r = r(x_0)/2$. Thus, we have

$$F(x) = F_1(x) + F_2(x), \quad F_1(x) = \int_{S_1(x_0)} K_0(x - \xi) f(\xi)\, d\xi$$

$$F_2(x) = \int_{B(x_{10}, r)} K_0(x_1 - s) f(s + i\,x_{20})\, ds, \tag{5.8.31}$$

$$x = x_1 + i\,x_{20}.$$

Clearly F_1 is analytic for x near x_0 and we can differentiate under the integral sign as often as is desired. Thus

$$F_{1,\alpha\beta\gamma}(x_0) = \int_{S_1(x_0)} K_{0,\alpha\beta\gamma}(x_0 - \xi)[f(\xi) - f(x_0)]\, d\xi + f(x_0) \cdot V_{,\alpha\beta\gamma}(x_0)$$

in which the second term vanishes, using Lemma 5.8.2. Now, from the form of our integrals, it follows that $|d\xi| \leq \nu$-dimensional element of area. From our construction it follows that

$$|d\xi| \leq (1 + h^2)^{\nu/2}\, d\xi_1$$

where $d\xi_1$ is the projection of the element on the real R_ν. Thus

$$|\nabla^3 F_1(x_0)| \leq Z_1 \cdot L \cdot \int_{R_\nu - B(x_{10}, r)} |\xi_1 - x_{10}|^{\mu - 1 - \nu}\, d\xi_1 \leq Z_2 \cdot L \cdot [\delta(x_0)]^{\mu - 1} \quad (L = \|f\|_1). \tag{5.8.32}$$

Now F_2 is just the potential of a complex-valued function. First we differentiate

$$F_{2,\alpha}(x) = \int_{B(x_{10},r)} -\frac{\partial}{\partial s^\alpha} K_0(x_1 - s) f(s + i x_{20})\, ds$$
$$= -\int_{\partial B(x_{10},r)} K_0(x_1 - s) f(s + i x_{20})\, ds'_\alpha +$$
$$+ \int_{B(x_{10},r)} K_0(x_1 - s) f_{,\alpha}(s + i x_{20})\, ds .$$

Differentiating twice more and using Theorem 2.6.5, we obtain

$$F_{2,\alpha\beta\gamma}(x_0) = -\int_{\partial B(x_{10},r)} K_{0,\beta\gamma}(x_{10} - s)[f(s + i x_{20}) - f(x_0)]\, ds'_\alpha +$$
$$+ C_{\beta\gamma} f_{,\alpha}(x_0) + \int_{B(x_{10},r)} K_{0,\beta\gamma}(x_{10} - s)[f_{,\alpha}(s + i x_{20}) - f_{,\alpha}(x_0)]\, ds .$$

Thus

$$(5.8.33) \qquad \begin{aligned} |\nabla^3 F_2(x_0)| &\le Z_3 L r^{\mu-1} + Z_4 L [\delta(x_0)]^{\mu-1} + Z_5 L r [\delta(x_0)]^{\mu-2} \\ &\le Z_6 L \cdot [\delta(x_0)]^{\mu-1} \end{aligned}$$

using Lemmas 5.8.1 and 5.8.3. The theorem follows from (5.8.32), (5.8.33), and Lemma 5.8.1 (c).

Theorem 5.8.5. *If H is harmonic and $H \in {}^*C^2_\mu(B_R)$, then $H \in {}^*C^2_\mu(B_{hR})$ for each $h < 1$ and*

$$\|H\|_{2h} \le C(\nu, \mu, h)\, \|H\|_2 .$$

Proof. From formula (2.4.4), it follows that

$$H(x) = -\int_{\partial B_R} [H(\xi)\, K_{0,\alpha}(x - \xi) + K_0(x - \xi)\, H_{R,\alpha}(\xi)]\, d\xi'_\alpha .$$

It is clear that H is analytic in any B_{hR} with $h < 1$. Differentiating, we obtain ($x = x_1 + i x_2$)

$$H_{,\beta\gamma\delta}(x) = -\int_{\partial B_R} \{K_{0,\alpha\beta\gamma\delta}(x - \xi)[H(\xi) - H(x_1) - \nabla H(x_1) \cdot (\xi - x_1) -$$
$$- \nabla^2 H(x_1) \cdot (\xi - x_1)^2/2!] + K_{0,\beta\gamma\delta}(x - \xi)[H_{,\alpha}(\xi) - H_{,\alpha}(x_1) -$$
$$- \nabla H_{,\alpha}(x_1) \cdot (\xi - x_1)]\}\, d\xi'_\alpha$$

since it follows from Theorem 2.6.5, differentiation, and analytic continuation that

$$\int_{\partial B_R} K_{0,\alpha\beta\gamma\delta}(x - \xi)\, d\xi'_\alpha = \int_{\partial B_R} K_{0,\alpha\beta\gamma\delta}(x - \xi) \cdot (\xi - x)\, d\xi'_\alpha = \ldots = 0$$

and also $\xi - x = \xi - x_1 + x_1 - x$, etc. Thus, using the properties of H for x real and the fact that $|\xi - x_1| \le |\xi - x| \le (1 + h^2)^{\frac{1}{2}} |\xi - x_1|$, we obtain

$$\begin{aligned} |\nabla^3 H(x)| &\le Z_1 L \int_{\partial B_R} |\xi - x_1|^{\mu-\nu}\, d\xi \le Z_2 L (R - |x_1|)^{\mu-1} \\ &\le Z_3 L [\delta(x)]^{\mu-1}, \quad L = \|H\|_2 \end{aligned}$$

from which the result follows.

We can now prove our interior analyticity theorem:

Theorem 5.8.6. *Suppose* $u \in C^2_\mu(\bar{B}_{R_0})$ *with* $u(0) = \nabla u(0) = \nabla^2 u(0) = 0$ *and satisfies* (5.8.5) *there, where* ψ *is analytic in the* (x, r)*-space near* $(0, 0)$ *and has the properties described there. Then* u *is analytic near* $x = 0$.

Proof. Let us choose $h < 1$. From Theorems 5.8.2 and 5.8.4 it follows that $\|H_R\|_{2h} \leq M_3$ uniformly for $R \leq R_0$. If we restrict ourselves to U_{1R} and U_{2R} in $*C^2_\mu(B_{hR})$ with norms $\leq M_3$, we see that (5.8.11) and (5.8.12) hold for complex x in B_{hR} so that if we take R_2, $0 < R_2 \leq R_1$, small enough T_R satisfies (5.8.9) with $\varepsilon(R) = 1/2$, considering it as an operator over $*C^2_\mu(B_{hR})$. Thus, there is a unique solution u_R of (5.8.10) with $\|u_R\|_{2h} \leq M_3$ and $\|u_R\|_2 \leq M_2$. These must coincide so that u_R also $\in *C^2_\mu(B_{hR})$. The result follows.

We now wish to consider a solution u of (5.8.1) where x_0 is a point on an analytic part of ∂G along which u vanishes. A simple analytic transformation of the x coordinates enables us to assume that $x_0 = 0$ and a part of $\bar{G}$ corresponds to $\bar{G}_{R_0}$ with a part of ∂G corresponding to σ_{R_0}. The substitutions made earlier lead to the equation before (5.8.5). Then an affine transformation, followed by a rotation of axes yields (5.8.5) on $\bar{G}_{R_0}$ where $u = 0$ on σ_{R_0} and $|\nabla u(0)| = |\nabla^2 u(0)| = 0$. We may assume that ψ has its previous properties. We write (5.8.6), where we make the following definitions; the extra difficulty in defining $P_R(f)$ arises from the fact that the hemisphere G_R^- is not a smooth domain.

Definition 5.8.1'. The *space* $*C^0_\mu(G_R)$ consists of all $f \in C^0_\mu(\bar{G}_R)$ with $f(0) = 0$, the norm $\|f\|_0$ being $h_\mu(f, G_R)$. The *space* $*C^2_\mu(G_R)$ consists of all $u \in C^2_\mu(\bar{G}_R)$ which vanish along σ_R and for which also $\nabla u(0) = \nabla^2 u(0) = 0$, the norm $\|u\|_2$ being $h_\mu(\nabla^2 u, G_R)$. For $f \in *C^0_\mu(G_R)$ we define $U = P_R(f)$ by the formula

$$U(x) = -Q(x) + \int_{B_{2R}} K_0(x-\xi) f_1(\xi)\, d\xi - 2 \int_{G_{2R}^-} K_0(x-\xi) f_1(\xi)\, d\xi \tag{5.8.34}$$

where Q is that quadratic polynomial chosen so that $\nabla U(0) = \nabla^2 U(0) = 0$, and f_1 is that extension of f to B_{2R} defined by the conditions

$$\begin{aligned} &f_1(x) = f(x), \quad x \in \bar{G}_R, \quad f_1(x^\nu, x'_\nu) = f_1(-x^\nu, x'_\nu), \quad x \in \bar{B}_{2R} \\ &f_1(r\xi) = f_1[(2R-r)\xi], \quad 0 \leq r \leq 2R, \quad \xi \in \partial B(0,1). \end{aligned} \tag{5.8.35}$$

Theorem 5.8.1'. *Theorem* 5.8.1 *holds with* B_R *replaced by* G_R.

Proof. As in the proof of Theorem 5.8.1, we find that $\Delta Q = 0$. Also since $U = 0$ along $x^\nu = 0$, it follows that Q is also. We need only prove the Hölder continuity of $\nabla^2 U$. Let $F_1(x)$ and $F_2(x)$ denote the first and second integrals, respectively, in (5.8.34). That $h_\mu(\nabla^2 F_1, B_R) \leq Z_1 h_\mu(f, B_R)$ follows from Theorem 2.6.7, Corollary 1. Obviously,

$h_\mu(f, B_R) = h_\mu(f_1, B_{2R})$. Differentiating F_2, we obtain

$$F_{2,\alpha\beta\gamma}(x) = -2\int_{G_{2R}^-} K_{0,\alpha\beta\gamma}(x-\xi)\,[f_1(\xi) - f_1(x)]\,d\xi + f_1(x)\,V_{,\alpha\beta\gamma}(x)$$

$$V(x) = -2\int_{G_{2R}^-} K_0(x-\xi)\,d\xi. \tag{5.8.36}$$

Obviously V is harmonic, and if α, say $,< \nu$, then

$$V_{,\alpha}(x) = 2\int_{\Sigma_{2R}^-} K_0(x-\xi)\,d\xi'_\alpha, \quad V_{,\alpha\beta\gamma}(x) = 2\int_{\Sigma_{2R}^-} K_{0,\beta\gamma}(x-\xi)\,d\xi'_\alpha. \tag{5.8.37}$$

Since V is harmonic, $V_{,\nu\nu\nu}$ can be found in terms of the other $V_{,\alpha\beta\gamma}$, and we find from (5.8.36) and (5.8.37) that

$$\begin{aligned}|\nabla^3 F_2(x)| &\le Z_1 L \int_{R_\nu - B(x, x^\nu)} |\xi - x|^{\mu-\nu-1}\,d\xi + Z_2 L R^\mu \int_{\Sigma_{2R}^-} |\xi - x|^{-\nu}\,dS(\xi)\\ &\le Z_3 L [\delta(x)]^{\mu-1}, \quad x \in \bar{G}_R,\end{aligned}$$

$\delta(x)$ being the distance of x from ∂G_R. The theorem follows from Theorem 2.6.6(b).

Again we define v_R and T_R by the formulas (5.8.8) and conclude easily as before that the following theorem holds.

Theorem 5.8.2′. *Theorem 5.8.2 holds with B_R replaced by G_R.*

Now it would be desirable to be able to introduce spaces of analytic functions as was done above and prove the analogs of the remaining theorems. However, it turns out to be more convenient to introduce first spaces of functions which are analytic only in $(x^1, \ldots, x^{\nu-1})$, keeping x^ν real. In this way, we shall be able to extend f to f_1 and to generalize easily the formulas (5.8.21). We first define our domains G_{hR} and our spaces of functions.

Definition 5.8.2′. We define B_{0hR} to be that part of B_{hR} for which x^ν is real, G_{hR} as the part of B_{0hR} where $x^\nu > 0$, and G_{hR}^- as the part of B_{0hR} where $x^\nu < 0$. The *space* $^*C_\mu^0(G_{hR})$ consists of all $f \in {}^*C_\mu^0(G_R)$ which can be extended to $\in C_\mu^0(\bar{G}_{hR})$ and to be analytic in $(x^1, \ldots, x^{\nu-1})$ for $x \in G_{hR}$, the norm $\|f\|_{0h}$ being $h_\mu(f, G_{hR})$. The space $^*C_\mu^2(G_{hR})$ consists of all $u \in {}^*C_\mu^2(G_R)$ which can be extended to $\in C_\mu^2(\bar{G}_{hR})$ and to be analytic in $(x^1, \ldots, x^{\nu-1})$ for $x \in G_{hR}$ with $u = 0$ for all $x \in \bar{G}_{hR}$ with $x^\nu = 0$, and $\nabla u(0) = \nabla^2 u(0) = 0$, the norm $\|u\|_{2h}$ being $h_\mu(\nabla^2 u, G_{hR})$.

Theorem 5.8.3′. *Suppose* (i) *that* $f_1 \in C_\mu^0(\bar{B}_{0hR})$ *and is analytic in* $(x^1, \ldots, x^{\nu-1})$ *for* $x \in B_{0hR}$; (ii) $x_0 \in G_{hR}$; (iii) $S_1 : \xi = \xi_1(s)$, $s \in \bar{B}_R$ *and* $S_1^* : \xi = \xi_1^*(s)$, $s \in \bar{G}_R^-$, *are admissible with respect to* $\mathfrak{F}_1(\xi; x_0) = K_0(x_0 - \xi) f_1(\xi)$; (iv) *that* S_1 *and* S_1^* *satisfy also the conditions* (5.8.20) *and also the conditions that*

$$\xi_{12}^\nu(s) = \xi_{12}^{*\nu}(s) = 0 \tag{5.8.38}$$

on their respective domains. Then the integrals of $\mathfrak{F}_1$ over S_1 and S_1^* exist. If S_2 and S_2^* satisfy their respective conditions, the integrals of $\mathfrak{F}_1$ over S_1

and S_2 *are equal as are the integrals of* $\mathfrak{F}_1$ *over* S_1^* *and* S_2^*. *Finally if, for each* $x \in G_{hR}$, $S(x)$ *and* $S^*(x)$ *satisfy all the conditions then the functions* F *and* F^* *defined by*

$$(5.8.39)\quad F(x) = \int_{S(x)} K_0(x-\xi) f_1(\xi)\, d\xi, \quad F^*(x) = \int_{S^*(x)} K_0(x-\xi) f_1(\xi)\, d\xi$$

are analytic in $(x^1, \ldots, x^{\nu-1})$ *for* x *on* G_{hR} *and of class* C^1 *in* x *and*

$$(5.8.40)\quad \begin{aligned} F_{,\alpha}(x) &= \int_{S(x)} K_{0,\alpha}(x-\xi) f_1(\xi)\, d\xi, \\ F^*_{,\alpha}(x) &= \int_{S^*(x)} K_{0,\alpha}(x-\xi) f_1(\xi)\, d\xi, \end{aligned} \qquad 1 \le \alpha \le \nu.$$

Proof. The statements concerning $F(x)$ and the integrals over S_1 and S_2 follows from the proof of Theorem 5.8.3, since the condition (5.8.38) guarantees that $\xi_t^\nu(s,t) \equiv 0$ so that the derivative of $\mathfrak{F}_1$ with respect to ξ^ν, which does not necessarily exist, is not involved anyway. An entirely similar analysis shows that the integral over S_1^* equals that over S_2^*: If we set

$$\xi^*(s,t) = s + i[(1-t)\xi_{12}^*(s) + t\,\xi_{22}^*(s)] \quad (\xi^{*\nu}(s,t) = s^\nu),$$
$$\varphi(t) = \int_{S^*(s,t)} K_0(x_0 - \xi) f_1(\xi)\, d\xi,$$

a repetition of the analysis in the proof of Theorem 5.8.3 yields

$$\varphi'(t) = \int_{\sigma_R} \sum_{\gamma=1}^{\nu-1} (-1)^{\gamma+\nu} \mathfrak{F}_1 \xi^{*\gamma} \frac{\partial \xi_\nu^{*\prime}}{\partial s_\nu'} ds_\nu' + \int_{\Sigma_{\bar R}} \sum_{\gamma=1}^{\nu-1} \sum_{\delta=1}^{\nu} (-1)^{\gamma+\delta} \mathfrak{F}_1 \xi_t^{*\gamma} \frac{\partial \xi_\gamma^{*\prime}}{\partial s_\delta'} ds_\delta'.$$

The first integral vanishes since $\xi^{*\nu} = s^\nu$ and the second vanishes since $\xi^*(s,t) = s$ on $\sum_{\bar R}$. The analyticity in the case F follows as before and that for F^* follows similarly; the formulas (5.8.40) are obtained in the course of the proof.

Definition 5.8.4′. For $f \in {}^*C_u^0(G_{hR})$, we define $U = P_R(f)$ by the formula

$$(5.8.41)\quad \begin{aligned} U(x) &= -Q(x) + F_1(x) + F_2(x) \\ F_1(x) &= \int_{B_{2R}-B_R} K_0(x-\xi) f_1(\xi)\, d\xi + \int_{S(x)} K_0(x-\xi) f_1(\xi)\, d\xi \\ F_2(x) &= \int_{G_{2R}^- - G_R^-} -2K_0(x-\xi) f_1(\xi)\, d\xi - 2\int_{S^*(x)} K_0(x-\xi) f_1(\xi)\, d\xi \end{aligned}$$

where f_1 is the extension of f to $\bar{B}_{0hR} \cup (\bar{B}_{2R} - B_R)$ which is uniquely defined by the *formulas* in (5.8.35) and Q is as usual.

Definition 5.8.5′. For $x = x_1 + i\, x_2 (x_2^\nu = 0)$ in B_{0hR}, we let $\delta_0(x)$ be the distance from ∂B_{0hR} (in the $2\nu - 1$ space) and $r(x)$ be the distance in the ν-plane $x_2 = \text{const.}$ of x from the intersection of that plane with ∂B_{hR}. For $x \in G_{hR}$, we define $\delta(x)$ as the distance of x from ∂G_{hR}.

Lemma 5.8.1′. (a) *B_{0hR} and G_{hR} are convex.*

(b) $r(x) = R - |x_1| - |x_2'|/h$, $\delta_0(x) = h\,r(x)/\sqrt{1+h^2}$

$$\delta(x) = \min[\delta_0(x), x^\nu], \quad x_2' = (x_2^1, \ldots, x_2^{\nu-1}).$$

(c) *Any two points z_1 and z_2 of G_{hR} can be joined by a path in G_{hR} as in Lemma 5.8.1. The same is true of B_{0hR}.*

The proof is like that of Lemma 5.8.1.

Lemma 5.8.2′. *Let $S^*(x)$ satisfy the conditions in Theorem 5.8.3′ and, for $x \in G_{hR}$, define*

$$V^*(x) = -2\int_{S^*(x)} K_0(x-\xi)\,d\xi - 2\int_{G_{2R}^- - G_R^-} K_0(x-\xi)\,d\xi. \tag{5.8.42}$$

Then V^* is analytic in x_ν' for $x \in G_{Rh}$, is harmonic there, and

$$|\nabla^3 V^*(x)| \le C(\nu, \mu, h) R^{-1}. \tag{5.8.43}$$

Proof. Choose $x_0 \in G_{hR}$. For x sufficiently near x_0, $V^*(x)$ is given by (5.8.42) with $S^*(x)$ replaced by $S^*(x_0)$ so that V^* is analytic in all x^α and its derivatives of any order can be found by differentiation under the integral sign. Thus V^* is harmonic. For real x_1

$$V_{,\alpha}^*(x_1) = -2\int_{G_{2R}^-} K_{0,\alpha}(x_1-\xi)\,d\xi = 2\int_{\Sigma_{2R}^-} K_0(x_1-\xi)\,d\xi_\alpha', \quad \alpha < \nu. \tag{5.8.44}$$

Since the right side of (5.8.44) is analytic, (5.8.44) must hold for x in G_{hR}. Hence

$$V_{,\alpha\beta\gamma}^*(x_0) = 2\int_{\Sigma_{2R}^-} K_{0,\beta\gamma}(x_0-\xi)\,d\xi_\alpha'$$

from which a bound (5.8.43) follows if $(\alpha, \beta, \gamma) \neq (\nu, \nu, \nu)$. But since V^* is harmonic, this derivative has a similar bound.

Lemma 5.8.3′. *If $f_1 \in C_\mu^0(\bar{B}_{0hR})$ and f_1 is analytic in x_ν' for all $x \in B_{0hR}$, then*

$$\begin{aligned} |f_{,\alpha}(x_0)| &\le C_1'(\nu, \mu, h)\cdot[\delta(x_0)]^{\mu-1}, \\ |f_{,\alpha\beta}(x_0)| &\le C_2'(\nu, \mu, h)\cdot[\delta(x_0)]^{\mu-2}, \end{aligned} \quad \alpha, \beta < \nu. \tag{5.8.45}$$

Proof. The proof of Lemma 5.8.3 with ν replaced by $\nu - 1$ demonstrates (5.8.45) with $\delta(x_0)$ replaced by the distance $d(x_0)$ in the plane $x^\nu = x_0^\nu$ from the intersection of that plane with ∂G_{hR}; obviously $d(x_0) \ge \delta(x_0)$, so the result follows.

Theorem 5.8.4′. *P_R is a bounded operator from ${}^*C_\mu^0(G_{hR})$ to ${}^*C_\mu^2(G_{hR})$ with bound independent of R.*

Proof. From the fact that F_1, ΔF_1, and f_1 are analytic in x_ν' for x on B_{0hR} and $\Delta F_1 = f_1$ for real x, it follows that this holds on B_{0hR}. It is also seen that $\Delta F_2 = 0$ on G_{hR}.

Choose $x_0 \in B_{0hR}$. For $x = x_1 + x_{20}$, $|x_1 - x_{10}| < r$, we can write

$$F_1(x) = F_{11}(x) + F_{12}(x), \quad F_{11}(x) = \int_{B(x_{10}, r)} K_0(x_1 - s)\,f_1(s + i x_{20}')\,ds, \tag{5.8.46}$$

$$F_{12}(x) = \int_{S_1(x_0)} K_0(x-\xi)\,f_1(\xi)\,d\xi + \int_{B_{2R}-B_R} K_0(x-\xi)\,f_1(\xi)\,d\xi.$$

As in the proof of Theorem 5.8.4, F_{12} is analytic in all x^α and

$$(5.8.47) \qquad |\nabla^3 F_{12}(x_0)| \le Z_1 \|f\|_0 \cdot [\delta(x_0)]^{\mu-1}.$$

In the case of F_{11}, we first choose $\alpha < \nu$ and differentiate to obtain

$$F_{11,\alpha}(x) = -\int_{\partial B(x_{10},r)} K_0(x_1 - s) f_1(s + i x'_{20})\, ds'_\alpha + \int_{B(x_{10},r)} K_0(x_1 - s) f_{1,\alpha}(s + i x'_{20})\, ds.$$

We can then estimate $\nabla^2 F_{11,\alpha}$ by

$$|\nabla^2 F_{11,\alpha}(x_0)| \le Z_2 \|f\|_0 [\delta(x_0)]^{\mu-1}$$

as was done in the proof of Theorem 5.8.4, using Lemma 5.8.3′ this time. Thus all second derivatives $F_{1,\alpha\beta}$ with $(\alpha, \beta) \neq (\nu, \nu)$ satisfy the desired Hölder condition. But then $F_{1,\nu\nu}$ does also since $\Delta F_1 = f_1$.

In the case of F_2, we see that

$$F_2(x_1 + i x_{20}) = \int_{S^*(x_0)} -2K_0(x - \xi) f_1(\xi)\, d\xi - 2\int_{G_{2R}^- - G_R^-} K_0(x - \xi) f_1(\xi)\, d\xi$$

for $x = x_1 + i x_{20}$ near x_0. Differentiation yields (see Lemma 5.8.2′)

$$\nabla^3 F_2(x_0) = f_1(x_0) \cdot \nabla^3 V^*(x_0) - 2\int_{S^*(x_0)} \nabla^3 K_0(x_0 - \xi)\, [f_1(\xi) - f_1(x_0)]\, d\xi -$$

$$(5.8.48) \qquad - 2\int_{G_{2R}^- - G_R^-} \nabla^3 K_0(x_0 - \xi)\, [f_1(\xi) - f_1(x_0)]\, d\xi.$$

The last two integrals can be estimated as in the proof of Theorem 5.8.1′ and the first term on the right in (5.8.48) has the estimate

$$|f_1(x_0) \cdot \nabla^3 V^*(x_0)| \le Z_3 \|f\|_1 \cdot R^{\mu-1} \le Z_3 \|f\|_1 \cdot [\delta(x_0)]^{\mu-1}$$

using Lemmas 5.8.1′ and 5.8.2′. The theorem follows.

Theorem 5.8.5′. *If H is harmonic on G_R and $H \in {}^*C^2_\mu(G_R)$, then $H \in {}^*C^2_\mu(G_{hR})$ and the inequality of Theorem 5.8.5 holds.*

Proof. This follows from Theorem 5.8.5 and the reflection principle.

Theorem 5.8.6′. *Suppose $u \in C^2_\mu(\bar{G}_{R_0})$ with $\nabla u(0) = \nabla^2 u(0) = 0$, u satisfies (5.8.5) on G_{R_0}, and vanishes along σ_R, where ψ is analytic near $(0, 0)$ and has the properties stated near that equation. Then u is analytic in G_R and can be continued analytically across σ_R.*

Proof. As in the proof of Theorem 5.8.6, one deduces that there is an R_2 which is such that $u \in {}^*C^2_\mu(G_{hR})$ for every R, $0 < R \le R_2$. From Theorem 5.8.6, we conclude that u is analytic in any such G_R. Now, let ν be replaced by $\nu + 1$ and let $x^{\nu+1} = y$. Then u satisfies on G_R an equation of the form

$$(5.8.49) \qquad u_{yy} = \varphi(y, x, u, u_{,\alpha}, u_y, u_{y,\alpha}, u_{,\alpha\beta}) \quad (\alpha, \beta = 1, \ldots, \nu)$$

in which φ is analytic near the origin; this is obtained by solving (5.8.5) for u_{yy}. If we now set

$$w^1 = u, \; w^2 = u_y, \; w^{2+\alpha} = u_{,\alpha}, \; \alpha = 1, \ldots, \nu$$

we see from (5.8.49) that the vector $w^1, \ldots, u^{\nu+2}$ satisfies a system of the Cauchy-Kovalevsky type

$$\frac{\partial w^j}{\partial y} = f^j(y, x, w, w_x), \quad j = 1, \ldots, P \quad (x = x^1, \ldots, x^\nu)$$

in which the f^j are analytic near the origin. Since the origin could correspond to any point on an analytic part of the boundary, we shall simply prove that the w^j are analytic at the origin. We already know that the w^j are analytic in (x, y) on G_R and in x for (x, y) on G_{hR}. Our method is to use the proof of the Cauchy-Kovalevsky theorem to show that a solution can be obtained with our given values $\varphi^j(x, y_0)$ which has a radius of convergence independent of y_0, if y_0 is small enough.

For each real $y_0 > 0$ but sufficiently small, we make the usual substitutions to prepare to prove the Cauchy-Kovalevsky theorem as it is proved in the cited book by GOURSAT, pp. 2—6:

$$(5.8.50) \qquad {}^*\omega^j(x, \eta; y_0) = w^j(x, y_0 + \eta) - \varphi^j(x; y_0).$$

This leads to equations

$$(5.8.51) \qquad \frac{\partial^* \omega^j}{\partial \eta} = f^j[x, y_0 + \eta, {}^*\omega + \varphi(x; y_0), {}^*\omega_x + \varphi_x(x, y_0)].$$

Subtracting off the constant term and making the final change

$$(5.8.52) \qquad \begin{aligned} \omega^j(x, \eta; y_0) &= {}^*\omega^j(x, \eta; y_0) - a^j(y_0)\,\eta \\ a^j(y_0) &= f^j[0, y_0, \varphi(0, y_0), \varphi_x(0, y_0)], \end{aligned}$$

we obtain the equations

$$(5.8.53) \qquad \begin{aligned} \frac{\partial \omega^j}{\partial \eta} = g^j(x, \eta, \omega, \omega_x; y_0) &= -a^j(y_0) + \\ &+ f^j[x, y, \omega + \varphi(x, y_0) + a(y_0)\,\eta, \omega_x + \varphi_x(x, y_0)]. \end{aligned}$$

Now the $f^j(x, y, w, p)$ are analytic with

$$|f^j(x, y, w, p)| \le M_1 \text{ for } |x|^\alpha \le r, \ |y| \le r, \ |w^j| \le r, \ |p^j_\alpha| \le r.$$

Moreover, we know that the φ^j are analytic in x and

$$|\varphi^j(x, y_0)| \le \frac{r}{3}, \quad |\varphi^j_{,\alpha}(x, y_0)| \le \frac{r}{3} \quad \text{if} \quad |x^\alpha| \le \delta, \quad 0 \le y_0 \le \delta.$$

Thus we see that there is a $\sigma > 0$ such that the g^j are analytic in (x, η, ω, π) for each small y_0 with

$$|g^j(x, y, \omega, \pi; y_0)| \le M_1 \quad \text{if} \quad |x^\alpha| \le \sigma, \ |\eta| \le \sigma, \ |\omega| \le \sigma, \ |\pi| \le \sigma, \\ 0 \le y_0 \le \sigma.$$

Consequently there are numbers M, $r > 0$ and $\varrho > 0$ such that each g^j is dominated by the function

$$G = \frac{M}{[1 - r^{-1}(\eta + x^1 + \cdots + x^\nu + \omega^1 + \cdots + \omega^P)] \cdot \left[1 - \varrho^{-1} \sum_{j,\alpha} \pi^j_\alpha\right]} - M.$$

The remainder of the proof as given in GOURSAT shows that there is a $\delta' > 0$ such that the $\omega^j(x, \eta; y_0)$ are analytic with

$$|\omega^j(x, \eta; y_0)| \leq \mu^j \text{ for } |x^\alpha| \leq \delta', \ |\eta| \leq \delta', \ 0 \leq y_0 \leq \delta'$$

where δ' and the μ^j are independent of y_0. Thus the w^j obtained as above by solving the differential equations have the same properties and must coincide with those obtained previously, for $y > 0$. Their analyticity at the origin follows.

5.9. Properties of the extremals; regular cases

In this section we complete the proofs of the theorems and remarks made in Chapter 1 concerning the differentiability of the solutions of variational problems and of the weak solutions of the quasi-linear differential equations mentioned in the introduction.

We begin by completing the proof of Theorem 1.11.1 which states in the case of the quasi-linear equations (1.10.13) and (1.10.14), where the A_i^α and B_i satisfy (1.10.7″) or (1.10.7‴) with $k \geq 2$, that the derivatives p_γ^i and the function $U = V^{k/2} \in H_2^1(D)$ for each $D \subset\subset G$ and the derivatives satisfy the differentiated equations (1.11.9) there. In Chapter 1, we introduced carefully the difference-quotient procedure and proved that the difference quotients

$$z_h(x) = h^{-1} \Delta z, \quad \Delta z = z(x + h\, e_\gamma) - z(x)$$

satisfied the equations (1.11.3) and that for each $D \subset D'_a \subset D' \subset G$,

$$\int_D A_h |\nabla z_h|^2 dx \leq (C_1 + C_2\, a^{-2}) \int_{D'} A_h(x)\, z_h^2(x)\, dx \tag{5.9.1}$$

where C_1 and C_2 are independent of h. Now, from the existence theory, it follows that each $z_{,\gamma} \in L_k$ with $k \geq 2$. Consequently, since

$$z_h(x) = \int_0^1 z_{,\gamma}(x + t\, h\, e_\gamma)\, dt,$$

it follows from Theorem 3.6.8 that $z_h \to z_{,\gamma}$ in $L_k(D')$ for each $D' \subset\subset G$. Moreover, by approximating strongly to z in $H_k^1(D)$ by $z_n \in C^1$ and by using the techniques of Theorem 3.6.8 and Lemma 3.4.2 we conclude that $A_h \to A$ in $L_{k/(k-2)}(D')$ if $k > 2$ and $A_h \equiv 1$ if $k = 2$. Thus, from the Hölder inequality we see that the right side of (5.9.1) tends to its expected limit as $h \to 0$. Next, since $A_h(x) \geq 1$, it follows that $z_h \rightharpoonup u$, say, in $H_2^1(D)$ for a subsequence of $h \to 0$. But from Theorem 3.4.4 (since $D \subset$ a strongly Lipschitz domain $\subset D'$, etc.) and the fact that $z_h \to p_\gamma$ in $L_k(D)$, it follows that $u = z_{,\gamma}$ so $z_{,\gamma} = p_\gamma \in H_2^1(D)$. Also, we see that $A_h(x) \to A(x) = V^{k-2}(x)$ a.e. and the coefficients $a_{hij}^{\alpha\beta}(x)$, $b_{hij}^\alpha(x)$, $e_{hi}^{\alpha\gamma}$, c_{hij}^α, d_{hij}, and f_{hi}^γ tend a.e. and boundedly to their expected limits as $h \to 0$, $P_h(x) \to V(x)$ a.e. and in $L_k(D')$. From the uniform boundedness of the left side of (5.9.1), it follows that $A_h^{\frac{1}{2}} a_{hij}^{\alpha\beta} z_{h,\beta}^j$, $A_h^{\frac{1}{2}} c_{hij}^{\alpha} z_{h,\alpha}^j$ tend weakly in $L_2(D)$ to certain quantities and $A_h^{\frac{1}{2}} b_h \cdot z_h$, $A_h^{\frac{1}{2}} P_h \cdot e_h^{\gamma}$, $A_h^{\frac{1}{2}} d_h \cdot z_h$, and $A_h^{\frac{1}{2}} f_h^{\gamma} \cdot P_h$ tend strongly in $L_2(D)$ to their expected limits as $h \to 0$ through a further subsequence. By multiplying the first two by bounded measurable functions and integrating, using the weak and strong convergence already established we see that the first two terms tend weakly to their expected limits. If $\zeta \in L\,i\,p_c(D)$, $A_h^{\frac{1}{2}} \nabla \zeta$ and $A_h^{\frac{1}{2}} \zeta$ tend strongly to their expected limits so that we may let $h \to 0$ to obtain (1.11.9) and (1.11.10).

Before proceeding further with the general theory, we now discuss the extremals of $I(z, G)$ in the case (1.10.8) with $\nu = k = 2$, N arbitrary. We have already seen in Theorem 4.3.1 that any minimizing function z satisfies a condition

$$(5.9.2) \quad \int_{B(x_0,r)} |\nabla z|^2 dx \le L^2 (r/R)^{2\mu}, \quad 0 \le r \le R, \quad B(x_0, R) \subset G, \quad \mu > 0$$

from which it follows that $z \in C_\mu^0(G)$. But now, if we consider a domain $D \subset D'_a \subset D' \subset\subset G$, we know that the difference quotient z_h satisfies the equations (1.11.3) in which $A_h \equiv 1$, the $a_{hij}^{\alpha\beta}$ satisfy (5.2.17), the b_h, c_h, and d_h satisfy (5.2.18), and e_h and f_h satisfy (5.4.8) uniformly in h and, as we have already seen above, each $z_h \in H_2^1(D)$ with norm uniformly bounded. It follows from Theorem 5.4.1, which has been seen in § 5.4 to generalize to systems, that the z_h satisfy a condition like (5.9.2) uniformly with G replaced by D, so that this is true of each $z_{,\gamma}$ which therefore $\in C_\mu^0(D)$. The whole analysis carries over for a solution z of (1.10.14) if the A_i^α and B_i satisfy (1.3.8″) with $k = \nu = 2$, N arbitrary. Once the z^i and p_γ^i are seen to $\in C_\mu^0(G)$, the equations (1.11.9) assume the form (5.1.1) (with $B(u, v)$, $C(u, v)$, and $L(v)$ given in (5.2.16) with coefficients e, and f Hölder continuous. Thus, in the case $N = 1$, we conclude from § 5.5 that $z \in C_\mu^2(G)$ and hence satisfies Euler's equation; then the theorems of § 5.5 or § 5.6 can be used to obtain the results stated in Chapter 1 that $z \in C_\mu^n$ if $f \in C_\mu^n$ or A_i^α and $B_i \in C_\mu^{n-1}$ in case $n = 2$ or if f and $f_p \in C_\mu^{n-1}$ or $A_i^\alpha \in C_\mu^{n-1}$ and $B_i \in C_\mu^{n-2}$ in case $n > 2$. The cases where $N > 1$ are treated in Chapter 6. Clearly $z \in C^\infty$ if $f \in C^\infty$ or A and $B \in C^\infty$. That z is analytic if f of A and B are analytic follows from the theorems in § 5.8 as generalized in § 6.7.

We now prove Theorem (1.11.1′) for which proof we need the following lemma due to Ladyzenskaya and Ural'tseva [2]:

Lemma 5.9.1. *Suppose the A_i^α and B_i satisfy (1.10.8″) with $k \ge \nu$ and that $z \in H_k^1(G) \cap C_\mu^0(\bar{G})$, $0 < \mu < 1$ and is a solution of (1.10.14). Then there are numbers R_1 and $C_1 > 0$ which depend only on ν, m, M, m_1,*

M_1, k, K, μ, and $h_\mu(z, \bar{G})$ *such that*

$$\int_{B(x_0,R)} V^k(x)\,\xi^2(x)\,dx \leq C_2 R^\mu \int_{B(x_0,R)} V^{k-2} |\nabla \xi|^2\,dx,$$

$$0 < R \leq R_1, \quad \xi \in H^1_{k0}[B(x_0, R)].$$

Proof. Suppose, first that $\xi \in Lip_c[B(x_0, R)]$, and define

$$\zeta^i(x) = \xi^2(x)\,[z^i(x) - z^i(x_0)]$$

this choice being allowed in (1.10.14) as one sees by an approximation. Using that ζ^i in (1.10.14), we obtain

$$\int_{B(x_0,R)} \{\xi^2 \cdot [p^i_\alpha A^\alpha_i + (z^i - z^i_0)\,B_i] + 2\xi(z^i - z^i_0)\xi_{,\alpha} A^\alpha_i\}\,dx = 0.$$

The hypotheses (1.3.8″) ($N = 1$) imply that

$$\begin{aligned} p^i_\alpha A^\alpha_i(x, z, p) &= p^i_\alpha A^\alpha_i(x, z, 0) + p^i_\alpha p^j_\beta \int_0^1 A^\alpha_{i p^j_\beta}(x, z, tp)\,dt \\ &\geq p^i_\alpha A^\alpha_i(x, z, 0) + m_1 |p|^2 \int_0^1 (1 + t^2 |p|^2)^{-1+k/2}\,dt \\ &\geq m_2 V^k - K_2, \quad m_2, K_2 = \varphi(\nu, m_1, M_1, K, k) \end{aligned}$$

$$|A(x, z, p)| \leq M_1 V^{k-1}, \quad |B(x, z, p)| \leq M_1 V^k.$$

It follows that

$$\int_{B(x_0,R)} V^k \xi^2\,dx \leq \int_{B(x_0,R)} \{Z_1 \xi^2 + Z_2 L R^\mu V^k \xi^2 + Z_3 L R^\mu V^{k-2} |\nabla\xi|^2\}\,dx$$

$$L = h_\mu(z, \bar{G}), \quad Z_i = Z_i(\nu, m_1, M_1, K, k)$$

from which the lemma follows easily, since

$$\int_{B(x_0,R)} \xi^2\,dx \leq 2^{-1} R^2 \int_{B(x_0,R)} |\nabla\xi|^2\,dx \leq 2^{-1} R^\mu \int_{B(x_0,R)} V^{k-2} |\nabla\xi|^2\,dx$$

by Poincaré's inequality, k being ≥ 2. The lemma follows easily for any $\xi \in H^1_{k0}[B(x_0, R)]$ by an approximation.

The proof of Theorem 1.11.1′ starts with the difference quotient procedure which leads to equations (1.11.3′) obtained from (1.11.3) by replacing b_h, c_h, d_h, and $f_h P_h$ by $b_h P_h$, $c_h P_h$, $d_h Q_h$, and $f_h Q_h$, respectively, where P_h is defined in (1.11.4) along with a_h and e_h and Q_h, b_h, c_h, d_h, and f_h are now defined by

$$A_h P_h b_h = \int_0^1 A_z\,dt, \quad A_h P_h c_h = \int_0^1 B_p\,dt,$$

$$A_h Q_h = \int_0^1 [1 + |z(x) + t\Delta z|^2 + |p(x) + t\Delta p|^2]^{k/2}\,dt, \tag{5.9.3}$$

$$A_h Q_h d_h = \int_0^1 B_z\,dt, \quad A_h Q_h f_h = \int_0^1 B_x\,dt.$$

From our hypothesis (verified for any extremal if $k > \nu$ and by any minimizing extremal if $k = \nu$), it follows that z satisfies (1.10.9) and is therefore $\in C^0_\mu(\bar{D})$ for any $D \subset\subset G$; so we may assume that G is already a subdomain of the domain of z and $z \in C^0_\mu(\bar{G})$. So, let us pick $B(x_0, R+a) \subset G_{h_0}$, $0 < a \leq R$, and define $\eta \in Lip_c[B(x_0, R+a)]$ with $\eta = 1$ on $\overline{B(x_0, R)}$ and define

$$\zeta = \eta Z, \quad Z = \eta z_h.$$

Substituting this ζ in the equations (1.11.3'), we obtain

$$\int\limits_{B(x_0, Ra)} A_h \{(\nabla Z + \nabla \eta \cdot z_h) \cdot [a_h \cdot (\nabla Z - \nabla \eta \cdot z_h) + b_h P_h Z + \eta e_h P_h] + $$
$$+ Z \cdot [c_h P_h \cdot (\nabla Z - \nabla \eta \cdot z_h) + d_h Q_h Z + f_h Q_h]\} dx = 0$$

from which we conclude in the usual manner (since $P_h^2 \leq Q_h$) that

$$\int\limits_{B(x_0, R+a)} A_h |\nabla Z|^2 dx \leq \int\limits_{B(x_0, R+a)} (Z_1 A_h Q_h Z^2 + Z_2 |\nabla \eta|^2 A_h z_h^2 + Z_3 A_h Q_h) dx$$
$$Z_i = Z_i(\nu, N, m_1, M_1, k, K).$$

Now, we observe that the function $z(x + h e_\gamma)$ $(= z + \Delta z)$ satisfies the hypotheses of Lemma 5.9.1 for the functions $*A$ and $*B$ defined by $*A(x, z, p) = A(x + h e_\gamma, z, p)$, etc., which obviously satisfy (1.10.8'') etc. Thus we may apply that lemma with $V(x)$ replaced by

$$V(x + h e_\gamma) = 1 + |z + \Delta z|^2 + |p + \Delta p|^2.$$

By using the lemma in conjunction with the inequalities

$$c[(1 + |p_1|^2)^r + (1 + |p_2|)^r]$$
$$(5.9.4) \qquad \leq \int_0^1 [(1 + |p_1| + t \Delta p|^2]^r dt \quad (\Delta p = p_2 - p_1, \text{ etc.})$$
$$\leq C[(1 + |p_1|^2)^r + (1 + |p_2|^2)^r] \quad (r = -1 + k/2 \text{ or } k/2)$$

we find, if $R + a \leq R_1(z(x + h e_\gamma) \text{ also} \in C^0_\mu[\overline{B(x_0, R+a)}])$, that

$$\int\limits_{B(x_0, R+a)} Z_1 A_h Q_h |Z|^2 dx \leq Z_4 (R+a)^\mu \int\limits_{B(x_0, R+a)} A_h |\nabla Z|^2 dx.$$

Thus, if $R + a \leq R_2$, depending on the same quantities as R_1, we obtain the inequality (1.11.8) with $D = B(x_0, R)$, $D' = B(x_0, R+a)$ and also

$$\int\limits_{B(x_0, R)} A_h Q_h z_h^2 dx \leq Z_5 a^{-2} R^\mu \int\limits_{B(x_0, R+a)} A_h Q_h dx.$$

Using this last inequality as well as the others we see that the rest of the proof can be carried through as before except that now

$$A_h^{\frac{1}{2}} b_h P_h \cdot z_h, \quad A_h^{\frac{1}{2}} c_h \cdot \nabla z_h, \quad A_h^{\frac{1}{2}} Q_h^{\frac{1}{2}} z_h$$

converge only weakly in $L_2[B(x_0, R)]$ to their limits and, for each $\zeta \in Lip_c[B(x_0, R)]$, $A_h^{\frac{1}{2}} \nabla \zeta$, $A_h^{\frac{1}{2}} P_h \zeta$, and $A_h^{\frac{1}{2}} Q_h^{\frac{1}{2}} \zeta$ tend strongly to their limits.

The proof of Theorem 1.11.1'' is deferred to the next section since a new idea is required.

We proceed to the proof of Theorem 1.11.2. Here we must assume that $N = 1$ (see Remark below). However we need assume only the hypotheses and conclusions of Theorems 1.11.1, 1.11.1', and 1.11.1''. From Lemma 5.3.1 and these results we conclude that we may substitute

$$\zeta = \omega V^{-\varepsilon} p_\gamma, \quad \omega \in Lip_c D, \quad \omega(x) \geq 0$$

in the equations (1.11.9) or (1.11.9') and sum on γ. Let us consider the case (1.10.7) in which $V = (1 + z^2 + |p|^2)^{\frac{1}{2}}$, the other case is treated similarly. Using the relations

$$V V_{,\alpha} = p_\gamma p_{\gamma,\alpha} + z p_\alpha, \quad p_\gamma p_\gamma = |p|^2 = V^2 - 1 - z^2,$$

$$|\nabla V|^2 \leq |\nabla p|^2 + |p|^2, \quad \zeta_{,\alpha} = V^{-\varepsilon}[\omega_{,\alpha} p_\gamma + \omega(p_{\gamma,\alpha} - \varepsilon V^{-1} p_\gamma V_{,\alpha})]$$

we find that V satisfies

$$\int_D V^{k-1-\varepsilon}\{\omega_{,\alpha}(a^{\alpha\beta} V_{,\beta} + 'B^\alpha V) + \omega('C^\alpha V_{,\alpha} + 'D V)\} dx \leq 0$$

where

$$'B^\alpha = V^{-2}[-a^{\alpha\beta} z p_\beta + b^\alpha(V^2 - 1 - z^2) + e^{\alpha\beta} V p_\gamma]$$

$$'C^\alpha = (b^\alpha + c^\alpha) - \varepsilon' B^\alpha$$

$$'D = -g + V^{-2}[-(b^\alpha + c^\alpha) z p_\alpha + d(V^2 - 1 - z^2) + V f^\gamma p_\gamma]$$

g and $\varepsilon > 0$ being chosen so that

$$a^{\alpha\beta} p_{\gamma,\alpha} p_{\gamma,\beta} + V e^{\alpha\gamma} p_{\gamma,\alpha} - \varepsilon a^{\alpha\beta} V_{,\alpha} V_{,\beta} + g V^2 \geq 0.$$

It is clear that g may be taken bounded so all the coefficients are bounded and the result (5.3.4) follows with

$$W = V^{k-\varepsilon} = U^\lambda, \quad \lambda = 2(k - \varepsilon)/k, \quad P \equiv 1.$$

In the case (1.10.8), where $V = (1 + |p|^2)^{\frac{1}{2}}$ we obtain (5.3.4) with $P = V$ and somewhat different coefficients B, C, and D.

Remark. The reason that we must assume $N = 1$ in this proof is that if we set $\zeta^i = \omega V^{-\varepsilon} p^i_\gamma$ above we obtain terms like

$$\omega_{,\alpha} V^{-\varepsilon} p^i_\gamma a^{\alpha\beta}_{ij} p^j_{\gamma,\beta}$$

which give $\omega_{,\alpha} V^{-\varepsilon} a^{\alpha\beta}(V V_{,\beta} - z p_\beta)$ in the case $N = 1$ but yield nothing tractable for $N > 1$.

The results above apply only to interior domains and, except for the cases $k = \nu = 2$, to cases where $N = 1$. However, if we have a weak solution of (1.10.14) for the general N which is known to $\in C^1_\mu(D)$, at least for $D \subset\subset G$, then the equations (1.11.9) have bounded coefficients $\in C^0_\mu$ and the theory of § 5.5, as generalized in § 6.4 allows us to conclude that the $p^i_\gamma \in C^1_\mu$ so the $z^i \in C^2_\mu$. Then repeated applications of the theorems of

§ 5.6 and 6.3 to Euler's equations allows us to conclude the higher differentiability, analyticity, etc. If $N = 1$, all the higher differentiability results follow if we have a weak solution known to be Lipschitz. Thus, in the case of an integrand of the form $f(p)$, we conclude the differentiability of the Lipschitz solutions found in § 4.2 and announced in §§ 1.7 and 1.10. And, as was stated in § 1.10, Ladyzenska and Ural'tseva have showed that any bounded weak solution of (1.10.14) in the case (1.10.8″) is differentiable, in the case $N = 1$. But, for the general system (1.10.14) for N arbitrary, it is possible to conclude all the higher differentiability of a weak solution if it is known to $\in C^1(D)$, $D \subset G$. The writer (MORREY [10]) showed this using the space $L_{2,\mu}$ of § 5.5 (see also NIRENBERG [1] for the two dimensional case). But it is more interesting and the techniques are results are more important using the L_p theory which has been highly developed for higher order equations by AGMON, AGMON, DOUGLIS, and NIRENBERG ([1], [2]), BROWDER ([1], [2]) and which has been presented in §§ 5.5 and 5.6 for a single second order equation. We shall present the theory as it applies to higher order equations in Chapter 6.

5.10. The extremals in the case $1 < k < 2$

In this section, we prove Theorem 1.11.1″. The only difficulty in this is due to the fact that it is no longer true that $A_h(x) \geq 1$ and the places where A_h is near zero do not necessarily occur at the places where z_h is large. So to avoid this difficulty, we introduce a new idea, namely to replace our minimum problem by that of minimizing $I(z, G)$ among those functions for which a certain other integral $J(z, G) \leq$ some number K. This same idea will be used again in § 9.5 in connection with the parametric problem. The writer feels that a further exploitation of this idea could lead to fruitful results.

Definition 5.10.1. In this section we define

$$J(z, G) = (1/2) \int_G |\nabla z|^2 dx. \tag{5.10.1}$$

For each $z^* \in H_2^1(G)$, we denote by z_0 the harmonic function in G such that $z_0 - z^* \in H_{20}^1(G)$ and we define $K_0 = J(z_0, G)$.

Theorem 5.10.1. *Suppose $z^* \in H_2^1(G)$. Then for each $K > K_0$, there exists a function z_K which minimizes $I(z, G)$ in the family F_K of all z such that $z - z^* \in H_{20}^1(G)$ and $J(z, G) \leq K$. The vector z_K satisfies the equation*

$$\int_G [\zeta^i_{,\alpha}(\mu p^i_\alpha + f_{p^i_\alpha}) + \zeta^i f_{z^i}]\, dx = 0, \quad \zeta \in H_{20}^1(G), \tag{5.10.2}$$

for some $\mu = \mu(K) \geq 0$. There exists a sequence $\{K_n\} \to +\infty$ such that $K_n \cdot \mu(K_n) \to 0$. If $z_K \neq z_0$, μ is unique.

Proof. The existence follows immediately from Theorem 1.9.1. That μ is unique if $z_K \neq z_0$ is evident. If $J(z_K, G) < K$, it is obvious that

(5.10.2) holds with $\mu = 0$. If $K > K_0$ and $J(z_K, G) = K$, then $z_K \neq z_0$ and is not harmonic so that there exists a function $\zeta_1 \in Lip_c(G)$ such that

$$\int_G \zeta^i_{1,\alpha} z^i_{K,\alpha}\, dx = 1. \tag{5.10.3}$$

It follows that $J(z_K + \lambda \zeta_1, G)$ is increasing with λ. Suppose that ζ satisfies

$$\int_G \zeta^i_{,\alpha} z^i_{K,\alpha}\, dx = 0 \tag{5.10.4}$$

and let us define

$$Z(x, \lambda) = z_K(x) + \lambda \zeta(x) - c\lambda^2 \zeta_1(x)$$

where c is a constant. We easily conclude that $J[Z(\cdot, \lambda), G] \leq K$ for all λ with $|\lambda| < \lambda_0$ if c is sufficiently large and positive. Since $J(z_K, G) = K$ and z_K minimizes $I(z, G)$ as above, we deduce that

$$\int_G (\zeta^i_{,\alpha} f_{p^i_\alpha} + \zeta^i f_{z^i})\, dx = 0 \tag{5.10.5}$$

for each $\zeta \in Lip_c(G)$ which satisfies (5.10.4).

Now, let ζ be any vector $\in Lip_c(G)$. We can write

$$\zeta = \zeta^* + \lambda \zeta_1 \quad \text{where} \quad \lambda = \int_G \zeta^i_{,\alpha} z^i_{K,\alpha}\, dx. \tag{5.10.6}$$

Then ζ^* satisfies (5.10.4). Accordingly (5.10.5) holds for $\zeta^* = \zeta - \lambda \zeta_1$ so that (5.10.2) holds with

$$\mu = -\int_G (\nabla \zeta_1 \cdot f_p + \zeta_1 \cdot f_z)\, dx \tag{5.10.7}$$

as one sees by substituting the value of λ from (5.10.6). Since $I(z, G) \geq I(z_K, G)$ whenever $J(z, G) < K$ and since $J(z_K + \lambda \zeta_1, G)$ is increasing, we conclude that

$$\begin{aligned} &\lambda^{-1}[I(z_K - \lambda \zeta_1, G) - I(z_K, G)] \geq 0 \quad \text{for } 0 < \lambda < \lambda_1, \\ &\lim_{\lambda \to 0^+} \lambda^{-1}[I(z_K - \lambda \zeta_1, G) - I(z_K, G)] = \mu \end{aligned} \tag{5.10.8}$$

using (5.10.7). Finally, let $\varphi(K) = I(z_K, G)$. Evidently φ is non-increasing and

$$\varphi(K + \Delta K) \leq I(z_K + \lambda \zeta_1, G), \quad \Delta K = J(z_K + \lambda \zeta_1, G) - J(z_K, G).$$

Since $\Delta K/\lambda \to 1$ as $\lambda \to 0$, we find that

$$\varphi'(K) = -\mu(K) \quad \text{for almost all } K.$$

Then, it follows that μ is summable on $[K_0, \infty)$ and

$$\int_{K_0}^{\infty} \mu(K)\, dK \leq I(z_0, G) - I(z_1, G)$$

z_1 being a minimizing function for $I(z, G)$.

We now prove Theorem 1.11.1''.

Proof of Theorem 1.11.1''. Suppose first that $z^* \in C^1(\Gamma)$ for some $\Gamma \supset \bar{G}$. Then, if $Z - z^* \in H^1_{k0}(G)$, there exists a sequence $\{z_n\}$, each z_n being in $C^1(\Gamma)$ with $(z_n - z^*)|_G \in C^1_c(G)$ such that $z_n \to Z$ in $H^1_k(G)$. Consequently $I(z_K, G) \to \inf I(Z, G)$ for $Z - z^* \in H^1_{k0}(G)$ so that $z_K \rightharpoonup$ some z as K runs through a subsequence of the sequence $\{K_n\}$ of Theorem 5.10.1 and z minimizes $I(Z, G)$ among all $Z \ni Z - z^* \in H^1_{k0}(G)$.

Now for K fixed and equal to one of the K_n above, we replace ζ *in the equation* 5.10.2 by $\zeta_h(x) = h^{-1}[\zeta(x - h\, e_\gamma) - \zeta(x)]$ and conclude that the difference quotient z_{Kh} satisfies

$$\begin{aligned}\int_{D'} \{\zeta^i_{.\alpha}[\mu\, z^i_{Kh,\alpha} + A_h(a^{\alpha\beta}_{ijh} z^j_{Kh,\beta} + b^\alpha_{ijh} z^j_{Kh} + e^{\alpha\gamma}_{jh} P_h)] + \\ + \zeta^i A_h(b^\alpha_{jih} z^j_{Kh,\alpha} + c_{ijh} z^j_{Kh} + f^\gamma_{ih} P_h)\}\, dx = 0. \end{aligned} \tag{5.10.9}$$

By putting $\zeta^i = \eta^2 z^i_{Kh}$, we find in the usual way that

$$\int_{D'} \eta^2(\mu + A_h)\,|\nabla z_{Kh}|^2 dx \le C \int_{D'} [|\nabla\eta|^2 (\mu + A_h)\, z_h^2 + \eta^2 A_h (z_h^2 + P_h^2)]\, dx.$$

Since $J(z_K, G) \le K$, we may let $h \to 0$ to obtain the limiting equations corresponding to 5.10.9) and the inequality

$$\begin{aligned}\int_D (\mu + V_K^{k-2})\,|\nabla p_K|^2 dx &\le C\, a^{-2} \int_{D'} (\mu\, V_K^2 + V_K^k)\, dx \\ &= 2Ca^{-2} K \cdot \mu(K) + C\, a^{-2} \int_{D'} V_K^k\, dx \quad (D \subset D'_a).\end{aligned} \tag{5.10.10}$$

The right side of (5.10.10) remains bounded as $K \to \infty$ in our subsequence, since $K \cdot \mu(K) \to 0$ and $I(z_K, G) \to \min$. Also

$$\begin{aligned}\int_D |\nabla p_K|^k dx &= \int_D V_K^h \cdot V_K^{-h} |\nabla p_K|^k dx \\ &\le \left(\int_D V_K^k dx\right)^{1-k/2} \left(\int_D V_K^{k-2} |\nabla p_K|^2 dx\right)^{k/2} \quad (h = k(2-k)/2).\end{aligned} \tag{5.10.11}$$

From this it follows that $p_K \rightharpoonup p$ in $H^1_k(D)$, that $V_K^{-1+k/2} p_K \rightharpoonup V^{-1+k/2} p$ in $H^1_2(D)$, that $U = V^{k/2} \in H^1_2(D)$, and that we can let $K \to \infty$ in the limiting equations $(h = 0)$ (5.10.9) to obtain the equation (1.11.9).

Finally, suppose merely that $z^* \in H^1_k(G)$. Let z_0 be a function minimizing $I(Z, G)$ among all $Z \ni Z - z^* \in H^1_k(G)$. Let $\{G_n\}$ be a sequence of domains such that $\bar{G}_n \subset G_{n+1}$ for each n and $G = \bigcup_n G_n$. For each n, there exists a sequence of functions $\{z^*_{np}\}$, each $z^*_{np} \in C^1(G_{n+1})$, such that $z^*_{np} \to z_0$ in $H^1_k(G_n)$ as $p \to \infty$. For each n and p, choose a function $z_{np} = \lim z_{Knp}$ as in the first part of this proof. For each n, we can choose a subsequence of the z_{np} which converges weakly in $H^1_k(G_n)$ to some z_n. This function z_n is minimizing for I on G_n and $z_n - z_0 \in H^1_{k0}(G_n)$. For each n, the function $Z_n = z_n$ in G_n and $= z_0$ in $G - G_n$ is minimizing for $I(z, G)$ and $Z_n - z^* \in H^1_{k0}(G)$. The bounds for the integrals of $|\nabla p_n|^k$, $V_n^{k-2} |\nabla p_n|^2$, etc., on each $D \subset\subset G$ are uniform, so that there exists a subsequence of the Z_n which converges weakly toward a function z of the type sought.

5.11. The theory of Ladyzenskaya and Ural'tseva[1]

In this section, we assume that z is a bounded weak solution of (1.10.14) with $N = 1$ in which the A^α and B satisfy (1.10.8″) with $k > 1$. Since we may write

$$p_\alpha A^\alpha(x, z, p) = p_\alpha A^\alpha(x, z, 0) + p_\alpha p_\beta \int_0^1 A^\alpha_{p_\beta}(x, z, t p)\, dt$$

it follows easily that the hypotheses (1.10.8″) imply the existence of numbers m_1 and M_1 such that

$$\begin{aligned} p_\alpha A^\alpha(x, z, p) &\geq m_1 |p|^k - M_1, \quad |A| \leq M_1 (1 + |p|^{k-1}) \\ |B| &\leq M_1 (1 + |p|^k). \end{aligned} \quad (k > 1) \tag{5.11.1}$$

The most difficult part of this analysis is that of showing that $z \in C^0_\mu(G)$ for some $\mu > 0$. For that purpose, it is convenient to introduce the following De Giorgi type space:

Definition 5.11.1. $z \in \mathfrak{B}_k(G; L; \gamma; \delta)$ iff (i) $z \in H^1_k(G)$, (ii) $|z(x)| \leq L$ on G, (iii) for each sphere $B(x_0, r) \subset G$, we have

$$\int_{B(x_0, r/2)} |\nabla z|^k dx \leq \gamma r^{\nu-k}, \tag{5.11.2}$$

(iv) for each h_1 for which $z(x) - h_1 \leq \delta$ in $B(x_0, r)$, we have

$$\int_{A(x_0, h_1, r - r\sigma)} |\nabla z|^k dx \leq \gamma |A(x_0, h_1, r)| \left\{1 + (\sigma r)^{-k} \max_{x \in A(x_0, h_1, r)} [z(x) - h_1]^k\right\}, \quad 0 < \sigma < 1, \tag{5.11.3}$$

(v) for each h_2 such that $h_2 - z(x) \leq \delta$ on $B(x_0, r)$, we have

$$\int_{B(x_0, h_2, r - r\sigma)} |\nabla z|^k dx \leq \gamma |B(x_0, h_2, r)| \left\{1 + (\sigma r)^{-k} \max_{x \in B(x_0, h_2, r)} [h_2 - z(x)]^k\right\}, \quad 0 < \sigma < 1.$$

Here $A(x_0, h, r) = B(x_0, r) \cap A(h)$ and $B(x_0, h, r) = B(x_0, r) \cap B(h)$ where $A(h)$ and $B(h)$ are respectively the sets where $z(x) > h$ and $z(x) < h$.

Theorem 5.11.1. *Suppose* $z \in \mathfrak{B}_k(G; L; \gamma; \delta)$, $x_0 \in G$, *and* a *is the distance of* x_0 *from* ∂G. *Then*

$$\text{osc.}\,[z, B(x_0, r)] \leq C_0 (r/a)^{\mu_0}, \quad 0 \leq r \leq a, \quad \mu_0 > 0$$

where C_0 *and* μ_0 *depend only on* k, L, γ, δ, *and* G.

[1] For a complete account of their work, see their new book (Ladyzenskaya and Ural'tseva [3]).

It is convenient first to prove the following lemmas:

Lemma 5.11.1. *Suppose* $w \in H_1^1[B(x_0, r)]$. *Then*

$$\left\{ \int_{B(x_0,r)} |w(x) - \bar{w}|^{\nu/(\nu-1)} dx \right\}^{(\nu-1)/\nu} \leq C(\nu) \int_{B(x_0,r)} |\nabla w(x)|\, dx$$

where $\bar{w}$ *denotes the average value of* w.

Proof. From the homogeneity of the situation, we may assume that $r = 1$ and $\bar{w} = 0$. From Theorem 3.6.5, it follows that

$$\int_{B(0,1)} |w(x)|\, dx \leq C_1(\nu) \int_{B(0,1)} |\nabla w(x)|\, dx .$$

From Theorem 3.5.4, we conclude that there is a bounded operator $\mathfrak{E}$ from $H_1^1[B(0,1)]$ to $H_{10}^1[B(0,2)]$ such that if $W = \mathfrak{E}\, w$, then

$$\int_{B(0,2)} |\nabla W(x)|\, dx \leq C_2 \int_{B(0,1)} [|\nabla w(x)| + |w(x)|]\, dx$$
$$\leq C_3 \int_{B(0,1)} |\nabla w(x)|\, dx, \quad W(x) = w(x), \quad x \in B(0, 1).$$

Then, from SOBOLEV's lemma (Theorem 3.5.3), we conclude that

$$\int_{B(0,1)} |w(x)|^{\nu/(\nu-1)} dx \leq \int_{B(0,2)} |W(x)|^{\nu/(\nu-1)} dx \leq \left(C_4 \int_{B(0,2)} |\nabla W(x)|\, dx \right)^{\nu/(\nu-1)}$$

from which the lemma follows.

Lemma 5.11.2 (DE GIORGI [1]). *There is a constant* $\beta = \beta(\nu)$ *such that if* $u \in H_1^1(G)$ *and* $B(x_0, r) \subset G$ *then*

$$(\lambda - h)\, \tau^{1-1/\nu}(h, \lambda; r) \leq \beta \int_{A(x_0,h,r) - A(x_0,\lambda,r)} |\nabla u(x)|\, dx \tag{5.11.4}$$
$$h < \lambda, \quad \tau(h, \lambda; r) = \min[|A(x_0, \lambda, r)|,\ \gamma_\nu r^\nu - |A(x_0, h, r)|].$$

Proof. Let us define

$$w(x) = \begin{cases} 0, & \text{if } x \in B(x_0, r) - A(x_0, h, r) \\ u(x) - h, & \text{if } x \in A(x_0, h, r) - A(x_0, \lambda, r) \\ \lambda - h, & \text{if } x \in A(x_0, \lambda, r) \end{cases}$$

Then $\nabla w(x) = 0$ unless $x \in A(x_0, h, r) - A(x_0, \lambda, r)$ when $\nabla w(x) = \nabla u(x)$ (almost everywhere). Applying Lemma 5.11.1 and using the definition of τ, we obtain

$$\tau \cdot [(\lambda - h - \bar{w})^t + \bar{w}^t] \leq \left\{ C \int_{B(x_0,r)} |\nabla w(x)|\, dx \right\}^t, \quad t = \nu/(\nu - 1)$$

from which the lemma follows easily.

Lemma 5.11.3. *For each* $\sigma \in (0,1)$, *there exists a* $\theta(\sigma, \gamma, k, \nu) > 0$ *such that if* $u \in \mathfrak{B}_k(G, L, \gamma, \delta)$, $B(x_0, \varrho) \subset G$, *and* $H = \max[u(x) - h] \leq \delta$ $(x \in A(x_0, h, \varrho))$, *then*

$$|A(x_0, h + H + \sigma^2 \varrho, \varrho - \sigma\varrho)| = 0 \quad \textit{whenever} \quad |A(x_0, h, \varrho)| \leq \theta \cdot \varrho^\nu$$

The corresponding result holds for the sets $B(x_0, h, \varrho)$.

Proof. Define

$$\varrho_j = \varrho - \sigma\varrho + 2^{-j}\sigma\varrho, \; h_j = h + \sigma H + \sigma^2\varrho - 2^{-j}(\sigma H + \sigma^2\varrho),$$
$$j = 0, 1, 2, \ldots$$

We shall show that $|A(x_0, h_j, \varrho_j)| \to 0$. We apply the estimate (5.11.3) with $\varrho = \varrho_j$ and $\varrho - \sigma\varrho = \varrho_{j+1}$ to obtain

$$\int\limits_{A(x_0, h_j, \varrho_{j+1})} |\nabla u|^k dx \leq \gamma \cdot |A(x_0, h_j, \varrho_j)| \cdot \{1 + (\varrho_j - \varrho_{j+1})^{-k} H^k\}. \tag{5.11.5}$$

We now estimate the left side of (5.11.5) using (5.11.4) and the Hölder inequality and assuming that $\theta \cdot \varrho^\nu \leq \gamma_\nu (\varrho - \sigma\varrho)^\nu/2$ to obtain

$$\int\limits_A |\nabla u|^k dx \geq |A|^{1-k}\Big(\int\limits_A |\nabla u|\, dx\Big)^k \geq |A|^{1-k}\Big(\int\limits_{A-A'} |\nabla u|\, dx\Big)^k$$
$$\geq \beta^{-k} |A''|^{1-k} (h_{j+1} - h_j)^k |A'|^{k(1-1/\nu)}. \tag{5.11.6}$$

($A = A(x_0, h_j, \varrho_{j+1})$, $A' = A(x_0, h_{j+1}, \varrho_{j+1})$, $A'' = A(x_0, h_j, \varrho_j)$; for $1 - k < 0$, $|A''| \geq |A|$, and $|A'| \leq |B(x_0, \varrho_{j+1})| - |A|$ since $|A(x_0, h_0, \varrho_0)| = |A(x_0, h, \varrho)| \leq \theta \cdot \varrho^\nu \leq |B(x_0, \varrho - \sigma\varrho)|/2$. From (5.11.6) and (5.11.5) and the fact that $|A'| \leq |A''|$, we conclude that

$$|A'| \leq \beta\gamma^{1/k} |A''|^{1+1/\nu} (h_{j+1} - h_j)^{-1} (\varrho_j - \varrho_{j+1})^{-1} \Big[H^k + \Big(\frac{\sigma\varrho}{2^{j+1}}\Big)^k\Big]^{1/k}. \tag{5.11.7}$$

We now prove by induction that

$$|A(x_0, h_j, \varrho_j)| \leq \theta \cdot 2^{-2\nu j} \varrho^\nu, \quad j = 0, 1, 2, \ldots \tag{5.11.8}$$

provided that we take

$$\theta = \min\Big\{\frac{1}{2}\varrho^{-\nu} |B(x_0, \varrho - \sigma\varrho)|, \quad \sigma^{2\nu}/(2^{2\nu+2}\beta\gamma^{1/k})^\nu\Big\}.$$

For $j = 0$, (5.11.8) is just the condition $|A(x_0, h, \varrho)| \leq \theta \cdot \varrho^\nu$. If we assume that (5.11.8) holds for some j, it follows from (5.11.7) and the definition of θ that it holds for $j + 1$.

Proof of the theorem. We consider the balls

$$B_j \equiv B(x_0, \varrho_j), \; \varrho_j = 4^{-j} \cdot a, \; j = 0, 1, 2, \ldots$$

We denote by ω_j the oscillation of u on B_j and set

$$C_1 = \max[2M/\varrho_0, 2^{s+1}], \quad (\varrho_0 = a),$$

where the constant s will be chosen below. Then we have

$$\omega_0 \leq C_1 \varrho_0.$$

Suppose for some j that we have the opposite inequality

$$\omega_j > C_1 \varrho_j.$$

Then

$$\omega_j > 2^{s+1} \varrho_j. \tag{5.11.9}$$

We shall show that s can be chosen so that (5.11.9) implies

(5.11.10) $$\omega_j \leq \eta\, \omega_{j-1} + \varrho_j, \quad \eta = 1 - \frac{1}{2^{s+2}} < 1.$$

To that end, we set

$$\begin{aligned} \omega &= \omega_{j-1}, \ \bar{\mu} = (\mu_1 + \mu_2)/2 \\ \mu_1 &= \max u(x), \ \mu_2 = \min u(x), \ x \in B_{j-1} \\ D_t &= A(x_0, \mu_1 - 2^{-t}\omega, 2\varrho_j) - A(x_0, \mu_1 - 2^{-t-1}\omega, 2\varrho_j), \\ t &= 1, \ldots, s. \end{aligned}$$

We assume that $|A(x_0, \bar{\mu}, 2\varrho_j)| \leq |B(x_0, 2\varrho_j)|/2$; if not, we must have $|B(x_0, \bar{\mu}, 2\varrho_j)| \leq |B(x_0, 2\varrho_j)|/2$.

We now apply Lemma 5.11.2 and the Hölder inequality to the integral of $|\nabla u|$ over D_t to obtain

(5.11.11) $$|A(x_0, \mu_1 - 2^{-t-1}\omega, 2\varrho_j)|^{k(1-1/\nu)} \cdot \left(\frac{\omega}{2^{t+1}}\right)^k \leq \beta^k |D_t|^{k-1} \int_{D_t} |\nabla u|^k\, dx, \quad t = 1, \ldots, s.$$

We estimate the integral on the right in (5.11.11) using (5.11.3), (5.11.2) in case $2^{-t}\omega > \delta$, and the facts that $|A(x_0, \bar{\mu}, 2\varrho_j)| \leq |B(x_0, 2\varrho_j)|/2$, $\bar{\mu} = \mu_1 - 2^{-1}\omega$, and $u(x) \leq \mu_1$ on $B(x_0, \varrho_{j-1})$. We obtain

(5.11.12) $$\int_{D_t} |\nabla u|^k\, dx \leq \begin{cases} \gamma\, \varrho_{j-1}^{\nu-k}, & 2^{-t}\omega > \delta \\ \gamma |A(x_0, \mu_1 - 2^{-t}\omega, 2\varrho_j)| \cdot [1 + (2\varrho_j)^{-k}(2^{-t}\omega)^k], & 2^{-t}\omega \leq \delta. \end{cases}$$

Using the fact that $|A(x_0, \mu_1 - 2^{-t}\omega, 2\varrho_j)| \leq |A(x_0, \mu_1 - 2^{-1}\omega, 2\varrho_j) \leq |B(x_0, 2\varrho_j)|/2$, (5.11.12) becomes

(5.11.13) $$\int_{D_t} |\nabla u|^k\, dx \leq \begin{cases} \gamma \cdot 2^{\nu-k} \varrho_j^{\nu-k}, & 2^{-t}\omega > \delta \\ K\gamma\, \varrho_j^{\nu-k}[\varrho_j^k + (2^{-t-1}\omega)^k], & 2^{-t}\omega \leq \delta, \quad K = 2^{\nu-1}\gamma_\nu. \end{cases}$$

But now

$$K[\varrho_j^k + (2^{-t-1}\omega)^k] > 2^{\nu-1}\gamma_\nu \cdot (\delta/2)^k \quad \text{if} \quad 2^{-t}\omega > \delta$$

so that we may write (5.11.13) in the form

(5.11.14) $$\int_{D_t} |\nabla u|^k\, dx \leq K_1\, \varrho_j^{\nu-k}[\varrho_j^k + (2^{-t-1}\omega)^k], \quad K_1 = K_1(\gamma, \delta, k, \nu).$$

Taking into account our assumption (5.11.9) and noting that $\omega = \omega_{j-1} \geq \omega_j > 2^{s+1}\varrho_j \geq 2^{t+1}\varrho_j$, we obtain, from (5.11.11) and (5.11.14),

$$\begin{aligned} &|A(x_0, \mu_1 - 2^{-t-1}\omega, 2\varrho_j)|^{k(1-1/\nu)}(2^{-t-1}\omega)^k \\ &\leq \beta^k |D_t|^{k-1} \cdot 2K_1 \cdot \varrho_j^{\nu-k}(2^{-t-1}\omega)^k. \end{aligned}$$

In addition, $|A(x_0, \mu_1 - 2^{-t-1}\omega, 2\varrho_j)| \geq |A(x_0, \mu_1 - 2^{-s-1}\omega, 2\varrho_j)$ so that

(5.11.15) $$|A(x_0, \mu_1 - 2^{-s-1}\omega, 2\varrho_j)|^{k(\nu-1)/\nu(k-1)} \leq (2K_1\beta^k \varrho_j^{\nu-k})^{1/(k-1)} \cdot |D_t|.$$

We sum these inequalities over t from 1 to s to obtain

$$s \cdot |A(x_0, \mu_1 - 2^{-s-1}\omega, 2\varrho_j)|^\tau \le (2K_1\beta^k \varrho_j^{\nu-k})^{1/(k-1)} |B(x_0, 2\varrho_j)|$$

(5.11.16) $$(\tau = k(\nu - 1)/\nu(k-1)).$$

If s is so large that

$$[s^{-1} \cdot (2K_1\beta^k)\gamma_\nu \cdot 2^\nu \cdot \varrho_j^{\nu+(\nu-k)/(k-1)}]^{1/\tau} < \theta\left(\frac{1}{2}\right)(2\varrho_j^\nu)$$

(the ϱ_j occur to the same power on both sides) then

$$|A(x_0, \mu_1 - 2^{-s-1}\omega, 2\varrho_j)| < \theta\left(\frac{1}{2}\right)(2\varrho_j^\nu)$$

If, in addition, $2^s > M/\delta$, then $2^{-s-1}\omega < \delta$ and it follows from Lemma 5.11.3 (with $h = \mu_1 - 2^{-s-1}\omega$, $\sigma = 1/2$, $\varrho = 2\varrho_j$, $H = 2^{-s-1}\omega$) that

$$|A(x_0, \mu_1 - 2^{-s-1}\omega + 2^{-s-2}\omega + 2^{-1}\varrho_j, \varrho_j)| = 0.$$

This means (since $z(x) \ge \mu_2$ on B_{j-1}) that

$$\omega_j \le \omega_{j-1} - 2^{-s-2}\omega_{j-1} + \varrho_j$$

which is the desired inequality (5.11.10).

Thus we have shown that, for any j,

(5.11.17) $$\omega_j \le C_1\varrho_j \text{ or } \omega_j \le \eta\,\omega_{j-1} + \varrho_j, \quad 0 < \eta < 1.$$

Hence we certainly obtain

$$\omega_j \le \eta\,\omega_{j-1} + (C_1 + 1)\,4^{-j}\varrho_0.$$

By induction, we find that

$$\omega_j \le \eta^j\{\omega_0 + (C_1+1)[(4\eta)^{-1} + (4\eta)^{-2} + \cdots + (4\eta)^{1-j}]\} \le C_2\eta^j,$$
$$C_2 = \omega_0 + (C_1+1)/(4\eta - 1)$$

since we must have $\eta > 3/4$. The theorem then follows from the argument in the last part of the proof of Theorem 5.3.3.

Theorem 5.11.2. *Suppose z is a solution of* (1.10.14) *on G with $N = 1$, where the A^α and B satisfy* (5.11.1), *in which $z(x) \le L$ on G and $z \in H_k^1(G)$. Then $z \in B_k(G; L; \gamma; \delta)$ for some numbers γ and $\delta > 0$ which depend only on ν, m, M, K, k, and L.*

Proof. First we set

$$\zeta(x) = e^{\lambda z(x)}\eta^k(x)$$

where η is defined as usual to equal 1 on $B(x_0, r/2)$ and to vanish near $\partial B(x_0, r)$ and in $G - B(x_0, r)$ and λ is a positive constant. We obtain

$$\int_{B(x_0,r)} e^{\lambda z}[\lambda\eta^k A^\alpha p_\alpha + k\eta^{k-1}A^\alpha\eta_{,\alpha} + B\eta^k]\,dx = 0.$$

Using (5.11.1), we conclude that

$$\int_{B(x_0,r)} e^{\lambda z}[\lambda m_1\eta^k|\nabla z|^k - \lambda M_1\eta^k - kM_1|\nabla\eta|\cdot\eta^{k-1}(1 + |\nabla z|^{k-1})$$

(5.11.18) $$- M_1\eta^k(1 + |\nabla z|^k)]\,dx \le 0.$$

Using Young's inequality,

$$a^{k-1}\, b = k^{-1}(k-1)\,\varepsilon\, a^k + k^{-1}\,\varepsilon^{1-k}\, b^k, \tag{5.11.19}$$

we conclude from (5.11.18) and (5.11.19) with $\varepsilon = 1$ that

$$\int\limits_{B(x_0,r)} e^{\lambda z}\{[\lambda m_1 - k M_1](\eta|\nabla z|)^k - (\lambda + k) M_1 \eta^k - 2M_1|\nabla\eta|^k\} dx \leq 0.$$

By choosing $\lambda = 2k\, M_1/m_1$, for instance, we find that

$$\int\limits_{B(x_0,r/2)} |\nabla z|^k dx \leq \gamma\, r^{\nu-k}$$

since $e^{\lambda z}$ is bounded above and below.

Next choose h, σ with $0 < \sigma < 1$, and $B(x_0, r) \subset G$, and define

$$\zeta(x) = \eta^k(x)\,\omega(x), \quad \omega(x) = \begin{cases} z(x) - h, & \text{if } z(x) \geq h, \\ 0, & \text{if } z(x) \leq h, \end{cases}$$

where $\eta(x) = 1$ on $B(x_0, r - \sigma r)$ and vanishes on and outside $\partial B(x_0, r)$. Since $\omega_{,\alpha}(x) = z_{,\alpha}(x)$ if $z(x) \geq h$, we obtain

$$\begin{aligned} 0 &= \int\limits_{B(x_0,r)} [\eta^k A^\alpha \omega_{,\alpha} + k\,\eta^{k-1}\,\omega\,\eta_{,\alpha} A^\alpha + \eta^k\,\omega\, B]\, dx \\ &\geq \int\limits_{A(x_0,h,r)} [\eta^k(m_1|\nabla\omega|^k - M_1) - k M_1\,\omega|\nabla\eta|\eta^{k-1}(|\nabla z|^{k-1} + 1) - \\ &\qquad - \eta^k\,\omega M_1(|\nabla z|^k + 1)]\, dx. \end{aligned} \tag{5.11.20}$$

Now, we choose h so that $\omega(x) \leq \delta = m_1/4\, M_1$ and use (5.11.19) with $\varepsilon = m_1/4\, M_1\,(k-1)$ and we conclude from (5.11.20) that

$$\int\limits_{A(x_0,h,r-r\sigma)} |\nabla z|^k dx \leq \gamma \int\limits_{A(x_0,h,r)} [1 + (\sigma r)^{-k}\,\omega^k(x)]\, dx$$

from which property (iv) follows. Property (v) is proved similarly.

For what follows it is convenient to note that the hypotheses (1.10.8″) imply (5.11.1) which, in turn, imply that

$$p_\alpha A^\alpha(x, z, p) \geq m_2 V^k - M_2, \quad |A| \leq M_2 V^{k-1}, \quad |B| \leq M_2 V^k \tag{5.11.21}$$

where m_2 and M_2 depend only on m, M, K, and k.

Lemma 5.11.4. *Suppose that the hypotheses of Theorem 5.11.2 hold with $k \geq 2$. Then there are numbers $\varkappa$, $0 < \varkappa < 1$, and C_1 which depend only on m_2, M_2, C_0, μ_0, and ν, such that*

$$\int\limits_{B(x_0,r)} V^k \xi^2(x)\, dx \leq C_1 (r/a)^{\mu_0} \int\limits_{B(x_0,r)} V^{k-2} |\nabla\xi|^2 dx,$$

$$0 \leq r \leq \varkappa a, \quad B(x_0, a) \subset G, \quad a \leq 1,$$

for any bounded $\xi \in H^1_{k0}[B(x_0, r)]$.

Proof. By approximations, it is easy to see that if $r < a$ we may set

$$\zeta(x) = [z(x) - z(x_0)]\,\xi^2(x)$$

in (1.10.14). From this substitution and Theorems 5.11.1 and 5.11.2, it follows that

$$\int_{B(x_0,r)} \{\xi^2[A^\alpha p_\alpha + (z-z_0)B] + 2\xi\,\xi_{,\alpha}(z-z_0)A^\alpha\}dx = 0$$

$$\geq \int_{B(x_0,r)} \{\xi^2[m_2 V^k - M_2 - 2C_0(r/a)^{\mu_0} M_2 V^k] - M_2 C_0 (r/a)^{\mu_0} V^{k-2} |\nabla\xi|^2 dx$$

from which the result follows easily, since

$$\int_{B(x_0,r)} \xi^2 dx \leq (r^2/2) \int_{B(x_0,r)} |\nabla\xi|^2 dx$$

by Poincaré's Inequality and $V^{k-2} \geq 1$.

As a consequence of this lemma and Theorems 5.11.1 and 5.11.2, we obtain the following useful theorem:

Theorem 5.11.3. *Suppose the hypotheses of Theorem 5.11.2 hold with $k \geq 2$. Then if $D \subset\subset G$, there are constants $R_0 > 0$, C_2 and C_3, depending only on ν, m, M, K, k, L, and D such that $z \in C^0_\mu(\bar{D})$ with*

$$h_\mu(z, \bar{D}) \leq C_2, \quad \int_{B(x_0,r)} V^k \xi^2 dx \leq C_3 R^{\mu_0} \int_{B(x_0,R)} V^{k-2} |\nabla\xi|^2 dx \quad R \leq R_0$$

for each bounded $\xi \in H^1_{k0}[B(x_0, R)]$, $B(x_0, R) \subset D$.

We can now prove the analog of Theorem 1.11.1.

Theorem 5.11.4. *Suppose that the hypotheses of Theorem 5.11.2 hold with $N = 1$, $k \geq 2$, and suppose $D \subset\subset G$. Then the p_γ and the function $U = V^{k/2} \in H^1_2(D)$ and they satisfy*

$$(1.11.9') \qquad \int_D V^{k-2}\{\zeta_{,\alpha}(a^{\alpha\beta} p_{\gamma,\beta} + b^\alpha V p_\gamma + e^{\alpha\gamma} V) + \zeta V(c^\alpha p_{\gamma,\alpha} + dV p_\gamma + f^\gamma V)\}dx = 0$$

$$(1.11.10) \qquad \int_D V^{k-2} |\nabla p|^2 dx \leq C(\nu, m, M, K, k, L, D) \int_G V^k dx.$$

In addition $V^{k+2} \in L_1(D)$.

Proof. We let D' be a domain such that $D \subset\subset D' \subset\subset G$ and apply the difference quotient procedure. This leads to the h-equations (1.11.3′) ((1.11.3) modified as described in § 5.9 in the proof of Theorem 1.11.1′); for $0 < |h| < h_0$, z_h is Hölder continuous.

Now, let us suppose that $R + a \leq R_0(\ldots, D')$, suppose $B_{R+a} \equiv B(x_0, R+a) \subset D'$, let $\eta = 1$ on B_R, etc., and set

$$\zeta = \eta Z_h, \quad Z_h = \eta z_h$$

in the equations (1.11.3′). The result is (see the notation of § 5.9).

$$\int_{B_{R+a}} A_h\{(\nabla Z_h + \nabla\eta \cdot z_h) \cdot [a_h \cdot (\nabla Z_h - \nabla\eta\, z_h) + b_h P_h z_h + \eta e_h P_h] + Z_h[c_h P_h \cdot (\nabla Z_h - \nabla\eta\, z_h) + d_h Q_h Z_h + f_h Q_h]\}dx = 0$$

from which we easily conclude as usual (since $P_h^2 \leq Q_h$)

$$(5.11.22)\quad \int\limits_{B_{R+a}} A_h |\nabla Z_h|^2 dx \leq \int\limits_{B_{R+a}} A_h (Z_1 Q_h Z_h^2 + Z_2 \eta^2 Q_h + Z_3 |\nabla \eta|^2 z_h^2)\, dx.$$

Now, by using the fact that $z(x + h e_\gamma)$ is a solution of (1.10.14) for the functions $A^\alpha(x + h e_\gamma, z, p)$ and $B(x + h e_\gamma, z, p)$ and using the inequalities (5.9.4), we see as in § 5.9 that

$$\int\limits_{B_{R+a}} A_h Q_h Z_h^2 dx \leq Z_4 \int\limits_{B_{R+a}} [V^k(x) + V^k(x + h e_\gamma)] Z_h^2(x)\, dx$$

$$\leq Z_4 C_3 (R + a)^{\mu_0} \int\limits_{B_{R+a}} [V^{k-2}(x) + V^{k-2}(x + h e_\gamma)] |\nabla Z_h|^2 dx$$

$$\leq Z_5 (R + a)^{\mu_0} \int\limits_{B_{R+a}} A_h |\nabla Z_h|^2 dx.$$

So, if we assume $Z_1 Z_5 (R + a)^{\mu_0} \leq 1/2$, we obtain

$$(5.11.23)\quad \int\limits_{B_{R+a}} A_h |\nabla Z_h|^2 dx \leq Z_6 \int\limits_{B_{R+a}} A_h (\eta^2 Q_h + |\nabla \eta|^2 z_h^2)\, dx$$

$$(5.11.24)\quad \int\limits_{B_{R+a}} A_h Q_h Z_h^2 dx \leq Z_7 \int\limits_{B_{R+a}} A_h (\eta^2 Q_h + |\nabla \eta|^2 z_h^2)\, dx.$$

We may now let $h \to 0$ in these inequalities and in (1.11.3′) to obtain the results.

Next we prove a preparatory lemma like Lemma 5.3.3.

Lemma 5.11.2. *Suppose the hypotheses and conclusion of Theorem 5.11.4 hold and suppose* $w = U^\tau = V^{k\tau/2} \in L_2(D)$ *for some* $\tau > 1$. *Then* $w \in H_2^1(\Delta)$ *for each* $\Delta \subset\subset D$ *and if* $B(x_0, R + a) \subset D$,

$$\int\limits_{B(x_0,R)} |\nabla w|^2 dx \leq C_4 \tau^\varrho a^{-2} \int\limits_{B(x_0,R+a)} w^2 dx, \quad 0 < a \leq R,$$

$$\varrho = 1 + 2\mu_0^{-1}, \quad C_4 = C_4(\nu, m, M, K, k, D);$$

C_4 does not depend on τ.

Proof. It follows from Theorem 5.11.4 and Lemma 5.3.1 with $\psi = V^{-1+k/2}$ and $\omega = \eta^2 V_L^\lambda p_\gamma$ that we may set

$$\zeta = \eta^2 V_L^\lambda p_\gamma, \quad \lambda = k(\tau - 1)$$

in the equations (1.11.9′), V_L being the truncated function for V (see (5.3.1)). Since $V_{L,\alpha} = 0$ when $V \geq V_L$, we see that

$$\zeta_{,\alpha} = V_L^\lambda (\eta^2 p_{\gamma,\alpha} + \eta^2 V^{-1} p_\gamma V_{L,\alpha} + 2\eta \eta_{,\alpha} p_\gamma)$$

$$V V_{,\alpha} = p_\gamma p_{\gamma,\alpha} \quad |\nabla V| \leq |\nabla p|.$$

Using these results we obtain

$$\int\limits_{B_{R+a}} V^{k-2} V_L^\lambda \eta^2 (|\nabla V|^2 + \lambda |\nabla V_L|^2)\, dx$$

$$\leq \int\limits_{B_{R+a}} (Z_1 V^{k+2} V_L^\lambda \eta^2 + Z_2 |\nabla \eta|^2 V^k V_L^\lambda)\, dx.$$

Now let

$$\omega_L = \eta V V_L^{\lambda/2}, \quad w_L = \eta V^{k/2} V_L^{\lambda/2}$$
$$\nabla \omega_L = V_L^{\lambda/2}(\eta \nabla V + V \nabla \eta + \lambda \eta \nabla V_L/2)$$
$$\nabla w_L = V^{-1+k/2} V_L^{\lambda/2}[(k\eta \nabla V/2] + (\lambda \eta \nabla V_L/2) + V \nabla \eta].$$

Then, since $k\tau = (\lambda + k)/k$, we conclude that

$$\int_{B_{R+a}} \eta^2 V^{k+2} V_L^\lambda dx = \int_{B_{R+a}} V^k \omega_L^2 dx$$
$$\leq C_3(R+a)^{\mu_0} \int_{B_{R+a}} \{Z_3 V^{k-2} V_L^\lambda \eta^2 \tau(|\nabla V|^2 + \lambda |\nabla V_L|^2) + $$
$$+ Z_4 |\nabla \eta|^2 V^k V_L^\lambda\} dx$$
$$\leq C_3 Z_1 Z_3 \tau (R+a)^{\mu_0} \int_{B_{R+a}} \eta^2 V^{k+2} V_L^\lambda dx +$$
$$+ C_3(R+a)^{\mu_0} \int_{B_{R+a}} (Z_2 Z_3 \tau + Z_4) |\nabla \eta|^2 V^k V_L^\lambda dx.$$

Let us now define α by

$$2C_3 Z_1 Z_3 \tau \alpha^{\mu_0} = 1, \quad \alpha \leq R_0,$$

replace a by $a/2$, set $R = a/2$, assume $B_a \subset G$ and suppose $a \leq \alpha$. Then

$$\int_{B(x_0,a)} |\nabla w_L|^2 dx \leq Z_4 \tau \int_{B(x_0,a)} |\nabla \eta|^2 V^k V_L^\lambda dx.$$

We may now let $L \to \infty$ and use the definition of η to conclude

$$\int_{B(x_0,a/2)} |\nabla w|^2 dx \leq Z_5 \tau a^{-2} \int_{B(x_0,a)} w^2 dx$$

which is exactly (5.3.16). The remainder of the proof is the same as that of the proof of Lemma 5.3.3; the different result for ϱ here is due to the different definition of α.

Theorem 5.11.5. *Suppose the hypotheses and conclusions of Theorem* 5.11.4 *hold. Then U is bounded on each $\Delta \subset\subset D$.*

Proof. The proof is exactly the same as that of Theorem 5.3.1.

Remarks. The higher differentiability results now follow as in the other cases. But we have assumed above that $k \geq 2$. LADYZENSKAYA and URAL'TSEVA have extended Lemma 5.11.4 to the cases $1 < k < 2$; it follows that Theorem 5.11.3 extends to those cases also. So the proof of Theorem 5.11.4 goes through as far as the equations (5.11.23) and (5.11.24). However, the writer has not found a way in these cases to show that

$$\int_{B_{R+a}} |\nabla \eta|^2 A_h z_h^2 dx$$

is uniformly bounded. Since the authors above (L and U) gave no hint as to why this is so, they may have overlooked this point. This is just the difficulty which caused the writer to resort to the argument in § 5.10 which does not apply in this case since we have assumed z bounded and would have to show that the z_K were uniformly bounded. However, once the results of Theorem 5.11.4 are shown to hold, then Theorem 5.11.5 will follow also.

5.12. A class of non-linear equations

In this section we present essentially the treatment of LERAY and LIONS of a theory of non-linear equations initiated by MINTY, BROWDER [3], and VISIK. Each of these latter three authors has written several papers on this subject; we have quoted a recent one of each author. The equations which can be solved by this method are in some respects more general than those discussed in the preceding three sections, since higher order equations and systems can be handled but not all of those previously discussed, namely those satisfying the conditions (1.10.8″), can be handled.

Lemma 5.12.1. *Suppose $u = f(x)$ is a mapping from R_m into itself such that*

$$\lim_{|x| \to +\infty} |x|^{-1} x \cdot f(x) = +\infty. \tag{5.12.1}$$

Then the range of f is the whole of R_m.

Proof. Let $u_0 \in R_m$ and define $f^*(x) = f(x) - u_0$. Then f^* satisfies (5.12.1). Consequently it is sufficient to prove that the range of any map satisfying (5.12.1) contains the origin. Using (5.12.1) we see that we may choose R large enough so that

$$|x|^{-1} x \cdot f(x) \geq 1 \quad \text{for} \quad |x| = R. \tag{5.12.2}$$

But from (5.12.2), it follows that the mapping

$$\omega = f(R\,\xi)/|f(R\,\xi)|, \quad |\xi| = 1$$

from $\partial B(0,1)$ into itself is homotopic to the identity. It follows that there is a solution x in $B(0, R)$ of $f(x) = 0$.

We begin with the statement and proof of the abstract theorem.

Theorem 5.12.1. *Suppose that V is a separable, reflexive, Banach space, V' is its dual* (the space of linear functionals), *and A is an (possibly non-linear) operator from the whole of V into V' which is continuous in the weak topology of V' on finite dimensional subspaces of V* (i.e. *if $x_n \to x_0$ in a finite dimensional subspace of V, then $A x_n \rightharpoonup$ (weak convergence) $A x_0$ in V'*). *We suppose further that* (see notation below)

$$\frac{(A(w), v)}{\|v\|} \to +\infty \quad \textit{as} \quad \|v\| \to +\infty \quad (\textit{coerciveness}). \tag{5.12.3}$$

Finally we assume the existence of an operator $\mathfrak{A}$ from $V \times V$ to V' which has the following properties:

(i) *$\mathfrak{A}(u, v)$ is defined everywhere and carries bounded sets into bounded sets and*

$$\mathfrak{A}(u, u) = A(u). \tag{5.12.4}$$

(ii) $\mathfrak{A}(u, v)$ *has the continuity properties of A as a function of v and*

(5.12.5) $(\mathfrak{A}(u, u) - \mathfrak{A}(u, v), u - v) \geq 0$ *for all* (u, v) (monotonicity).

(iii) *If* $u_n \rightharpoonup u$ in V *and*

(5.12.6) $$(\mathfrak{A}(u_n, u_n) - \mathfrak{A}(u_n, u), u_n - u) \to 0$$

then $\mathfrak{A}(u_n, v) \rightharpoonup \mathfrak{A}(u, v)$ *for each* $v \in V$.

(iv) *If* $u_n \rightharpoonup u$ *and* $\mathfrak{A}(u_n, v) \rightharpoonup v^*$, *then* $(\mathfrak{A}(u_n, v), u_n) \to (v_*, u)$.
Then the range of A is the whole of V'.

Notation. If $v \in V$ and $v' \in V'$, then (v', v) means $v'(v)$.

Proof. Let $\{w_j\}$ be a basis for V and let V_m be the space spanned by $w_1, \ldots, w_m$. The equation

(5.12.7) $$(A(u), v) = (f, v), \quad u, v \in V_m$$

is reduced to a mapping

$$f^j = A^j(x_1, \ldots, x_m), \quad t = 1, \ldots, m$$

and the condition (5.12.3) is reduced to (5.12.1) by the substitution

$$u = \sum_{j=1}^{m} x_j w_j, \quad v = \sum_{j=1}^{m} y_j w_j, \quad (f, v) = \sum_{j=1}^{m} f^j y_j,$$

etc. Thus, for each m, there is a solution u_m of (5.12.7). Since

$$(A(u_m), u_m) = (f, u_m) \leq \|f\|_* \cdot \|u_m\|$$

it follows from (5.12.3) that $\|u_m\| \leq M$ and hence that $\|A(u_m)\|_* \leq M_*$ (using (i)) for some M and M^* and all m. We may therefore extract a subsequence $\{u_n\}$ such that

(5.12.8) $$u_n \rightharpoonup u \text{ in } V, \quad A(u_n) \rightharpoonup \chi \text{ in } V', \quad \mathfrak{A}(u_n, u) \rightharpoonup \psi \text{ in } V'.$$

Since (5.12.7) holds with $u = u_n$ for all v in V_n, it follows that

(5.12.9) $$\chi = f$$

since $(f, v) = (A(u_n), v) \to (\chi, v)$ for $v \in U V_m$ which is dense in V.

We next note that

$$\lim_{n\to\infty}(\mathfrak{A}(u_n, u_n) - \mathfrak{A}(u_n, u), u_n - u) = \lim(A(u_n), u_n)$$

(5.12.10) $$-\lim(A(u_n), u) - \lim(\mathfrak{A}(u_n, u), u_n) + \lim(\mathfrak{A}(u_n, u), u)$$
$$= \lim(f, u_n) - (\chi, u) - (\psi, u) + (\psi, u) = 0$$

on account of (5.12.8), (5.12.9), and hypothesis (iv).

Now, from this result and (iii) it follows that

$$\mathfrak{A}(u_n, v) \rightharpoonup \mathfrak{A}(u, v) \text{ for each } v \in V.$$

Then as in (5.12.10), we conclude that

$$\lim_{n\to\infty}(\mathfrak{A}(u_n, u_n) - \mathfrak{A}(u_n, v), u_n - v) = \lim(A(u_n), u_n) -$$
$$- \lim(A(u_n), v) - \lim(\mathfrak{A}(u_n, v), u_n) + \lim(\mathfrak{A}(u_n, v), v)$$
$$= (\chi, u) - (\chi, v) - (\mathfrak{A}(u, v), u) + (\mathfrak{A}(u, v), v).$$

Thus

$$(\chi - \mathfrak{A}(u, v), u - v) \geq 0 \text{ for all } v. \tag{5.12.11}$$

If, in (5.12.11), we set $v = u - \xi w$, $w \in V$, $\xi > 0$, we obtain

$$(\chi - \mathfrak{A}(u, u - \xi w), w) \geq 0 \text{ for all } w \in V.$$

By letting $\xi \to 0^+$ and using the continuity along lines we obtain

$$(\chi - \mathfrak{A}(u, u), w) \geq 0 \text{ for all } w \in V$$

from which it follows at once that

$$A(u) = \mathfrak{A}(u, u) = \chi = f.$$

We now present the existence theorem of VIŠIK essentially as extended by LERAY and LIONS. We have allowed u to be a vector. The theorem concerns equations of the form

$$(A\, u, w) \equiv \int_G \sum_{i=1}^{N} \sum_{|\alpha| \leq m_i} A_i^\alpha(x, \delta\, u, D^m u) D^\alpha w^i\, dx = (f, w) \tag{5.12.12}$$

where $u = (u^1, \ldots, u^N)$ and $w = (w^1, \ldots, w^N)$ range over a reflexive Banach space $V = V_1 \times \cdots \times V_N$ in which

$$H_{p0}^{m_i}(G) \subset V_i \subset H_p^{m_i}(G), \quad m_i \geq 1, \quad i = 1, \ldots, N,$$

and, as in Theorem 5.12.1, (f, w) means $f(w)$, f being a linear functional on V; here $\delta\, u$ denotes the tensor $\{D^\alpha u^i\}$ where $0 \leq |\alpha| \leq m_i - 1$ and $D^m u$ denotes the tensor $\{D^\alpha u^i\}$ with $|\alpha| = m_i$. It can be shown, as in §§ 1.3 and 1.10 that if the vector u minimizes the integral

$$\int_G F(x, \delta\, u, D^m u)\, dx$$

among all vectors u in a space such as V, then u satisfies an equation of the form (5.12.12) where

$$A_i^\alpha(x, \delta\, u, D^m u) = \partial F / \partial p_\alpha^i \quad (p_\alpha^i = D^\alpha u^i).$$

Thus the existence theorem is an alternative method to that of the calculus of variations.

We shall first prove the theorem under somewhat simplified assumptions concerning the A_i^α; we shall indicate later how our simplified assumptions may be relaxed somewhat. We shall assume that G is strongly Lipschitz (see § 3.4) unless all the $V_i = H_{p0}^{m_i}(G)$ in which case we shall allow G to be merely bounded. Concerning the A_i^α we shall assume.

(a) Each A_i^α is measurable in (x, ξ, η) and continuous in (ξ, η) for almost all x on G; here $\xi = \{\xi_\alpha^i\}$ where $|\alpha| \leq m_i - 1$ and $\eta = \{\eta_\alpha^i\}$ where $|\alpha| = m_i$.

(b) We assume that for all (x, ξ, η), we have

$$|A_i(x, \xi, \eta)| \leq C\left\{1 + \sum_j \left[\sum_{|\beta| < m_j} |\xi_\beta^j|^{p-1} + \sum_{|\beta| = m_j} |\eta_\beta^j|^{p-1}\right]\right\}. \tag{5.12.13}$$

(c) We assume a strong monotonicity condition:

$$\sum_i \sum_{|\alpha|=m_i} [A_i^\alpha(x, \xi, \eta_2) - A_i^\alpha(x, \xi, \eta_1)] \cdot (\eta_{2\alpha}^i - \eta_{1\alpha}^i)$$

$$\geq \varrho(|\eta_2 - \eta_1|, R, S, x) \quad \text{if} \quad |\xi| \leq R, |\eta_1| \leq S, x \in G \tag{5.12.14}$$

where $\varrho(r, R, S, x) > 0$ whenever $r > 0$ and is non-decreasing in r.

Theorem 5.12.2. *We assume that G and the A_i^α satisfy the conditions above and we assume that the operator A defined in* (5.12.12) *satisfies the coerciveness condition* (5.12.3). *Then the equation* (5.12.12) *has a solution $u \in V$ for each f in V'.*

Proof. Let us define the functional $\mathfrak{A}(u, v)$ by

$$(\mathfrak{A}(u, v), w) = \int_G \sum_i \Big\{ \sum_{|\alpha|=m_i} A_i^\alpha[x, \delta u(x), D^m v(x)] D^\alpha w^i(x) +$$

$$+ \sum_{|\alpha|<m_i} A_i^\alpha[x, \delta u(x), D^m u(x)] D^\alpha w^i(x) \Big\} dx. \tag{5.2.15}$$

It is clear that the conditions on the A_i^α and G ensure that $\mathfrak{A}(u, v)$ is defined for all (u, v), $\mathfrak{A}$ carries bounded sets in $V \times V$ into bounded sets in V', and if $u_n \to u$ and $v_n \to v$ in a manifold of finite dimensionality in $V \times V$, then $\mathfrak{A}(u_n, v_n) \rightharpoonup \mathfrak{A}(u, v)$. Clearly $\mathfrak{A}(u, u) = A(u)$. The condition (5.12.5) follows immediately from the condition (5.12.14). We need only to verify conditions (iii) and (iv) of Theorem 5.12.1; our theorem will then follow from that theorem.

To prove (iii), let us suppose that $u_n \rightharpoonup u$ and that (5.12.6) holds. Let v and $w \in V$ and let $\{r\}$ be any subsequence of $\{n\}$. Since $D^\alpha u_n^i \to D^\alpha u^i$ in $L_p(G)$ if $|\alpha| < m_i$ (Rellich's theorem, Theorem 3.4.4), it follows that we may choose a subsequence $\{s\}$ of $\{r\}$ such that $\delta u_s(x) \to \delta u(x)$ for almost every x. Thus, for such x, $\delta u_s(x)$ is bounded with respect to s. Thus, the condition (5.12.6) together with (5.12.14) implies that

$$\lim_{s\to\infty} \int_G \varrho[|D^m(u_s - u)|, R(x), |D^m u(x)|, x] \, dx = 0$$

$R(x)$ being a common bound for $|\delta u_s(x)|$ and $|\delta u(x)|$. It follows that we may extract a further subsequence $\{t\}$ such that we also have $D^\alpha u_t^i(x) \to D^\alpha u^i(x)$ almost everywhere if $|\alpha| = m_i$. Thus, if we define

$$f_{ni}^\alpha(x) = \begin{cases} A_i^\alpha[x, \delta u_n(x), D^m v(x)], & \text{if } |\alpha| = m_i \\ A_i^\alpha[x, \delta u_n(x), D^m u_n(x)], & \text{if } |\alpha| < m_i \end{cases} \tag{5.12.16}$$

f_i^α being defined similarly, then we see that $f_{ti}^\alpha(x) - f_i^\alpha(x) \to 0$ almost everywhere. From (5.12.13) we conclude that

$$|A_i^\alpha(x, \xi, \eta)|^{p'} \leq C'\{1 + |\xi|^p + |\eta|^p\}, \quad p' = p/(p-1). \tag{5.12.17}$$

From (5.12.17) and the uniform boundedness of $\|u_n\|$, it follows that $\|f_{ni}^\alpha\|_{p'}$ are uniformly bounded. Also, since v is a fixed vector and $\delta u_t(x) \to \delta u(x)$ in $L_p(G)$ we see that the set functions

$$\int_e |f_{ti}^\alpha(x) - f_i^\alpha(x)|^{p'} dx, \quad |\alpha| = m_i$$

are uniformly absolutely continuous. Thus $f_{ti}^{\alpha} \to f_i^{\alpha}$ in $L_{p'}(G)$ if $|\alpha| = m_i$. Since $|D^\alpha w^i| \in L_q(G)$ for some $q > p$ if $|\alpha| < m_i$, it follows that

$$(\mathfrak{A}(u_t, v) - \mathfrak{A}(u, v), w) \to 0, \quad w \in V.$$

Since $\{r\}$ was any subsequence the result follows.

To prove (iv), we assume that $v \in V$, $u_n \rightharpoonup u$ in V, and $\mathfrak{A}(u_n, v) \rightharpoonup v_*$ in V'. Again, let $\{r\}$ be any subsequence of $\{n\}$. Since $u_n \rightharpoonup u$ it follows as above that the norms $\|f_{ni}^{\alpha}\|_{p'}$ are uniformly bounded. Thus we may extract a further subsequence $\{s\}$ of $\{r\}$ such that $\delta u_s \to \delta u$ in $L_p(G)$, $\delta u_s(x) \to \delta u(x)$ almost everywhere, and $f_{s\alpha}^i \rightharpoonup$ some functions f_α^i in $L_{p'}(G)$. As in the preceding paragraph it follows that $f_{si}^{\alpha} \to f_i^{\alpha}$ in $L_{p'}(G)$ if $|\alpha| = m^i$ where f_i^{α} has its previous significance if $|\alpha| = m_i$. Thus v_* is given by

$$(v_*, w) = \int_G \sum_i \sum_{|\alpha| \le m_i} f_i^{\alpha}(x)\, D^\alpha w^i(x)\, dx.$$

From the strong convergence of f_{si}^{α} to f_i^{α} and the weak convergence of $D^\alpha u_s^i$ to $D^\alpha u^i$ if $|\alpha| = m_i$ and from the weak convergence of f_{si}^{α} to f_i^{α} and the strong convergence of $D^\alpha u_s^i$ to $D^\alpha u^i$ when $|\alpha| < m_i$, the desired result in (iv) holds for the subsequence $\{s\}$. But since $\{r\}$ was arbitrary, the desired result follows for the whole sequence $\{n\}$.

Remark 1. The reader can easily verify that the theorem still holds if the condition (5.12.13) on the A_i^{α} is replaced by

$$|A_i^{\alpha}(x, \xi, \eta)| \le C\left\{K(x) + \sum_j \left[\sum_{|\beta| < m_j} |\xi_\beta^j|^{r(j,\beta)} + \sum_{|\beta| = m_j} |\eta_\beta^j|^{p-1}\right]\right\}, \quad |\alpha| = m_i,$$

$$|A_i^{\alpha}(x, \xi, \eta)| \le C\left\{L(x) + \sum_j \left[\sum_{|\beta| < m_j} |\xi_\beta^j|^{s(j,\beta)} + \sum_{|\beta| = m_j} |\eta_\beta^j|^{t(j,\beta)}\right]\right\}, \quad |\alpha| < m_i,$$

$$\begin{aligned} (5.12.18) \qquad & r^{-1}(j, \beta) > (p-1)^{-1}[1 - \nu^{-1}(m_j - |\beta|)p], \\ & t(j, \beta) < (p-1) + \nu^{-1} p(m_i - |\alpha|) \\ & s^{-1}(j, \beta) > \frac{\nu - p(m_j - |\beta|)}{\nu(p-1) + p(m_i - |\alpha|)}, \; r(j,\beta),\; s(j,\beta),\; t(j,\beta) > 1, \\ & K \in L_{p'},\; L \in L_{q(i,\alpha)},\; q(i, \alpha) > 1, \\ & q^{-1}(i, \alpha) > 1 - p^{-1} + \nu^{-1}(m_i - |\alpha|). \end{aligned}$$

The proof is essentially the same but makes stronger use of Sobolev's lemma.

It is instructive to specialize to the case where all the $m_i = 1$ and the A_i^{α} are differentiable with respect to all arguments. Then the condition (5.12.13) is implied by the first inequality in (1.10.7″). The last inequality in (1.10.7″) implies that

$$\begin{aligned}
&\sum_{i,\alpha} (\eta^i_{2\alpha} - \eta^i_{1\alpha})\,[A^\alpha_i(x, \xi, \eta_2) - A^\alpha_i(x, \xi, \eta_1)] \\
(5.12.19)\qquad &= \sum (\eta^i_{2\alpha} - \eta^i_{1\alpha})\,(\eta^j_{2\beta} - \eta^j_{1\beta}) \int_0^1 A^\alpha_{i\eta^j_\beta}[x, \xi, \eta_1 + t(\eta_2 - \eta_1)]\,dt \\
&\geq m_1\,|\eta_2 - \eta_1|^2 \int_0^1 [1 + |\xi|^2 + |\eta_1 + t\,(\eta_2 - \eta_1)|^2]^{-1+p/2}\,dt.
\end{aligned}$$

If $p \geq 2$, the condition (5.12.14) holds with $\varrho = m_1 r^2$. In case $1 < p < 2$, let us define

$$\varrho(r, R, S) = \inf I(s, \xi, \eta_1) \quad \text{for} \quad |s| \geq r,\ |\xi| \leq R,\ |\eta_1| \leq S,$$

$$I(s, \xi, \eta_1) = m_1 |s|^2 \int_0^1 [1 + |\xi|^2 + |\eta_1 + t\,s|^2]^{-1+p/2}\,dt.$$

If $\varrho(r, R, S) = 0$ for some R, S, and $r > 0$, $\exists$ $\{\xi_n\}$, $\{\eta_{1n}\}$, and $\{s_n\}$ such that $|\xi_n| \leq R$, $|\eta_{1n}| \leq S$, and $|s_n| \geq r$ and $I(s_n, \xi_n, \eta_{1n}) \to 0$. If, for some subseequence, $|s_n|$ remains bounded, a compacness argument shows that $I \not\to 0$. Thus we must have $|s_n| \to \infty$. But

$$[1 + |\xi_n|^2 + |\eta_{1n} + t\,s_n|^2] \leq 1 + R^2 + 2S^2 + 2|s_n|^2.$$

Consequently we see that $I(s_n, \xi_n, \eta_{1n}) \to +\infty$, a contradiction. Thus (5.12.14) holds in this case also. Clearly this monotonicity condition, then, is implied by the very strong ellipticity of the system as assumed in (1.10.7″).

Using (5.12.19) with $\eta_2 = \eta$ and $\eta_1 = 0$ and using the facts that

$$\begin{aligned}
t^{p-2}(1 + |\xi|^2 + |\eta|^2)^{-1+p/2} &\leq (1 + |\xi|^2 + t^2|\eta|^2)^{-1+p/2} \\
&\leq (1 + |\xi|^2 + |\eta|^2)^{-1+p/2}
\end{aligned}$$

we conclude that

$$\begin{aligned}
&\sum_i \sum_{|\alpha|=1} \eta^i_\alpha A^\alpha_i(x, \xi, \eta) + \xi^i A_i(x, \xi, \eta) \\
(5.12.20)\qquad &\geq m_2|\eta|^2 V^{p-2} - M_1|\eta| \cdot (1 + |\xi|^2)^{(p-1)/2} - M_1|\xi| \cdot V^{p-1} \\
&\geq m_3 V^p - M_3(1 + |\xi|^2)^{p/2}, \quad V = (1 + |\xi|^2 + |\eta|^2)^{1/2}.
\end{aligned}$$

In case all the $V_i = H^{m_i}_{p0}(G)$ (Dirichlet problem) then (5.12.20) guarantees the coerciveness condition (5.12.3) *provided the diameter of G is sufficiently small* (Poincaré's inequality). The conditions (1.10.7‴) and (1.10.8″) are sufficient for the theorem provided that $p > \nu$ and the coerciveness condition holds. But, in these latter cases the coerciveness condition is more of a restriction than it is in the case of assumptions (1.10.7″). The coerciveness condition corresponds roughly to the condition that $f(x, z, p) \geq m\,V^p - K$ for all (x, z, p).

Chapter 6

Regularity theorems for the solutions of general elliptic systems and boundary value problems

6.1. Introduction

In the preceding chapter, we developed rather completely the regularity properties of weak and strong solutions of certain first order variational problems and second order differential equations involving only one unknown function. Complete results concerning the solutions of the corresponding problems involving several unknown functions were obtained only in the case where $\nu = 2$; these results were obtained by the writer before the war and were described in Chapter 1. In the preceding chapter some first differentiability results were obtained for certain systems of equations but these results did not imply the continuity of the first derivatives. In 1952, the writer presented a paper (MORREY [10]) at the Arden House Conference on Partial Differential Equations, in which it was proved that any vector solution of class C^1 of a regular variational problem of class C^n_μ, $n \geq 2$, was also of class C^n_μ. These results still leave a gap in the theory for systems, which can only be filled by an extension of the DE GIORGI-NASH results, developed in § 5.3, to systems or some entirely new device. But, in proving the C^n_μ differentiability results for systems of equations, the writer was forced to use the very important formulas of F. JOHN ([1], [3]) for fundamental solutions and to use methods which are appropriate for the discussion of elliptic systems of higher order. Hence the inclusion of this chapter in this book.

Our presentation will be a modification of that to be found in the two important papers of AGMON, DOUGLIS, and NIRENBERG ([1] and [2]) although some ideas essentially due to BROWDER ([1], [2]) have been used. A brief but rather comprehensive treatment of the topics in this chapter is to be found in the last chapter of the recent book by HÖRMANDER [1]; he includes some topics such as existence theorems on the "index" of a problem, etc., not included here. Many people, particularly among the Russians, have contributed to the development of this theory and have proved supplementary results not stated here. Many additional papers are referred to in the papers to which we refer in this chapter. The theorems proved in this chapter will enable us to conclude the differentiability of the solutions of the higher dimensional PLATEAU problem discussed in Chapter 10. They will suggest differentiability hypotheses on the functions $A_i^\alpha(x, \xi, \eta)$ in § 5.12 which, together with an a priori assumption that each $u^j \in C^{m_j}$, will imply further differentiability properties of the u^j.

We shall begin by studying linear systems of the form

$$(6.1.1) \qquad L_{jk}(x, D)\, u^k(x) = f_j(x), \quad j = 1, \ldots, N, \quad x \in G,$$

and shall be concerned also with the solutions of (6.1.1) subject to boundary conditions of the form

$$(6.1.2) \qquad B_{rk}(x, D)\, u^k(x) = g_r(x), \quad r = 1, \ldots, m, \quad x \text{ on } \partial G,$$

where m is determined below. It is assumed that there are integers $s_1, \ldots, s_N$ and $t_1, \ldots, t_N$ such that each operator L_{jk} is of order $s_j + t_k$; since we may add an integer τ to all the t_j if we subtract it from all the s_j, we may as well assume that $\max s_j = 0$. We allow complex-valued functions throughout and allow the coefficients in the operators to be complex. We require the system (6.1.1) to be elliptic in the sense of the following definition:

Definition 6.1.1. The system (6.1.1) is said to be *elliptic* if and only if the determinant $L(x, \lambda)$ of the characteristic polynomials $L'_{jk}(x, \lambda)$ is not zero for any real non-zero λ; here $L'_{jk}(x, D)$ is the *principal part* of the operator L_{jk}, i.e. the part of order exactly $s_j + t_k$.

It is to be noted that L is a homogeneous polynomial of degree

$$(6.1.3) \qquad P = \sum_{j=1}^{N} (s_j + t_j).$$

It is assumed that if $s_j + t_k < 0$ for some (j, k), then $L_{jk} \equiv 0$. From the ellipticity condition, it follows that for each k one of the operators $L'_{jk} \not\equiv 0$, so that, for each k, $\max(s_j + t_k) = t_k \geq 0$.

Systems of this sort were introduced by DOUGLIS and NIRENBERG. They showed that if $P = 0$ in (6.1.3) then the equations (6.1.1) may be solved for the u^k in terms of the f_j and their derivatives. In fact, with our normalization, $\max s_j = 0$, it turns out that if several $t_k = 0$, then the corresponding u^k may be found in terms of the remaining u^k, f_j, and their derivatives. This is easy to see as follows: (1) By relabelling the u^k, we may assume that $t_1 \leq t_2 \leq \cdots \leq t_N$. (2) By reordering the equations, we may assume $s_1 \leq s_2 \leq \cdots \leq s_N = 0$. If $t_1 = t_2 = \cdots = t_r = 0 < t_{r+1} \leq \cdots \leq t_N$, we must have $s_{N-\varrho+1} = \cdots = s_N = 0$, where $\varrho \geq r$, and the ellipticity guarantees that the matrix of functions $L_{jk}(x)$ with $N - \varrho + 1 \leq j \leq N$ and $1 \leq k \leq r$ has rank r. DOUGLIS and NIRENBERG then showed that if these u^k are eliminated from the remaining equations, the reduced system is still elliptic. So it is sufficient to consider reduced systems and we restrict ourselves to such systems; we present such results in § 6.2. To sum up, we assume

$$(6.1.4) \qquad \max s_j = 0, \quad t_k \geq 1, \quad k = 1, \ldots, N, \quad P > 0.$$

In order to study boundary value problems, we impose the following conditions of proper ellipticity (sometimes called the root condition) and uniform ellipticity:

Definition 6.1.2. The system (6.1.1) is said *to be properly elliptic* or *to satisfy the root condition* or *to satisfy the supplementary condition* if and only if P is even, say $P = 2m$, and, for each pair ξ and ξ' of linearly independent ν-vectors the equation

$$L(x, \xi + z\xi') = 0 \tag{6.1.5}$$

has m roots with positive imaginary part and m with negative imaginary part. We say that the system is *uniformly elliptic* on a set S if and only if there is a number M such that

$$|L(x, \lambda)| \geq M^{-1} > 0, \quad |\lambda| = 1, \quad x \in S \quad (\lambda \text{ real})$$

and the absolute values of all the coefficients in the operators are $\leq M$ for x on S.

Lemma 6.1.1. (a) *The conditions of ellipticity and proper ellipticity are invariant under transformations of the independent variables.*

(b) *If* $\nu > 2$, *any elliptic system* (6.1.1) *is properly elliptic.*

Proof. (a) This can be easily verified by the reader.

(b) Suppose, for some x_0, ξ_0, and ξ_0', that z_0 is a root of (6.1.5) corresponding to the pair (ξ_0, ξ_0'), i.e. $L(x_0, \xi_0 + z_0\xi_0') = 0$. Then obviously, $-z_0$ is a root corresponding to $(\xi_0, -\xi_0')$. Since, if $\nu > 2$, the pair (ξ_0, ξ_0') can be deformed continuously into $(\xi_0, -\xi_0')$, always keeping the vectors linearly independent, and since no root of (6.1.5) is real, and since the roots of (6.1.5) (x_0 fixed) vary continuously with (ξ, ξ'), it follows that, for any fixed x_0, ξ, ξ', the number of roots of (6.1.5) with positive imaginary part equals the number with negative imaginary part.

The result (b) is not necessarily true if $\nu = 2$ as is seen by replacing the system (6.1.1) by a single operator L, where

$$L_1 = \frac{\partial}{\partial x} + i\frac{\partial}{\partial y} \quad \text{or} \quad L_2 = \left(\frac{\partial}{\partial x} + i\frac{\partial}{\partial y}\right)^2.$$

The reader can easily verify that $L_2 u = 0$ for any u of the form

$$u = (1 - x^2 - y^2) f(z), \quad z = x + yi$$

where f is any holomorphic function. Thus the DIRICHLET problem for the operator L_2 fails very seriously to have a unique solution. Another example of this sort was given by BICADZE. The operators L_1 and L_2 are not properly elliptic. An example is given in § 6.5 of a properly elliptic operator L for which the Dirichlet boundary condition is complementing (see Definition 6.1.3 below) but which is such that every complex λ is an eigenvalue for L. The following system is seen to be properly elliptic and is even strongly elliptic as defined in § 6.5:

$$\begin{aligned} u_{xx} - u_{yy} \qquad & + v_{xyyy} = f \\ -u_{xyyy} \qquad & (D_x^2 - D_y^2)^3 v = g. \end{aligned} \tag{6.1.6}$$

The determinant $L(\lambda, \mu)$ $(\lambda^1 = \lambda, \lambda^2 = \mu)$ is

$$\begin{vmatrix} \lambda^2 - \mu^2 & \lambda \mu^3 \\ -\lambda \mu^3 & (\lambda^2 - \mu^2)^3 \end{vmatrix} = (\lambda^2 - \mu^2)^4 + \lambda^2 \mu^6 .$$

We now consider the boundary conditions (6.1.2); we shall assume that the portion of the boundary ∂G on which we consider these conditions is sufficiently smooth, the degree of smoothness being specified in each case being considered. First of all, the m in (6.1.2) will be precisely $P/2$. We require also that there are integers $h_1, \ldots, h_m$, some of which may be negative, such that the order of B_{rk} is $\leq t_k - h_r$; we let h_0 be the largest of 0 and the $-h_r$. If $t_k - h_r < 0$, we assume that $B_{rk} \equiv 0$; otherwise we let B'_{rk} denote the principal part of B_{rk}. At any point x_0 of ∂G, we let n denote the unit normal at x_0 and ξ any (real) vector tangent to ∂G at x_0. Let $z_s^+(x_0, \xi)$, $s = 1, \ldots, m$, be the roots of $L(x_0, \xi + z\, n) = 0$ with positive imaginary part; these exist since we assume that the system (6.1.1) is properly elliptic. Define

$$L_0^+(x_0, \xi; z) = \prod_{s=1}^{m} [z - z_s^+(x_0, \xi)] \tag{6.1.7}$$

and let $\|L^{jk}(x_0, \xi + z\, n)\|$ be the matrix adjoint to $\|L_{jk}(x_0, \xi + z\, n)\|$.

Definition 6.1.3. For any $x_0 \in G$ and any real vector ξ tangent to ∂G at x_0, let us regard $L_0^+(x_0, \xi; z)$ and the elements of the matrix $\| Q_{rk}(x_0, \xi; z)\|$ as polynomials in z, where

$$Q_{rk}(x_0, \xi; z) = \sum_{j=1}^{N} B'_{rj}(x_0, \xi + z\, n)\, L^{jk}(x_0, \xi + z\, n) .$$

The system (6.1.2) of boundary operators is said to satisfy *the complementing condition* (*with respect to the system* (6.1.1)) if and only if the rows of the Q matrix are linearly independent modulo $L_0^+(x_0, \xi; z)$, that is, the polynomial

$$\sum_{r=1}^{m} C_r\, Q_{rk}(x_0, \xi; z) \equiv 0 \ (\operatorname{mod} L_0^+)$$

only if the C_r are all 0.

In the system (6.1.1) the maximum order of any derivative of a particular u^k which occurs is t_k. It is to be noticed that some (or all) of the h_r are allowed to be negative, in which case derivatives of u^k of order higher than t_k may occur among the boundary conditions. As an example, we might have

$$L\, u \equiv \Delta u \equiv (D^2_{x^1} + D^2_{x^2} + D^2_{x^3})\, u , \quad \xi = (\xi_1, \xi_2, 0) , \quad n = (0, 0, 1) ,$$

$$\begin{aligned} B u = {} & a\, D^3_{x^3} u + (b_1\, D_{x^1} + b_2\, D_{x^2} + b_3)\, D^2_{x^3} u + \\ & + (c_{11}\, D^2_{x^1} + 2c_{12}\, D_{x^1} D_{x^2} + c_{22}\, D^2_{x^2} + 2c_{13} D_{x^1} + 2c_{23} D_{x^2} + c_{33}) \times \\ & \times D_{x^3} u + P_3(D_{x^1}, D_{x^2}) , \end{aligned}$$

P_3 being a cubic polynomial. Then

$$L_0^+(\xi, z) = z - i\sqrt{\xi_1^2 + \xi_2^2}$$
$$B(\xi + z\,n) = a\,z^3 + (b_1\,\xi_1 + b_2\,\xi_2 + b_3)\,z^2 + (c_{11}\,\xi_1^2 + \cdots)\,z + + P_3(\xi_1, \xi_2)$$

and the complementing condition is merely that $B(\xi + z\,n)$ not be divisible by L_0^+ (considered as polynomials in z) for any $\xi \neq 0$.

Notations. In this chapter, we shall obtain estimates for the solutions of various problems which involve constants which do not depend on the particular solution. We shall often denote such constants simply by C but C's occurring in different places may denote different constants. On the other hand, we shall sometimes distinguish different constants by using subscripts; however, we do not guarantee that C_2, for example, always denotes the same constant. We shall frequently use $Z_1, Z_2, \ldots$ to denote similar constants. Our notation here is not consistent, but the situation is usually clear from the context. Estimates for solutions of boundary value problems in a half-space where the operators L_{jk} and B_{rk} have constant coefficients and coincide with their principal parts will involve constants which depend only on a common bound for these coefficients, for M (Def. 6.1.2), and for I^{-1} where

$$I = \inf_{\xi}\{\max|Q^\alpha|\}, \quad |\xi| = 1, \quad \xi\cdot n = 0,$$

the Q^α being the various $m \times m$ determinants in the matrix Q of (6.3.19), below, which is related to the Q-matrix above. Such constants will be said to *depend only on E*; interior estimates will not depend on I. Constants which depend also on bounds and/or moduli of continuity of *all* the coefficients and possibly on the domain G will be said to *depend only on E and E'*. If a constant depends on other quantities such as h, q, μ, etc., we may write $C = C(h, q, \mu)$, etc. In § 6.3, we shall replace ν by $\nu + 1$, let $x = (x^1, \ldots, x^\nu)$, and $y = x^{\nu+1}$; we then often let $X = (x, y)$. We say that *a vector function has support in* $G_R \cup \sigma_R$ if and only if it vanishes outside G_R and on and near $\sum_R$; *it need not vanish on* σ_R. If $\alpha = (\alpha_1, \ldots, \alpha_\nu)$ is a multi-index, we define

$$\alpha! = (a_1!)\,(\alpha_2!)\ldots(\alpha_\nu!), \quad x^\alpha = (x^1)^{\alpha_1}\ldots(x^\nu)^{\alpha_\nu}, \quad \text{etc.}$$

Definition 6.1.4. A function or vector function φ is said to be *essentially homogeneous of degree s* if and only if (1) φ is positively homogeneous of degree s or (2) s is an integer ≥ 0 and

$$\varphi(x) = \varphi_1(x)\log|x| + \varphi_2(x)$$

where φ_1 is a *homogeneous polynomial* of degree s and φ_2 is positively homogeneous of degree s.

Remark. If $\varphi \in C^1(R_\nu - \{0\})$ and is essentially homogeneous of degree s, then any first partial derivative is essentially homogeneous of degree $s - 1$.

In § 6.2, we introduce certain fundamental solutions of equations (6.1.1) with constant coefficients and present interior estimates for the solutions of systems (6.1.1) with variable coefficients; the principal results are stated in Theorems 6.2.5 and 6.2.6. In § 6.3, we introduce the "Poisson kernels" of AGMON, DOUGLIS, and NIRENBERG ([1], [2]) and VOLEVICH and introduce certain solutions of certain boundary value problems for a half space in which the L_{jk} and B_{rk} have constant coefficients and coincide with their principal parts. These results are used to obtain the principal results on higher differentiability, given in Theorem 6.3.7, and on "coerciveness" as given in Theorem 6.3.9. In § 6.4, we first obtain results on local coerciveness and higher differentiability on the interior for "weak solutions" of equations (6.1.1) and then obtain similar results at the boundary for weak solutions of (6.1.1) which satisfy the boundary conditions in a certain weak sense. In § 6.5 we treat the Dirichlet problem for the "strongly-elliptic" systems introduced by NIRENBERG [2]. In § 6.6, we extend the analyticity results for linear systems given in the recent paper by MORREY and NIRENBERG to apply to solutions satisfying analytic *general* boundary conditions along an analytic part of the boundary; the developments are like those of § 5.7. In § 6.7, we extend the results of the author (MORREY [12]) and A. FRIEDMAN ([1]) concerning the analyticity of the solutions of *non-linear* systems to the case of general non-linear analytic boundary conditions; the developments are like those of § 5.8. Finally, in § 6.8, we obtain results on higher differentiability of the solutions and weak solutions of certain non-linear problems and then prove a perturbation theorem generalizing that (Theorem 12.6) in the large paper of AGMON, DOUGLIS, and NIRENBERG [1]. Our estimates are of two types, the L_p type involving L_p estimates of the solutions and their derivatives and the Schauder type involving estimates of Hölder norms.

We have not attempted to present the latest developments in the theory of elliptic differential equations nor have we attempted to trace the development of the subject. But a great many references are to be found in the recent books and expository papers by HÖRMANDER [1], LIONS [2], MIRANDA [2], ROSENBLOOM, BROWDER [1], [2], GÅRDING [2], STAMPACCHIA [1], NIRENBERG [3], and the author (MORREY [15], [18]).

We have not presented an account of what might be called "natural" or "variational" boundary conditions. We present only special cases of such problems in Chapters 7 and 8. We have in mind here the regularity at the boundary and the boundary conditions satisfied by the solutions u of equations of the form (6.4.1) where we require that these hold for all v in some linear manifold V and perhaps require u to satisfy (possibly in a weak sense) fewer than m boundary conditions. STAMPACCHIA [2] has shown that if $u (N = 1) \in H_2^1(G)$ and satisfies an equation of the form

(5.2.1) for all v in some V such that $H^1_{20}(G) \subset V \subset H^1_2(G)$ and if ∂G satisfies a mild degree of smoothness, then $u \in C^0_\mu(\bar{G})$. For higher order equations this approach to boundary values was formulated by VIŠIK and SOBOLEV [1]. Quite a few results along these lines have been obtained recently by SCHECHTER. No doubt other workers in the field, such as AGMON, BROWDER, NIRENBERG, etc., have contributed results but the author is not familiar with them.

6.2. Interior estimates for general elliptic systems

In this section, we consider interior estimates for the solutions of the general elliptic systems of the form (6.1.1). We shall be able to include a simple proof of the L_p estimates. Our method is essentially that of §§ 5.5 and 5.6. We begin by writing the equations in the form

$$L_{0jk} u^k = f_j - (L_{jk} - L_{0jk}) u^k, \tag{6.2.1}$$

where L_{0jk} denotes that operator with constant coefficients obtained from the *principal part* of L_{jk} by evaluating the coefficients at the point x_0. Since these equations with constant coefficients cannot, in general, be reduced to a simpler form by affine transformations of the x variables, we can only assume x_0 to be the origin and have to consider the general case. Moreover, for such systems, there is generally no simple law of reflection, so the boundary neighborhoods require the more complicated treatment given in the next section.

We shall assume that the coefficients in L_{jk}, and the functions u^k and f_j satisfy the following minimum conditions:

Minimum conditions. *In case $s_j = 0$, we require the coefficients in the principal part of each L_{jk} to be continuous and the others to be bounded and measurable. If $s_j < 0$, we require merely that all the coefficients $\in C_1^{-1-s_j}(\bar{G})$. We shall assume that $m > 0$, m being the total order of the system. We assume that $f_j \in H_q^{-s_j}(D)$ and $H_q^{t_k}(D)$ for some $q > 1$ and each $D \subset\subset G$. We do not assume that the system is properly elliptic.*

Remarks. The reader will see that some of the results can be generalized by relaxing somewhat the boundedness requirements as was done in §§ 5.5 and 5.6. However, an attempt to state the most general theorem possible would add greatly to the complications without adding much interest. After reading those sections and this one, the reader will be able to formulate and prove whatever results he may need in this line.

As in §§ 5.5, 5.6, we shall first assume that the vector u has support in B_R and satisfies (6.1.1) there. Next, for each sufficiently small R, we alter the coefficients of the L_{jk} outside B_R thus obtaining new operators L_{Rjk}, the coefficients of which reduce to those of L_{0jk} on, near, and outside of ∂B_{2R}, the requisite continuity properties being preserved. Then we define $P_{2R}(f)$ to be a particular solution U_R on B_{2R} of the equations

$$L_{0jk}\, U_R^k = f_j. \tag{6.2.2}$$

Next, we extend u to be zero on $B_{2R} - B_R$ and write $u = U_R + H_R$ where $u_R = P_{2R}[L_0(u)]$ so that H_R is a solution of the homogeneous equations (6.2.2). It is then shown that H_R is a polynomial of bounded degree with $\|H\|^t \le C\,\|u\|^t$ in a convenient space. If we regard H_R as known, then u_R satisfies an equation of the form

$$\begin{aligned} &u_R - T_R\, u_R = v_R, \quad v_R = P_{2R}(f) - P_{2R}[(L_R - L_0)\, H_R]^1 \\ &T_R\, u_R = P_{2R}[-(L_R - L_0)\, u_R]^1 \quad (L_0\, u = f - (L_R - L_0)\, u). \end{aligned} \tag{6.2.3}$$

If R is small enough, $\|T_R\| \le 1/2$ and it follows that

$$\|u\|^t \le C_1\,\|f\|^r + C_2(R)\,\|u\|^0,$$

where the norms are specified below. In case f and the coefficients have additional smoothness properties, it is shown that u then possesses additional smoothness properties. These results are sufficient for interior boundedness theorems and, together with the results in the next section, are sufficient for the estimates in the large including the boundary.

We now set about defining the operator P_{2R}. Suppose that we let $L_0(\lambda)$ be the determinant of the matrix $L_{0jk}(\lambda)$ of characteristic polynomials, let $L_0^{kl}(\lambda)$ be the cofactor of $L_{0lk}(\lambda)$, and let $L_0(D)$ and $L_0^{kl}(D)$ be the corresponding operators. Then a solution of (6.2.2) can be obtained by setting

$$U^k = L_0^{kl}(D)\, F_l \quad \text{where} \quad L_0\, F_l = f_l \tag{6.2.4}$$

and our problem is reduced to solving the second equation in (6.2.4). We now do that heuristically by Fourier transforms. Introducing the transforms by their usual formulas

$$\hat{F}(y) = (2\pi)^{-\tau} \int_{R_\nu} e^{-i x\cdot y}\, F(x)\, dx, \quad (\tau = \nu/2),$$

we find formally

$$i^m\, L_0(y)\, \hat{F}(y) = \hat{f}(y), \quad \hat{F}(y) = (-i)^m\, L_0^{-1}(y)\, \hat{f}(y)^2.$$

Taking the inverse Fourier transforms, we find formally

$$\begin{aligned} F(x) &= \int_{R_\nu} K(x-\xi)\, f(\xi)\, d\xi \\ K(x) &= (2\pi)^{-\nu}(-i)^m \int_{R_\nu} e^{i x\cdot y}\, L_0^{-1}(y)\, dy. \end{aligned} \tag{6.2.5}$$

---

[1] Of course $(L_R - L_0)\, U$ denotes the vector F defined by

$$F_j = L_{Rjk}\, U^k - L_{0jk}\, U^k$$

[2] Recall that m is the total order of the system and may be odd if $\nu = 2$.

If we evaluate the integral for K by introducing polar coordinates in the y space, we obtain

$$K(x) = (2\pi)^{-\nu}(-i)^m \int_\Sigma L_0^{-1}(\sigma) \left\{ \int_0^\infty r^{\nu-1-m} e^{ir(x\cdot\sigma)} dr \right\} d\Sigma. \tag{6.2.6}$$

If $Im\,(x \cdot \sigma)$ were >0, the integral in the brace would converge to

$$J_{\nu-1-m}(x\cdot\sigma), \quad J_h(w) = h!(-iw)^{-h-1}, \quad h \geq 0, \quad m \leq \nu - 1. \tag{6.2.7}$$

In case $m > \nu - 1$, the integral would diverge. But we notice that (formally)

$$J_h^{(p)}(w) = i^p J_{h+p}(w). \tag{6.2.8}$$

If we define

$$\begin{aligned} J_h(w) &= \frac{(iw)^{-h-1}}{(-h-1)!}[-\log(-iw) + C_{-h-1}], \quad h < 0 \\ C_0 &= 0, \quad C_t = 1 + 2^{-1} + \cdots + t^{-1} \quad \text{if} \quad t \geq 1, \end{aligned} \tag{6.2.9}$$

we see that (6.2.8) holds for all integers h and p; for the log, we take the principal log and cut the w plane along the negative imaginary axis. Then, instead of defining K by (6.2.6) with the brace replaced by $J_{\nu-1-m}$, we choose P as a positive integer such that

$$\nu - 1 - 2P < 0 \tag{6.2.10}$$

and define

$$\begin{aligned} K_P^*(x) &= (2\pi)^{-\nu}(-i)^{m+2P} \int_\Sigma L_0^{-1}(\sigma) J_{\nu-1-m-2P}(\sigma\cdot x)\, d\Sigma \\ K(x) &= \Delta^P K_P^*(x), \quad M_P(x) = L_0 K_P^*(x). \end{aligned} \tag{6.2.11}$$

This function K will be seen in Theorem 6.2.1 below to be the desired function. We first prove the following lemma:

Lemma 6.2.1. *If y_0 is a real vector $\neq 0$, there are functions $c_\alpha^\gamma(y)$, analytic for complex y near y_0, such that the transformation*

$$\zeta^\gamma = c_\alpha^\gamma(y)\, \lambda^\alpha \tag{6.2.12}$$

is a rotation of axes for each real y near y_0 and such that

$$y \cdot \lambda = |y|\, \zeta^1.$$

Proof. Let $\eta^\alpha = y^\alpha/|y|$. Making a *fixed* rotation of axes, we may assume $\eta_0 = (1, 0, \ldots, 0)$. If, in (6.2.12), α denotes the row, we define

$$c_\alpha^1(y) = \eta^\alpha, \quad \alpha = 1, \ldots, \nu$$

and then notice that there is a unique way to complete the matrix $c_\alpha^\gamma(y)$ in such a way that the matrix is orthogonal, $c_\alpha^\gamma(y) = 0$ for $\alpha > \gamma \geq 2$, and each determinant in the upper left hand corner is positive. The c_α^γ are analytic.

Theorem 6.2.1. (a) *The functions K_P^*, K, and M_P are analytic for all complex $x = x_1 + i\,x_2 \neq 0$ with $|x_2| < h_1\,|x_1|$, where h_1 depends only on the ellipticity bound E.*

(b) K_P^* *and* K *are essentially homogeneous of degree* $m + 2P - \nu$ *and* $m - \nu$, *respectively, and satisfy*

$$K_P^*(-x) = (-1)^m K_P^*(x), \quad K(-x) = (-1)^m K(x). \tag{6.2.13}$$

(c) $\Delta^{P-1} M_P(x) = K_0(x)$ [1] *if* $\nu > 2$; if $\nu = 2$, $\Delta^{P-1} M_P(x) = K_0(x) + const.$

(d) If $f \in C_\mu^0(\bar{B}_R)$ *and* F *is defined by*

$$F(x) = \int_{B_R} K(x-\xi) f(\xi)\, d\xi, \tag{6.2.14}$$

then $F \in C_\mu^m(\bar{B}_R)$ *with* $h_\mu(\nabla^m F, B_R) \le C\, h_\mu(f, B_R)$ *and if* $f \in C_{\mu c}^k(B_R)$, $F \in C_\mu^{m+k}(\bar{B}_R)$ *with* $h_\mu(\nabla^{m+k} F, B_R) \le C\, h_\mu(\nabla^k f, B_R)$, *where* C *depends only on* E *and on* μ. *Moreover* $L_0 F = f$.

(e) If $f \in L_q(B_R)$ *for some* $q > 1$ *and* F *is defined by* (6.2.14), *then* $F \in H_q^m B_R)$ *with* $\|\nabla^m F\|_{qR}^0 \le C \cdot \|f\|_{qR}^0$ *and if* $f \in H_{0q}^k(B_R)$ *then* $F \in H_q^{m+k}(B_R)$ *with'* $\|\nabla^m F\|_{q,R}^k \le C_1 \|\nabla^k f\|_{qR}^0$ *where* C *depends only on* q *and* E *and* C_1 *depends on* q, E, *and* k.

Proof. We begin by using carefully the definition of $J_h(w)$ and breaking up the integral for K_P^* into integrals over $\sum^+$ and over $\sum^-$ where $\sigma \cdot x > 0$ and < 0, respectively. This yields

$$K_P^*(x) = C_P \int_{\Sigma^+} L_0^{-1}(\sigma) \cdot (\sigma \cdot x)^h [-\log|\sigma \cdot x| + C_h + i\pi/2]\, d\textstyle\sum +$$
$$+ C_P \int_{\Sigma^-} L_0^{-1}(\sigma) \cdot (\sigma \cdot x)^h [-\log|\sigma \cdot x| + C_h - i\pi/2]\, d\textstyle\sum,$$
$$C_P = 1/h!\,(2\pi i)^\nu \quad (h = 2P + m - \nu).$$

Now, since L_0 is of degree m, $L_0^{-1}(\sigma) \cdot (\sigma \cdot x)^h$ is even or odd according as ν is even or odd. Thus

$$K_P^*(x) = \begin{cases} 2C_P \int_{\Sigma^+} L_0^{-1}(\sigma) \cdot (\sigma \cdot x)^h [-\log|\sigma \cdot x| + C_h]\, d\sum, & \nu \text{ even} \\ i\pi C_P \int_{\Sigma^+} L_0^{-1}(\sigma) \cdot (\sigma \cdot x)^h\, d\sum & , \nu \text{ odd.} \end{cases} \tag{6.2.15}$$

The properties of K_P^* in (b) follow from this representation and the analyticity of K_P^* follows by making the rotation

$$\sigma^\alpha = c_\gamma^\alpha(x)\, \tau^\gamma \tag{6.2.16}$$

of Lemma 6.2.1, which is analytic in x near an $x_0 \ne 0$. When this is done, $\sum^+$ is just the subset of $\sum$ where $\tau^1 > 0$ and

$$\sigma = \sigma(x, \tau), \quad \sigma \cdot x = |x|\, \tau^1.$$

To prove (c) we note that

$$M_P(x) = (2\pi)^{-\nu} (-1)^P \int_{\Sigma} J_{\nu-1-2P}(\sigma \cdot x)\, d\textstyle\sum.$$

---

[1] K_0 is the kernel for Poisson's equation, as defined in (2.4.2).

We shall prove (c) in the case $\nu = 2k$; the proof for ν odd is similar. Then we notice that

$$\Delta^{P-k} M_P(x) = (2\pi)^{-\nu}(-1)^k \int_{\Sigma} J_{-1}(\sigma \cdot x)\, d\Sigma$$
$$= A \log|x| + C, \quad A = (-1)^{k-1}(2\pi)^{-\nu}\Gamma_\nu$$

and C is a constant. If $k = 1$, so $\nu = 2$, this is the result, since $\Gamma_\nu = 2\pi$. If $k > 1$, we find by taking Δ^{k-1} of this, that

$$\Delta^{P-1} M_P(x) = -(2\pi)^{-\nu}\Gamma_\nu \cdot 2 \cdot 4^{k-2}(k-2)!\,(k-1)!\, r^{2-\nu}$$

which is seen to reduce to $K_0(x)$; one uses the formula

$$\Gamma_\nu = \nu\gamma_\nu, \quad \Gamma_\nu = \gamma_{\nu-1} \cdot 2\int_0^{\pi/2} \cos^\nu\theta\, d\theta.$$

If $f \in C_\mu^0(\bar{B}_R)$, F is given by (6.2.14), and F_P^* is given by

$$F_P^*(x) = \int_{B_R} K_P^*(x-\xi) f(\xi)\, d\xi \tag{6.2.17}$$

it follows that

$$L_0 F_P^*(x) = \int_{B_R} M_P(x-\xi) f(\xi)\, d\xi, \quad \Delta^P F_P^*(x) = F(x)$$
$$L_0 \Delta^{P-1} F_P^*(x) = \int_{B_R} K_0(x-\xi) f(\xi)\, d\xi$$

so that it follows from Theorem 2.6.7 that $F \in C_\mu^m(B_R)$, satisfies $L_0 F = f$, and has the stated bound. If $f \in C_{\mu c}^k(B_R)$, we see by integrating by parts successively that $\nabla^j F$ is given by (6.2.14) in terms of $\nabla^j f$, $j \leq k$. Thus the remaining results in (d) follow. The results in (e) follow from these, the usual approximations, and Theorem 3.4.2.

We now define our spaces and norms.

Definition 6.2.1. For each integer $h \geq 0$, we define two pairs of spaces, ${}^*H_{q0}^h(B_R)$ and ${}^{**}H_q^h(B_R)$ of vectors having components in various H_q^t-spaces ($q > 1$), and ${}^*C_{\mu 0}^h(B_R)$ and ${}^{**}C_\mu^h(B_R)$ $(0 < \mu < 1)$ of vectors having components in various $C_\mu^t(\bar{B}_R)$ spaces with respective norms defined as follows:

$$\begin{aligned} {}^*\|f\|_{qR}^h &= \sum_j{}'\|f_j\|_{qR}^{h-s_j}, & {}^{**}\|u\|_{qR}^h &= \sum_k{}'\|u^k\|_{qR}^{t_k+h}, \\ {}^*|||f|||_{\mu R}^h &= \sum_j{}'|||f_j|||_{\mu R}^{h-s_j}, & {}^{**}|||u|||_{\mu R}^h &= \sum_k{}'|||u^k|||_{\mu R}^{t_k+h} \end{aligned} \tag{6.2.18}$$

where we define the norms

$$\begin{aligned} {}'\|\varphi\|_{qR}^t &= \sum_{s=0}^t R^{-s}\|\nabla^{t-s}\varphi\|_{qR}^0 \\ {}'|||\varphi|||_{\mu R}^t &= \sum_{s=0}^t [R^{-s} h_\mu(\nabla^{t-s}\varphi) + R^{-s-\mu}|||\nabla^{t-s}\varphi|||_R] \\ |||\varphi|||_R &= \max_{x\in\bar{B}_R}|\varphi(x)|. \end{aligned} \tag{6.2.19}$$

We now define our operator $P_{Rh_0}(f)$ depending on the integer h_0 defined in § 6.1 in connection with the boundary value problem.

Definition 6.2.2. If $f \in {}^* H^h_{q0}(B_R)$ for some $h \geq h_0 \geq 0$ and $q > 1$ or $f \in$ some ${}^*C^h_{\mu 0}(B_R)$ for some $h \geq h_0$ and μ, $0 < \mu < 1$, we define $P_{Rh_0}(f) = U$ where

$$(6.2.20)\qquad \begin{aligned} U^k &= L_0^{kl}[F_l - F_l^*], \quad F_l = \int_{B_R} K(x-\xi) f_l(\xi)\, d\xi \\ F_l^*(x) &= \sum_{s \leq \varrho} \frac{1}{s!} \nabla^s F_l(0) \cdot x^s, \quad \varrho = m + h_0 - s_l - \nu \end{aligned}$$

(the sum being 0 if $\varrho < 0$).

The following lemma simplifies the handling of the derivatives of low order.

Lemma 6.2.2. (a) *If $u \in H^t_p(B_R)$ $p \geq q \geq 1$, $\varepsilon > 0$, and s is an integer with $0 < s \leq t$, then*

$$\|\nabla^{t-s} u\|^0_{qR} \leq \varepsilon R^s \|\nabla^t u\|^0_{pR} + C_1 R^{s-t-\varrho} \|u\|^0_{qR},$$
$$C_1 = C_1(\varepsilon, p, q, \nu, s, t), \quad \varrho = \nu(p-q)/pq.$$

Thus

$$'\|u\|^t_{qR} \leq C_2(\|\nabla^t u\|^0_{pR} + R^{-t-\varrho}\|u\|^0_{qR}), \quad C_2 = C_2(p, q, \nu, t).$$

(b) *If $u \in C^t_\mu(\bar{B}_R)$, $0 < \mu \leq 1$, $\varepsilon > 0$, and $q \geq 1$, then*

$$h_\mu(\nabla^{t-s} u, B_R) \leq \varepsilon R^s h_\mu(\nabla^t u, B_R) + C_3 R^{s-t-\sigma-\mu}\|u\|^0_{qR}, \quad 0 < s \leq t,$$
$$\|\nabla^{t-s} u\|_R \leq \varepsilon R^{s+\mu} h_\mu(\nabla^t u, B_R) + C_4 R^{s-t-\sigma}\|u\|^0_{qR}, \quad 0 \leq s \leq t;$$
$$C_3 = C_3(\varepsilon, q, \mu, \nu, s, t), \quad C_4 = C_4(\varepsilon, q, \mu, \nu, s, t), \quad \sigma = \nu/q.$$

Thus

$$'\|u\|_R \leq C_5(q, \mu, \nu, t)\,[h_\mu(\nabla^t u, B_R) + R^{-t-\sigma-\mu}\|u\|^0_{qR}].$$

(c) *The same results hold with B_R replaced by G_R.*

Proof. In each part, the last statement follows from the first. From considerations of homogeneity, it follows that it is sufficient to prove the first statement in each part for the case $R = 1$. If (a) were not true on B_1, there would exist a sequence $\{u_n\}$ with $'\|u_n\|^t_p = 1$ such that $u_n \rightharpoonup u$ in $H^t_p(B_1)$ but

$$\|\nabla^{t-s} u_n\|^0_{p1} > \varepsilon \|\nabla^t u_n\|^0_{p1} + n\|u_n\|^0_{q1}.$$

Then $u = 0$ and (since $s > 0$) $\nabla^{t-s} u_n \to 0$ in $L_p(B_1)$. But then $\nabla^t u_n \to 0$ (strongly) so that $u_n \to 0$ in $H^t_p(B_1)$, contradicting the fact that $'\|u_n\|^t_{p1} = 1$.

Suppose one of the first inequalities in (b) fails to hold on B_1 for some $s > 0$. Then there exists a sequence $\{u_n\}$ such that $'|\!\|u_n|\!\|^t_1 = 1$, $\nabla^{t-s} u_n$ converges uniformly to $\nabla^{t-s} u$ for some u, $0 \leq s \leq t$, and

$$h_\mu(\nabla^{t-s} u_n) > \varepsilon h_\mu(\nabla^t u_n, B_1) + n\|u_n\|^0_{q1}.$$

It then follows that $u \equiv 0$ and it is easy to see that $h_\mu(\nabla^{t-s} u_n, B_1) \to 0$ if $s > 0$. But then $h_\mu(\nabla^t u_n, B_1) \to 0$ so $'|\!\|u_n|\!\|^t_{\mu 1} \to 0$, a contradiction.

The other inequalities in (b) are proved similarly. Part (c) is proved in the same way.

Theorem 6.2.2. *P_{Rh_0} is a bounded operator from $^*H^h_{p0}(B_R)$ to $^{**}H^h_p(B_R)$ and from $^*C^h_{\mu 0}(B_R)$ to $^{**}C^h_\mu(B_R)$ with bound independent of R if $h \geq h_0 \geq 0$. The bounds depend only on h, h_0, p, or μ, and E. Moreover if $U = P_{Rh_0}(f)$, then U satisfies the equations*

$$L_{0jk}\, U^k(x) = f_j(x) - f_j^*(x)$$

where f_j^* is the Maclaurin expansion of f_j out to (and including) the terms of degree $h_0 - s_j - \nu$.

Proof. By integrating by parts we see that

$$F_l(x) - F_l^*(x) = \int_{B_R} \sum_{|\alpha|=h_0-s_l} C_\alpha K_\alpha(x,\xi)\, D^\alpha f_l(\xi)\, d\xi, \quad C_\alpha = \frac{|\alpha|!}{\alpha_1! \dots \alpha_\nu!}$$

$$K_\alpha(x,\xi) = K_\alpha(x-\xi) - \sum_{0 \leq s \leq \varrho} \frac{1}{s!} \nabla^s K_\alpha(-\xi) \cdot x^s, \quad \varrho = m + h_0 - s_l - \nu$$

$$K_\alpha(x) = D^\alpha K^*_{|\alpha|}(x), \ \sum_{|\alpha|=h_0-s_l} C_\alpha D^\alpha K_\alpha(x) = K(x). \tag{6.2.21}$$

Since (see Theorem 6.2.1) K_α is essentially homogeneous of degree ϱ, we have

$$K_\alpha(y) = K_{1\alpha}(y) \log|y| + K_{2\alpha}(y), \quad \nabla^s K_\alpha(y) = \nabla^s K_{1\alpha}(y) \cdot \log|y| + K_{2s\alpha}(y)$$

where $K_{1\alpha}$ is a homogeneous *polynomial* of degree ϱ and $K_{2\alpha}$ and $K_{2s\alpha}$ are positively homogeneous of degrees ϱ and $\varrho - s$ respectively. Thus, if we set $x = R\,y$, $\xi = R\,\eta$, $y, \eta \in B_1$, then

$$\begin{aligned} K_\alpha(Ry, R\eta) &= [K_\alpha(y-\eta) - \sum_{s \leq \varrho} \frac{1}{s!} \nabla^s K_\alpha(-\eta) \cdot y^s] \cdot R^\varrho + \\ &+ R^\varrho \log R \cdot [K_{1\alpha}(y-\eta) - \sum_{s \leq \varrho} \frac{1}{s!} \nabla^s K_{1\alpha}(-\eta) \cdot y^s] \end{aligned} \tag{6.2.22}$$

and the second term vanishes. Thus

$$\int_{B_R} |K_\alpha(x,\xi)|\, d\xi \leq C\, R^{\varrho+\nu}, \ \int_{B_R} |K_\alpha(x,\xi)|\, dx \leq C\, R^{\varrho+\nu}, \ x, \xi \in B_R. \tag{6.2.23}$$

The analysis above assumes that $\varrho \geq 0$. If $\varrho < 0$, the results (6.2.23) follow from Lemma 3.4.3. Using the Hölder inequality, we find that

$$\begin{aligned} &|F_l(x) - F_l^*(x)|^q \\ &\leq (C\, R^{\varrho+\nu})^{q-1} \int_{B_R} \sum_{|\alpha|=h_0-s_l} C_\alpha |K_\alpha(x,\xi)|\, |\nabla^{h_0-s_l} f_l(\xi)|^q\, d\xi \end{aligned} \tag{6.2.24}$$

so that we can say that

$$\|F_l - F_l^*\|_q^0 \leq Z_1\, R^{\varrho+\nu} \|\nabla^{h_0-s_l} f_l\|_q^0 \leq Z_2\, R^{\varrho+\nu+h-h_0} \|\nabla^{h-s_l} f_l\|_q^0.$$

The fact that $F_l \in H^{m+h-s_l}$ with $\|\nabla^{m+h-s_l} F_l\|_q^0 \leq Z_3 \|\nabla^{h-s_l} f_l\|_q^0$ follows from Theorem 6.2.1. Using Lemma 6.2.2 we obtain the first result. The Hölder results are obtained similarly. The last statement follows from Theorem 6.2.1.

It is convenient to introduce the following terminology:

Definition 6.2.3. The operator L (i.e. the matrix $\{L_{jk}\}$) is said to *satisfy the h-conditions* ($h \geq 0$) on a set $\bar{G}$ if and only if

(i) if $s_j = h = 0$, the coefficients of the highest order derivatives in the operators L_{jk} are continuous on $\bar{G}$, the others being bounded and measurable there; and

(ii) if $h - s_j > 0$, the coefficients in $L_{jk} \in C_1^{h-s_j-1}(\bar{G})$. The operator is said to satisfy the $h - \mu$ conditions on $\bar{G}$ if and only if the coefficients in the operators $L_{jk} \in C_\mu^{h-s_j}(\bar{G})$.

Given the operator L defined on $\bar{B}_A$, we define L_R by

$$L_{Rjk} = L_{0jk} + \varphi(R^{-1}|x|)\,(L_{jk} - L_{0jk}), \quad 0 < R \leq A/2, \tag{6.2.25}$$

where L_{0jk} is defined near (6.2.1), and φ is a fixed function $\in C^\infty(R_1)$ with $\varphi(s) = 1$ for $s \leq 1$, $\varphi(s) = 0$ for $s \geq 7/4$, and $\varphi'(s) \leq 0$ for $1 \leq s \leq 2$.

Theorem 6.2.3. (a) *Suppose that L satisfies the h-conditions on* $\bar{B}_A$ *and* L_R *is defined as in* (6.2.25). *Then* L_R *satisfies the h-conditions on* $\bar{B}_{2R}$ *and* $(L_R - L_0)$ *is a bounded operator from* $^{**}H_p^h(B_{2R})$ *into* $^{*}H_{p0}^h(B_{2R})$ *with bound* $\varepsilon(R)$, *which depends on E and E', and which* $\to 0$ *as* $R \to 0$.

(b) *If L also satisfies the h-μ-conditions* ($0 < \mu < 1$) *on* $\bar{B}_A$, *then so does* L_R *and* $L_R - L_0$ *is a bounded operator from* $^{**}C_\mu^h(\bar{B}_{2R})$ *into* $^{*}C_{\mu 0}^h(\bar{B}_{2R})$ *with bound of the form* $C R^\mu$ *where C depends only on E and E'*.

(c) *If in* (a) *or* (b) *above* $h \geq 1$, *then the bound of the operator* $L_R - L_0$ *is* $\leq CR$ *where C depends only on E and E'*.

Proof. The proof of (a) is similar to but simpler than that of (b) and the proof of (c) is similar. We sketch the proof of (b). Suppose $u \in {}^{**}C_\mu^h(\bar{B}_{2R})$ and $f = (L_R - L_0)\,u$. Then

$$f_j(x) = \sum_{|\alpha| = s_j + t_k} [a_{Rjk}^\alpha(x) - a_{0jk}^\alpha]\, D^\alpha u^k(x) + \sum_{|\alpha| < s_j + t_k} a_{Rjk}^\alpha(x)\, D^\alpha u^k(x)$$

$$a_{Rjk}^\alpha(x) - a_{0jk}^\alpha = \varphi(R^{-1}|x|)\,[a_{jk}^\alpha(x) - a_{jk}^\alpha(0)] \quad \text{if} \quad |\alpha| = s_j + t_k$$

$$a_{Rjk}^\alpha(x) = \varphi(R^{-1}|x|)\, a_{jk}^\alpha(x) \quad \text{if} \quad |\alpha| < s_j + t_k$$

the a_{jk}^α being the coefficients in L_{jk}. If $|\beta| = h - s_j (\geq 0)$, $D^\beta f_j$ is a sum of terms of the form

$$D^\gamma \varphi_R \cdot D^\delta (a_{jk}^\alpha - a_{0jk}^\alpha) \cdot D^\theta u^k \quad \text{and} \quad D^\varkappa \varphi_R \cdot D^\lambda a_{jk}^\alpha \cdot D^\varrho u^k,$$

$$\varphi_R(x) = \varphi(R^{-1}|x|),$$

where γ, δ, etc., are multi-indices and

$$|\gamma| + |\delta| + |\theta| = t_k + h, \quad |\gamma| + |\delta| \leq h - s_j,$$

$$|\varkappa| + |\lambda| + |\varrho| < t_k + h, \quad |\varkappa| + |\lambda| \leq h - s_j.$$

The derivatives of the a_{jk}^α and their Hölder constants are bounded independently of R while we easily see from the definition of $'|||u|||_{\mu R}^h$ and the form of φ_R that

$$|||D^\gamma \varphi_R|||_{2R} \le C(|\gamma|) R^{-|\gamma|}, \quad h_\mu[D^\gamma \varphi_R, \bar{B}_{2R}] \le C(|\gamma|, \mu) R^{-|\gamma|-\mu},$$

$$|||D^\theta u^k|||_{2R} \le (2R)^{t_k+h-|\theta|+\mu} K, \quad h_\mu[D^\theta u^k, B_{2R}] \le (2R)^{t_k+h-|\theta|} \cdot K$$

$$|||a - a_0|||_{2R} \le C \cdot R^\sigma, \quad \sigma = \mu \text{ if } h - s_j = 0; \quad \sigma = 1 \text{ if } h - s_j > 0,$$

$$(K = {}^{**}|||u|||^h_{\mu R}). \tag{6.2.26}$$

It follows easily that $h_\mu(\nabla^{h-s_j} f_j B_{2R}) \le C K$ and the rest follows since the f_j vanish on $B_{2R} - B_{7R/4}$.

Definition 6.2.4. If L satisfies the h-conditions on $\bar{B}_A$ and $R \le A/2$, we define the operator T_R on the spaces ${}^{**}H^h_p(B_{2R})$ and ${}^{**}C^h_\mu(\bar{B}_{2R})$ by the condition that

$$T_R u = -P_{2R}[(L_R - L_0) u].$$

From Theorems 6.2.2 and 6.2.3, we immediately conclude the following theorem:

Theorem 6.2.4. *If L satisfies the h-conditions on $\bar{B}_A$, then T_R is a bounded operator on ${}^{**}H^h_p(B_{2R})$ for each $p > 1$ and $R \le A/2$ with bound $\varepsilon(R, p)$ where ε depends also on E and E' but $\to 0$ as $R \to 0$. If L also satisfies the $h - \mu$ conditions, then T_R is bounded on ${}^{**}C^h_\mu(\bar{B}_{2R})$ with bound $\varepsilon(R, \mu)$ of the form $C(\mu, E, E') \cdot R^\sigma$, $\sigma = \mu$ or 1 as in* (6.2.26).

Remark. T_R depends on the integer h_0 which we may, of course, regard as fixed.

We can now prove the following interior differentiability theorem:

Theorem 6.2.5. *Suppose that L satisfies the h-conditions on the bounded domain G, $u^k \in H^{t_k+h_0}_q(G)$ for some $q > 1$, and the $f_j = L_{jk} u^k \in H^{h-s_j}_p(D)$ for each $D \subset\subset G$ and some $p \ge q$ and $h \ge h_0$. Then $u^k \in H^{t_k+h}_p(D)$ for each $D \subset\subset G$.*
If, also, L satisfies the $h - \mu$ conditions on G and the $f_j \in C^{h-s_j}_\mu(\bar{D})$ for each $D \subset\subset G$, then $u^k \in C^{t_k+h}_\mu(D)$ for each such D. If L and $f \in C^\infty(G)$, then $u \in C^\infty(G)$.

Proof. The last statement follows from the preceding ones. To prove the first statement suppose $D \subset\subset G$ and suppose $D \subset\subset \Delta \subset\subset G$. Then the $f_j \in H^{h-s_j}_p(\Delta)$. We first assume $h \ge 1 + h_0$ and shall show that $u^k \in H^{t_k+h}_q(D)$. With each point x_0 of $\bar{D}$ is associated an $R > 0$ such that $B(x_0, \overline{2R}) \subset \Delta$ and the bounds of T_R on the spaces ${}^{**}H^0_q[B(x_0, 2R)]$ and ${}^{**}H^1_q[B(x_0, 2R)]$ are both $\le 1/2$; we take $h_0 = 0$ for this proof. We can find a C^∞ partition of unity $\zeta_1, \ldots, \zeta_S$ defined on a domain $\Gamma \supset \bar{D}$ and such that each ζ_s has support in some one of the $B(x_0, R)$. Let $u^k_s = \zeta_s u^k$; we shall show that each $u_s \in H^{t_k+1}_q(G)$.

To do this, we suppose that the support of u_s is in B_R. Then, clearly $u_s \in {}^{**}H^0_q(B_{2R})$ and

$$L_{jk} u^k_s = \zeta_s f_j + M_{sjk} u^k = f_{js} \quad (f_j = L_{jk} u^k) \tag{6.2.27}$$

and $f_s \in {}^{*}H^1_q(B_{2R})$ since the operator M_{sjk} is of order $\le s_j + t_k - 1$.

Now, write $u_s = u_{sR} + H_{sR}$ where $u_{sR} = P_{2R}[L_{0s} u_s]$. If $u_s \in C_c^\infty(B_{2R})$, then

$$L_{0s}^{kl} \int_{B_{2R}} K(x-\xi) L_{0slm} u_s^m(\xi)\, d\xi = L_{0s}^{kl} F_{ls}(x)$$
$$= L_{0s}^{kl} L_{0slm} \int_{B_{2R}} K(x-\xi) u_s^m(\xi)\, d\xi = L_{0s} \int_{B_{2R}} K(x-\xi) u_s^k(\xi)\, d\xi = u_s^k(x).$$

Thus, from the definition of P_{2R}, we see that H_R^s is a polynomial of degree $\le t_k - \nu$. Thus u_{sR} satisfies

$$u_{sR} = P_{2R}(f_s) - P_{2R}[(L_{Rs} - L_{0s}) u_{sR}] - P_{2R}[(L_{Rs} - L_{0s}) H_{sR}], \quad \text{or}$$
$$u_{sR} - T_{sR} u_{sR} = v_{sR} = P_{2R}(f_s) - P_{2R}[(L_{Rs} - L_{0s}) H_{sR}].$$

Since $\|T_{sR}\| \le 1/2$, it follows that u_{sR} and hence $u_s \in {}^{**}H_q^1(B_{2R})$. Accordingly $u \in {}^{**}H_q^1(\Delta)$ for some $\Delta \supset D$. The argument may be repeated choosing R smaller if necessary. In a finite number of steps we find that $u \in {}^{**}H_q^h(D)$.

To raise the exponent from q to p, we proceed similarly but choosing R so that $\|T_R\| \le 1/2$ in ${}^{**}H_{q_1}^h(B_{2R})$ where $q_1 = \nu q/(\nu - q)$, the Sobolev exponent corresponding to q. Then, from (6.2.27), we see that $f_s \in {}^{*}H_{q_1}^h(B_{2R})$. A finite number of repetitions accomplishes our purpose. In case L satisfies the $h - \mu$ conditions and $f_j \in C_\mu^{h-s_j}$, we may take p so large that $u^k \in C_\mu^{t_k+h-1}$ and then $f_s \in {}^{*}C_\mu^h$ and we may conclude that $u^k \in C_\mu^{t_k+h}$ by taking R small enough.

For later use, we need the following lemma and theorem:

Lemma 6.2.3. *There exists constants $C_1(\nu, s, t, p, q)$ and $C_2(\eta, s, t, \mu, q)$ such that*

$$'\| H \|_{pR}^t \le C_1 R^{-t-\varrho} \| H \|_{qR}^0, \quad '\|| H \||_{\mu R}^t \le C_2 R^{-t-\sigma-\mu} \| H \|_{qR}^0,$$
$$\varrho = \nu(p-q)/pq, \quad \sigma = \nu/q, \quad p > q$$

for all polynomials H of degree $\le s$.

Proof. Let n be the larger of $s+1$ and $t+1$. Then, from Lemma 6.2.2 with $\varepsilon = 1$, t replaced by n, and s replaced by $r = n - t$, we see that (since $\nabla^n H \equiv 0$)

$$\| \nabla^t H \|_{pR}^0 \le C_1' R^{-t-\varrho} \| H \|_q^0, \quad h_\mu(\nabla^t H) \le C_2' R^{-t-\sigma-\mu} \| H \|_{qR}^0.$$

The lemma follows from Lemma 6.2.2.

Theorem 6.2.6. *Suppose L satisfies the h-conditions on $\bar{B}_A$. Then there exist constants $R_2 > 0$ and C_2 depending only on ν, h, q, E, and E' such that*

$$^{**}\| u \|_{qR}^h \le C_2 [{}^{*}\| L u \|_{qR}^h + {}^{*\prime}\| u \|_{1R}^0], \quad {}^{*\prime}\| u \|_{1R}^0 = \sum_k R^{-t_k-h-\varrho} \| u^k \|_{1R}^0$$
$$0 < R \le R_2, \quad u^k \in H_{q0}^{t_k+h}(B_R), \quad \varrho = \nu(q-1)/q.$$

Also there exist constants $R_3 > 0$ and C_3, depending on ν, h, μ $(0 < \mu < 1)$ E, and E' such that if L satisfies the $\mu - h$ conditions on $\bar{B}_A$, then

$$^{**}\|| u \||_{\mu R}^h \le C_3 [{}^{*}\|| L u \||_{\mu R}^h + {}^{*\prime}\|| u \||_{1R}^0], \quad {}^{*\prime}\|| u \||_{1R}^0 = R^{\varrho-\nu-\mu}\, {}^{*\prime}\| u \|_{1R}^0$$
$$0 < R \le R_3, \quad u^k \in C_{\mu c}^{t_k+h}(\bar{B}_R).$$

Proof. We prove the second, the proof of the first is similar. We omit the subscripts μR. We confine ourselves to R so small that $\| T_R \| \leq 1/2$ in the space $^{**}C^h_\mu(\bar{B}_{2R})$. Suppose $u^k \in C^{t_k+h}_{\mu c}(\bar{B}_R)$, define L_R, P_{2R}, and T_R as above (with some fixed $h_0 \leq h$); we assume $u = 0$ on $\bar{B}_{2R} - \bar{B}_R$. Let $u_R = P_{2R}(L_0 u)$. Then, as was seen in the proof of Theorem 6.2.5, $u^k = u^k_R + H^k_R$, where H^k_R is a polynomial of degree $\leq t_k - \nu$, and

$$(6.2.28) \quad u_R - T_R u_R = P_R(f) - P_R[(L_R - L_0) H_R], \quad f = L_R u \equiv L u.$$

Using the definition of u_R, we obtain (Z_i independent of R)

$$^{**}|||u_R|||^h \leq Z_1^{**} |||u|||^h, \qquad ^{**}|||H_R|||^h \leq Z_2^{**} |||u|||^h.$$

Using (6.2.28), theorems 6.2.2, 6.2.3, and 6.2.4 and the fact that $\| T_R \| \leq 1/2$, we find further that

$$(6.2.29) \qquad ^{**}|||u_R|||^h_{\mu R} \leq Z_3^* |||f|||^h_{\mu R} + \varepsilon_1(R) ^{**}|||u|||^h_{\mu R}.$$

From the definition of the norm, we conclude that

$$(6.2.30) \quad \begin{aligned} &^{*'}|||u_R|||^0_{1R} \leq Z_4 Z_3^* |||f|||^h_{\mu R} + Z_4 \varepsilon_1(R) ^{**}|||u|||^h_{\mu R} \qquad (Z_4 = \gamma_\nu^{-1}) \\ &\therefore ^{*'}|||H_R|||^0_{1R} \leq ^{*'}|||u|||^0_{1R} + ^{*'}|||u_R|||^0_{1R}. \end{aligned}$$

From (6.2.30), (6.2.29), and Lemma 6.2.3, we conclude that

$$^{**}|||H_R|||^h_{\mu R} \leq Z_5^{*'} |||u|||^0_{1R} + Z_6^* |||f|||^h_{\mu R} + \varepsilon_2(R) ^{**}|||u|||^h_{\mu R}$$

$$^{**}|||u|||^h_{\mu R} \leq Z_5^{*'} |||u|||^0_{1R} + Z_6^* |||f|||^h_{\mu R} + \varepsilon_3(R) ^{**}|||u|||^h_{\mu R}$$

from which the result follows.

6.3. Estimates near the boundary; coerciveness

We now carry out the boundary estimates for the boundary problem described in § 6.1, namely

$$(6.3.1) \qquad L_{jk} u^k = f_j \text{ in } G, \; B_{rk} u^k = g_r \text{ on } \partial G.$$

We shall begin by considering functions defined on hemispheres G_A instead of full spheres B_A but otherwise shall try to imitate the developments of the preceding section. For functions u having support in $G_R \cup \sigma_R$ we obtain estimates for u in terms of those for f and g as defined in (6.3.1).

The fact that the coefficients in the boundary conditions are also variable causes an extra complication which we take care of in the following way: We define the operator $P_R(f, g)$ for the *pair* (f, g) to be a certain solution of the equations

$$(6.3.2) \qquad L_{0jk} U^k = f_j \text{ on } R^+\!: x^{\nu+1} > 0, \; B_{0rk} U^k = g_r \text{ on } \sigma$$

where the f_j and g_r have support on $G_{2R} \cup \sigma_{2R}$ and σ_{2R}, respectively. Then if u is given with support in $G_R \cup \sigma_R$, we alter the coefficients of L and also B outside G_R to obtain operators L_R and B_{Rrk} as in the preceding section and then write $u = u_R + H_R$ where

$$u_R = P_R[f - (L_R - L_0)\, u, g - (B_R - B_0)\, u]. \tag{6.3.3}$$

It turns out again that H_R is a polynomial of bounded degree. We define

$$T_R(\varphi) = P_R[-(L_R - L_0)\,\varphi, -(B_R - B_0)\,\varphi] \tag{6.3.4}$$

and the equation (6.3.1) becomes

$$u_R - T_R\, u_R = v_R = P_R[f - (L_R - L_0)\, H_R, g - (B_R - B_0)\, H_R]. \tag{6.3.5}$$

We define

$$P_R(f, g) = U_R + V_R - W_R, \quad U_R = P_R(0, g), \quad W_R = P_R(0, \gamma),$$
$$\gamma_r = B_{0rk}\, V_R^k \tag{6.3.6}$$

and V_R is defined in a way very similar to the $P_{h0,2R}(\tilde{f})$ of the preceding section, $\tilde{f}$ being a smooth extension of f to B_{2R}; $P_R(0, g)$ and $P_R(0, \gamma)$ are defined by means of the "Poisson kernels" introduced by Agmon, Douglis, and Nirenberg ([1], [2]). We write P_R instead of P_{2R} in this section, but still work on G_{2R}. In order to obtain the L_p results, it is shown how to express the derivatives of highest order by means of singular integrals of Calderon-Zygmund type; this is an idea similar to one used by Browder [1], [2]. We shall replace ν by $\nu + 1$ and $x^{\nu+1}$ by y and shall write the operators $L_{0jk}(D)$ and $B_{0rk}(D)$ in the forms $L_{0jk}(D_x, D_y)$ and $B_{0rk}(D_x, D_y)$ where $x = (x^1, \ldots, x^\nu)$.

We now develop formulas for the Poisson kernels by attempting to solve the equations

$$L_{0jk}(D_x, D_y)\, u^k = 0 \text{ on } R^+, \tag{6.3.7}$$
$$B_{0rk}(D_x, D_y)\, u^k(x, 0) = g_r(x) \text{ on } \sigma,$$

formally by introducing the Fourier transforms $\hat{u}^k$ and $\hat{g}_r$ with respect to x defined by

$$\hat{u}^k(\lambda, y) = (2\pi)^{-\tau} \int_{R_\nu} e^{-i\lambda\cdot x}\, u^k(x, y)\, dx, \quad (\tau = \nu/2) \tag{6.3.8}$$

with a similar formula for $\hat{g}_r$. This yields the equations

$$L_{0jk}(i\,\lambda, D_y)\, \hat{u}^k(\lambda, y) = 0, \quad y > 0 \tag{6.3.9}$$
$$B_{0rk}(i\,\lambda, D_y)\, \hat{u}^k(\lambda, 0) = g_r(\lambda), \quad y = 0.$$

It is convenient to normalize these equations by setting

$$\hat{u}^k(\lambda, y) = (i\,|\lambda|)^{-t_k}\, U^k(\lambda, |\lambda|\, y) \tag{6.3.10}$$
$$\hat{g}_r(\lambda) = (i\,|\lambda|)^{-h_r}\, G_r(\lambda).$$

Then the equations (6.3.9) in $u = |\lambda|\, y$ become

$$L_{0jk}(\sigma, -i\, D_u)\, U^k(\lambda, u) = 0, \qquad u > 0, \tag{6.3.11}$$

$$B_{0rk}(\sigma, -i\, D_u)\, U^k(\lambda, 0) = G_r(\lambda) \qquad (\sigma = |\lambda|^{-1}\,\lambda,\ |\sigma| = 1). \tag{6.3.12}$$

We begin the solution of these equations by stating the following well known lemma:

Lemma 6.3.1. (a) *Every solution U of a single ordinary differential equation (with constant coefficients) of the form*

$$L(-i\,D)\,U = 0,\ L(z) = a_0 \prod_{j=1}^{n} (z - r_j)^{m_j}$$

is of the form

$$U(u) = \sum_{j=1}^{n} P_j(u)\, e^{i r_j u}$$

where each P_j is a polynomial of degree $\leq m_j - 1$. Moreover, every function U of this form is a solution.

(b) *If $s \neq$ any r_j, the equation*

$$L(-i\,D)\,U = P(u)\, e^{i s u} \quad (P \text{ a polynomial})$$

has a unique solution U of the form $Q(u)\, e^{i s u}$, where $Q(u)$ is a polynomial of the same degree as P. If $s = r_j$, the equation has a unique solution U of the form $u^{m_j} Q(u)\, e^{i r_j u}$ where Q is a polynomial of the same degree as P.

Definition 6.3.1. A function of the form $P(u)\, e^{i s u}$, (P a polynomial) where $I m\ s > 0$ is said to be *exponentially decaying.*

We now prove the following fundamental theorem of AGMON, DOUGLIS, and NIRENBERG ([1], [2]):

Theorem 6.3.1. *We suppose that the matrix of operators L_{0jk} is properly elliptic. Then there are fixed rectangles R^+ in the upper half-plane and R^- in the lower half-plane such that R^+ contains all the zeros with positive imaginary part of $L_0(\sigma, z)$ for each σ with $|\sigma| = 1$ and R^- contains all those with negative imaginary parts. The operators B_{0rk} satisfy the complementing condition with respect to the L_{0jk} if and only if there are exponentially decaying solutions U_s of the homogeneous equations in (6.3.11) which satisfy*

$$B_{0rk}(\sigma, -i\,D_u)\, U_s^k(\sigma, 0) = \delta_{rs}, \quad r, s = 1, \ldots, m. \tag{6.3.13}$$

If they exist, these functions are unique and are analytic near $|\sigma| = 1$. If we introduce the functions $\hat{U}_s^k(\sigma, z)$ by the formulas

$$\hat{U}_s^k(\sigma, z) = (2\pi)^{-1} \int_0^\infty e^{-i u z}\, U_s^k(\sigma, u)\, du \tag{6.3.14}$$

we see that the $\hat{U}_s^k$ are analytic in σ near $|\sigma| = 1$, are rational functions of z having denominator $L_0^+(\sigma, z)$, defined as in (6.1.6), and we have

$$U_s^k(\sigma, u) = \int_{\partial R^+} e^{i u z}\, \hat{U}_s^k(\sigma, z)\, dz. \tag{6.3.15}$$

Finally there are polynomials $\hat{P}_{js}(\sigma, z)$ in z such that

$$L_{0jk}(\sigma, z)\, \hat{U}_s^k(\sigma, z) = \hat{P}_{js}(\sigma, z). \tag{6.3.16}$$

Moreover, the $\hat{U}_s^k$ satisfy the conditions.

$$\int_{\partial R} B_{0rk}(\sigma, z)\, \hat{U}_s^k(\sigma, z)\, dz = \delta_{rs}, \quad r, s = 1, \ldots, m. \tag{6.3.17}$$

Proof. Since the zeros of $L_0(\sigma, z)$ vary continuously with σ, the set $|\sigma| = 1$ is compact, and no zero is real, the existence of the rectangles R^+ and R^- follows.

Until the last paragraph of this proof, we assume that σ is a constant and suppress it. Let us consider the matrix $L_{0jk}(z)$ of characteristic polynomials. By applying the elementary operations of (a) interchanging two *rows* and (b) adding to one row another row multiplied by a polynomial, one can replace $\| L_{0jk}(z) \|$ by a triangular matrix $\| \tilde{L}_{jk}(z) \|$ in which $\tilde{L}_{jk}(z) \equiv 0$ if $j > k$. Clearly

$$L_0(z) = \pm \prod_{j=1}^{N} \tilde{L}_{jj}(z).$$

If we factor $\tilde{L}_{jj}(z)$ into $\tilde{L}_{jj}^+(z) \cdot \tilde{L}_{jj}^-(z)$, then, clearly

$$L_0^+(z) = \pm \prod_{j=1}^{N} \tilde{L}_{jj}^+(z), \quad m = m_j,$$

m_j being the degree of $\tilde{L}_{jj}^+$. Moreover, the system

$$\tilde{L}_{jk}(-i\,D)\,U^k = 0 \tag{6.3.18}$$

is equivalent to the homogeneous system (6.3.11).

Next we see that every exponentially decaying solution of (6.3.18) (and hence (6.3.11)) can be obtained as follows: First let U^N be any exponentially decaying solution of $\tilde{L}_{NN}(-i\,D)\,U^N = 0$. There are m_N linearly independent solutions U^N (Lemma 6.3.1). With each given such U^N, we associate the particular solutions of Lemma 6.3.1 (b) for the equations $\tilde{L}_{N-1,N-1}\,U^{N-1} = -\tilde{L}_{N-1,N}\,U^N$, $\tilde{L}_{N-2,N-2}\,U^{N-2} = -\tilde{L}_{N-2,N-1}\,U^{N-1} - \tilde{L}_{N-2,N}\,U^N$, etc., in turn. Next, take $U^N = 0$, let U^{N-1} be any exponentially decaying solution of $L_{N-1,N-1}\,U^{N-1} = 0$, and determine the other U^k as the particular solutions obtained by solving the other equations in turn. There are m_{N-1} such solutions. This process may be continued yielding $m = m_1 + \cdots + m_N$ linearly independent exponentially decaying vector solutions of (6.3.11).

Thus the totality of these vector solutions is an m-dimensional complex vector space V. If $U \in V$, each expression $B_{0rk}(-i\,D)\,U^k(0)$ is a linear functional $F_r(U)$. Thus the transformation $X_r = F_r(U)$ is a $1-1$ transformation from V to C_m the range of the transformation is C_m. To show this, we set

$$L_{0+}(z) = z^m + a_1 z^{m-1} + \cdots + a_m$$

and define

$$L_{0\beta}^+(z) = z^\beta + a_1 z^{\beta-1} + \cdots + a_\beta, \quad \beta = 0, \ldots, m-1.$$

We notice that if R^+ is a rectangle enclosing the roots of L_0^+, then

$$\frac{1}{2\pi i} \int_{\partial R^+} \frac{z^\gamma L_{0,m-1-\beta}^+(z)}{L_0^+(z)}\, dz = \delta_\beta^\gamma, \; \beta, \gamma = 0, \ldots, m-1.$$

For, if $\gamma < \beta$, the degree of the numerator is $\leq m - 2$ and the result follows by replacing ∂R^+ by a large circle; if $\gamma = \beta$, the result follows similarly; if $\beta < \gamma \leq m - 1$, the numerator differs from $z^{\gamma-\beta-1} L_0^+(z)$ by a polynomial of degree $\leq \gamma - 1 \leq m - 2$. Next, we write

$$B_{0rk}(z) L_0^{kl}(z) = \sum_{\beta=1}^{m-1} q_r^{l\beta} z^\beta \pmod{L_0^+}; \tag{6.3.19}$$

we consider the matrix $Q = (q_r^{l\beta})$ as having m rows (indexed with r) and $m N$ columns. The complementing condition is equivalent to the statement that Q is of rank m. Thus, if X_r are any numbers, there are constants $c_{l\beta}$ such that

$$c_{l\beta} q_r^{l\beta} = X_r. \tag{6.3.20}$$

With these $c_{l\beta}$, we define

$$U^k(u) = \frac{1}{2\pi i} \int_{\partial R^+} c_{l\beta} \frac{L_0^{kl}(z)}{L_0^+(z)} L_{m-\beta-1}^+(z) e^{iuz} dz. \tag{6.3.21}$$

Then

$$B_{0rk}(-iD) U^k(0) = \frac{1}{2\pi i} \int_{\partial R^+} c_{l\beta} B_{0rk}(z) L_0^{kl}(z) \frac{L_{0,m-\beta-1}^+(z)}{L_0^+(z)} dz$$

$$= \frac{1}{2\pi i} \int_{\partial R^+} c_{l\beta} q_r^{l\gamma} z^\gamma \frac{L_{0,m-1-\beta}^+(z)}{L_0^+(z)} dz = c_{l\beta} q_r^{l\beta} = X_r.$$

It is clear from (6.3.21) that the vector U^k is an exponentially decaying solution of the homogeneous equations. It is unique in spite of the fact that the $c_{l\beta}$ satisfying (6.3.20) are not unique.

Thus, for each σ, there is a unique solution $U_s^k(\sigma, u)$ of (6.3.11) and (6.3.13). Since $L_0(\sigma, z)$[1] and the L_{0jk} and L_0^{kl} and B_{0rk} are analytic in σ, it is easy to see that we may choose, for each s, functions $c_{sl\beta}(\sigma)$ which are analytic in σ near any given σ_0 which satisfy (6.3.20) with $X_r = \delta_{rs}$. Thus $U_s^k(\sigma, z)$ is analytic in σ also. If we introduce $\hat{U}_s^k$ by (6.3.14), it is clear from (6.3.21) that

$$\hat{U}_s^k(\sigma, z) = (-i) c_{sl\beta}(\sigma) [L_0^+(\sigma, z)]^{-1} L_0^{kl}(\sigma, z) L_{m-1-\beta}^+(\sigma, z)$$

---

[1] To see that $L_0^+(\sigma, z)$ is analytic in σ, we note that

$$c_n(\sigma) = \int_{\partial R^+} z^n [L_0^+(\sigma, z)]^{-1} L_0^{+\prime}(\sigma, z) dz = \int_{\partial R^+} z^n L_0^{-1} L_0' dz$$

so that each c_n is analytic in σ and

$$(L_0^+)^{-1} L_0^{+\prime} = m z^{-1} + \sum_{n=1}^{\infty} c_n(\sigma) z^{-n-1}$$

where the series converges uniformly for $|z| \geq$ some A, independent of σ; of course R^+ is a fixed rectangle in the upper half-plane which contains all roots of $L_0^+(\sigma, z)$ for all σ with $|\sigma| = 1$.

where, near any σ_0, the $c_{sl\beta}(\sigma)$ satisfy (6.3.20) with $X_r = \delta_{rs}$. The statements about the $\hat{U}_s^k$ follow.

Thus, from this theorem it follows that the equations (6.3.11) and (6.3.12) have a unique exponentially decaying solution which is given by

$$U^k(\lambda, u) = \sum_s G_s(\lambda)\, U_s^k(\sigma, u), \quad U_s^k = \int_{\partial R^+} e^{iuz}\, \hat{U}_s^k(\sigma, z)\, dz$$

where R^+ is a fixed rectangle in the upper half-plane which contains all the roots of $L_{0+}(\sigma, z)$ for each σ with $|\sigma| = 1$. If we now use the substitutions (6.3.10) and formally apply the inverse Fourier transform, we obtain

(6.3.22) $$u^k(x, y) = \int_{R^\nu} L^{kr}(x - \xi, y)\, g_r(\xi)\, d\xi, \quad \text{where}$$

(6.3.23) $$L^{kr}(x, y) = C^{kr} \int_{R^\nu} |\lambda|^{-t_k + h_r}\, d\lambda \int_{\partial R^+} e^{i(\lambda \cdot x + |\lambda| yz)}\, \hat{U}_s^k(\sigma, z)\, dz$$

$$C^{kr} = (2\pi)^{-\nu}(-i)^{t_k - h_r}.$$

The integral in (6.3.23) is divergent, but if we introduce polar coordinates (s, σ) with $s = |\lambda|$ and $|\sigma| = 1$, we obtain

$$L^{kr}(x, y) = C^{kr} \int_\Sigma d\Sigma \int_{\partial R^+} \hat{U}_r^k(\sigma, z) \left[\int_0^\infty s^{\nu - 1 - t_k + h_r}\, e^{is(\sigma \cdot x + yz)}\, ds\right] dz.$$

By replacing the bracket by $J_{\nu - 1 - t_k + h_r}(\sigma \cdot x + yz)$ where the $J_h(w)$ are defined in (6.2.7) and (6.2.9), we obtain

(6.3.24) $$L^{kr}(x, y) = C^{kr} \int_\Sigma d\Sigma \int_{\partial R^+} J_{\nu - 1 - t_k + h_r}(\sigma \cdot x + yz)\, \hat{U}_r^k(\sigma, z)\, dz.$$

It is convenient also to have the following integrated kernels.

Definition 6.3.2. We define, for large P,

$$^*L_P^{kr}(x, y) = (-1)^P C^{kr} \int_\Sigma d\Sigma \int_{\partial R^+} J_{\nu - 1 - t_k + h_r - 2P}(\sigma \cdot x + yz)\, \hat{U}_r^k(\sigma, z)\, dz,$$

$$^*L_{P\alpha}^{kr}(x, y) = D_x^\alpha\, {}^*L_{P + |\alpha|}^{kr}(x, y), \quad L_\alpha^{kr}(x, y) = \Delta_x^P\, {}^*L_{P\alpha}^{kr}(x, y).$$

(6.3.25)

The following lemma is very useful:

Lemma 6.3.2. *Suppose φ is essentially homogeneous of degree $n \geq 0$ and $\varphi \in C^\infty(R_{\nu+1} - \{0, 0\})$ and suppose*

(6.3.26) $$\varphi(X, \Xi) = \varphi(X - \Xi) - \sum_{s \leq n} \frac{1}{s!} \nabla^s \varphi(-\Xi) \cdot X^s.$$

Then

$$|D_X^\beta \varphi(X, \Xi)| \leq C_A |X|^{n+1-|\beta|} |\Xi|^{-1} \quad \text{if} \quad 0 \leq |X| \leq A|\Xi|, \quad |\beta| \leq n, \quad A < 1$$

$$|D_X^\beta \varphi(X, \Xi)| = |D^\beta \varphi(X - \Xi)| \leq C_\beta |X - \Xi|^{n - |\beta|} \quad \text{if} \quad |\beta| > n,$$

where C_A, and C_β depend also on bounds for the derivatives of φ on $\partial B(0,1)$.

Proof. The last statement is evident. Moreover, if $|\beta| \leq n$, then $D_X^\beta \varphi(X, \Xi)$ is obtained from $D^\beta \varphi(X - \Xi)$ by the formula corresponding to (6.3.26). So it is sufficient to prove the first statement only for $|\beta| = 0$. We write

$$\Xi = S\xi, \quad |\xi| = 1, \quad \varphi(X) = \varphi_1(X) \log|X| + \varphi_2(X)$$

where φ_1 is a polynomial of degree n and φ_2 is positively homogeneous of degree n. Then we have

$$D^\alpha \varphi(X) = D^\alpha \varphi_1(X) \log|X| + \varphi_{2\alpha}(X) \tag{6.3.27}$$

where $\varphi_{2\alpha}$ is positively homogeneous of degree $n - |\alpha|$. Thus

$$\begin{aligned}\varphi(X, S\xi) &= S^n \varphi_1(S^{-1}X - \xi)\,[\log S + \log|S^{-1}X - \xi|] + S^n \varphi_2(S^{-1}X - \xi)\\ &\quad - S^n \sum_{|\alpha| \leq n} \frac{(S^{-1}X)^\alpha}{\alpha!}[D^\alpha \varphi_1(-\xi) \log S + \varphi_{2\alpha}(-\xi)]\\ &= S^n \left[\varphi(S^{-1}X - \xi) - \sum_{|\alpha| \leq n} \frac{(S^{-1}X)^\alpha}{\alpha!} D^\alpha \varphi(-\xi)\right],\\ &\qquad \alpha! = (\alpha_1!) \ldots (\alpha_\nu!),\end{aligned}$$

since, from (6.3.27) it follows that $D^\alpha \varphi(-\xi) = \varphi_{2\alpha}(-\xi)$. The result follows since $|S^{-1}X| \leq A < 1$.

Theorem 6.3.2. (a) *The functions* $^*L_P^{kr}$, $^*L_{P\alpha}^{kr}$, L^{kr}, *and* L_α^{kr} *are analytic for all real* $(x, y) \neq (0, 0)$ *with* $y \geq 0$ *and are essentially homogeneous of degrees* $2P + t_k - h_r - \nu$, $|\alpha| + 2P + t_k - h_r - \nu$, $t_k - h_r - \nu$, *and* $|\alpha| + t_k - h_r - \nu$, *respectively.*

(b) *For each* r, P, *and* α, *the functions above satisfy*

$$\begin{gathered}\sum_k L_{0jk}\,{}^*L_P^{kr} = \sum_k L_{0jk}\,{}^*L_{P\alpha}^{kr} = \sum_k L_{0jk} L^{kr} = \sum_k L_{0jk} L_\alpha^{kr} = 0, \quad y > 0\\ \sum_{|\alpha|=t} C_\alpha D_x^\alpha\,{}^*L_{P\alpha}^{kr} = {}^*L_P^{kr}, \quad \sum_{|\alpha|=t} C_\alpha D_x^\alpha L_\alpha^{kr} = L^{kr}, \quad y > 0\\ \sum_k B_{0tk}\,{}^*L_P^{kr}(x, 0) = \delta_t^r M_P(x), \quad C_\alpha = \frac{|\alpha|!}{(\alpha_1!) \ldots (\alpha_\nu!)}\end{gathered} \tag{6.3.28}$$

where M_P *is defined in* (6.2.11); *it is assumed that* P *is large.*

Proof. The analyticity for $y > 0$ is evident from (6.3.25) and the definitions of the J_h given in (6.2.7) and (6.2.9). The essential homogeneity follows from an argument simpler than that given in the first part of the proof of Theorem 6.2.1. Since $|\sigma| = 1$, it follows immediately from (6.3.25) (and (6.2.8)) that

$$\Delta_x^{|\alpha|}\,{}^*L_{P+|\alpha|}^{kr}(x, y) = {}^*L_P^{kr}(x, y), \quad y > 0.$$

Thus the first formula in the second line of (6.3.28) follows. From (6.3.25) if follows that

$$\begin{aligned}\sum_k L_{0jk}\,{}^*L_P^{kr}(x, y) &= C_{jP}^r \int_\Sigma d\Sigma \int_{\partial R^+} J_\varrho(\sigma \cdot x + yz) \sum_k L_{0jk}(\sigma, z) \hat{U}_r^k(\sigma, z)\, dz\\ &= 0, \quad \varrho = \nu - 1 + s_j + h_r - 2P, \quad y > 0, \quad C_{jP}^r = (2\pi)^{-\nu}(-1)^P i^{s_j + h_r}\end{aligned}$$

on account of (6.3.16). Likewise

$$\sum_k B_{0tk} {}^*L_P^{kr}(x, y) + {}'C_{tP}^r \int_\Sigma d\Sigma \int_{\partial R^+} J_\lambda(\sigma \cdot x + y z) \sum_k B_{0tk}(\sigma, z) \hat{U}_r^k(\sigma, z)\, dz$$

$${}'C_{tP}^r = (-1)^P (2\pi)^{-\nu} i^{h_r - h_t}, \quad \lambda = \nu - 1 + h_r - h_t - 2P.$$

If P is large, we may let $y \to 0^+$ and obtain

$$\sum_k B_{0tk} {}^*L_P^{kr}(x, 0) = (-1)^P (2\pi)^{-\nu} \delta_t^r \int_\Sigma J_{\nu-1-2P}(\sigma \cdot x)\, d\Sigma$$

from which the last line of (6.3.28) follows, using (6.2.11). The remaining formulas in (6.3.28) follows easily.

To prove the analyticity of ${}^*L_P^{kr}$ at a point $(x_0, 0)$, $x_0 \neq 0$, we make the rotation

$$\sigma = \sigma(x, \tau) : \sigma^\alpha = c_\gamma^\alpha(|x|^{-1} \cdot x)\, \tau^\gamma, \quad \sigma \cdot x = |x| \cdot \tau^1$$

of Lemma 6.2.1 for x near x_0. Then we set

$$\tau^1 = \cos\varphi, \quad \tau^{1+\alpha} = \omega^\alpha \sin\varphi, \quad |\omega| = 1,$$

$${}^*N_P^{kr}(x, \varphi, z) = \sin^{\nu-2}\varphi \int_{|\omega|=1} \hat{U}_r^k[\sigma(x, \tau), z]\, d\omega.$$

Then ${}^*N_P^{kr}$ is analytic in its arguments for φ near the segment $[0, \pi]$ in the φ-plane, x complex but near x_0, and z outside and on ∂R^+. Also, for a suitable integer n and constant C,

$${}^*L_P^{kr}(x, y) = C \int_0^\pi \left\{ \int_{\partial R^+} J_n(|x| \cos\varphi + y z)\, {}^*N_P^{kr}(x, \varphi, z)\, dz \right\} d\varphi, \qquad y > 0.$$

We conclude the analyticity at $(x_0, 0)$ by noticing that the segment $[0, \pi]$ may be replaced by a nearby arc from 0 to π along which $Im \cos\varphi > 0$ except at the end points.

Theorem 6.3.3. *Suppose $h \geq h_0 \geq 0$ and $h_0 + h_r \geq 1$ for each r and suppose that the g_r satisfy the following conditions for some L:*

(i) $g_r \in C_\mu^{h+h_r}(\bar{G}_A)$ *with* $h_\mu(\nabla^{h+h_r} g_r, \bar{G}_A) \leq L$, *for every* A.

(ii) $$|\nabla^t g_r(X)| \leq \begin{cases} L R^{h+h_r+\mu-t}, & 0 \leq t \leq h + h_r, \quad |X| \leq 2R, \quad y \geq 0, \\ L R^{h+h_r+\mu-t} (X/2R)^{\varrho-t} [1 + \log|X/2R|], & 0 \leq t \leq \varrho, \\ L R^{h+h_r+\mu-t} (X/2R)^{\varrho-t}, & \varrho < t \leq h + h_r \end{cases}$$

where $\varrho = h_0 + h_r - \nu - 1$ and $|X| \geq 2R$, $y \geq 0$ in the last two lines of (ii). *Suppose also that u is defined by*

$$u^k(x, y) = \int_\sigma \sum_r \sum_{|\alpha|=q_r} C_\alpha L_\alpha^{kr}(x, y; \xi)\, D^\alpha g_r(\xi)\, d\xi, \quad q_r = h_0 + h_r - 1$$

$$L_\alpha^{kr}(x, y; \xi) = L_\alpha^{kr}(x - \xi, y) - \sum_{|\beta| \leq \tau} D_X^\beta L_\alpha^{kr}(-\xi, 0) \frac{X^\beta}{\beta!}, \tag{6.3.29}$$

$$\tau = |\alpha| + t_k - h_r - \nu = t_k + h_0 - \nu - 1.$$

Then $u^k \in C_\mu^{t_k+h}(\bar{G}_A)$ *for every* A *and*

$$\sum_k{}' \| u^k \|_{\mu,2R}^{t_k+h} \le C L. \tag{6.3.30}$$

If the g_r *satisfy* (ii) *only for* $0 \le t \le h + h_r - 1$, *then* $u^k \in C_\mu^{t_k+h-1}(\bar{G}_A)$ *for any* A *and* (6.3.30) *holds with* h *replaced by* $h - 1$. *In either case* u *is a solution of the boundary value problem* (6.3.7) *with* g_r *replaced by* $g_r - g_r^*$ *where* g_r^* *is the expansion of* g_r *about* 0 *out to and including the terms of degree* ϱ. *Of course if* $\varrho < 0$, *the conditions for* $0 \le t \le \varrho$ *are to be omitted.*

Proof. We may extend the $L^{kr}(x, y)$ to be of class C^∞ on the whole of $R_{\nu+1}$ preserving the essential homogeneity and the analyticity along $y = 0$, $x \ne 0$. From Lemma 6.3.2 we conclude that

$$|D_X^\beta L_\alpha^{kr}(x, y; \xi)| \le \begin{cases} C \cdot |X|^{t_k+h_0-\nu-|\beta|} \cdot |\xi|^{-1} & \text{if } |\beta| \le \tau,\ |X| \le A|\xi|,\ A < 1 \\ C \cdot |X - \xi|^{t_k+h_0-\nu-1-|\beta|} & \text{if } |\beta| > \tau,\ \tau = t_k + h_0 - \nu - 1. \end{cases}$$

Thus we see that $u^k \in C^\infty$ (analytic in fact) for $y > 0$ and we may differentiate under the integral sign as often as desired to conclude that u is a solution of the equations (6.3.1) with L_{jk} replaced by L_{0jk} and the f_j being possible polynomials of degrees $\le h_0 - s_j - \nu - 1$; the convergence of the integral (6.3.29) at ∞ is guaranteed by the fact that $|\alpha| = \tau + \nu$.

Next, we note that any derivative of the form $D_u^{h-h_0+2} D_X^{\tau+\nu} u^k$ ($h - h_0 + 1$ of the x-derivatives can be shifted onto g_r) is of the form

$$\int_\sigma C D_x \Gamma(x - \xi, y) \gamma(\xi)\, d\xi, \quad \gamma = D^\delta g_r, \quad |\delta| = h + h_r, \tag{6.3.31}$$

where Γ is positively homogeneous of degree $-\nu$. Since $D_x \Gamma(x - \xi, y)$ is absolutely integrable over σ with value 0 (if $y > 0$), we may replace $\gamma(\xi)$ by $\gamma(\xi) - \gamma(x)$ in (6.3.31). Thus we find that

$$\begin{aligned} &|D_x^{h-h_0+2} D_X^{t_k+h_0-1} u^k(x, y)| \\ &\le C L \int_\sigma [|x - \xi|^2 + y^2]^{-(\nu+1)/2} \cdot |\xi - x|^\mu\, d\xi \le C L\, y^{\mu-1}. \end{aligned} \tag{6.3.32}$$

By differentiating the $j-th$ equation in (6.3.1) (as modified above) $h_0 - s_j$ times with respect to y and using the ellipticity, we see that we can solve for the $D_y^{t_k+h_0} u^k$ in terms of the other derivatives $D_X^{t_k+h_0} u^k$. By performing suitable differentiations we obtain the derivatives $D_x^{h-h_0+1} D_X^{t_k+h_0} u^k$ in terms of the $D_x^{h-h_0+2} D_X^{t_k+h_0-1} u^k$, and so on. Thus we obtain a bound (6.3.32) for $\nabla^{t_k+h+1} u^k$. From Theorem 2.6.6 it then follows that

$$h_\mu(\nabla^{t_k+h} u^k, \bar{G}_A) \le C L \quad \text{for all } A. \tag{6.3.33}$$

We now wish to show that u is a solution of the boundary problem (6.3.7) except for the polynomials g_r^*. Let $\varphi \in C^\infty(R_1)$, $\varphi(t) = 1$ for $t \le 1$, $\varphi(t) = 0$ for $t \ge 2$, $\varphi'(t) \le 0$ for $1 \le t \le 2$. If, for large n, we define

$$g_{nr}(\xi) = \varphi[(n R)^{-1} |\xi|] \cdot g_r(\xi)$$

we see that each $\{g_{nr}\}$ and its derivatives of order $\leq h + h_r$ are equicontinuous, converge uniformly on any ball to those of g_r, and the g_{nr} satisfy conditions (ii), at least with the log term added. Thus the corresponding u_n^k and all their derivatives of order $\leq t_k + h$ converge uniformly on any $\bar{G}_A$ to u^k and its derivatives. Then, for each n, $u_n = u_{0n} - \tilde{u}_n$ where u_{0n} is defined by (6.3.29) with $L_\alpha^{kr}(x, y; \xi)$ replaced by $L_\alpha^{kr}(x - \xi, y)$ and $\tilde{u}_n$ is the Maclaurin expansion of u_{0n} out to terms of degree τ. It follows from Theorem 6.3.2 that u_{0n} satisfies the homogeneous equations (6.3.7). Since $L_{0jk} \tilde{u}_n^k$ is just the corresponding Maclaurin expansion of $L_{0jk} u_{0n}^k$, it follows that $\tilde{u}_n$ also satisfies (6.3.7). Hence u satisfies (6.3.7).

Let us define

$$u^{*k}(x, y) = \int_\sigma \sum_r \sum_{|\alpha| = s_r} C_\alpha L_{P\alpha}^{*kr}(x, y; \xi) D^\alpha g_r(\xi)\, d\xi$$

$$L_{P\alpha}^{*kr}(x, y; \xi) = L_{P\alpha}^{*kr}(x-\xi, y) - \sum_{|\beta| \leq \tau'} \frac{X^\beta}{\beta!} D^\alpha L_{P\alpha}^{*kr}(-\xi, 0),$$

$$\tau' = t_k + h_0 - \nu - 1 + 2P.$$

For each n, we find by integrating by parts, etc., that

$$u_n^{*k} = U_n^{*k} - V_n^{*k}, \quad u_n^k = U_n^k - V_n^k, \quad \Delta_x^P u^{*k} = u^k, \quad \text{etc.,}$$

$$U_n^{*k}(x, y) = \int_\sigma \sum_r L_P^{*kr}(x - \xi, y) g_{nr}(\xi)\, d\xi \tag{6.3.34}$$

where V_n^k and V_n^{*k} are the Maclaurin expansions of U_n^k and U_n^{*k} out to terms of degree $t_k + h_0 - \nu - 1$ and $t_k + h_0 - \nu - 1 + 2P$, respectively, and U_n^k has the same formula with L_P^{*kr} replaced by L^{kr}. By using Theorem 6.3.2(b) with P large, we find that

$$B_{0rk} U_n^{*k}(x, 0) = \int_\sigma M_P(x - \xi) g_{nr}(\xi)\, d\xi$$

$$B_{0rk} U_n^k(x, 0) = g_{nr}(x).$$

Since V_n^k is a polynomial of degree $\leq t + h_0 - \nu - 1$, $B_{0rk} V_n^k$ is a polynomial of degree $\leq h_0 + h_r - \nu - 1 = q_r - \nu$. Hence

$$B_{0rk} u_n^k(x, 0) = g_{nr}(x) - \sum_{|\beta| \leq q_r - \nu} \frac{x^\beta}{\beta!} D^\beta g_{nr}(0). \tag{6.3.35}$$

Since $\nabla^t u^k(0, 0) = 0$ for $0 \leq t \leq t_k + h_0 - \nu - 1$, the bound (6.3.30) will follow from (6.3.33) and bounds which we now establish for $\nabla^t u^k(x, y)$ with $t = t_k + h_0 - \nu$. From (6.3.29) it follows that

$$\nabla^t u^k(x, y) = \int_\sigma \sum_r \sum_{|\alpha| = q_r} C_\alpha \nabla^t L_\alpha^{kr}(x - \xi, y) D^\alpha g_r(\xi)\, d\xi, \quad t = t_k + h_0 - \nu.$$

Using the bounds (ii) for $D^\alpha g_r$ ($|\alpha| = h_0 + h_r - 1$, $\nabla^t L_\alpha^{kr}$ pos. hom degree -1) we obtain

$$\begin{aligned}|\nabla^t u^k(X)| &\le \int_{\sigma_{3R}} C[|x-\xi|^2+y^2]^{-1/2}\cdot L\cdot R^{h-h_0+\mu+1}\,d\xi + \\ &+ \int_{R_\nu-\sigma_{3R}} C[|x-\xi|^2+y^2]^{-1/2}\cdot L\cdot R^{h-h_0+\mu+1}(|\xi|/R)^{-\nu}\,d\xi \\ &\le CLR^{h-h_0+\mu+\nu}, \quad |X|\le 2R.\end{aligned}$$

The second result on differentiability is proved similarly.

Corollary. *Suppose $h \ge h_0 \ge 0$, $h_0 + h_r \ge 1$ for each r, and suppose that $g_r \in C_{\mu c}^{h+h_r-1}(\sigma_{2R})$. Suppose also that u is given by (6.3.29). Then (6.3.30) holds with h replaced by $h-1$. If $g_r \in C_{\mu c}^{h+h_r}(\sigma_{2R})$, then (6.3.30) holds. In either case, u is a solution of the boundary value problem (6.3.7) with g_r replaced by $g_r - g_r^*$ where g_r^* is the polynomial in (6.3.35).*

Proof. For $u = U - V$ where $U = \Delta_\alpha^P U^*$, U^* is given by (6.3.34) and V^k is the Taylor expansion of U^k out to terms of degree $t_k + h_0 - \nu - - 1$.

We now develop a representation of u^k in which the highest order derivatives $D^{t_k+h} u^k$ are expressed as integrals of Calderon-Zygmund type.

Theorem 6.3.4. *Suppose $h \ge h_0 \ge 0$, $q > 1$, and $h_0 + h_r \ge 1$ for each r and suppose the g_r satisfy the following conditions:*

(i) *$g_r \in H^{h+h_r}(G_A)$ with $\|\nabla^{h+h_r} g_r\|_{q,A}^0 \le L$ for every A;*

(ii) *$'\|g_r\|_{q,3R}^{h+h_r} \le L$;*

(iii) *if $\varrho = h_0 + h_r - \nu - 1 \ge 0$, then $\nabla^t g_r$ is continuous on any $\bar G_A$ with $0 \le t \le \varrho$; and*

(iv) *for $|X| \ge 3R$ we have*

$$|\nabla^t g_r(X)| \le \begin{cases} L\cdot R^{h+h_r-(\nu+1)/q-t}(|X|/2R)^{\varrho-t}(1+\log|X/2R|), & \text{if } 0\le t\le\varrho \\ L\cdot R^{h+h_r-(\nu+1)/q-t}(|X|/2R)^{\varrho-t}, & \text{if } \varrho< t\le h+h_r.\end{cases}$$

Suppose, further that u is given by (6.3.29). Then $u^k \in H_q^{t_k+h}(G_A)$ for any A and

$$\sum_k {}'\|u^k\|_{q,2R}^{t_k+h} \le CL, \quad \sum_k \|\nabla^{t_k+h} u^k\|_{q,A}^0 \le L \quad \textit{for every } A.$$

Proof. We begin by defining

$$(6.3.36)\qquad \begin{aligned} M_\alpha^{kr\varepsilon}(x,y) &= D_y L_{\alpha+\varepsilon}^{kr}(x,y) \quad (x,y)\neq(0,0), \quad y\ge 0 \\ N_\alpha^{kr}(x,y) &= -D_{x^\varepsilon} L_{\alpha+\varepsilon}^{kr}(x,y) \quad |\varepsilon| = 1 \quad (\varepsilon = 1,\ldots,\nu)\end{aligned}$$

and where we extend $L_{\alpha+\varepsilon}^{kr}$ to belong to C^∞ on $R_{\nu+1} - \{0,0\}$ and to be essentially homogeneous of degree $t_k + h_0 - \nu$. We then define

$$\tilde u^k(x,y) = U^k(x,y) - U^{*k}(x,y),\ v^k(x,y) = T\tilde u^k(x,y),\ (q_r = h_0 + h_r - 1)$$

$$\begin{aligned} U^k(x,y) &= \int_{R_{\nu+1}^+} \sum_r \sum_{|\alpha|=q_r} C_\alpha \Big\{ \sum_{|\varepsilon|=1} M_\alpha^{kr\varepsilon}(X-\Xi) D_\xi^{\alpha+\varepsilon} g_r(\xi,\eta) + \\ &+ N_\alpha^{kr}(X-\Xi) D_\xi^\alpha D_\eta g_r(\xi,\eta)\Big\}\,d\xi\,d\eta, \quad T = D_X^t D_x^\beta, \end{aligned}$$

$$(6.3.37)\qquad |\beta| = h - h_0, \quad t = t_k + h_0 - 1.$$

U^{*k} being the Taylor expansion of U^k out to terms of degree $t_k + h_0 - \nu - 1$.

By using mollifiers and the methods of the proof of Theorem 6.3.3, we may approximate to g_r by functions $\in C_c(R^+_{\nu+1})$. For such functions we may integrate by parts in (6.3.37) if $t_k + h_0 > 1$ and conclude that $\tilde{u}^k(x, y) = u^k(x, y)$, since

$$N^{kr}_\alpha(x, y) = -L^{kr}_\alpha(x, y), \quad y \geq 0, \quad (x, y) \neq (0, 0)$$
$$M^{kr\varepsilon}_{\alpha,\varepsilon}(x, y) + N^{kr}_{\alpha y}(x, y) = 0, \quad (x, y) \neq (0, 0).$$

In case $t_k + h_0 = 1$, we may obtain the same result by first removing a small sphere $B(x, y; \varrho)$ from G_R, integrating by parts, letting $\varrho \to 0$, and using the result

$$\int_{\partial B(0,1)} [x^\varepsilon M^{kr\varepsilon}_\alpha(x, y) + y N^{kr}_\alpha(x, y)]\, dS = 0;$$

since this integral depends only on the values of the $L^{kr}_{\alpha+\varepsilon}$ for (x, y) near $\partial B(0, 1)$, its vanishing is seen to follow from (6.3.36) by altering the L' s near $(0, 0)$ so that they $\in C^\infty[\overline{B(0, 1)}]$. Then v^k has the form

$$v^k(x, y) = \int_{R^+_{\nu+1}} \sum_r \sum_{|\alpha| = h_0 + h_r - 1} C_\alpha \Big\{ \sum_{|\varepsilon|=1} D^t_X M^{kr\varepsilon}_\alpha(X - \Xi) D^{\alpha+\varepsilon+\beta}_\xi g_r(\Xi) +$$
$$+ D^t_X N^{kr}_\alpha(X - \Xi) D^{\alpha+\beta}_\xi D_\eta g_r(\Xi)\Big\}\, d\Xi$$

which is of the form 2.6.1. The results concerning the derivatives $D^{h-h_0}_x D^{t_k+h_0}_X u^k$ then follow from the Theorems of §§ 2.6 and 2.7. The same results for the other derivatives of order $t_k + h$ are found by differentiating the equations (6.3.7) as in the proof of Theorem 6.3.3.

From (6.3.37) we see that $u^k\,(= \tilde{u}^k)$ is a sum of terms of the form

$$\omega(x) = \int_{R^+_{\nu+1}} M(X, \Xi) D^\beta g_r(\Xi)\, d\Xi, \quad |\beta| = h_0 + h_r$$
$$M(X, \Xi) = M(X - \Xi) - \sum_{|\gamma| \leq \varrho} \frac{X^\gamma}{\gamma!} D^\gamma M(-\Xi), \quad \varrho = t_k + h_0 - \nu - 1$$

and M is essentially homogeneous of degree $t_k + h_0 - \nu - 1$. By the method of proof of Theorem 6.2.2 we conclude that (recall that ν has been replaced by $\nu + 1$)

$$(6.3.38) \qquad \int_{G_{3R}} |M(X, \Xi)|\, dX \leq C R^{t_k+h_0}, \quad \int_{G_{3R}} |M(X, \Xi)|\, d\Xi \leq C R^{t_k+h_0}.$$

Letting $\omega = \omega_1 + \omega_2$ where ω_1 is the integral over G_{3R}, we obtain

$$(6.3.39) \qquad |\omega_1(X)|^q \leq (C R^{t_k+h_0})^{q-1} \int_{G_{3R}} |M(X, \Xi)| \cdot |D^\beta g_r(\Xi)|^q\, d\Xi$$

using the Hölder inequality. From (ii), we conclude that

$$(6.3.40) \qquad \| D^\beta g_r \|^0_q \leq C R^{h-h_0}{}' \| g_r \|^{h+h_r}_{q,3R} \leq C L R^{h-h_0}.$$

By integrating (6.3.39) and using (6.3.40), we obtain

$$\| \omega_1 \|^0_{q,2R} \leq C L R^{t_k+h}.$$

Using Lemma 6.3.2 and (iv), we see that if $|X| \leq 2R$,

$$|\omega_2(X)| \leq C \int\limits_{R^+_{\nu+1}-G_{3R}} |\Xi|^{-1} \cdot |X|^{t_k+h_0-\nu} \cdot L \cdot R^{\nu+1+h-h_0-(\nu+1)/q} |\Xi|^{-\nu-1} d\Xi$$

$$\leq C L R^{t_k+h+1-(\nu+1)/q} \int\limits_{R^+_{\nu+1}-G_{3R}} |\Xi|^{-\nu-2} d\Xi \leq C L R^{t_k+h-(\nu+1)/q}.$$

Thus, also

$$\| \omega_2 \|^0_{q,2R} \leq C L R^{t_k+h}.$$

Corollary. *Suppose $h \geq h_0 \geq 0$, $h_0 + h_r \geq 1$ for each r, and suppose that $g_r \in H^{h+h_r}_q(G_{2R})$ and has support in $G_{2R} \cup \sigma_{2R}$. Suppose also that u is given by (6.3.29). Then $u^k \in H^{t_k+h}_q(G_A)$ for any A and u satisfies the conclusions of the theorem with L replaced by $\sum_r{}' \| g_r \|^{h+h_r}_{q,2R}$. Moreover u is a solution of the boundary value problem (6.3.7) with g_r replaced by $g_r - g^*_r$ where g^*_r is the polynomial in (6.3.35).*

Theorem 6.3.5: *Suppose $h \geq h_0 \geq 0$, $h_0 + h_r \geq 1$ for each r, $u^k \in H^{t_k+h}_q(G_A)$ for each A and each k, each $u^k \in C^\infty$ for $|X| \geq R_0$ and $y \geq 0$ for some $R_0 > 0$, and*

$$\text{(6.3.41)} \qquad L_{0jk} u^k(x, y) = f^*_j, \quad y > 0, \quad B_{0rk} u^k = g^*_r \quad \text{on} \quad y = 0,$$

$$\text{(6.3.42)} \qquad |D^p_X u^k(X)| \leq C_p(1 + |X|)^{t_k+h_0-\nu-1-p+1/2}, \quad p = 0, 1, 2, \ldots, \quad |X| \geq R_0, \quad y \geq 0,$$

where f^*_j and g^*_r are polynomials of degree $\leq h_0 - s_j - \nu - 1$ and $h_0 + h_r - \nu - 1$, respectively. Then u^k is a polynomial of degree $\leq t_k + h_0 - \nu - 1$.

Proof. Let φ be a mollifier in the x-variables and let $u^k_\varrho(x, y)$ be the mollified functions:

$$u^k_\varrho(x, y) = \int\limits_{B(x,\varrho)} \varphi^*_\varrho(x - \xi) u^k(\xi, y) d\xi = \int\limits_{R_\nu} \varphi^*_\varrho(x - \xi) u^k(\xi, y) d\xi$$

$$= \int\limits_R^y \int\limits_{B(x,\varrho)} \varphi^*_\varrho(x - \xi) u^k_y(\xi, \eta) d\xi \, d\eta.$$

It is easy to see that all derivatives of u_ϱ of order $\leq t^k + h - 1$ are continuous in (x, y) and C^∞ in x for $y \geq 0$ and the u^k_ϱ clearly satisfy equations like (6.3.41) and (6.3.12) if $\varrho > 0$ (the polynomials are altered but are of the same degrees). By repeatedly using the equation (6.3.41) and its derivatives, we conclude that each $u^k_\varrho \in C^\infty(\bar{R}^+_{\nu+1})$ if $\varrho > 0$. Since u^k_ϱ converges uniformly with all derivatives to u^k if $|X| \leq R_0$ and $y \leq 0$, it is sufficient to prove the theorem for $u \in C^\infty(\bar{R}^+_{\nu+1})$.

Let $v^k = D^t_x u^k$ where $t \geq \max(t_k + h_0, h_0 - s_j, h_0 + h_r)$. Then v^k satisfies (6.3.41) with f^*_j and $g^*_r \equiv 0$ and, from (6.3.42) we conclude that

$$|v^k(X)| \leq C(1 + |X|)^{-\nu-1/2}, \quad X \in R^+_{\nu+1}.$$

Now, if we let $\hat{v}^k(\lambda, y)$ be the Fourier transform of v_k as defined in (6.3.8) and introduce $V^k(\lambda, u)$ by (6.3.10), i.e. $V^k(\lambda, u) = (i|\lambda|)^{t_k}\hat{v}^k(\lambda, u/|u|)$, we see that $|V^k(\lambda, u)| \leq C|\lambda|^{t_k}$ for all (λ, u) and satisfies

$$L_{0jk}(\sigma, -iD_u)V^k(\lambda, u) = 0, \quad B_{0rk}(\sigma, -iD_u)V^k(\lambda, 0) = 0, \quad r = 1, \ldots, m.$$

Since V^k is bounded for $u \geq 0$ and λ fixed, it must be exponentially decaying and hence $\equiv 0$ by Theorem 6.3.1. Thus all the $D_x^t u^k \equiv 0$ and we conclude from this and (6.3.42) that

$$u^k(x, y) = \sum_{0 \leq |\beta| \leq p_k} c_\beta^k(y)x^\beta, \quad p_k \leq \tau_k = t_k + h_0 - \nu - 1.$$

By applying the operators $D_y^\tau L_0$ for $\tau \geq \tau_k - 2m$ we see that each u^k satisfies $D_y^\tau L_0 u^k(x, y) = 0$. Of course, if $p_k < 0$, $u^k \equiv 0$. Otherwise, since $D_y^\tau L_0 u^k = 0$, it follows that $D_y^{\tau+2m} c_\beta^k(y) = 0$ for $|\beta| = p_k$, this being the coefficient of x^β in $D_y^\tau L_0 u^k(x, y)$. Thus, using (6.3.42), we conclude that the $c_\beta^k(y)$ are polynomials of degree $\leq \tau_k - p_k$. Next, suppose that $0 \leq p < p_k$ and that all the $c_\beta^k(y)$ with $|\beta| > p$ are polynomials of degree $\tau_k - |\beta|$. Then, since $D_y^\tau L_0 D_y^{\tau_k - p} u^k(x, y) = 0$ and $D_y^{\tau_k - p} u^k(x, y)$ contains no terms with $|\beta| > p$ we see as above that $D_y^{\tau + \tau_k + 2m - p} c_\beta^k(y) = 0$ if $|\beta| = p$. Thus such c_β^k are polynomials which, by (6.3.42) must be of degree $\leq \tau_k - p$. The fact that each u^k is a polynomial of degree $\leq \tau_k$ now follows by induction.

Definition 6.3.3. For $h \geq 0$ we define the spaces $^*H_q^h(G_R)$, $^*C_\mu^h(\bar{G}_R)$, $^{**}H_q^h(G_R)$, and $^{**}C_\mu^h(\bar{G}_R)$ just as they were defined in Definition 6.2.1 with B_R replaced by G_R; if $f \in {}^*H_q^h$ or $^*C_\mu^h$ we require $\nabla^t f_j$ to vanish on $\sum_R$ but not necessarily along σ_R, $t = 0, \ldots, h - s_j - 1$.

Definition 6.3.4. If $h_0 \geq 0$, $h \geq h_0 - 1$, and $h_0 + h_r \geq 1$ for every r, we define the space $^*C_\mu^h(\sigma_R)$ to consist of vectors g with $g_r \in C_{\mu 0}^{h+h_r}(\sigma_R)$ with norm

$$^{*\prime}|||g|||_{\mu R}^h = \sum_r{}' |||g_r|||_{\mu R}^{h+h_r}.$$

If $h \geq h_0$, we define the space $^{*\prime}H_q^h(\sigma_R)$ to consist of all vectors g such that $g_r(x) = \tilde{g}_r(x, 0)$ where $\tilde{g}_r \in H_q^{h+h_r}(G_R)$ and $\nabla^t \tilde{g}_r = 0$ along $\sum_R$ for $0 \leq t \leq h + h_r - 1$ with norm

$$^{*\prime}\|g\|_{qR}^h = \inf_{\tilde{g}} \sum_r{}' \|\tilde{g}_r\|_{qR}^{h+h_r}, \quad \tilde{g}_r(x, 0) = g_r(x).$$

Lemma 6.3.3. (Lions) *There are bounded extension operators τ from $^*H_q^h(G_R)$ into $^*H_q^h(B_R)$ and from $^*C_\mu^h(\bar{G}_R)$ into $^*C_\mu^h(\bar{B}_R)$.*

Proof. Let $t = \max h - s_j$. We may define $\tau f = F$, where $F(x, y) = f(x, y)$ if $(x, y) \in G_R$, $F(x, y) = 0$ if $y \geq 0$, $(x, y) \notin G_R$,

$$F(x, y) = \sum_{s=1}^{t+1} C_s^t F(x, -sy), \quad y < 0$$

where the C_s^t are uniquely defined by the conditions

$$\sum_{s=1}^{t+1} (-s)^u C_s^t = 1, \quad u = ,0 \ldots, t.$$

Definition 6.3.5. Suppose $h \geq h_0 \geq 0$ and $h_0 + h_r \geq 1$ for every r. For f in $^*C_\mu^h(\bar{G}_R)$ and g in $^{*\prime}C_\mu^h(\bar{\sigma}_R)$ or for f in $^*H_q^h(G_R)$ and $g \in {}^{*\prime}H_q^h(\sigma_R)$ we define

$$P_R(f, g) = U_R + V_R - W_R = P_{1R}(g) + P_{2R}(f) + P_{3R}(f)$$

where U_R is defined by (6.3.29) in terms of g,

$$\text{(6.3.43)}\quad \begin{gathered} V_R^k = L_0^{kl}(F_l - F_l^*), \quad F_l(X) = \int_{B_{2R}} K(X - \Xi)\tilde{f}_l(\Xi)\, d\Xi, \quad \tilde{f} = \tau f \\ F_l^*(X) = \sum_{|\alpha| \leq \varrho} \frac{X^\alpha}{\alpha!} D^\alpha F_l(0), \quad \varrho = 2m + h_0 - s_l - \nu - 1, \end{gathered}$$

τ being the extension operator of Lemma 6.3.3, and W_R is defined by (6.3.29) with $g(x)$ replaced by $\tilde{\gamma}(x, 0)$ where

$$\tilde{\gamma}_r(X) = B_{0rk} V_R^k(X) \equiv \tilde{\gamma}_{r0}(X) - \tilde{\gamma}_r^*(X), \quad \tilde{\gamma}_r^*(X) = B_{0rk} L_0^{kl} F_l^*(X).$$

Theorem 6.3.6. *P_R is a bounded operator from $^*H_q^h(G_R)\, x^{*\prime}H_q^h(\sigma_R)$ to $^{**}H_q^h(G_R)$ and from $^*C_\mu^h(\bar{G}_R)\, x^{*\prime}C_\mu^h(\bar{\sigma}_R)$ to $^{**}C_\mu^h(\bar{G}_R)$ with bound independent of R. If $u = P_R(f, g)$, then*

$$\text{(6.3.44)}\quad L_{0jk} u^k = f_j - f_j^* \text{ in } R_{\nu+1}^+ \text{ and } B_{0rk} u^k = g_r - g_r^* \text{ on } \sigma,$$

f_j^* and g_r^* being Maclaurin expansions of f_j and g_r out to the terms of degree $h_0 - s_j - \nu - 1$ and $h_0 + h_r - \nu - 1$ if these are ≥ 0, being 0 otherwise. The second result holds if $h = h_0 - 1$ if $h_0 - 1 \geq 0$.

Proof. From the two corollaries it follows that P_{1R} is bounded from $^{*\prime}H_q^h(\sigma_R)$ to $^{**}H_q^h(G_R)$ and from $^{*\prime}C_\mu^h(\sigma_R)$ to $^{**}C_\mu^h(\bar{G}_R)$ and that $U_R = P_{1R}(g)$ is a solution of (6.3.44) with $f_j \equiv f_j^* = 0$. Now V_R^k coincides with the function U^k in (6.2.20). If $f \in {}^*C_\mu^h(\bar{G}_R)$, the norm of f in $^*C_\mu^{h_0}(\bar{G}_R) \leq C \cdot R^{h-h_0}$ times its norm in $^*C_\mu^h(\bar{G}_R)$. We see by integrating by parts $h_0 - s_l$ times that

$$\begin{gathered} F_l(X) = \int_{B_{2R}} \sum_{|\alpha| = h_0 - s_l} C_\alpha K_\alpha(X - \Xi) D^\alpha \tilde{f}_l(\Xi)\, d\Xi \\ K_\alpha(X) = D_X^{\alpha*} K_{|\alpha|}(X) \end{gathered}$$

(see (6.2.11)) so that $F_l \in C_\mu^{2m+h-s_l}(\bar{B}_A)$ for any A. Thus, we see that $\tilde{\gamma}_r \in C_\mu^{h+h_r}(\bar{B}_A)$ for any A and that $\tilde{\gamma}_{r0}$ is a sum of terms of the form

$$\begin{gathered} \int_{B_{2R}} \Gamma(X - \Xi) f(\Xi)\, d\Xi, \quad f = D_X^\alpha \tilde{f}_l, \quad |\alpha| = h_0 - s_l, \\ f \in C_{\mu 0}^{h-h_0}(\bar{B}_{2R}), \quad \Gamma \text{ ess. hom. degree } h_0 + h_r - \nu - 1. \end{gathered}$$

We see also from (6.3.43) that $\tilde{\gamma}_r^*$ is the Maclaurin expansion of $\tilde{\gamma}_{r0}$ out to and including the terms in X^t, $t = h_0 + h_r - \nu - 1$. Accordingly the Maclaurin expansion of $\tilde{\gamma}_r$ out to terms of degree t vanishes. From this representation it is easily seen that the $\tilde{\gamma}_r$ satisfy the conditions on the g_r in Theorem 6.3.3. Thus in this case P_{2R} and P_{3R} are seen to be bounded operators from $^*C_\mu^h(\bar{G}_{2R})$ to $^{**}C_\mu^h(\bar{G}_{2R})$ and from Theorem 6.3.3, it follows that $v_R = V_R - W_R$ satisfies (6.3.44) with $g_r = g_r^* = 0$. The case $h = h_0 - 1 \geq 0$ follows by inspection in the $*C_\mu^h$ case.

In case $f \in *H_q^h(G_{2R})$, it follows from Theorem 6.2.2 and its proof that the $\tilde{\gamma}_r$ satisfy condition (i) for the g_r in Theorem 6.3.4. Condition (iii) follows from condition (ii) and the relevant SOBOLEV lemmas (§ 3.5). For almost all X in $R_{\nu+1}^+$, we see that $\nabla^t \tilde{\gamma}(X)$ is a sum of terms of the form

$$\omega(X) = \int_{B_{2R}} M(X, \Xi) f_0(\Xi)\, d\Xi, \quad f_0(\Xi) = D^\alpha f_l(\Xi), \quad |\alpha| = h_0 - s_l,$$

$$M(X, \Xi) = M(X - \Xi) - \sum_{|\delta| \leq \varrho - t} \frac{X^\delta}{\delta!} D^\delta M(-\Xi) \tag{6.3.45}$$

where M is essentially homogeneous of degree $\varrho - t$ and $\varrho = h_0 + h_r - \nu - 1$. From this it follows from Lemma 6.3.2 that

$$\int_{B_{3R}} |M(X, \Xi)|\, dX \leq C R^{h_0 + h_r - t}, \quad \int_{B_{3R}} |M(X, \Xi)|\, d\Xi \leq C R^{h_0 + h_r - t},$$
$$(\text{if } h_0 + h_r - t \geq 0).$$

Using the method of proof of Theorem 6.2.2 and the facts that

$$\|f_0\|_q^0 \leq C R^{h - h_0} \sum_l{}' \|\tilde{f}_l\|_{q\,3R}^{h - s_l} \leq C R^{h - h_0}\, {}^*\|f\|_{q, 2R}^h \tag{6.3.46}$$

we obtain the result (ii) for the $\tilde{\gamma}_r$ by setting $t = 0$, $L = C\,{}^*\|f\|_{q,2R}^h$ above and using Lemma 6.2.2. Now, if $\varrho - t < 0$, $|X| \geq 3R$, $|\Xi| \leq 2R$, we see that

$$|M(X, \Xi)| \leq C |X|^{\varrho - t}, \quad \int_{B_{2R}} |f_0(\Xi)|\, d\Xi \leq C R^{(\nu+1)(1 - 1/q)} \|f_0\|_{q, 2R}^0 \tag{6.3.47}$$

so that the bound (iv) holds in this case. If $\varrho - t \geq 0$, it follows from the essential homogeneity that

$$|M(X, \Xi)| \leq C |X|^{\varrho - t} [1 + \log |X/2R|] \quad (|X| \geq 3R,\ |\Xi| \leq 2R)$$

from which the bound (ii) again follows as in (6.3.47). The remaining results follow easily from (6.3.45), (6.3.46), and (6.3.47).

Definition 6.3.6. Suppose h_0 is the smallest integer such that $h_0 \geq 0$ and $h_0 + h_r \geq 1$ for every r. If $h \geq h_0$, we say that the operators B_{rk} satisfy the h-conditions in $\overline{B(x_0, A)}$ if and only if the coefficients all $\in C_1^{h + h_r - 1}[\overline{B(x_0, A)}]$; they satisfy the $h - \mu$ conditions there if and only if the coefficients all $\in C_\mu^{h + h_r}[\overline{B(x_0, A)}]$; we allow the $h - \mu$ conditions if we merely have $h \geq h_0 - 1$ provided that $h \geq 0$.

We can now imitate the developments at the end of the preceding section. We first prove the following theorem like Theorem 6.2.5.

Theorem 6.3.7. *Suppose* $h \geq h_0 \geq 0$ *and* $h_0 + h_r \geq 1$ *for every* r. *Suppose* G *is of class* $C_1^{t_0 + h - 1}$, $t_0 = \max t_k$, *and suppose the operators* $L = \{L_{jk}\}$ *and* $B = \{B_{rk}\}$ *satisfy the h-conditions on a domain* $\Gamma \supset \bar{G}$.

Suppose that $u^k \in H_q^{t_k+h_0}(G)$, $f_j \in H_p^{h-s_j}(G)$, *and* $g_r = \tilde{g}_r$ *on* ∂G, *where* $\tilde{g}_r \in H_p^{h+h_r}(G)$ *and*

$$p \geq q, \; f_j = L_{jk} u^k, \; g_r = B_{rk} u^k. \tag{6.3.48}$$

Then $u^k \in H_p^{t_k+h}(G)$. *If* G *is of class* $C_\mu^{t_0+h}$, L *and* B *satisfy the* $h-\mu$ *conditions on a domain* $\Gamma \supset \bar{G}$, *and* $u^k \in H_q^{t_k+h_0}(G)$, $f_j \in C_\mu^{h-s_j}(\bar{G})$, *and* $g_r \in C_\mu^{h+h_r}(\partial G)$, *then* $u^k \in C_\mu^{t_k+h}$. *If* G, f_j, g_r, L, *and* $B \in C^\infty$, *so does* u^k. *If* $h \geq h_0$, $f_j \in C_\mu^{h+s_j}(\bar{G})$, $g_r \in C_\mu^{h+h_r}(\partial G)$, $h_0 \geq 1$, $u^k \in C_\mu^{t_k+h_0-1}(\bar{G})$, *then* $u^k \in C_\mu^{t_k+h}$.

Proof. The proof is similar to that of Theorem 6.2.5 from which the interior results follow. But now, also, with each x_0 on ∂G is associated an $R > 0$ and a mapping of class $C_1^{t_0+h-1}$ (in the first case) of a neighborhood of x_0 onto G_{2R}; R can be taken so small that the transformation T_R defined in (6.3.4) (in terms of the coordinates on G_{2R}) has bound $\leq 1/2$ in a given desired space. Suppose $\zeta \in C^\infty(\bar{G})$ and has support in the neighborhood of x_0 corresponding to $G_R \cup \sigma_R$ and let U be the transform of ζu. Then, as in (6.2.27), we see that

$$\tilde{L}_{jk} U^k = \tilde{\zeta} \tilde{f}_j + \tilde{M}_{jk} \tilde{u}^k = \tilde{f}_j^*$$
$$\tilde{B}_{rk} U^k = \tilde{\zeta} \tilde{g}_r + \tilde{N}_{rk} \tilde{u}^k = \tilde{g}_r^*$$

where the tildas denote the transformed functions and the operators $\tilde{M}_{jk}$ and $\tilde{N}_{jk}$ involve only derivatives of lower order. Since these are smoother than those of highest order, the proof proceeds as before. The basic fact is that H_R is again a polynomial on account of Theorem 6.3.5.

Next we prove a local theorem like Theorem 6.2.6.

Theorem 6.3.8. *Suppose* B *and* L *satisfy the* h*-conditions* ($h \geq h_0$) *on* $\bar{B}_A$. *Then there are constants* $R_2 > 0$ *and* C_2 *depending only on* ν, h, q, E, *and* E' *such that*

$$^{**}\| u \|_{q,R}^h \leq C_2 [^*\| L u \|_{qR}^h + {}^{*\prime}\| B u \|_{qR}^h + {}^{**\prime}\| u \|_{1R}^0], \quad 0 < R \leq R_2,$$
$$u^k \in H_q^{t_k+h}(G_R), \quad {}^{**\prime}\| u \|_{1R}^0 = \sum_k R^{-t_k-h-\varrho} \| u^k \|_{1R}^0, \quad \varrho = (\nu+1)(q-1)/q$$

whenever u *also has support in* $G_R \cup \sigma_R$. *Also there exist constants* $R_3 > 0$ *and* C_3, *depending only on* ν, h, μ $(0 < \mu < 1)$ *and the quantities above such that if* L *and* B *satisfy the* $h-\mu$ *conditions on* $\bar{B}_A$, *then*

$$^{**}||| u |||_{\mu R}^h \leq C_3 [^* ||| L u |||_{\mu R}^h + {}^{*\prime}||| B u |||_{\mu R}^h + R^{\varrho-\nu-1-\mu} \, {}^{*\prime}\| u \|_{1R}^0],$$
$$0 < R \leq R_3, \text{ supt } u \subset \bar{G}_R \cup \sigma_R.$$

This last result holds if $h = h_0 - 1$ *if* $h_0 - 1 \geq 0$.

Proof. The proof is exactly analogous to that of Theorem 6.2.6. The differences involve only P_R: we let $u = u_R + H_R$,

$$u_R = P_R(L_0 u, B_0 u), \quad T_R \varphi = -P_R[(L_R - L_0)\varphi, (B_R - B_0)\varphi]$$

and then we have

$$u_R - T_R u_R = P_R(f, g) - P_R[(L_R - L_0) H_R, (B_R - B_0) H_R,]$$
$$f = L_R u \equiv L u, \quad g = B_R u \equiv B u.$$

We conclude this section with the following well-known "coerciveness" inequality.

Theorem 6.3.9. *Suppose $h \geq h_0 \geq 0$, $h_0 + h_r \geq 1$ for every r. Suppose G is of class $C_1^{t_0+h-1}$, $t_0 = \max t_k$, and suppose L and B satisfy the h-conditions on a domain $\Gamma \supset G$. Then*

(6.3.49)
$$\sum_k \| u^k \|_{q,G}^{t_k+h} \leq C\Big[\sum_j \| f_j \|_{q,G}^{h-s_j} + \sum_r \| \tilde{g}_r \|_{q,G}^{h+h_r} + \sum_k \| u^k \|_{1,G}^{0}\Big],$$
$$u^k \in H_q^{t_k+h}(G), \quad \tilde{g}_r \in H_q^{h+h_r}(G), \quad \tilde{g}_r = g_r \quad \text{on} \quad \partial G,$$

where C depends only on v, h, q, G, E and E', and f_j and g_r are given by (6.3.48). In case G is of class $C_\mu^{t_0+h}$ and L and B satisfy the $h - \mu$ conditions on Γ, then a similar inequality holds for the various $\|| \ \||$-norms. This last result holds if $h = h_0 - 1$ if $h_0 - 1 \geq 0$.

Proof. We prove the first statement, the proof of the second is similar. G can be covered by a finite number of neighborhoods $\mathfrak{N}_P$ each of which is either a sphere $B(P, R)$ with $B(P, 2R) \subset G$ or is the image of $G_R \cup \sigma_R$ under a mapping τ of class $C_1^{t_0+h-0}$ ($t_0 = \max t_k$) which maps $\bar{B}_A$ onto a neighborhood $\supset \mathfrak{N}_p$ with $A \geq 2R$. Since the transformed operators under such mappings still satisfy the h-conditions, it is clear that we may choose each $R \leq$ the number R_2 in Theorems 6.2.6 or 6.3.8. And there is a partition of unity $\zeta_1, \ldots, \zeta_s$ of class $C_1^{t_0+h-1}(\bar{G})$ each ζ_s having support in some one such neighborhood.

Now suppose there is no such constant. Then $\exists$ a sequence $\{u_n\}$ such that the left side of (6.3.49) is unity for each n, $u_n^k \rightharpoonup u^k$ in $H_q^{t_k+h}(G)$ and the bracket on the right in (6.3.49) $\to 0$. Then we must have $u = 0$ and hence $u_n^k \to 0$ in $H_q^{t_k+h-1}(G)$. Now let $u_{ns} = \zeta_s u_n$. Then $u_{ns}^k \to 0$ in $H_q^{t_k+h-1}(G)$ and u_{ns} satisfies

(6.3.50) $L_{jk} u_{ns} = \zeta_s f_{nj} + M_{jks} u_n, \; B_{rk} u_{ns} = \zeta_s g_{nr} + N_{rks} u_n$

where the operators M_{jks} and N_{rks} involve only derivatives of lower order. Clearly, then, the right sides of (6.3.50) $\to 0$ strongly in the respective spaces. Thus, from Theorems 6.2.6 and 6.3.8, it follows that each $u_{ns}^k \to 0$ strongly in $H_q^{t_k+h}(G)$. But since $u_n^k = \sum u_{ns}^k$, this contradicts our assumption that the left side of (6.3.49) is 1 for each n.

6.4. Weak solutions

In this section, we consider solutions of equations of the form

(6.4.1)
$$\int_G \sum_{j=1} \sum_{|\alpha| \leq m_j} D^\alpha v_j \Big[\sum_{k=1}^N \sum_{|\beta| \leq \varrho_{jk}} a_{jk}^{\alpha\beta} D^\beta u^k - f_j^\alpha\Big] dx = 0,$$
$$v^j \in C_c^\infty(G), \quad \varrho_{jk} = t_k + s_j - m_j$$

where we shall assume that the s_j, t_k, and m_j satisfy

$$\min(s_j + t_k) \geq 1, \quad \min(t_k + s_j - m_j) \geq 0, \quad \min m_j \geq 0, \qquad (6.4.2)$$
$$\max s_j = 0,$$

and where we also assume that there exists an integer h_0 *which may be negative* such that

$$\min(t_k + h_0) \geq 0, \quad \min(h_0 - s_j + m_j) \geq 0. \qquad (6.4.3)$$

It follows that

$$2m = \sum_{j=1}^{N} (s_j + t_j) \geq N, \quad m_j \leq s_j + t_j \leq 2m. \qquad (6.4.4)$$

We note that if the u^k, $a_{jk}^{\alpha\beta}$, and the f_j^α are sufficiently differentiable, then the u^k satisfy the equations (6.1.1) where

$$L_{jk} u^k = \sum_{|\alpha| \leq m_j} \sum_{|\beta| \leq \varrho_{jk}} (-1)^{|\alpha|} D^\alpha a_{jk}^{\alpha\beta} D^\beta u^k, \quad f_j = \sum_{|\alpha| \leq m_j} (-1)^{|\alpha|} D^\alpha f_j^\alpha. \qquad (6.4.5)$$

For each x_0 we define the operators L_{0jk} by

$$L_{0jk} u^k = \sum_{|\alpha| = m_j} \sum_{|\beta| = \varrho_{jk}} (-1)^{|\alpha|} a_{jk}^{\alpha\beta}(x_0) D^{\alpha+\beta} u^k \qquad (6.4.6)$$

and then define the operators L_0 and L_0^{kl} as in §§ 6.1—6.3. We assume that the operator L_0 is elliptic and shall denote the collection of numbers s_j, t_k, m_j, h_0, ν, and a *bound* for the coefficients in (6.4.6) and for $L_0(\lambda)^{-1}$ for λ real with $|\lambda| = 1$ by E. In the case of variable coefficients, E' shall stand for the moduli of continuity of the coefficients in (6.4.6) and *bounds* for their derivatives and for the other $a_{jk}^{\alpha\beta}$ and their derivatives as required.

We notice that the equations

$$\int_G [v_{,\alpha}(a^{\alpha\beta} u_{,\beta} + b^\alpha u + f^\alpha) + v(c^\beta u_{,\beta} + du + f)]\, dx = 0$$

$$\int_G [v_{,\alpha\beta}(a^{\alpha\beta} u - f^{\alpha\beta}) + v_{,\alpha}(b^\alpha u - f^\alpha) + v(c u - f)\, q\, dx = 0$$

are special cases of (6.4.1) (here α and β are single indices).

As in § 6.2, we begin by considering solutions u having support in a small ball B_R and define the coefficients $a_{Rjk}^{\alpha\beta}$ by

$$a_{Rjk}^{\alpha\beta}(x) - a_{0jk}^{\alpha\beta} = \varphi(R^{-1}|x|)\,[a_{jk}^{\alpha\beta}(x) - a_{jk}^{\alpha\beta}(x_0)]$$
$$a_{0jk}^{\alpha\beta} = \begin{cases} a_{jk}^{\alpha\beta}(x_0), & \text{if } |\alpha| = m_j, \quad |\beta| = \varrho_{jk}, \\ 0, & \text{otherwise} \end{cases} \qquad (6.4.7)$$

where φ is defined in (6.2.25).

Definition 6.4.1. If $h \geq h_0$, we say that *the a' s satisfy the h-conditions* on a set Γ if and only if

(i) if $h = h_0$ and $s_j = 0$, the $a_{jk}^{\alpha\beta}$ with $|\alpha| = m_j$ and $|\beta| = \varrho_{jk}$ are continuous; and

(ii) if $h - s_j + |\alpha| > 0$ then the $a_{jk}^{\alpha\beta} \in C_1^{h-s_j+|\alpha|-1}(\Gamma)$; otherwise the $a_{jk}^{\alpha\beta}$ are bounded and measurable.

The a's satisfy the $h - \mu$-conditions on Γ if and only if the coefficients $a_{jk}^{\alpha\beta}$ with $|\alpha| = m_j$ and $|\beta| = \varrho_{jk} \in C_\mu^0(\bar{\Gamma})$ at least, those for which $h - s_j + |\alpha| > 0$ belonging to $C_\mu^{h-s_j+|\alpha|}(\bar{\Gamma})$, the remaining a's belonging to $C_\mu^0(\bar{\Gamma})$ at least.

Definition 6.4.2. For $h \geq h_0$, we define the spaces $^{**}H_q^h(B_{2R})$ and $^{**}C_\mu^h(\bar{B}_{2R})$ of vectors u just as they were defined in §§ 6.2 and 6.3 exept that now h may be ≤ 0. We define the space $^*H_q^h(B_{2R})$ to consist of those vectors $f = \{f_j^\alpha\}$ such that

$$f_j^\alpha \begin{cases} \in H_{q0}^{h-s_j+|\alpha|}(B_{2R}), & \text{if} \quad h - s_j + |\alpha| > 0, \\ \in L_q(B_{2R}) & , \text{if} \quad h - s_j + |\alpha| \leq 0, \end{cases}$$

$$^*\|f\|_{q,2R}^h = \sum_{h-s_j+|\alpha|>0} {}' \|f_j^\alpha\|_{q,2R}^{h-s_j+|\alpha|} + \sum_{h-s_j+|\alpha|\leq 0} R^{(h-s_j+|\alpha|)} \|f_j^\alpha\|_{q,2R}^0.$$

We define the space $^*C^h(\bar{B}_{2R})$ to consist of those vectors f such that

$$f_j^\alpha \begin{cases} \in C_{\mu 0}^{h-s_j+|\alpha|}(\bar{B}_{2R}), & \text{if} \quad h - s_j + |\alpha| > 0 \\ \in C_{\mu 0}^0(\bar{B}_{2R}) & , \text{if} \quad h - s_j + |\alpha| \leq 0 \end{cases}$$

$$^*|||f|||_{\mu,2R}^h = \sum_{h-s_j+|\alpha|>0} {}' |||f_j^\alpha|||_{\mu,2R}^{h-s_j+|\alpha|} + \sum_{h-s_j+|\alpha|\leq 0} R^{(h-s_j+|\alpha|)} {}' |||f_j^\alpha|||_{\mu,2R}^0.$$

Definition 6.4.3. If $h \geq h_0$ and $f \in {}^*H_q^h(B_{2R})$ or $^*C_\mu^h(\bar{B}_{2R})$, we define $P_{h_0 R}(f) = U_R$ where

$$\begin{aligned} U_R^k &= \sum_l \sum_{|\alpha|\leq m_l} L_0^{kl\prime}(-1)^{|\alpha|} D^\alpha (F_l^\alpha - F_l^{*\alpha}) \\ F_l^\alpha(x) &= \int_{B_{2R}} K(x-\xi) f_l^\alpha(\xi)\, d\xi, \end{aligned} \tag{6.4.8}$$

and $F_l^{*\alpha}$ is the Maclaurin expansion of F_l^α out to terms of degree σ_l^α where $\sigma_l^\alpha = 2m - \nu - 1$ if $h_0 - s_l + |\alpha| \leq 0$ and $\sigma_l^\alpha = 2m + h_0 - s_l + |\alpha| - \nu - 1$ if $h_0 - s_l + |\alpha| > 0$.

Using the methods and results of § 6.2 and approximations to the f_j^α by smooth functions we conclude the following theorem:

Theorem 6.4.1. *$P_{h_0 R}$ is a bounded operator from $^*H_q^h(B_{2R})$ into $^{**}H_q^h(B_{2R})$ and from $^*C_\mu^h(\bar{B}_{2R})$ into $^{**}C_\mu^h(\bar{B}_{2R})$ with bound independent of R if $q > 1$, $0 < \mu < 1$, and $h \geq h_0$. If $U_R = P_{h_0 R}(f)$, then U_R satisfies (6.4.1) with $G = B_{2R}$ and $a_{jk}^{\alpha\beta}$ replaced by $a_{0jk}^{\alpha\beta}$ and f_j^α replaced by $f_j^\alpha - f_j^{*\alpha}$ where $f_j^{*\alpha}$ is the expansion of f_j^α out to terms of degree $h_0 - s_j + |\alpha| - \nu - 1$ (if this is ≥ 0). Moreover if u has support in B_{2R} and satisfies the same equations with the $f_j^{*\alpha} \equiv 0$, then $u^k - U_R^k \equiv 0$ or is a polynomial of degree $\leq \max(t_k + h_0 - \nu - 1, t_k + s_l - \nu - 1)$. It is understood that the $a_{0jk}^{\alpha\beta}$ are defined in (6.4.7).*

The last statement follows since the mollified functions u_ϱ^k satisfy the same equations as does u with the f_j^α replaced by their mollified functions $f_{j\varrho}^\alpha$; of course ϱ is assumed so small that u_ϱ has support in B_{2R} (we may work in $B_{2R+\delta}$ with $\varrho < \delta$). Then

$$L_0^{kl}\Big[\int_{B_{2R}} K(x-\xi)\, L_{0lm}\, u_\varrho^m(\xi)\, d\xi\Big] = L_0^{kl} L_{0lm} \int_{B_{2R}} K(x-\xi)\, u_\varrho^m(\xi)\, d\xi$$

$$= L_0 \int_{B_{2R}} K(x-\xi)\, u_\varrho^k(\xi)\, d\xi = u_\varrho^k(x), \quad L_{0lm}\, u_\varrho^m = f_{l\varrho} = \sum_{|\alpha|\le m_j} (-1)^{|\alpha|} D^\alpha f_{j\varrho}^\alpha$$

(6.4.9)

since u_ϱ has compact support. Using (6.4.9), integrating by parts and letting $\varrho \to 0$, we obtain

$$u^k(x) = L_0^{kl} \sum_{|\alpha|\le m_j} (-1)^{|\alpha|} D^\alpha F_l^\alpha(x) \tag{6.4.10}$$

where F_l^α is defined in (6.4.8). The result follows by comparing (6.4.10) and (6.4.8) and calculating the degrees of the polynomials $L_0^{kl}(-1)^{|\alpha|} D^\alpha F_l^{*\alpha}$.

Remark. We could have defined $P_{h_0 R}(f)$ by defining $F_l^{*\alpha}$ as the expansion of F_l^α out to terms of degree $2m - \nu - 1$. This would give a definition of P_R which is independent of h_0 and, moreover U_R^k would satisfy the equations without polynomials $f_l^{*\alpha}$. However, the presence of the h_0 and the polynomials $f_j^{*\alpha}$ is not a serious drawback and their presence is necessary for the treatment of boundary values.

Definition 6.4.4. We define the coefficients a_R as in (6.4.7) and define T_R on the spaces $^{**}H_q^h(B_{2R})$ and $^{**}C_\mu^h(\bar{B}_{2R})$ by

$$T_R u_R = U_R = P_{h_0 R}(f_R), \quad f_{jR}^\alpha = -(a_{Rjk}^{\alpha\beta} - a_{0jk}^{\alpha\beta})\, D^\alpha u_R^k. \tag{6.4.11}$$

Remark. T_R obviously depends also on h_0.

Then if u has support in B_R and satisfies (6.4.1) and if we define $u_R = P_{h_0 R}(f_R)$ where f_R is defined in (6.4.11) with u_R replaced by u, we see from Theorem 6.4.1 that

$$u = u_R + H_R \tag{6.4.12}$$

where H_R^k is a polynomial of degree $\le \max(t_k + h_0 - \nu - 1, t_k - \nu - 1)$ and the equations (6.4.1) are equivalent to

$$u_R - T_R u_R = v_R = P_{h_0 R}(f) - P_{h_0 R}[(a_R^\beta - a_0^\beta) \cdot D^\beta H_R]. \tag{6.4.13}$$

The remaining developments of § 6.2 can be repeated to yield the following results:

Theorem 6.4.2. *If $h \ge h_0$ and the a' s satisfy the h-conditions on $\bar{B}_A$, then T_R is a bounded operator on $^{**}H_p^h(B_{2R})$ for each $p > 1$ and each $R \le A/2$ with bound $\varepsilon(R, p)$ where ε also depends on h, h_0, E, and E' but $\varepsilon(R, p) \to 0$ as $R \to 0$. If the a' s also satisfy the $h - \mu$-conditions, then T_R is bounded on $^{**}C_\mu^h(\bar{B}_{2R})$ with $\varepsilon(R, \mu)$ of the form $C(\mu, h, h_0, E, E') \times R^\sigma$, $\sigma = \mu$ or 1 according as $h = h_0$ or $h > h_0$.*

Theorem 6.4.3. *Suppose the a' s satisfy the h-conditions on G, $u^k \in H_q^{t_k+h_0}(G)$ for some $q > 1$, the $f_j^\alpha \in H_p^{h-s_j+|\alpha|}(D)$ if $h - s_j + |\alpha| > 0$ and $f_j^\alpha \in L_p(D)$ if $h - s_j + |\alpha| \leq 0$ for each $D \subset\subset G$ and some $h \geq h_0$ and $p \geq q$, and u satisfies* (6.4.1) *on G. Then $u^k \in H_p^{t_k+h}(D)$ for each $D \subset\subset G$. If, also, the a' s satisfy the $h - \mu$ conditions on G and the $f_j^\alpha \in C^{h-s_j+|\alpha|}(D)$ if $h - s_j + |\alpha| > 0$ and $f_j^\alpha \in C_\mu^0(D)$ if $h - s_j + |\alpha| \leq 0$, then $u^k \in C_\mu^{t_k+h}(D)$ for each $D \subset\subset G$.*

The proof of this is just like that of Theorem 6.2.5. The equations replacing (6.2.27) for u_s are of the form

$$\int_{B_{2R}} \sum_j \sum_{|\alpha| \leq m_j} D^\alpha v^j \left[\sum_k \sum_{|\beta| \leq \varrho_{jk}} a_{Rjk}^{\alpha\beta} D^\beta u_s^k - g_{js}^\alpha\right] dx = 0, \quad v \in C_c^\infty(B_{2R})$$

(6.4.14)
$$g_{js}^\alpha = \zeta_s f_j^\alpha + k_{js}^\alpha,$$

where if $|\alpha| = m_j$ the k_{js}^α involve only derivatives of u^k of order $< \varrho_{jk}$. The k_{js}^α are not uniquely determined but a definite set can be obtained by replacing v^j in (6.4.1) by $\zeta_s v^j$ and writing

$$D^\alpha(\zeta_s v^j) = \zeta_s D^\alpha v^j + [D^\alpha(\zeta_s v^j) - \zeta_s D^\alpha v^j]$$

and then shifting the ζ_s factor to the bracket in (6.4.1) to obtain

$$\begin{aligned}\zeta_s &\left[\sum_k \sum_{|\beta|} a_{Rjk}^{\alpha\beta} D^\beta u^k - f_j^\alpha\right] \\ &= \sum_k \sum_{|\beta|} a_{Rjk}^{\alpha\beta} \{D^\beta u_s^k - [D^\beta(\zeta_s u^k) - \zeta_s D^\beta u^k]\} - \zeta_s f_j^\alpha.\end{aligned}$$

Thus the equations (6.4.14) must be equivalent to

$$\begin{aligned}\int_{B_{2R}} &\Big\{\sum_j \sum_\alpha D^\alpha v^j \Big[\sum_k \sum_\beta a_{Rjk}^{\alpha\beta} D^\beta u_s - \zeta_s f_j^\alpha - \sum_k \sum_\beta a_{Rjk}^{\alpha\beta} \times \\ &\times [D^\beta(\zeta_s u^k) - \zeta_s D^\beta u^k] + \sum_j \sum_\alpha [D^\alpha(\zeta_s v^j) - \zeta_s D^\alpha v^j] \times \\ &\times \Big[\sum_k \sum_\beta a_{Rjk}^{\alpha\beta} D^\beta u^k - f_j^\alpha\Big]\Big\} dx = 0, \quad v \in C_c^\infty(B_{2R}).\end{aligned}$$

Theorem 6.4.4. *Suppose the a' s satisfy the h-conditions on $\bar{B}_A$. Then there are constants $R_2 > 0$ and C_2, depending only on h_0, h, q, E, and E' such that if u is a solution of* (6.4.1) *with $u^k \in H_{q0}^{t_k+h}(B_R)$,* and $f \in {}^*H_q^h(B_R)$, *then*

$$^{**}\|u\|_{qR}^h \leq C_2 [^*\|f\|_{qR}^h + {}^{**'}\|u\|_{1R}^0], \quad {}^{**'}\|u\|_{1R}^0 = \sum_k R^{-t_k-h-\varrho} \|u^k\|_{1R}^0,$$
$$0 < R \leq R_2, \quad \varrho = \nu(q-1)/q.$$

Also there exist constants C_3 and R_3, depending only on h_0, h, μ ($0 < \mu < 1$), E, and E' such that if u is a solution of (6.4.1), *$u \in {}^{**}C_\mu^h(\bar{B}_R)$ and has support in B_R and if $f \in {}^*C_0^h(\bar{B}_R)$, then*

$$^{**}|||u|||_{\mu R}^h \leq C_3 [^*|||f|||_{\mu R}^h + {}^{**'}|||u|||_{1R}^0], \quad {}^{**'}|||u|||_{1R}^0 = R^{\varrho-\nu-\mu} \cdot {}^{**'}\|u\|_{1R}^0.$$

Remarks. It is clear that we may allow the f_j^α with $h - s_j + |\alpha| < 0$ to be distributions in some properly defined space $H_q^{h-s_j+|\alpha|}$; such spaces have been defined and used extensively in the study of differential equations (see HORMANDER [1], A. FRIEDMAN [2], LIONS [2], LIONS and MAGENES, and many others; see Chapter 1). We could say that $\varphi \in H_q^{-t}$ if and only if $\varphi = \Delta^t \varphi$ where $\varphi \in H_q^t$; no doubt spaces C_μ^{-t} could be defined similarly. Moreover the coefficients $a_{jk}^{\alpha\beta}$ may be allowed to be somewhat more general as in §§ 5.5 and 5.6.

We now wish to consider weak solutions satisfying general boundary conditions in some weak sense and would like to obtain global results like those in § 6.3. We shall consider boundary conditions of the form

$$\int_{\partial G} \sum_{r,k} \sum_{|\gamma| \le p_r} D^\gamma \zeta^r [B_{rk\gamma} u^k - g_{r\gamma}]\, dS = 0, \quad \zeta \in C^\infty(\bar{G}) \tag{6.4.15}$$

where the operators $B_{rk\gamma}$ are of order $\le t_k - h_r - p_r$ and where, if (x, y) are boundary coordinates the equations (6.4.15) reduce to

$$\int_{\sigma_{2R}} \sum_{r,k} \sum_{|\gamma| \le p_r} D_x^\gamma \zeta^r(x) [B_{rk\gamma}(x, D_x, D_y) u^k(x, 0) - g_{r\gamma}(x)]\, dx = 0$$

$$\zeta^r \in C_c^\infty(\bar{\sigma}_{2R}). \tag{6.4.16}$$

We assume that the integers p_r and h_0 satisfy

$$\min p_r \ge 0, \quad \min h_0 + h_r + p_r \ge 1 \tag{6.4.17}$$

as well as (6.4.3). *We assume that s_j, t_k and m_j continue to satisfy* (6.4.2). We define $B_{0rk\gamma} = 0$ if $|\gamma| < p_r$; if $|\gamma| = p_r$, we define $B_{0rk\gamma}$ as that operator of order $t_k - h_r - p_r$ obtained by replacing the coefficients of the principal part of $B_{rk\gamma}$ by their values at the origin. Then we define

$$B_{0rk} = \sum_{|\gamma| = p_r} (-1)^{p_r} D_x^\gamma B_{0rk\gamma}. \tag{6.4.18}$$

We suppose that for each point on the part of ∂G in which we are interested, that the operator B_{0rk} above and the operator L_0 satisfy the root and complementing conditions of § 6.1.

Remark. If $B_{rk}(x, y; D_x, D_y)$ is any operator of order $t_k - h_r$ with sufficiently differentiable coefficients of degree $\le t_k - h_r - p_r$ in D_y, it can be written in the form

$$\sum_{|\gamma| \le p_r} (-1)^{|\gamma|} D_x^\gamma B_{rk\gamma}(x, y; D_x, D_y)$$

in which the operators $B_{rk\gamma}$ are of order $\le t_k - h_r - p_r$. This justifies the restriction of our attention to boundary conditions of the form (6.4.16).

Definition 6.4.5. We define the space ${}^{*\prime}H_q^h(\sigma_{2R})$ $(h \ge h_0,\ q > 1)$ to consist of those vectors $g_{r\gamma}$ for which there exists a $\tilde{g}_{r\gamma} \in H_{q0}^{h+h_r+|\gamma|}(B_{2R})$ if $h + h_r + |\gamma| \ge 1$ and $H_{q0}^1(B_{2R})$ if $h + h_r + |\gamma| \le 1$ such that $\tilde{g}_{r\gamma}(x, 0) = g_{r\gamma}(x)$ on σ_{2R}, and define the norm by

$${}^{*\prime}\|g\|_{q,2R}^h = \inf \Big\{ \sum_{h+h_r+|\gamma| \ge 1} {}'\|g_{r\gamma}\|_{q,2R}^{h+h_r+|\gamma|} + \sum_{h+h_r+|\gamma| < 1} R^{h+h_r+|\gamma|-1}\, {}'\|g_{r\gamma}\|_{q,2R}^1 \Big\}.$$

We define the space $*'C_\mu^h(\bar{\sigma}_{2R})$ to consist of those $g_{r\gamma} \in C_{0\mu}^{h+h_r+|\gamma|}(\bar{\sigma}_{2R})$ in case $h + h_r + |\gamma| \geq 1$ and $g_{r\gamma} \in C_{0\mu}^1(\bar{\sigma}_{2R})$ otherwise, with norm

$$*'|||g|||_{\mu,2R}^h = \sum_{h+h_r+|\gamma|\geq 1} {}'|||g_{r\gamma}|||_{\mu,2R}^{h+h_r+|\gamma|} + \sum_{h+h_r+|\gamma|<1} R^{h+h_r+|\gamma|-1}\, {}'|||g_{r\gamma}|||_{\mu,2R}^1 .$$

We define the space $*'\tilde{C}_\mu^h(\bar{\sigma}_{2R})$ in the same way as $*'C_\mu^h(\bar{\sigma}_{2R})$ but replacing 1 by 0.

Definition 6.4.6. For g in one of the spaces above, we define $P_{h_0R}(0, g) = U_R$, where

$$U_R^k(X) = \sum_r \sum_{|\gamma|\leq p_r} (-1)^{|\gamma|} D_x^\gamma[u_{0\gamma}^{kr}(X) - u_\gamma^{*kr}(X)],$$

$$u_{0\gamma}^{kr}(X) = \int_{\sigma_{2R}} L^{kr}(x-\xi, y)\, g_{r\gamma}(\xi)\, d\xi \quad (r \text{ not summed}),$$

u_γ^{*kr} is the Maclaurin expansion of $u_{0\gamma}^{kr}$ out to terms of degree $t_k - h_r - \nu$, and the L^{kr} are defined in (6.3.24).

Theorem 6.4.5. *P_{h_0R} is independent of h_0 and is a bounded operator from $*'H_q^h(\sigma_{2R})$ into $**H_q^h(G_{2R})$ and from either $*'C_\mu^h(\bar{\sigma}_{2R})$ or $*'\tilde{C}_\mu^h(\bar{\sigma}_{2R})$ into $**C_\mu^h(\bar{G}_{2R})$ with bound independent of R. In any case U_R is a solution of the homogeneous equations* (6.3.7) *for $y > 0$ and satisfies the boundary* conditions (6.4.16) *with $B_{rk\gamma}$ replaced by $B_{0rk\gamma}$.*

Proof. Using the methods of proof and the results of Theorem 6.3.3 we conclude that P_{h_0R} is a bounded operator as stated. For $y > 0$, we see from Theorem 6.3.2 that

$$\sum_k L_{0jk}\, u_{0\gamma}^{kr}(X) = 0, \quad j = 1, \ldots, N \tag{6.4.19}$$

for each fixed r and γ. Now L_{0jk} is an operator of order $t_k + s_j$ and so the degree of $\sum_k L_{0jk}\, u^{*kr}$ is $\leq -s_j - h_r - \nu$. If this is negative,

$$\sum_k L_{0jk}\, u_\gamma^{*kr}(X) = 0. \tag{6.4.20}$$

If it is positive, then $t_k - h_r - \nu > t_k + s_j$ so that

$$\sum_k L_{0jk}[u_{0\gamma}^{kr}(X) - u_\gamma^{*kr}(X)]$$

vanishes like X^ϱ where $\varrho = 1 - s_j - h_r - \nu$. In this case (6.4.20) follows on account of (6.4.19).

Now, suppose the $g_{r\gamma} \in C_c^\infty(\bar{\sigma}_{2R})$. Then from Theorem 6.3.3 and its proof we conclude that

$$\sum_k B_{0sk}\, u_{0\gamma}^{kr}(x, 0) = \begin{cases} 0 \quad , & \text{if} \quad s \neq r \\ g_{r\gamma}(x), & \text{if} \quad s = r. \end{cases} \tag{6.4.21}$$

Also $\sum_k B_{0sk}\, u_\gamma^{*kr}(X)$ is of degree $h_s - h_r - \nu$. If this is negative,

$$\sum_k B_{0sk}\, u_\gamma^{*kr}(X) = 0. \tag{6.4.22}$$

If it is positive we conclude (6.4.22) using (6.4.21) and the fact that $[u_{0\gamma}^{kr}(X) - u_{\gamma}^{*kr}(X)]$ vanishes at the origin to the power $1 + h_s - h_r - \nu$.

We wish now to define $P_R(f, 0)$. By virtue of Lemma 6.3.3, we may as well assume that f is already extended to B_{2R} and we then use the norms defined in Definition 6.4.2.

Definition 6.4.7. For f in the space $^*C_\mu^h(\bar{G}_{2R})$ or $^*H_\mu^h(G_{2R})$, $h \geq h_0$, we define (we omit the subscript R)

$$P_R(f, 0) = \sum_{l=1}^{N} \sum_{|\varepsilon| \leq m_l} (V^{l\varepsilon} - W^{l\varepsilon}),$$

$$V^{kl\varepsilon}(X) = L_0^{kt}(F_t^{l\varepsilon} - F_t^{l\varepsilon *}),$$

$$F_t^{l\varepsilon}(X) = \delta_t^l \cdot (-1)^{|\varepsilon|} D^\varepsilon \mathfrak{F}_l^\varepsilon(X), \quad (l, \varepsilon \text{ not summed})$$

$$\mathfrak{F}_l^\varepsilon(X) = \int_{B_{2R}} K(X - \Xi) \tilde{f}_l^\varepsilon(\Xi)\, d\Xi, \quad \tilde{f} = \tau f$$

$$W^{kl\varepsilon}(X) = \sum_r \sum_{|\gamma| = p_r} (-1)^{|\gamma|} D_x^\gamma \int_\sigma \sum_{|\beta| = q_{rl}} C_\beta L_\beta^{kr}(x, y; \xi) D_\xi^\beta g_{r\gamma}(\xi, 0)\, d\xi$$

$$g_{r\gamma}^{l\varepsilon}(X) = B_{0rk\gamma} V^{kl\varepsilon}(X),$$

$$q_{rl} = \max(h + h_r + p_r - 1, h_r + p_r + s_l - |\varepsilon| - 1)$$

where $F_l^{l\varepsilon *}$ is the Maclaurin expansion of $F_l^{l\varepsilon}$ out to the terms of degree $2m - |\varepsilon| - \nu - 1$ (if this is ≥ 0, $F_l^{l\varepsilon *} \equiv 0$ otherwise).

Theorem 6.4.6. *$P_R(f, 0)$ is a bounded operator from $^*C_\mu^h(\bar{G}_{2R})$ into $^{**}C_\mu^h(\bar{G}_{2R})$ and from $^*H_q^h(G_{2R})$ into $^{**}H_q^h(G_{2R})$ with bound depending on E and not R. If $Y = P_R(f, 0)$, then Y satisfies the equations (6.4.1) with a replaced by a_0 and satisfies the boundary conditions (6.4.16) with $B_{rk\gamma}$ replaced by $B_{0rk\gamma}$ and $g_{r\gamma}$ by 0.*

Proof. By virtue of Lemma 6.3.3, we may as well assume that $f \in {}^*C_\mu^h(\bar{B}_{2R})$ or $^*H_q^h(B_{2R})$. It is easy to see, using the carefully constructed definitions of the norms that the mapping via P_R from f to V is bounded as stated. Let us define V_0, V^*, g_0, and g^* by

$$V = V_0 - V^*, \quad g = g_0 - g^*$$

$$V^{kl\varepsilon *} = L_0^{kt} F_t^{l\varepsilon *}, \quad g_{r\gamma}^{l\varepsilon *} = B_{0rk\gamma} V^{kl\varepsilon *}.$$

Then we see easily by checking the degrees of the polynomials that the vector $V^{l\varepsilon *}$ satisfies the homogeneous equations (6.3.7). If, for some (l, ε), $f_l^\varepsilon \in C_c^\infty(B_{2R})$ it follows that $V^{l\varepsilon}$ satisfies the equations (6.3.1) with a replaced by a_0 and

$$f_t^{l\varepsilon} = (-1)^{|\varepsilon|} D^\varepsilon f_l^\varepsilon \cdot \delta_t^l \quad (l, \varepsilon \text{ not summed}).$$

By virtue of our hypotheses (6.4.17), we see that

$$h + h_r + p_r - 1 \geq h_0 + h_r + p_r - 1 \geq 0$$

$$h_r + p_r + s_l - |\varepsilon| - 1 = (h + h_r + p_r - 1) - (h - s_l + |\varepsilon|)$$

and $q_{rl} \geq 0$ in all cases. Also, in all cases, $g_{r\gamma}^{l\varepsilon}$ behaves at ∞ like a function which is essentially homogeneous of degree $h_r + p_r + s_l - |\varepsilon| - \nu - 1$ and $g_{r\gamma}^{l\varepsilon *}$ is a polynomial of degree $\leq$ this. Since this $\leq q_{rl} - \nu$ in all cases, the polynomial $g_{r\gamma}^{l\varepsilon *}$ may be neglected in the expression for W if desired. Since $L_\beta^{kr}(x, y; \xi)$ is, for a fixed ξ, of the form $G_0^k(X) - G^{k*}(X)$, where G_0 satisfies the homogeneous equations (6.3.7) and G_0^* is a Maclaurin expansion of G_0, it follows that W satisfies the homogeneous equations (6.3.7) for $y > 0$ where it is analytic; the convergence at ∞ of the integrals follows since L_β^{kr} has degree -1 in $|\xi|$ and $D_\xi^\beta g_{r\gamma}^{l\varepsilon}$ has degree $\leq -\nu$ in $|\xi|$ at ∞. A proof like that in the first part of Theorem 6.3.3 shows that the mapping from f to W via $P_R(f, 0)$ is bounded as stated; one sees that $g_{r\gamma}^{l\varepsilon} \in C_u^{q_{rl}+1}(\bar{B}_A)$ for every A with the expected bounds.

Now for some (l, ε), let us assume that $f_l^\varepsilon \in C_c^\infty(\bar{B}_{2R})$. Then the $g_{r\gamma}^{l\varepsilon} \in C^\infty(\bar{B}_A)$ for every A and $W^{l\varepsilon}$ can be seen to $\in C^\infty(\overline{R_{\nu+1}^+})$. Let us approximate to $g_{r\gamma}^{l\varepsilon}$ by $g_{r\gamma n}^{l\varepsilon} \in C_c^\infty(R_{\nu+1})$ using the device in the proof of Theorem 6.3.3. Then the derivatives $D^t W_n^{l\varepsilon}$ converge as desired in $\overline{R_{\nu+1}^+}$ to $\nabla^t W^{l\varepsilon}$ and we have (omitting (l, ε))

$$W_n^k(X) = W_{n0}^k(X) - W_n^{k*}(X), \quad g_{rn} = \sum_{|\gamma|=p_r} (-1)^{|\gamma|} D_x^{|\gamma|} g_{r\gamma n}$$

$$W_{n0}^k(X) = \int_\sigma L^{kr}(x-\xi, y)\, g_{rn}(\xi, 0)\, d\xi,$$

and $W_n^{kl\varepsilon *}$ is the Maclaurin expansion of $W_{n0}^{kl\varepsilon}$ out to the terms of degree $t_k - \nu - 1 + \max(h, s_l - |\varepsilon|) = t_k + q_{rl} - h_r - p_r - \nu$. By the method of proof of Theorem 6.3.3, we conclude that

$$B_{0rk} W_{n0}^k(\xi, 0) = g_{rn}(\xi, 0).$$

Thus, by passing to the limit, we see that

$$B_{0rk} W^{kl\varepsilon}(\xi, 0) = g_r^{l\varepsilon}(\xi, 0) - g_r^{l\varepsilon *}(\xi, 0)$$

where $g_r^{l\varepsilon *}(X)$ is the Maclaurin expansion of $g_r^{l\varepsilon}(X)$ out to the terms of degree $q_{rl} - p_r - \nu$. Since the polynomials $g_{r\gamma}^{l\varepsilon *}$ could be neglected in the formula for W, we must have $g_r^{l\varepsilon *} \equiv 0$. The results follow.

From here on the developments parallel those of § 6.3. The theorem corresponding to Theorem 6.3.5 is:

Theorem 6.4.7. *Suppose $u^k \in H_q^{t_k+h}(G_A)$ for each A and each k, $u^k \in C^\infty$ for $|X| \leq$ some R_0 and $y \geq 0$, and suppose u satisfies (6.4.1) with a replaced by a_0 and f_j^α replaced by $f_j^{*\alpha}$ and satisfies (6.4.16) with $B_{rk\gamma}$ replaced by $B_{0rk\gamma}$ and $g_{r\gamma}$ replaced by $g_{r\gamma}^*$, where $f_j^{*\alpha}$ and $g_{r\gamma}^*$ are polynomials of degrees $\leq h_0 - s_j + |\alpha| - \nu - 1$ and $h_0 + h_r + p_r - \nu - 1$, respectively. Suppose also that u satisfies*

$$|D_X^p u^k(X)| \leq C_p(1 + |X|)^{t_k+h_0-\nu-1-p+1/2}, \quad p = 0, 1, \ldots, |X| \geq R_0.$$

Then each u^k is a polynomial of degree $\leq t_k + h_0 - \nu - 1$.

Remark. Since if $u_R = P_R(f, g)$, then u_R satisfies (6.4.1) with a replaced by a_0 and satisfies (6.4.16) with $B_{rk\gamma}$ replaced by $B_{0rk\gamma}$ with no residual polynomials f^* and g^*, we need Theorem 6.4.7 only for the cases $f^* = g^* = 0$.

By paralleling the previous developments we may conclude the following results:

Theorem 6.4.8. *Suppose h_0 is the smallest integer satisfying all the conditions of this section, suppose $h \geq h_0$, suppose the a' s and the coefficients in the $B_{rk\gamma}$ satisfy the h-conditions on a domain $\Gamma \supset \bar{G}$, suppose that for each point x_0 of $\bar{G}$ the operator L_0 is properly elliptic and that L_0 and the operators B_{0rk} satisfy the root and complementing condition for each x_0 on ∂G, suppose G is of class $C_1^{t_0+h-1}$ $(t_0 = \max t_k)$, and suppose u satisfies (6.4.1) and (6.4.15) where $f_j^\alpha \in H_q^\varrho(G)$, $\varrho = \max(0, h - s_j + |\alpha|)$, $g_{r\gamma} \in H_q^\tau(G)$, $\tau = \max(1, h + h_r + |\gamma|)$, and $u^k \in H_q^{t_k+h}(G)$. Then*

$$\sum_k \| u^k \|_{qG}^{t_k+h} \leq C \Big[\sum_{j,\alpha} \| f_j^\alpha \|_{qG}^\varrho + \sum_{r,\gamma} \| g_{r\gamma} \|_{qG}^\tau + \sum_k \| u^k \|_{1G}^0 \Big]$$

where C depends only on h, q, G, E, and E'. If the a' s and the b' s satisfy the $h-\mu$-conditions on $\Gamma \supset \bar{G}$, G is of class $C_\mu^{t_k+h}$, $u^k \in C_\mu^{t_k+h}(\bar{G})$, $f_j^\alpha \in C_\mu^\varrho(\bar{G})$, and $g_{r\gamma} \in C_\mu^\tau(\bar{G})$, $\tau = \max(0, h + h_r + |\gamma|$. Then

$$\sum_k ||| u^k |||_{\mu G}^{t_k+h} \leq C \Big[\sum_{j,\alpha} ||| f_j |||_{\mu G}^\varrho + \sum_{r,\gamma} ||| g_{r\gamma} |||_{\mu G}^\tau + \sum_k \| u^k \|_{1G}^0 \Big]$$

where C depends only on h, μ, G, E, and E'.

If G, f_j^α, $g_{r\gamma}$, and the a' s and b' s satisfy the conditions in the first statement for some $h' \geq h$ and $q' \geq q$ and $u^k \in H_q^{t_k+h}(G)$, then $u^k \in H_{q'}^{t_k+h'}(G)$. If G, f_j^α, $g_{r\gamma}$, and the a' s and b' s satisfy the (Hölder) conditions in the second statement with some $h' \geq h$ and if $u^k \in H^{t_k+h}(G)$, then $u^k \in C_\mu^{t_k+h'}(\bar{G})$. If G, f_j^α, $g_{r\gamma}$, and the a's and b' s $\in C^\infty(\bar{G})$ and $u^k \in H_q^{t_k+h}(G)$, then $u^k \in C^\infty(\bar{G})$

6.5. The existence theory for the Dirichlet problem for strongly elliptic system

It is clear that the results of § 6.3 imply that for every λ, the set of solutions of the homogeneous boundary value problem

$$(6.5.1) \qquad L_{jk} u^k + (-1)^{s_j} \lambda u^j = 0 \quad \text{on } G, \quad B_{rk} u^k = 0 \quad \text{on } \partial G,$$

form a manifold of finite dimensionality, provided the L_{jk} and B_{rk} satisfy the supplementary and complementing conditions of § 6.1. It is evident (since for one equation we regard L^{kl} as the identity) that the Dirichlet boundary conditions are always complementary for a single equation of order $2m$ which satisfies the supplementary condition. However the following example due to Seeley and communicated orally to the author (essentially) by F. Browder shows that every complex number λ may be an eigenvalue: Letting (r, θ) be polar coordinates in the plane, we consider the operator

$$L = e^{-2i\theta}\left[\frac{\partial^2}{\partial\theta^2} + \frac{\partial^2}{\partial r^2} + I\right]$$

on the annulus $\pi < r < 2\pi$. For each complex λ, the function

$$U(r, \theta; \lambda) = J_0(w) \sin r, w = \lambda^{1/2} e^{i\theta}$$

is an eigen-function for the given eigenvalue λ, J_0 being the Bessel function. That is, U satisfies

$$LU - \lambda U = 0.$$

In this section we present the existence theory for the Dirichlet problem for a strongly elliptic system, as defined by NIRENBERG [2]; in this problem, the eigenvalues are isolated. The developments parallel those in §§ 5.2 and 5.6.

In discussing strongly elliptic systems, it is convenient to change the notation somewhat from that used in §§ 6.1—6.4. We assume that the operator L_{jk} is of order $s_j + s_k$ where we also assume

$$(6.5.2) \qquad s_j \geq 1, \quad j = 1, \ldots, N; \quad \text{thus} \quad m = s_1 + \cdots + s_N \geq N.$$

In the previous notation t_k was the maximum order of any of the L_{jk} for fixed k. Thus if we define $s = \max s_j$, then

$$(6.5.3) \qquad s = \max s_j, \quad t_k = s + s_k, \quad \text{old} \quad s_j = -s + \text{new } s_j,$$

where t_k refers to the previous notation and the s_j are new.

Definition 6.5.1. If we let

$$L_{jk}(x, D) = \sum_{|\alpha| \leq s_j + s_k} a_{jk}^{\alpha}(x) D^{\alpha}, \quad L'_{jk} = \sum_{|\alpha| = s_j + s_k} a_{jk}^{\alpha} D^{\alpha},$$

L'_{jk} denoting the *principal part* of L_{jk}, then the system is said to be *strongly elliptic* if and only if

$$(6.5.4) \qquad \begin{gathered} Re \sum_{|\alpha| = s_j + s_k} a_{jk}^{\alpha}(x) \lambda_\alpha \xi^j \bar{\xi}^k \geq M^{-1} \sum_{j=1}^{N} |\lambda|^{2s_j} |\xi^j|^2, \quad M > 0 \\ x \in \bar{G}, \quad \lambda \text{ real}, \quad \xi \text{ complex}. \end{gathered}$$

A vector u will be said to satisfy 0 *Dirichlet data* on ∂G if and only if

$$(6.5.5) \qquad \nabla^{r-1} u^k = 0, \quad r = 1, \ldots, s_k \text{ on } \partial G.$$

In order to fit these special boundary value problems into our general framework, we replace the index r by the pair (j, r); then the operators B_{rk}, in (x, y) coordinates, are

$$(6.5.6) \qquad B_{jr;k} = \begin{cases} 0 \quad , & \text{if } j \neq k \\ D_y^{r-1}, & r = 1, \ldots, s_j \quad \text{if } j = k. \end{cases}$$

Since, in the old notation, B_{rk} was of order $t_k - h_r$, it follows that

$$(6.5.7) \qquad h_{jr} = s + s_j + 1 - r.$$

In order to prove our existence theorems, we begin as in § 5.2 by assuming, in addition to (6.5.4), that

$$a_{jk}^{\alpha} \in C_1^{|\alpha|-s_k-1}(\bar{G}), \quad |\alpha| \geq s_k + 1, \tag{6.5.8}$$
$$a_{jk}^{\alpha} \text{ bounded and measurable, all } \alpha, j, k.$$

In that case, it is possible to integrate by parts to eliminate derivatives of u^k of order higher than s_k and thus obtain

$$\begin{aligned} &\int_G \sum_j (-1)^{s_j} \bar{v}^j \sum_k L_{jk} u^k \, dx \equiv B(u, v) \\ &= \int_G \sum_{j,k} \sum_{|\alpha| \leq s_j} \sum_{|\beta| \leq s_k} A_{jk}^{\alpha\beta}(x) \, D^\alpha \bar{v}^j \, D^\beta u^k \, dx, \quad v^j \in H_{20}^{s_j}(G). \end{aligned} \tag{6.5.9}$$

The integrations by parts are not unique but we shall assume that they are done in some fixed way, obtaining the definite formula (6.5.9) for $B(u, v)$. However these integrations are performed, it follows easily that

$$\sum_{j,k} \sum_{|\alpha| = s_j} \sum_{|\beta| = s_k} A_{jk}^{\alpha\beta}(x) \, \lambda_\alpha \lambda_\beta \xi^j \bar{\xi}^k = \sum_{j,k} \sum_{|\gamma| = s_j + s_k} a_{jk}^{\gamma} \lambda_\gamma \xi^j \bar{\xi}^k. \tag{6.5.10}$$

With these assumptions, the equations

$$L_{jk} u^k + (-1)^{s_j} u^j = f_j, \quad B_{kr,l} u^l = 0 \quad \text{on } \partial G \tag{6.5.11}$$

may be solved in the weak sense by solving the equations

$$\begin{aligned} &B(u, v) + \lambda \, C(u, v) = L(v) \\ &C(u, v) = \int_G \sum_j u^j \bar{v}^j \, dx, \quad L(v) = \int_G \sum_j (-1)^{s_j} f_j \bar{v}^j \, dx \end{aligned} \tag{6.5.12}$$

in which $B(u, v)$ is defined in (6.5.9) and G need only be bounded.

As in § 5.2 we first prove the following theorem:

Theorem 6.5.1 (Gårding's inequality). *We suppose that G is bounded and the coefficients $A_{jk}^{\alpha\beta}$ are bounded and measurable and that those whith $|\alpha| = s_j$ and $|\beta| = s_k$ are continuous and satisfy*

$$\begin{aligned} &Re \sum_{j,k} \sum_{|\alpha| = s_j} \sum_{|\beta| = s_k} A_{jk}^{\alpha\beta}(x) \, \lambda_\alpha \lambda_\beta \xi^j \bar{\xi}^k \geq M^{-1} \sum_j |\lambda|^{2 s_j} |\xi^j|^2, \\ &M > 0, \quad \lambda \text{ real}, \quad \xi \text{ complex}. \end{aligned} \tag{6.5.13}$$

Then there exist constants M_1 and λ_0, depending only on ν, M, the s_j, G, bounds for the coefficients, and the moduli of continuity of those $A_{jk}^{\alpha\beta}$ for which $|\alpha| = s_j$ and $|\beta| = s_k$, such that

$$\begin{aligned} &|B(u, v)| \leq M_1 \cdot \|u\| \cdot \|v\|, \quad u, v \in \mathfrak{H}_0 \\ &Re \, B(u, u) \geq (m_1/2) \cdot \|u\|^2 - \lambda_0 \, C(u, u), \quad m_1 = M^{-1} \end{aligned} \tag{6.5.14}$$

where the space $\mathfrak{H}_0$ is the space of vectors u in which $u^j \in H_{20}^{s_j}(G)$ and

$$(u, v) = \int_G \sum_j \sum_{|\alpha| = s_j} C_\alpha \, D^\alpha u^j \, D^\alpha \bar{v}^j \, dx, \quad C_\alpha = |\alpha|!/\alpha!. \tag{6.5.15}$$

Proof. It is sufficient to prove this for $u, v \in C_c^\infty(G)$. The existence of M_1 follows immediately from Poincaré's inequality (Theorem 3.2.1). To prove the second inequality, let $\varepsilon > 0$. $\bar{G}$ can be covered by a finite number of spheres $B(x_i, r_i)$ such that $|A_{jk}^{\alpha\beta}(x) - A_{jk}^{\alpha\beta}(x_i)| < \varepsilon$ whenever $x \in \overline{B(x_i, r_i)}$ and $|\alpha| = s_j$ and $|\beta| = s_k$. Clearly there is a sequence $\{\zeta_t\}$, $t = 1, \ldots, S$, in which each $\zeta_t \in C_c^\infty(R_\nu)$ and has support in some one $B(x_i, r_i)$ and $\zeta_1^2 + \cdots + \zeta_S^2 \equiv 1$ on $\bar{G}$. Then, as in the proof of the corresponding inequality at the end of § 5.2, we see that

$$Re\, B(u, u) = Re\, B'(u, u) + Re \int_G \sum_t \sum_{|\alpha|=s_j} \sum_{|\beta|=s_k} \zeta_t^2 A_{jk}^{\alpha\beta} D^\alpha \bar{u}^j D^\beta u^k \, dx$$

$$= Re\, B''(u, v) + Re \int_G \sum_t \sum_{|\alpha|=s_j} \sum_{|\beta|=s_k} A_{jk}^{\alpha\beta} D^\alpha u_t^j D^\beta \bar{u}_t^k \, dx, \quad u_t^j = \zeta_t u^j \tag{6.5.16}$$

where $B''(u, v)$ is a form like $B(u, v)$ in which, however, $''A_{jk}^{\alpha\beta} \equiv 0$ whenever $|\alpha| = s_j$ and $|\beta| = s_k$. But now the last term in (6.5.16)

$$\begin{aligned} &= Re \int_G \sum_t \sum_{|\alpha|=s_j} \sum_{|\beta|=s_k} A_{jk}^{\alpha\beta}(x_t) D^\alpha \bar{u}_t^j D^\beta u_t^k \, dx \\ &+ Re \int_G \sum_t \sum_{|\alpha|=s_j} \sum_{|\beta|=s_k} [A_{jk}^{\alpha\beta}(x) - A_{jk}^{\alpha\beta}(x_t)] D^\alpha \bar{u}_t^j D^\beta u_t^k \, dx . \end{aligned} \tag{6.5.17}$$

Clearly the absolute value of the second term in (6.5.17)

$$\leq Z_1(\nu, s_1, \ldots, s_N, N) \cdot \varepsilon \cdot \sum_t \| u_t \|^2 . \tag{6.5.18}$$

If we introduce the Fourier transforms $\hat{u}_t^j$ of u_t^j by their usual formulas (see between (6.2.4) and (6.2.5)), the PLANCHEREL theorem shows that the first term in (6.5.17)

$$= \int_{R_\nu} \sum_t Re \sum_{|\alpha|=s_j} \sum_{|\beta|=s_k} A_{jk}^{\alpha\beta} y^\alpha y^\beta \bar{\hat{u}}_t^j \hat{u}_t^k \, dy \geq m_1 \sum_t \| u_t \|^2 . \tag{6.5.19}$$

Now, using the fact that $\nabla^{s_j} u_t^j = \zeta_t \nabla^{s_j} u^j + M_t^j u^j$, where the M_t^j are of order $s_j - 1$, and using (6.5.16) — (6.5.19), we see that

$$Re\, B(u, u) \geq (m_1 - Z_1 \varepsilon) \| u \|^2 - Re\, B'''(u, u)$$

where B''' is a form like B''.

Almost exactly as in § 5.2, we conclude the following two theorems:

Theorem 6.5.2. *Suppose the transformation U is defined on $\mathfrak{H}_0$* (see Theorem 6.5.1) *by the condition that*

$$C(u, v) = (U u, v)_{\mathfrak{H}_0}, \quad v \in \mathfrak{H}_0 . \tag{6.5.20}$$

Then U is completely continuous.

Theorem 6.5.3. *If λ is not in a set $\mathfrak{C}$ which has no limit points in the plane, the equation* (6.5.12), *with $L(v)$ defined by the more general expression*

$$L(v) = \int_G \sum_j \sum_{|\alpha| \leq s_j} f_j^\alpha D^\alpha \bar{v}^j \, dx , \tag{6.5.21}$$

has a unique solution u in $\mathfrak{H}_0$ for each set of f_j^α in $L_2(G)$. If $\lambda \in \mathfrak{C}$, the manifold of solutions u of the equation with $L = 0$ has a finite non-zero dimension. It is assumed that the coefficients satisfy the conditions of Theorem 6.5.1.

Now we notice that the equation (6.5.12) with $L(v)$ given by (6.5.21) is of the form discussed in § 6.4 with

(6.5.22) $$h = -s, \quad m_j = s_j, \quad h_{jr} = s + s_j + 1 - r, \quad p_{jr} = 0.$$

Consequently we may conclude that if the a^α_{jk} satisfy (6.5.8) the $f^\alpha_j \in H^{s-s_j+|\alpha|}_2(G)$ and G is of class C^{2s-1}_1, then the $u^j \in H^{s+s_j}_2(G) \cap H^{s_j}_{20}(G)$ and the integrations by parts yielding (6.5.9) may really be performed (in reverse) to yield the following theorem:

Theorem 6.5.4. *Suppose the a^α_{jk} satisfy* (6.5.4) *and* (6.5.8) *and G is of class C^{2s-1}_1. Then if λ is not in a set $\mathfrak{C}$ which has no limit points in the plane, the equations*

(6.5.23) $$L_{jk}\, u^k + (-1)^{s_j} u^j = f_j \quad \text{on } G$$

have a unique solution u with $u^j \in H^{s+s_j}_2(G) \cap H^{s_j}_0(G)$ for each f for which each $f^j \in H^{s-s_j}_2(G)$. If $\lambda \in \mathfrak{C}$, the manifold of solutions of the homogeneous equations has a finite nonzero dimension.

Of course, higher differentiability follows from more assumptions about the coefficients and the f_j. We shall not state these since it is possible to eliminate many of the assumptions made in (6.5.8) about the coefficients, as was done in § 5.6. We prove below a theorem (Theorem 6.5.6) corresponding to Theorem 5.6.4, but only for systems in which $s_1 = s_2 = \cdots = s_N = m > 0$; the more general systems introduce difficulties. We must first prove the following theorem which is strongly suggested by the results above and has been proved in AGMON-DOUGLIS-NIRENBERG [2].

Theorem 6.5.5. *If the system* (6.1.1) *is strongly elliptic, it satisfies the supplementary condition and the Dirichlet boundary conditions are complementing.*

Proof. That the matrix L_{jk} satisfies the supplementary conditions follows from the fact that the set S of coefficients a^α_{jk} satisfying (6.5.4) is convex and contains the set

(6.5.24) $$\begin{aligned} &a^\alpha_{jk} = 0 \quad \text{if } j \neq k \text{ or if } j = k \text{ and } \alpha \neq 2\beta \text{ with } |\beta| = s_j \\ &a^{2\beta}_{jj} = C_\beta = s_j!/\beta!, \quad (\beta! = \beta_1! \ldots \beta_\nu!); \end{aligned}$$

obviously the coefficients above correspond to the system

$$\Delta^{s_j} u^j = f_j, \quad j = 1, \ldots, N.$$

Clearly the determinant $L(\lambda; x)$ of the matrix $\| a^\alpha_{jk} \lambda_\alpha \|$ cannot vanish for any real λ or there would exist a complex ξ so that

$$a^\alpha_{jk} \lambda_\alpha \xi^j = 0, \quad k = 1, \ldots, N,$$

contradicting (6.5.4). The rest of the supplementary condition requires proof only if $\nu = 2$. Then if σ and τ are orthogonal unit vectors then the roots z of $L(\sigma + z\tau; x) = 0$ are never real and vary continuously with the coefficients as they vary over S.

To show that the Dirichlet conditions are complementing, it is sufficient, according to Theorem 6.3.1 to show the existence of exponentially decaying vector solutions U_{lt}^k of

(6.5.25)
$$L_{0jk}(\sigma, -i D_u)\, U_{ls}^k(\sigma, u) = 0, \quad D^{r-1}\, U_{lt}^k(\sigma, 0) = \delta_l^k\, \delta_t^r, \quad |\sigma| = 1,$$

where, here, we are using the (x, y) notation of § 6.3. These equations made non-homogeneous are ordinary differential equations of the form

(6.5.26)
$$\sum_{k=1}^{N} \sum_{p=0}^{s_j+s_k} b_{jkp}(\sigma)\,(-i\, D_u)^p\, U^k(\sigma, u) = f_j(\sigma, u)$$

where (6.5.4) implies that

(6.5.27)
$$Re \sum_{j,k} \left[\sum_{p=0}^{s_j+s_k} b_{jkp}(\sigma)\, \mu^p\right] \xi^j\, \bar{\xi}^k \geq m_1 \sum_j (1 + \mu^2)^{s_j}\, |\xi^j|^2, \quad (|\sigma| = 1), \quad m_1 = M^{-1}.$$

Let us now define H^0 as the set of all vectors U (henceforth we suppress σ) such that $U^k \in C^{s_k-1}$ with $D^{s_k-1}\, U^k$ absolutely continuous and satisfying

(6.5.28)
$$D_u^p\, U^k(0) = 0, \quad p = 0, \ldots, s_k - 1$$

and with norm defined by

(6.5.29)
$$\|U\|^2 = \int_0^\infty \sum_j \sum_{p=0}^{s_j} \binom{s_j}{p} |D^p\, U^j|^2\, du < \infty.$$

For U and V having finite norm, we define

(6.5.30)
$$B(U, V) = \int_0^\infty \sum_{j,k} \Big\{\overline{V^j} \sum_{p=0}^{s_k} b_{jkp}\,(-i\, D)^p\, U^k + \sum_{p=s_k+1}^{s_k+s_j} b_{jkp}(-i)^p\,(-1)^{p-s_k}\, D^{s_k} U^k\, D^{p-s_k}\, \overline{V^j}\Big\}\, du;$$

we also define $B_a^b(U, V)$ similarly with $[0, \infty)$ replaced by $[a, b]$.

Now, let us introduce the Fourier transforms $\hat{U}$ and $\hat{V}$ of U and V in H_0. Then, for such U and V, we see that

(6.5.31)
$$B(U, V) = \int_{-\infty}^{\infty} \sum_{j,k} \sum_{p=0}^{s_j+s_k} b_{jkp}\, z^p\, \hat{U}^k(z)\, \hat{\bar{V}}^j(z)\, dz.$$

From (6.5.27) and (6.5.31), we see that B satisfies the hypotheses of B_0 in the Lemma of Lax and Milgram (Theorem 5.2.2). Consequently if $f_j \in C^{s-s_j}[0, \infty)$ and vanishes for $u \geq 1$, say, there exists a unique U in H_0 such that

(6.5.32)
$$B(U, V) = \int_0^\infty \sum_j f_j\, \overline{V^j}\, du, \quad V \in H_0.$$

The usual difference quotient procedure shows that $U^k \in C^{s+s_k}$ and satisfies (6.5.28) and (6.5.26). So, to find U^k_{lt}, let ω^k_{lt} be any function $\in C^{s+s_k}[0, \infty)$ and vanishing for $u \geq 1$ and satisfying

$$D^{r-1}\,\omega^k_{lt}(0) = \delta^k_l\,\delta^r_t, \quad r = 1, \ldots, s_k.$$

Then, let U be the solution above with

$$f_j = \sum_k \sum_{p=0}^{s_j+s_k} b_{jkp}(-i\,D)^p\,\omega^k_{lt}.$$

Then we may define

$$U^k_{lt} = \omega^k_{lt} - U^k.$$

The following theorem and its proof are essentially due to NIRENBERG (see AGMON-DOUGLIS-NIRENBERG [1], p. 693):

Theorem 6.5.6. *Suppose that $G \in C^{2s-1}_1$, that $s_j = s$ for every j, that the a^α_{jk} with $|\alpha| = 2s \in C^0(\bar{G})$, the other coefficients being bounded and measurable. Then the conclusions of Theorem 6.5.4 hold, In fact, there exist real numbers λ_0 and C, which depend only on ν, G, E, and E' such that*

$$\sum_k (\|u^k\|^{2s}_{2,G})^2 \leq C \sum_j \Big(\|\sum_k L_{jk}\,u^k + (-1)^s\,\lambda\,u^j\|^0_{2,G}\Big)^2, \quad \lambda \geq \lambda_0,$$
$$u^k \in H^{2s}_2(G) \cap H^s_{20}(G), \quad \lambda \text{ real}.$$

Proof. We first prove the last statement. The proof of the first is like that of Theorem 5.6.5.

For any given $\eta_1 > 0$, each point x_0 is in a neighborhood or boundary neighborhood in which $|a(x) - a(x_0)| < \eta_1$ where $a(x) = \{a_{jk}(x)\}$ with $|\alpha| = 2s$. We choose a sequence ζ_t, $t = 1, \ldots, T$, such that each $\zeta_t \in C^\infty(\bar{G})$ with spt ζ_t in some one such neighborhood and $\zeta_1^2 + \cdots + \zeta_T^2 = 1$. We let $u^k_t = \zeta_t\,u^k$ and note that

$$\zeta_t\,L_{jk}\,u^k = L'_{jk}\,u^k_t + M_{jkt}\,u^k$$

where L'_{jk} is the principal part of L_{jk} and M_{jkt} is of order $< 2s$. Then, using (f, g) to denote the L_2 inner product, we obtain

$$\begin{aligned}\sum_j (-1)^s (L_{jk}\,u^k, u^j) &= \sum_{j,t} (-1)^s (\zeta_t\,L_{jk}\,u^k, \zeta_t\,u^j)\\ &= \sum_{j,t} [(-1)^s (L'_{jk}\,u^k_t, u^j_t) + (M'_{jkt}\,u^k, u^j)]\end{aligned} \tag{6.5.33}$$

where M'_{jkt} is of order $< 2s$. But

$$\begin{aligned}\sum_j (-1)^s (L'_{jk}\,u^k_t, u^j_t) &= \sum_j (-1)^s (L'_{jkt}\,u^k_t, u^j_t) +\\ &+ \sum_j (-1)^s\,(L'_{jk}\,u^k_t - L'_{jkt}\,u^k_t, u^j_t),\end{aligned} \tag{6.5.34}$$

where L'_{jkt} is the operator with the constant coefficients $a^\alpha_{jk}(x_t)$ where $|\alpha| = 2s$ and x_t is in the small support of ζ_t. Since for each t (cf. (6.5.9), (6.5.17), and (6.5.19))

$$\begin{aligned}Re \sum_j (-1)^s (L'_{jkt}\,u^k_t, u^j_t) &= Re \int_{G\,j,k,\alpha,\beta} A^{\alpha\beta}_{jk}(x_t)\,D^\alpha\,\bar{u}^j_t\,D^\beta\,u^k_t\,dx\\ &\geq m_1(\|u_t\|^s)^2 \geq 0 \quad ((\|\varphi\|^s)^2 = \sum_j (\|\varphi^j\|^s)^2),\end{aligned} \tag{6.5.35}$$

we see, using (6.5.33)—(6.5.35) that

$$\begin{aligned} &Re \sum_j (-1)^s \Big(\sum_k L_{jk} u^k, u^j\Big) \geq -Z_1 \eta_1 \|u\|^{2s} \cdot \|u\|^0 - \\ (6.5.36) \qquad &- C(\eta_1) \sum_{j,k} \|u^k\|^{2s-1} \|u^j\|^0 \geq -Z_2 \eta_1 \|u\|^{2s} \|u\|^0 - Z_3(\|u\|^0)^2. \end{aligned}$$

But now, for real λ,

$$\begin{aligned} &\sum_j \Big(\Big\|\sum_k L_{jk} u^k + (-1)^s \lambda u^j\Big\|^0\Big)^2 = \sum_j \Big(\Big\|\sum_k L_{jk} u^k\Big\|^0\Big)^2 + \\ (6.5.37) \qquad &+ \lambda^2 (\|u\|^0)^2 + 2\lambda\, Re(-1)^s \sum_j \Big(\sum_k L_{jk} u^k, u^j\Big). \end{aligned}$$

From Theorem 6.5.5, it follows that the boundary value problem satisfies the conditions of § 6.3, so that

$$(6.5.38) \qquad (\|u\|^{2s})^2 \leq C_1 \Big[\sum_j (\|L_{jk} u^k\|^0)^2 + (\|u\|^0)^2\Big].$$

Using (6.5.36) and (6.5.38), we see that (6.5.37) becomes

$$\begin{aligned} \sum_j \|L_{jk} u^k + (-1)^s \lambda u^j\|^2 &\geq \sum_j (\|L_{jk} u^k\|^0)^2 + \lambda^2 (\|u\|^0)^2 - \\ &\quad - 2\lambda Z_2 \eta_1 \|u\|^{2s} \cdot \|u\|^0 - 2\lambda Z_3 (\|u\|^0)^2 \\ &\geq (1 - Z_2 \eta_1 C_1) \sum_j (\|L_{jk} u^k\|^0)^2 + \\ &\quad + [\lambda^2 (1 - Z_2 \eta_1) - 2\lambda Z_3 - Z_2 \eta_1 C_1] (\|u\|^0)^2 \\ &\geq \frac{1}{2} \Big[\sum_j (\|L_{jk} u^k\|^0)^2 + \lambda^2 (\|u\|^0)^2\Big] \quad \text{if } \lambda \geq \text{some } \lambda_1. \end{aligned}$$

The result follows easily from this and (6.5.38).

Higher differentiability results follow from § 6.3.

6.6. The anlyticity of the solutions of analytic systems of linear elliptic equations

In this section we carry over the methods and results of § 5.7 to apply to systems of the type (6.1.1) on the interior and to those of the type studied in § 6.3 at the boundary. The developments follow those of MORREY-NIRENBERG except that we treat general boundary conditions.

For our results on the interior, we consider solutions on a sphere B_R. Since we already know that the solutions $\in C^\infty$, we shall assume that $u \in C^\infty(\bar{B}_R)$. We define

$$\begin{aligned} &e_p(f, B_r) = \Big[\int_{B_r} \sum_j |\nabla^{-s_j+p} f_j(x)|^2 dx\Big]^{1/2}, \; p \geq s_0 = \min s_j \; (\leq 0), \\ (6.6.1) \qquad &d_p(u, B_r) = \Big[\int_{B_r} \sum_k |\nabla^{t_k+p} u(x)|^2 dx\Big]^{1/2}, \; p \geq -t, \; t = \max t_k \end{aligned}$$

where, in general, $\nabla^q \varphi = 0$ *if* $q < 0$. With $[p!]$ given by (5.7.6), we define

$$\begin{aligned} &M_{R,p}(f) = [p!]^{-1} \sup_{R/2 \leq r < R} (R-r)^{t+p} e_p(f, B_r), \quad p \geq s_0 \quad (\geq -t) \\ (6.6.2) \qquad &N_{R,p}(u) = [p!]^{-1} \sup_{R/2 \leq r < R} (R-r)^{t+p} d_p(u, B_r), \quad p \geq -t. \end{aligned}$$

As in § 5.7, we shall show that

$$N_{R,p}(u) \leq M\, L^p, \quad p \geq -t \tag{6.6.3}$$

for some constants M and L.

We begin by studying equations of the form

$$L^0_{jk}\, u^k = f_j, \quad j = 1, \ldots, N, \tag{6.6.4}$$

where the L^0_{jk} are operators with constant coefficients and involving derivatives only of order $s_j + t_k$. We use the notation of § 6.2 until we start considering boundary value problems. Since, for each j, there must be at least one non-zero operator L^0_{jk}, we must have

$$s_0 + t \geq 0. \tag{6.6.5}$$

Lemma 6.6.1. *Suppose $u \in C^\infty_c(B_R)$ and satisfies* (6.6.4) *on B_R. Then there is a constant $K_1(\nu, E)$ such that*

$$d_0(u, B_R) \leq K_1\, e_0(f, B_R).$$

Proof. Multiplying both sides of (6.6.4) by $K(x - \xi)$ (see (6.2.11)) and integrating with respect to ξ we find that

$$L^0_{jk} U^k(x) = F_j(x), \quad U^k(x) = \int_{B_R} K(x-\xi)\, u^k(\xi)\, d\xi, \tag{6.6.6}$$
$$F_j(x) = \int_{B_R} K(x-\xi)\, f_j(\xi)\, d\xi.$$

Operating on both sides of (6.6.6) with L^{lj}_0 and summing, we obtain

$$u^l(x) = L^{lj}_0 F_j(x)$$

from which the lemma follows, using Theorem 6.2.1.

Lemma 6.6.2. *Suppose $u \in C^\infty(\bar{B}_R)$ and satisfies* (6.6.4) *on B_R. Then there is a constant $K_2(\nu, E)$ such that, for $0 < r < r + \delta < R$, $r > \delta$,*

$$d_0(u, B_r) \leq K_2 \left\{ e_0(f, B_{r+\delta}) + \sum_{q=1}^{-s_0} \delta^{-q}\, e_{-q}(f, B_{r+\delta}) + \sum_{q=1}^{t} \delta^{-q}\, d_{-q}(u, B_{r+\delta}) \right\}. \tag{6.6.7}$$

Proof. Let φ be befined as in (6.2.25) et seq. and let

$$\zeta(x) = \varphi[\delta^{-1}(|x| - r + \delta)], \quad U^k(x) = \zeta(x)\, u^k(x). \tag{6.6.8}$$

Then U has support on $B_{r+\delta} \subset B_R$ and satisfies

$$L^0_{jk} U^k = F_j = \zeta f_j + \sum_k \sum_{q=1}^{s_j + t_k} \nabla^q \zeta \cdot L_{jkq}\, u^k \tag{6.6.9}$$

for appropriate operators L_{jkq}. Moreover

$$\nabla^{-s_j} F_j = \zeta \nabla^{-s_j} f_j + \sum_{q=1}^{-s_j} \nabla^q \zeta \cdot m_{jq} f_j + \sum_k \sum_{q=1}^{t_k} \nabla^q \zeta \cdot M_{jkq}\, u^k \tag{6.6.10}$$

where m_{jq} and M_{jkq} are appropriate operators with constant coefficients of orders $-s_j - q$ and $t_k - q$, respectively, being zero if these integers are negative. Since

$$|\nabla^q \zeta(x)| \le Z_q(\nu, \varphi) \cdot \delta^{-q}, \tag{6.6.11}$$

the lemma follows by applying Lemma 6.6.1 to the vector U and using (6.6.10) and (6.6.11).

Lemma 6.6.3. *Suppose* $u \in C^\infty(\bar{B}_R)$ *and satisfies* (6.6.4) *on* B_R. *Then there is a constant* $K_3(\nu, F)$ *such that*

$$N_{R,p}(u) \le K_3 \left\{ M_{R,p}(f) + \sum_{q=1}^{-s_0} M_{R,p-q}(f) + \sum_{q=1}^{t} N_{R,p-q}(u) \right\}, \quad p > 0. \tag{6.6.12}$$

Proof. Suppose r is any number with $R/2 \le r < R$ and let

$$\delta = (R-r)/(p+1) \quad \text{so} \quad R - r - \delta = \left[1 - \frac{1}{p+1}\right] \cdot (R - r). \tag{6.6.13}$$

Then, if $p > 0$, $\nabla^p u$ also satisfies (6.6.4) with f replaced by $\nabla^p f$, so that Lemma 6.6.2 holds with all the indices increased by p. Multiplying both sides of this result by $(p!)^{-1} \cdot (R-r)^{t+p}$ and using (6. 6. 13), we obtain

$$\begin{aligned}
&(p!)^{-1}(R-r)^{t+p} d_p(u, B_r) \\
&\le (p!)^{-1} K_2 \left\{ e_p(f, B_{r+\delta})(R - r - \delta)^{t+p}\left(1 - \frac{1}{p+1}\right)^{-t-p} + \right. \\
&+ \sum_{q=1}^{-s_0} (p+1)^q e_{p-q}(f, B_{r+\delta}) \cdot (R - r - \delta)^{t+p-q}\left(1 - \frac{1}{p+1}\right)^{-t-p+q} + \\
&\left. + \sum_{q=1}^{t} (p+1)^q d_{p-q}(u, B_{r+\delta})(R - r - \delta)^{t+p-q}\left(1 - \frac{1}{p+1}\right)^{-t-p+q} \right\}
\end{aligned}$$

from which the result follows by taking the sup.

We turn now to the general analytic system

$$L_{jk} u^k = f_j \tag{6.6.14}$$

which we write in the form

$$L^0_{jk} u^k = f_j + (L^0_{jk} - L_{jk}) u^k = F_j \tag{6.6.15}$$

where L^0_{jk} is the operator whose coefficients are just those of the highest order derivatives evaluated at the origin. We set

$$\nabla^{-s_j+\lambda} F_j = \nabla^{-s_j+\lambda} f_j + \sum_{q=0}^{t+\lambda} a^{\lambda q}_{jk}(x) \cdot \nabla^{t_k+\lambda-q} u^k, \quad s_0 \le \lambda \le 0, \tag{6.6.16}$$

where the $a^{\lambda q}_{jk}$, to be abbreviated by $a^{\lambda q}$, are appropriate analytic tensors. (We recall that $\nabla^p \varphi = 0$ if $p < 0$.) Accordingly, there are numbers $A \ge 2$, L, and R_0, such that

$$\left.\begin{aligned}
&\Big[\sum_j |\nabla^{-s_j+p} f_j|^2\Big]^{1/2} \le [p!] L A^p, \quad p \ge s_0 \\
&|\nabla^p a^{\lambda q}| \le p! L A^p, \quad p \ge 0 \\
&a^{00}(0) = 0, \quad |a^{00}(x)| \le L A \cdot |x|,
\end{aligned}\right\} \text{ in } B_{R_0}. \tag{6.6.17}$$

Applying the vector inequality to (6.6.16) and using Lemma 5.7.4, we obtain

$$\left[\sum_j |\nabla^{-s_j+p} F_j|^2\right]^{1/2} \leq \left[\sum_j |\nabla^{-s_j+p} f_j|^2\right]^{1/2} + $$
$$+\sum_{q=0}^{t} \sum_{\varkappa=0}^{p} \binom{p}{\varkappa} |\nabla^{p-\varkappa} a^{0q}| \cdot \left[\sum_k |\nabla^{t_k+\varkappa-q} u^k|^2\right]^{1/2}, \quad p \geq 0$$
$$\left[\sum_j |\nabla^{-s_j+p} F_j|^2\right]^{1/2} \leq \left[\sum_j |\nabla^{-s_j+p} f_j|^2\right]^{1/2} + $$
$$+\sum_{q=1}^{t+p} |a^{pq}(x)| \left[\sum_k \nabla^{t_k+p-q} u^k|^2\right]^{1/2}, \quad p \leq 0.$$

Using (6.6.17) and separating out the term where $q = 0$ and $\varkappa = p$, in case $p \geq 0$, we obtain for $0 < r < R \leq R_0$ (setting $\tau = \nu/2$),

$$\begin{aligned} e_p(F, B_r) \leq {} & p!\, L A^p \gamma_\nu^{1/2} r^\tau + \\ & +\sum_{\varkappa=0}^{p-1} \binom{p}{\varkappa} (p-\varkappa)!\, L A^{p-\varkappa} d_\varkappa(u, B_r) + L A R \cdot d_p(u, B_r) + \\ & +\sum_{\lambda=1}^{t} \sum_{\varkappa=0}^{p} \binom{p}{\varkappa} \cdot (p-\varkappa)!\, L A^{p-\varkappa} d_{\varkappa-\lambda}(u, B_r), \quad p \geq 0, \end{aligned} \tag{6.6.18}$$
$$e_p(F, B_r) \leq [p!]\, L A^p \gamma_\nu^{1/2} r^\tau + \sum_{\lambda=0}^{t+p} L [(p-\varkappa)!]\, d_{p-\lambda}(u, B_r), \; p < 0.$$

Thus, from the definitions (6.6.2), we find that

$$\begin{aligned} M_{R,p}(F) \leq {} & \gamma_\nu^{1/2} L (A R)^p R^{t+\tau} + \sum_{\varkappa=0}^{p-1} L (A R)^{p-\varkappa} N_{R,\varkappa}(u) + \\ & + L A R \cdot N_{R,p}(u) + \sum_{\lambda=1}^{t} \sum_{\varkappa=0}^{p} \frac{[\varkappa-\lambda]!}{\varkappa!} L (A R)^{p-\varkappa} R^\lambda N_{R,\varkappa-\lambda}(u) \\ & \qquad p \geq 0, \end{aligned} \tag{6.6.19}$$
$$M_{R,p}(F) \leq \gamma_\nu^{1/2} L A^p R^{p+t+\tau} + \sum_{\lambda=0}^{t+p} L R^\lambda N_{R,p-\lambda}(u), \quad s_0 \leq p < 0.$$

We now apply Lemma 6.6.3. If we take R so small that

$$K_3 L A R \leq 1/2, \tag{6.6.20}$$

we see that we can solve for $N_{R,p}(u)$ in terms of the non-homogeneous term and the $N_{R,q}(u)$ with $q < p$. The result is, for $p > 0$,

$$\begin{aligned} N_{R,p}(u) \leq 2K_3 \Big\{ & \gamma_\nu^{1/2} L R^{t+\tau} \sum_{q=0}^{-s_0} (A R)^{p-q} + \sum_{\varkappa=0}^{p-1} L (A R)^{p-\varkappa} N_{R,\varkappa}(u) + \\ & + \sum_{\lambda=1}^{t} \sum_{\varkappa=0}^{p} L (A R)^{p-\varkappa} R^\lambda N_{R,\varkappa-\lambda}(u) + \sum_{q=1}^{p} \sum_{\lambda=0}^{t} \sum_{\varkappa=0}^{p-q} L (A R)^{p-q-\varkappa} R^\lambda N_{R,\varkappa-\lambda}(u) + \\ & + \sum_{q=p+1}^{-s_0} \sum_{\lambda=0}^{t} L R^\lambda N_{R,p-q-\lambda}(u) + \sum_{q=1}^{t} N_{R,p-q}(u) + L A R \sum_{q=1}^{p} N_{R,p-q}(u) \Big\} \\ & 0 < p < -s_0. \end{aligned} \tag{6.6.21}$$

In case $p \geq -s_0$ (which is the general case), the double sum for $p+1 \leq q \leq -s_0$ is missing and q ranges from 1 to $-s_0$ in the triple sum. We note that

$$\sum_{q=0}^{-s_0} \gamma_\nu^{1/2} L (A R)^{p-q} R^{t+\tau} = \gamma_\nu^{1/2} L A^{s_0} R^{t+s_0+\tau} \sum_{\mu=0}^{-s_0} (A R)^{p+\mu}$$
$$\leq 2^{1-p} \gamma_\nu^{1/2} L A^{s_0} R^{t+s_0+\tau} \quad \text{if} \quad AR \leq 1/2.$$

Thus, if we assume that

(6.6.22) $\qquad N_{R,q}(u) \leq M R^{t+s_0+\tau} \cdot P^q, \quad -t \leq q < p, \quad p \geq -s_0$

we see that

$$N_{R,p}(u) \leq 2 K_3 R^{t+s_0+\tau} \Big\{ 2^{1-p} \gamma_\nu^{1/2} L A^{s_0} + \sum_{\varkappa=0}^{p-1} M L\, 2^{\varkappa-p} P^\varkappa +$$
$$+ \sum_{\lambda=1}^{t} \sum_{\varkappa=0}^{p} L A^{-\lambda}\, 2^{\varkappa-p-\lambda} M P^{\varkappa-\lambda} + \sum_{q=1}^{-s_0} \sum_{\lambda=0}^{t} \sum_{\varkappa=0}^{p-q} L A^{-\lambda}\, 2^{q+\varkappa-p-\lambda} M P^{\varkappa-\lambda} +$$
$$+ \sum_{q=1}^{t} M P^{p-q} + 2^{-1} L \sum_{q=1}^{p} P^{p-q} \Big\} \leq M \cdot R^{t+s_0+\tau} P^p \quad (AR \leq 1/2)$$

(6.6.23)

if M and P are chosen so that (6.6.22) holds for $-t \leq p \leq -s_0 - 1$; it is easy to see that P may be chosen > 1 and large enough so that the coefficient of M in the brace is $\leq P^p/4K_3$ and then M may be taken so large that the remaining terms in the brace are $\leq M\, P^p/4K_3$. We therefore conclude the following theorem:

Theorem 6.6.1. *If u is a solution of* (6.6.14) *and f_j and the coefficients in the operators L_{jk} are analytic at x_0, then u is analytic at x_0.*

We now consider the analyticity at the boundary. Using an analytic change of independent variables, we may transform a boundary neighborhood along an analytic portion of the boundary into G_{R_0}, the part of the boundary corresponding to σ_{R_0}. We begin by considering solutions u of (6.6.4) which satisfy the boundary conditions

(6.6.24) $\qquad B_{0rk}\, u^k = g_r$ on σ_R, $R \leq R_0$,

where B_{0rk} has its significance in § 6.3. We assume that

(6.6.25) $\qquad h_0 \geq 0,\ h_0 + h_r \geq 1$ for each r,

h_0 and h_r being the numbers in § 6.3. We assume that the boundary conditions and equations satisfy the conditions in § 6.3 and we use the (x, y) and other notations of that section. Instead of (6.6.1) and (6.6.2), we define

$$e_p(f, G_r) = \begin{cases} \sum_j \| \nabla^{h_0 - s_j} \nabla_x^p f_j \|_{G_r}, & p \geq 0, \\ \sum_j \| \nabla^{h_0 - s_j + p} f_j \|_{G_r}, & 0 \geq p \geq s_0 - h_0, \end{cases}$$

(6.6.26) $$d_p(u, G_r) = \begin{cases} \sum_k \| \nabla^{t_k + h_0} \nabla_x^p u^k \|_{G_r}, & p \geq 0, \\ \sum_k \| \nabla^{t_k + h_0 + p} u^k \|_{G_r}, & 0 \geq p \geq -t - h_0, \end{cases}$$

$$c_p(\tilde{g}, G_r) = \begin{cases} \sum_s \| \nabla^{h_0 + h_s} \nabla_x^p \tilde{g}_s \|_{G_r}, & p \geq 0, \\ \sum_s \| \nabla^{h_0 + h_s + p} \tilde{g}_s \|_{G_r}, & 0 \geq p \geq -h_0 - h,\ h = \max h_s, \end{cases}$$

the norms being the L_2 norms. Since all functions considered $\in C^\infty(\bar{G}_r)$, we may take $\tilde{g}_s \in C^\infty(\bar{G}_r)$. We then define

$$(6.6.27)\quad \begin{aligned} N_{R,p}(u) &= \sup_{R/2 \le r < R} [p!]^{-1}(R-r)^{t+h_0+p} d_p(u, G_r), \quad p \ge -t-h_0 \\ M_{R,p}(f) &= \sup_{R/2 \le r < R} [p!]^{-1}(R-r)^{t+h_0+p} e_p(f, G_r), \quad p \ge s_0 - h_0 \\ Q_{R,p}(\tilde{g}) &= \sup_{R/2 \le r < R} [p!]^{-1}(R-r)^{t+h_0+p} c_p(\tilde{g}, G_r), \quad p \ge -h_0 - h \end{aligned}$$

We shall first establish the analog of (6.6.22).

Lemma 6.6.1′. *Suppose u, f, and $\tilde{g} \in C^\infty(\bar{G}_R)$ and have support on $G_R \cup \sigma_R$, suppose $g \in C_c^\infty(\sigma_R)$, suppose $\tilde{g}_s(x, 0) = g_s(x)$, suppose u satisfies (6.6.4) on G_R and (6.6.24) on σ_R. Then there is a constant $K_1'(\nu, E)$ such that*

$$(6.6.28)\qquad d_0(u, B_R) \le K_1'[e_0(f, B_R) + c_0(\tilde{g}, B_R)].$$

Proof. We note that u, f, $\tilde{g}$, and g satisfy the hypotheses on any G_A with $A \ge R$ and vanish outside G_R in $R_{\nu+1}^+$. Let $U = P_R(f, g)$. Then, from Theorems 6.3.6 and 6.3.5, it follows that $u^k - U^k$ is a polynomial of degree $\le t_k + h_0 - \nu - 1$ and hence that (6.6.28) holds, since only the highest order derivatives of the u^k are involved.

Lemma 6.6.2′. *Suppose u, f, and $\tilde{g} \in C^\infty(\bar{G}_R)$, suppose $g \in C^\infty(\bar{\sigma}_R)$ and $\tilde{g}(x, 0) = g(x)$ on $\bar{\sigma}_R$, suppose u satisfies (6.6.4) on G_R, and suppose u satisfies (6.6.24) on σ_R. Then there is a constant $K_2'(\nu, E)$ such that*

$$(6.6.29)\quad \begin{aligned} d_0(u, B_r) \le K_2' \Big\{ & e_0(f, G_{r+\delta}) + \sum_{q=1}^{h_0-s_0} \delta^{-q} e_{-q}(f, G_{r+\delta}) + c_0(\tilde{g}, G_{r+\delta}) + \\ & + \sum_{q=1}^{h_0+h} \delta^{-q} c_{-q}(\tilde{g}, G_{r+\delta}) + \sum_{q=1}^{t+h_0} \delta^{-q} d_{-q}(u, G_{r+\delta}) \Big\}, \\ & 0 < r < r+\delta < R, \quad r > \delta. \end{aligned}$$

Proof. Define ζ and U^k by (6.6.8). Then U satisfies (6.6.9) and

$$(6.6.30)\qquad B_{0sk} U^k = \zeta g_s + \sum_{q=1}^{t_k - h_s} \nabla^q \zeta \cdot B_{skq} u^k \equiv \gamma_s \quad \text{on} \quad \sigma_R.$$

If we define $\tilde{\gamma}_s$ by (6.6.30) with g_s replaced by $\tilde{g}_s$, we see that $\tilde{\gamma}_s(x, 0) = \gamma_s(x)$ and, in fact $U, F, \tilde{\gamma}$, and γ satisfy the hypotheses of Lemma 6.6.1′. Just as in the proof of Lemma 6.6.2 we find that

$$e_0(F, B_{r+\delta}) \le e_0(f, G_{r+\delta}) + Z_1 \left[\sum_{q=1}^{h_0-s_0} \delta^{-q} e_{-q}(f, G_{r+\delta}) + \sum_{q=1}^{t+h_0} \delta^{-q} d_{-q}(u, G_{r+\delta}) \right]$$

$$c_0(\tilde{\gamma}, B_{r+\delta}) \le c_0(\tilde{g}, G_{r+\delta}) + Z_2 \left[\sum_{q=1}^{h_0+h} \delta^{-q} c_{-q}(\tilde{g}, G_{r+\delta}) + \sum_{q=1}^{t+h_0} \delta^{-q} d_{-q}(u, G_{r+\delta}) \right].$$

The lemma then follows from Lemma 6.6.1′.

By differentiating in the tangential directions only and proceeding as in the proof of Lemma 6.6.3, we prove the following lemma:

Lemma 6.6.3'. *Suppose u, f, $\tilde{g}$, and g satisfy the hypotheses of Lemma 6.6.2' on $\bar{G}_R$. Then there is a constant $K_3'(\nu, E)$ such that*

$$N_{R,p}(u) \leq K_3 \Big\{ M_{R,p}(f) + \sum_{q=1}^{h_0-s_0} M_{R,p-q}(f) + Q_{R,p}(\tilde{g}) + \\ + \sum_{q=1}^{h_0+h} Q_{R,p-q}(\tilde{g}) + \sum_{q=1}^{t+h_0} N_{R,p-q}(u) \Big\}, \quad p > 0.$$

Remark. In this proof, we use the facts that

$$d_p(u, G_r) = d_0(\nabla_x^p u, G_r) \text{ if } p \geq 0, \; e_{-q}(\nabla_x^p f, G_r) \leq e_{p-q}(f, G_r), \; p, q \geq 0,$$
$$c_{-q}(\nabla_x^p \tilde{g}, G_r) \leq c_{p-q}(\tilde{g}, G_r), \quad p, q \geq 0.$$

To handle the general analytic system, we write it in the form

$$(6.6.31) \quad L_{jk}^0 u^k = F_j, \quad B_{0sk} u^k = \tilde{\gamma}_s = \tilde{g}_s + (B_{0sk} - B_{sk}) u^k \text{ on } \sigma_R,$$

where F_j is defined as before, in (6.6.15) and $\tilde{g}_s$ is analytic on $\bar{G}_R$ with $\tilde{g}_s(x, 0) = g_s(x)$. As before, we have

$$\nabla^{h_0-s_j+\lambda} F_j = \nabla^{h_0-s_j+\lambda} f_j + \sum_{q=0}^{t+h_0} a_{jk}^{\lambda q}(x, y) \nabla^{t_k+h_0+\lambda-q} u^k(x, y), \quad 0 \geq \lambda \geq s_0 - h_0$$

$$\nabla^{h_0-h_s+\lambda} \tilde{\gamma}_s = \nabla^{h_0+h_s+\lambda} \tilde{g}_s + \sum_{q=0}^{t+h_0} b_{sk}^{\lambda q}(x, y) \nabla^{t_k+h_0+\lambda-q} u^k(x, y), \quad 0 \geq \lambda \geq -h_0 - h$$

$$a^{00}(0) = b^{00}(0) = 0, \quad |a^{00}(x, y)| \leq LA \sqrt{|x|^2 + y^2},$$
$$|b^{00}(x, 0)| \leq LA \sqrt{|x|^2 + y^2},$$

with bounds like those in (6.6.17) for $\nabla^p f_j$, $\nabla^p a^{\lambda q}$, $\nabla_x^p b^{\lambda q}$, $\nabla^p \tilde{g}_s$. Using the previous analysis and the idea of the remark above, we obtain the inequalities (6.6.18) and (6.6.19) with t replaced by $t + h_0$ and we obtain the corresponding inequalities for $Q_{R,p}(g)$. As before, if we take $R (\leq R_0)$ so small that $2K_3 LAR \leq 1/2$ this time, we can obtain an inequality like (6.6.21) with t replaced by $t + h_0$ and terms on the right coming from the analysis of $\tilde{\gamma}$. An analysis like the previous one demonstrates the existence of numbers $R > 0$, M, and P such that

$$(6.6.32) \quad N_{R,p}(u) \leq M R^{t+s_0+\tau} P^p, \quad p \geq -t - h_0, \quad 0 < R \leq R_1.$$

Now, in order to prove our analyticity theorem, we must obtain bounds like (6.6.32) for *all* the derivatives of u. For this purpose we define

$$(6.6.33) \quad \begin{aligned} d_{p,q}(u, G_r) &= \sum_k \| D_y^{t_k+h_0+q} \nabla_x^p u^k \|_{G_r}, \quad p \geq 0, \quad q \geq -t - h_0 \\ N_{R,p,q}(u) &= \sup_{R/2 \leq r < R} [(p+q)!]^{-1} (R - r)^{t+h_0+p+q} d_{p,q}(u, G_r). \end{aligned}$$

Using the idea of the remark above, we conclude that

$$(6.6.34) \quad N_{R,p,q}(u) \leq N_{R,p+q}(u) \text{ if } q \leq 0 \text{ and } p \geq 0.$$

We shall show that there exist numbers R_2, $0 < R_2 \leq R_1$, $\bar{M}$, $\bar{P} \geq 1$, and $\theta \leq 1/2$, R_2, $\bar{P}$, and θ depending only on the bounds for the coefficients, f, and $\tilde{g}$ and their derivatives, such that

$$N_{R,p,q}(u) \leq \bar{M}\bar{P}^{p+q}\theta^p R^{t+s_0+\tau}, \quad p \geq 0, \quad q \geq -t-h_0, \quad 0 < R \leq R_2. \tag{6.6.35}$$

If we differentiate the $j-th$ equation $-s_j+\sigma$ times with respect to y, $0 \leq \sigma \leq h_0$, the ellipticity implies that we can solve for $D_y^{t_k+\sigma} u^k$ in terms of the other derivatives, obtaining

$$D_y^{t_k+\sigma} u^k = f^{\sigma k} + \sum_{\varrho=1}^{t}\sum_{\omega=0}^{\varrho} b_{\varrho l\omega}^{\sigma k}(x,y) D_y^{t_l+\sigma-\varrho}\nabla_x^{\omega} u^l, \quad 0 \leq \sigma \leq h_0 \tag{6.6.36}$$

where $f^{\sigma k}$ and $b_{\varrho l\omega}^{\sigma k}$ are suitable functions analytic near $(0,0)$. We may assume that the f's and b's satisfy

$$\begin{aligned} &\sum_k |D_y^q \nabla_x^p f^{\sigma k}| \leq L\,p!\,q!\,A^{p+q} \\ &\sum_k \max_{(x,y,l)} |D_y^q \nabla_x^p b_{\varrho l\omega}^{\sigma k}(x,y)| \leq L\,p!\,q!\,A^{p+q} \end{aligned} \tag{6.6.37}$$

for each set (σ,ϱ,ω). Differentiating (6.6.36) with $\sigma = h_0$ and using (6.6.37) and Lemma 5.7.4, we obtain, for $p \geq 0$ and $q \geq 0$,

$$\begin{aligned} \sum_k |D_y^{t_k+h_0+q}\nabla_x^p u^k| \leq{}& L\cdot p!\,q!\,A^{p+q} \\ & \sum_{\varrho=1}^{t}\sum_{\omega=0}^{\varrho}\sum_{\lambda=0}^{p}\sum_{\mu=0}^{q}\binom{p}{\lambda}\binom{q}{\mu}(p-\lambda)!\,(q-\mu)!\,L\,A^{p+q-\lambda-\mu}\times \\ & \times\sum_l |D_y^{t_l+h_0+\mu-\varrho}\nabla_x^{\omega+\lambda} u^l|. \end{aligned} \tag{6.6.38}$$

From (6.6.38) and (6.6.33), we obtain, for $p \geq 0$, $q \geq 0$,

$$\begin{aligned} N_{R,p,q}(u) \leq{}& L\frac{p!\,q!}{(p+q)!}A^{p+q}R^{t+h_0+\tau+p+q} + \\ & + \sum_{\varrho,\omega,\lambda,\mu} B_{\varrho\omega\lambda\mu}^{pq} L\,A^{p+q-\lambda-\mu} N_{R,\omega+\lambda,\mu-\varrho}(u)\,R^{p+q-\lambda-\mu+\varrho-\omega}, \\ B_{\varrho\omega\lambda\mu}^{pq} ={}& \frac{p!\,q!}{(p+q)!}\cdot\frac{[(\lambda+\mu+\omega-\varrho)!]}{\lambda!\,\mu!} \leq 1. \end{aligned} \tag{6.6.39}$$

Now, from (6.6.34), it follows that (6.6.35) holds for all $q \leq 0$ and all $p \geq 0$, provided merely that

$$0 < R \leq R_1 \quad\text{and}\quad M P^{p+q} \leq \bar{M}(\bar{P}\theta)^{p+q}\theta^{-q} \quad\text{for all}\quad p \geq 0, \quad 0 \geq q \geq -t-h_0. \tag{6.6.40}$$

Then, from (6.6.39), we see that if (6.6.35) holds for all (p', q') with $p' \geq 0$ and $-t-h_0 \leq q' < q\ (> 0)$, then it will hold also for all (p', q') with $q' \leq q$, if, for example, we choose

$$A R_2 = \tfrac{1}{2}, \quad \bar{P}\theta \geq P, \quad \bar{P}\theta \geq 1, \quad \theta \leq 1, \quad \bar{M} \geq M, \quad 2^{p+q}\bar{M} \geq 2L\cdot(2A)^{-h_0+s_0},$$
$$\bar{P} \geq 1, \quad A\,\bar{P}\theta \geq 1, \quad\text{and}\quad \theta \leq 1/16\,L.$$

Since A was supposed ≥ 2, these conditions may be satisfied.

6.7. The analyticity of the solutions of analytic nonlinear elliptic-systems

In this section we extend the results of § 5.8 to general elliptic systems; the analyticity at the boundary is proved for the systems and boundary conditions discussed in § 6.3.

As usual, we consider analyticity on the interior first. We consider solutions u $(= u^1, \ldots, u^N)$ of systems of the form

$$\varphi_j(x, Du) = 0, \quad t = 1, \ldots, N \tag{6.7.1}$$

in which the φ_j are analytic for all values of their arguments near the values of those arguments along the solution at x_0. The system is elliptic along the solution in the sense that the linear equations of variation

$$L_{jk}(x, D)\, v^k \equiv \frac{d}{d\lambda} \varphi_j[x, Du + \lambda D v]_{\lambda=0} = 0 \tag{6.7.2}$$

form a system of the type discussed in § 6.2. We shall use the notations of that section.

In (6.7.1), we may suppose that x_0 is the origin. In addition, we make the substitution

$$u^k = v^k + p^k, \quad k = 1, \ldots, N \tag{6.7.3}$$

where p^k is that polynomial of degree $\leq t_k$ such that

$$\nabla^r P^k(0) = \nabla^r u^k(0), \quad 0 \leq r \leq t_k. \tag{6.7.4}$$

We therefore assume that $x_0 = 0$ and $D^\alpha u^k(0) = 0$ for $|\alpha| \leq t_k$ in (6.7.1). If we expand the φ_j about the origin, we may write (6.7.1) in the form

$$L^0_{jk} u^k = M^0_{jk} u^k + \psi_j(x, Du) \tag{6.7.5}$$

where the operators L^0_{jk} have constant coefficients and are zero or of order $s_j + t_k$, the M^0_{jk} have constant coefficients and are of lower order, and the ψ_j are the remainders. Since the derivatives $D^\alpha u^k(0) = 0$ if $|\alpha| \leq t_k$, we see by differentiating the $j-th$ equation in (6.7.5) up to $-s_j$ times, that the Taylor expansion of ψ_j begins with a homogeneous polynomial in x of degree $1 - s_j$, linear terms in the $D^\alpha u^k$ with coefficients homogeneous and linear in x, and quadratic terms in the $D^\alpha u^k$ with constant coefficients.

Definition 6.7.1. We define the spaces $^*C_\mu(\bar{B}_R)$ of vectors f for which $f_j \in C_\mu^{-s_j}(\bar{B}_R)$ with $\nabla^r f_j(0) = 0$ for $0 \leq r \leq -s_j$ and $^{**}C_\mu(\bar{B}_R)$ of vectors u for which $u^k \in C_\mu^{t_k}(\bar{B}_R)$ with $\nabla^r u^k(0) = 0$ for $0 \leq r \leq t_k$; we define the norms by

$$^*\|f\|_{\mu,R} = \sum_j h_\mu(\nabla^{-s_j} f_j, \bar{B}_R), \quad ^{**}\|u\|_{\mu,R} = \sum_k h_\mu(\nabla^{t_k} u^k, \bar{B}_R). \tag{6.7.6}$$

Definition 6.7.2. For $f \in {}^*C_\mu(\bar{B}_R)$, we define $P_R(f) = U$, where

$$U^k = u^k - P^k, \quad u^k = L_0^{kl} F_l, \quad F_l(x) = \int_{B_R} K(x - \xi) f_l(\xi)\, d\xi \tag{6.7.7}$$

where P^k is the polynomial in (6.7.4). We are here supposing that the order m of the operator L_0 is >0; otherwise L_0 is just a constant and we set $F_l = L_0^{-1} f_l$.

In order to show that P_R is a bounded operator from $^*C_\mu(\bar{B}_R)$ we need some more refined results than those in Theorem 6.2.1(d).

Theorem 6.7.1. *P_R is a bounded operator from $^*C_\mu(\bar{B}_R)$ to $^{**}C_\mu(\bar{B}_R)$ with bound independent of R (dependent only on ν, μ, E). Moreover, if $U = P_R(f)$ and $m > 0$, then U satisfies*

$$L_{jk}^0 U^k = f_j. \tag{6.7.8}$$

Proof. We extend f_j to $\in C_{\mu c}^{-s_j}(\bar{B}_{2R})$ (properly; cf. Lemma 6.3.3) and write

$$F_j(x) = F_{j1}(x) - F_{j2}(x), \quad F_{j1}(x) = \int_{B_{2R}} K(x-\xi) f_j(\xi)\, d\xi$$

$$F_{j2}(x) = \int_{B_{2R}-B_R} K(x-\xi) f_j(\xi)\, d\xi. \tag{6.7.9}$$

From Theorem 6.2.1, we conclude that

$$F_{1j} \in C_\mu^{m-s_j}(\bar{B}_R), \quad h_\mu(\nabla^{m-s_j} F_{1j}, \bar{B}_R) \le Z_1 h_\mu(\nabla^{-s_j} f_j, \bar{B}_R). \tag{6.7.10}$$

By differentiating F_{j2} and using Theorem 2.6.5, we obtain

$$D^{m+1-s_j} F_{j2}(x) = \int_{B_{2R}-B_R} D\Delta(x-\xi)\, [D^{-s_j} f_j(\xi) - D^{-s_j} f_j(x)]\, d\xi,$$

$$\Delta(y) = D^m K(y), \quad \Delta(-y) = \Delta(y), \quad x \in B_R. \tag{6.7.11}$$

From (6.7.11), we see easily that

$$|\nabla^{m+1-s_j} F_{j2}(x)| \le Z_2 h_\mu(\nabla^{-s_j} f_j, \bar{B}_R) \cdot (R - |x|)^{\mu-1}. \tag{6.7.12}$$

The result (6.7.10) for F_{j2} follows as usual from Theorem 2.6.6. Thus, if we define u by (6.7.7), we see that $u^k \in C_\mu^{t_k}(\bar{B}_R)$ and u satisfies (6.7.8). Thus U^k satisfies

$$L_{jk}^0 U^k = f_j + P_j$$

where P_j is a polynomial of degree $\le -s_j$. But by differentiating and setting $x = 0$, we find that $P_j \equiv 0$. The theorem follows.

Now, as in § 5.8, we assume that u is a solution of (6.7.5) and that $u \in {}^{**}C_\mu(B_{R_0})$, $R_0 > 0$. For $0 < R \le R_0$, we write

$$u = u_R + H_R, \quad u_R = P_R[M u + \psi(x, Du)]. \tag{6.7.13}$$

If we define

$$v_R = P_R[M H_R + \psi(x, D H_R)]$$

$$T_R(u_R, H_R) = P_R[M u_R + \psi(x, D u_R + D H_R) - \psi(x, D H_R)], \tag{6.7.14}$$

then the equation for u_R becomes

$$u_R - T_R(u_R, H_R) = v_R \tag{6.7.15}$$

and H_R is seen to satisfy (6.7.8) *with $f_j = 0$ on B_R.*

Now, if u is a solution on $\bar{B}_{R_0}$, then it is clear that $^{**}\|u\|_R \leq M_1$ independently of R. The proof of the following theorem is identical with that of Theorem 5.8.2:

Theorem 6.7.2. *Suppose u is a solution $\in {}^{**}C_\mu(\bar{B}_{R_0})$ of (6.6.5) and suppose u_R, v_R, H_R, and T_R are defined as above. Then $^{**}\|u_R\|_R$, $^{**}\|v_R\|_R$, and $^{**}\|H_R\|_R$ are uniformly bounded by some number M_2 for $0 < R \leq R_0$, H_R satisfies (6.7.8) with $f = 0$ on B_R, and*

$$T_R(0; H_R) = 0, \quad \lim_{R\to 0} {}^{**}\|v_R\|_R = \lim_{R\to 0} {}^{**}\|u_R\|_R = 0,$$

$$^{**}\|T_R(U_{1R}; H_R) - T_R(U_{2R}; H_R)\|_R \leq \varepsilon(R)\ {}^{**}\|U_{1R} - U_{2R}\|_R,$$

$$\lim_{R\to 0} \varepsilon(R) = 0, \quad \text{if} \quad {}^{**}\|H_R\|_R,\ {}^{**}\|U_{1R}\|_{1R},\ {}^{**}\|U_{2R}\|_R \leq M_2.$$

Thus, if $0 < R \leq R_1$, u_R is the only solution of (6.7.15) for which $^{**}\|u_R\| \leq M_2$, provided $^{**}\|H_R\|_R \leq M_2$, H_R being supposed given.

Now, as in § 5.8, we introduce the following spaces of analytic functions; we use the notations of that section.

Definition 6.7.3. The *space* $^*C_{\mu h}(\bar{B}_{hR})$ consists of all $f \in {}^*C_\mu(\bar{B}_R)$ which can be extended to $\bar{B}_{hR}$ so that each $f_j \in C_\mu^{-s_j}(\bar{B}_{hR})$ and is holomorphic in B_{hR}. The *space* $^{**}C_{\mu h}(\bar{B}_{hR})$ consists of all $u \in {}^{**}C_\mu(\bar{B}_R)$ which can be extended to $\bar{B}_{hR}$ so that each $u^k \in C_\mu^{t_k}(\bar{B}_{hR})$ and is holomorphic on B_{hR}. We define the norms by

$$(6.7.16) \qquad ^*\|f\|_{hR} = \sum_j h_\mu(\nabla^{-s_j} f_j, \bar{B}_{hR}), \qquad ^{**}\|u\|_h = \sum_k h_\mu(\nabla^{t_k} u^k, \bar{B}_{hR}).$$

The following theorem is proved by imitating the proof of Theorem 5.8.3:

Theorem 6.7.3. *Suppose $0 < h < h_1$, h_1 being the number in Theorem 6.2.1(a), suppose $f \in {}^*C_{\mu h}(\bar{B}_{hR})$, suppose $x_0 \in B_{hR}$, suppose $S : \xi = \xi(s)$, $s \in \bar{B}_R$, is admissible with respect to the function $\mathfrak{F}(\xi; x_0) = K(x_0 - \xi) f(\xi)$, and ξ satisfies (5.8.20). Then the integral (5.8.18) exists. If ξ and ξ^* are both admissible and satisfy (5.8.20), the integrals have the same value. Finally, if for each $x \in B_{hR}$, $S(x) : \xi = \xi(s; x)$ is admissible with respect to $\mathfrak{F}(\xi; x)$ and satisfies (5.8.20) for each x, then the function F defined by*

$$(6.7.17) \qquad F(x) = \int_{S(x)} K(x - \xi) f(\xi)\, d\xi$$

is analytic on B_{hR} and

$$(6.7.18) \qquad L_0 F(x) = f(x), \quad D^\alpha F(x) = \int_{S(x)} D^\alpha K(x - \xi) f(\xi)\, d\xi, \quad |\alpha| < m.$$

Definition 6.7.4. For $f \in {}^*C_{h\mu}(B_{hR})$, we define $P_R(f) = U$, where U is defined by (6.7.7) with F_l replaced by its extension

$$(6.7.19) \qquad F_l(x) = \int_{S(x)} K(x - \xi) f_l(\xi)\, d\xi.$$

Theorem 6.7.4. *P_R is a bounded operator from $*C_{h\mu}(\bar{B}_{hR})$ to $**C_{h\mu}(B_{hR})$ with bound depending only on μ, h, and E and not on R.*

Proof. Let $f \in *C_{h\mu}(\bar{B}_{hR})$ and suppose f is extended to $\bar{B}_{hR}$ as in the definition and is also extended so that each $f_j \in C_\mu^{-s_j}(B_{2R})$ and is zero near $\sum_{2R}$, the norm of f being increased at most by a factor $C(E)$; the possibility of this is proved as in Lemma 6.3.3 using r instead of x^ν. We then write

$$F_l(x) = F_{l1}(x) - F_{l2}(x), \quad F_{l2}(x) = \int_{B_{2R}-B_R} K(x-\xi) f_l(\xi)\, d\xi.$$

Clearly F_{l2} is analytic on B_{hR} and

$$\nabla^{m+1-s_l} F_{l2}(x) = \int_{B_{2R}-B_R} \nabla^{1-s_l} \Delta(x-\xi)\left[f_l(\xi) - \sum_{|\alpha|=0}^{-s_l} \frac{1}{\alpha!} D^\alpha f_l(x)\cdot(\xi-x)^\alpha\right] d\xi,$$

$$\Delta(y) = \nabla^m K(y), \quad x \in B_{hR}$$

for, since Δ satisfies (2.6.9) and (2.6.10), we see by induction and differentiation using Theorem 2.6.5 that

$$\int_{B_{2R}-B_R} \nabla^{1-s_l}\Delta(x-\xi)\cdot(\xi-x)^\alpha\, d\xi \equiv 0, \quad 0 \le |\alpha| \le -s_l. \tag{6.7.20}$$

Thus

$$|\nabla^{m+1-s_l} F_{l2}(x_0)| \le Z(\mu, h, E)\, h_\mu(\nabla^{-s_l} f_l, \bar{B}_{hR})\cdot[r(x_0)]^{\mu-1}, \quad x_0 \in B_{hR}. \tag{6.7.21}$$

Since the derivatives up to the order $-s_j$ of the f_j join up continuously across ∂B_R, we find that

$$\begin{aligned} D^{m-1-s_l} F_{1l}(x) &= \int_{S(x)} D^{m-1} K(x-\xi) D^{-s_l} f_l(\xi)\, d\xi + \\ &+ \int_{B_{2R}-B_R} D^{m-1} K(x-\xi) D^{-s_l} f_l(\xi)\, d\xi = \varphi_1(x) + \varphi_2(x), \end{aligned} \tag{6.7.22}$$

$$\varphi_2(x) = \int_{B(x_{10}, r)} D^{m-1} K(x_1 - s) D^{-s_l} f_l(s + i x_{20})\, ds,$$

$$x = x_1 + i x_{20}, \quad x_0 = x_{10} + i x_{20} \,\varepsilon\, B_{hR}, \quad r = r(x_0)/2.$$

where $S(x) = S_1(x_0) \cup S_2(x_0)$, $S_1(x_0)$ is the surface in Definition 5.8.5, and $S_2 = \{x = s + i x_{20} \mid s \in B(x_{10}, r)\}$. In case $m = 1$ and $-s_l > 0$, this has to be proved by removing a sphere $B(x_1, \varrho)$ as is done for φ_2 below. For x near x_0, we may find $D^2\varphi_1(x)$ by differentiating under the integral sign to get

$$\begin{aligned} D^2\varphi_1(x) &= \int_{S_1(x_0)} D\Delta(x-\xi)\,[D^{-s_l} f_l(\xi) - D^{-s_l} f_l(x)]\, d\xi + \\ &+ \int_{B_{2R}-B_R} D\Delta(x-\xi)\,[D^{-s_l} f_l(\xi) - D^{-s_l} f_l(x)]\, d\xi \end{aligned}$$

since, by replacing $S_1(x_0)$ by a surface near it but containing a thin annulus of points $\xi = s + i x_{20}$, $s \in B(x_{10}, r + \varepsilon) - B(x_0, r)$, we can prove that

$$\int_{S_1(x_0)} D_\alpha \Delta(x-\xi)\,d\xi + \int_{B_{2R}-B_R} D_\alpha \Delta(x-\xi)\,d\xi$$
$$= \int_{\partial B_{2R}} \Delta(x-\xi)\,d\xi'_\alpha - \int_{\partial B(x_{10}, r)} \Delta(x_1 - s)\,ds'_\alpha = 0, \quad x_1 \in B(x_{10}, r).$$

Using Theorem 2.6.5 and analytic continuation. Thus

$$|\nabla^2 \varphi_1(x_0)| \le Z(\mu, h, E)\cdot[r(x_0)]^{-1}. \tag{6.7.23}$$

Now, for points $x = x_1 + i x_{20}$, $x_1 \in B(x_{10}, r)$ we write

$$\varphi_{2\varrho}(x) = \int_{B(x_{10}, r)-B(x_1,\varrho)} \Gamma(x_1 - s) f(s + i x_{20})\,ds, \tag{6.7.24}$$
$$\Gamma(y) = D^{m-1} K(y), \quad f(\xi) = D^{-s_l} f_l(\xi).$$

Differentiating (6.7.24) with respect to x^α, integrating by parts, and letting $\varrho \to 0$, we obtain

$$D_\alpha \varphi_2(x) = -\int_{\partial B(x_{10}, r)} \Gamma(x_1 - s) f(s + i x_{20})\,ds\,\alpha' + \tag{6.7.25}$$
$$+ \int_{B(x_{10}, r)} \Gamma(x_1 - s) f_{,\alpha}(s + i x_{20})\,ds.$$

Since these are real integrals of a complex-valued function f, we may differentiate (6.7.25) using Theorems 2.6.2 and 2.6.5 to obtain

$$D_\beta D_\alpha \varphi_2(x) = -\int_{\partial B(x_{10}, r)} \Gamma_{,\beta}(x_1 - s)\,[f(s + i x_{20}) - f(x_1 - i x_{20})]\,ds'_\alpha +$$
$$+ C_\beta f_{,\alpha}(x_1 + i x_{20}) + \int_{B(x_{10}, r)} \Gamma_{,\beta}(x_1 - s)\,[f_{,\alpha}(s + i x_{20}) - f_{,\alpha}(x_1 + i x_{20})]\,ds$$
$$C_\beta = -\int_{\partial B(0,1)} \Gamma(-s)\,ds'_\beta$$

since $\Gamma_{,\beta}$ can be taken as a Δ in Theorem 2.6.5. From this and Lemma 5.8.3 and the fact that $\delta(x_0) \le r(x_0)$, we conclude that

$$|\nabla^2 \varphi_2(x)| \le Z(\mu, h, E)\cdot[\delta(x_0)]^{\mu-1}. \tag{6.7.26}$$

The result follows from (6.7.21), (6.7.22), (6.7.32), and (6.7.25).

Theorem 6.7.5. *Suppose $H \in {}^{**}C_\mu(\bar{B}_R)$ and satisfies (6.7.8) with $f = 0$. Then $H \in {}^{**}C_{h\mu}(\bar{B}_{hR})$ and*

$${}^{**}\|H\|_{hR} \le C(\mu, h, E)\cdot {}^{**}\|H\|_R.$$

Proof. We may define V to be an extension of H to B_{2R} to have support in B_{2R} and so that ${}^{**}\|V\|_R \le Z_1(\mu, \nu)\cdot {}^{**}\|H\|_R$. As in the proof of Lemma 6.6.1, we conclude that

$$H^l(x) = L_0^{lj} F_j(x), \quad F_j(x) = \int_{B_{2R}-B_R} K(x-\xi) f_j(\xi)\,d\xi, \tag{6.7.27}$$
$$f_j(x) = L_{jk}^0 V^k(x), \qquad x \in B_{2R}.$$

From (6.7.27), we immediately conclude that H is holomorphic on B_{hR}. Moreover

$$\nabla^{m+1-s_j} F_j(x) = \int_{B_{2R}-B_R} \nabla^{m+1} K(x-\xi)\, [\nabla^{-s_j} f_j(\xi) - \nabla^{-s_j} f_j(x)]\, d\xi$$

so that

$$\begin{aligned} |\nabla^{m+1-s_j} F(x)| &\leq Z_2(\mu, h, E) \cdot h_\mu(\nabla^{-s_j} f_j, \bar{B}_{2R}) \cdot [\delta(x)]^{\mu-1} \\ &\leq Z_3(\mu, h, E) \sum_k h_\mu(\nabla^{t_k} H, B_R) \cdot [\delta(x)]^{\mu-1}, \quad x \in B_{hR}. \end{aligned}$$

The result follows easily from this.

The proof of the following interior analyticity theorem is essentially like that of Theorem 5.8.6; one verifies the conclusions of Theorem 6.7.2 for the spaces $^{**}C_{h\mu}(\bar{B}_{hR})$.

Theorem 6.7.6. *Suppose $u \in {}^{**}C_\mu(B_R)$ for some $R > 0$ and satisfies (6.7.5) there where L^0_{jk}, M_{jk}, and the ψ_j have the properties as described above. Then u is analytic near $x = 0$.*

We now wish to prove analyticity along an analytic portion of the boundary of the solutions of a system of the form (6.7.1) which satisfy non-linear analytic boundary conditions of the form

$$\chi_r(X, Du) = 0, \quad r = 1, \ldots, m \tag{6.7.28}$$

where we now use the notation of § 6.3; the order of L_0 is now $2m$. We assume that the χ_r contain derivatives of u^k of order $\leq t_k - h_r$, where some of the h_r may be negative, and we assume that the linearized boundary conditions

$$\frac{\partial}{\partial \lambda} \chi_r(X, Du + \lambda Dv|_{\lambda=0} = 0 \tag{6.7.29}$$

satisfy the conditions of §§ 6.1 and 6.3. We assume that h_0 is the smallest integer such that

$$h_0 \geq 0, \quad h_0 + h_r \geq 1, \quad r = 1, \ldots, m, \tag{6.7.30}$$

and we assume that our solution u is such that $u^k \in C_u^{t_k+h_0}(\mathfrak{N})$ in a neighborhood on $\bar{G}$ of a point X_0 on ∂G, assumed analytic near X_0.

We begin by making an analytic change of independent variables which carries x_0 into the origin, $\mathfrak{N}$ onto $G_{R_0} \cup \sigma_{R_0}$, and $\partial G \cap \mathfrak{N}$ onto σ_{R_0}; we then use the (x, y) notation of § 6.3. In the resulting equations (6.7.1) and (6.7.28), we make the substitutions (6.7.3) where P^k is that polynomial of degree $t_k + h_0$ such that (6.7.4) holds for all r, $0 \leq r \leq t_k + h_0$. If we then expand the new φ_j and χ_r about the origin in the (x, y, p) space, we obtain the system (6.7.5) and the boundary conditions

$$B_{0rk} u^k = C_{0rk} u^k + \omega_r(x, 0, Du) \tag{6.7.31}$$

where B_{0rk} has constant coefficients and contains all and only derivatives $D^\alpha u^k$ of order $t_k - h_r$, C_{0rk} has constant coefficients but is of lower order, and the Taylor expansion of ω_r begins with a homogeneous polynomial in (x, y) of degree $h_0 + h_r + 1$, linear terms in the $D^\alpha u^k$ with coefficients homogeneous in (x, y) of degree ≥ 1, and second powers of the derivatives; the expansion of ψ_j has its previous form except that the homogeneous polynomial in (x, y) has degree $h_0 + 1 - s_j$.

The following lemma can be proved by the method of proof of Lemma 6.3.3:

Lemma 6.7.1. *There exist bounded extension operators τ_{1jR} from $C^j_\mu(\bar{G}_R)$ to $C^j_\mu(\bar{G}_{2R})$, τ_{2jR} from $C^j_\mu(\bar{B}_R)$ to $C^j_\mu(\bar{B}_{2R})$, τ_{3jR} from $C^j_\mu(\bar{G}_R)$ to $C^j_\mu(\bar{B}_R)$, and τ_{4jR} from $C^j_\mu(\bar{\sigma}_R)$ to $C^j_\mu(\bar{\sigma}_{2R})$, with bounds independent of R if the norms are defined using h_μ, and which are such that $\tau_{1jR}(f)$ vanishes near $\sum^+_{2R}$ and $\tau_{2jR}(f)$ vanishes near ∂B_{2R}. Thus*

$$\tau_{2j,2R}[\tau_{3jR}(f)] \in C^j_{\mu c}(\bar{B}_{2R}) \quad \text{if} \quad f \in C^j_\mu(\bar{G}_R).$$

The construction in Lemma 6.3.3 is to be replaced by

$$F(x, y) = \sum_{s=1}^{j+1} C^j_s F\left(x, -\frac{s y}{j+1}\right), \quad y < 0 \tag{6.7.32}$$

to yield τ_{3jR}.

Definition 6.7.1'. We define the space $*C_\mu(\bar{G}_R)$ as the space of vectors f such that $f_j \in C^{h_0 - s_j}_\mu(\bar{G}_R)$ with $\nabla^\varrho f_j(0, 0) = 0$ for $0 \leq \varrho \leq h_0 - s_j$, the norm $*\|f\|_R$ being $\sum_j h_\mu(\nabla^{h_0 - s_j} f_j, \bar{G}_R)$. We define the space $*'C_\mu(\bar{\sigma}_R)$ as the space of vectors g such that $g_r \in C^{h_0+h_r}_\mu(\bar{\sigma}_R)$ with $\nabla^\varrho g_r(0) = 0$ for $0 \leq \varrho \leq h_0 + h_r$ and norm $*'\|g\| = \sum_r h_\mu(\nabla^{h_0+h_r} g_r, \bar{\sigma}_R)$. We define the space $**C^\mu(\bar{G}_R)$ as the space of vectors u such that $u^k \in C^{t_k+h_0}_\mu(\bar{G}_R)$, $\nabla^\varrho u^k(0,0) = 0$ for $0 \leq \varrho \leq t_k + h_0$, and norm $**\|u\| = \sum_k h_\mu(\nabla^{t_k+h_0} u^k, \bar{G}_R)$. For $f \in *C_\mu(\bar{G}_R)$ and $g \in *'C(\bar{\sigma}_R)$, we define $P_R(f, g)$ as follows: Let

$$\tilde{f}_j = \tau_{2j,2R}[\tau_{3jR} f_j], \quad \tilde{g}_r = \tau_{4\sigma_r R} g_r, \quad \sigma_r = h_0 + h_r.$$

We then define $P_R(f, g) = u$, where

$$u^k = U^k + V^k - W^k - P^k$$

where U^k is given in terms $\tilde{g}$ by (6.3.29), V^k is given by (6.3.43), W^k is defined by (6.3.29) with g_r replaced by

$$\tilde{\gamma}_r(x) = B_{0rk} V^k(x, 0), \text{ (all } x);$$

we assume, of course, that $\tilde{g}_r(x) = 0$ outside σ_{2R} and, as usual, P^k is chosen so that $\nabla^\varrho u^k(0,0) = 0$ for $0 \leq \varrho \leq t_k + h_0$.

Since $P_R(f, g)$ as we have defined it differs from the $P_R(\tilde{f}, \tilde{g})$ of Definition 6.3.5 by a polynomial of degree $\leq t_k + h_0$, the following theorem follows from our normalization at the origin and Theorem 6.3.6:

Theorem 6.7.1'. *P_R is a bounded operator from $*C_\mu(\bar{G}_R) \times *'C_\mu(\bar{\sigma}_R)$ into $**C_\mu(\bar{G}_R)$ with bound depending only on (μ, E). Moreover, if $U = P_R(f, g)$ then U satisfies (6.7.8) with f replaced by $\tilde{f}$ everywhere on $R^+_{\nu+1}$*

and also satisfies

(6.7.33) $$B_{0rk} U^k(x, 0) = \tilde{g}_r(x), \quad \text{all} \quad x.$$

Now we assume that u is a solution of (6.7.5) on $\bar{G}_R$ and (6.7.31) on σ_R and that $u \in {}^{**}C_\mu(\bar{G}_{R_0})$, $R_0 > 0$. For $0 < R \leq R_0$, we write

$$u = u_R + H_R, \quad u_R = P_R[M_0 u + \psi(x, y, D u), \ C_0 u + \omega(x, D u)].$$

If we define

$$v_R = P_R[M_0 H_R + \psi(x, y, D H_R), \ C_0 H_R + \omega(x, D H_R)]$$

(6.7.34) $$T_R(u_R, H_R) = P_R[M_0 u_R + \psi(x, y, D u_R + D H_R) - \\ - \psi(x, y, D H_R), \ C_0 u_R + \omega(x, D u_R + D H_R) - \omega(x, D H_R)]$$

then the equation for u_R reduces to (6.7.15). An analysis like that in the proof of Theorem 5.8.2 demonstrates the following theorem:

Theorem 6.7.2'. *The conclusions of Theorem* 6.7.2 *hold with* B_R *replaced by* G_R. *In addition* H_R *satisfies* (6.7.33) *with* $\tilde{g}_r = 0$ *on* σ_R.

As in § 5.8, we now introduce certain spaces of functions which are analytic in the x-variables only.

Definition 6.7.3'. We define B_{0hR} (as in § 5.8) to be the part of B_{hR} where y is real, G_{hR} as the part of B_{0hR} where $y > 0$, and $G_{\bar{h}R}^-$ the part of B_{0hR} where $y < 0$. The *space* ${}^*C_{h\mu}(\bar{G}_{hR})$ consists of all vectors $f \in {}^*C_\mu(\bar{G}_R)$ which can be extended to $\bar{G}_{hR}$ so that $f_j \in C_\mu^{h_0 - s_j}(\bar{G}_{hR})$ and f_j is analytic in x for all (x, y) in G_{hR}. The *space* ${}^{*\prime}C_{h\mu}(\bar{G}_{hR})$ consists of all $g \in {}^*C_\mu(\bar{\sigma}_R)$ such that g_r can be extended to $\bar{\sigma}_{hR}$ so that $g_r \in C_\mu^{h_0+h_r}(\bar{\sigma}_{hR})$ and g_r is analytic in σ_{hR}; here σ_{hR} is the part of B_{0hR} where $y = 0$. The *space* ${}^{**}C_{h\mu}(\bar{G}_{hR})$ consists of all $u \in {}^{**}C_\mu(\bar{G}_R)$ which can be extended to $\bar{G}_{hR}$ so that $u^k \in C_\mu^{t_k+h_0}(\bar{G}_{hR})$ and u^k is analytic in x for (x, y) in G_{hR}. We define the norms as usual:

$$^*\|f\|_{hR} = \sum_j h_\mu[\nabla^{h_0 - s_j} f_j, \bar{G}_{hR}], \quad \text{etc.}$$

We wish now to extend the definition of the operator $P_R(f, g)$ to the space ${}^*C_{h\mu}(\bar{G}_{hR}) \times {}^{*\prime}C_{h\mu}(\bar{G}_{hR})$ and wish to show then that P_R is a bounded operator from that space to ${}^{**}C_{h\mu}(\bar{G}_{hR})$. We consider the separate terms in the definition 6.7.1'.

Theorem 6.7.3' (a). *Suppose* $g \in {}^{*\prime}C_{h\mu}(\bar{G}_{hR})$, $\tilde{g}_r = \tau_{4s_rR}\, g_r$ *where* $s_r = h_0 + h_r$, *and we define*

$$u^k = U^k - P^k, \quad U^k(x, y) = \int_{\sigma_{2R}} \sum_r \sum_{|\alpha| = s_r} C_\alpha L^{kr}(x - \xi, y) D^\alpha \tilde{g}_r(\xi, 0)\, d\xi$$

where $C_\alpha = |\alpha|!/\alpha!$ *and* P^k *is the usual polynomial. Then* $u \in {}^{**}C_{h\mu}(\bar{G}_{hR})$, *satisfies the homogeneous equations* (6.7.8) on $R_{\nu+1}^+$, *satisfies the boundary conditions* (6.7.33) for all real x and *all complex* x *on* $\bar{\sigma}_{hR}$, *and finally*

(6.7.35) $$^{**}\|u\|_h \leq C(\mu, h, E) \cdot {}^{*\prime}\|g\| \quad \textit{provided} \quad 0 < h \leq h_2(E) \leq h_1(E).$$

Proof. From Theorem 6.3.2, we conclude that the $L_\alpha^{kr}(x, y)$ are analytic for all $(x, y) \neq 0$ such that y is real and ≥ 0 and $x = x_1 + i x_2$ is complex with $|x_2| \leq h\sqrt{|y|^2 + |x_1|^2}$ if $0 < h \leq h_2(E)$. For such h, we extend g_r to $\bar{\sigma}_{hR}$ and then extend U^k to G_{hR} by the formula

$$U^k(x, y) = \int_{S(x,y)} \mathfrak{F}^k(x, y; \xi)\, d\xi + \int_{\sigma_{2R}-\sigma_R} \mathfrak{F}^k(x, y; \xi)\, d\xi,$$

$$\mathfrak{F}^k(x, y; \xi) = \sum_r \sum_{|\alpha|=s_r} C_\alpha L_\alpha^{kr}(x - \xi, y)\, D^\alpha g_r(\xi), \quad D_\alpha = |\alpha|!/\alpha!$$

where

$$S(x, y) : \xi = s + i\,\xi_2(s), \quad s \in \sigma_R,$$

is admissible. We have seen in § 5.8 that such integrals are independent of the path $S(x, y)$ as long as it is admissible so that U^k is analytic. Near any point (x_0, y_0), we may choose a fixed surface containing a disc $\xi = s + i\,x_{20}$, $|s - x_{10}| \leq \varrho$, independent of (x, y). We may calculate the derivatives of U^k by differentiating in the real directions, so that we would find that a derivative $D_x D^{t_k+h_0} U^k(x, y)$ is a sum of terms of the form

$$C\left[\int_{S(x,y)} D_x \Gamma(x - \xi, y)\, \gamma(\xi)\, d\xi + \int_{\sigma-\sigma_R} D_x \Gamma(x - \xi, y)\, \gamma(\xi)\, d\xi\right]$$

where $\gamma(\xi) = D^{s_r} g_r$, $\Gamma(x, y) = D^{t_k+h_0} L_\alpha^{kr}(x, y)$ so that Γ is homogeneous of degree $-\nu$. Now, if $y > 0$, we find, as in the proof of Theorem 6.3.3 and by using analytic continuation, that

$$\int_{S(x,y)} D_x \Gamma(x - \xi, y)\, d\xi + \int_{\sigma-\sigma_R} D_x \Gamma(x - \xi, y)\, d\xi = 0, \quad (x, y) \in G_{hR}$$

so that we may replace $\gamma(\xi)$ above by $\gamma(\xi) - \gamma(x)$ and thus obtain a bound like (6.3.23) for all such derivatives. Using the differential equations we get a similar bound for $D_y^{t_k+h_0+1}$ and hence conclude the inequality (6.7.35) and thus the theorem.

Theorem 6.7.3′ (b). *Suppose $f \in {}^*C_{h\mu}(\bar{G}_R)$ and we define V^k by (6.3.43) where $\tilde{f}_j = \tau_{2\lambda_j 2R}\, \tau_{3\lambda_j 2R} f_j$, $\lambda_j = h_0 - s_j$. Then $V^k \in C_\mu^{t_k+h_0}(\bar{B}_A)$ for any A, V^k is analytic in x for $(x, y) \in B_{0hR}$, $V^k \in C_\mu^{t_k+h_0}(\bar{B}_{0hR})$, $h_\mu(\nabla^{t_k+h_0} V^k, \bar{B}_{0hR}) \leq C(\mu, h, E)\, {}^*\|f\|_h$, and $\tilde{\gamma}_r(X) \equiv B_{0rk} V^k(X)$ satisfies the conditions at ∞ in Theorem 6.3.3 and the analyticy conditions in Theorem 6.7.3′(a). Thus $W^k \in C_\mu^{t_k+h_0}(\bar{G}_A)$ for any A, W^k it is analytic in x for $(x, y) \in G_{hR}$, $W^k \in C_\mu^{t_k+h_0}(\bar{G}_A)$, and $h_\mu(\nabla^{t_k+h_0} W^k, \bar{G}_{0hR}) \leq C(\mu, h, E)\, {}^*\|f\|_h$. If $v^k = V^k - W^k - P^k$, P^k being the usual normalizing polynomial, then v satisfies (6.7.8) with f_j replaced by $\tilde{f}_j$ on $R_{\nu+1}^+ \cup G_{hR}$ and satisfies (6.7.33) with $\tilde{g}_r = 0$ on $\sigma \cup \sigma_{hR}$.*

Proof. The results for real (x, y) follow from Theorem 6.3.6. We may extend $\tilde{f}_j$ to $R_{\nu+1} \cup \bar{B}_{0hR}$ by defining $\tilde{f}_j$ as above on B_{2R}, defining $\tilde{f}_j(x, y) = 0$ on $R_{\nu+1} - B_{2R}$, extending f_j to $\bar{G}_{hR}$ as in the definition, and then extending $\tilde{f}_j$ to $\bar{B}_{0hR}$ using the formulas (6.7.32). It is clear that the extended $\tilde{f}_j$ is analytic in x for (x, y) on B_{0hR}. We may then extend F_j to $R_{\nu+1} \cup B_{0hR}$ by defining

$$F_j(x, y) = \int_{S(x,y)} \mathfrak{F}_j(x, y; \xi, \eta)\, d\xi\, d\eta + \int_{B_{2R}-B_R} \mathfrak{F}_j(x, y; \xi, \eta)\, d\xi\, d\eta,$$

$$\mathfrak{F}_j(x, y; \xi, \eta) = K(x - \xi, y - \eta)\, f_j(\xi, \eta), \quad (x, y) \in \bar{B}_{0hR},$$

where

$$\text{(6.7.36)} \quad S(x, y) : \xi = s + i\,\xi_2(s, t; x, y), \ \eta = t \text{ (real)}, \ (s, t) \in B_R$$

is admissible for $\mathfrak{F}_j$. An analysis like that in the proof of Theorem 5.8.3 shows that F_j is analytic in x for $(x, y) \in B_{0hR}$ and that we may differentiate F_j $2m + h_0 - s_j - 1$ times, shifting $h_0 - s_j$ of the differentiations onto the $\tilde{f}_j$. The result is

$$\text{(6.7.37)} \quad \varphi(x, y) = \int_{S(x,y)} \Gamma(x - \xi, y - \eta)\tilde{f}(\xi, \eta)\, d\xi\, d\eta + \int_{B_{2R}-B_R} \Gamma(X - \Xi)\tilde{f}(\Xi)\, d\Xi$$

where $\tilde{f} \in C^0_\mu(\bar{B}_{2R} \cup \bar{B}_{0hR})$ and Γ is homogeneous of degree $-\nu$.

Near a point (x_0, y_0) in B_{0hR}, we may take $S(x, y)$ as the union of a fixed surface S_1, given by (6.7.36) with $(s, t) \in B_R - B(x_{10}, y_0; r)$ and the disc $\xi = s + i\,x_{20}$, $\eta = t$ with $(s, t) \in B(x_{10}, y_0; r)$. Then, for points (x, y) with $x = x_1 + i\,x_{20}$ and $(x_1, y) \in B(x_{10}, y_0; r)$, (6.7.37) becomes

$$\varphi(x, y) = \varphi_1(x, y) + \varphi_2(x, y),$$

$$\text{(6.7.38)} \quad \varphi_1(X) = \int_{S_1} \Gamma(X - \Xi)\tilde{f}(\Xi)\, d\Xi + \int_{B_{2R}-B_R} \Gamma(X - \Xi)\tilde{f}(\Xi)\, d\Xi,$$

$$\varphi_2(x, y) = \int_{B(x_{10}, y_0; r)} \Gamma(x_1 - s, y - t)\tilde{f}(s + i\,x_{20}, t)\, ds\, dt.$$

An analysis exactly like that applied to the φ_1 in the proof of Theorem 6.7.4 shows that

$$\text{(6.7.39)} \qquad |\nabla^2 \varphi_1(X)| \le Z(\mu, h, E) \cdot h_\mu(f, \bar{B}_{0hR}) \cdot [\delta(X)]^{\mu-1}$$

$\delta(X)$ being the distance from X to ∂B_{0hR} where B_{0hR} is considered as a set in $R_{2\nu+1}$. We may differentiate φ_2 with respect to an x^α obtaining the formula analogous to (6.7.28). This may then be differentiated with respect to y or an x^β so that we obtain a bound (6.7.39) for $|\nabla_x \nabla \varphi_1(X)|$. It then follows that any derivative $D^{t_k+h_0-s_j-1} \in C^0_\mu(\bar{B}_{0hR})$ with the desired bound. The results for V then follow from the differential equations. The proof of the results for W differs from the proof of Theorem 6.7.3(a) only in the consideration of the convergence at ∞, since $\tilde{\gamma}$ does not have compact support.

Thus we obtain our first desired result.

Theorem 6.7.4′. *P_R is a bounded operator from $^*C_{h\mu}(\bar{G}_{hR}) \times {}^{*\prime}(C_{h\mu}\bar{\sigma}_{hR})$ into $^{**}C_{h\mu}(\bar{G}_{hR})$ with bound $C(\mu, h, E)$.*

Theorem 6.7.5'. *Suppose* $H \in {}^{**}C_\mu(\bar{G}_R)$, *satisfies* (6.7.8) *with the* $f_j = 0$ *on* G_R, *and satisfies* (6.7.33) *with* $\tilde{g}_r = 0$ *on* σ_R. *Then* $H \in {}^{**}C_{h\mu}(\bar{G}_{hR})$ *and*

$$(6.7.40) \qquad {}^{**}\|H\|_{hR} \leq C(\mu, h, E)\, {}^{**}\|H\|_R \text{ if } 0 < h \leq h_2 \leq h_1,$$

h_2 *being the number in Theorem* 6.7.3'(a).

Proof. Let $\tilde{H}^k = \tau_{1,\, t_k + h_0,\, R} H^k$ and define $\tilde{H}^k(x, y) = 0$ elsewhere on $R^+_{\nu+1}$ and define

$$(6.7.41) \qquad f_j(x, y) = L_{0jk}\, \tilde{H}^k, \quad \tilde{g}_r(x) = B_{0rk}\, \tilde{H}^k(x, 0).$$

Then $\tilde{g}_r = 0$ on σ_R and on $\sigma - \sigma_{2R}$ and $f_j = 0$ on G_R and on $R^+_{\nu+1} - G_{2R}$. We now define $\tilde{f} = \tau' f$ where φ is defined as usual ($= 1$ for $s \leq 1$, $= 0$ for $s \geq 7/4$, etc.)

$$f_j(X) = \varphi(R^{-1}|X|)\, f'_j(X)$$

$$f'_j(x, y) = \sum_{t=1}^{\tau_j+1} C'_t\, f_j[x, -(\tau_j + 1)^{-1} t\, y]$$

where the C'_j are the unique constants such that

$$\sum_{t=1}^{\tau_j+1} (-1)^l (\tau_j + 1)^{-l} t^l C'_t = 1, \quad l = 0, \ldots, \tau_j = h_0 - s_j.$$

Then τ' is a bounded extension operator and we see that $\tilde{f}_j(X) = 0$ for $X \in B_R$ and in $R_{\nu+1} - B_{2R}$. Then we define V by (6.3.43) and then define W as indicated immediately below and define U by (6.3.29) in terms of g. Then U, V, and W have the requisite differentiability bounds everywhere, V is analytic in B_{hR}, U is analytic in $G_{Rh} \cup \sigma_{Rh}$, if h is small enough, and the $\tilde{\gamma}_r$ are analytic in B_{hR}. From a minor extension of Theorem 6.7.3'(a), it follows that W is analytic in x on $G_{Rh} \cup \sigma_{Rh}$. Finally, from Theorem 6.3.6, it follows that $u = U + V - W$ satisfies (6.3.44) with $f^* = g^* = 0$ since f and g are 0 near 0 (g^*_r could be a polynomial vanishing on σ). From Theorem 6.3.5, it follows that $\tilde{H} - u$ is a polynomial of low degree. The results follow.

Theorem 6.7.6'. *Suppose* $u \in {}^{**}C_\mu(\bar{G}_{R0})$ *and satisfies* (6.7.5) *on* G_R *and* (6.7.31) *along* σ_R, *where the* ψ_j *and* ω_r *satisfy the analyticity conditions specified near the respective equations. Then* u *is analytic at the origin.*

Proof. Suppose $0 < h \leq h_2(E) \leq h_1(E)$. From Theorem 6.7.2' and 6.7.5', it follows that ${}^{**}\|H_R\|_{h,R} \leq M_3$ *for* $0 < R \leq R_2$. The theorems above show that the conclusions of Theorem 6.7.2' hold for the spaces ${}^{**}C_{h\mu}(\bar{G}_{hR})$. Thus if $0 < R \leq R_2 \leq R_1$, u_R is the only solution of (6.7.15) with ${}^{**}\|u_R\|_h \leq M_3$. Thus the previous u_R must $\in {}^{**}C_{h\mu}(\bar{G}_{hR})$. Thus, this is true of u also. The ellipticity shows that we may, after differentiating the $j-th$ equation in (6.7.5) $-s_j$ times, solve for $D_y^{t_k} u^k$ in terms of the other derivatives. Introducing the various derivatives of the u^k as new variables w^j, we see that the $w^j(x, y)$ are analytic in x for (x, y) on G_{hR} and satisfy a system of the form (5.8.49). The analyticity at $(0, 0)$ follows from our previous discussion of that system in the proof of Theorem 5.8.6'.

6.8. The differentiability of the solutions of non-linear elliptic systems; weak solutions; a perturbation theorem

In this section we first obtain results concerning the differentiability of the solutions or weak solutions of non-linear elliptic systems which satisfy general non-linear boundary conditions. We assume that the equations of variation, as defined in (6.7.2), are properly elliptic along the solution and the linearized boundary conditions, as defined in (6.7.29) satisfy the complementing condition. We make the corresponding assumptions regarding the non-linear systems analogous to those in § 6.4. We conclude the section with a perturbation theorem for such non-linear boundary value problems.

We shall obtain our results on the interior first and shall make use of the spaces $*H^h_{p0}(B_R)$, etc., defined in Definition 6.2.1, on balls B_R interior to our domain.

Theorem 6.8.1. *Suppose that $u^k \in C^{t_k}(G)$ and satisfies the equations*

$$\varphi_j(x, D u) = 0, \quad j = 1 - \ldots, N, \tag{6.8.1}$$

on G, where $\varphi_j(x, p) \in C^{h-s_j}(\mathfrak{N})$, $h \geq 1$, and $\mathfrak{N}$ is an open set in (x, p)-space containing all the points $(x, D u)$; we assume that the equations of variation are properly elliptic. Then $u^k \in H_p^{t_k+h}(D)$ for each $p > 1$ and each $D \subset\subset G$ and the derivatives $D^\delta u^k$, $|\delta| \leq h$, satisfy the corresponding differentiated equations (almost everywhere if $|\delta| = h$). If, also, the $\varphi_j \in C_\mu^{h-s_j}(\mathfrak{N})$ then the $u^k \in C_\mu^{t_k+h}(D)$ for each $D \subset\subset G$. If the φ_j are of class C^∞ (analytic) on $\mathfrak{N}$, then the u^k are of class C^∞ (analytic) on G.

Proof. We first prove the first statement for $h = 1$. Let x_0 be any point of G and let $\overline{B(x_0, A)} \subset G$; we hereafter assume that $x_0 = 0$. Let σ_0 be a positive number $< A/2$, let γ be a positive integer $\leq \nu$, let e_γ be the unit vector along the x^γ axis, and define

$$u^k_\sigma(x) = \sigma^{-1}[u^k(x + \sigma e_\gamma) - u^k(x)], \quad |\sigma| < \sigma_0.$$

Then the u^k_σ satisfy the equations

$$\sum_{k=1}^{N} \sum_{|\beta| \leq s_j + t_k} a^{\sigma\beta}_{jk}(x)\, D^\beta u^k_\sigma(x) = f_{j\sigma}(x), \quad (\sigma \text{ not summed})$$

$$a^{\sigma\beta}_{jk}(x) = \int_0^1 \varphi_{j p^k_\beta}\{x + \sigma t e_\gamma, p(x) + t[p(x + \sigma e_\gamma) - p(x)]\}\, dt \tag{6.8.2}$$

$$f_{j\sigma}(x) = -\int_0^1 \varphi_{j x^\gamma}\{\text{same}\}\, dt, \quad (p = \{p^k\}).$$

Clearly $a_j^\sigma \to a_j (a^\sigma = \{a_{jk}^{\sigma\beta}\}$, etc.) and $f_{\sigma j} \to f_j$ in C^{-s_j}, where

$$a_{jk}^\beta(x) = \varphi_{j p_\beta^k}[x, Du(x)], \quad f_j(x) = -\varphi_{j x^\gamma}. \tag{6.8.3}$$

So the equations (6.8.2) are properly elliptic and the constants C_2 and R_2 in Theorem 6.2.6 are independent of σ. So, choose $R \leq \min(R_2, A/2)$ and let $\zeta_R \in C_c^\infty(\bar{B}_R)$ with $\zeta_R = 1$ on $B_{R/2}$ and define $U_\sigma^k = \zeta\, u_\sigma^k$. Multiplying (6.8.2) by ζ, we see that U_σ satisfies

$$\sum_k L_{jk}^\sigma U_\sigma^k = F_{j\sigma} = \zeta_R f_{j\sigma} + \sum_k M_{jk}^\sigma u_\sigma^k \tag{6.8.4}$$

where the operators M_{jk}^σ (which involve the derivatives of ζ_R) are of order $\leq s_j + t_k - 1$ and the L_{jk}^σ have the $a_{jk}^{\alpha\beta}$ as coefficients.

It follows that the $F_{j\sigma}$ converge in C^{-s_j} to certain F_j which are expressible in terms of x and the $D^\beta u^k$, $|\beta| \leq t_k$. From Theorem 6.2.6 it follows that the norms of U_σ are uniformly bounded in $^{**}H_p^0(B_R)$ for any given p, if R is small enough. Thus a subsequence of values of $\sigma \to 0$ exists such that $U_\sigma \rightharpoonup$ in $^{**}H_p^0(B_R)$ to some U. By writing (6.8.4) in the form

$$\sum_k L_{jk} U_\sigma^k = F_{j\sigma} - G_{j\sigma}, \quad G_{j\sigma} = \sum_k (L_{jk}^\sigma - L_{jk})\, U_\sigma^k \tag{6.8.5}$$

and noting that $G_{j\sigma} \to 0$ in $H_{p0}^{-s_j}(B_R)$, we see that $U_\sigma \to U$ in $^{**}H_p^0(B_R)$ and that U satisfies the limiting equations. Thus $u_{,\gamma}^k \in H_p^{t_k}(B_{R/2})$ and satisfies the differentiated equations on $B_{R/2}$. Since x_0 was arbitrary the first result follows for $h = 1$.

The remaining results follow from repeated applications of Theorem 6.2.5 since the $u_{,\gamma}^k$ satisfy.

$$L_{jk} u_{,\gamma}^k(x) \equiv \sum_{k,\beta} a_{jk}^\beta(x)\, D^\beta u_{,\gamma}^k = f_{j\gamma}(x) \equiv -\varphi_{j x^\gamma}(x, D u). \tag{6.8.6}$$

Since $u^k \in H^{t_k+1}(D)$ for each $p > 1$, it follows that $u^k \in C_{\mu'}^{t_k}(D)$ for each μ', $0 < \mu' < 1$. Thus, if the $\varphi_j \in C_\mu^{1-s_j}(\mathfrak{N})$ for some μ, $0 < \mu < 1$, it follows first the a_{jk}^β and f_j in (6.8.3) $\in C_{\mu\mu'}^{-s_j}$ so that $u^k \in C_{\mu\mu'}^{t_k+1}$. Then the a' s and f' s $\in C_\mu^{-s_j}$ so that $u^k \in C_\mu^{t_k+1}$. If, now the $\varphi_j \in C_\mu^{2-s_j}$, it will follow that the a' s and f' s $\in C^{1-s_j}$ so that $u^k \in H_p^{t_k+2}$ for each $p > 1$. If $\varphi_j \in C_\mu^{2-s}$, it follows as before that $u^k \in C_\mu^{t_k+2}$. The argument may be repeated.

The following theorem is proved in a similar way using Theorem 6.3.8 for boundary neighborhoods.

Theorem 6.8.2. *Suppose that h_0 is the smallest integer satisfying the conditions*

$$h_0 \geq 0, \quad h_0 + h_r \geq 1 \quad \textit{for each } r.$$

Suppose $u^k \in C^{t_k+h_0}(\bar{G})$ and the u^k satisfy (6.8.1) *on G and*

$$\chi_r(x, D u) = 0 \quad \textit{on } \partial G. \tag{6.8.7}$$

Suppose, for some $h \geq h_0$, that G is bounded and of class C^{t_0+h}, $t_0 = \max t_k$, $\varphi_j \in C^{h-s_j}(\mathfrak{N})$, and $\chi_r \in C^{h+h_r}(\mathfrak{N}')$, where $\mathfrak{N}$ and $\mathfrak{N}'$ are appropriate *neighborhoods in the* (x, p)*-space containing all the points* $[x, D\,u(x)]$ *for* $x \in \bar{G}$. *We assume that the linearized equations are properly elliptic and the linearized boundary conditions satisfy the corresponding complementing conditions on* ∂G. *Then* $u^k \in H_p^{t_k+h}(G)$ *for each* $p > 1$ *and the* $D^\delta u^k$ *satisfy the differentiated equations on* G. *If, also, the* $\varphi_j \in C_\mu^{h-s_j}(\mathfrak{N})$ *and* $\chi_r \in C_\mu^{h+h_r}(\mathfrak{N})$, *then* $u^k \in C_\mu^{t_k+h}(\bar{G})$. *Corresponding results hold in the* C^∞ *and analytic cases. If* $h_0 \geq 1$, G *is of class* $C_\mu^{t_0+h_0}$, $u^k \in C_\mu^{t_k+h_0-1}(\bar{G})$, $\varphi_j \in C_\mu^{h_0-s_j}(\mathfrak{N})$, *and the* $\chi_r \in C_\mu^{h_0+h_r}(\mathfrak{N})$, *then* $u^k \in C_\mu^{t_k+h_0}(\bar{G})$.

Proof. We prove the first conclusion for $h = 1$. The interior results have been proved in Theorem 6.8.1. So, let x_0 be a boundary point and map a neighborhood of x_0 onto G_A in the usual way. Let e_γ be the unit vector in the (new) x^γ direction, assumed tangential. If we define the u_σ^k as before, we see that they satisfy (6.8.2) and the boundary conditions

$$B_{rk}^\sigma u_\sigma^k \equiv \sum_{k,\beta} b_{rk}^{\sigma\beta}(x)\, D^\beta u_\sigma^k = g_r^\sigma, \quad x \in \sigma_{A/2}$$

where $b_{rk}^{\sigma\beta}$ and g_r^σ are defined in terms of the $\chi_{r\,p_\beta^k}$ and $-\chi_{r\,x^\gamma}$ as in (6.8.2). From Theorem 6.3.8 and the convergence in $C^{h_0+h_r}$ of b_r^σ and g_r^σ to b_r and g_r, respectively, we conclude that we may find an $R_2 > 0$ such that the bound in (6.3.49) holds on G_R, independently of σ if $0 < R \leq R_2$. Then we define ζ_R and U_σ^k as before, we may repeat the argument of the preceding proof to show that $U_\sigma^k \to u_{,\gamma}^k$ in $H^{t_k+h_0}(G_R)$. Using the differential equations, we find that the normal derivative $u_{,\nu}^k$ also $H_p^{t_k+h_0}(G_R)$. The further differentiability results are obtained by repeated application of Theorem 6.3.7. The last statement is proved in the same way.

The equations analogous to (6.4.1) are of the form

$$(6.8.8) \qquad \int_G \sum_{j=1}^N \sum_{|\alpha| \leq m_j} D^\alpha v^j \cdot \varphi_j^\alpha(x, D\,u)\, dx = 0, \quad v \in C_c^\infty(G)$$

where we shall assume that s_j, t_k, m_j, h_0, and ϱ_{jk} satisfy the conditions in (6.4.1), (6.4.2), (6.4.3), and (6.4.4) and φ_j^α involves derivatives $D^\beta u^k$ where $|\beta| \leq \varrho_{jk}$. We assume that each φ_j^α with $|\alpha| = m_j \in C^1$ at least and that the system of operators in (6.4.6) is properly elliptic for each set of numbers

$$(6.8.9) \qquad a_{jk0}^{\alpha\beta} = \varphi_{j p_\beta^k}^\alpha(x_0, p_0), \quad |\alpha| = m_j, \quad |\beta| = \varrho_{jk}$$

for (x_0, p_0) in a neighborhood $\mathfrak{N}$ of all points $[x, D\,u(x)]$. The theorem analogous to Theorem 6.8.1 may be stated as follows:

Theorem 6.8.3. *Suppose that* s_j, t_k, m_j, ϱ_{jk}, h_0, *and the* φ_j^α *satisfy the conditions stated above and suppose* $u^k \in C^{t_k+h_0}(G)$ *and satisfies equations* (6.8.8) *on* G. *Suppose, for some* $h \geq h_0 + 1$, *that the* $\varphi_j^\alpha \in C^\tau(\mathfrak{N})$ *where* $\tau = \max(0, h - s_j + |\alpha|)$ *and* $\mathfrak{N}$ *has its usual significance. Then* $u^k \in H_p^{t_k+h}(D)$ *for each* $D \subset\subset G$. *If the* $\varphi_j^\alpha \in C_\mu^\tau(\mathfrak{N})$, $0 < \mu < 1$, *then* $u^k \in C_\mu^{t_k+h}(D)$ *for each* $D \subset\subset G$.

Sketch of proof. We prove the first statement for $h = h_0 + 1$. The second statement follows from the theory of § 6.4. The results for larger h follow by repeating the argument.

We begin by replacing v^j in (6.8.8) by $v^j_\sigma = \sigma^{-1}[v^j(x - \sigma e_\gamma) - v^j(x)]$, e_γ having its usual significance; we assume that $v \in C^\infty_c(G)$ and that $|\sigma|$ is sufficiently small. By proceeding as in § 1.11, we see that the u^k_σ satisfy equations of the form (6.4.1) where the $a^{\sigma\alpha\beta}_{jk}$ and $f^{\sigma\alpha}_j$ are defined by formulas like (6.8.2) for those (j, α) for which $\varphi^\alpha_j \in C^\tau(\mathfrak{N})$ for some $\tau \geq 1$, namely those for which $h_0 - s_j + |\alpha| \geq 0$ (for each j, there is at least one such α since $h_0 - s_j + m_j \geq 0$). For the remaining (j, α), for which the φ^α_j are not differentiable, we write

$$\int_G D^\alpha v^j_\sigma(x)\cdot\varphi^\alpha_j[x, u(x)]\,dx = \int_G \varphi^\alpha_j[x, u(x)]\cdot\left\{-\int_0^1 D^\alpha v^j_{,\gamma}(x - t\sigma e_\gamma)\,dt\right\}dx$$

$$= \int_G D^\alpha v^j_{,\gamma}(x)[-f^{\sigma\alpha}_j(x)]\,dx,\quad f^{\sigma\alpha}_j(x) = \int_0^1 \varphi^\alpha_j[x + t\sigma e_\gamma, u(x + t\sigma e_\gamma)]\,dt.$$

(6.8.10)

Since any φ^α_j with $|\alpha| = m_j$ is differentiable, we see that the u^k_σ satisfy equations of the form (6.4.1) where the $a^{\sigma\alpha\beta}_{jk} = 0$ and the $f^{\sigma\alpha}_j$ are given in (6.8.10) if $h_0 - s_j + |\alpha| < 0$. The usual argument can be carried through using Theorem 6.4.4.

The boundary conditions analogous to (6.4.15) are:

$$\text{(6.8.11)}\qquad \int_{\partial G}\sum_r\sum_{|\gamma|\leq p_r} D^\gamma\zeta^r\,\chi_{r\gamma}[x, D\,u(x)]\,dx = 0,\quad \zeta\in C^\infty(\bar G)$$

where the $\chi_{r\gamma}$ involve derivatives $D^\beta u^k$ with $|\beta| \leq t_k - h_r - p_r$ and are such that if ζ has support on a boundary patch mapped on G_R, then (6.8.11) reduces to

$$\text{(6.8.12)}\qquad \int_{\sigma R}\sum_r\sum_{|\gamma|\leq p_r} D^\gamma_x\zeta^r\cdot\chi_{r\gamma}[x, 0; D\,u(x, 0)]\,dx = 0$$

if we use the (x, y) notation of § 6.3. We shall require that the $\chi_{r\gamma}$ with $|\gamma| = p_r$ shall be differentiable and that the operators

$$\text{(6.8.13)}\qquad \begin{aligned} B_{0rk}u^k &= \sum_{|\gamma|=p_r}\ \sum_{|\delta|=q_r}(-1)^{|\gamma|}b^\delta_{rk\gamma0}D^\gamma_x D^\delta_X u^k,\quad ,q_r = t_k - h_r - p_r\\ b^\delta_{rk\gamma0} &= \chi_{r\gamma p^k_\delta}[x_0, p_0],\quad (x_0, p_0)\in\mathfrak{N}',\quad (X = (x, y)),\end{aligned}$$

$\mathfrak{N}'$ being a neighborhood of all the $[X, D\,u(X)]$, $X \in \bar G$, satisfy the complementing condition with respect to the L_{0jk}. We assume that s_j, t_k, m_j, ϱ_{jk} and h_0 satisfy their previous conditions and that h_0, h_r, and p_r satisfy the conditions

$$p_r \geq 0,\quad h_0 + h_r + p_k \geq 1,\quad t_k - h_r - p_r \geq 0$$

and we assume that h_0 (probably negative) is the smallest integer satisfying all these conditions.

Using the methods of proof outlined above and in § 6.4, we can prove the following theorem:

Theorem 6.8.4. *Suppose that the* s_j, t_k, ϱ_{jk}, h_0, h_r, p_r, φ_j^α, *and* $\chi_{r\gamma}$ *satisfy all the conditions above, suppose that the* $u^k \in C^{t_k+h_0}(\bar{G})$ *and satisfy the equations* (6.8.8) *on* G *and* (6.8.11) *on* ∂G, *suppose that, for some* $h \geq h_0 + 1$, G *is of class* C^{t_0+h}, $\varphi_j^\alpha \in C^\tau(\mathfrak{N})$ *and* $\chi_{r\gamma} \in C^\omega(\mathfrak{N}')$, where $\tau = \max(0, h - s_j + |\alpha|)$, $\omega = \max(1, h + h_r + |\gamma|)$, and $\mathfrak{N}$ *and* $\mathfrak{N}'$ *are neighborhoods of the sets* $\{[x, D\,u(x)], x \in \bar{G}\}$ *in the appropriate* (x, p)-*spaces. Then* $u^k \in H_p^{t_k+h}(G)$ *for each* $p > 1$. *If, also,* G *is of class* $C_\mu^{t_0+h}$, $\varphi_j^\alpha \in C_\mu^\tau(\mathfrak{N})$, *and* $\chi_{r\gamma} \in C_\mu^\omega(\mathfrak{N}')$, *then* $u^k \in C_\mu^{t_k+h}(\bar{G})$. If $h_0 \geq 1$, $u^k \in C_\mu^{t_k+h_0-1}$, G *is of class* $C_\mu^{t_0+h_0}$, $\varphi_j^\alpha \in C_\mu^\varkappa(\mathfrak{N})$, *and* $\chi_{r\gamma} \in C_\mu^\lambda(\mathfrak{N}')$, *where* $\varkappa = \max(0, h_0 - s_j + |\alpha|)$ *and* $\lambda = \max(1, h_0 + h_r + |\gamma|)$, *then* $u^k \in C_\mu^{t_k+h_0}(\bar{G})$.

We now wish to prove the perturbation theorem of AGMON, DOUGLIS, and NIRENBERG ([1], Theorem 12.6). We first prove the following standard lemma:

Lemma 6.8.1. *Suppose the* L_{jk} *and* B_{rk} *satisfy the* $h_0 - \mu$ *conditions*[1] *on a domain* $\bar{G}$ *of class* $C_\mu^{t_0+h_0}$ *and suppose the system* L *is properly elliptic and the boundary operators* B_{rk} *satisfy the complementing condition on* ∂G. *Suppose also that the problem*

$$L_{jk}\, u^k = 0 \text{ on } G, \quad B_{rk}\, u^k = 0 \text{ on } \partial G \tag{6.8.14}$$

has a unique solution. Then there is a constant C, *independent of* u, *such that if* $L_{jk}\, u^k = f_j$ *on* G *and* $B_{rk}\, u^k = g_r$ *on* ∂G, *then*

$$\sum_k |||u^k|||_\mu^{t_k+h_0} \leq C\Big[\sum_j |||f_j|||_\mu^{h_0-s_j} + \sum_r |||\tilde{g}_r|||_\mu^{h_0+h_r}\Big] \tag{6.8.15}$$

where g_r *is any function* $\in C_\mu^{h_0+h_r}(\bar{G})$ *such that* $\tilde{g}_r = g_r$ *on* ∂G *and*

$$|||\varphi|||_\mu^t = h_\mu(\nabla^t \varphi) + \sum_{\tau=0}^{t} \max|\nabla^\tau \varphi(x)|. \tag{6.8.16}$$

Proof. Suppose this is not true. Then there is a sequence u_n such that the left side of (6.8.15) equals 1 for each n and the right side tends to zero. From the equi-continuity, we may assume then that $L_{jk}\, u_n^k \to 0$ and $B_{rk}\, u_n^k \to 0$ uniformly. But, from Theorem 6.3.9 and the negation of (6.8.15), we conclude also that $u_n \to u$ uniformly and $u \neq 0$ and also $\in C_\mu^{t_k+h_0}(\bar{G})$ and u satisfies (6.8.14). This contradicts the hypothesis.

Let us now suppose that the numbers s_j, t_k, m_j, ϱ_{jk}, h_0, and h_r satisfy the condition in Theorems 6.8.1 and 6.8.2, that G is of class $C_\mu^{t_0+h_0+2}$, that $u^k \in C_\mu^{t_k+h_0+2}(\bar{G})$, and that $\varphi_j^{(0)} \in C_\mu^{h_0+s-s_j}$ and $\chi_r^{(0)} \in C_\mu^{h_0+h_r+2}$ in their arguments, and that the u_0^k satisfy the equations (6.8.1) on G and (6.8.7) on ∂G. These respective equations of variation, as defined in (6.7.2)

[1] See Definitions 6.2.3 and 6.3.6.

and (6.7.29), reduce to

$$
\begin{aligned}
&L_{jk}\, v^k = 0 \text{ on } G, \quad B_{rk}\, v^k = 0 \text{ on } \partial G \\
&L_{jk}\, v^k \equiv \sum_k \sum_{|\beta| \le \tau} a_{jk}^{\beta}(x)\, D^{\beta} v^k, \quad \tau = s_j + t_k \\
&B_{rk}\, v^k = \sum_k \sum_{|\beta| \le \lambda} b_{rk}^{\beta}(x)\, D^{\beta} v^k, \quad \lambda = t_k - h_r \\
&a_{jk}^{\beta}(x) = \varphi_{j p_{\beta}^k}^{(0)}[x, D\, u_0(x)], \quad b_{rk}^{\beta}(x) = \chi_{r p_{\beta}^k}^{(0)}[x, D\, u_0(x)].
\end{aligned} \tag{6.8.17}
$$

We wish to find a solution $u = u_0 + v$ of the equations

$$\varphi_j^{(0)}(x, D\, u) = f_j(x) \text{ in } G, \quad \chi_r^0(x, D\, u) = g_r(x) \text{ on } \partial G, \tag{6.8.18}$$

for all f and g with $^*|||f|||_{\mu}^0$ and $^{*\prime}|||g|||_{\mu}^0$ sufficiently small, where we define

$$^*|||f|||_{\mu}^0 = \sum_j |||f_j|||_{\mu}^{h_0 - s_j}, \quad ^{*\prime}|||g|||_{\mu}^0 = \inf \sum_r |||\tilde{g}_r|||_{\mu}^{h_0 + h_r}, \tag{6.8.19}$$

the norms on the right having been defined in (6.8.16). Suppose we define

$$\varphi_j(x, p) = \varphi_j^{(0)}[x, D\, u_0(x) + p], \quad \chi_r(x, p) = \chi_r^{(0)}[x, D\, u_0(x) + p]. \tag{6.8.20}$$

Since we have assumed that $u_0 \in C_{\mu}^{t_k + h_0 + 2}(\bar{G})$, it follows that $\varphi_j \in C_{\mu}^{h_0 + 2 - s_j}$ and $\chi_r \in C_{\mu}^{h_0 + h_r + 2}$ in its arguments and

$$\varphi_j(x, 0) \equiv 0, \quad \chi_r(x, 0) \equiv 0. \tag{6.8.21}$$

Now, let us define ψ_j and ω_r by

$$
\begin{aligned}
&\varphi_j(x, p) - \sum_k \sum_{|\beta| \le \tau} \varphi_{j p_{\beta}^k}(x, 0)\, p_{\beta}^k = \psi_j(x, p), \quad \tau = s_j + t_k \\
&\chi_r(x, p) - \sum_k \sum_{|\beta| \le \lambda} \chi_{r p_{\beta}^k}(x, 0)\, p_{\beta}^k = \omega_r(x, p), \quad \lambda = t_k - h_r.
\end{aligned} \tag{6.8.22}
$$

Then the ψ_j and $\psi_{jp} \in C_{\mu}^{h_0 + 1 - s_j}$ and the χ_r and $\chi_{rp} \in C_{u}^{h_0 + h_r + 1}$ in their arguments and

$$\psi_j(x, 0) = \psi_{jp}(x, 0) = 0, \quad \omega_r(x, 0) = \omega_{rp}(x, 0). \tag{6.8.23}$$

The equations (6.8.18) then become, in turn,

$$\varphi_j(x, D\, v) = f_j(x) \text{ in } G, \quad \chi_r(x, D\, v) = g_r(x) \text{ on } \partial G, \tag{6.8.24}$$

$$
\begin{aligned}
L_{jk}\, v^k &= f_j(x) - \psi_j[x, D\, v(x)] \text{ on } G \\
B_{rk}\, v^k &= g_r(x) - \omega_r[x, D\, v(x)] \text{ on } \partial G
\end{aligned} \tag{6.8.25}
$$

where L_{jk} and B_{rk} are exactly the operators in (6.8.17).

We now state the theorem.

Theorem 6.8.5. *Suppose* (1) *that all the numbers* s_j, t_k, $t_0 = \max t_k$, m_j, ϱ_{jk}, h_0, *and* h_r *satisfy the conditions in Theorems* 6.8.1 *and* 6.8.2; (2) *that* G *is of class* $C_{\mu}^{t_0 + h_0 + 2}$, (3) *that* $u^k \in C_{\mu}^{t_k + h_0}(\bar{G})$, (4) *that* $\varphi_j^{(0)} \in C_{\mu}^{h_0 + 2 - s_j}$ *and* $\chi_r^{(0)} \in C_{\mu}^{h_0 + h_r + 2}$ *in their arguments in* $\mathfrak{N}$ *and* $\mathfrak{N}'$, (5) *that the* u^k *satisfy* (6.8.1) *and* (6.8.7) *with* φ_j *replaced by* $\varphi_j^{(0)}$ *and* χ_r *by* $\chi_r^{(0)}$, (6) *that the* L-*system in* (6.8.17) *is properly elliptic,* (7) *that the* B_{rk} *satisfy the complementing condition on* ∂G, *and* (8) *that the homogeneous system* (6.8.17) *has only the zero solution.*

Then $v^k \in C_\mu^{t_k+h_0+2}(\bar{G})$ *and there exist numbers* A *and* $B > 0$ *such that if the norms of* f *and* g*, as defined in* (6.8.19), *are* $< A$*, then the equations* (6.8.18) *have a unique solution* v *of norm* $< B$*. If* $h_0 \geq 1$*, all the differentiability requirements may be reduced by* 1 *in terms of* h_0 *(*G *of class* $C_\mu^{t_0+h_0+1}$, etc.) *and we conclude that* $v^k \in C_\mu^{t_k+h_0+1}(\bar{G})$.

Proof. The additional differentiability of v follows from Theorem 6.8.2. Using this differentiability, we introduce the functions φ_j, χ_r, ψ_j, and ω_r as in (6.8.20) and (6.8.22) and we conclude (6.8.23). From Lemma 6.8.1, it follows that the problem

$$(6.8.26) \qquad L_{jk}\, w^k = f_j \text{ on } G, \quad B_{rk}\, w^k = g_r \text{ on } \partial G$$

has a unique solution which we denote by $P(f, g) = w$ in which P is a bounded operator from ${}^*C_\mu^0 \times {}^{*\prime}C_\mu^0$ (norms in (6.8.19)) into ${}^{**}C_\mu^0$ which consists of all vectors $u \ni u^k \in C_\mu^{t_k+h_0}(\bar{G})$ with

$${}^{**}|||u|||_\mu^0 = \sum_k |||u^k|||_\mu^{t_k+h_0}.$$

Then the equations (6.8.18) and (6.8.25) are equivalent to

$$(6.8.27) \qquad v + T\, v = w, \quad w = P(f, g), \quad T\, v = P[\psi, \omega].$$

We wish to show that

$${}^{**}|||T\, v_1 - T\, v_2|||_\mu^0 \leq \frac{1}{2}\, {}^{**}|||v_1 - v_2|||_\mu^0 \quad \text{if} \quad {}^{**}|||v_i|||_\mu^0 \leq L, \quad i = 1, 2$$

(6.8.28)

where L is sufficiently small.

Let

$$(6.8.29) \quad f_{ji}^*(x) = \psi_j[x, D\, v_i(x)], \quad g_{ri}^*(x) = \omega_r[x, D\, v_i(x)], \quad i = 1, 2,$$

and let us suppose that

$$(6.8.30) \qquad {}^{**}|||v_i|||_\mu^0 \leq L, \quad {}^{**}|||v_1 - v_2|||_\mu^0 = R.$$

Then, from (6.8.23), we see that

$$(6.8.31) \qquad |f_{j1}^*(x) - f_{j2}^*(x)| \leq C\, L \cdot R, \quad x \in \bar{G}.$$

Now, letting the p_β^k be arranged in a single sequence $p^1, \ldots, p^t$, we see that

$$f_j^*(x) \equiv f_{j1}^*(x) - f_{j2}^*(x) = \sum_{i=1}^t A_{ji}(x) \cdot [p_1^i(x) - p_2^i(x)],$$

$$(6.8.32) \quad A_{ji}(x) = \int_0^1 \psi_{j\, p^i}[x, p(x, t')]\, dt', \quad p(x, t') = p_2(x) + t'[p_1(x) - p_2(x)].$$

In case $h_0 - s_j = 0$, so that $h_0 = -s_j = 0$, all we need to do is show that $h_\mu(f_j^*, G) \leq C\, L^\mu R$. From (6.8.32), (6.8.23), and the fact that ψ_j and $\psi_{j\, p} \in C_\mu^1$, we easily find that

$$|A_{ji}(x)| \leq C\, L, \quad |A_{ji}(x_2) - A_{ji}(x_1)| \leq C\, L^\mu |x_2 - x_1|^\mu \quad \text{if} \quad L \leq 1.$$

The desired result in this case follows.

More generally, a derivative of order $\tau = h_0 - s_j\,(> 0)$ of $f^*_{j1} - f^*_{j2} \equiv f^*_j$ is a sum of products of a derivative of order q of an A_{ji} and one of order $\tau - q$ of a $(p^i_1 - p^i_2)$, $0 \leq q \leq \tau$. Any derivative of order q of an A_{ji} is the sum of terms of the form

$$\begin{gathered} \varrho = \int_0^1 \psi_{ji\alpha K}[x, p(x,t')] \cdot (D^{\gamma_1} p^{k_1}) \ldots (D^{\gamma_m} p^{k_m})\, dt' \\ K = (k_1, \ldots, k_m), \quad m \geq 0, \quad |\alpha| + |\gamma_1| + \cdots + |\gamma_m| = q, \\ \gamma_\omega > 0, \quad \omega = 1, \ldots, m \text{ if } m > 0, \quad \psi_{ji\alpha K} = D^\alpha_x D_{p^{k_1}} \ldots D_{p^{k_m}} \psi_{j\,p^i} \end{gathered} \tag{6.8.33}$$

where α and the γ_ω are multi-indices. If $m \geq 2$ or $q < \tau$, it is easy to see that every such term $\varrho \in C^1$ and that

$$h_\mu(\varrho) \leq \max_{x \in \bar{G}} |\nabla \varrho(x)| \leq CL, \quad |\varrho(x)| \leq CL,$$

the last holding even if $q = \tau$. If $q = \tau$, the only non-differentiable terms (assuming $\tau > 0$) ϱ are of the form

$$\varrho = \int_0^1 \psi_{j\,p^i\,p^k}[x, p(x,t)]\, D^\beta p^k(x,t)\, dt, \quad |\beta| = \tau.$$

Let ϱ be such a term and abbreviate the integrand $\psi\, Q$. Then

$$\begin{aligned} \varrho(x_2) - \varrho(x_1) = \int_0^1 \psi(x_2, t')\,[Q(x_2,t') - Q(x_1,t')] + \\ + Q(x_1,t')\,[\psi(x_2,t') - \psi(x_1,t')]\, dt'. \end{aligned}$$

Using the definitions of $p(x,t')$ and $Q(x,t') = D^\beta p$, and the further differentiability of ψ_j, we obtain

$$|\varrho(x_2) - \varrho(x_1)| \leq CL\, |x_2 - x_1|^\mu.$$

The other f^*_j and the correspondingly defined $g^*_r = g^*_{r1} - g^*_{r2}$ can be handled in the same way. Thus if $^{**}|||v_i|||^0_\mu \leq L$, we conclude from these results and the boundedness of P that

$$^{**}|||T v_1 - T v_2|||^0_\mu \leq C L^{\mu\, **}|||v_1 - v_2|||^0_\mu,$$

if L is small enough. The result follows.

The following special case of a variant of Theorem 6.8.5 can be extended to systems:

Theorem 6.8.6. *Suppose G is of class C^2_μ, the $A^\alpha(x,z,p) \in C^3_\mu$, $B(x,z,p) \in C^2_\mu$, $z_0 \in C^2_\mu(\bar{G})$, z_0 satisfies the equations* (1.10.13) ($N = 1$), *and the equation of variation*

$$\begin{gathered} L(z;\zeta) \equiv \frac{\partial}{\partial x^\alpha}(a^{\alpha\beta}\zeta_{,\beta} + b^\alpha \zeta) - (c^\alpha \zeta_{,\alpha} + d\zeta) = 0 \text{ with } z = z_0, \\ a^{\alpha\beta} = A^\alpha_{p_\beta}[x, z(x), \nabla z(x)], \quad b^\alpha = A^\alpha_z, \quad c^\alpha = B_{p_\alpha}, \quad d = B_z \end{gathered} \tag{6.8.34}$$

has only the zero solution if $\zeta = 0$ on ∂G. Then there are constants C, $\varrho_0 > 0$, and $\sigma_0 > 0$ such that the boundary problem

$$L(z;\zeta) = 0, \quad \zeta = 1 \text{ on } \partial G \tag{6.8.35}$$

has a unique solution $\zeta \in C^2_\mu(\bar{G})$ *provided* $\|z - z_0\| \leq \sigma_0$. *If we denote this solution by* $\mathfrak{F}(z)$, *then we have*

$$\|\mathfrak{F}(z_2) - \mathfrak{F}(z_1)\| \leq C\|z_2 - z_1\| \quad \text{if} \quad \|z_1 - z_2\| \leq \sigma_0 (\|\varphi\| = |||\varphi|||^2_\mu). \tag{6.8.36}$$

Thus there exists a unique function Z *from* $[-\varrho_0, \varrho_0]$ *into* $C^2_\mu(\bar{G})$ *which* $\in C^1[-\varrho_0, \varrho_0]$ *and satisfies*

$$\frac{dZ}{d\varrho} = \mathfrak{F}(Z), \quad Z(0) = z_0. \tag{6.8.37}$$

Proof. From Lemma 6.8.1 it follows that there is a constant C_1 such that if $f \in C^0_\mu(\bar{G})$ and $g \in C^2_\mu(\bar{G})$, there is a unique solution of the problem

$$L(z_0; \zeta) = f \text{ on } G, \quad \zeta = g \text{ on } \partial G \tag{6.8.38}$$

and that

$$\|\zeta\| \leq C_1(|||f|||^0_\mu + \|g\|) \quad (\|\varphi\| = |||\varphi|||^2_\mu). \tag{6.8.39}$$

So we set $\zeta = \zeta_0 + \omega$ in (6.8.35) where ζ_0 is the solution of the problem (6.8.35) with $z = z_0$. Then we write (6.8.35) in the form

$$L(z_0; \omega) = -[L(z; \omega) - L(z_0; \omega)] - [L(z; \zeta_0) - L(z_0; \zeta_0)]. \tag{6.8.40}$$

From this, (6.8.38), and (6.3.39), we find that

$$\|\omega\| \leq C_2\|z - z_0\|(\|\omega\| + \|\zeta_0\|). \tag{6.8.41}$$

Thus, if $C_2\|z - z_0\| \leq 1/2$, we see that

$$\|\omega\| \leq 2C_2\|z - z_0\| \cdot \|\zeta_0\|, \quad \|\zeta\| \leq \|\zeta_0\| \cdot (1 + 2C_2\|z - z_0\|). \tag{6.8.42}$$

Now, if we have $\zeta_k = \mathfrak{F}(z_k)$, $\|z_k - z_0\| \leq \sigma_0$, $k = 1, 2$, then we have $L(z_2, \zeta_2) = L(z_1, \zeta_1) = 0$, $\zeta_2 - \zeta_1 = 0$ on ∂G, and

$$L(z_2, \zeta_2 - \zeta_1) = -[L(z_2; \zeta_1) - L(z_1; \zeta_1)].$$

An analysis like that in (6.3.40)—(6.8.42) shows that

$$\|\zeta_2 - \zeta_1\| \leq C_3\|z_2 - z_1\| \cdot \|\zeta_1\| \leq C_4\|z_2 - z_1\|.$$

Thus $\mathfrak{F}(Z)$ satisfies a Lipschitz condition so the theorem follows.

Theorem 6.8.7. *Suppose that* G, *the* A^α, B, *and* z_0 *satisfy all the hypotheses of Theorem* 6.8.6 *except the last. If* $x_0 \in G$, *there exists an* $R > 0$ *such that the problem* (6.8.35) *with* $z = z_0$ *and* $G = B(x_0, R)$ *has a unique solution* ζ *and* $\zeta(x) > 0$ *in* $\overline{B(x_0, R)}$.

Proof. We notice from (6.8.34) that

$$L(z_0; \zeta) = a^{\alpha\beta}(x)\zeta_{,\alpha\beta} + {}'b^\alpha(x)\zeta_{,\alpha} + {}'c(x)\zeta$$

where the $a^{\alpha\beta}$ and ${}'c \in C^1_\mu(\bar{G})$ and ${}'b^\alpha \in C^0_\mu(\bar{G})$ and the coefficients depend on z_0, of course. Now for each R such that $\overline{B(x_0, R)} \subset G$, we define

$$\eta_R(y) = \zeta_R(x_0 + Ry) = 1 + \omega_R(y), \quad y \in \overline{B(0,1)}.$$

Then the desired ω_R satisfies the conditions

$$A_0^{\alpha\beta}\,\omega_{R,\alpha\beta} + (A_R^{\alpha\beta} - A_0^{\alpha\beta})\,\omega_{R,\alpha\beta} + R\,B_R^{\alpha}\omega_{R,\alpha} + R^2\,C_R\,\omega_R + R^2\,C_R = 0,$$
$$\omega_R = 0 \text{ on } \partial B(0,1), \quad A_R^{\alpha\beta}(y) = a^{\alpha\beta}(x_0 + R\,y), \text{ etc.}, \tag{6.8.43}$$
$$A_0^{\alpha\beta} = A_R^{\alpha\beta}(0) = a^{\alpha\beta}(x_0).$$

From (6.8.43), we conclude that

$$\|\omega_R\| \le (Z_1\,R + Z_2\,R^{1+\mu} + Z_3\,R^2 + Z_4\,R^3)\,\|\omega_R\| + R^2\,Z_5.$$

The result follows.

Chapter 7

A variational method in the theory of harmonic integrals

7.1. Introduction

In this chapter, we show how one can apply the variational method to the study of the theory of harmonic integrals. In his first paper on the subject, W. V. D. HODGE [1] used a variational method in this theory to study certain boundary value problems for forms defined on domains in Euclidean space (using Cartesian coordinates). But, in order to carry his theory over to compact Riemannian manifolds, he and subsequent writers found it expedient to employ methods involving integral equations (see HODGE [2], KODAIRA, DE RHAM-KODAIRA, and references therein). More recently, MILGRAM and ROSENBLOOM ([1], [2]) and GAFFNEY ([1], [2]), have treated certain problems by their "heat equation" method involving parabolic equations. The variational method was applied to general compact Riemannian manifolds by MORREY and EELLS and to such manifolds with boundary by MORREY [11]. A closely related method was employed concurrently by FRIEDRICHS [3] in both cases. Certain boundary value problems had been discussed previously by DUFF and SPENCER and by CONNOR ([1], [2]).

One of the principal results which was obtained early is now known as the "Kodaira decomposition" and states that

$$\mathfrak{L}_2 = \mathfrak{H} \oplus \mathfrak{C} \oplus \mathfrak{D} \quad \text{or} \quad \mathfrak{L}_2^r = \mathfrak{H}^r \oplus \mathfrak{C}^r \oplus \mathfrak{D}^r \tag{7.1.1}$$

where $\mathfrak{L}_2^r$ denotes the space of forms of degree r_s which are either all even or all odd, which have components in $\mathfrak{L}_2$, the inner product being defined in § 7.2; $\mathfrak{H}^r$ denotes the space of harmonic fields in $\mathfrak{L}_2^r$, and $\mathfrak{C}$ and $\mathfrak{D}$ are the respective closures of the spaces $\{\delta\alpha\}$ and $\{d\beta\}$ where α and β are smooth (see § 7.2 for definitions). In case all the forms involved are smooth (which is the case if M and the given form in $\mathfrak{L}_2^r$ are smooth—see below) this decomposition leads to the principal theorem of HODGE which is that *there is a unique harmonic field in each cohomology class of closed forms* (i.e. those ω with $d\omega = 0$) *on* M. This statement is equivalent to the statement: *Suppose M is of class C^∞, $\omega \in C^\infty(M)$, and $d\omega = 0$; then there is a unique harmonic field H ($\in C^\infty(M)$) such that*

$$\omega = H + d\varphi \quad \textit{for some} \quad \varphi \in C^\infty(M). \tag{7.1.2}$$

The writers mentioned above, except for EELLS, FRIEDRICHS, and MORREY, were not especially concerned either with the differentiability of M and of the forms or with the nature of the elements in the spaces $\mathfrak{C}$ and $\mathfrak{D}$ above. Not only does the variational method yield a very simple proof of the decomposition in (7.1.1), the finite dimensionality of $\mathfrak{H}$ (in the case where M is without boundary), and the fact that in the general case in (7.1.1) where $c \in \mathfrak{C}$ and $d \in \mathfrak{D}$, there exist forms α and β in the SOBOLEV spaces $H_2^1(M)$ such that

$$\delta\alpha = c, \quad d\beta = d, \quad d\alpha = \delta\beta = 0, \tag{7.1.3}$$

but also shows that the general results hold for manifolds M of class C_1^1 on which the metric tensor is merely Lipschitz. The success of the method depends on (a) the use of the SOBOLEV spaces $H_p^1(M)$, $p \geq 2$, (b) the recent results concerning the differentiability of "weak solutions" of elliptic differential equations (see Chapter 5), and (c) an inequality due, in the case of forms, to GAFFNEY [1] and in the case of a single differential equation of higher order to GÅRDING [1]. The results in (b) above lead to very exhaustive results concerning the differentiability of various solution forms. However, some additional differentiability is obtained by using the special character of the equations (see Theorem 7.4.1).

In the case of a compact manifold with boundary, it turns out that the SOBOLEV space $H_2^1(M)$ splits into the two closed linear subspaces $H_2^{1+}(M)$ and $H_2^{1-}(M)$ of forms ω for which $n\,\omega = 0$ and $t\,\omega = 0$, respectively, on the boundary $b\,M$, $n\,\omega$ and $t\,\omega$ being the normal and tangential parts of ω (see § 7.5). The GAFFNEY inequality holds, on each of these spaces separately and this, together with the boundary differentiability theory of Chapter 5, makes it possible to carry over the preceding theory to each of these partial spaces. By using these two spaces and certain potentials defined therein, it is possible to prove first a decomposition like (7.1.1) in which $\alpha \in H_2^{1+}$ and $\beta \in H_2^{1-}$. In this decomposition however, $\mathfrak{H}$ may have infinite dimensionality (see § 7.7). After this is shown a great variety of boundary value problems treated by DUFF and SPENCER and CONNOR ([1], [2]) are treated.

It was pointed out that FRIEDRICHS was working on these problems at about the same time as were EELLS and the author. After a number of conferences, it seemed as though there was enough difference between our results as well as our methods to warrant publishing both versions and this was done. The version presented here is mainly due to MORREY and EELLS and MORREY [11]. In §§7.2—7.4, we consider manifolds without boundary and in §§ 7.5—7.8, we consider manifolds with boundary.

We do not assume that the manifold is orientable; there are completely parallel theories involving only "even forms" or only "odd forms" (see DE RHAM-KODAIRA). We shall assume that all the forms considered are even or else that they are all odd.

7.2. Fundamentals; the Gaffney-Gårding inequality

Definition 7.2.1. We adopt the usual definition of a *Riemannian manifold of class* C^k (with or without boundary) and dimension ν and define those *of class* C^k_μ, $0 < \mu \leq 1$, in the obvious way: any two admissible coordinate systems are related by a transformation of class C^k_μ, and if θ is one such with domain $G \subset R_\nu$, the induced components $g_{ij}(x) \in C^{k-1}_\mu$ on G.

Assumption. *We assume that any manifold M is compact and of class at least C^1_1.**

We shall be concerned with exterior differential forms of degree r $(0 \leq r \leq \nu)$ on a manifold M; we call these simply *r-forms*. In the domain of any coordinate system, such a form may be represented as follows:

$$\omega = \sum_{1 \leq i_1 < \cdots < i_r \leq \nu} \omega_{i_1 \ldots i_r} \, dx^{i_1} \wedge \ldots \wedge dx^{i_r} \tag{7.2.1}$$

where $\omega_{i_1 \ldots i_r}$ are the *components* of ω in that coordinate system and $\wedge$ denotes the exterior product. If two coordinate systems with coordinates (x) and $(\bar{x})$ overlap, the relation

$$\omega_{i_1 \ldots i_r}(\bar{x}) = \varepsilon \sum_{j_1 < \cdots < j_r} \omega_{j_1 \ldots j_r}[x(\bar{x})] \frac{\partial(x^{j_1}, \ldots, x^{j_r})}{\partial(\bar{x}^{i_1}, \ldots, \bar{x}^{i_r})},$$

$$\varepsilon = \begin{cases} +1 & \text{for even forms} \\ J/|J| & \text{for odd forms}, \quad J = \dfrac{\partial(x^1, \ldots, x^\nu)}{\partial(\bar{x}^1, \ldots, \bar{x}^\nu)}, \end{cases} \tag{7.2.2}$$

holds between the components. The notation in (7.2.1) is often abbreviated to

$$\omega = \sum \omega_I \, dx^I \tag{7.2.3}$$

where I [or other capital letter, correspondingly] denotes a sequence in which $1 \leq i_1 < i_2 < \cdots < i_r \leq \nu$.

The relation (7.2.2) allows us to define $\mathfrak{C}_2$ as follows:

Definition 7.2.2. An r-form ω on M will be said to be *in* $\mathfrak{L}_p$ *(resp.* H^1_p*)* $\Leftrightarrow$ its components are in L_p (resp. H^1_p). If ω and η are r-forms of the same kind in $\mathfrak{L}_2$, we define their *inner product* by

$$(\omega, \eta) = \int_M F(P; \omega, \eta) \, dS(P) \tag{7.2.4}$$

* Actually the reader will observe that many of the results have proper generalizations to manifolds of class H^2_p where $p > \nu$. Such manifolds are of class C^1_μ with $\mu = 1 - \nu/p$.

where in any admissible coordinate system

$$F(P;\,\omega,\eta) = \sum_{I,K} g^{IK}(x)\,\omega_I(x)\,\eta_K(x), \quad dS(P) = \Gamma(x)\,|\,dx^1 \dots dx^\nu\,|,$$

(7.2.5)
$$g^{IK} = \begin{vmatrix} g^{i_1 k_1} \dots g^{i_1 k_r} \\ g^{i_r k_1} \dots g^{i_r k_r} \end{vmatrix}, \quad \Gamma(x) = [g(x)]^{1/2},$$

and g is the determinant of the g_{ij} (and the g^{ij} matrix is the inverse of the g_{ij} matrix). It is easy to see that the value at P of F is independent of the coordinate system provided that ω and η are of the same kind.

The following theorem is well-known and evident:

Theorem 7.2.1. *For each r, $0 \le r \le \nu$, the totality of even (resp. odd) r-forms in $\mathfrak{L}_2^r$ (with equivalent forms identified) forms a real Hilbert space with inner product given by* (7.2.4).

Forms ω are often regarded as alternating tensors and written in the form

(7.2.6)
$$\omega = \frac{1}{r!} \sum_{i_1 \dots i_r = 1}^{\nu} \omega_{i_1 \dots i_r}\, dx^{i_1} \wedge \dots \wedge dx^{i_r}.$$

When this is done the components are defined for all values of the indices from 1 to ν and are antisymmetric in the indices, a component being zero if two indices have the same value. In this case, ω transforms like an ordinary covariant tensor, except for the factor ε in (7.2.2) and $F(P;\,\omega,\eta)$ has the form

(7.2.7)
$$F(P;\,\omega,\eta) = \frac{1}{r!} \sum_{\substack{i_1,\dots,i_r \\ k_1,\dots,k_r}} g^{i_1 k_1} \dots g^{i_r k_r}\, \omega_{i_1 \dots i_r}\, \eta_{k_1 \dots k_r}$$

where the i's and k's all run independently from 1 to ν. We shall not often use this form but it is very useful in computations (see the proof of Theorem 7.2.4 below and §§ 7.5, 8.3, etc.).

We now wish to introduce an inner product into the space H_2^1.

Definition 7.2.3. Let $\mathfrak{U} = (U_1, \dots, U_Q)$ be a finite open covering of M by coordinate patches θ_q with domains G_q (LIPSCHITZ and in R_ν) and ranges U_q. If ω and $\eta \in H_2^1$ (see Def. 7.2.2), we define

(7.2.8)
$$((\omega,\eta))_{\mathfrak{U}} = (\omega,\eta) + \sum_{q=1}^{Q} \int_{G_q} \sum_I{}' \sum_{\alpha=1}^{\nu} \omega_{I x^\alpha}^{(q)}\, \eta_{I x^\alpha}^{(q)}\, dx,$$

where the $\omega_I^{(q)}$ and $\eta_I^{(q)}$ are the components of ω and η with respect to θ_q. Then we define the corresponding norm

(7.2.9)
$$\|\omega\|_{\mathfrak{U}} = ((\omega,\omega))_{\mathfrak{U}}^{1/2}.$$

Theorem 7.2.2. *For each coordinate cover $\mathfrak{U}$ and each r, the space of r-forms in H_2^1 forms a real Hilbert space H_2^{1r} with inner product given by* (7.2.8). *Any two such inner products are topologically equivalent.*

This is evident.

Definition 7.2.4. If ω is an r-form $\in H_2^1$ with $r < \nu$, we define its *exterior derivative* $d\omega$ as the $r+1$—form defined by

$$d\omega = \sum_{I,\alpha} \omega_{I x^\alpha}\, dx^\alpha \wedge dx^I. \tag{7.2.10}$$

If $r = \nu$, we define $d\omega = 0$; we have assumed ω given by (7.2.1) or (7.2.3), of course.

Remark. We note that if we wish to write

$$d\omega = \sum_J (d\omega)_J\, dx^J$$

then we must have

$$(d\omega)_J = \sum_{\gamma=1}^{r+1} (-1)^{\gamma-1}\, \partial\omega_{J'_\gamma}/\partial x^{j_\gamma} \tag{7.2.11}$$

$$(J'_\gamma = j_1, \ldots, j_{\gamma-1}, j_{\gamma+1}, \ldots, j_{r+1}).$$

This definition is set up so that the following version of Stokes' theorem holds.

Theorem 2.3.7 (Stokes' theorem). *If $\mathfrak{b}$ is an oriented manifold of class C^1 in M having dimension $r+1$ and boundary $b\,\mathfrak{b}$ of class C^1 and if ω is even and of class C^1 on $\mathfrak{b}$, then*

$$\int_{\mathfrak{b}} d\omega = \int_{b\mathfrak{b}} \omega \tag{7.2.12}$$

where $b\,\mathfrak{b}$ has its usual orientation induced by that of $\mathfrak{b}$.

We do not need this theorem and therefore omit its proof.

Definition 7.2.5. If ω is an r-form $\in H_2^1$ given by (7.2.1) and $r \geq 1$, we define its co-differential $\delta\,\omega$ by the condition that

$$(\omega, d\varphi) = (\delta\,\omega, \varphi) \tag{7.2.13}$$

for every $\varphi \in H_2^{1,r-1}$. If ω is a 0-form, we define $\delta\,\omega = 0$.

Theorem 7.2.4. *If $\omega \in H_2^1$ and is given by (7.2.5) with $r \geq 1$, then*

$$(\delta\omega) = \frac{1}{(r-1)!} \sum_{k_1 \ldots k_{r-1}} (\delta\omega)_{k_1 \ldots k_{r-1}}\, dx^{k_1} \wedge \ldots \wedge dx^{k_{r-1}},$$

$$-(\delta\omega)_{k_1\ldots k_{r-1}} = g^{\alpha\beta}\, \omega_{\beta k_1 \ldots k_{r-1} x^\alpha} + (g^{\alpha\beta}_{x^\alpha} + \Gamma^{-1} g^{\alpha\beta}\, \Gamma_{x^\alpha})\, \omega_{\beta k_1 \ldots k_{r-1}} +$$

$$+ \sum_{\delta=1}^{r-1} \sum_l g_{k_\delta l}\, g^{tl}_{x^\alpha}\, g^{\alpha\beta}\, \omega_{\beta k_1 \ldots k_{\delta-1} t k_{\delta+1} \ldots k_{r-1}}. \tag{7.2.14}$$

If ω is even (resp. odd), then $d\,\omega$ and $\delta\,\omega$ are even (resp. odd).

Proof. We may choose φ in (7.2.13) to have support in a coordinate patch with domain G; we assume φ given in the form (7.2.6). Then, from (7.2.4), (7.2.7), and (7.2.11), we obtain

$$(\omega, d\varphi) = \frac{1}{r!} \int_G \sum_{(i),(j)} g^{i_1 j_1} \ldots g^{i_r j_r}\, \omega_{(i)} \left[\sum_{\gamma=1}^{r} (-1)^{\gamma-1}\, \varphi_{(j'_\gamma) x^{j_\gamma}} \right] \Gamma\, dx$$

$$= -\frac{1}{r!} \int_G \sum_{\nu=1}^{r} (-1)^{\gamma-1} \sum_{(i),(j)} \varphi_{(j'_\gamma)} \left\{ \Gamma^{-1} \frac{\partial}{\partial x^{j_\gamma}} (\Gamma g^{i_1 j_1} \ldots g^{i_r j_r}\, \omega_{(i)}) \right\} \Gamma\, dx.$$

$$(i) = i_1 \ldots i_r, \quad (j'_\gamma) = j_1 \ldots j_{\gamma-1} j_{\gamma+1} \ldots j_r, \quad \text{etc.} \tag{7.2.15}$$

If for each γ we now choose $(l_1, \ldots, l_{r-1}) = (j'_\gamma)$ and set $j_\gamma = \alpha$, the last integral in (7.2.15) becomes

$$-\frac{1}{r!}\int_G \sum_{\gamma=1}^{r} (-1)^{\gamma-1} \sum_{(i),(l)} \varphi_{(l)} \times$$

$$\times\left\{\Gamma^{-1}\frac{\partial}{\partial x^\alpha}\left(\Gamma g^{i_\gamma\alpha} g^{i_1 l_1}\ldots g^{i_{\gamma-1}l_{\gamma-1}} g^{i_{\gamma+1}l_\gamma}\ldots g^{i_r l_{r-1}}\,\omega_{(i)}\right)\right\}\Gamma\,dx. \tag{7.2.16}$$

If we now set $i_\gamma = \beta$, $k_1 \cdots k_{r-1} = (i'_\gamma)$ and use the anti-symmetry in the indices, the integral (7.2.16) becomes

$$-\frac{1}{(r-1)!}\int_G \sum_{(k)(l)} \varphi_{(l)}\left\{\Gamma^{-1}\frac{\partial}{\partial x^\alpha}\left(\Gamma g^{k_1 l_1}\ldots g^{k_{r-1}l_{r-1}} g^{\alpha\beta}\,\omega_{\beta k_1\ldots k_{r-1}}\right)\right\}\Gamma\,dx$$

which must, for all $\varphi_{(l)}$, be equal to

$$\frac{1}{(r-1)!}\int_G \sum_{(k)(l)} g^{k_1 l_1}\ldots g^{k_{r-1}l_{r-1}}(\delta\omega)_{(k)}\,\varphi_{(l)}\,\Gamma\,dx.$$

By equating coefficients of $\varphi_{(l)}$, multiplying both sides by $g_{l_1 s_1}\cdots g_{l_{r-1}s_{r-1}}$ and summing over (l), we obtain

$$(\delta\omega)_{(s)} = -\Big\{g^{\alpha\beta}\,\omega_{\alpha s_1\ldots s_{r-1}x^\beta} + \left(g^{\alpha\beta}_{x^\alpha} + \Gamma^{-1} g^{\alpha\beta}\Gamma_{x^\alpha}\right)\omega_{\alpha s_1\ldots s_{r-1}} +$$

$$+\sum_{\delta=1}^{r-1} g_{s_\delta l_\delta}\, g^{k_\delta l_\delta}_{x^\alpha}\, g^{\alpha\beta}\,\omega_{\beta s_1\ldots s_{\delta-1}k_\delta s_{\delta+1}\ldots s_{r-1}}\Big\}$$

from which the result (7.2.14) follows. The reader may verify the last statement.

Remark. We notice that if $\omega_{i_1\ldots i_r}$ is anti-symmetric in the indices (i), $(\delta\omega)_{k_1\ldots k_{r-1}}$, as defined by (7.2.14) is also anti-symmetric in $k_1\ldots k_{r-1}$, and $(d\omega)_{j_1\ldots j_{r+1}}$ as defined in (7.2.11) (with J'_γ replaced by (j'_γ)) is anti-symmetric in $j_1\ldots j_{r+1}$. Thus we may write $d\omega$ or $\delta\omega$ in either form (7.2.3) or (7.2.6).

Definition 7.2.6. We define the *Dirichlet integral* by

$$D(\omega) = (d\omega, d\omega) + (\delta\omega, \delta\omega). \tag{7.2.17}$$

Theorem 7.2.5. *d is a bounded operator from the whole of H_2^{1r} into $\mathfrak{L}_2^{r+1}$, and δ is a bounded operator from the whole of H_2^{1r} into $\mathfrak{L}_2^{r-1}$. $D(\omega)$ is a lower semi-continuous function with respect to weak convergence in H_2^{1r}. If ω_k tends weakly to ω_0 in H_2^{1r} on M, then ω_k tends strongly to ω_0 in $\mathfrak{L}_2^r$ on M.*

Proof. This is clear from (7.2.11) in the case of d and from (7.2.14) in the case of δ, since the g_{ij} are at least Lipschitz and have bounded first derivatives. Now if ω_k tends weakly in H_2^1 to ω, $d\omega_k$ and $\delta\omega_k$ tend weakly in $\mathfrak{L}_2$ to $d\omega$ and $\delta\omega$, whence the last statement about $D(\omega)$ follows from the lower-semicontinuity of the norm with respect to weak convergence. The last statement is an application of Theorem 3.4.4.

Essentially the following lemma has been proved by GAFFNEY [1]:

Lemma 7.2.1. *Take any $\varepsilon > 0$ and r such that $0 \leq r \leq \nu$. Given any point $x_0 \in M$ there is a coordinate system θ mapping the open ball $B(0, \varrho)$ of radius ϱ in E_ν onto a neighborhood U of x_0 and a constant l such that*

$$D(\omega) \geq (1 - \varepsilon) \int_{B(0,\varrho)} \sum_{I,\alpha} (\omega_{Ix^\alpha})^2 \, dx - l(\omega, \omega) \tag{7.2.18}$$

for any r-form $\omega \in H_2^{1r}$ whose support is in U.

Proof. Let us set $B_\varrho = B(0, \varrho)$. Then for any ω whose support is in U we can write

$$D(\omega) = \int_{B_\varrho} \sum_{IJ} [a^{IJ\alpha\beta} \omega_{Ix^\alpha} \omega_{Jx^\beta} + 2 b^{IJ\alpha} \omega_{Ix^\alpha} \omega_J + c^{IJ} \omega_I \omega_J] \, dx \tag{7.2.19}$$

and

$$a^{IJ\alpha\beta}(x) = a^{JI\beta\alpha}(x), \quad c^{IJ}(x) = c^{JI}(x),$$

where the a's are combinations of the g_{ij} only and the b's and c's are similar combinations of the g_{ij} and their first partial derivatives. Consequently it follows from our assumption that the a's are Lipschitz and the b's and c's bounded and measurable at least.

We begin by choosing a fixed coordinate system mapping 0 into x_0 and B_ϱ onto a neighborhood of x_0, with $g_{ij}(0) = \delta_{ij}$. Since the a's are LIPSCHITZ, we may choose ϱ so small that

$$\int_{B_\varrho} a^{IJ\alpha\beta}(x) \, \omega_{Ix^\alpha} \omega_{Jx^\beta} \, dx \geq D_0(\omega) - \frac{\varepsilon}{2} \int_{B_\varrho} \sum_{I,\alpha} (\omega_{Ix^\alpha})^2 \, dx, \tag{7.2.20}$$

where $D_0(\omega)$ is the Dirichlet integral in the euclidean case with cartesian coordinates. Since the b's and c's are bounded and since for any $\eta > 0$ we have

$$|2\alpha\beta| \leq \eta \alpha^2 + \eta^{-1} \beta^2,$$

it follows that

$$D(\omega) \geq D_0(\omega) - \varepsilon \int_{B_\varrho} \sum_{I,\alpha} (\omega_{Ix^\alpha})^2 \, dx - l(\omega, \omega).$$

Now suppose that the ω_I are of class C^2 on B_ϱ and vanish on and near S_ϱ. Then by using the formula (7.2.13), we may write

$$D_0(\omega) = (\delta_0 \, d\omega + d\delta_0 \, \omega, \omega)_0.$$

Using the formulas (7.2.11), (7.2.14), (7.2.4), and (7.2.5) with $g^{\alpha\beta} \equiv \delta^{\alpha\beta}$ we find that

$$(\delta_0 \, d\omega + d\,\delta_0 \, \omega)_0 = -\sum_I \int_{B_\varrho} \omega_I \Delta_0 \omega_I \, dx = \int_{B_\varrho} \sum_{I,\alpha} (\omega_{Ix^\alpha})^2 \, dx = D_0(\omega). \tag{7.2.21}$$

Using approximations, the last equality in (7.2.21) holds for all forms ω in H_2^1 with the stated properties.

Theorem 7.2.6. *For each $r = 0, \ldots, n$ and coordinate covering $\mathfrak{U}$ of M, there exist constants $K_{\mathfrak{U}} > 0$ and $L_{\mathfrak{U}}$ such that*

$$D(\omega) \geq K_{\mathfrak{U}}((\omega, \omega))_{\mathfrak{U}} - L_{\mathfrak{U}}(\omega, \omega) \tag{7.2.22}$$

for every $\omega \in H_2^{1r}$.

Proof. From Theorem 7.2.2 it is sufficient to prove this for some particular $\mathfrak{U}$. Let $\mathfrak{U} = (U_1, \ldots, U_Q)$ be an open covering of M by coordinate patches such that each point $x \in M$ is in some U_k satisfying (7.2.18) with $\varepsilon = \frac{1}{2}$, say. Let $G_1, \ldots, G_Q$ be the domains in E^n such that $U_k = \theta_k(G_k)$ for all k. There exists a finite sequence $\varphi_1, \ldots, \varphi_S$ of Lipschitz functions on M, each of which has support interior to some U_q, and such that

$$\sum_{s=1}^{S} \varphi_s(x) = 1 \tag{7.2.23}$$

for all $x \in M$.

Now if (7.2.22) were false for the $\mathfrak{U}$ just described, there would exist a sequence $\{\omega_p\}$ of r-forms in H_2^{1r} such that $(D\,\omega_p)$ and (ω_p, ω_p) were uniformly bounded but $\|\omega_p\|_{\mathfrak{U}} \to \infty$. Then, for some s, q, and some subsequence, still called ω_p, we would have

$$\int_{G_q} \sum_{I,\alpha} (\varphi_s\, \omega_I^{(q)})_{x^\alpha}^2\, dx \to \infty,$$

where φ_s has support in U_q, since

$$\|\omega_p\|_{\mathfrak{U}} \leq \sum_{s=1}^{S} \|\varphi_s\, \omega_p\|_{\mathfrak{U}}$$

and

$$\|\varphi_s \omega\|_{\mathfrak{U}}^2 = (\varphi_s \omega_p, \varphi_s \omega_p) + \sum_{q=1}^{Q} \int_{G_q} \sum_{I,\alpha} (\varphi_s\, \omega_I^{(q)})_{x^\alpha}^2\, dx.$$

But it is easy to see that $D(\varphi_s\, \omega_p)$ and $(\varphi_s\, \omega_p, \varphi_s\, \omega_p)$ are uniformly bounded. From our choice of neighborhoods we have reached a contradiction with the fact that

$$D(\varphi_s\, \omega_p) \geq \frac{1}{2} \int_{G_q} \sum_{I,\alpha} (\varphi_s\, \omega_I^{(q)})_{x^\alpha}^2\, dx.$$

7.3. The variational method

We begin with the following lemma.

Lemma 7.3.1. *Let $\mathfrak{M}$ be any closed linear manifold in the space $\mathfrak{L}_2^r$ of r-forms on M. Then either there is no form ω of $\mathfrak{M}$ which is in H_2^{1r} or there is a form ω_0 in $\mathfrak{M} \cap H_2^{1r}$ with $(\omega_0, \omega_0) = 1$ which minimizes $D(\omega)$ among all such forms.*

Proof. If $\mathfrak{M}$ contains no form in H_2^{1r}, there is nothing to prove. Otherwise let $\{\omega_k\}$ be a minimizing sequence, i.e. one such that $(\omega_k, \omega_k) = 1$ and $\omega_k \in \mathfrak{M} \cap H_2^{1r}$ for $k = 1, 2, \ldots$, and such that $D(\omega_k)$ approaches its infimum for all $\omega \in \mathfrak{M} \cap H_2^{1r}$. From Theorem 7.2.6 it follows that the $((\omega_k, \omega_k))_{\mathfrak{U}}$ are uniformly bounded. Accordingly, a subsequence, still called $\{\omega_k\}$, exists which converges weakly in H_2^{1r} to some form ω_0. But from Theorem 7.2.5. ω_k tends strongly in $\mathfrak{L}_2^r$ to ω_0 and $D(\omega)$ is lower-semicontinuous with respect to weak convergence in H_2^{1r}. The proof of the lemma is now complete.

Definition 7.3.1. *A harmonic field ω on M* is a form in H_2^1 on M for which $d\,\omega = \delta\,\omega = 0$ almost everywhere. We will let $\mathfrak{H}^r$ denote the linear manifold of harmonic fields on M of degree r.

Theorem 7.3.1. *For each $r = 0, \ldots, \nu$ (= dim M) the linear manifold $\mathfrak{H}^r$ is finite dimensional.*

Proof. The H_2^1-forms are dense in $\mathfrak{L}_2^r$, since the Lipschitz forms are. Let $\mathfrak{M}_1 = \mathfrak{L}_2^r$. There is a form ω_1 in $\mathfrak{M}_1 \cap H_2^{1r}$ which minimizes $D(\omega)$ among all such forms with $(\omega, \omega) = 1$. Let $\mathfrak{M}_2$ be the closed linear manifold in $\mathfrak{L}_2^r$ orthogonal to ω_1, and let ω_2 be the corresponding minimizing form in $\mathfrak{M}_2$. By continuing this process, we may determine successive minimizing forms $\omega_1, \omega_2, \omega_3, \ldots$, each satisfying $(\omega_k, \omega_k) = 1$ and being orthogonal to all the preceding ones.

Now if $D(\omega_1) > 0$, there are no harmonic fields $\neq 0$ since $D(\omega_1) \leq D(\omega_2) \leq \cdots$. On the other hand, suppose $D(\omega_k) = 0$ for all values of k. Then by Theorem 7.2.6, $((\omega_k, \omega_k))_{\mathfrak{U}}$ is uniformly bounded in k, whence a subsequence $\{\omega_p\}$ converges weakly in H_2^{1r} and hence strongly in $\mathfrak{L}_2^r$ to some form ω_0 in H_2^{1r}. This is impossible since the ω_k form an orthonormal system in $\mathfrak{L}_2^r$.

Theorem 7.3.2. *For each coordinate covering $\mathfrak{U}$ of M there is a constant λ_0 such that*

$$D(\omega) \geq \lambda_0 ((\omega, \omega))_{\mathfrak{U}} \tag{7.3.1}$$

for any ω in H_2^{1r} which is orthogonal to $\mathfrak{H}^r$.

Proof. For, let ω_0 be that form in H_2^{1r} (there is one since each harmonic field is in H_2^1) which minimizes $D(\omega)$ among all ω in H_2^{1r} with $(\omega, \omega) = 1$ and ω orthogonal to $\mathfrak{H}^r$. Then clearly $D(\omega_0) > 0$ and by homogeneity

$$D(\omega) \geq D(\omega_0)\,(\omega, \omega) \tag{7.3.2}$$

for all ω in H_2^{1r} and orthogonal to $\mathfrak{H}^r$. By Theorem 7.2.6 we see that

$$K_{\mathfrak{U}} ((\omega, \omega))_{\mathfrak{U}} \leq \{1 + L_{\mathfrak{U}}/D(\omega_0)\}\, D(\omega), \tag{7.3.3}$$

from which (7.3.1) follows.

Theorem 7.3.3. *Suppose ω_0 is any form in $\mathfrak{L}_2^r$ and orthogonal to $\mathfrak{H}^r$. Then there is a unique form Ω_0 in H_2^{1r} and orthogonal to $\mathfrak{H}^r$ such that*

$$(d\Omega_0, d\zeta) + (\delta\Omega_0, \delta\zeta) = (\omega_0, \zeta) \tag{7.3.4}$$

for every ζ in H_2^{1r}. Moreover, the transformation from ω_0 to Ω_0 is a bounded linear transformation from $\mathfrak{L}_2^r$ into H_2^{1r}.

Proof. We see from Theorem 7.3.2 that

$$\begin{aligned} I(\omega) &= D(\omega) - 2(\omega, \omega_0) \\ &\geq \lambda_0 ((\omega, \omega))_{\text{u}} - (\lambda_0/2)(\omega, \omega) - (2/\lambda_0)(\omega_0, \omega_0) \\ &\geq (\lambda_0/2)((\omega, \omega))_{\text{u}} - (2/\lambda_0)(\omega_0, \omega_0) \end{aligned} \tag{7.3.5}$$

for all ω in H_2^{1r} and orthogonal to $\mathfrak{H}^r$. Also, since (ω, ω_0) is a (continuous) linear functional on H_2^{1r}, it follows that $I(\omega)$ is lower-semicontinuous with respect to weak convergence in H_2^{1r}. From (7.3.5) we see that $((\omega_k, \omega_k))_{\text{u}}$ is uniformly bounded in any minimizing sequence; the existence of the minimizing form Ω_0 is established as usual. Since $\omega = 0$ is allowed we have

$$I(\Omega_0) \leq 0, \ (\lambda_0/2)((\Omega_0, \Omega_0))_{\text{u}} \leq (2/\lambda_0)(\omega_0, \omega_0), \tag{7.3.6}$$

from which the last statement will follow when the others are established.

Now let ζ be any form in H_2^{1r} and orthogonal to $\mathfrak{H}^r$. Then

$$\begin{aligned} I(\Omega_0 + \lambda\zeta) = I(\Omega_0) + 2\lambda[(d\Omega_0, d\zeta) + (\delta\Omega_0, \delta\zeta) - \\ - (\omega_0, \zeta)] + \lambda^2 D(\zeta). \end{aligned} \tag{7.3.7}$$

Since $I(\Omega_0 + \lambda\zeta) \geq I(\Omega_0)$ for all λ, we see that (7.3.4) must hold for all ζ orthogonal to $\mathfrak{H}^r$ and hence for all ζ in H_2^{1r} since any such can be writtem uniquely in the form $\zeta = H + \zeta_0$, where ζ_0 is orthogonal to $\mathfrak{H}^r$ and $dH = \delta H = 0$. If Ω_1 also satisfied (7.3.4) for all ζ in H_2^{1r}, we would have

$$(d\Omega_0 - d\Omega_1, d\zeta) + (\delta\Omega_0 - \delta\Omega_1, \delta\zeta) = 0 \tag{7.3.8}$$

for all ζ in H_2^{1r}, in particular for $\zeta = \Omega_0 - \Omega_1$, from which it would follow that $\Omega_0 - \Omega_1$ was harmonic. But then $\Omega_0 - \Omega_1 = 0$, since both are orthogonal to $\mathfrak{H}^r$.

Definition 7.3.2. The form Ω_0 of Theorem 7.3.3 is called the *potential* of ω_0.

We observe that if all the forms in (7.3.4) were sufficiently differentiable, then (7.3.4) would imply that

$$\Delta\Omega_0 \equiv d\delta\,\Omega_0 + \delta\, d\Omega_0 = \omega_0. \tag{7.3.9}$$

7.4. The decomposition theorem. Final results for compact manifolds without boundary

The defining equation (7.3.4) for potentials is a special case of the equations

$$(d\omega - \varphi, d\zeta) + (\delta\omega - \psi, \delta\zeta) - (\eta, \zeta) = 0 \tag{7.4.1}$$

for all ζ in H_2^1. We will have occasion below to use these more general equations. Referring to the defining formulas for d and δ, for any ζ in H_2^1 whose support is in a single coordinate patch, we see that (7.4.1) is of the form

$$\int_G \sum_{IJ} \{\zeta_{Ix^\alpha} [a^{IJ\alpha\beta}(x)\, \omega_{Jx^\beta} + b^{IJ\alpha}(x)\, \omega_J + e^{I\alpha}(x)] +$$

$$(7.4.2) \qquad + \zeta_I [b^{JI\alpha}(x)\, \omega_{Jx^\alpha} + c^{IJ}(x)\, \omega_J + f^I(x)]\} dx = 0$$

for all ζ in H_{20}^1; the symmetry properties of the coefficients and their dependence on the g_{ij} and their derivatives have been noted in (7.2.19). The e' s and f' s are linear combinations of the φ' s, ψ' s, and η' s with the coefficients of the φ' s and ψ' s in the e' s and those of the η' s in the f' s not involving the derivatives of the g_{ij}; those of the φ' s and ψ' s in the f' s possibly do involve such derivatives. Thus if the manifold M is of class C_μ^k (at least C_1^1), then the g_{ij} and hence the a' s and the coefficients in the e' s and of the η' s in the f' s are of class C_μ^{k-1}; the b' s and c' s and the coefficients of the φ' s and ψ' s in the f' s are of class C_μ^{k-2} (bounded and measurable if $k = \mu = 1$). Moreover, if $G = B(0, R)$ and the $g_{ij}(0) = \delta_{ij}$, then the a' s reduce to what they are in E^n.

By repeating the argument at the end of the proof of Lemma 7.2.1 and assuming that ζ vanishes on and near ∂G, we see that we may add certain constants to the a' s which do not affect the value of (7.4.2) for any such ζ, whence

$$(7.4.3) \qquad a^{IJ\alpha\beta}(0) = \delta_{j_1}^{i_1} \ldots \delta_{j_r}^{i_r} \delta^{\alpha\beta}.$$

Thus the equations (7.4.2) are of the type discussed in §§ 5.2 and 5.5.

From the results of §§ 5.2 and 5.5 we may draw the following conclusions about the harmonic fields and potentials discussed in § 7.3. Corresponding results hold for the more general equations (7.4.1):

(i) *If M is of class C_1^1 and ω is an r-form in $\mathfrak{L}_2$ on M, then the potential Ω of ω is in H_2^1 on M if $r \geq 1$; Ω and its derivatives are in H_2^1 in any coordinate system if $r = 0$. If $\omega \in \mathfrak{L}_p$ with $p \geq 2$, then $\Omega \in H_p^1$ and in C_μ^0 if $p > \nu$, $\mu = 1 - \nu/p$.*

(ii) *If M is of class C_μ^2, where $0 < \mu < 1$ and $\omega \in C_\mu^0$, then $\Omega \in C_\mu^1$ if $r \geq 1$; $\Omega \in C_\mu^2$ if $r = 0$.*

(iii) *If M is of class C_1^2 and $\omega \in \mathfrak{L}_2$, then the components of Ω and their derivatives in any coordinate system are in H_2^1. If $\omega \in \mathfrak{L}_p$, then Ω and its derivatives are in H_p^1; $\Omega \in C_\mu^1$ if $p > \nu$ and $\mu = 1 - \nu/p$.*

(iv) *If M is of class C_μ^k, $0 < \mu < 1$ and $k \geq 3$ and $\omega \in C_\mu^{k-3}$, then $\Omega \in C_\mu^{k-1}$; if $\omega \in C_\mu^{k-2}$, then $\Omega \in C_\mu^k$ if $r = 0$.*

(v) *If M and ω are analytic or C^∞, then Ω is analytic or C^∞, respectively.*

(vi) *If ω is a harmonic field and M is of class C_μ^k, then ω is of class C_μ^{k-1}* (C_μ^k for 0-forms); *if M is analytic or C^∞, then ω is also.*

We begin with a lemma which follows at once from the relations $d\,d = \delta\,\delta = 0$ whenever M and the forms are of class C^2.

Lemma 7.4.1. *Suppose that α and β are in H_2^1 on M. Then*

$$(\delta\alpha, d\beta) = 0 \quad \text{and} \quad (\delta\alpha, \beta) = (\alpha, d\beta).$$

Proof. Suppose $\mathfrak{U} = (U_1, \ldots, U_Q)$ is an open covering of M by coordinate patches. There is a sequence $\varphi_1, \ldots, \varphi_S$ of Lipschitz functions such that $\varphi_s \geq 0$, $\varphi_1 + \cdots + \varphi_S = 1$ on M and such that the support of $\varphi_s + \varphi_t$ is contained in some U_q whenever the supports of φ_s and φ_t intersect. If we write $\alpha_s = \varphi_s\,\alpha$ and $\beta_t = \varphi_t\,\beta$, then

$$(\delta\alpha, d\beta) = \sum_{s,t} (\delta\alpha_s, d\beta_t), \quad (\delta\alpha, \beta) = \sum_{s,t} (\delta\alpha_s, \beta_t),$$

the sums being over all ordered pairs (s, t) such that the supports of φ_s and φ_t intersect (and hence both lie in some U_q). Thus the proofs of both parts are reduced to the case where both have support interior to some U_q. But then in the domain G_q, we may approximate to the components of α and β strongly in H_2^1 by C^2 functions; we may approximate to the g_{ij} by functions g_{pij} of class C_μ^3 in such a way that the g_{pij} satisfy the same Lipschitz condition as g_{ij}, and the respective first derivatives converge almost everywhere to those of g_{ij}. The result follows easily.

Theorem 7.4.1. *Suppose M is of class C_1^1, ω is in $\mathfrak{L}_2$ on M, and Ω is the potential of ω. Then*

(i) *$\alpha = d\Omega$ and $\beta = \delta\Omega$ are in H_2^1 and satisfy*

$$\omega = \delta\alpha + d\beta = \delta(d\Omega) + d(\delta\Omega), \qquad (7.4.4)$$
$$d\alpha = \delta\beta = 0.$$

(ii) *If $\omega \in H_2^1$, then α and β are the potentials of $d\omega$ and $\delta\omega$, respectively, so that $\delta\alpha$ and $d\beta$ are also in H_2^1.*

(iii) *If $\omega \in \mathfrak{L}_p (p \geq 2)$, then Ω, $d\Omega$, and $\delta\Omega$ are in H_p^1 and in C_μ^0 if $p > \nu$ and $\mu = 1 - \nu/p$.*

(iv) *If M is of class C_μ^k with $0 < \mu < 1$ and $\omega \in C_\mu^{k-2}$, then Ω, $d\Omega$, and $\delta\Omega$ are in C_μ^{k-1} with $k \geq 2$.*

Proof. We shall prove that (7.4.4) holds in some neighborhood of each point of M and will also show that each point x of M is in some neighborhood U such that

$$(d\alpha, d\zeta) + (\delta\,\alpha, \delta\,\zeta) - (d\omega, \zeta) = 0 \qquad (7.4.5)$$
$$(d\beta, d\zeta) + (\delta\,\beta, \delta\,\zeta) - (\delta\,\omega, \zeta) = 0$$

for all ζ in H_2^1 with support in U if ω is in H_2^1, or

$$(d\alpha, d\zeta) + (\delta\,\alpha - \omega, \delta\,\zeta) = 0 \qquad (7.4.6)$$
$$(d\beta - \omega, d\zeta) + (\delta\,\beta, \delta\,\zeta) = 0$$

for such ζ if ω is merely in $\mathfrak{L}_2$. Since M can be covered by a finite number of such neighborhoods and any $\zeta \in H_2^1$ may be written as the sum of a finite number of ζ_s, each of which has support in some one such neighborhood, (7.4.5) and (7.4.6) hold for all ζ in H_2^1 on M. The regularity results follow from equations (7.4.5) and (7.4.6), the results for systems of §§ 5.2, 5.5 and 6.4, and the remarks at the beginning of this section.

Choose $P_0 \in M$ and an admissible coordinate system θ with domain $G = B(0, R)$ such that $g_{ij}(0) = \delta_{ij}$. Then the equations (7.3.4), when expressed in terms of the components Ω_I (and certain integrals of jacobians which have constant coefficients and which vanish if the components $\in H^1_{20}$ are added (cf. § 7.2)) take the form (5.2.2), where $B(u, v)$, $B_1(u, v)$, and $B_2(u, v)$ are defined in (5.2.16) and the coefficients satisfy (5.2.17) and (5.2.18) (with $\mu_1 = 2$ since our coefficients are bounded). Let us approximate the components ω_I, assumed in $H^1_2(G)$, and g_{ij} by functions ω_{pI} and $g_{pij} \in C^\infty(\bar{G})$ so that $\omega_{pI} \to \omega_I$ in $H^1_2(G)$ and the $g_{pij} \to g_{ij}$ uniformly so that all the g_{pij} satisfy the same Lipschitz condition as the g_{ij} and their derivatives converge almost everywhere to those of the g_{ij}. It is clear from Theorem 5.2.1 that we may assume that R is so small that

$$B_p(u, u) \geq m_1 (\| u \|^1_2)^2, \quad u \in H^1_{20}(G), \quad m_1 = m/2,$$

for every p (and degree r) and that (5.2.6) is satisfied for all p. Then, by writing $\Omega_{pI} = \Omega_I + \eta_{pI}$, $\eta_{pI} \in H^1_{20}(G)$, we see using Theorems 5.2.1 and 5.2.2 that, for each p, there is a unique solution vector Ω_{pI} of the equations corresponding to (7.3.4) with ω_0 replaced by ω_p and which coincides on ∂G with Ω_I. From the interior regularity theorem (§ 5.2, C^∞ case) it follows that each $\Omega_{pI} \in C^\infty(G)$. Also, from Theorems 5.2.1 and 5.2.7, we see that $\Omega_{pI} \to \Omega_I$ in $H^1_2(G)$. It follows that the vectors $\alpha_{pJ} = (d\Omega_p)_J$ and $\beta_{pK} = (\delta \Omega_p)_K$ converge strongly in $\mathfrak{L}_2(G)$ to α_J and β_K. For each p it is easy to see that the vectors α_{pJ} and β_{pK} satisfy the equations corresponding to (7.4.5) and (7.4.6) with the ω_I replaced by the ω_{pI}. From Theorems 5.2.1, 5.2.2, 5.2.5, and 5.2.7, it follows that the $\alpha_{pJ} \to \alpha_J$ and $\beta_{pK} \to \beta_K$ strongly in $H^1_2(\Gamma)$ for each Γ with $\bar{\Gamma} \subset G$. Thus α_J and $\beta_K \in H^1_2(\Gamma)$ for such Γ and satisfy the equations corresponding to (7.4.5) and (7.4.6). If the ω_I are merely in $\mathfrak{L}_2(G)$, we can still approximate to ω_I strongly in $\mathfrak{L}_2(G)$ by the ω_{pI}. In this case we still have the strong convergence of α_{pJ} to α_J and β_{pK} to β_K in $H^1_2(\Gamma)$ for $\bar{\Gamma} \in G$ so that α and $\beta \in H^1_2$ and satisfy (7.4.6). In case $\omega \in H^1_2$, the equations (7.4.5) are the defining equations which show that α and β are the respective potentials of $d\omega$ and $\delta\omega$, so that $\delta\alpha$, $d\alpha$, $\delta\beta$, and $d\beta$ all $\in H^1_2$.

We next prove a decomposition theorem for r-forms in $\mathfrak{L}_2$ on M which strengthens the result of K. Kodaira mentioned in our Introduction.

Theorem 7.4.2. *If $\omega \in \mathfrak{L}^r_2$ on M, then there are H^1_2 forms H, α, and β, where H is a harmonic field, such that*

$$\omega = H + \delta\alpha + d\beta \tag{7.4.7}$$

$$d\alpha = \delta\beta = 0, \quad \alpha = d\Omega, \quad \beta = \delta\Omega,$$

where Ω is the potential of $\omega - H$. The three forms in (7.4.7) are mutually orthogonal in $\mathfrak{L}_2^r$, and the totality of the elements in each is a closed linear manifold. If H_1, α_1, β_1 are any forms in H_2^1, where H_1 is a harmonic field and

$$\omega = H_1 + \delta\,\alpha_1 + d\beta_1, \tag{7.4.8}$$

then $H_1 = H$, $\delta\,\alpha_1 = \delta\,\alpha$, and $d\beta_1 = d\beta$.

Proof. Since $\mathfrak{H}^r$ is finite dimensional, there is a unique $H \in \mathfrak{H}^r$ which is nearest ω. Then $\omega - H$ is orthogonal to $\mathfrak{H}^r$, whence its potential Ω exists and has the properties mentioned in Theorem 7.4.1. Thus (7.4.7) defines a decomposition which is easily seen to have the desired type, using Lemma 7.4.1. Moreover, suppose $\{\alpha_p\}$ is a sequence of H_2^1 forms such that $d\alpha_p = 0$ for each p and $\delta\,\alpha_p \to$ some η strongly in $\mathfrak{L}_2$, each $\alpha_p = d\Omega_p$ being orthogonal to $\mathfrak{H}^r$; then we see that

$$D\,(\alpha_p - \alpha_q) = (\delta\,\alpha_p - \delta\,\alpha_q, \delta\,\alpha_p - \delta\,\alpha_q) \to 0;$$

it follows that the α_p form a Cauchy sequence in H_2^1, using Theorem 7.3.2. Hence the $\alpha_p \to$ some α strongly in H_2^1 with $d\alpha = 0$, $\delta\,\alpha = \eta$. A similar proof holds for the manifold of $d\beta$' s.

Next, suppose that H_1, α_1 and β_1 are H_2^1 forms satisfying (7.4.8), where H_1 is a harmonic field. From Lemma 7.4.1 we see that the three components are mutually orthogonal, from which we immediately conclude that $H_1 = H$ and $\delta\,(\alpha - \alpha_1) = d\,(\beta_1 - \beta)$. Since these last two are orthogonal by Lemma 7.4.1 they must both be zero.

Theorem 7.4.3. *If $\omega \in H_2^1$ on M, then in the decomposition of Theorem 7.4.2 we have $\delta\,\alpha$ and $d\beta$ in H_2^1 also.*

Proof. This follows at once from Theorem 7.4.1, since H is always in H_2^1.

Theorems 7.4.2 and 7.4.3 lead immediately to the following theorem:

Theorem 7.4.4. *Each cohomology class of forms in H_2^1 contains a unique harmonic field.*

Proof. For if $d\omega = 0$, then $\delta\alpha$ must be closed and hence zero in (7.4.7).

Remarks. The further differentiability properties of the forms $\delta\alpha$ and $d\beta$ in (7.4.7) follow from Theorem 7.4.1 and the other regularity properties stated at the beginning of the section.

In his book, Geometric Integration Theory, H. Whitney [2] considered a class of differential forms (the "flat" forms) analogous to those described in the following definition.

Definition 7.4.1. An r-form ω in $\mathfrak{L}_2$ on M is said to be $\mathfrak{L}_2$-flat $\Leftrightarrow$ there exists an $\mathfrak{L}_2$ form η and a sequence $\{\omega_p\}$ in H_2^1 such that $\omega_p \to \omega$ and $d\omega_p \to \eta$ strongly in $\mathfrak{L}_2$ on M.

Theorem 7.4.5. *An r-form ω in $\mathfrak{L}_2$ on M is $\mathfrak{L}_2$—flat if and only if $\delta\,\alpha \in H_2^1$ in the decomposition (7.4.7).*

Proof. Let ω be $\mathfrak{L}_2$—flat, and let η and $\{\omega_p\}$ be as in Definition 7.4.1. If we write (7.4.4) for each p, we have H_p, $\delta\,\alpha_p$, and $d\beta_p$ in H_2^1, where α_p is the potential of $d\omega_p$. It follows that α is the potential of η, whence $\delta\,\alpha$ is in H_2^1.

Conversely, suppose $\delta\,\alpha$ is in H_2^1. We may approximate to ω strongly in $\mathfrak{L}_2$ by $\{\omega'_p\}$ with each ω'_p in H_2^1. Then if we write (7.4.4) for each p, the forms β'_p and $d\beta'_p$ are in H_2^1 and $d\beta'_p$ tend strongly in $\mathfrak{L}_2$ to $d\beta$. Now, if we define ω_p by

$$\omega_p = H + \delta\,\alpha + d\beta'_p$$

we see that $\{\omega_p\}$ is a sequence as in Definition 7.4.1; it follows that ω is $\mathfrak{L}_2$-flat.

7.5. Manifolds with boundary

Definitions 7.5.1. For a ν-dimensional manifold M with boundary $b\,M$ of class $C_\mu^k (0 \le \mu \le 1)$ (C^∞, analytic) we adopt the standard definition: each point of $M \cup b\,M$ is contained in some set $\mathfrak{N}$, open on $M \cup b\,M$, which is either the homeomorphic image of the unit ball in R_ν or of the part of it for which $x^\nu \le 0$ in which latter case, the points where $x^\nu = 0$ correspond to $\mathfrak{N} \cap b\,M$; any two coordinate systems are related by a transformation of class C_μ^k (C^∞, analytic). Any coordinate system with a Lipschitz domain G is *admissible* if it is related in this way to one of the preferred ones and $\mathfrak{N} \cap b\,M$ corresponds to a part of ∂G. We still allow even and odd forms and define the spaces $\mathfrak{L}_p$ and H_p^1 and the inner products (ω, η) and $((\omega, \eta))_{\text{u}}$ and their corresponding norms as before. The differential operators d and δ are defined as before.

Assumption. *We assume M compact and connected and of class at least C_1^1.*

Theorem 7.5.1. *The spaces $\mathfrak{L}_2$ and H_2^1 of forms of a given kind and degree are Hilbert spaces. The operators d and δ are bounded operators from H_2^1 to $\mathfrak{L}_2$, preserving even-ness or odd-ness, and $D(\omega)$ is lower-semicontinuous with respect to weak convergence in H_2^1. Finally, if $\omega_n \rightharpoonup \omega$ in H_p^1, then $\omega_n \to \omega$ in $\mathfrak{L}_p$.*

In order to define boundary values and the normal and tangential parts of a form on $b\,M$, it is necessary to define admissible boundary coordinate systems.

Definition 7.5.2. An *admissible boundary coordinate system* on a manifold M with boundary $b\,M$ of class C_μ^k is a coordinate system of class C_μ^k which maps its (Lipschitz) domain $G \cup \sigma$ onto a boundary neighborhood $\mathfrak{N}$ in such a way that σ, the part of ∂G on $x^\nu = 0$, is not empty and open and is mapped onto $\mathfrak{N} \cap b\,M$, and such that the metric is of the form

$$ds^2 = \sum_{\gamma,\delta=1}^{\nu-1} g_{\gamma\delta}(x'_\nu, 0)\, dx^\gamma\, dx^\delta + (dx^\nu)^2 \quad \text{on } \sigma. \tag{7.5.1}$$

Lemma 7.5.1. (a) *If M ($\cup\, bM$) is of class C^k_μ, each point P_0 of bM is in the range of an admissible boundary coordinate system.*

(b) *If $k \geq 2$, each point P_0 of $b\,M$ is in the image of a coordinate system of class C^{k-1}_μ which satisfies all the other conditions of Definition 7.5.1 and is such that the metric takes the form* (7.5.1) *with $(x'_\nu, 0)$ replaced by any x in G. If M is of class C^∞ (analytic) this coordinate system may be taken to be of class C^∞ (analytic) and hence admissible.*

(c) *If (x) and (y) are overlapping admissible boundary coordinate sytems (or two systems as in* (b)), *then,*

$$y^\alpha_{,\nu}(x'_\nu, 0) = y^\nu_{,\alpha}(x'_\nu, 0) = 0, \quad \alpha < \nu, \quad y^\nu_{,\nu}(x'_\nu, 0) = 1. \tag{7.5.2}$$

Proof. Let (x) be any coordinate system of class C^k_μ which maps $G \cup \sigma$ onto a boundary neighborhood $\mathfrak{N}$ of P_0 in the usual manner. Suppose (ξ) is another similar coordinate system and let $G_{\alpha\beta}$ and $g_{\alpha\beta}$ be the metric tensors of the (ξ) and (x) systems, respectively. Then we want $G_{\nu\alpha} = \delta_{\nu\alpha}$, at least if $\xi^\nu = 0$, so that we want

$$g_{\gamma\varepsilon} \frac{\partial x^\gamma}{\partial \xi^\alpha} \frac{\partial x^\varepsilon}{\partial \xi^\nu} = \delta_{\alpha\nu} \tag{7.5.3}$$

at least on $\xi^\nu = 0$. If we multiply (7.5.3) by $\partial \xi^\alpha / \partial x^\theta$ and sum we obtain

$$g_{\theta\varepsilon} \frac{\partial x^\varepsilon}{\partial \xi^\nu} = \frac{\partial \xi^\nu}{\partial x^\theta} = \tau\, \delta^\nu_\theta, \quad \tau = \frac{\partial \xi^\nu}{\partial x^\nu}, \tag{7.5.4}$$

since, of course, $\xi^\nu = 0$ when $x^\nu = 0$. Multiplying (7.5.4) by $g^{\lambda\theta}$ and summing, we obtain

$$\frac{\partial x^\lambda}{\partial \xi^\nu} = \tau g^{\lambda\nu} \tag{7.5.5}$$

where we find, by multiplying by $\partial \xi^\nu / \partial x^\lambda$, using (7.5.4) and summing on λ, that

$$\tau^2 g^{\nu\nu} = 1. \tag{7.5.6}$$

A coordinate system as in (b) may be obtained for ξ in a 1-sided neighborhood of σ by solving the differential equations (7.5.5) with initial conditions

$$x^\lambda(\xi'_\nu, 0) = \xi^\lambda, \quad \lambda < \nu, \quad x^\nu(\xi'_\nu, 0) = 0. \tag{7.5.7}$$

If the (x) system already satisfies (7.5.1) it is clear that

$$\tau = 1, \quad \frac{\partial x^\lambda}{\partial \xi^\nu} = \delta_{\lambda\nu}, \quad \frac{\partial x^\nu}{\partial \xi^\alpha} = \delta^\nu_\alpha \quad \text{on} \quad \xi^\nu = 0 \tag{7.5.8}$$

which proves (b).

In order to obtain an *admissible* coordinate system (ξ), it is necessary only to define functions x^λ *of class* C^k_μ on a 1-sided neighborhood of σ which satisfy (7.5.5), (7.5.6), and (7.5.7) along σ. This may be done by defining

$$x^\lambda(\xi'_\nu, \xi^\nu) = \xi^\lambda + \xi^\nu \int\limits_{B(\xi'_\nu, \varrho)} \varphi^*_\varrho(\eta'_\nu - \xi'_\nu) f^\lambda(\eta'_\nu)\, d\eta'_\nu, \quad \varrho = -\xi^\nu > 0$$

$$(7.5.9) \quad x^\nu(\xi'_\nu, \xi^\nu) = \xi^\nu \int\limits_{B(\xi'_\nu, \varrho)} \varphi^*_\varrho(\eta'_\nu - \xi'_\nu) f^\nu(\eta'_\nu)\, d\eta'_\nu,$$

$$f^\lambda(\eta'_\nu) = \tau(\eta'_\nu, 0)\, g^{\lambda\nu}(\eta'_\nu, 0)$$

for instance, where $B(\xi'_\nu, -\xi^\nu)$ denotes the $(\nu - 1)$-dimensional ball with center at ξ'_ν and radius $-\xi^\nu > 0$, φ is a mollifier in $R_{\nu-1}$ and $\varphi^*_\varrho(y) = \varrho^{1-\nu}\varphi(\varrho^{-1}y)$.

Definition 7.5.2. Suppose $\omega \in H^1_2$ on M. We say that the *tangential part* $t\,\omega$ *or the normal part* $n\,\omega$ *of* ω *vanishes on* $b\,M$ if and only if

$$(7.5.10) \qquad \omega_{i_1 \ldots i_r}(x'_\nu, 0) = 0 \quad \text{if all} \quad i_\gamma < \nu, \quad \text{or}$$

$$(7.5.11) \qquad \omega_{i_1 \ldots i_r}(x'_\nu, 0) = 0 \quad \text{if some} \quad i_\gamma = \nu,$$

respectively, x being an admissible boundary coordinate system.

Remark. These are invariantly defined by virtue of (7.5.2).

Lemma 7.5.2. (a) *If* $\omega_n \rightharpoonup \omega$ *in* H^1_2 *and* $t\,\omega_n(n\,\omega_n)$ *vanishes on* bM *for each* n, *then* $t\,\omega$ $(n\,\omega)$ *vanishes on* $b\,M$.

(b) *If* M *is of class* C^k *with* $k \geq 2$ *and* $\omega \in C^1$ *and* $t\,\omega = 0$ $(n\,\omega = 0)$ *on* $b\,M$, *then* $t\,d\omega = 0\,(n\,\delta\,\omega = 0)$ *on* $b\,M$.

Proof. (a) follows from the theorems in Chapter 3 on H^1_2 functions. Using a partition of unity, one sees that it is sufficient to prove (b) for forms ω having support in the range of some admissible boundary coordinate system. For such an ω, one sees immediately from (7.5.10) and (7.2.11) that if (7.5.10) holds, then all the $(d\omega)_{j_1 \ldots j_{r+1}}$ with all $j_\gamma < \nu$ satisfy (7.5.10). Also, if (7.5.11) holds we see from (7.2.14) and the facts that $g_{\nu\alpha} = g^{\nu\alpha} = \delta_{\nu\alpha}$ in an admissible boundary coordinate system, that $(\delta\,\omega)_{i_1 \ldots i_{r-1}} = 0$ if any $i_\gamma = \nu$. This proves the results.

Definition 7.5.3. If φ and ψ are given by (7.2.1), we define

$$(7.5.12) \qquad \langle \varphi, \psi \rangle (P) = F(P; \varphi, \psi)$$

where $F(P; \varphi, \psi)$ is defined in (7.2.5).

Lemma 7.5.3. *Suppose* α *and* β *are* r *and* $(r - 1)$*-forms respectively, in* H^1_2 *which are of the same kind. Then*

$$(7.5.13) \qquad (\alpha, d\beta) = (\delta\,\alpha, \beta) + \int\limits_{bM} \langle (-1)^{r-1} n\,\alpha, t\,\beta \rangle\, dS(P).$$

If either $n\,\alpha$ *or* $t\,\beta = 0$ *on* $b\,M$, *then* $(\delta\,\alpha, d\beta) = 0$.

Proof. We prove (7.5.13) first. There is a partition of unity ζ_s, $s = 1, \ldots, S$, such that if the supports of ζ_s and ζ_t intersect, their union is in the range of some one coordinate patch. Thus, by setting $\alpha_s = \zeta_s\,\alpha$, $\beta_t = \zeta_t\,\beta$, we see that it is sufficient to prove the theorem for the case that α and β both have their supports in the range $\mathfrak{N}$ of some one coordinate patch. If $\mathfrak{N} \subset M^{(0)}$, (7.5.13) follows from § 7.2 with no boundary integral. So suppose $\mathfrak{N}$ is a boundary neighborhood and (x) is an admissible boundary coordinate system with domain $G_R \cup \sigma_R$ (in this chapter $x^\nu < 0$ on G_R). We may approximate the $g_{\gamma\delta}$ and and the components of α and β by smooth functions (Theorems (3.1.3) and (3.4.1) on $\bar{G}_R$. Then by looking at (7.2.15) we obtain

$$(\alpha, d\beta) = (\delta\alpha, \beta) + \frac{1}{r!}\int_{\sigma_R} \sum_{\substack{\gamma\,(i)\,(j)\\ j_\gamma=\nu}} \sum (-1)^{\gamma-1} g^{i_1 j_1} \dots g^{i_r j_r} \alpha_{(i)} \beta_{(j'_\gamma)} \Gamma \, dS(x).$$

Now if for each γ we set $j_\gamma = \nu$ and hence $i_\gamma = \nu$, since $g^{\nu\alpha} = 0$ if $\alpha < \nu$, move the index $i_\gamma \,(= \nu)$ to the first subscript of α, replace i'_γ by $k_1 \dots k_{r-1}$ and j'_γ by $l_1 \dots l_{r-1}$, we obtain

$$(\alpha, d\beta) = (\delta\alpha, \beta) + \frac{1}{(r-1)!}\int_{\sigma_R} \sum_{(k)(l)} g^{k_1 l_1} \dots g^{k_{r-1} l_{r-1}} \alpha_{\nu k_1 \dots k_{r-1}} \times$$
$$\times \beta_{l_1 \dots l_{r-1}} \, dS(P)$$

from which (7.5.13) follows since all the $l_1 \dots l_{r-1}$ are $< \nu$, since $j_\gamma = \nu$. Clearly $dS(P) = \Gamma(x)\, dS(x)$ since $g_{\nu\alpha}(x) = \delta_{\nu\alpha}$ for x on σ_R.

To prove the second result we may approximate as above and, since $d^2\varphi = \delta^2\varphi = 0$, we see that

$$(\delta\alpha, d\beta) = \pm \int_{bM} \langle n\,\delta\alpha, t\beta\rangle \, dS(P) = 0$$

if α, β and the metric are smooth, since it follows from Lemma 7.5.2 that $n\,\delta\alpha = 0$ if $n\,\alpha = 0$ on $b\,M$. The result follows by passing to the limit.

The following corollary will be useful later:

Corollary. *If M is of class C^k with $k \geq 3$ and $\omega \in C^2(\bar{M})$, then*

$$(d\omega, d\psi) + (\delta\omega, \delta\psi) = (\Delta\omega, \psi) + \int_{bM} (-1)^r \{\langle n\, d\omega, t\zeta\rangle + \langle t\,\delta\omega, n\,\zeta\rangle\} \, dS,$$

(7.5.14) $$\Delta = d\delta + \delta d.$$

It is now important to develop a formula for $D_0(\omega)$, where

(7.5.15) $$D_0(\omega) = (d\omega, d\omega)_0 + (\delta_0\,\omega, \delta_0\,\omega)_0,$$

δ_0 and the inner products being formed using the Euclidean metric, ω having its support in G_R (perhaps $\neq 0$ on σ_R) and being smooth. From (7.2.11) and (7.2.14), we obtain

(7.5.16)
$$(d\omega)_{m_1 \dots m_{r+1}} = \sum_{\gamma=1}^{r+1} (-1)^{\gamma-1} \omega_{(m'_\gamma),\, x^{m_\gamma}}$$
$$(\delta_0\,\omega)_{s_1 \dots s_{r-1}} = -\sum_\alpha \omega_{\alpha s_1 \dots s_{r-1},\, x^\alpha}.$$

We substitute the expressions (7.5.16) into (7.5.15) and form the inner products using the form (7.2.7) and the Euclidean metric in which $g^{ij} = \delta^{ij}$, etc. We obtain

(7.5.17) $$(d\omega, d\omega)_0 = \frac{1}{(r+1)!}\int_{G_R} \sum_{\gamma,\delta=1}^{r+1} \sum_{(m)} (-1)^{\gamma+\delta} \omega_{(m'_\gamma),\, x^{m_\gamma}}\, \omega_{(m'_\gamma),\, x^{m_\delta}} \, dx.$$

We break the sum on γ and δ into the sum where $\gamma = \delta$, that where $\gamma < \delta$ and that where $\gamma > \delta$. In the first sum, we replace m'_γ by $i_1 \ldots i_r$ and call $m_\gamma = \alpha$. In the sum where $\gamma < \delta$, we move the index m_δ to the first place in m'_γ and move m_γ into the first place of m'_δ, using the anti-symmetry of the indices; we then replace $m'_{\gamma\delta}$ (in which both m_γ and m_δ have been omitted) by $s_1 \ldots s_{r-1}$ and call $m_\delta = \beta$ and $m_\gamma = \alpha$. We do the corresponding thing for the sum where $\gamma > \delta$. Then we combine these two sums with the integral for $(\delta_0\, \omega, \delta_0\, \omega)_0$. The result is

$$D_0(\omega) = \frac{1}{r!} \int_{G_R} \sum_{\alpha, i_1 \ldots i_r} \omega^2_{i_1 \ldots i_r x^\alpha}\, dx + \frac{1}{(r-1)!} \int_{G_R} \sum_{s_1 \ldots s_{r-1}} \sum_{\alpha,\beta} (\omega_{\alpha s_1 \ldots s_{r-1} x^\alpha} \times$$

$$\times\, \omega_{\beta s_1 \ldots s_{r-1} x^\beta} - \omega_{\alpha s_1 \ldots s_{r-1} x^\beta}\, \omega_{\beta s_1 \ldots s_{r-1} x^\alpha})\, dx$$

for smooth forms, from which we obtain

$$(7.5.18) \quad D_0(\omega) = \int_{G_R} \sum_{I\alpha} \omega^2_{I x^\alpha}\, dx + \int_{\sigma_R} \sum_{T\beta} (\omega_{\nu T}\, \omega_{\beta T x^\beta} - \omega_{\nu T x^\beta}\, \omega_{\beta T})\, dS$$

$$T: 1 \leq s_1 < \cdots < s_{r-1} < \nu$$

for any form $\omega \in C^2$ on $\bar{G}_R$ which vanishes near the spherical surface of G_R.

Lemma 7.5.4. (a) *If ω (considered as a set of functions) $\in H^1_2$ on G_R, ω is zero on and near the spherical part of the surface of G_R, and if either $t\,\omega = 0$ or $n\,\omega = 0$, there exists a sequence ω_p of similar forms of class C^∞ on $\bar{G}_R$ which converge strongly in H^1_2 on G_R to ω and such that $t\,\omega_p = 0$ or $n\,\omega_p = 0$ (respectively) for each p.*

(b) *If ω satisfies the hypotheses of* (a), *then*

$$(7.5.19) \qquad D_0(\omega) = \int_{G_R} \sum_{(i)\alpha} (\omega_{(i) x^\alpha})^2\, dx.$$

Proof. Clearly (b) follows from (a) and (7.5.18). Also, since the condition $t\,\omega = 0$ is just the same as saying that the ω' s with $i_r < \nu$ vanish on σ_R *and* $n\,\omega = 0$ is the same as saying that those with $i_r = \nu$ vanish on σ_R, part (a) is just reduced to proving the theorem for functions. If the function ω is not required to be zero on σ_R, we extend ω to the whole of B_R by $\omega(-x^\nu, x'_\nu) = \omega(x^\nu, x'_\nu)$ and then note that the h mollified functions with respect to a spherically symmetric mollifier of ω are of class C^∞ and have support $\subset B_R$ if h is small enough; these tend strongly in H^1_2 to ω on B_R. If ω is zero on σ_R, we begin by extending ω to B_R by $\omega(-x^\nu, x'_\nu) = -\omega(x^\nu, x'_\nu)$ and then proceeding as above; we note that the mollified functions vanish on σ_R.

Lemma 7.5.5 (GAFFNEY [1]). *With each point P of M and each $\varepsilon > 0$ is associated an admissible coordinate system $\mathfrak{C}$ with domain G and range $\mathfrak{U}$ and a constant l such that*

$$D(\omega) \geq (1 - \varepsilon) \int_G \sum_{i,\alpha} (\omega_{i x^\alpha})^2\, dx - l(\omega, \omega)$$

for any form $\omega \in H_2^1$ *with support on* $\mathfrak{U}$ *and either* $t\,\omega = 0$ *or* $n\,\omega = 0$ *on* $b\,M$[1].

Proof. This has been proved for interior points in Lemma 7.2.1. If P is a boundary point, it is clear that we may choose an admissible boundary coordinate system with domain $G_R \cup \sigma_R$ which carries the origin into P and $G_R \cup \sigma_R$ into a neighborhood of P in which $g_{ij}(0) = \delta_{ij}$. Then, exactly as in the proof for interior points, we conclude that we may choose R so small that

$$D(\omega) \geq D_0(\omega) - \varepsilon \int_{G_R} \sum_{i,\alpha} (\omega_{i\,x^\alpha})^2\, dx - l(\omega, \omega)$$

for some l and all $\omega \in H_2^1$ with support in $\mathfrak{N}$, $D_0(\omega)$ having its significance in Lemma 7.5.4. The result follows from that lemma.

The following theorem follows from the lemma above in exactly the same way as in the case of Theorem 7.2.6.

Theorem 7.5.1. *For each finite system* $\mathfrak{N}$ *of admissible coordinate systems whose ranges cover* M, *there are constants* k *and* l *such that*

$$D(\omega) \geq k\,\|\omega\|^2 - l(\omega, \omega) \quad (k > 0)$$

for any form in H_2^1 *with either* $t\,\omega = 0$ *or* $n\,\omega = 0$ *on* $b\,M$, *the norm being that corresponding to* $\mathfrak{N}$.

7.6. Differentiability at the boundary

In this section, we discuss the regularity at the boundary of the solutions of the equations (7.4.1). The resulting systems (7.4.2) for boundary neighborhoods are not quite the same as those in Chapters 5 and 6 because we are requiring only that *some* (normal part or tangential part) of the components vanish on $b\,M$. However, the general ideas are the same as those in §§ 5.2, 5.5 and 5.6. On account of equation (7.5.18) and Lemma 7.5.5 and its proof, it follows that if $P_0 \in b\,M$, there is an admissible boundary coordinate system with domain G_R in which the origin corresponds to P_0 and $g_{\alpha\beta}(0) = \delta_{\alpha\beta}$, and if either $n\,\omega$ or $t\,\varphi = 0$ on $b\,M$, then the corresponding boundary integral to that in (7.5.18) (which then vanishes) may be subtracted off and the equations (7.4.1) take the form

$$\int_{G_R} [v^i_{,\alpha}(a^{\alpha\beta}_{ij} u^j_{,\beta} + b^\alpha_{ij} u^j + e^\alpha_i) + v^i(c^\alpha_{ij} u^j_{,\alpha} + d_{ij} u^j + f_i)]\, dx = 0$$

$$(7.6.1) \qquad v \in H^{1*}_{20}, \quad a^{\alpha\beta}_{ij}(0) = \delta_{ij}\delta^{\alpha\beta}, \quad i, j = 1, \ldots, N,$$

where H^{1*}_{20} denotes the set of all vectors $v \in H_2^1$ such that

$$(7.6.2) \qquad \text{all } v^i = 0 \text{ on } \Sigma_R^-, \quad v^i = 0 \text{ on } \sigma_R \text{ for } i = 1, \ldots, k \quad (0 \leq k \leq n).$$

[1] Here as elsewhere if ω has support in a boundary neighborhood $\mathfrak{N}$, it is not required to vanish on $b\,M \cap \mathfrak{N}$.

k being given. We have already noted that the a's are combinations of the $g_{\alpha\beta}$ whereas the b's, c's and d's involve also the derivatives of the g's; the e_i^α are combinations of the g's and the non-homogeneous terms in equations (7.4.1) and the f_i involve these and the derivatives of the g's.

We shall obtain results concerning the differentiability of u by obtaining such results for $U = \zeta u$ where ζ is a 0-form $\in C^\infty(\bar{G}_R)$ which has support on $G_R \cup \sigma_R$; results for interior neighborhoods have already been obtained in § 7.4. Assuming that v has support in $G_R \cup \sigma_R$ and that $v \in H_{20}^{1*}$, we replace v^i by ζv^i in (7.6.1) and find that U satisfies (7.6.1) with e and f replaced by E and F, respectively, where

$$\begin{aligned} E_i^\alpha &= \zeta e_i^\alpha - a_{ij}^{\alpha\beta} \zeta_{,\beta} u^j \\ F_i &= \zeta f_i + \zeta_{,\alpha} (a_{ij}^{\alpha\beta} u^j_{,\beta} + b_{ij}^\alpha u^j - c_{ij}^\alpha u^j + e_i^\alpha). \end{aligned} \tag{7.6.3}$$

So we begin by considering solutions u of (7.6.1) which have support on $G_R \cup \sigma_R$ and which $\in H_{20}^{1*}$.

We begin by altering the coefficients outside G_R as in (5.5.3) to obtain new coefficients $a_{ijR}^{\alpha\beta}$, etc. Then, if $u \in H_{20}^{1*}$, is a solution of (7.6.1), and has support in $G_R \cup \sigma_R$, it is also a solution of

$$\int_{G_{2R}} [v^i_{,\alpha} (a_{ijR}^{\alpha\beta} u^j_{,\beta} + b_{ijR}^\alpha u^j + e_i^\alpha) + v^i (c_{ijR}^\alpha u^j_{,\alpha} + d_{ijR} u^j + f_i)]\, dx = 0, \qquad v \in H_{20}^{1*}. \tag{7.6.4}$$

We then write, as in (5.5.5),

$$\begin{aligned} u &= u_R + H_R, \\ u_R &= Q_{2R}[(a_R - a_0) \cdot \nabla u + b_R \cdot u + e] + \\ &\quad + P_{2R}[c_R \cdot \nabla u + d_R \cdot u + f] \end{aligned} \tag{7.6.5}$$

where we now wish to define the vectors $U = Q_{2R}(e)$ and $V = P_{2R}(f)$ to satisfy the boundary conditions

$$U^i(x'_\nu, 0) = V^i(x'_\nu, 0) = 0, \quad i = 1, \ldots, k. \tag{7.6.6}$$

So we define Q_{2R} and P_{2R} as follows:

Definition 7.6.1. We define the vectors $U = Q_{2R}(e)$ and $V = P_{2R}(f)$ as follows: If $i \leq k$ we define $U^i(x)$ and $V^i(x)$ as in Definition 5.5.1′ and Equations (5.5.19), (5.5.20), and (5.5.21). If $i > k$, we define $P_{2R}(f)$ as the restriction to G_{2R} of the potential of $\tilde{f}$ over B_{2R}, where $\tilde{f}$ is defined. by extending f to B_{2R} by positive reflection:

$$\tilde{f}(x'_\nu, -x^\nu) = \tilde{f}(x'_\nu, x^\nu), \tag{7.6.7}$$

with the usual modification if $\nu = 2$. If $i > k$, we define $\tilde{e}$ by positive reflection and then define

$$\begin{aligned} U^i(x) &= \sum_{\alpha=1}^{\nu-1} \int_{B_{2R}} - K_{0,\alpha}(x - \xi)\, \tilde{e}_i^\alpha(\xi)\, d\xi + W^i_{,\nu}(x), \\ W^i(x) &= -\int_{G_{2R}} [K_0(x - \xi) - K_0(x' - \xi)]\, e^\nu(\xi)\, d\xi, \\ x' &= (x'_\nu, -x^\nu) \quad \text{if} \quad x = (x'_\nu, x^\nu). \end{aligned} \tag{7.6.8}$$

Lemma 7.6.1. (a) *If* $U = Q_{2R}(e)$ *and* $V = P_{2R}(f)$, $e \in L_2(G_{2R})$, $f \in L_p(G_{2R})$, $p \geq 2\nu/(\nu+2)$, $p > 1$, *then* U^i *and* V^i *vanish along* σ_{2R} *if* $i \leq k$ *and* U^i *and* V^i *are solutions of*

$$\int_{G_{2R}} v_{,\alpha}(\delta^{\alpha\beta} U^i_{,\beta} + e^\alpha_i)\,dx = 0, \quad \int_{G_{2R}} (v_{,\alpha}\delta^{\alpha\beta} V^i_{,\beta} + v f_i)\,dx = 0 \quad v \in H^1_{20}(G_{2R}). \tag{7.6.9}$$

If $i > k$, U^i *and* V^i *are restrictions to* G_{2R} *of functions which are defined on* R_ν *and satisfy* (7.6.7); *they are solutions of* (7.6.9) *for all* $v \in H^1_2(G_{2R})$ *which vanish on* Σ^-_{2R}.

(b) *If* $H \in H^1_2(G_{2R})$, *vanishes along* σ_{2R} *and satisfies*

$$\int_{G_{2R}} v_{,\alpha}\delta^{\alpha\beta} H_{,\beta}\,dx = 0, \quad v \in H^1_{20}(G_{2R}), \tag{7.6.10}$$

then H *is harmonic on* B_{2R} *if extended by negative reflection. If* $H \in H^1_2(G_{2R})$ *and satisfies* (7.6.10) *for all* $v \in H^1_2(G_{2R})$ *which vanish along* Σ^-_{2R}, *then* H *is harmonic if extended to* B_{2R} *by positive reflection.*

Proof. (a) The results concerning U^i and V^i for $i \leq k$ were proved in Theorem 5.5.1'. If the $e^\alpha_i \in C^0_\mu$ and have support in $G_{2R} \cup \sigma_{2R}$, then the $\tilde{e}^\alpha_i \in C^0_\mu(\bar{B}_{2R})$ and have support in B_{2R}. Since $K_0(x-\xi) - K_0(x'-\xi)$ is the Green's function for the lower half space, it follows that W^i, as defined in (7.6.8) $\in C^2_\mu$ on each closed half-space (but not necessarily on the whole space), this can be proved by writing

$$W^i(x) = -\int_{B_{2R}} K_0(x-\xi)\,\tilde{e}^\nu(\xi)\,d\xi + 2\int_{G^+_{2R}} K_0(x-\xi)\,\tilde{e}^\nu(\xi)\,d\xi$$

and using the method of proof in Theorem 5.5.1'. It follows that $U^i \in C^1_\mu(\bar{G}_{2R})$. If we approximate uniformly (with the same Hölder condition) to the e^α_i by smooth functions e^α_{ni} and (7.6.9) holds for each n, it will hold in the limit. Then

$$\int_{G_{2R}} v_{,\alpha}(\delta^{\alpha\beta} U^i_{n,\beta} + e^\alpha_{ni})\,dx = \int_{\sigma_{2R}} v(U^i_{n,\nu} + e^\nu_{ni})\,dS - \int_{G_{2R}} v(\Delta U^i_n + e^\alpha_{ni,\alpha})\,dx. \tag{7.6.11}$$

By integrating by parts, we obtain

$$U^i_n(x) = \int_{B_{2R}} -K_0(x-\xi)\sum_{\alpha=1}^{\nu-1} e^\alpha_{ni,\alpha}(\xi)\,d\xi + W^i_{n,\nu}(x).$$

From Chapter 2, we conclude that

$$\Delta W^i_n = -e^\nu_{ni}(x), \qquad \Delta U^i(x) = -\sum_{\alpha=1}^{\nu} e^\alpha_{ni,\alpha}(x)$$

$$U^i_{n,\nu}(x) = W^i_{n,\nu\nu}(x) \to \Delta W^i_n \quad \text{as} \quad x^\nu \to 0$$

since $W^i = 0$ along σ_{2R}. Thus the integral on the left in (7.6.11) vanishes for each n. If $f_i \in C^0_\mu(\bar{G}_{2R})$, then $\tilde{f}_i \in C^0_\mu(\bar{B}_{2R})$ if $i > k$ and so $\Delta V^i = f_i$ and $V^i_{,\nu} = 0$ along σ_{2R} since V^i is even with respect to x^ν.

In part (b), it follows from (7.6.10) and the interior differentiability theorems of Chapter 5 that H is harmonic on G_{2R}. If H vanishes along σ_{2R} and $\tilde{H}$ is the extension of H to B_{2R} by negative reflection (see (5.5.19)), then $\tilde{H}$ is easily seen to $\in H_2^1(B_{2R})$ by using the absolute continuity properties of § 3.1. If, in (7.6.10), we replace G_{2R} by B_{2R} and let $v \in C_C^\infty(B_{2R})$, we see that

$$\int_{B_{2R}} v_{,\alpha} \delta^{\alpha\beta} \tilde{H}_{,\beta}\, dx = \int_{G_{2R}} \tilde{v}_{,\alpha} \delta^{\alpha\beta} H_{,\beta}\, dx = 0, \tag{7.6.12}$$
$$\tilde{v}(x'_\nu, x^\nu) = v(x'_\nu, x^\nu) - v(x'_\nu, -x^\nu) \in H_{20}^1(G_{2R}).$$

If we do not assume that H vanishes along σ_{2R} but do assume that (7.6.10) holds for all v which vanish along $\sum_{2R}^-$, we obtain a result like (7.6.12) if we let $\tilde{H}$ be the extension by positive reflection, let $v \in C_C^\infty(B_{2R})$ and define $\tilde{v}$ by

$$\tilde{v}(x'_\nu, x^\nu) = v(x'_\nu, x^\nu) + v(x'_\nu, -x^\nu).$$

The developments in proving regularity at the boundary now proceed as in § 5.5. We define the norms and spaces as in Definition 5.5.1 except that the spaces $C_{\mu 0}^0(\bar{G}_{2R})$ will consist of those *vectors* in $C_\mu^0(\bar{G}_{2R})$ which vanish along $\sum_{2R}^-$ (the spherical part of ∂G_{2R}; recall that, in this chapter G_{2R} is the *lower* hemisphere); we do not require them to vanish along σ_{2R}. However, the range spaces $H_q^{1*}(G_{2R})$ and $C_\mu^{1*}(\bar{G}_{2R})$ of the operators Q_{2R} and P_{2R} are spaces of *vectors* which satisfy the boundary conditions in Lemma 7.6.1. With that understanding we obtain the result.

Theorem 7.6.1. *The results of Theorem* 5.5.1 *hold.*

In our case, the coefficients a_R are Lipschitz and the b_R, c_R, and d_R (in (7.6.4)) are bounded and measurable if the manifold M is of class C_1^1 and, more generally the $a_R \in C_\mu^{k-1}$ and the b_R, c_R, and $d_R \in C_\mu^{k-2}$ if M is of class C_μ^k, $k \geq 2$, and so in these cases satisfy the H_q^1 conditions in Definition 5.5.2 for any q. *We note again that if M is of class H_p^2 with $p > \nu$, then the coefficients satisfy the H_q^1-conditions of Definition* 5.5.2 *with $q \leq p$.* So if we write the equation (7.6.5) in the form

$$\begin{aligned} u_R - T_R u_R &= v_R + w_R, \quad v_R = Q_{2R}(e) + P_{2R}(f) \\ T_R u_R &= Q_{2R}[(a_R - a_0) \cdot \nabla u_R + b_R u_R] + P_{2R}[c_R \cdot \nabla u_R + d_R \cdot u_R] \\ w_R &= Q_{2R}[(a_R - a_0) \cdot \nabla H_R + b_R H_R] + P_{2R}[c_R \cdot \nabla H_R + d_R \cdot u_R], \end{aligned} \tag{7.6.13}$$

we obtain the theorem

Theorem 7.6.2. *The results of Theorem* 5.5.2 *carry over to our case.*

Now, we notice, if our given data e and f vanish outside G_{2R} and R_ν^-, we may carry over the analysis with R replaced by any A; we then notice that $Q_{2A}(e) = Q_{2R}(e)$ and $P_{2R}(e) = P_{2A}(e)$ for any A and we may let $A \to \infty$. If u vanishes outside G_R, we then see that the harmonic vector

$H_{2A} = H_{2R}$ can be extended to the whole space where it is bounded so that we have.

Theorem 7.6.3. *If* $u \in H_2^1(G_R)$, *vanishes on* $\sum_R^-$, *and satisfies* (7.6.4), *and if* H_R *is defined in* (7.6.5), *then* $H_R = const.$ *if* $\nu = 2$ *and* $H_R = 0$ *if* $\nu > 2$. *If* $\nu = 2$ *and* $i \leq k$, $H_R^i = 0$. *If* $\nu = 2$ *and* $i > k$, *then* H_R^i *is the average value over* B_{2R} *of the ordinary logarithmic potential of* f_i^*, *where* f_i^* *is the extension to* B_{2R} *by positive reflection of* $c_{Rij}^{\alpha} u_{,\alpha}^j + d_{Rij} u^j + f_i$.

Remark. One must recall the definition given in § 5.5 of $P_{2R}(f)$ when $\nu = 2$.

Thus, if e and $f \in L_q(G_{2R})$ for some $q > 2$, for example, R is small enough, depending on q, u vanishes outside G_{2R}, $u \in H_2^1(G_{2R})$, $u^i = 0$ along σ_{2R} if $i \leq k$, and u satisfies (7.6.1), then $u \in H_q^1(G_{2R})$ and

$$'\|u\|_{q,2R}^1 \leq C(\|e\|_q^0 + R\|f\|_q^0).$$

Corresponding results hold if e and $f \in C_{\mu 0}^0(\bar{G}_{2R})$. But now, suppose that the components u are components of a form ω which satisfies equations (7.4.1) in the large on M and suppose we assume that the forms φ, ψ, and $\eta \in \mathfrak{L}_q$ for some $q > 2$. We assume that $n\,\omega = n\,\zeta = 0$ for all forms ω and ζ considered or else that $t\,\omega = t\,\zeta = 0$. Then we know that $\omega \in H_2^1$ at any rate and from our interior results $\omega \in H_q^1$ on any D with $D \subset\subset M$. In any boundary neighborhood, then $U = \zeta u$ satisfies (7.6.1) with e and f replaced by E and F as given by (7.6.3). Now the terms ζe_i^{α}, ζf_i, and $\zeta_{,\alpha} e_i^{\alpha} \in \mathfrak{L}_q$ but the second term in E and the terms $\nabla\zeta \cdot (b - c) \cdot u$ in $F \in \mathfrak{L}_s$, where $s = 2\nu/(\nu - 2)$ and the term $\nabla\zeta \cdot a \cdot \nabla u$ merely $\in \mathfrak{L}_2$. If $q \leq s$, we conclude from the theorems above that each $U \in H_q^1$; if $q > s$, we conclude that $U \in H_s^1$. Since this is true in each sufficiently small boundary patch, we conclude that $\omega \in H_q^1$ if $q \leq s$ or to H_s^1 otherwise. In this latter case, then we conclude that $u \in \mathfrak{L}_{s'}$, where $s' = \nu s/(\nu - s)$. We may repeat the argument (with smaller patches if necessary) to conclude that $\omega \in H_q^1$ if $q \leq s'$ or in $H_{s'}^1$ otherwise. Clearly any q may be reached in a finite number of steps. If M is of class C_{μ}^2, $0 < \mu < 1$, and if φ, ψ, and $\eta \in C_{\mu}^0$, we first show that $\omega \in H_q^1$ for some $q > \nu$ which implies that $u \in C_{\mu}^0$, so that E and $F \in C_{\mu}^0$ so that $\omega \in C_{\mu}^1$. The higher differentiability results can be obtained as in §§ 5.5, 5.6.

7.7. Potentials; the decomposition theorem

In this and the next section, we shall assume that all of our forms are of the same kind, completely parallel theories being obtained for each kind.

Definition 7.7.1. We define the *closed linear manifolds* $\mathfrak{P}_2^+$ and $\mathfrak{P}_2^-$ (see Theorem 7.5.1) of H_2^1 as the totality of forms in H_2^1 for which $n\,\omega = 0$ and $t\,\omega = 0$ on $b\,M$ respectively.

Just as in Section 7.3, we obtain the following result:

Lemma 7.7.1. *Let $\mathfrak{M}$ be any closed linear manifold of $\mathfrak{L}_2$ such that $\mathfrak{M} \cap \mathfrak{P}_2^+ (\mathfrak{P}_2^-)$ is not empty. Then there exists a form ω in $\mathfrak{M} \cap \mathfrak{P}_2^+ (\mathfrak{P}_2^-)$ which minimizes $D(\omega)$ among all such forms with $(\omega, \omega) = 1$.*

Theorem 7.7.1. *The manifold $\mathfrak{H}^+(\mathfrak{H}^-)$ of harmonic fields in $\mathfrak{P}_2^+ (\mathfrak{P}_2^-)$ is finite dimensional.*

Theorem 7.7.2. *If $\omega \in \mathfrak{P}_2^+ (\mathfrak{P}_2^-)$ and is $\mathfrak{L}_2$-orthogonal to $\mathfrak{H}^+(\mathfrak{H}^-)$, then there are positive constants λ^+ and λ^- for each $\mathfrak{N}$ such that*

$$D(\omega) \geq \lambda^+ \|\omega\|^2 (\lambda^- \|\omega\|^2).$$

Theorem 7.7.3. *If $\eta \in \mathfrak{L}_2$ and is $\mathfrak{L}_2$-orthogonal to $\mathfrak{H}^+(\mathfrak{H}^-)$, there is a unique form $\Omega^+(\Omega^-)$ in $\mathfrak{P}_2^+ \perp \mathfrak{H}^+ (\mathfrak{P}_2^- \perp \mathfrak{H}^-)$ such that*

$$(d\Omega^+, d\zeta) + (\delta\Omega^+, \delta\zeta) = (\eta, \zeta), \quad \zeta \in \mathfrak{P}_2^+ (\mathfrak{P}_2^-), \tag{7.7.1}$$

Definition 7.7.2. The functions Ω^+ and Ω^- are called the plus-potential and minus-potential of η, respectively.

The defining equations (7.7.1) for the potentials are a special case of the more general equations

$$(d\omega - \varphi, d\zeta) + (\delta\omega - \psi, \delta\zeta) - (\eta, \zeta) = 0, \quad \zeta \in \mathfrak{P}_2^+ \text{ or } \mathfrak{P}_2^- \tag{7.7.2}$$

which were discussed in Section 7.4. The differentiability results for such equations on the interior of M follow from the discussion there given. But now, suppose we select a point P on bM and choose an admissible boundary coordinate system with domain G_R and range a boundary neighborhood $\mathfrak{U}$ of P such that $g_{ij}(0) = \delta_{ij}$. In such a system the conditions $n\omega = n\zeta = 0$ for $\mathfrak{P}_2^+$ and $t\omega = t\zeta = 0$ for $\mathfrak{P}_2^-$ correspond under a proper ordering of the sets I and J to the equations (7.6.1). Accordingly we see that if the support of ζ is confined to $\mathfrak{U}$, the system (7.7.2) reduces to the system (7.6.1) discussed in Section 7.6. From the theorems of Section 7.6 we may conclude that the differentiability results for the plus and minus potentials stated near the beginning of Section 7.4 for potentials, hold right up to the boundary. We now extend these results as in Section 7.4 and summarize as follows:

Theorem 7.7.4. *Suppose $\omega \in \mathfrak{L}_2 \ominus \mathfrak{H}^+ (\mathfrak{L}_2 \ominus \mathfrak{H}^-)$ and Ω is its plus (minus)-potential*

(i) *If M is of class C_1^1, then Ω, $d\Omega$, and $\delta\Omega$ are in $\mathfrak{P}_2^+ (\mathfrak{P}_2^-)$.*

(ii) *If M is of class C_1^1 and ω is in $\mathfrak{L}_p$ with $p > \nu$, then Ω, $d\Omega$, and $\delta\Omega$ are in $\mathfrak{P}_p^+ (\mathfrak{P}_p^-)$ and C_μ^0 if $\mu = 1 - \nu/p$.*

(iii) *If M is of class C_μ^k and $\omega \in C_\mu^{k-2}$ $(k \geq 2,\ 0 < \mu < 1)$, then Ω, $d\Omega$, and $\delta\Omega \in C_\mu^{k-1}$. If $k \geq 3$ and $\omega \in C_\mu^{k-3}$, then $\Omega \in C_\mu^{k-1}$.*

(iv) *If M and ω are of class C^∞ or analytic, then so is Ω*

(v) *If Ω and ω are 0-forms, then Ω has an additional degree of differentiability in all cases above except the second half of* (iii).

In all cases, if we set $\alpha = d\Omega$ *and* $\beta = \delta\Omega$,

$$(7.7.3)\qquad \delta\alpha + d\beta = \delta(d\Omega) + d(\delta\Omega) = \omega, \quad d\alpha = \delta\beta = 0;$$

$$(d\alpha, d\zeta) + (\delta\alpha - \omega, \delta\zeta) = (d\beta - \omega, d\zeta) + (\delta\beta, \delta\zeta) = 0, \quad \zeta \in \mathfrak{P}_2^+ (\mathfrak{P}_2^-).$$
(7.7.4)

Proof. The results for Ω follows directly from the discussion above and Section 7.6. The proof of the results for $d\Omega$ and $\delta\Omega$ is like that of Theorem 7.4.1 where it is already done for the interior of M. We choose a boundary point P and an admissible boundary coordinate system of the type described in the preceding paragraph and approximate (if necessary) to ω and the g_{ij} by smooth functions. For each of the approximating functions Ω, we see from formula (7.5.14) that $\Delta\Omega = \omega$ and

$$(7.7.5)\qquad n\, d\Omega = 0 \ \text{ if } \ \Omega \in \mathfrak{P}_2^+ \ \text{ and } \ t\,\delta\Omega = 0 \text{ if } \Omega \in \mathfrak{P}_2^-$$

since, in the integral over $b\,M$, $t\,\zeta$ is arbitrary if $\zeta \in \mathfrak{P}_2^+$ and $n\,\zeta$ is arbitrary if $\zeta \in \mathfrak{P}_2^-$. From Lemma 7.5.2 we see that $t\,d\varphi = 0$ whenever $t\,\varphi = 0$ and φ and M are differentiable. From Lemma 7.5.2 it follows also that

$$(7.7.6)\qquad n\,\varphi = 0 \rightarrow n\,\delta\varphi = 0.$$

Hence, from this and (7.7.5), we see that both α and $\beta \in \mathfrak{P}_2^+ (\mathfrak{P}_2^-)$ if ω and $\Omega \in \mathfrak{P}_2^+ (\mathfrak{P}_2^-)$ at each stage of the approximation, that α and β satisfy (7.7.3) and hence (7.7.4) using Lemma 7.5.3. The approximation may then be carried through as before on each G_r with $r < R$. Since a finite number of the smaller boundary neighborhoods cover $b\,M$, the results (7.7.4) for all ζ in $\mathfrak{P}_2^+ (\mathfrak{P}_2^-)$ follow and the differentiability of α and β now follow from Section 7.6.

Remark. Except in the case of zero forms Ω, the individual derivatives of the individual components of Ω *do not*, in general have the same differentiability properties as do $d\Omega$ and $\delta\Omega$ (the coordinate transformations will not allow it).

The following two theorems are useful and important:

Theorem 7.7.5. *Suppose* $\eta \in \mathfrak{P}_2 = H_2^1$, H^+ *and* H^- *are its projections in* $\mathfrak{H}^+$ *and* $\mathfrak{H}^-$ *and* Ω^+ *and* Ω^- *are the plus and minus potentials of* $\eta - H^+$ *and* $\eta - H^-$, *respectively, and* $\alpha^\pm = d\Omega^\pm$, $\beta^\pm = \delta\Omega^\pm$. *Then*

(i) α^+ *is the plus potential of* $d\eta$ *and* β^- *is the minus potential of* $\delta\eta$ *and* $d\eta \in \mathfrak{L}_2 \ominus \mathfrak{H}^+$ *and* $\delta\eta \in \mathfrak{L}_2 \ominus \mathfrak{H}^-$.

(ii) *If* $\eta \in \mathfrak{P}_2^+$, *then* β^+ *is the plus potential* $\delta\eta \in \mathfrak{L}_2 \ominus \mathfrak{H}^+$.

(iii) *If* $\eta \in \mathfrak{P}_2^-$, *then* α^- *is the minus potential of* $d\eta \in \mathfrak{L}_2 \ominus \mathfrak{H}^-$.

Proof. These results follow from Theorem 7.7.4, equation (7.7.4) and Lemma 7.5.3.

Theorem 7.7.6. (i) *If* $\eta^+ \in \mathfrak{P}_2^+$ *and* $\eta^- \in \mathfrak{P}_2^-$, *there are unique forms* α *and* β *where* $\alpha \in \mathfrak{P}_2^+$ *and is* $\mathfrak{L}_2$*-orthogonal to* $\mathfrak{H}^+$ *and* $\beta \in \mathfrak{P}_2^-$ *and is* $\mathfrak{L}_2$*-orthogonal to* $\mathfrak{H}^-$ *such that* $\delta\alpha = \delta\eta^+$ $d\alpha = 0$, $d\beta = d\eta^-$, $\delta\beta = 0$.

(ii) *If* $\eta \in H_2^1$, *there are unique forms* $\gamma \in \mathfrak{P}_2^+ \cap (\mathfrak{L}_2 \ominus \mathfrak{H}^+)$ *and* ε *in* $\mathfrak{P}_2^- \cap (\mathfrak{L}_2 \ominus \mathfrak{H}^-)$ *such that* $d\gamma = d\eta$, $\delta\gamma = 0$, $\delta\varepsilon = \delta\eta$, $d\varepsilon = 0$.

Proof. The uniqueness is evident. To prove (i), let Ω^+ and Ω^- be the respective plus and minus potentials of $\eta^+ - H^+$ and $\eta^- - H^-$ and let $\Gamma = \delta\Omega^+$ and $E = d\Omega^-$. From Theorem 7.7.5, we see that Γ is the plus potential of $\delta\eta^+$ and E is the minus potential of $d\eta^-$. Then, from Theorem 7.7.4 we conclude that $\alpha = d\Gamma$ and $\beta = \delta E$ have the desired properties. To prove (ii), let Ω^+ and Ω^- be the respective plus and minus potentials of $\eta - H^+$ and $\eta - H^-$ and let

$$A = d\Omega^+, \quad B = \delta\Omega^-, \quad \gamma = \delta A, \quad \varepsilon = dB$$

and (ii) follows from Theorem 7.7.5 (i).

Definition 7.7.3. We define the *linear sets* $\mathfrak{C}$ *and* $\mathfrak{D}$ *as* the sets of all forms of the form $\delta\alpha$ and $d\beta$, where $\alpha \in \mathfrak{P}_2^+$ and $\beta \in \mathfrak{P}_2^-$, respectively.

We now can prove an analog for the Kodaira decomposition theorem for manifolds with boundary.

Theorem 7.7.7. *The sets* $\mathfrak{C}$ *and* $\mathfrak{D}$ *and the set* $\mathfrak{H}$ *of all harmonic fields in* $\mathfrak{L}_2$ *on* M *are closed linear manifolds in* $\mathfrak{L}_2$ *and*

$$\mathfrak{L}_2 = \mathfrak{C} \oplus \mathfrak{D} \oplus \mathfrak{H}. \tag{7.7.7}$$

Moreover, if $\omega \in H_2^1$, *its* $\mathfrak{L}_2$ *projections* γ, ε, *and* H *on* $\mathfrak{C}$, $\mathfrak{D}$, *and* $\mathfrak{H}$ *belong to* $\mathfrak{P}_2^+$, $\mathfrak{P}_2^-$, *and* H_2^1, *respectively, and* $\delta\gamma = d\varepsilon = 0$.

Proof. That $\mathfrak{C}$ and $\mathfrak{D}$ are closed linear manifolds follows immediately from Theorem 7.7.6 and 7.7.2 and that $\mathfrak{H}$ is also follows from the interior regularity theorems of §§ 5.5. Using Lemma 7.5.3, we see that $\mathfrak{C}$ and $\mathfrak{D}$ are orthogonal and that $\mathfrak{C}$ and $\mathfrak{D}$ are both orthogonal to $\mathfrak{H} \cap H_2^1$ and, in fact, if $H \in H_2^1 \cap (\mathfrak{L}_2 \ominus \mathfrak{C} \ominus \mathfrak{D})$, then $H \in \mathfrak{H}$ ($\mathfrak{P}_2^+$ and $\mathfrak{P}_2^-$ are both everywhere dense in $\mathfrak{L}_2$).

Now, suppose $\eta \in H_2^1$ and let γ and ε be its projections on $\mathfrak{C}$ and $\mathfrak{D}$, respectively. Using Theorem 7.7.6 we conclude the existence of unique forms α and β in $\mathfrak{P}_2^+ \cap (\mathfrak{L}_2 \ominus \mathfrak{H}^+)$ and $\mathfrak{P}_2^- \cap (\mathfrak{L}_2 \ominus \mathfrak{H}^-)$ respectively, such that

$$\delta\alpha = \gamma, \quad d\alpha = 0, \quad \delta\beta = 0, \quad d\beta = \varepsilon. \tag{7.7.8}$$

Since γ and ε are the projections of η on $\mathfrak{C}$ and $\mathfrak{D}$, we see from (7.7.8) that α and β satisfy

$$\begin{aligned}(d\alpha, d\zeta^+) + (\delta\alpha, \delta\zeta^+) - (\eta, \delta\zeta^+) &= (d\alpha, d\zeta^+) + (\delta\alpha, \delta\zeta^+) - \\ &\quad - (d\eta, \zeta^+) = 0 \\ (d\beta, d\zeta^-) + (\delta\beta, \delta\zeta^-) - (\eta, d\zeta^-) &= (d\beta, d\zeta^-) + \\ &\quad + (\delta\beta, \delta\zeta^-) - (\delta\eta, \zeta^-) = 0\end{aligned} \tag{7.7.9}$$

for all ζ^+ in $\mathfrak{P}_2^+$ and ζ^- in $\mathfrak{P}_2^-$. Thus α is the plus potential of $d\eta$ and β is the minus potential of $\delta\eta$. The results follow from Theorems 7.7.3 and 7.7.5.

The following theorem contains further information concerning the decomposition (7.7.7).

Theorem 7.7.8. *Suppose $\omega \in \mathfrak{L}_2$ and γ, ε, and H are its projections on $\mathfrak{C}$, $\mathfrak{D}$, and $\mathfrak{H}$, respectively, and suppose Ω^+ and Ω^- are the plus and minus potentials of $\omega - H$, respectively. Then*

(8.7.10) $\quad \gamma = \delta\alpha, \quad \varepsilon = d\beta, \quad \alpha = d\Omega^+, \quad \beta = \delta\Omega^-, \quad d\alpha = \delta\beta = 0.$

If $\omega \in H_2^1$, then α and β are the plus and minus potentials of $d\omega$ and $\delta\omega$, respectively. We have the following differentiability results on the closure of M:

(i) *If M is of class C_1^1 and $\omega \in \mathfrak{L}_p (p \geq 2)$, then γ, ε, and $H \in \mathfrak{L}_p$ and α, β, Ω^+, and $\Omega^- \in H_p^1$; if also $\omega \in H_p^1$, then γ, ε, and $H \in H_p^1$ with γ, δ and H in C_μ^0 in case $\mu = 1 - \nu/p$ and $p > \nu$.*

(ii) *If M is of class C_μ^k with $k \geq 2$ and $0 < \mu < 1$ and if $\omega \in C_\mu^{k-2}$, then H, γ, and $\varepsilon \in C_\mu^{k-2}$ and α, β, Ω^+, and $\Omega^- \in C_\mu^{k-1}$; if also, $\omega \in C_\mu^{k-1}$, then H, γ, and $\varepsilon \in C_\mu^{k-1}$.*

(iii) *If M and ω are C^∞ or analytic, so are α, β, γ, ε, Ω^+, and Ω^-.*

(iv) *In the case of zero forms, H is a constant, and $\varepsilon = 0$. If $\omega \in H_2^1$ with $d\omega = 0$ or if $\omega \in \mathfrak{L}_2$ and $\omega = d\eta$ where $\eta \in H_2^1$, then $\gamma = 0$; if $\omega \in H_2^1$ and $\delta\omega = 0$ or if $\omega \in \mathfrak{L}_2$ and $\omega = \delta\eta$ where $\eta \in H_2^1$ then $\varepsilon = 0$.*

Proof. Suppose, first, that $\omega \in H_2^1$. If we then define α, β, γ, ε, Ω^+, and Ω^- by (7.7.10), the results follow from (7.7.8) and (7.7.9). In case ω is merely in $\mathfrak{L}_2$, we use the left sides of (7.7.9) and approximate, using the Theorems of § 7.6 and Theorem 7.7.2. The regularity results and the last statement follow from the facts that α and β are the respective potentials of $d\omega$ and $\delta\omega$, since $dH = \delta H = 0$, in case $\omega \in H_2^1$. The last results for ω merely in $\mathfrak{L}_2$ follow from Lemma 7.5.3.

We may now prove a slightly strengthened form of an inequality due to Friedrichs [3]:

Theorem 7.7.9. *There is a $\lambda > 0$ such that if $\omega \in H_2^1 \perp \mathfrak{H}$, then*

$$D(\omega) \geq \lambda \|\omega\|^2.$$

Proof. For if $\omega \in H_2^1 \perp \mathfrak{H}$, then

(7.7.11) $\quad \omega = \gamma + \varepsilon, \quad \delta\gamma = d\varepsilon = 0, \quad \gamma \in \mathfrak{P}_2^+, \quad \varepsilon \in \mathfrak{P}_2^-, \quad (\gamma, \varepsilon) = 0.$

Hence from (7.7.9) and Theorem 7.7.2, we see that

$$D(\omega) = D(\gamma) + D(\varepsilon) \geq \lambda^+ \|\gamma\|^2 + \lambda^- \|\varepsilon\|^2 \geq \lambda [2\|\gamma\|^2 + 2\|\varepsilon\|^2]$$
$$\geq \lambda \|\omega\|^2, \quad \lambda = \min[\lambda^+/2, \lambda^-/2].$$

The following theorem completes the analogy with the case for a compact manifold without boundary.

Theorem 7.7.10. *If $\eta \in \mathfrak{L}_2$ and is $\mathfrak{L}_2$-orthogonal to $\mathfrak{H}$, there is a unique form Ω in H_2^1 and $\mathfrak{L}_2$-orthogonal to $\mathfrak{H}$ such that*

(7.7.12) $\quad (d\Omega, d\zeta) + (\delta\Omega, \delta\zeta) = (\eta, \zeta), \quad \zeta \in H_2^1.$

Moreover,

(7.7.13) $$d\Omega = d\Omega^+ \quad \text{and} \quad \delta\Omega = \delta\Omega^-.$$

Ω^+ and Ω^- being the respective plus and minus potentials of η. The differentiability properties of Ω are the same as those in Theorem 7.7.4.

Proof. The proof that Ω exists in H_2^1 and is unique is just like that of Theorem 7.7.3. Obviously $\eta \in \mathfrak{L}_2 \ominus \mathfrak{H}^+$ and $\mathfrak{L}_2 \ominus \mathfrak{H}^-$ so that its plus and minus potentials exist. Accordingly we have, for example,

(7.7.14) $$(d\Omega - d\Omega^+, d\zeta) + (\delta\Omega - \delta\Omega^+, \delta\zeta) = 0 \quad \text{for all} \quad \zeta \in \mathfrak{P}_2^+.$$

But, from Theorem 7.7.6, we may find a $\zeta \in \mathfrak{P}_2^+$ such that

(7.7.15) $$d\zeta = d\Omega - d\Omega^+, \quad \delta\zeta = 0.$$

Using (7.7.14) and (7.7.15) and a similar argument for Ω^-, we derive (7.7.13).

Now, let us consider the decomposition (7.7.7) for Ω, Ω^+, and Ω^-. Using (7.7.13) and the last statement in Theorem 7.7.8 also, we obtain

(7.7.16) $$\begin{aligned} &\Omega = \Gamma + E, \quad \Omega^\pm = H^\pm + \Gamma^\pm + E^\pm, \quad \Omega = \Omega^\pm + K^\pm, \\ &dK^+ = \delta K^- = 0, \quad K^+ = H_1^+ + E_1^+, \quad K^- = H_1^- + \Gamma_1^- \end{aligned}$$

where the Γ' $s \in \mathfrak{C}$, the E' $s \in \mathfrak{D}$, and the H' $s \in \mathfrak{H}$. From (7.7.16) and the uniqueness of the decomposition, we obtain

$$H^\pm + H_1^\pm = 0, \quad \Gamma + E = \Gamma^+ + (E^+ + E_1^+) = (\Gamma^- + \Gamma_1^-) + E^-$$

(7.7.17) $$\Gamma = \Gamma^+, \quad E = E^-.$$

The differentiability properties of $d\Omega$ and $\delta\Omega$ follow from (7.7.13) and Theorem 7.7.4 and those for Ω follow from (7.7.16) and (7.7.17) and Theorems 7.7.4 and 7.7.9.

Remark. We cannot conclude the differentiability of Ω directly from (7.7.12) and the theorems of § 7.6 since the equations (7.7.12) are not the same as (7.6.1) since the boundary integral corresponding to the first term in (7.5.18) is not necessarily zero in this case.

Definition 7.7.4. The form Ω in Theorem 7.7.10 is called the *potential of η*.

Important Remark. All the differentiability properties at either interior or boundary points are local; all differentiability results extend immediately to cases where the given hypotheses hold in some coordinate patch, the conclusions then holding in that patch.

7.8. Boundary value problems

In this section, we derive briefly the results concerning boundary value problems for harmonic forms and fields which have been obtained by other methods by Duff and Spencer and Connor [1]. The differentiability results on the interior have been obtained in § 7.4; the corresponding results on the boundary depend on the given boundary values as well as on the differentiability of M and are stated below.

The following theorem is seen (from their proofs) to be equivalent to Theorems 3 and 4 of the paper by DUFF and SPENCER (pp. 150, 151):

Theorem 7.8.1. (a) *If ω is any closed form in H_2^1, there is a unique harmonic field H such that* (r = *degree of* ω).

$$\omega = H + d\beta, \quad tH = t\omega, \beta, d\beta \in \mathfrak{P}_2^- (t\beta = t\,d\beta = 0), \quad 0 \leq r \leq n-1.$$

(b) *If ω is any co-closed form in H_2^1, there is a unique harmonic field H such that*

$$\omega = H + \delta\alpha, \quad nH = n\omega, \alpha, \delta \in \mathfrak{P}_2^+ (n\alpha = n\,\delta\alpha = 0), \quad 1 \leq r \leq n.$$

In either case, the differentiability results are as follows:

(i) *If M is of* class C_1^1 *and $\omega \in H_p^1$, $p \geq 2$, then H is also and ω and $H \in C_\mu^0$ on $\bar{M}$ if $p > \nu$ and $\mu = 1 - \nu/p$. If $r = 0$, ω and H are constants.*

(ii) *If M is of class C_μ^k (C^∞, analytic) and ω is of class C_μ^{k-1} (C^∞, C^ω), $k \geq 2$, $0 < \mu < 1$, then H is of class C_μ^{k-1} (C^∞, analytic).*

Proof. If $d\omega = 0$, then from Theorems 7.7.7 and 7.7.8, we see that the term $\delta\alpha = 0$ in the decomposition which proves (a); (b) is proved similarly. The differentiability results follow from Theorem 7.7.8.

The next theorem is a refinement of Theorem 2 of the paper of DUFF and SPENCER.

Theorem 7.8.2. *If η is any form in H_2^1, there is a form ω in H_2^1 such that $t\omega = t\eta$ and $d\omega$ is a harmonic field. If, also, $\eta = \delta\chi$ for some χ in H_2^1, then there is a unique ω of the form $\omega = \delta\xi$ with ξ in H_2^1 which satisfies the conditions above.*

Proof. Let H^- be the projection of $d\eta$ on $\mathfrak{H}^-$, let α be the minus potential of $d\eta - H^-$, and let

$$\omega = \eta - \gamma, \quad \gamma = \delta\alpha, \quad \varepsilon = d\alpha. \tag{7.8.1}$$

Then γ and ε are in $\mathfrak{P}_2^-$ and from (7.7.3) ($\alpha = \Omega$), we have

$$\begin{aligned} t\gamma = 0 = t\,\varepsilon, \quad d\gamma + \delta\varepsilon = d\eta - H^-, \\ d\omega = d\eta - d\gamma = H^- + \delta\varepsilon. \end{aligned} \tag{7.8.2}$$

From (7.8.2) and the last statement in Theorem 7.7.8, we see that

$$d\omega \in (\mathfrak{H} \oplus \mathfrak{D}) \cap (\mathfrak{H} \oplus \mathfrak{C}) = \mathfrak{H}.$$

Now, suppose $\eta = \delta\chi$ for some χ in H_2^1. Then $\omega = \delta(\chi - \alpha)$ from (7.8.1). Suppose ω_1 also satisfies all these conditions. Then

$$t(\omega - \omega_1) = 0, \quad \omega - \omega_1 = \delta\nu (\nu = \chi - \alpha - \xi_1), \quad d(\omega - \omega_1) \in \mathfrak{H}.$$

But then, from the definitions of $\mathfrak{C}$ and $\mathfrak{D}$ and Theorem 7.7.8, we obtain $\omega - \omega_1 \in \mathfrak{P}_2^-$, $\therefore d(\omega - \omega_1) \in \mathfrak{D} \cap \mathfrak{H}$, $\therefore d(\omega - \omega_1) = 0$, $\therefore \omega - \omega_1 \in \mathfrak{H} \oplus \mathfrak{D}$, $\omega - \omega_1 = \delta\nu \in \mathfrak{H} \oplus \mathfrak{C}$, $\therefore \omega - \omega_1 \in \mathfrak{H}^-$, $\omega - \omega_1 \in \mathfrak{L}_2 - \mathfrak{H}^-$ (Theorem 7.7.5 (i)),

$$\therefore \omega - \omega_1 = 0.$$

We now consider boundary value problems for harmonic forms as distinct from harmonic fields. We begin by defining

$$H^1_{20} = \mathfrak{P}_2^+ \cap \mathfrak{P}_2^-, \quad \mathfrak{H}_0 = \mathfrak{H}^+ \cap \mathfrak{H}^- = \mathfrak{H} \cap H^1_{20}.$$

Then H^1_{20} consists of all H^1_2 forms ω with $t\,\omega = n\,\omega = 0$.

Theorem 7.8.3. *$\mathfrak{H}_0$ consists of the element* 0.

Proof. Obviously if $H \in \mathfrak{H}_0$, then $H \in H^1_{20}$ and minimizes $D(H)$ among all forms in H^1_{20}. We may clearly find a Riemannian manifold M' of the same class as M such that $\bar{M} \subset M'$. If we extend H to vanish on $M' - M$, then H satisfies the same conditions on M' and hence satisfies

$$(7.8.3) \qquad (dH, d\zeta) + (\delta H, \delta\zeta) = 0$$

for all ζ in H^1_2 and hence in $\mathfrak{P}_2^+$ or $\mathfrak{P}_2^-$ and so has the interior differentiability properties on M' stated in §§ 5.5 and 5.6. The theorem follows from the unique continuation theorem of ARONSZAJN et al.

Definition 7.8.1. A form K is *harmonic* on M if and only if K, dK, and $\delta K \in H^1_2$ on any domain interior to M with $\delta\, dK + d\,\delta K = 0$ there.

Theorem 7.8.4. *If $\omega \in H^1_2$, there exists a unique harmonic form K in H^1_2 such that $t\,K = t\,\omega$, $n\,K = n\,\omega$. The differentiability results for K are the same as those in Theorem* 7.8.1 *except that in the case of zero-forms, $K \in C^k_\mu$ if $\omega \in C^k_\mu$.*

Proof. Write $K = \omega + \eta$ and minimize

$$D(\omega + \eta) = D(\eta) + 2(d\eta, d\omega) + 2(\delta\eta, \delta\omega) + D(\omega)$$

among all η in $H^1_{20} \cap (\mathfrak{L}_2 \ominus \mathfrak{H}_0)$. Then $D(\eta) \geq \lambda\,\|\eta\|^2$ so that the minimizing function exists as usual. Then K is easily seen to satisfy (7.9.3) for all ζ in H^1_{20} so that K is harmonic on the interior of M (using the differentiability results of § 7.4). Since $\eta \in H^1_{20} \cap (\mathfrak{L}_2 \ominus \mathfrak{H}_0)$, the equivalent equation

$$(7.8.4) \qquad (d\eta + d\omega, d\zeta) + (\delta\eta + \delta\omega, \delta\zeta) = 0$$

for all $\zeta \in H^1_{20}$ is of the form (7.6.1) on boundary coordinate patches. The differentiability results follow from § 7.6. The uniqueness follows from Theorem 7.8.3.

More general theorems were proved in the authors paper [11]. But the statements of the results are too long to be reproduced here.

Chapter 8

The $\bar{\partial}$-Neumann problem on strongly pseudo-convex manifolds

8.1. Introduction

In this chapter we present a simplification of the recent solution due to J. J. KOHN ([1], [2]) of the so-called $\bar{\partial}$-NEUMANN problem introduced by GARABEDIAN and SPENCER for complex exterior differential forms on a compact complex-analytic manifold with strongly pseudo-convex boundary. The problem in its present form was investigated by D. C. SPENCER and J. J. KOHN by means of integral equations. The present author [13] solved this problem for the special cases of 0-forms and $\bar{z}$ — 1-forms (i.e. forms of the types (0,0) and (0,1) in our current notation) on certain "tubular" manifolds and used those results to prove that any compact real-analytic manifold can be analytically embedded in a Euclidean space of sufficiently high dimension. Unfortunately there is an error in that paper which is corrected in § 8.2 by using the results of KOHN presented in this chapter. These results apply to forms of arbitrary type (p, q) and the solution forms are shown to be of class C^∞ on the closed manifold provided the metric, boundary, and non-homogeneous term $\in C^\infty$ there. Recently HÖRMANDER [2] has extended these results using L_2-methods and certain weight functions. He was able to demonstrate existence (in the sense treated in § 8.4 below) of forms of type (p, q) in cases where the Levi form ((1.2) below) either has at least $q+1$ negative eigenvalues or at least $n-q$ positive eigenvalues. This is a much less restrictive condition on $b\,M$ than our condition of pseudo-convexity. Finally, KOHN and L. NIRENBERG have obtained a still greater simplification of this theory. I am presenting here essentially their simplification but confined to the $\bar{\partial}$-NEUMANN problem, whereas they have applied their method to a more general class of problems.

In his recent papers, KOHN sketched applications of his results (a) to the study of the $\bar{\partial}$-cohomology theory, (b) to the study of deformations of complex structures, and (c) to obtain a new proof of the result of NIRENBERG and NEWLANDER which showed that a complex analytic structure could be introduced on an integrable almost-complex manifold. However, part of the interest in this problem to those working in partial differential equations lies in the fact that the problem is not a regular boundary value problem in the sense of AGMON-DOUGLIS-NIRENBERG ([1] and [2]), BROWDER ([1], [2]) LOPATINSKY, etc.). We shall give an example below after we have introduced the notations and sketched the results; we shall also show the connection with the $\bar{\partial}$-cohomology.

We assume that $\bar{M} = M \cup b\,M$ is a compact complex-analytic manifold having boundary $b\,M$ of class C^∞. We assume that we are given a hermitian metric

$$(8.1.1) \qquad ds^2 = g_{\alpha\beta}\, dz^\alpha\, d\bar{z}^\beta \;{}^1 \quad (g_{\beta\alpha} = \bar{g}_{\alpha\beta},\ \alpha, \beta = 1, \ldots, \nu)$$

which is of class C^∞ on $\bar{M}$. We suppose that the function $r \in C^\infty(\bar{M})$ and equals the negative of the geodesic distance to $b\,M$ for points within a distance $-s_0$ of $b\,M$, $s_0 < 0$. It is clear that there exists a slightly larger such manifold $\bar{M}'$ such that $\bar{M} \subset M'$ and that the metric r can be ex-

[1] Repeated Greek indices are summed from 1 to ν.

tended to $\in C^\infty(\bar{M}')$ so that r is the geodesic distance from $b\,M$ on $M' - \bar{M}$. The strong pseudo-convexity of the boundary $b\,M$ implies that there is a constant $c_0 > 0$ such that at any point P_0 on $b\,M$ (where $r = 0$) we have

$$r_{z^\beta \bar{z}^\nu}\, T^\beta\, \bar{T}^\gamma \geq c_0\, g_{\beta\gamma}\, T^\beta\, \bar{T}^\gamma \tag{8.1.2}$$

for all complex vectors $(T^1, \ldots, T^\nu)$ such that

$$r_{z^\beta}\, T^\beta = 0. \tag{8.1.3}$$

If f is any other real function of class C^2 near $b\,M$ such that $\nabla f \neq 0$ and $f = 0$ on $b\,M$, and $f < 0$ on M near $b\,M$, then the positiveness of the form $f_{z^\beta \bar{z}^\nu}\, T^\beta\, \bar{T}^\gamma$ for T such that $f_{z^\beta}\, T^\beta = 0$ follows. In the above and throughout this chapter we assume that the operators $\partial/\partial z^\alpha$ and $\partial/\partial \bar{z}^\alpha$ are defined by

$$\frac{\partial}{\partial z^\alpha} = \frac{1}{2}\left(\frac{\partial}{\partial x^\alpha} - i\,\frac{\partial}{\partial y^\alpha}\right), \quad \frac{\partial}{\partial \bar{z}^\alpha} = \frac{1}{2}\left(\frac{\partial}{\partial x^\alpha} + i\,\frac{\partial}{\partial y^\alpha}\right), \tag{8.1.4}$$
$$z^\alpha = x^\alpha + i\,y^\alpha, \quad \bar{z}^\alpha = x^\alpha - i\,y^\alpha.$$

We let $\mathfrak{A}$ denote the set of all exterior differential forms of class $C^\infty(\bar{M})$ (i.e. C^∞ on $\bar{M}$) and denote by $\mathfrak{A}^{p,q}$ the set of all those which are of type (p, q), i.e. which can be expressed in any local analytic coordinate system in the form

$$\varphi = \sum_{\substack{i_1 < \cdots < i_p \\ j_1 < \ldots < j_q}} \varphi_{i_1 \ldots i_p j_1 \ldots j_q}\, d z^{i_1} \wedge \ldots \wedge d z^{i_p} \wedge d\bar{z}^{j_1} \wedge \ldots \wedge d\bar{z}^{j_q}. \tag{8.1.5}$$

We abbreviate the notation to

$$\varphi = \sum \varphi_{IJ}\, d z^I \wedge d\bar{z}^J; \tag{8.1.6}$$

when we use this notation I and J will always stand for increasing sequences as in (8.1.5). However, we shall often wish to have the $\varphi_{I, j_1 \ldots j_q}$ defined for *all* sequences of indices $j_1 \ldots j_q$; in this case, we assume that the φ's are defined so as to be antisymmetric in the j-indices. We shall at times wish to do the same with the I indices and shall sometimes write $\varphi_{I,\alpha R}$ where $R = (r_1, \ldots, r_{q-1})$ with $r_1 < \cdots < r_{q-1}$ and α runs from 1 to ν independently of R.

We shall wish to consider M (or $\bar{M}$) as a real manifold with metric given in $(x, y) = (x^1, \ldots, x^\nu, y^1, \ldots, y^\nu)$ coordinates by (1.1) which becomes, on setting $d z^\alpha = d x^\alpha + i\,d y^\alpha$ and $d\bar{z}^\beta = d x^\beta - i\,d y^\beta$,

$$g_{1\alpha\beta}(d x^\alpha\, d x^\beta + d y^\alpha\, d y^\beta) + 2 g_{2\alpha\beta}\, d x^\alpha\, d y^\beta, \tag{8.1.7}$$
$$g_{\alpha\beta} = g_{1\alpha\beta} + i\,g_{2\alpha\beta}, \quad g_{1\beta\alpha} = g_{1\alpha\beta}, \quad g_{2\beta\alpha} = -g_{2\alpha\beta}.$$

Then the dual $*\varphi$ would be defined by first expressing φ in terms of real differentials $d x^\alpha$ and $d y^\alpha$ and then taking the ordinary real dual of the real and imaginary parts. This procedure introduces a factor 2^{p+q} into the customary inner product

$$(\varphi, \psi) = \int_M \varphi \wedge *\bar{\psi} \tag{8.1.8}$$

of two forms of the same type. Along with most workers in this field we omit these factors. The space $\mathfrak{L}^{p,q}$ is the completion of the space $\mathfrak{A}^{pq}$ using the inner product (8.1.8) and $\mathfrak{L}$ is just the HILBERT space sum of all the $\mathfrak{L}^{pq}$. For two forms of the same type, it is convenient to define the point function $\langle\varphi, \psi\rangle$ by

$$\langle\varphi, \psi\rangle \, dM = \varphi \wedge {}^*\bar{\psi}, \quad \langle\varphi, \varphi\rangle = |\varphi|^2$$

where dM is the element of volume on M. The formulas for (φ, ψ), dM, and $\langle\varphi, \psi\rangle$ are

$$(\varphi, \psi) = \int_M \langle\varphi, \psi\rangle \, dM \tag{8.1.9}$$

where in any analytic coordinate system

$$\begin{aligned} dM &= \Gamma(x, y) \, dx \, dy \\ \langle\varphi, \psi\rangle &= \sum g^{KI} g^{JL} \varphi_{IJ} \bar{\psi}_{KL} \end{aligned} \tag{8.1.10}$$

if φ is given by (8.1.6) and ψ is correspondingly defined. Here Γ is the $\nu \times \nu$ determinant of the $g_{\alpha\beta}$, g^{KI} is the $p \times p$ determinant of the $g^{k_\gamma i_\delta}$ and g^{JL} is the $q \times q$ determinant of the $g^{j_\alpha l_\beta}$. If we use the anti-symmetry of the φ's and ψ's in all their indices, we may write

$$\langle\varphi, \psi\rangle = \frac{1}{p!\,q!} g^{k_1 i_1} \ldots g^{k_p i_p} g^{j_1 l_1} \ldots g^{j_q l_q} \varphi_{i_1 \ldots i_p j_1 \ldots j_q} \times \\ \times \bar{\psi}_{k_1 \ldots k_p l_1 \ldots l_q}.$$

For forms in $\mathfrak{A}$ we define the operator $\bar{\partial}$ as follows: If φ is of type (p, q) with $q = \nu$, we define $\bar{\partial}\varphi = 0$; if $q < \nu$ and φ is given by (8.1.6), we define

$$\bar{\partial}\varphi = \sum \varphi_{IJ\bar{z}^\alpha} \, d\bar{z}^\alpha \wedge dz^I \wedge d\bar{z}^J. \tag{8.1.11}$$

For forms in $\mathfrak{A}$ we define $\mathfrak{b}\varphi$ as follows: If $\varphi \in \mathfrak{A}^{p,q}$ and $q = 0$, we define $\mathfrak{b}\varphi = 0$; otherwise we define $\mathfrak{b}\varphi$ as that form of type $(p, q-1)$ such that

$$(\mathfrak{b}\varphi, \psi) = (\varphi, \bar{\partial}\psi) \tag{8.1.12}$$

for every ψ in $\mathfrak{A}^{p,q-1}$ with compact support in M. As is seen by integrating by parts (see § 3), this leads to a formula of the form

$$(\mathfrak{b}\varphi)_{IR} = (-1)^{p+1} g^{\alpha\beta} (\varphi_{I,\alpha R z^\beta} + A^{ST}_{IR\beta} \varphi_{S,\alpha T}), \tag{8.1.13}$$

for suitable functions $A^{ST}_{IR\beta}$, and to the general formula

$$(\varphi, \bar{\partial}\psi) = (\mathfrak{b}\varphi, \psi) + \int_{bM} \langle\omega, \psi\rangle \, dS \tag{8.1.14}$$

where dS is the invariant surface element on $b\,M$ and

$$\omega = \nu\,\varphi, \quad \omega_{IR} = (-1)^p g^{\alpha\beta} \varphi_{I,\alpha R} r_{z^\beta}. \tag{8.1.15}$$

From (8.1.15), we may also derive the formula

$$(\bar{\partial}\varphi, \psi) = (\varphi, \mathfrak{b}\psi) + \int_{bM} \langle\varphi, \nu\,\psi\rangle \, dS. \tag{8.1.16}$$

We let $\mathfrak{A}_0$ denote the subset of φ in $\mathfrak{A}$ for which $\nu\,\varphi = 0$ on $b\,M$.

It follows immediately from (8.1.11) and the antisymmetry of the exterior product that

$$\bar{\partial}\,\bar{\partial}\varphi = 0, \quad \varphi \in \mathfrak{A}. \tag{8.1.17}$$

From (8.1.14) and (8.1.16) if follows that if φ and $\psi \in \mathfrak{A}$, then

$$\begin{aligned}(\mathfrak{d}\varphi, \bar{\partial}\psi) &= (\varphi, \bar{\partial}\,\bar{\partial}\psi) - \int_{bM} \langle \nu\,\varphi, \bar{\partial}\psi\rangle\, dS \\ &= (\mathfrak{d}\,\mathfrak{d}\varphi, \psi) + \int_{bM} \langle \nu\,\mathfrak{d}\varphi, \psi\rangle\, dS.\end{aligned} \tag{8.1.18}$$

By first letting ψ be arbitrary with compact support in M and then letting it be arbitrary we find that

$$\mathfrak{d}\,\mathfrak{d}\varphi = 0 \quad \text{on } M \text{ and } \quad \nu\,\mathfrak{d}\varphi = 0 \quad \text{on } b\,M \text{ if } \quad \varphi \in \mathfrak{A}_0. \tag{8.1.19}$$

The $\bar{\partial}$-NEUMANN problem is to show the existence and regularity of the solutions of the complex Poisson equation

$$\Box\,\varphi \equiv \mathfrak{d}\,\bar{\partial}\varphi + \bar{\partial}\,\mathfrak{d}\,\varphi = \omega \tag{8.1.20}$$

subject to the boundary conditions

$$\nu\,\varphi = \nu\,\bar{\partial}\varphi = 0 \quad \text{on } bM. \tag{8.1.21}$$

This boundary value problem is seen to arise formally from the variational problem of minimizing the integral

$$d(\varphi, \varphi) - 2Re(\omega, \varphi), \quad d(\varphi, \psi) = (\bar{\partial}\varphi, \bar{\partial}\psi) + (\mathfrak{d}\varphi, \mathfrak{d}\psi) \tag{8.1.22}$$

among all $\varphi \in \mathfrak{A}_0$ ($\nu\,\varphi = 0$ on $b\,M$). If ω and $\varphi \in C^\infty(\bar{M})$, we see that φ satisfies

$$\begin{aligned}&(\bar{\partial}\varphi, \bar{\partial}\psi) + (\mathfrak{d}\varphi, \mathfrak{d}\psi) - (\omega, \psi) = 0 \\ &= (\Box\,\varphi - \omega, \psi) + \int_{bM} \{\langle \nu\,\bar{\partial}\varphi, \psi\rangle - \langle \delta\varphi, \nu\,\psi\rangle\}\, dS\end{aligned} \tag{8.1.23}$$

for all $\psi \in \mathfrak{A}_0$; the second line follows from (8.1.14) and (8.1.16). Since $\nu\,\psi = 0$, we see from (8.1.23) (and known formulas-see end of § 3) that the condition $\nu\,\bar{\partial}\varphi = 0$ on $b\,M$ is a natural boundary condition.

8.2. Results. Examples. The analytic embedding theorem

In order to get a complete picture of the results, we define

$$D(\varphi, \psi) = d(\varphi, \psi) + (\varphi, \psi) \quad (\varphi, \psi \in \mathfrak{A}_0) \tag{8.2.1}$$

where $d(\varphi, \psi)$ was defined in (8.1.22). *We define $\mathfrak{D}$ as the closure of $\mathfrak{A}_0$ with respect to the norm corresponding to the inner product D.* The following preliminary theorem is proved in § 8.4:

Theorem. *If $\varphi \in \mathfrak{D}$, then $\varphi \in H_2^1(M_s)$ for each $s < 0$ and if $\psi \in \mathfrak{D}$ and we form $\bar{\partial}\varphi$, $\mathfrak{d}\varphi$, $\bar{\partial}\psi$ and $\mathfrak{d}\psi$ as in (8.1.11) and (8.1.13) using the strong derivatives then the inner product $D(\varphi, \psi)$ is still given by (8.2.1). Moreover $\nu\,\varphi \in H_{20}^1(M)$. Finally, if $\varphi \in H_2^1(M)$ and $\nu\,\varphi \in H_{20}^1(M)$, then $\varphi \in \mathfrak{D}$.*

Then we define the operator L as follows: $\varphi \in \mathfrak{D}(L) \Leftrightarrow \varphi \in \mathfrak{D}$ and there exists an α in $\mathfrak{L}$ such that

$$d(\varphi, \psi) = (\alpha, \psi), \quad \psi \in \mathfrak{D};$$

if $\varphi \in \mathfrak{D}(L)$, we define $L\varphi = \alpha$. We define $\mathfrak{H}$ as the set of all φ in $\mathfrak{D}$ such that $\bar{\partial}\varphi = \mathfrak{b}\varphi = 0$. We define $\mathfrak{D}^{pq} = \mathfrak{D} \cap \mathfrak{L}^{pq}$, $\mathfrak{H}^{pq} = \mathfrak{H} \cap \mathfrak{L}^{pq}$. In §§ 8.4 and 8.6, we prove the following principal theorem:

Theorem. (i) *L is self adjoint, $\mathfrak{R}(L) = \mathfrak{L} \ominus \mathfrak{H}$, and $\mathfrak{H}$ is closed*

(ii) *If $\omega \in \mathfrak{L} \ominus \mathfrak{H}$, $\exists$ a unique $\varphi \in \mathfrak{D}(L) \cap (\mathfrak{L} \ominus \mathfrak{H}) \ni L\varphi = \omega$.*

(iii) *If we define $N\omega = 0$ for $\omega \in \mathfrak{H}$ and $N\omega$ as the solution in* (ii) *if $\omega \in \mathfrak{L} \ominus \mathfrak{H}$, then N is completely continuous.*

(iv) *If $q \geq 1$, $\mathfrak{H}^{pq}$ is finite-dimensional.*

(v) *If $\omega \in \mathfrak{A}$, then $N\omega \in \mathfrak{A}$.*

(vi) *If $q \geq 1$, $\mathfrak{H}^{pq} \subset \mathfrak{A}_0^{pq}$.*

Parts (i)—(iv) are proved, for forms of type (p, q) with $q \geq 1$, in § 8.4. Only parts of the results are proved there for forms of type $(p, 0)$; these are treated completely in § 8.6. The smoothness results in (v) and (vi) are proved in § 8.6. The principal tools used are (a) the important formula (8.3.15) for integration by parts, and (b) the "s-norms" introduced in §§ 8.5 and 8.6.

Before proceeding, we introduce some additional notations: The manifold $M_s = M(s)$ for $s < 0$ consists of all points P on M for which $r(P) \leq s$. An analytic coordinate patch with domain G and range $\mathfrak{N}$ is said to be *tangential* at some point P_0 of $bM \Leftrightarrow$ a part g of bG contains the origin and corresponds, under the mapping from G to $\mathfrak{N}$, to $\mathfrak{N} \cap bM$, the origin corresponding *to* P_0, and at the origin $g_{\alpha\beta}(0) = \delta_{\alpha\beta}$ and the exterior normal to M at P_0 corresponds to the positive y^ν axis (i.e. $r_{y^\nu}(0,0) = 1$). In case $\tau = (\tau^1, \ldots, \tau^r)$ or $(\tau_1, \ldots, \tau_r)$ is a set of indices, τ'_ν denotes the set $(\tau^1, \ldots, \tau^{\nu-1}, \tau^{\nu+1}, \ldots, \tau^r)$. If $\alpha = (\alpha_1, \ldots, \alpha_\nu)$ is a sequence of non-negative integers, then D^α means $D_{x^1}^{\alpha_1} \ldots D_{a^\nu}^{\alpha_\nu}$. If φ is a vector function $\nabla\varphi$ denotes its gradient.

Next, we give an example to illustrate the fact that the $\bar{\partial}$-Neumann problem is not regular, except in the case where $q = \nu$ when it reduces to the DIRICHLET problem since $\bar{\partial}\varphi \equiv 0$ and $\nu\varphi = 0$ on $bM \Leftrightarrow \varphi = 0$ on bM in that case. In the case $q = 0$, the problem is obviously not regular since the null space is just the space of holomorphic forms $\mathfrak{H}^{p0}$. To show that the problem is not regular for $1 \leq q \leq \nu - 1$, we take, as an example, $\nu = 2$, M the unit ball in R_4, the metric Euclidean, and set

$$\varphi = \varphi_1\, d\bar{z}^1 + \varphi_2\, d\bar{z}^2, \quad \varphi_1 = -\bar{z}^2\Phi, \quad \varphi_2 = \bar{z}^1\Phi,$$

$$\Phi = \frac{3 - 2r^2}{6} A(z), \quad r^2 = z^1\bar{z}^1 + z^2\bar{z}^2,$$

where $A(z) \in H_2^1(M)$ and is holomorphic on $M^{(0)}$ but is not in $H_2^2(M)$.

Then

$$r \cdot \nu \varphi = \bar{z}^1 \varphi_1 + \bar{z}^2 \varphi_2 \equiv 0, \quad \bar{\partial} \varphi = \omega \, d\bar{z}^1 \wedge d\bar{z}^2$$

$$\omega = \varphi_{2\bar{z}^1} - \varphi_{1\bar{z}^2} = \bar{z}^1 \Phi_{\bar{z}^1} + \bar{z}^2 \varphi_{\bar{z}^2} + 2\Phi = (1 - r^2) A(z) \in H^1_{20}(M)$$

$$\Delta \varphi_1 = -\bar{z}^2 \Delta \Phi - \Phi_{z^2}, \quad \Delta \varphi_2 = \bar{z}^1 \Delta \Phi + \Phi_{z^1}$$

$$\Delta \Phi = -\frac{1}{3}(z^1 A_{z^1} + z^2 A_{z^2} + 2A) \in L_2(M).$$

It follows easily that $\varphi \in \mathfrak{D}(L)$ but φ does not $\in H^2_2(M)$ as it would if the problem were regular.

We now prove a theorem indicating the connection with the $\bar{\partial}$-cohomology theory:

Theorem. *If $\varphi \in \mathfrak{A}^{pq}$ with $q \geq 1$ and if $\bar{\partial}\varphi = 0$, there is a harmonic field $\varphi_0 (\in \mathfrak{H}^{pq})$ such that $\varphi - \varphi_0 = \bar{\partial}\Phi$ for some $\Phi \in \mathfrak{A}^{p,q-1}$.*

Proof. First of all, suppose $f \in C^\infty$ on R_1, and we define

$$\Phi_1 = f(r)(\nu\varphi), \quad f(0) = 0, \quad f^1(0) = 4, \quad f(r) = 0$$

for $r \leq s_0 < 0$ for some such s_0. A computation something like that in (8.4.11) and (8.4.12) below shows that

$$\nu(\varphi - \bar{\partial}\Phi_1) = 0 \quad \text{on } bM.$$

Hence we may assume that $\varphi \in \mathfrak{A}^{pq}_0$. Then, let $\Phi = N\,\mathfrak{b}\varphi$. Then it follows from Theorem 8.6.3 that Φ, $\bar{\partial}\Phi$, and $\mathfrak{t}\,\Phi$ all $\in \mathfrak{A}_0$, so that $\bar{\partial}\mathfrak{b}\Phi$ is orthogonal to $\mathfrak{b}(\varphi - \bar{\partial}\Phi)$ (see (8.1.14)—(8.1.16)) and also equal to it and hence zero. Thus (since $\bar{\partial}\varphi = 0$)

$$\mathfrak{b}(\varphi - \bar{\partial}\Phi) = \bar{\partial}(\varphi - \bar{\partial}\Phi) = 0 \quad \text{or} \quad \varphi - \bar{\partial}\Phi \in \mathfrak{H}^{pq}.$$

The following analog of the KODAIRA decomposition theorem is of some interest:

Theorem. *$\mathfrak{L} = \mathfrak{H} \oplus \mathfrak{D}' \oplus \mathfrak{C}$ where $\mathfrak{H}$ has its usual significance, $\mathfrak{D}'$ is the totality of forms of the form $\bar{\partial}\varphi$ for some $\varphi \in \mathfrak{D}$, and $\mathfrak{C}$ is that of forms of the form $\mathfrak{b}\psi$ for ψ in $\mathfrak{D}$. If the given form in $\mathfrak{L} \in \mathfrak{A}$, then its projections on $\mathfrak{H}$, $\mathfrak{D}$, and $\mathfrak{C}$ also $\in \mathfrak{A}$.*

Proof. It is clear that if $h \in \mathfrak{H}$ and φ and $\Psi \in \mathfrak{D}$, then the forms h, $\bar{\partial}\varphi$, and $\mathfrak{b}\Psi$ (see (8.1.14)—(8.1.16) and approximation) are mutually orthogonal. Since $\mathfrak{H}^{p0}$ just consists of the holomorphic forms of degree p and the $\mathfrak{H}^{pq}$ with $q > 0$ are finite-dimensional, it follows that $\mathfrak{H}$ is closed. If $\omega \in \mathfrak{L} \ominus \mathfrak{H}$, let $\Phi = N\omega$, $\varphi = \mathfrak{b}\Phi$, and $\psi = \bar{\partial}\Phi$. From Theorem 8.6.3, it follows that φ and $\psi \in \mathfrak{D} \cap (\mathfrak{L} \ominus \mathfrak{H})$ and that

$$(8.2.2) \qquad \bar{\partial}\varphi + \mathfrak{b}\psi = \omega \quad \text{and} \quad \mathfrak{b}\varphi = \bar{\partial}\psi = 0.$$

It is clear from the first statement above that the decomposition is unique. That $\mathfrak{C}$ and $\mathfrak{D}'$ are closed follows from our principal results and the particular choice of φ and ψ in (8.2.2).

In case $\omega \in \mathfrak{A}$, we first subtract off the form ω_1 defined (different notation unfortunately) in the proof of Theorem 8.4.4, so that $\omega_0 = \omega - \omega_1 \in \mathfrak{A}_0$. If we then define Φ_0, φ_0, and ψ_0 as above in terms of ω_0, we see that (8.2.2) holds for ω_0 and that φ_0 and $\psi_0 \in \mathfrak{A}_0$.

We can now present a simplification and correction to the proof given in the paper, MORREY [13], mentioned above of the possibility of embedding analytically a real-analytic abstract manifold in Euclidean space. The error in that paper was in the proof of Theorem C of that paper which was given in §§ 8—11. The embedding theorem was proved by BOCHNER for compact manifolds in 1937 ([1]) assuming the existence of an analytic metric; this result was extended by MALGRANGE in 1957 to the case of non-compact manifolds. The result of MORREY [13] was generalized to manifolds with a countable topology by GRAUERT using methods of the theory of functions of several complex variables and some results of REMMERT which had not been published at that time.

We now outline our method of proof. First of all, on account of BOCHNER's result, it is sufficient to show the following:

Theorem A. *With each point P_0 of the given real analytic compact manifold M_0 there are associated ν functions w_γ which are analytic over the whole of M_0 and have linearly independent gradients at P_0.*

For the gradients will remain linearly independent in some neighborhood of P_0 and thus M_0 can be covered by neighborhoods $\mathfrak{R}_q$, $q = 1, \ldots, Q$, where the functions $w_{q\gamma}$, $\gamma = 1, \ldots, \nu$ are analytic over M_0 and have linearly independent gradients over $\mathfrak{R}_q$, $q = 1, \ldots, Q$. The mapping $w_{q\gamma} = w_{q\gamma}(P)$ maps M_0 analytically into Euclidean space of Q^ν dimensions; the mapping may not be $1-1$ in the large but is locally $1-1$ and non-singular and the Euclidean metric induces an analytic metric on M_0.

To prove Theorem A, we first embed M_0 in an open complex extension M (see MORREY ([13], § 2, SHUTRICK, or WHITNEY-BRUHAT where this embedding is discussed for manifolds with countable topology). Let P_0 be any point on M_0, let τ_0 be a complex-analytic coordinate patch with domain G_0 containing the origin and range $\mathfrak{R}_0$ containing P_0, in which P_0 and the origin correspond and the part of G_0 in R^ν (i.e. for which $y = 0$) corresponds to $\mathfrak{R}_0 \cap M_0$, and choose a Hermitian metric (8.1.1) which is of class $C^\infty(M)$, which is real on M_0 (i.e. $g_{\alpha\beta}$ is real on M_0), and for which we have

$$g_{\alpha\beta}(x, y) = \delta_{\alpha\beta} \quad (x, y) \text{ on } G_0, \tag{8.2.3}$$

with respect to the coordinate system τ_0.

For points P on M near M_0, we define $r'(P)$ to be the geodesic distance from P to M_0. It is easily shown (see MORREY [13], § 3) that the function $K(P) = [r'(P)]^2/2$ is of class C^∞ in a neighborhood of M_0

including all points where $r'(P) \leq R_0$. We define M_R as the complex analytic manifold $r'(P) \leq R$. It is easy to compute that ($g_{\alpha\beta}(x, 0)$ is real)

(8.2.4) $$4K_{z^\alpha \bar{z}^\beta}(x,0)\, T^\alpha \bar{T}^\beta = K_{y^\alpha y^\beta}(x,0)\, T^\alpha \bar{T}^\beta = g_{\alpha\beta}(x,0)\, T^\alpha \bar{T}^\beta$$

for any complex-analytic patch which carries the points $(x, 0)$ into M_0. Thus if $0 < R \leq R_1$, $b\,M_R$ is regular and of class C^∞ and M_R is strongly pseudo-convex; the function $r(P)$ used in this paper reduces to $r'(P) - R$ on M_R and the pseudo-convexity follows from (8.2.4). In fact, since $r(P) = r'(P) - R$ and $K(P) = [r'(P)]^2/2$ on M_R, we see by following through the proof of Theorem 5.8 (of the cited paper) as far as equation (5.18), that we obtain

$$d_R(\varphi) = I_R(\varphi) + \int_{bM(R)} \sum g^{KI} g^{ST} g^{\alpha\beta} g^{\gamma\delta} r_{z^\beta \bar{z}^\gamma}\, \varphi_{I\alpha S}\, \varphi_{K\delta T}\, dS(P)$$
$$\geq -C_1(\varphi,\varphi)_R + C_2 R^{-1} \int_{bM(R)} |\varphi|^2\, dS, \quad \varphi \in \mathfrak{D}(M_R)$$

(cf. 8.3.15 below). Thus we obtain

(8.2.5) $$\int_{bM(R)} |\varphi|^2\, dS \leq C R [d_R(\varphi) + (\varphi,\varphi)_R], \quad 0 < R \leq R_1 \leq R_0, \quad \varphi \in \mathfrak{D}^{pq}(M_R),$$

where C is independent of R.

We now sketch the proof, given in § 7 of the cited paper (MORREY [13]), of the important inequality

(8.2.6) $$(\varphi,\varphi)_R \leq C R^2 d_R(\varphi), \quad 0 < R \leq R_2 \leq R_1, \quad \varphi \in \mathfrak{D}^{pq}(M_R), \quad q \geq 1.$$

Incidentally, this shows that $\mathfrak{H}^{pq}(M_R)$ *consists only of the zero element if* $q \geq 1$ and R is small enough. We conclude first that there is an R_3, $0 < R_3 \leq R_1$, such that $\overline{M(R_3)}$ can be covered by a finite number of neighborhoods $\mathfrak{N}_t^*$ each of which $\subset \mathfrak{N}_t$, the range of a complex analytic coordinate patch τ_t of the type of τ_0 (i.e. real on M_0) with domain G_t, and is the range of *a* "quasi-geodesic" non-analytic (but C^∞) coordinate system τ_t^* with domain of the form $G_{0t} \times \overline{B(0, R_3)}$, where G_{0t} is a domain of class C^∞ in real ν-space R^ν and $\bar{G}_{0t} \subset G_t \cap R^\nu$. If $P \in \mathfrak{N}_t^*$, its quasi-geodesic coordinates (ξ, η) with respect to τ_t^* are determined as follows: There is a unique geodesic through P which is orthogonal to M_0 at some point of M_0. Let its equations be

$$x^\alpha = x^\alpha(r), \quad y^\alpha = y^\alpha(r), \quad 0 \leq r \leq r'(P),$$

in the (x, y) coordinates of τ_t. We define

$$\xi^\alpha(P) = x^\alpha(0), \quad \zeta^\alpha(P) = {}'y^\alpha(0) \quad ({}'y = dy/dr).$$

Since the metric is real along M_0, we see that

$$ds^2 = g_{1\alpha\beta}(x,0)\,(dx^\alpha dx^\beta + dy^\alpha dy^\beta),$$
$$= g_{1\alpha\beta}(x,0)\,(d\xi^\alpha d\xi^\beta + d\zeta^\alpha d\zeta^\beta), \quad \text{along } M_0.$$

By using the Lagrange method of reducing a quadratic form to a sum of squares, we introduce the η^α by

$$\zeta^\alpha = d^\alpha_\gamma(\xi)\,\eta^\gamma$$

where the $d^\alpha_\gamma \in C^\infty$ and the matrix is non-singular so that

(8.2.7) $$ds^2 = g_{1\alpha\beta}(\xi, 0)\,d\xi^\alpha\,d\xi^\beta + \delta_{\alpha\beta}\,d\eta^\alpha\,d\eta^\beta \quad \text{along } M_0.$$

From the construction, it follows that

(8.2.8) $$K(P) = |\eta(P)|^2/2, \quad r'(P) = |\eta(P)|.$$

It is shown in MORREY [13] 3, that the coordinates are C^∞.

With each t and each R, $0 < R \le R_3$, we define an analytic manifold $\bar{M}_{tR}$ as follows: Let $\bar{\mathfrak{N}}_{tR} = \tau_t^*[\bar{G}_{0t} \times \overline{B(0, R)}]$, and let $\bar{G}_{tR} = \tau_t^{-1}(\mathfrak{N}_{tR})$; clearly $\bar{G}_{0t} = \bar{G}_{tR} \cap R^\nu$. Choose positive numbers a_t and A_t such that if we define

$$F_t = \overline{F_t^{(0)}}, \quad F_t^{(0)}: |x^\alpha| < A_t, \quad \alpha = 1, \ldots, \nu$$

then

$$\bar{G}_{tR_3} \subset F_t^{(0)} \times B(0, a_t/2), \quad G_{0t} \subset F_t^{(0)}.$$

Extend the metric $g_{t\alpha\beta}$ to be of class C^∞ for all (x, y), to be periodic of period $2A_t$ in each x^α to be real if $y = 0$, and so that $g_{t\alpha\beta}(x, y) = \delta_{\alpha\beta}$ for all x on and near ∂F_t and all y with $|y| \ge 3a_t/4$. Then, if R_3 is small enough, the quasi-geodesic (ξ, η) coordinates can be extended to all (ξ, η) with $|\eta| \le a_t$ to be periodic of period A_t in each ξ^α. We then let M_{tR} be the set of all (x, y) corresponding to the (ξ, η) with $|\eta| \le R$, any two points (x_1, y) and (x_2, y) where each $x_2^\alpha - x_1^\alpha = 2A_t n^\alpha$, n^α an integer, being identified.

Now, suppose $\varphi \in \mathfrak{D}^{pq}(M_R)$, $0 < R \le R_3$. Let $\{\zeta_s\}$, $s = 1, \ldots, S$ be a partition of unity such that each $\zeta_s \in C^\infty(\bar{M}_{R_3})$ and has support in $\bar{M}_{R_3} \cap \bar{\mathfrak{N}}_t^*$ for some t (ζ_s need not vanish on $b\,M_{R_3}$) and let $\varphi_s = \zeta_s\,\varphi$. Then, by approximating φ by smooth forms, as we may on account of our principal results, we see that each $\varphi_s \in \mathfrak{D}^{pq}(M_R)$. But now, we may associate each φ_s with a form φ_{st} on M_{tR} by defining the components of φ_{st} on $\bar{G}_{tR}$ to agree with those of φ_s there and to vanish elsewhere on M_{tR}. Then, clearly,

$$d_R(\varphi_s) = d_{tR}(\varphi_{st}), \quad (\varphi_s, \varphi_s)_R = (\varphi_{st}, \varphi_{st})_{tR},$$
$$d_R(\varphi_s) \le C[d_R(\varphi) + (\varphi, \varphi)_R]$$

since the ζ_s do not depend on R. Accordingly, it is sufficient to prove (8.2.6) for forms $\in \mathfrak{D}^{pq}(M_{tR})$, since, if this is done, we would have

$$(\varphi, \varphi)_R^{1/2} \le \sum_{s=1}^{S} (\varphi_s, \varphi_s)_R^{1/2} \le C \cdot S \cdot R \cdot [d_R(\varphi) + (\varphi, \varphi)_R]^{1/2}$$

from which the result follows easily if $0 < R \le R_2 \le R_3$.

So we consider some M_{tR} and drop the t. We first prove (8.2.6) for forms $\varphi_0 \in H^1_{20}(M_R)$ if $0 < R \le R_3 \le R_1$.

We shall hereafter denote the components φ_{IJ} simply by φ^j. Then

$$(8.2.9)\qquad \begin{aligned}(\varphi_0, \varphi_0) &= \int_0^R \int_{bM(r)} \langle\varphi_0, \varphi_0\rangle\, dS\, dr \\ &\le C \int_0^R r^{\nu-1}\, dr \int_F d\xi \int_\Sigma \sum_j |\varphi_0^j(r, \xi, \theta)|^2\, d\textstyle\sum(\theta),\end{aligned}$$

where $\sum = bB(0,1)$ and θ denotes coordinates on $bB(0,1)$, (ξ, η) being the quasi-geodesic coordinates, and (r, θ) being polar coordinates in the η-space. Since $\varphi_0(R, \xi, \theta) = 0$, we have

$$(8.2.10)\qquad \begin{aligned}\varphi_0^j(r, \xi, \theta) &= -\int_r^R \varphi_{0r}^j(s, \xi, \theta)\, ds \\ |\varphi_0^j(r, \xi, \theta)|^2 &\le (R - r) \int_r^R |\varphi_{0r}^j(s, \xi, \theta)|^2\, ds\end{aligned}$$

Substituting (8.2.10) into (8.2.9) and using the conditions $s \ge r$, we obtain

$$\begin{aligned}(\varphi_0, \varphi_0)_R \le CR^2 \cdot ((\varphi_0, \varphi_0))_R &= CR^2((\varphi_0, \varphi_0))_{R_3} \le CR^2\, d_{R_3}(\varphi_0) \\ &= CR^2\, d_R(\varphi_0)\end{aligned}$$

(for the definition of the strong norm $((\varphi, \varphi))^{1/2}$ in H_2^1, see § 8.4) if we define $\varphi_0 = 0$ for $R \le r \le R_3$ and use Theorem 8.4.1 for $M(R_3)$. This is (8.2.6) for φ_0.

From this, it follows that if $\varphi \in \mathfrak{D}$, $\exists$ a unique $\varphi_0 \in H_{20}^1(M_R)$ which minimizes the integral $d_R(\varphi_0 - \varphi)$ among all such φ_0 if $0 < R \le R_3$. It follows that $\varphi_0 - \varphi$ is a harmonic form H and that

$$(8.2.11)\qquad d_R(\varphi) = d_R(\varphi_0) + d_R(H)$$

To prove (8.2.6) for a harmonic H, we shall prove for any harmonic H, whether in $\mathfrak{D}^{pq}(M_R)$ or not, that

$$(8.2.12)\qquad (H, H)_R \le C \cdot R \int_{bM(R)} \langle H, H\rangle\, dS \quad \text{(for } R \text{ small enough)}$$

from which (8.2.6) follows, using (8.2.5). We may assume $H \in C^\infty(\bar{M}_R)$. To prove (8.2.12), we see as in § 5, that the real and imaginary parts H^j of the components satisfy a system of differential equations of the form

$$(8.2.13)\qquad \begin{aligned}a^{\alpha\beta} H^j_{\xi^\alpha\xi^\beta} + 2b^{\alpha\beta} H^j_{\xi^\alpha\eta^\beta} + c^{\alpha\beta} H^j_{\eta^\alpha\eta^\beta} + 2d_k^{j\alpha} H^k_{\xi^\alpha} + \\ + 2e_k^{j\alpha} H^k_{\eta^\alpha} + f_k^j H^k = 0\end{aligned}$$

in terms of the quasi-geodesic coordinates, where

$$(8.2.14)\qquad a^{\alpha\beta}(\xi, 0) = g_1^{\alpha\beta}(\xi, 0), \quad b^{\alpha\beta}(\xi, 0) = 0, \quad c^{\alpha\beta}(\xi, 0) = \delta^{\alpha\beta}.$$

Let us now take spherical coordinates in the η-space as above. Then the equations (8.2.13) are seen to be equivalent to

$$(8.2.15)\qquad \begin{aligned}H^j_{rr} + (\nu - 1) r^{-1} H^j_r + r^{-2} \Delta_{2\theta} H^j + 2c^\gamma H^j_{r\theta^\gamma} + r^{-1} C^{\gamma\delta} H^j_{\theta^\gamma\theta^\delta} + \\ + 2r B^\alpha H^j_{r\xi^\alpha} + 2B^{\alpha\gamma} H^j_{\xi^\alpha\theta^\gamma} + A^{\alpha\beta} H^j_{\xi^\alpha\xi^\beta} + 2D_k^{j\alpha} H^k_{\xi^\alpha} + \\ + E_k^j H^k_r + 2r^{-1} E_k^{j\gamma} H^k_{\theta^\gamma} + F_k^j H^k = 0\end{aligned}$$

where $\Delta_{2\theta}$ denotes the Beltrami operator on the unit sphere and all the coefficients $\in C^\infty$ in (r, ξ, θ). We define the positive form Q by

$$Q = \sum_{j=1}^{N} [(H_r^j)^2 + r^{-2} |\nabla_\theta H^j|^2 + 2 C^\gamma H_r H_{\theta\gamma} + r^{-1} C^{\gamma\delta} H_{\theta\gamma}^j H_{\theta\delta}^j + + 2r B^\alpha H_{\xi^\alpha}^j H_r^j + 2 B^{\alpha\gamma} H_{\xi^\alpha}^j H_{\theta\gamma}^j + A^{\alpha\beta} H_{\xi^\alpha}^j H_{\xi^\beta}^j] \tag{8.2.16}$$

$$F(s) = \int_{F\times\Sigma} \sum_j [H^j(s, \xi, \theta)]^2 \, d\Sigma(\theta). \tag{8.2.17}$$

Then we see easily that

$$F'(s) = 2 \int_{F\times\Sigma} \sum_j H^j(s, \xi, \theta)\, H_r^j(s, \xi, \theta)\, d\xi\, d\Sigma(\theta)$$

$$F''(s) = 2 \int_{F\times\Sigma} \sum_j H^j H_{rr}^j + (H_r^j)^2]\, d\xi\, d\Sigma \tag{8.2.18}$$

$$F'(0) = 2 \int_{F\times\Sigma} \sum_j H^j(\xi, 0)\, H_r^j(0, \xi, \theta)\, d\xi\, d\Sigma = 0.$$

Using (8.2.15) to eliminate the H_{rr}^j, integrating by parts with respect to the ξ^α and θ^γ, and using the fact that

$$\int_\Sigma u \Delta_{2\theta} u\, d\Sigma = -\int_\Sigma |\nabla_\theta u|^2\, d\Sigma$$

we find that

$$F''(s) = 2 \int_{F\times\Sigma} \Big\{ Q + \sum_j [-(\nu - 1) r^{-1} H^j H_r^j + 2 C_{\theta\gamma}^\gamma H^j H_r^j + + r^{-1} C_{\theta\delta}^{\gamma\delta} H^j H_{\theta\gamma}^j + 2r B_{\xi^\alpha}^\alpha H^j H_r^j + 2 B_{\theta\gamma}^{\alpha\gamma} H^j H_{\xi^\alpha}^j + + A_{\xi^\alpha}^{\alpha\beta} H^j H_{\xi^\beta}^j] - 2 D_k^{j\alpha} H^j H_{\xi^\alpha}^k - \text{etc.} \Big\}\, d\xi\, d\Sigma(\theta). \tag{8.2.19}$$

Using the positivity of Q and the simple device $|2ab| \le \varepsilon a^2 + \varepsilon^{-1} b^2$, we conclude from (8.2.19) that

$$F''(s) \ge -(\nu - 1)\, s^{-1} F'(s) - \lambda^2 F(s)$$

where λ depends only on bounds for the coefficients and their derivatives (i.e. on the metric). Thus if R is small enough $(\le \lambda^{-1} \varrho_1)$

$$F(s) \le 2F(R), \quad 0 \le s \le R. \tag{8.2.20}$$

thus

$$(H, H)_R \le C \int_0^R r^{\nu-1} F(r)\, dr \le C R^\nu F(R) \le C R \int_{bM(R)} \langle H, H \rangle\, dS$$

as desired.

The inequality (8.2.6) states that the constant C_3 in Theorem 8.4.5 can be replaced by $C R^2$ for the manifolds M_R. Thus, from Theorem 8.6.4 we conclude the following theorem:

Theorem B′. *If $w \in \mathfrak{D}(L_R^{0,0})$ on M_R and $R \le R_2$, then*

$$\|N_R L_R w\| \le C_5 R^2 \|L_R w\| \tag{8.2.21}$$

where $\| \ \|$ denotes the norm in $\mathfrak{L}^{(0,0)}$.

We now show how to prove Theorem A using Theorem B'. We first construct, for each R, functions $w_{2R\gamma}$, $\gamma = 1, \ldots,$ which $\in \mathfrak{D}(L_R^{0,0})$, are of class $C^\infty(\overline{M}_R)$, which are analytic at least in the ball $B(P_0, R)$ with

$$w_{2R\gamma z^\beta}(0) = \delta_{\gamma\beta} \quad \text{(with respect to } \tau_0)$$
$$(8.2.22) \qquad |L_R\, w_{2R\gamma}| \leq Z_1 R^h \quad (\text{on } \overline{M}_R)$$
$$h = [\nu/2], \quad 0 < R \leq R_4 \leq R_2, \quad Z_1 \text{ independent of } R.$$

To do this we first define $w_{1\gamma} = z^\gamma$ in $\overline{B(P_0, R_2)}$ with respect to τ_0, extend $w_{1\gamma}$ to be of class C^∞ on $M_0 \cup \overline{B(P_0, R_2)}$. We then extend $w_{1\gamma}$ into some M_{R_2} using Whitney's extension theorem, assigning the various derivatives of order $\leq h + 1$ with respect to the y^α in each complex-analytic patch, real on M_0, in such a way that the Cauchy-Riemann equations and all their derivative of order $\leq h$ hold along M_0. Thus the second condition in (8.2.22) is satisfied. The functions $w_{2R\gamma}$ are constructed to $\in \mathfrak{D}(L_R^{0,0})$ by a method like that in the proof of Theorem 8.4.4 which retains the second condition in (8.2.22). Then if we set $w_{3R\gamma} = N_R\, L_R\, w_{2R\gamma}$, we see from Theorem B' that

$$\|w_{3R\gamma}\| \leq C R^2 \|L_R\, w_{2R\gamma}\| \leq Z_2 R^{h+2+\nu/2}.$$

But also $w_{4R\gamma} = w_{2R\gamma} - w_{3R\gamma}$ is analytic on M_R, so $w_{3R\gamma}$ is analytic on $B(P_0, R)$. From the inequalities of § 2.2, it follows that

$$|\nabla w_{3R\gamma}(0)| \leq Z_3 R^k, \quad k = 1 + (\nu/2) + [\nu/2] - \nu = \frac{1}{2} \text{ or } 1.$$

Hence if R is small enough, the gradients of the $w_{3R\gamma}$ are so small that those of the $w_{4R\gamma}(0)$ are linearly independent at 0.

8.3. Some important formulas

In case $\varphi \in \mathfrak{A}^{pq}$, $\bar{\partial}\varphi$ was defined to be 0 if $q = \nu$ and was defined in (8.1.11) otherwise. Starting from that definition, we obtain

$$\begin{aligned}
\bar{\partial}\varphi &= \frac{(-1)^p}{q!} \sum_I \sum_{j_1 \ldots j_q} \sum_\alpha \varphi_{I, j_1 \ldots j_q \bar{z}^\alpha}\, dz^I \wedge d\bar{z}^\alpha \wedge d\bar{z}^{j_1} \wedge \ldots \wedge d\bar{z}^{j_q} \\
&= \frac{(-1)^p}{(q+1)!} \sum_I \sum_{(j)\alpha} \varphi_{I, j_1 \ldots j_q \bar{z}^\alpha} \Big\{ dz^I \wedge d\bar{z}^\alpha \wedge d\bar{z}^{j_1} \wedge \ldots \wedge d\bar{z}^{j_q} + \\
&\quad + \sum_{\delta=1}^{q} (-1)^\delta dz^I \wedge d\bar{z}^{j_1} \wedge \ldots \wedge d\bar{z}^{j_\delta} \wedge d\bar{z}^\alpha \wedge d\bar{z}^{j_{\delta+1}} \wedge \ldots \wedge d\bar{z}^{j_q} \Big\} \\
&= (-1)^p \sum_{I,M} \sum_{\gamma=1}^{q+1} (-1)^{\gamma-1} \varphi_{I, M'_\gamma \bar{z}^{m_\gamma}}\, dz^I \wedge d\bar{z}^M \; (|M| = q+1, q < \nu).
\end{aligned}$$

(8.3.1)

The form (8.3.1) has the advantage that the coefficients are antisymmetric in all the indices $m_1, \ldots, m_{q+1}$.

We shall be concerned with boundary integrals over the manifolds $b\, M_s$. We note that the function r defined in § 8.7 is a real-valued function of class $C^\infty(\overline{M})$ such that $|\nabla r|$ (as measured on M) $= 1$ near

$b\,M$, $r = 0$ on $b\,M$, and $r < 0$ on M near $b\,M$. If we let $d\,M(P)$ denote the volume element and $dS(P)$ denote the surface element along some $b\,M_s$ at P, then

$$\begin{aligned} dM(P) &= \Gamma(x,y)\,dx\,dy, \quad dM(P) = dS(P)\,dr(P) \\ dS(P) &= |\nabla r(x,y)|^{-1}\,\Gamma(x,y)\,dS(x,y), \end{aligned} \tag{8.3.2}$$

(x, y) corresponding to P in some coordinate patch; here $|\nabla r(x, y)|$ denotes the gradient of r with respect to the coordinates (x, y) and $dS(x, y)$ denotes the surface element of the surface of integration in the (x, y)-space. Let G be the domain of an analytic coordinate patch having range $\mathfrak{R}$ such that $\mathfrak{R} \cap b\,M$ is not empty and let g be the part of $b\,G$ which corresponds to $\mathfrak{R} \cap b\,M$. Then, if φ or ψ vanishes on and near $b\,G - g$, we note that

$$\begin{aligned} \int_G \varphi\,\psi_{z^\beta}\,\Gamma(x,y)\,dx\,dy &= \int_g \varphi\,\psi\cdot r_{z^\beta}\cdot|\nabla r(x,y)|^{-1}\,\Gamma(x,y)\,dS(x,y) - \\ &\quad - \int_G \psi\left[\Gamma^{-1}\frac{\partial}{\partial z^\beta}\,\Gamma\,\varphi\right]\cdot\Gamma\,dx\,dy \end{aligned} \tag{8.3.3}$$

and the integral over g can be expressed in the form

$$\int_{\mathfrak{R}\cap bM} \varphi\,\psi\,r_{z^\beta}\,dS(P). \tag{8.3.4}$$

Next, suppose that $\varphi \in \mathfrak{A}^{pq}$, $\psi \in \mathfrak{A}^{p,q-1}$, $q \geq 1$, φ is given by (8.1.6) and ψ is given by a similar formula. Then, using (8.3.1), (8.1.10), and the antisymmetry in the indices, we obtain

$$\begin{aligned} \langle\varphi, \bar\partial\psi\rangle &= \frac{1}{q!}\sum_{I,K}\sum_{(j),(m)} g^{KI}\,g^{j_1 m_1}\cdots g^{j_q m_q}\times \\ &\quad\times\sum_{\gamma=1}^{q}(-1)^{p+\gamma-1}\,\varphi_{I,j_1\ldots j_q}\,\bar\psi_{K,m'_\gamma z^{m_\gamma}} \\ &= \sum_{I,K}\sum_{R,L}\sum_{\alpha,\beta}(-1)^p\,\bar\psi_{KLz^\beta}\,g^{KI}\,g^{RL}\,g^{\alpha\beta}\,\varphi_{I,\alpha R}. \end{aligned} \tag{8.3.5}$$

If we cover M with coordinate patches and use a partition of unity $\{\zeta_s\}$, each with support in one patch, let $\varphi_s = \zeta_s\,\varphi$, integrate by parts using (8.3.3), (8.3.4), and (8.3.5), and add up, we obtain

$$(\varphi, \bar\partial\psi) = \int_{bM}\sum_{I,K}\sum_{R,L} g^{KI}\,g^{RL}\,\omega_{IR}\,\bar\psi_{KL}\,dS(P) + (\mathfrak{d}\,\varphi, \psi) \tag{8.3.6}$$

where $\omega = \nu\,\varphi$ is given by (8.1.15) and $\mathfrak{d}\varphi$ is given by (8.1.13) where the $A^{ST}_{IR\beta}$ are determined so that

$$\begin{aligned} (\mathfrak{d}\,\varphi, \psi) &= \int_M \sum g^{KI}\,g^{RL}(\mathfrak{d}\,\varphi)_{IR}\,\bar\psi_{KL}\,dM(P) \\ &= (-1)^{p+1}\int_M \bar\psi_{KL}\left\{\Gamma^{-1}\frac{\partial}{\partial z^\beta}\,(\Gamma\,g^{KI}\,g^{RL}\,g^{\alpha\beta}\,\varphi_{I,\alpha R})\right\}dM(P) \end{aligned} \tag{8.3.7}$$

for all $\psi \in \mathfrak{A}^{p,q-1}$.

Now, let φ and $\psi \in \mathfrak{A}^{pq}$; we wish to develop an important formula for

$$d(\varphi, \psi) = (\mathfrak{b}\varphi, \mathfrak{b}\psi) + (\bar{\partial}\varphi, \bar{\partial}\psi). \tag{8.3.8}$$

Clearly, we may write φ and ψ each as sums of forms each having compact support and such that if the supports of $\varphi^{(u)}$ and $\psi^{(v)}$ intersect, their union lies in one coordinate patch. So we assume φ and ψ have supports in one patch. Suppose φ and ψ are given by (8.1.6), $q < \nu$, and $\varrho = \bar{\partial}\varphi$, $\sigma = \bar{\partial}\psi$; then ϱ and σ are given by formulas like (8.3.1), so

$$\begin{aligned}(\bar{\partial}\varphi, \bar{\partial}\psi) = \frac{1}{(q+1)!} \int_M \sum_{K,I} \sum_{\substack{m_1\ldots m_{q+1}=1\\ r_1\ldots r_{q+1}=1}}^{\nu}{}' \sum_{\gamma,\delta=1}^{q+1} (-1)^{\gamma+\delta} g^{KI} \times \\ \times g^{m_1 r_1} \ldots g^{m_{q+1} r_{q+1}} \varphi_{I m'_\gamma \bar{z}^{m_\gamma}} \bar{\psi}_{K, r'_\delta z^{r_\delta}} \, dM = I_1 + I_2\end{aligned} \tag{8.3.9}$$

where I_1 is the part of the sum where $\delta = \gamma$ und I_2 is the remainder. We obtain[1]

$$I_1 = \int \sum g^{KI} g^{JL} g^{\alpha\beta} \varphi_{IJ\bar{z}^\alpha} \bar{\psi}_{KLz^\beta} \, dM. \tag{8.3.10}$$

Using the antisymmetry of the indices we see that

$$\varphi_{I, m'_\gamma \bar{z}^{m_\gamma}} \bar{\psi}_{K, r'_\delta z^{r_\delta}} = (-1)^{\gamma+\delta-1} \varphi_{I, m_\delta m'_{\gamma\delta} \bar{z}^{m_\gamma}} \bar{\psi}_{K, r_\gamma r'_{\gamma\delta} z^{r_\delta}}$$

where $m'_{\gamma\delta}$ denotes the m sequence with m_γ and m_δ both omitted, etc. Thus, we obtain

$$I_2 = -\int_M \sum g^{KI} g^{ST} g^{\alpha\beta} g^{\gamma\delta} \varphi_{I,\gamma S \bar{z}^\alpha} \psi_{K,\beta T z^\delta} \, dM. \tag{8.3.11}$$

Using (8.3.7), we thus obtain (interchanging (α, γ) and (β, δ) in (8.3.11))

$$\begin{aligned} d(\varphi, \psi) = I_1 + \int \sum g^{KI} g^{ST} g^{\alpha\beta} g^{\gamma\delta} [- \varphi_{I,\alpha S \bar{z}^\gamma} \bar{\psi}_{K,\delta T z^\beta} + \\ + (\varphi_{I,\alpha S z^\beta} + \sum A^{UV}_{IS\beta} \varphi_{U,\alpha V}) (\bar{\psi}_{K,\delta T \bar{z}^\gamma} + \\ + \sum A^{WX}_{KT\gamma} \psi_{W,\delta X})] \, \Gamma \, dx \, dy = I_1 + I_3. \end{aligned} \tag{8.3.12}$$

Now we define the forms $'\chi$ and $'\omega \in \mathfrak{A}^{p,q-1}$ by

$$\begin{aligned} '\chi &= \sum{}' \, '\chi_{IS} \, dz^I \wedge d\bar{z}^S = (-1)^p \nu \varphi, \\ '\omega &= (-1)^p \nu \psi = \sum{}' \, '\omega_{KT} \, dz^K \wedge d\bar{z}^T \\ '\chi_{IS} &= g^{\alpha\beta} r_{z^\beta} \varphi_{I,\alpha S}, \quad '\omega_{KT} = g^{\delta\gamma} r_{z^\gamma} \bar{\psi}_{K\delta T}. \end{aligned} \tag{8.3.13}$$

Next, we note that

$$\begin{aligned} 2(\varphi_{I,\alpha S z^\gamma} \bar{\psi}_{K,\delta T \bar{z}^\gamma} - \varphi_{I,\alpha S \bar{z}^\gamma} \bar{\psi}_{K,\delta T z^\beta}) = \frac{\partial}{\partial z^\beta} (\varphi_{I,\alpha S} \bar{\psi}_{K,\delta T \bar{z}^\gamma} - \\ - \varphi_{I,\alpha S \bar{z}^\gamma} \bar{\psi}_{K,\delta T}) + \frac{\partial}{\partial \bar{z}^\gamma} (\varphi_{I,\alpha S z^\beta} \bar{\psi}_{K,\delta T} - \varphi_{I,\alpha S} \bar{\psi}_{K,\delta T z^\beta}) \\ g^{\alpha\beta} r_{z^\beta} \varphi_{I,\alpha S \bar{z}^\gamma} = {'\chi}_{IS\bar{z}^\gamma} - (g^{\alpha\beta} r_{z^\beta \bar{z}^\gamma} + g^{\alpha\beta}_{\bar{z}^\gamma} r_{z^\beta}) \varphi_{I,\alpha S} \\ g^{\gamma\delta} r_{\bar{z}^\gamma} \bar{\psi}_{K,\delta T z^\beta} = {'\bar{\omega}}_{KTz^\beta} - (g^{\gamma\delta} r_{z^\beta \bar{z}^\gamma} + g^{\gamma\delta}_{z^\beta} r_{\bar{z}^\gamma}) \bar{\psi}_{K,\delta T}. \end{aligned} \tag{8.3.14}$$

---

[1] Strictly speaking, the integrand in I_1 is not invariant under changes of coordinates; however, the final result in (8.3.15) is invariant.

We then integrate I_3 by parts until there are no terms like $\varphi_{U,\alpha V z^\beta}$ and $\psi_{W,\delta X \bar{z}^\gamma}$ in the remaining integral over M. The result is of the form

$$\begin{aligned} d(\varphi,\psi) = \int_M \sum \{ & g^{KI} g^{JL} g^{\alpha\beta} \varphi_{IJ\bar{z}^\alpha} \bar{\psi}_{KLz^\beta} + C^{IJ,KL\beta} \varphi_{IJ} \bar{\psi}_{KLz^\beta} + \\ & + \bar{C}^{KL,IJ\alpha} \varphi_{IJ\bar{z}^\alpha} \bar{\psi}_{KL} + D^{IJ,KL} \varphi_{IJ} \bar{\psi}_{KL}\} \, dM + \\ & + \frac{1}{2} \int_{bM} \sum g^{KI} g^{ST} \{ '\chi_{IS} [\varepsilon (\bar{\mathfrak{d}} \psi)_{KT} + \bar{B}_{KT}^{UV\delta} \bar{\psi}_{U,\delta V}] + \\ & + '\bar{\omega}_{KT} [\varepsilon (\mathfrak{d} \varphi)_{IS} + B_{IS}^{UV\alpha} \varphi_{U,\alpha V}] - g^{\gamma\delta} {}'\chi_{IS\bar{z}^\gamma} \bar{\psi}_{K,\delta T} - \\ & - g^{\alpha\beta} {}'\bar{\omega}_{KTz^\beta} \varphi_{I,\alpha S} + 2 r_{z^\beta \bar{z}^\gamma} g^{\alpha\beta} g^{\gamma\delta} \varphi_{I,\alpha S} \bar{\psi}_{K,\delta T}\} \, dS \end{aligned}$$

(8.3.15) $\qquad (\varepsilon = (-1)^{p+1})$.

In the case $q = \nu$, $\bar{\partial}\varphi = \bar{\partial}\psi = 0$; a special computation leads to the result (8.3.15) in this case. In the case $q = 0$, $\mathfrak{d}\varphi = \mathfrak{d}\psi = 0 = '\chi = '\omega$ and (8.3.15) holds without a boundary integral. In the case $q = \nu$ it follows that $\nu\,\varphi = 0$ on $b\,M \Leftrightarrow$ the components of $\varphi = 0$ on $b\,M$.

From the result (8.3.6) it follows that $\varphi \in \mathfrak{A} \cap \mathfrak{D} \Leftrightarrow \nu\,\varphi = 0$ on $b\,M$. But now if φ and $\psi \in \mathfrak{A} \cap \mathfrak{D}$, the boundary integral in (8.3.15), is seen to reduce to that of the last term since $'\chi$ and $'\omega$ vanish on $b\,M$ and hence

(8.3.16) $\qquad '\chi_{IS\bar{z}^\gamma} = \lambda_{IS} r_{\bar{z}^\gamma} \quad$ and $\quad '\omega_{KTz^\beta} = \bar{\mu}_{KT} r_{z^\beta} \quad$ on $b\,M$

for suitable functions λ_{IS} and μ_{KT}. From our hypothesis of pseudo-convexity of $b\,M$, it follows that

$$r_{z^\beta \bar{z}^\gamma} \tau^\beta \bar{\tau}^\gamma \geq C\, g_{\beta\gamma} \tau^\beta \bar{\tau}^\gamma, \quad C > 0, \quad \text{whenever} \quad r_{z^\beta} \tau^\beta = 0.$$

Consequently, if $\psi = \varphi$ and $\varphi \in \mathfrak{A} \cap \mathfrak{D}$, then

$$\begin{aligned} & \int_{bM} g^{KI} g^{ST} r_{z^\beta \bar{z}^\gamma} g^{\alpha\beta} g^{\gamma\delta} \varphi_{I,\alpha S} \bar{\varphi}_{K\delta T} \, dS \\ & \geq C \int_{bM} g^{KI} g^{ST} g^{\alpha\delta} \varphi_{I,\alpha S} \bar{\varphi}_{K\delta T} \, dS = C \int_{bM} |\varphi|^2 \, ds. \end{aligned} \tag{8.3.17}$$

Now, for forms φ having support in the range $\mathfrak{N}$ of some coordinate patch, it is sometimes desirable to introduce a new basis $\zeta^1, \ldots, \zeta^\nu$, $\bar{\zeta}^1, \ldots, \bar{\zeta}^\nu$ for the 1-forms, given by

$$\begin{aligned} \zeta^\alpha &= c_\gamma^\alpha \, dz^\gamma, & dz^\gamma &= d_\alpha^\gamma \zeta^\alpha \\ \bar{\zeta}^\alpha &= \bar{c}_\gamma^\alpha \, d\bar{z}^\gamma, & d\bar{z}^\gamma &= \bar{d}_\alpha^\gamma \bar{\zeta}^\alpha \end{aligned} \tag{8.3.18}$$

and to introduce the corresponding differential operators

$$\begin{aligned} u_{,\gamma} &= d_\gamma^\alpha u_{z^\alpha}, & u_{z^\alpha} &= c_\alpha^\gamma u_{,\gamma}, \\ u_{.\bar{\gamma}} &= \bar{d}_\gamma^\alpha u_{\bar{z}^\alpha}, & u_{\bar{z}^\alpha} &= \bar{c}_\alpha^\gamma u_{.\bar{\gamma}} \end{aligned} \tag{8.3.19}$$

where of course the matrices (c_γ^α) and (d_γ^α) are inverses of one another as are the matrices $(\bar{c}_\gamma^\alpha)$ and $(\bar{d}_\alpha^\gamma)$. The exterior multiplication allows us to express any form in terms of the ζ^γ and $\bar\zeta^\gamma$:

$$(8.3.20)\quad \begin{aligned}\varphi &= \frac{1}{p!\,q!}\sum_{\substack{i_1\ldots i_p\\ j_1\ldots j_q}} \tilde\varphi_{i_1\ldots i_p j_1\ldots j_q} dz^{i_1}\wedge\ldots\wedge dz^{i_p}\wedge d\bar z^{j_1}\wedge\ldots d\bar z^{j_q}\\ &= \frac{1}{p!\,q!}\sum \tilde\varphi_{i_1\ldots i_p j_1\ldots j_q} d^{i_1}_{k_1}\ldots d^{i_p}_{k_p}\bar d^{j_1}_{l_1}\ldots \bar d^{j_q}_{l_q}\zeta^{k_1}\wedge\ldots\wedge\zeta^{k_p}\\ &\qquad \bar\zeta^{l_1}\wedge\ldots\wedge\bar\zeta^{l_q}.\end{aligned}$$

In case the bases are introduced so that

$$(8.3.21)\quad g^{\alpha\beta}\bar c^\gamma_\alpha c^\delta_\beta = \Delta^{\gamma\delta} \quad\text{and}\quad g_{\alpha\beta}d^\alpha_\gamma \bar d^\beta_\delta = \Delta_{\gamma\delta},\quad \gamma,\delta = 1,\ldots,\nu$$

we see that

$$(8.3.22)\quad \begin{aligned}&\langle\varphi,\psi\rangle = \sum_{I,J}\varphi_{IJ}\bar\psi_{IJ}, \quad\text{if}\\ &\varphi = \sum_{I,J}\varphi_{IJ}\zeta^I\wedge\bar\zeta^J,\quad \psi = \sum_{I,J}\psi_{IJ}\zeta^I\wedge\bar\zeta^J.\end{aligned}$$

We call a basis in which the c^γ_α satisfy (8.2.21) an *orthogonal basis*. Such bases were used by Kohn ([1], [2]).

In terms of such a basis, we see that

$$(8.3.23)\quad \begin{aligned}\bar\partial\varphi &= \sum \chi_{IM}\zeta^I\wedge\bar\zeta^M,\quad \mathfrak{d}\varphi = \sum \varrho_{IR}\zeta^I\wedge\bar\zeta^R\\ \chi_{IM} &= \sum_{\gamma=1}^{q+1}(-1)^{p+\gamma-1}\varphi_{I,M'_\gamma,\bar m_\gamma} + \sum B^{KT}_{IM}\varphi_{KT}\\ \varrho_{IR} &= (-1)^{p+1}\sum_\alpha \varphi_{I,\alpha R,\alpha} + \sum A^{UV\alpha}_{IR}\varphi_{U,\alpha V}\end{aligned}$$

where the A's and B's are suitable C^∞ functions.

Such bases are more useful in boundary neighborhoods $\mathfrak{N}$ (in which $\mathfrak{N}\cap bM$ is not empty). In case σ is a *tangential analytic coordinate patch* with domain $G\cup g$ and range $\mathfrak{N}$, we may, by taking a smaller $\mathfrak{N}$ if necessary, choose an orthogonal basis ζ such that

$$(8.3.24)\quad 2i\,g^{\alpha\beta}r_{z\beta} = \bar d^\alpha_\nu.$$

It is also possible, choosing $\mathfrak{N}$ smaller if need be, to introduce non-analytic boundary coordinates (t, r), $t = (t^1,\ldots,t^{2\nu-1})$ of class C^∞ which range over some $G_R\cup\sigma_R$ and which are such that the metric takes the form

$$(8.3.25)\quad ds^2 = \sum_{\lambda,\mu=1}^{2\nu-1} a_{\lambda\mu}(t,r)\,dt^\lambda dt^\mu + dr^2,\quad a_{\lambda\mu}(0,0) = \delta_{\lambda\mu}.$$

Now, since the basis is complex-orthogonal, each $u_{,\gamma} = ('D^\gamma u - i''D^\gamma u)/2$ and $u_{,\bar\gamma} = ('D^\gamma u + i''D^\gamma u)/2$, where $'D^\gamma$ and $''D^\gamma$ are real operators which are, in fact, directional derivatives along real unit vectors e'_γ and e''_γ in which all 2ν are mutually orthogonal and $e''_\nu = \nabla r$. Thus, in terms of the (t, r) coordinates

$$(8.3.26)\quad \begin{aligned}u_{,\gamma} &= \sum_\lambda e^\lambda_\gamma u_{t^\lambda},\quad \gamma<\nu \quad (e^\lambda_\gamma\ \text{complex});\\ u_{,\nu} &= \sum_\lambda e^\lambda_\nu u_{t^\lambda} - (i/2)\,u_r \quad (e^\lambda_\nu\ \text{real});\\ u_{,\bar\gamma} &= \sum_\lambda \bar e^\lambda_\gamma u_{t^\lambda},\quad \gamma<\nu;\quad u_{,\bar\nu} = \sum_\lambda e^\lambda_\nu u_{t^\lambda} + (i/2)\,u_r.\end{aligned}$$

Finally, by using the relations implied by (8.3.20) between the components with respect to the $(dz^\alpha, d\bar{z}^\alpha)$ basis and the orthogonal $(\zeta^\alpha, \bar{\zeta}^\alpha)$ basis, we find easily that

$$\nu\varphi = \sum \omega_{IR}\zeta^I \wedge \bar{\zeta}^R, \quad \omega_{IR} = (-1)^p(-i/2)\,\varphi_{I,\nu R} \tag{8.3.27}$$

provided the $(\zeta^\alpha, \bar{\zeta}^\alpha)$ basis satisfies (8.3.24) on the boundary neighborhood $\mathfrak{N}$.

To see that (8.1.23) for all $\psi \in \mathfrak{A}_0$ implies that $\nu\,\bar{\partial}\varphi = 0$ on bM in case $\varphi \in \mathfrak{A}_0$, we note that if ψ has support in a boundary coordinate patch, $\omega = \nu\,\bar{\partial}\varphi$, and $(\zeta, \bar{\zeta})$ is an orthogonal basis satisfying (8.3.24), and we write

$$\omega = \sum \omega_{IJ}\zeta^I \wedge \bar{\zeta}^J, \quad \psi = \sum \psi_{IJ}\zeta^I \wedge \bar{\zeta}^J,$$

then

$$\int_{bM} \langle \omega, \psi\rangle\, ds = \int_{bM} \sum_{IJ} \omega_{IJ}\bar{\psi}_{IJ}\, ds.$$

If (8.1.23) holds, it follows that all the $\omega_{IJ} = 0$ unless $j_q = \nu$. But, from (8.3.27), we conclude that

$$\omega_{IJ} = (-1)^p\,(-i/2)\,(\bar{\partial}\varphi)_{I,\nu J}$$

and so ω_{IJ} automatically $= 0$ if $j_q = \nu$.

8.4. The Hilbert space results

In this section we prove the first four principal results stated in § 8.2 for forms of type (p, q) with $q \geq 1$. Only a few results are proved for $(p, 0)$ forms here, they are treated fully in § 8.6. We begin with preliminary results including those stated there.

Let $\bar{M}$ be covered by the ranges $\mathfrak{N}_s$ of analytic coordinate patches τ_s with domains G_s, where each τ_s can be extended to a domain $\Gamma_s \supset \bar{G}_s$. For each s, suppose $\zeta_s = 1$ on G_s, $\zeta_s \in C^\infty(\Gamma_s)$, ζ_s has support on Γ_s, and satisfies $0 \leq \zeta_s \leq 1$ on Γ_s. For forms of a definite type (p, q), we define

$$\begin{aligned} ((\varphi, \psi)) &= (\varphi, \psi) + \sum_s \int_{\Gamma_s} \zeta_s \sum_{IJ\alpha} (\varphi^{(s)}_{IJ\,x^\alpha}\bar{\psi}^{(s)}_{IJ\,x^\alpha} + \varphi^{(s)}_{IJ\,y^\alpha}\bar{\psi}^{(s)}_{IJ\,y^\alpha})\,dx\,dy, \\ ((\varphi, \psi))_{\bar{z}} &= (\varphi, \psi) + \sum_s \int_{\Gamma_s} \zeta_s \sum_{IJ\alpha} \varphi^{(s)}_{IJ\,\bar{z}^\alpha}\bar{\psi}^{(s)}_{IJ\,\bar{z}^\alpha}\,dx\,dy \end{aligned} \tag{8.4.1}$$

where $\varphi^{(s)}_{IJ}$ and $\psi^{(s)}_{IJ}$ are the components of φ and ψ with respect to τ_s. For composite forms, we define these inner products by summing over (p, q). Evidently the space H^1_2 is the closure of $\mathfrak{A}$ under the strong norm defined by the inner product $((\varphi, \psi))$.

Theorem 8.4.1. $H^1_{20} \subset \mathfrak{D}$. *If* $\varphi \in H^1_{20}$, *then*

$$((\varphi, \varphi)) \leq C((\varphi, \varphi))_{\bar{z}}. \tag{8.4.2}$$

If $\varphi \in \mathfrak{D}$, *then* $\nu\varphi \in H^1_{20}$ *and*

$$((\varphi, \varphi))_{\bar{z}} \leq C\,D(\varphi, \varphi), \quad ((\nu\varphi, \nu\varphi)) \leq C(\nu\varphi, \nu\varphi)_{\bar{z}} \leq C\,D(\varphi, \varphi). \tag{8.4.3}$$

Proof. The first statment is obvious. If (8.4.2) were not true, there would exist a sequence $\{\varphi_n\} \ni \varphi_n \in C_c^\infty(M)$ with

$$((\varphi_n, \varphi_n)) = 1 \quad \text{and} \quad ((\varphi_n, \varphi_n))_{\bar{z}} \to 0 \qquad (8.4.4)$$
$$\varphi_n \rightharpoonup \varphi \quad \text{in} \quad H^1_{20}.$$

Then $\varphi_n \to \varphi$ in $\mathfrak{L}$ so $\varphi = 0$. Also, for each s

$$\int_{\Gamma_s} \zeta_s \sum_{IJ\alpha} |\varphi^{(s)}_{nIJ\bar{z}^\alpha}|^2 \, dx\, dy \to 0. \qquad (8.4.5)$$

But, if we write $\varphi = \varphi_1 + i\varphi_2$ and assume $\varphi \in C_c^\infty(\Gamma)$, then

$$\int_\Gamma \zeta |\varphi_{\bar{z}^\alpha}|^2 \, dx\, dy = \int_\Gamma \zeta [(\varphi_{1x^\alpha} - \varphi_{2y^\alpha})^2 + (\varphi_{2x^\alpha} + \varphi_{1y^\alpha})^2] \, dx\, dy$$
$$= \int_\Gamma \zeta(|\varphi_{x^\alpha}|^2 + |\varphi_{y^\alpha}|^2)\, dx\, dy + \int_\Gamma [\zeta_{x^\alpha}(\varphi_1 \varphi_{2y^\alpha} - \varphi_2 \varphi_{1y^\alpha}) + \qquad (8.4.6)$$
$$+ \zeta_{y^\alpha}(\varphi_{1x^\alpha}\varphi_2 - \varphi_{2x^\alpha}\varphi_1)]\, dx\, dy.$$

Applying the result (8.4.6) to (8.4.5) and using the strong convergence of φ_n to 0 in $\mathfrak{L}$, we find that $((\varphi_n, \varphi_n)) \to 0$ which contradicts (8.4.4).

If $\varphi \in \mathfrak{A}_0^{pq}$ with $q \geq 1$, $\nu\varphi \in H^1_{20}$ since it $\in \mathfrak{A}$ and vanishes on bM. The result (8.4.3) for such φ follows from (8.3.15) with $\psi = \varphi$ and $\omega' = \chi' = 0$ on bM and from (8.3.17), from which we obtain

$$d(\varphi, \varphi) \geq I(\varphi, \varphi) + c_1 \int_{bM} |\varphi|^2 \, dS, \quad I(\varphi, \varphi) \geq ((\varphi, \varphi))_{\bar{z}} - C(\varphi, \varphi), \quad c_1 > 0,$$

$I(\varphi, \varphi)$ being the integral over M in (8.3.15). Moreover if $\omega = \nu\varphi$, then

$$((\omega, \omega)) \leq C((\omega, \omega))_{\bar{z}} \leq C((\varphi, \varphi))_{\bar{z}} \leq C D(\varphi, \varphi).$$

The results for φ in $\mathfrak{D}$ follow by an easy limit process. If $q = 0$, then $D(\varphi, \varphi)$ and $((\varphi, \varphi))_{\bar{z}}$ yield equivalent norms.

Before proving the next theorem, we conclude from (8.1.13) and (8.3.1) and the definition of exterior multiplication that

$$\bar{\partial}(\eta\varphi) = \eta\bar{\partial}\varphi + \bar{\partial}\eta \wedge \varphi, \quad \mathfrak{d}(\eta\varphi) = \eta\,\mathfrak{d}\varphi - \omega \qquad (8.4.7)$$
$$\omega_{IR} = (-1)^p g^{\alpha\beta}\eta_{z^\beta}\varphi_{I,\alpha R}, \quad \eta \in \mathfrak{A}^{00}.$$

From (8.4.7) and (8.1.15) we see that if $\eta = f(r)$, then

$$\omega = f'(r)\,\nu\varphi. \qquad (8.4.8)$$

Theorem 8.4.2. *If $\varphi \in \mathfrak{D}$, then $\varphi \in H^1_2(M_s)$ for each $s < 0$. If each $\varphi_n \in \mathfrak{A}_0$ and $\varphi_n \to \varphi$, $\bar{\partial}\varphi_n \to \psi$, and $\mathfrak{d}\varphi_n \to \chi$, then $\psi = \bar{\partial}\varphi$* and *$\chi = \mathfrak{d}\varphi$ where $\bar{\partial}\varphi$ and $\mathfrak{d}\varphi$ are computed from (8.3.1) and (8.1.13), respectively, the derivatives being the strong derivatives. Finally*

$$((\varphi, \varphi))_s \leq C_1 s^{-2} D(\varphi, \varphi), \quad s_0 \leq s < 0.$$

Here $((\ ))_s$ denotes the strong inner product on $M(s)$.

Proof. For each s, choose $\eta \in C_c^\infty(M)$ with $\eta = 1$ on M_s and $\eta = f(r)$ for $s \leq r \leq 0$. Then if φ_n is a sequence as above, $\eta\varphi_n \in H^1_{20}$ for each n

and, from Theorem 8.4.1, we see that

$$((\eta\,\varphi_n, \eta\,\varphi_n)) \le C((\eta\,\varphi_n, \eta\,\varphi_n))_{\bar z} \le C\,D(\eta\,\varphi_n, \eta\,\varphi_n)$$
$$\le C(\eta)\cdot D(\varphi_n, \varphi_n).$$

Thus we see that $\eta\,\varphi_n \to \eta\,\varphi$ in H^1_{20} so that the results follow since we may assume that $|f'(r)| \le 2|s|^{-1}$.

Theorem 8.4.3. (a) *L is self-adjoint.*

(b) *$\mathfrak{H}$ is closed.*

(c) *$\varphi \in \mathfrak{L} \ominus \mathfrak{R}(L) \Leftrightarrow \varphi \in \mathfrak{D}(L)$ and $L(\varphi) = 0 \Leftrightarrow \varphi \in \mathfrak{H}$.*

Proof. (a) Since $D(\varphi, \varphi) \ge (\varphi, \varphi)$, it follows that $L + I$ is $1-1$ and $(L+I)^{-1}$ is defined everywhere and has bound ≤ 1. Suppose ω and $\eta \in \mathfrak{L}$ and let $\varphi = (L+I)^{-1}\,\omega$ and $\psi = (L+I)^{-1}\,\eta$. Then φ and $\psi \in \mathfrak{D}$ and we see from the definitions of L and D that

$$((L+I)^{-1}\,\omega, \eta) = (\varphi, (L+I)\,\psi) = D(\varphi, \psi) = ((L+I)\,\varphi, \psi)$$
$$= (\omega, (L+I)^{-1}\,\eta).$$

Thus $(L+I)^{-1}$ is self-adjoint so that $L + I$ and L are also.

(b) Suppose $\varphi_n \in \mathfrak{H}$ for each n and $\varphi_n \to \varphi$ in $\mathfrak{L}$. Since $\bar\partial\varphi_n \to 0$ and $\mathfrak{b}\varphi_n \to 0$, it follows that $\varphi \in \mathfrak{D}$ and $\bar\partial\varphi = \mathfrak{b}\varphi = 0$.

(c) The first statement follows from HILBERT space theory. The second statement follows since $\varphi \in \mathfrak{D}(L)$ with $L\,\varphi = 0 \Leftrightarrow$

$$d(\varphi, \psi) = 0 \text{ for each } \psi \in \mathfrak{D}.$$

Theorem 8.4.4. *Suppose $\varphi \in \mathfrak{D}^{pq}$ and $q \ge 1$. Then*

$$\int_{bM(s)} |\varphi|^2\,dS \le C_2\,D(\varphi, \varphi), \quad s_0 \le s \le 0, \quad s_0 < 0. \tag{8.4.9}$$

Proof. It is sufficient to prove this for $\varphi \in \mathfrak{A}_0$. Let $\omega = \nu\varphi$. From Theorem 8.4.1, we conclude that $\omega \in H^1_{20}$ and

$$((\omega, \omega)) \le C\,D(\varphi, \varphi). \tag{8.4.10}$$

Next, let us define

$$\psi = \varphi - 4\,\varrho, \quad \varrho_{I, j_1, \ldots, j_q} = (-1)^p \sum_{\gamma=1}^{q} (-1)^{\gamma-1}\, r_{\bar z j_\gamma}\, \omega_{I, j'_\gamma}. \tag{8.4.11}$$

Then

$$\begin{aligned}(-1)^p g^{\alpha\beta} r_{z\beta}\, \psi_{I,\alpha S} = \omega_{IS} - 4\cdot g^{\alpha\beta} r_{z\beta} \Big[r_{\bar z\alpha}\, \omega_{IS} + \\ + \sum_{\gamma=1}^{q-1} (-1)^\gamma\, r_{\bar z s_\gamma I, \alpha s'_\gamma}\Big] = 0 \quad (\text{near } bM)\end{aligned} \tag{8.4.12}$$

since

$$(-1)^p g^{\alpha\beta} r_{z\beta}\, \omega_{I,\alpha s'_\gamma} = g^{\alpha\beta} g^{\varepsilon\theta} r_{z\beta} r_{z\theta}\, \varphi_{I, \varepsilon\alpha s'_\gamma} = 0$$

on account of the antisymmetry of φ in the indices ε and α. Thus $\nu\,\psi = 0$ near bM. Hence for $|s|$ sufficiently small, (8.3.15) and (8.3.17) with $\varphi = \psi$ yields

$$\begin{aligned} d_s(\psi, \psi) &= I_s(\psi, \psi) + \int_{bM(s)} \sum g^{KI} g^{ST} r_{z\beta\bar z\gamma}\, g^{\alpha\beta} g^{\gamma\delta}\, \psi_{I,\alpha S}\, \bar\psi_{K,\delta T}\, dS \\ &\ge c_2((\psi, \psi))_{\bar z s} - C(\psi, \psi)_s + c\int_{bM(s)} |\psi|^2\, ds, \quad c > 0, \quad c_2 > 0. \end{aligned} \tag{8.4.13}$$

From (8.4.10) and (8.4.11), one sees easily that $\varrho \in H^1_{20}$ and

$$((\varrho, \varrho)) \leq C\, D(\varphi, \varphi). \tag{8.4.14}$$

Since $\varrho = 0$ on $b\,M$, we see that

$$\int_{bM(s)} |\varrho(s, P)|^2 \, dS(P) \leq 2 \int_{bM(s)} |\varrho(s, P)|^2 \, d\Sigma(P) \tag{8.4.15}$$
$$\leq C\,|s| \cdot \int_{M-M(s)} |\varrho_r(s, P)|^2 \, dM \leq C\,|s| \cdot D(\varphi, \varphi), \quad s_1 \leq s \leq 0, \quad s_1 < 0.$$

The result follows from (8.4.11), (8.4.13), (8.4.14), and (8.4.15).

Theorem 8.4.5. *Suppose each $\varphi_n \in \mathfrak{D}^{pq}$ with $q \geq 1$ and suppose $\varphi_n \rightharpoonup \varphi$, $\bar{\partial}\varphi_n \rightharpoonup \psi$, and $\mathfrak{d}\varphi_n \rightharpoonup \chi$ in $\mathfrak{L}$. Then $\varphi_n \to \varphi$ in $\mathfrak{L}$, $\varphi \in \mathfrak{D}$, $\psi = \bar{\partial}\varphi$, and $\chi = \mathfrak{d}\varphi$. Also $\mathfrak{H}^{pq}$ has finite dimensionality. If $\varphi \in \mathfrak{L}^{pq} \ominus \mathfrak{H}^{pq}$, then*

$$(\varphi, \varphi) \leq C_3\, d(\varphi, \varphi).$$

Proof. Clearly $\varphi \in \mathfrak{D}$ with $\bar{\partial}\varphi = \psi$ and $\mathfrak{d}\varphi = \chi$, since the manifold of triples $(\varphi, \bar{\partial}\varphi, \mathfrak{d}\varphi)$ is closed in $\mathfrak{L} \oplus \mathfrak{L} \oplus \mathfrak{L}$. To prove the strong convergence, let $\{r\}$ be any subsequence of $\{n\}$. From Theorems 8.4.1, 8.4.2, and 3.4.4, it follows that there is a further subsequence $\{s\}$ such that $\varphi_s \to$ some φ', which must be φ, on each M_t with $t < 0$. Now, choose $\varepsilon > 0$. From Theorem 8.4.4 it follows that there is a $t_2 < 0$ and so small that

$$\int_{M-M(t_2)} |\varphi_s|^2 \, dM \leq C_2\,|t_2| \cdot D(\varphi_s, \varphi_s) < (\varepsilon/3)^2. \tag{8.4.16}$$

From the weak convergence, it follows that (8.4.16) holds also for φ. Since $\varphi_s \to \varphi$ on $M(t_2)$, we conclude that

$$\int_{M(t_2)} |\varphi - \varphi_s|^2 \, dM < (\varepsilon/3)^2, \quad s > s_0. \tag{8.4.17}$$

It follows that $\varphi_s \to \varphi$ on M. It follows that $\varphi_n \to \varphi$ on M.

From this, it follows easily that if $\mathfrak{M}$ is any closed linear manifold in $\mathfrak{L}^{pq}$, there is a form $\varphi \in \mathfrak{D}^{pq}$ which minimizes $d(\varphi, \varphi)$ among all $\varphi \in \mathfrak{D}^{pq} \cap \mathfrak{M}$ for which $(\varphi, \varphi) = 1$. We may let $\mathfrak{M}_1 = \mathfrak{L}^{pq}$ and φ_1 a minimizing form in $\mathfrak{M}_1$, $\mathfrak{M}_2$ be those forms in $\mathfrak{L}^{pq}$ orthogonal to φ_1 and φ_2 a minimizing form in $\mathfrak{M}_2$, etc. Clearly $0 \leq d(\varphi_1, \varphi_1) \leq d(\varphi_2, \varphi_2) \leq \cdots$. Now, suppose all the $d(\varphi_k, \varphi_k) = 0$. Since each $(\varphi_k, \varphi_k) = 1$, we may extract a subsequence $\{\varphi_n\}$ such that $\varphi_n \rightharpoonup \varphi$, $\bar{\partial}\varphi \to \psi\ (= 0)$, and $\mathfrak{d}\varphi_n \to \chi\ (= 0)$. But then $\varphi_n \to \varphi$. But this is impossible since the φ_n form a normal orthogonal set. Hence $\mathfrak{H}^{pq}$ has finite dimensionality and

$$d(\varphi, \varphi) \geq C(\varphi, \varphi), \quad C > 0, \quad \text{if } \varphi \in \mathfrak{D}^{pq} \cap (\mathfrak{L}^{pq} \ominus \mathfrak{H}^{pq})$$

from which the last result follows.

We can now complete the proofs of our principal results (i)—(iv) for forms in $\mathfrak{L}^{pq}$ with $q \geq 1$:

Theorem 8.4.6. (a) *If* $q \geq 1$, $\mathfrak{R}(L^{pq}) = \mathfrak{L}^{pq} \ominus \mathfrak{H}^{pq}$.

(b) *If* $\omega \in \mathfrak{L}^{pq} \ominus \mathfrak{H}^{pq}$ *and* $q \geq 1$, *there exists a unique solution* $\varphi \in \mathfrak{D}(L^{pq}) \cap (\mathfrak{L}^{pq} \ominus \mathfrak{H}^{pq})$ *of* $L^{pq}\varphi = \omega$.

(c) *If we define* $N^{pq}\omega$ *as this solution if* $\omega \in \mathfrak{L}^{pq} \ominus \mathfrak{H}^{pq}$ *and* $N^{pq}\omega = 0$ *if* $\omega \in \mathfrak{H}^{pq}$, *then* N^{pq} *is completely continuous* ($q \geq 1$).

Proof. By virtue of Theorem 8.4.5, it follows that there is a unique φ in $\mathfrak{D}^{pq} \cap (\mathfrak{L}^{pq} \ominus \mathfrak{H}^{pq})$ which minimizes

$$d(\varphi, \varphi) - 2Re(\omega, \varphi)$$

among all such φ. If $\omega \in \mathfrak{L}^{pq} \ominus \mathfrak{H}^{pq}$, then φ satisfies

$$d(\varphi, \psi) = (\omega, \psi), \quad \psi \in \mathfrak{D} \tag{8.4.18}$$

so that $\varphi \in \mathfrak{D}(L^{pq})$ and $L\varphi = \omega$. From (8.4.18) and Theorem 8.4.5, we conclude that

$$\|\varphi\|^2 \leq C_3 d(\varphi, \varphi) = C_3(\omega, \varphi) \leq C_3 \|\omega\| \cdot \|\varphi\|,$$

so that

$$\|\varphi\| \leq C_3 \|\omega\|, \quad d(\varphi, \varphi) \leq C_3 \|\omega\|^2$$

so N is bounded. Finally, suppose $\omega_n \rightharpoonup \omega$ in $\mathfrak{L}^{pq} \ominus \mathfrak{H}^{pq}$. Then, if $\varphi_n = N\omega_n$ and $\varphi = N\omega$, $\varphi_n \rightharpoonup \varphi$, $\bar{\partial}\varphi_n \rightharpoonup \bar{\partial}\varphi$, and $\mathfrak{d}\varphi_n \rightharpoonup \mathfrak{d}\varphi$, so that $\varphi_n \to \varphi$ by Theorem 8.4.5. The complete continuity of N^{pq} follows.

8.5. The local analysis

In this section, we consider an equation of the form

$$\begin{gathered} u_{rr}(t, r) + \sum_{\alpha,\beta=1}^{2\nu-1} a^{\alpha\beta}(t, r)\, u_{,\alpha\beta}(t, r) \\ = \sum_{\alpha,\beta} f^{\alpha\beta}_{,\alpha\beta}(t, r) + \sum_{\alpha} f_{,\alpha}(t, r) + f_r(t, r) + g(t, r) \\ (t = t^1, \ldots, t^{2\nu-1},\ \varphi_{,\alpha} \equiv \partial\varphi/\partial t^\alpha, \text{ etc.}) \end{gathered} \tag{8.5.1}$$

in which the coefficients are real and of class C^∞ for all (t, r) with $-R \leq r \leq 0$ and are periodic of period $2R$ in each t^α and we assume that this is true of each $f^{\alpha\beta}$, f^α, f, and g. Actually, we assume that the $a^{\alpha\beta}$ depend on R (as do the other functions, and satisfy

$$\begin{gathered} |\nabla^p \varepsilon^{\alpha\beta}(t, r)| \leq K_p R^{1-p}, \quad p = 0, 1, 2, \ldots \\ \varepsilon^{\alpha\beta}(0,0) = 0, \quad \varepsilon^{\alpha\beta}(t, r) = \delta^{\alpha\beta} - a^{\alpha\beta}(t, r). \end{gathered} \tag{8.5.2}$$

Notations. In the remainder of this chapter σ_R denotes the hypercube $|t^\alpha| < R$, $\alpha = 1, \ldots, 2\nu - 1$, and G_R denotes the cell $t \in \sigma_R$, $-R < r < 0$.

We shall be interested in functions $f^{\alpha\beta}$, f^α, f, g, and u which $\in C^\infty(\bar{G}_R)$ and vanish near $r = -R$ and $t^\alpha = \pm R$ and we wish to obtain bounds for certain norms which we now define: We define

$$\begin{aligned} (|||u|||_{sR})^2 = (2R)^{2\nu-1} \int_{-R}^{0} \sum_m [&(1 + |m|^2)^s |u_m(r)|^2 + \\ &+ (1 + |m|^2)^{s-1} R^2 |u'_m(r)|^2]\, dr \end{aligned} \tag{8.5.3}$$

for any real s, where here and below

(8.5.4)
$$u_m(r) = (2R)^{1-2\nu} \int_{\sigma_R} u(t,r)\, e^{-i\pi m\cdot t/R}\, dt, \quad m = (m_1 \ldots m_{2\nu-1}),$$
$$\Big(u(t,r) = \sum_m u_m(r)\, e^{i\pi m\cdot t/R}\Big).$$

The following lemma is an immediate consequence of the definition:

Lemma 8.5.1. $|||u|||^2_{s+h} = \sum_{|\alpha|\le s} C_\alpha |||D^\alpha u|||^2_h \cdot (R/\pi)^{2|\alpha|}$

$$\alpha = (\alpha_1, \ldots, \alpha_{2\nu-1}), \quad |\alpha| = \alpha_1 + \cdots + \alpha_{2\nu-1}, \quad C_\alpha = \frac{|\alpha|!}{\alpha_1! \ldots \alpha_{2\nu-1}!},$$

s being an integer and h being any real number. Moreover the norm $|||u|||_{s,R}$ increases with s.

Theorem 8.5.1. *Suppose $H \in C^\infty$ for $-R \le r \le 0$, is periodic of period $2R$ in each t^α, vanishes for $r = -R$, and is harmonic for $-R < r < 0$. Then*

(8.5.5)
$$|||H|||^2_{1/2,R} \le R\, C(\nu) \int_{\sigma_R} |H(t,0)|^2\, dt.$$

Proof. Expand H into a Fourier series as in (8.5.4). Then the $H_m \in C^\infty[-R, 0]$ and satisfy

$$H''_m(r) - \pi^2 |m|^2 R^{-2} H_m(r) = 0, \quad H_m(-R) = 0.$$

Thus we see that

$$H_0(r) = c_0 R^{-1}(r+R)$$
$$H_m(r) = c_m (\sinh \pi |m|)^{-1} \sinh \pi |m| (r+R)/R, \quad m \ne 0$$
$$\int_{\sigma_R} |H(t,0)|^2\, dt = \sum |c_m|^2 \cdot (2R)^{2\nu-1}.$$

The theorem follows from a straightforward computation.

Let us define the "prime-norm" $'|||u|||_{sR}$ by

$$'|||u|||^2_{sR} = (2R)^{2\nu-1} \int_{-R}^{0} \sum_m (1 + |m|^2)^s |u_m(r)|^2\, dr.$$

Then

$$|||u|||^2_{sR} = {}'|||u|||^2_{sR} + R^2\, '|||u_r|||^2_{s-1,R}.$$

Theorem 8.5.2. *Suppose that $a(t,r)$ and $u(t,r) \in C^\infty$ for $-R \le r \le 0$ and are periodic with period $2R$ in each t^α. Then, for each real s*

$$'|||au|||_{sR} \le A_0\, '|||u|||_{sR} + C(a,s)\, '|||u|||_{s-1,R}$$

(8.5.6)
$$A_0 = \max_{-R\le r\le 0} \sum_n |a_n(r)|, \quad C(a,s) = \max_{-R\le r\le 0} \sum_n C_1(n,s)\, |a_n(r)|$$
$$C_1(n,s) = 2^{|s-1|/2} |s| \cdot |n| \left[1 + (1+|n|^2)^{|s-1|/2}\right].$$

A corresponding result holds for the full s-norm.

Proof. Expanding a and u in Fourier series, we obtain

$$(8.5.7)\qquad \begin{aligned}(1+|m|^2)^{s/2}(a\,u)_m(r) &= \sum_n a_n(r)\, v_{n,m}(r),\\ v_{n,m}(r) &= (1+|m|^2)^{s/2}\, u_{m-n}(r)\end{aligned}$$

and think of the $a_n(r)$ as scalars and the $v_{n,m}$ (and others below) as components of a vector function v_n. We now write

$$v_{n,m}(r) = V_{n,m}(r) + W_{n,m}(r), \quad V_{n,m} = (1+|m-n|^2)^{s/2}\, u_{m-n}(r)$$
$$W_{n,m}(r) = [(1+|m|^2)^{s/2} - (1+|m-n|^2)^{s/2}]\, u_{m-n}(r).$$
(8.5.8)

Using the inequalities

$$|a^s - b^s| \le |s| \cdot (a^{s-1} + b^{s-1})\, |a-b|, \quad a>0, \quad b>0,$$
$$2^{-1}(1+|n|^2)^{-1}(1+|m-n|^2) \le 1+|m|^2 \le 2(1+|n|^2)(1+|m-n|^2),$$

we find that

$$(8.5.9)\qquad \begin{aligned}|W_{n,m}(r)| &\le C_2(n,s)\cdot[1+|m-n|^2]^{(s-1)/2}\,|u_{m-n}(r)|,\\ |C_2(n,s)| &\le C_1(n,s).\end{aligned}$$

The result follows easily from (8.5.7), (8.5.8), (8.5.9), the definition of the prime norm, and the fact that $'|||V_n|||_{sR} = '|||u|||_{sR}$ for each n.

Theorem 8.5.3. *Suppose that u, v, f, and $g \in C^\infty$ for $-R \le r \le 0$, all are periodic with period $2R$ in the t^α, u and v vanish for $r = -R$ and 0, and*

$$\Delta u = f, \quad \Delta v = g_r.$$

Then there are absolute constants C_3 and C_4 such that

$$|||u|||_{sR} \le C_3 R^2\, '|||f|||_{s-2,R}, \quad |||v|||_{sR} \le C_4 R \cdot '|||g|||_{sR}.$$

Proof. The u_m satisfy the conditions

$$(8.5.10)\qquad u''_m(r) - \pi^2 |m|^2 R^{-2} u_m(r) = f_m(r), \quad u_m(-R) = u_m(0) = 0$$

Multiplying both sides of (8.5.10) by $-R^2(2R)^{2\nu-1}(1+|m|^2)^{s-1}\bar{u}_m(r)$ and integrating by parts yields

$$(8.5.11)\qquad \begin{aligned}&(2R)^{2\nu-1}\int_{-R}^{0}\sum_m (1+|m|^2)^{s-1}[\pi^2|m|^2|u_m|^2 + R^2|u'_m|^2]\,dr\\ &= -R^2(2R)^{2\nu-1}\int_{-R}^{0}\sum_m (1+|m|^2)^{s-1} f_m(r)\,\bar{u}_m(r)\,dr.\end{aligned}$$

Since $u_m(-R) = u_m(0) = 0$, one easily concludes that

$$(8.5.12)\qquad \begin{aligned}&(2R)^{2\nu-1}\int_{-R}^{0}\sum_m (1+|m|^2)^{s-1}|u_m(r)|^2\,dr\\ &\le Z_1 R^2 (2R)^{2\nu-1}\int_{-R}^{0}\sum_m (1+|m|^2)^{s-1}|u'_m(r)|^2\,dr.\end{aligned}$$

Combining (8.5.11) and (8.5.12) yields

$$(8.5.13)\qquad |||u|||_{sR}^2 \leq Z_1 R^2 (2R)^{2\nu-1} \left| \int_{-R}^{0} \sum_m (1 + |m|^2)^{s-1} f_m(r)\, \bar{u}_m(r)\, dr \right|$$

from which the first result follows using the Schwarz inequality. In the case of v, we obtain the result (8.5.13) with f_m replaced by g'_m. The second result then follows by integrating by parts.

Theorem 8.5.4. *Suppose that u and the functions $\varepsilon_R^{\alpha\beta}$, $f^{\alpha\beta}$, f^α, f, and $g \in C^\infty$ for $-R \leq r \leq 0$, are periodic of period $2R$ in each t^ν, and vanish along $r = -R$. Suppose also that the $\varepsilon_R^{\alpha\beta}$ satisfy (8.5.2) and that u satisfies (8.5.1). Then there is a number R_0, depending only on ν and the K_p and a constant C, depending only on ν such that*

$$|||u|||_{1/2,R} \leq C \Big\{ \sum_{\alpha,\beta}{}' |||f^{\alpha\beta}|||_{1/2,R} + R \sum_\alpha{}' |||f^\alpha|||_{-1/2,R} + $$
$$+ R'|||f|||_{-1/2,R} + R^2\,'|||g|||_{-3/2,1} + \Big[R \int_{\sigma_R} |u(t,0)|^2\, dt \Big]^{1/2} \Big\},$$
$$0 < R \leq R_0.$$

Proof. Let H be the unique harmonic function of Theorem 8.5.1 for which $H(t,0) = u(t,0)$ and let $U = u - H$. Then U satisfies all the conditions with different $f^{\alpha\beta}$, etc., and also vanishes along $r = 0$. We may write the equation (8.5.1) for U in the form

$$\begin{aligned} \Delta U &= \varepsilon_R^{\alpha\beta} U_{,\alpha\beta} + \varepsilon_R^{\alpha\beta} H_{,\alpha\beta} + f^{\alpha\beta}_{,\alpha\beta} + f^\alpha_{,\alpha} + f_r + g \\ &= (f^{\alpha\beta} + \varepsilon_R^{\alpha\beta} U + \varepsilon_R^{\alpha\beta} H)_{,\alpha\beta} + (f^\alpha - 2\varepsilon^{\alpha\beta}_{R,\beta} U - 2\varepsilon^{\alpha\beta}_{R,\beta} H)_{,\alpha} + f_r + \\ &\quad + g + \varepsilon^{\alpha\beta}_{R,\alpha\beta} U + \varepsilon^{\alpha\beta}_{R,\alpha\beta} H = F^{\alpha\beta}_{,\alpha\beta} + F^\alpha_{,\alpha} + F_r + G. \end{aligned}$$

Clearly

$$(8.5.14)\qquad \begin{gathered} U = V^{\alpha\beta}_{,\alpha\beta} + V^\alpha_{,\alpha} + V + W, \\ \Delta V^{\alpha\beta} = F^{\alpha\beta}, \quad \Delta V^\alpha = F^\alpha, \quad \Delta V = F_r, \quad \Delta W = G \end{gathered}$$

and $V^{\alpha\beta}$, V^α, V, and W all vanish along $r = -R$ and $r = 0$.

Now, from Theorem 8.5.2 we conclude that

$$(8.5.15)\qquad '|||\varepsilon_R^{\alpha\beta} U|||_{1/2} \leq A_0^{\alpha\beta}\,'|||U|||_{1/2} + C^{\alpha\beta}(a,s)\,'|||U|||_{-1/2}$$

where $A_0^{\alpha\beta}$ and $C^{\alpha\beta}(a,s)$ are the constants of Theorem 8.5.2 for $\varepsilon_R^{\alpha\beta} = a^{\alpha\beta}$ and where we have left out the subscript R on the norms. But, from (8.5.2), (8.5.4), and differentiation, we see that

$$(8.5.16)\qquad \begin{gathered} \sum_m |m|^{2p} |\varepsilon^{\alpha\beta}_{Rm}(r)|^2 \leq Z_p^2 K_p^2 R^2 \\ , \; p = 0, 1, 2, \ldots \\ \sum_m |m|^{2p} |'\varepsilon^{\alpha\beta}_{Rm}(r)|^2 \leq Z_p^2 K_p^2. \end{gathered}$$

Thus, from (8.5.15) and (8.5.16), we see that

$$(8.5.17)\qquad '|||\varepsilon_R^{\alpha\beta} V|||_{1/2,R} \leq A^{\alpha\beta} \cdot R \cdot '|||V|||_{1/2,R}, \quad V = U \text{ or } H.$$

where $A^{\alpha\beta}$ depends only on the K_p. From (8.5.14), (8.5.17) and Theorem 8.5.3 and Lemma 8.5.1, it follows that

$$\begin{aligned} ||| U |||_{1/2,R} \leq C_1 \Big\{ \sum_{\alpha,\beta} [' ||| f^{\alpha\beta} |||_{1/2} + R A^{\alpha\beta} (' ||| U |||_{1/2} + ' ||| H |||_{1/2})] + \\ (8.5.18) \qquad + R \sum_{\alpha} [' ||| f^{\alpha} |||_{-1/2} + A^{\alpha} (' ||| U |||_{-1/2} + ' ||| H |||_{-1/2})] + \\ + R ' ||| f |||_{-1/2} + R^2 [' ||| g |||_{-3/2} + R^{-1} A (' ||| U |||_{-3/2} + ' ||| H |||_{-3/2})] \Big\} \end{aligned}$$

where $C_1 = C_1(\nu)$ and the $A^{\alpha\beta}$, A^{α}, and A depend only on the K_p. The result follows from (8.5.18), Theorem 8.5.1, and the fact that $u = U + H$.

A similar analysis can be carried through for functions $u \in C^\infty(\overline{C}_R)$ where C_R is the full cube

$$C_R : |t^\alpha| < R, \quad \alpha = 1, \ldots, 2\nu.$$

We are interested in the solutions of the equations

$$(8.5.1)' \qquad \sum_{\alpha,\beta=1}^{2\nu} a^{\alpha\beta}(t)\, u_{,\alpha\beta} = \sum_{\alpha,\beta} f^{\alpha\beta}_{,\alpha\beta} + \sum_{\alpha} f_{,\alpha} + f,$$

where we have replaced r by $t^{2\nu}$ and the coefficients and functions $\in C^\infty$ and are periodic of period $2R$ in each t^α and the $a^{\alpha\beta}$ satisfy (8.5.2). We define the s-norms by

$$||| u |||^2_{s,R} = (2R)^{2\nu} \sum_m (1 + |m|^2)^s |u_m|^2,$$

$$u(t) = \sum_m u_m e^{i\pi m\cdot t/R}, \quad (m = m_1, \ldots, m_{2\nu}).$$

Theorem 8.5.5. *Lemma* 8.5.1 *holds if* $\alpha = (\alpha_1, \ldots, \alpha_{2\nu})$. *Theorem* 8.5.2 *carries over without change. In Theorem* 8.5.3 *we need retain only the case* $\Delta u = f$. *In Theorem* 8.5.4 *we need consider only the cases where* u *satisfies*

$$\Delta u = \varepsilon^{\alpha\beta} u_{,\alpha\beta} + f^{\alpha\beta}_{,\alpha\beta} + f^{\alpha}_{,\alpha} + g,$$

where $\alpha, \beta = 1, \ldots, 2\nu$, in which case the result is

$$||| u |||_{1/2} \leq C \Big[\sum_{\alpha,\beta} ' ||| f^{\alpha\beta} |||_{1/2} + R \sum_{\alpha} ' ||| f^{\alpha} |||_{-1/2} + R^2 ' ||| g |||_{-3/2} \Big], \quad 0 \leq R \leq R_0.$$

8.6. The smoothness results

Until further notice, we shall be concerned with forms of a given type (p, q) in which $q \geq 1$. We begin by defining the s-norms for forms $\varphi \in \mathfrak{A}^{pq}$. Each interior point P of M is in a neighborhood $\mathfrak{N}_P$ which is the image of a cube C_R with respect to a holomorphic coordinate patch τ. Each boundary point P of M is in a neighborhood $\mathfrak{N}_P$ on $M \cup bM$ which is the image of $G_R \cap \sigma_R$ under a non-analytic (t, r) boundary coordinate system of the type described at the end of § 8.3. Let us choose a covering $\mathfrak{U}$ of $M \cup bM$ by means of such neighborhoods $\mathfrak{N}_1, \ldots, \mathfrak{N}_T$ and choose a partition of unity $\{\eta_t\}$, $t = 1, \ldots, T$, of class C^∞, each having support in the corresponding $\mathfrak{N}_t$. In each $\mathfrak{N}_t$, let us choose an orthogonal basis $(\zeta, \bar{\zeta})$ as in § 8.3 which satisfies (8.3.24) if $\mathfrak{N}_t$ is a boundary neighborhood. We then define

$$\||| \varphi \|||_s^2 = \sum_t \sum_{IJ} \||| \eta_t \varphi_{IJ}^{(t)} \|||_s^2 \tag{8.6.1}$$

where the $\varphi_{IJ}^{(t)}$ are the components of φ with respect to the basis $(\zeta, \bar{\zeta})$ in $\mathfrak{N}_t$. Of course the norm depends on the $\mathfrak{N}_t$, the coordinate patches τ_t, and the bases $(\zeta_t, \bar{\zeta}_t)$ but any two such norms, for a fixed s, are topologically equivalent. We choose one fixed set of $\mathfrak{N}_t$, η_t, and $(\zeta_t, \bar{\zeta}_t)$. Then we have the following fact:

Lemma 8.6.1. *$\||| \varphi \|||_s$ is non-decreasing in s and if $\eta \in \mathfrak{A}^{0,0}$, $\||| \eta\varphi \|||_s \leq C(\eta) \||| \varphi \|||_s$.*

We begin by proving the fundamental inequality:

Theorem 8.6.1. *If $\varphi \in \mathfrak{A}_0^{p\,q}$ (and $q \geq 1$), then*

$$\||| \varphi \|||_{1/2}^2 \leq C\, D(\varphi, \varphi), \tag{8.6.2}$$

where C depends on M and the metric and the choices of $\mathfrak{N}_t$, η_t, and $(\zeta_t, \bar{\zeta}_t)$.

Proof. It is clear, from Theorem 8.4.1. that, for a given φ in $\mathfrak{A}_0^{p\,q}$, there is a unique $\varphi_0 \in H_{20}^1$ which minimizes $D(\tilde{\varphi} - \varphi, \tilde{\varphi} - \varphi)$ among all $\tilde{\varphi}$ in H_{20}^1. Since this just leads to the Dirichlet problem, φ_0 and hence $H = \varphi - \varphi_0 \in \mathfrak{A}_0$. Clearly, also, φ_0 satisfies

$$D(\varphi_0 - \varphi, \psi) \equiv -D(H, \psi) = 0, \quad \psi \in H_{20}^1 \tag{8.6.3}$$

so that we have

$$(L + I)\, H = 0, \quad D(\varphi, \varphi) = D(\varphi_0, \varphi_0) + D(H, H). \tag{8.6.4}$$

From Theorem 8.4.1, the definitions, and Lemma 8.6.1 we conclude that

$$\||| \varphi_0 \|||_{1/2}^2 \leq \||| \varphi_0 \|||_1^2 \leq Z_1 \cdot ((\varphi_0, \varphi_0)) \leq Z_2\, D(\varphi_0, \varphi_0) \leq Z_2\, D(\varphi, \varphi). \tag{8.6.5}$$

Thus we need only prove (8.6.2) for $H \in \mathfrak{A}_0$ and satisfying (8.6.4).

From the form of the integral in (8.3.15) it follows that the components $\tilde{H}_{IJ}$ of H in any holomorphic coordinate system satisfy equations of the form

$$g^{\alpha\beta}\, \tilde{H}_{IJ\, \bar{z}^\alpha z^\beta} + (\text{lower order terms in all the } \tilde{H}_{ST}) = 0. \tag{8.6.6}$$

The components H_{IJ} with respect to an orthogonal basis, being linear combinations of the $\tilde{H}_{IJ}$ (with C^∞ variable coefficients), satisfy similar equations. If we multiply by 4, write $g^{\alpha\beta} = g_1^{\alpha\beta} + i\, g_2^{\alpha\beta}$, and express the derivatives in terms of those with respect to x^α and y^α, the equations (8.6.6) take the form

$$g_1^{\alpha\beta}\, u^j_{x^\alpha x^\beta} + g_2^{\alpha\beta}\, u^j_{x^\alpha y^\beta} + g_2^{\beta\alpha}\, u^j_{y^\alpha x^\beta} + g_1^{\alpha\beta}\, u^j_{y^\alpha y^\beta} + \cdots = 0 \tag{8.6.7}$$

where we have denoted the components by u^j.

Now with each boundary point P we choose a (t, r) boundary coordinate patch τ_P from $G_A \cup \sigma_A$ to $M \cup b\,M$ so that the origin corresponds to P and the metric has the form

$$ds^2 = a_{\alpha\beta}(t, r)\, dt^\alpha\, dt^\beta + (dr)^2, \quad a_{\alpha\beta}(0,0) = \delta_{\alpha\beta}. \tag{8.6.8}$$

We also choose a $(\zeta, \bar{\zeta})$ orthogonal basis satisfying (8.3.24). The equations (8.6.7) then take the form

$$\text{(8.6.9)} \qquad \begin{aligned} u^j_{rr} + a^{\alpha\beta}(t,r)\, u^j_{,\alpha\beta} &= b^\alpha_{jk}\, u^k_{,\alpha} + b_{jk}\, u^k_r + c_{jk}\, u^k \\ a^{\alpha\beta}(0,0) &= \delta^{\alpha\beta}. \end{aligned}$$

It is clear that the mappings may be chosen so that A and the bounds for the coefficients and all their derivatives are independent of P Having chosen τ_P, we define

$$\text{(8.6.10)} \qquad \begin{aligned} a^{\alpha\beta}_R(t,r) &= \delta^{\alpha\beta} + h_R(t,r)\,[a^{\alpha\beta}(t,r) - \delta^{\alpha\beta}], \quad (t,r) \in \bar{G}_R, \\ h_R(t,r) &= h[2R^{-1} r] \prod_{\gamma=1}^{2\nu-1} h[2R^{-1} t^\gamma], \end{aligned}$$

where $h \in C^\infty(R_1)$, $h(t) = 1$ for $|t| \leq 5/4$, $h(t) = 0$ for $|t| \geq 7/4$, $0 \leq h(t) \leq 1$ for all t; then we extend $a^{\alpha\beta}_R$ to be periodic of period $2R$ in each t^γ. We define b^α_{jkR}, etc., similarly. Then the $a^{\alpha\beta}_R$ satisfy (8.5.2) with K_p independent of P. Then we can choose an R_1, $0 < R_1 < c\, R_0$ (Theorem 8.5.4) and a finite covering of the strip $-R_1 \leq r \leq 0$ by the parts of the $\mathfrak{N}_P$ corresponding to $G_{R/2}$. In like manner, we may cover $\bar{M}_{-R_1}$ by the parts corresponding to $C_{R/2}$ of holomorphic coordinate patches with $0 < R \leq R_0$ so the $a^{\alpha\beta}_R$ satisfy (8.5.2).

Now, suppose H is harmonic and in $\mathfrak{A}_0$ and let η have support in one of these small neighborhoods, say a boundary one. Then if H^j are the components of H, with respect to $(\zeta, \bar{\zeta})$ and $u^j = \eta H^j$, the u^j satisfy (since the $a^{\alpha\beta} = a^{\alpha\beta}_R$ in $G_{R/2}$)

$$\begin{aligned} &u^j_{rr} + a^{\alpha\beta}_R\, u^j_{,\alpha\beta} = f^\alpha_{j,\alpha} + f_{jr} + g_j \\ &f^\alpha_j = (\eta\, b^\alpha_{jkR} + 2\delta_{jk}\, a^{\alpha\beta}_R\, \eta_{,\beta})\, H^k, \quad f_j = (\eta\, b_{jkR} + 2\delta_{jk}\, \eta_r)\, H^k \\ &g_j = [\eta\, c_{jkR} + \delta_{jk}(\eta_{rr} + a^{\alpha\beta}_R\, \eta_{,\alpha\beta}) - (\eta\, b^\alpha_{jkR} + 2\delta_{jk}\, a^{\alpha\beta}_R\, \eta_{,\beta})_{,\alpha} - \\ &\qquad - (\eta\, b_{jkR} + 2\delta_{jk}\, \eta_r)_r]\, H^k. \end{aligned}$$

From this and Theorem 8.4.4 with $\varphi = H$, we conclude that

$$\sum_{j,\alpha}{}' |||f^\alpha_j|||_0^2 + \sum_j{}' |||f_j|||_0^2 + {}' |||g_R|||_0^2 + \sum_j \int_{\sigma_R} |u^j|^2\, dt \leq C\, D(H,H),$$

where C depends on η (possibly R), and the K_p and the norms may be taken on $\mathfrak{M}$ or on G_R as desired. Thus, from Theorem 8.5.4, it follows that

$$|||\eta\, H|||^2_{1/2,\mathfrak{M}} \leq C\, D(H,H).$$

A similar result holds for the inner neighborhoods. If the η_t form a partition of unity, the result follows.

We have already seen that $(L+I)^{-1}$ is a bounded operator defined everywhere on $\mathfrak{L}^{pq}$. We shall prove the following theorem concerning the solutions φ of

$$\text{(8.6.11)} \qquad (L+I)\,\varphi = \omega, \quad \varphi \in \mathfrak{D}(L), \quad \text{or} \quad \varphi = (L+I)^{-1}\,\omega.$$

We let $\mathfrak{H}^s$ denote the closure of $\mathfrak{A}$ with respect to the s-norm.

Theorem 8.6.2. *If* $\omega \in \mathfrak{L}^{pq}$, *then* $\varphi \in \mathfrak{H}^{1/2}$ *and*

$$\|\!|\varphi|\!\|_{1/2} \leq C\,\|\omega\|. \tag{8.6.12}$$

If $\omega \in {}'\mathfrak{H}^{s-1/2}$ *for an integer* $s \geq 1$, *then* $\varphi \in \mathfrak{H}^{s+1/2}$ *and*

$$\|\!|\varphi|\!\|_{s+1/2} \leq C(s)\,'\|\!|\omega|\!\|_{s-1/2}. \tag{8.6.13}$$

If $\omega \in \mathfrak{A}$, *then* $\varphi \in \mathfrak{A}_0$.

Proof. We shall show first that if $\omega \in \mathfrak{A}$, then $\varphi \in \mathfrak{A}_0$ and all the inequalities (8.6.12) and (8.6.13) hold. The first statements then follow by approximations. To do this, we assume $\omega \in \mathfrak{A}^{pq}$ and for each $\varepsilon > 0$, we let φ_ε be the function minimizing

$$D_\varepsilon(\varphi,\varphi) - 2Re(\omega,\varphi), \quad \varphi \in \mathfrak{A}_0, \quad D_\varepsilon(\varphi,\psi) \equiv \varepsilon((\varphi,\psi)) + D(\varphi,\psi). \tag{8.6.14}$$

This minimum exists and is unique and $\in \mathfrak{A}_0$ as is seen from the theory of § 6.4 (as long as $\varepsilon > 0$). It satisfies

$$D_\varepsilon(\varphi_\varepsilon,\psi) = (\omega,\psi), \quad \psi \in \mathfrak{A}_0. \tag{8.6.15}$$

We shall show by induction that if φ_ε satisfies (8.6.15), then

$$\varepsilon\|\!|\varphi_\varepsilon|\!\|^2_{s+1} + \|\!|\varphi_\varepsilon|\!\|^2_{s+1/2} \leq C(s)\,'\|\!|\omega|\!\|^2_{s-1/2}, \quad s = 0, 1, \ldots \tag{8.6.16}$$

We note first that (8.6.16) holds for $s = 0$, since

$$\begin{aligned} &\varepsilon\|\!|\varphi_\varepsilon|\!\|^2_1 + \|\!|\varphi_\varepsilon|\!\|^2_{1/2} \leq C\,D_\varepsilon(\varphi_\varepsilon,\varphi_\varepsilon) = C(\omega,\varphi_\varepsilon) \\ &\leq C\,'\|\!|\omega|\!\|_{-1/2}\cdot{}'\|\!|\varphi_\varepsilon|\!\|_{1/2} \leq \frac{1}{2}\|\!|\varphi_\varepsilon|\!\|^2_{1/2} + C\,'\|\!|\omega|\!\|^2_{-1/2} \end{aligned} \tag{8.6.17}$$

(using Theorem 8.6.1) from which the result follows. From the definitions in (8.6.1) and from Lemma 8.5.1, we find that

$$\begin{aligned} \varepsilon\|\!|\varphi|\!\|^2_{s+1} + \|\!|\varphi|\!\|^2_{s+1/2} &= \sum_n \sum_{IJ} [\varepsilon\|\!|\eta_n\varphi^{(n)}_{IJ}|\!\|^2_{s+1} + \|\!|\eta_n\varphi^{(n)}_{IJ}|\!\|^2_{s+1/2}] + \\ &+ \sum_n \sum_{IJ} \sum_{|\alpha|\leq s} C_\alpha[\varepsilon\|\!|D^\alpha_t\eta_n\varphi^{(n)}_{IJ}|\!\|^2_1 + \|\!|D^\alpha_t\eta_n\varphi^{(n)}_{IJ}|\!\|^2_{1/2}](R/\pi)^{2|\alpha|} \end{aligned} \tag{8.6.18}$$

where n indexes the coordinate patches and the norms on the right in (8.6.18) are those in G_R as defined in § 8.5. Our method of showing this will be to use the first inequality in (8.6.17) for the tangential derivatives $D^\alpha(\eta_n\varphi_\varepsilon)$ and then use results like (8.6.15) for these derivatives.

We begin by considering what happens in one of the coordinate patches which we take as a boundary patch, the treatment for interior patches being similar and simpler. We first replace the subscripts IJ, etc., by a single superscript. Then, if φ or ψ has support in the range of a coordinate system with domain G $(= G_R)$,

$$D_\varepsilon(\varphi,\psi) = \int_G [A^{\alpha\beta}_{ij}\varphi^i_{,\alpha}\bar{\psi}^j_{,\beta} + B^\alpha_{ij}\varphi^i_{,\alpha}\bar{\psi}^j + B^\alpha_{ji}\varphi^i\bar{\psi}^j_{,\alpha} + C_{ij}\varphi^i\bar{\psi}^j]\,dt \tag{8.6.19}$$

where we have put $r = t^{2\nu}$ and D_ε is hermitian symmetric, i.e.

$$A^{\beta\alpha}_{ji} = \bar{A}^{\alpha\beta}_{ij}, \quad C_{ji} = \bar{C}_{ij}. \tag{8.6.20}$$

Next we prove by induction that (if φ_ε and $\psi = 0$ outside our patch)

$$(8.6.21)\quad D_\varepsilon(D^\lambda \varphi_\varepsilon, \psi) = (-1)^{|\lambda|} D_\varepsilon(\varphi_\varepsilon, D^\lambda \psi) + \sum_{\mu<\lambda} D^\lambda_{\varepsilon\mu}(D^\mu \varphi_\varepsilon, \varphi),$$

where all the $D^\lambda_{\varepsilon\mu}$ are given by hermitian symmetric integrals like D_ε, where D^λ and D^μ are tangential derivatives, and where

$$(8.6.22)\quad \mu < \lambda \text{ means } 0 \le \mu_\alpha \le \lambda_\alpha,\ \alpha = 1, \ldots, 2\nu - 1 \text{ and } |\mu| < |\lambda|.$$

(8.6.21) is evident for $|\lambda| = 1$. To complete the induction, we assume (8.6.21) and use that for $|\varrho| = 1$ to obtain

$$(8.6.23)\quad D_\varepsilon(D^{\varrho+\lambda} \varphi, \psi) = -D_\varepsilon(D^\lambda \varphi, D^\varrho \psi) + D^\varrho_{\varepsilon 0}(D^\lambda \varphi, \psi) \quad \varphi = \varphi_\varepsilon.$$

Applying (8.6.21) to the first term on the right in (8.6.23), we obtain

$$(8.6.24)\quad \begin{aligned} D_\varepsilon(D^{\varrho+\lambda} \varphi, \psi) &= (-1)^{|\lambda|+1} D_\varepsilon(\varphi, D^{\lambda+\varrho} \psi) - \\ &- \sum_{\mu<\lambda} D^\lambda_{\varepsilon u}(D^\mu \varphi, D^\varrho \psi) + D^\varrho_{\varepsilon 0}(D^\lambda \varphi, \psi). \end{aligned}$$

The result for $\lambda + \varrho$ follows by applying (8.6.21) for $|\varrho| = 1$ backwards to the terms in the middle sum on the right in (8.6.24).

Now, it is easy to see that we may choose a real function η of class C^∞ and vanishing outside our patch (see (8.6.10)) such that

$$(8.6.25)\quad |\nabla \eta(t)|^2 \le C \eta(t), \quad t \in G.$$

We now assume that φ ($\equiv \varphi_\varepsilon$) satisfies (8.6.15) and then set

$$(8.6.26)\quad \psi = \eta^2 D^\lambda \varphi, \quad \lambda \text{ given}$$

in (8.6.21). By using the hermitian symmetry, the reader may verify that

$$(8.6.27)\quad \begin{aligned} Re\, D_\varepsilon(D^\lambda \varphi, \eta^2 D^\lambda \varphi) &= D_\varepsilon(\eta D^\lambda \varphi, \eta D^\lambda \varphi) - \int_G A^{\alpha\beta}_{ij} \eta_{,\alpha} \eta_{,\beta} \Phi^i \bar{\Phi}^j\, dt, \\ (-1)^{|\lambda|} D_\varepsilon(\varphi, D^\lambda \psi) &= (-1)^{|\lambda|}(\omega, D^\lambda \psi) = (D^\lambda \omega, \eta^2 D^\lambda \varphi), \\ \Phi^i &= D^\lambda \varphi^i, \quad \varphi = \varphi_\varepsilon, \end{aligned}$$

using (8.6.15). In order to handle the various integrals $D^\lambda_{\varepsilon u}$, we let I denote any hermitian symmetric integral with coefficients $a^{\alpha\beta}_{ij}$, b^α_{ij}, c_{ij}. We notice that, for any vectors ϱ and σ,

$$(8.6.28)\quad \begin{aligned} I(u, \eta v) &= I(\eta u, v) + \sum_j [(d_j u, v^j) - (u^j, d_j v)] \\ d_j u &= a^{\alpha\beta}_{ij} \eta_{,\beta} u^i_{,\alpha} + \bar{b}^\alpha_{ji} \eta_{,\alpha} u^i. \end{aligned}$$

We also note that

$$(8.6.29)\quad d_j \eta u = \eta d_j u + a^{\alpha\beta}_{ij} \eta_{,\alpha} \eta_{,\beta} u^i.$$

Now, suppose ϱ is a multi-index with $|\varrho| = 1$. Then, by applying (8.6.28) twice and (8.6.29) once, we see that

$$(8.6.30)\quad I(\psi, \eta^2 \psi_{,\varrho}) = I(\eta^2 \psi, \psi_\varrho) + 2\sum_j [(d_j \psi, \eta \psi^j_{,\varrho}) - (\psi^j, \eta d_j \psi_{,\varrho})].$$

Integrating $I(\eta^2 \psi, \psi_{,\varrho})$ by parts and transposing one integral, we obtain

$$(8.6.31)\quad 2 Re\, I(\psi, \eta^2 \psi_{,\varrho}) = Re\{-2 I(\eta \eta_{,\varrho} \psi, \psi) - I_\varrho(\eta^2 \psi, \psi) + {}^*\}$$

where * denotes the sum in (8.6.30) and I_ϱ denotes the integral with coefficients $a^{\alpha\beta}_{ij,\varrho}$, etc. Applying (8.6.28), (8.6.29), etc., we obtain

$$\begin{aligned} 2Re\, I(\psi, \eta^2 \psi_{,\varrho}) &= -I_\varrho(\eta\psi, \eta\psi) + L_\varrho(\psi,\psi) + \\ &+ 2Re\Big\{-I(\eta_{,\varrho}\psi, \eta\psi) + \sum_j [(d_j \eta_{,\varrho}\psi, \psi^j) - (\eta_{,\varrho}\psi^j, d_j\psi)] + \\ &+ \sum_j [(d_j\psi, \eta\psi^j_{,\varrho}) - (\psi^j, \eta\, d_j \psi_{,\varrho})] \end{aligned} \tag{8.6.32}$$

where L_ϱ is an integral like that in (8.6.27) with coefficients $a^{\alpha\beta}_{ij,\varrho}$.

We now complete our proof of (8.6.16) by induction on s: We suppose it is true for all $s \le k$ and consider the case $s = k+1$. Suppose $|\lambda| = k+1$. From (8.6.17), (8.6.21) with $\psi = \eta^2 D^\lambda \varphi$, and from (8.6.27), we conclude that

$$\begin{aligned} &\varepsilon |||\eta D^\lambda \varphi|||_1^2 + |||\eta D^\lambda \varphi|||_{1/2}^2 \le C\, D_\varepsilon(\eta D^\lambda \varphi, \eta D^\lambda \varphi), \\ &D_\varepsilon(\eta D^\lambda \varphi, \eta D^\lambda \varphi) = \int_G A^{\alpha\beta}_{ij}\, \eta_{,\alpha}\, \eta_{,\beta}\, D^\lambda \varphi^i D^\lambda \bar{\varphi}^j\, dt + \\ &+ Re\Big\{(\eta D^\lambda \omega, \eta D^\lambda \varphi) + \sum_{\mu<\lambda} D^{(\lambda)}_{\varepsilon\mu}(D^\mu \varphi, \eta^2 D^\lambda \varphi)\Big\}, \quad \varphi = \varphi_\varepsilon. \end{aligned}$$

Using (8.6.25), (8.6.32) and the facts that

$$|(f,g)| \le {}'|||f|||_\tau \cdot {}'|||g|||_{-\tau}, \quad {}'|||\nabla u|||_\tau \le C R^{-1} |||u|||_{\tau+1}, \quad \text{for all } \tau, \tag{8.6.33}$$

where ∇u is the vector function $(u_{,\alpha})$, $\alpha = 1, \ldots, 2\nu$, $t^{2\nu} = r$, we find that (taking all norms on $\mathfrak{M}$ and allowing the C' s to depend on R)

$$\begin{aligned} &\Big|\int_G A^{\alpha\beta}_{ij}\, \eta_{,\alpha}\, \eta_{,\beta}\, D^\lambda \varphi^i D^\lambda \bar{\varphi}^j\, dt\Big| \le C\, |||D^\lambda \varphi|||_{-1/2} \cdot |||\eta D^\lambda \varphi|||_{1/2} \\ &|(\eta D^\lambda \omega, \eta D^\lambda \varphi)| \le C\, |||D^\lambda \omega|||_{-1/2}\, |||\eta D^\lambda \varphi|||_{1/2}. \end{aligned} \tag{8.6.34}$$

Regarding the integrals $D^\lambda_{\varepsilon\mu}$ with $|\mu| \le |\lambda| - 2$ as sums of appropriate inner products one can easily obtain the bound

$$\begin{aligned} |Re\, D^\lambda_{\varepsilon\mu}(D^\mu \varphi_\varepsilon, \eta^2 D^\lambda \varphi_\varepsilon)| &\le C[|||D^\mu \varphi_\varepsilon|||_{3/2} \cdot |||\eta D^\lambda \varphi_\varepsilon|||_{1/2} + \\ &+ |||D^\mu \varphi_\varepsilon|||_{1/2} \cdot |||\eta D^\lambda \varphi_\varepsilon|||_{-1/2}], \quad |\lambda - \mu| \ge 2. \end{aligned} \tag{8.6.35}$$

One may proceed in much the same way for those integrals with $|\lambda - \mu| = 1$; however, one starts with (8.6.32). We obtain, for example

$$\begin{aligned} |I(\eta_\varrho D^\mu \varphi, \eta D^\mu \varphi)| &\le C\, '|||\nabla \eta_\varrho D^\mu \varphi|||_{-1/2} \cdot {}'|||\nabla \eta D^\mu \varphi|||_{1/2} + \cdots \\ &\le C\, |||\eta_\varrho D^\mu \varphi|||_{1/2} \cdot |||\eta D^\lambda \varphi|||_{1/2} + \cdots \quad (\varphi = \varphi_\varepsilon) \\ |(D^\mu \varphi^j, \eta\, d_j D^\lambda \varphi)| &\le C\, |||D^\mu \varphi|||_{1/2} \cdot |||\eta D^\lambda \varphi|||_{1/2} + \cdots \end{aligned}$$

where the dots denote terms of lower order. The other terms can be handled similarly. By using the common device that

$$|2ab| \le \xi a^2 + \xi^{-1} b^2, \quad \xi > 0, \ a, b \text{ real}, \ \xi \text{ small}$$

and by summing over all λ with $|\lambda| = k+1$ we find that

$$\begin{aligned} \sum_{|\lambda|=k+1} C_\lambda [\varepsilon |||\eta D^\lambda \varphi|||_1^2 + |||\eta D^\lambda \varphi|||_{1/2}^2] &\le C \sum_{|\lambda|=k+1} C_\lambda\, |||\eta D^\lambda \omega|||_{-1/2}^2 + \\ &+ C \sum_{|\mu| \le k} |||D^\mu \varphi|||_{1/2}^2. \end{aligned}$$

Then by summing over the patches and adding in the lower order terms according to (8.6.1) and Lemma 8.5.1 we obtain (8.6.16) for $s = k + 1$. This completes the induction.

By repeated use of the differential equations (8.6.9) we find in each patch that *all* the derivatives of the φ_ε^j satisfy similar inequalities and (8.6.12) and (8.6.13) hold for φ_ε, independently of ε. Thus as $\varepsilon \to 0$ it follows easily that $\varphi_\varepsilon \to \varphi$ together with all its derivatives. Thus $\varphi \in \mathfrak{A}_0$ if $\omega \in \mathfrak{A}$ and (8.6.12) and (8.6.13) hold.

We can now prove our desired smoothness theorem:

Theorem 8.6.3. *If* $\varphi \in \mathfrak{D}(L^{pq})$ *and* $q \geq 1$, *then* φ, $\bar\partial\varphi$, *and* $\mathfrak{b}\varphi \in \mathfrak{D}$. *If, also,* $L^{pq}\varphi \in \mathfrak{A}$, *then* φ, $\bar\partial\varphi$, *and* $\mathfrak{b}\varphi \in \mathfrak{A}_0$, $\bar\partial\varphi \in \mathfrak{D}(L^{p,q+1})$ *and*

(8.6.36) $$L^{p,q+1}\bar\partial\varphi = \bar\partial L^{pq}\varphi, \quad L^{p,q-1}\mathfrak{b}\varphi = \mathfrak{b}L^{pq}\varphi.$$

If $\varphi \in \mathfrak{H}^{pq}$ *and* $q \geq 1$, *then* $\varphi \in \mathfrak{A}_0$.

Proof. We first prove the smoothness results. Let $L^{pq}\varphi \equiv L\varphi = \omega \in \mathfrak{A}$. Then $\varphi \in \mathfrak{L}^{pq}$ and

$$(L + I)\varphi = \omega + \varphi.$$

Hence φ and therefore $\omega + \varphi \in \mathfrak{H}^{1/2}$ by Theorem 8.6.2. Using induction and Theorem 8.6.2 repeatedly we see that $\varphi \in \mathfrak{H}^{s+1/2}$ for every s. Using the differential equations, we find that all the derivatives of $\varphi \in \mathfrak{H}^{s+1/2}$ so $\varphi \in \mathfrak{A}_0$ (Sobolev lemma, Chapter 3). That $\bar\partial\varphi \in \mathfrak{A}_0$ follows from the natural boundary condition (8.1.23). That $\mathfrak{b}\varphi \in \mathfrak{A}_0$ follows from (8.1.19). Since $L\psi = \bar\partial\mathfrak{b}\psi + \mathfrak{b}\bar\partial\psi$ for ψ and $\bar\partial\psi \in \mathfrak{A}_0$, (8.6.36) follows easily from the facts that $\bar\partial\bar\partial\psi = 0$ and $\mathfrak{b}\,\mathfrak{b}\psi = 0$ for $\psi \in \mathfrak{A}$.

Now, suppose $\omega \in \mathfrak{L}^{pq} \ominus \mathfrak{H}^{pq}$ and $L\varphi = \omega$. Let $\omega_n \in \mathfrak{A}^{pq} \cap (\mathfrak{L}^{pq} \ominus \mathfrak{H}^{pq})$, $\varphi_n \in \mathfrak{D}(L)$, $L\varphi_n = \omega_n$, and suppose $\omega_n \to \omega$ in $\mathfrak{L}^{pq}$. For each n,

$$\bar\partial(\bar\partial\varphi_n) = 0, \quad \mathfrak{b}(\mathfrak{b}\varphi_n) = 0, \quad \mathfrak{b}\bar\partial\varphi_n + \bar\partial\mathfrak{b}\varphi_n = \omega_n,$$
$$(\mathfrak{b}\bar\partial\varphi_n, \bar\partial\psi) = (\bar\partial\mathfrak{b}\varphi_n, \mathfrak{b}\chi) = 0, \quad \psi \in \mathfrak{A}^{p,q-1}, \quad \chi \in \mathfrak{A}_0^{p,q+1}$$
$$\bar\partial\varphi_n \in L^{p,q+1} \ominus \mathfrak{H}^{p,q+1}, \quad \mathfrak{b}\varphi_n \in \mathfrak{L}^{p,q-1} \ominus \mathfrak{H}^{p,q-1}.$$

Thus $\partial\varphi_n$ and $\mathfrak{b}\varphi_n$ satisfy the equations

$$(\bar\partial\bar\partial\varphi_n, \bar\partial\psi) + (\mathfrak{b}\bar\partial\varphi_n - \omega_n, \mathfrak{b}\psi) = 0, \quad \psi \in \mathfrak{A}_0^{p,q+1}$$
$$(\bar\partial\mathfrak{b}\varphi_n - \omega_n, \bar\partial\chi) + (\mathfrak{b}\,\mathfrak{b}\varphi_n, \mathfrak{b}\chi) = 0, \quad \chi \in \mathfrak{A}_0^{p,q-1}.$$

It follows that

$$d(\bar\partial\varphi_n, \bar\partial\varphi_n) + d(\mathfrak{b}\varphi_n, \mathfrak{b}\varphi_n) = (\omega_n, \omega_n)$$

from which the remaining statements follow (by taking $\varphi_n = N\omega_n$ so that $\varphi_n \to \varphi$).

We can now complete the proofs of our principal results in the cases where $q = 0$.

Theorem 8.6.4. (a) $\Re(L^{p\,0}) = \mathfrak{L}^{p\,0} \ominus \mathfrak{H}^{p\,0}$.

(b) *If* $\omega \in \mathfrak{L}^{p\,0} \ominus \mathfrak{H}^{p\,0}$, *there is a unique* $\varphi \in \mathfrak{D}(L^{p\,0}) \cap [\mathfrak{L}^{p\,0} \ominus \mathfrak{H}^{p\,0}]$ *such that* $L^{p\,0}\varphi = \omega$.

(c) *If we define* $N^{p\,0}\omega$ *as this solution when* $\omega \in \mathfrak{L}^{p\,0} \ominus \mathfrak{H}^{p\,0}$ *and* $N^{p\,0}\omega = 0$ *when* $\omega \in \mathfrak{H}^{p\,0}$, *then* $N^{p\,0}$ *is completely continuous. Moreover*

$$\|N^{p\,0}\| \le C_3,$$

C_3 *being the constant in Theorem 8.4.5.*

(d) *If* $\omega \in \mathfrak{A}^{p\,0}$, *then* $N^{p\,0}\omega \in \mathfrak{A}^{p\,0}$.

Proof. Suppose $\omega \in \mathfrak{A}^{p\,0}$. Then $\bar{\partial}\omega \in \mathfrak{A}^{p\,1}$ and if $\psi \in \mathfrak{H}^{p\,1}$,

$$(\bar{\partial}\omega, \psi) = \int_{bM} \langle \omega, \nu\,\psi \rangle \, dS = 0$$

so $\bar{\partial}\omega \in \mathfrak{L}^{p\,1} \ominus \mathfrak{H}^{p\,1}$. So let $\Phi = N\,\bar{\partial}\omega$. Then $\bar{\partial}\Phi \in \mathfrak{D}(L^{p\,2}) \cap (\mathfrak{L}^{p\,2} \ominus \mathfrak{H}^{p\,2})$ (as above) and $L^{p\,2}\,\bar{\partial}\Phi = \bar{\partial}\bar{\partial}\omega = 0$ so $\bar{\partial}\Phi = 0$ and hence

$$(8.6.37) \qquad \bar{\partial}(\mathfrak{b}\Phi - \omega) = 0, \quad \mathfrak{b}\Phi - \omega \in \mathfrak{H}^{p\,0}.$$

On the other hand if $\omega \in \mathfrak{L}^{p\,0} \ominus \mathfrak{H}^{p\,0}$ and $\psi \in \mathfrak{H}^{p\,0}$, then

$$(8.6.38) \qquad (\mathfrak{b}\Phi - \omega, \psi)_s = (\Phi, \bar{\partial}\psi)_s - (\omega, \psi)_s - \int_{bM_s} \langle \nu\Phi, \psi \rangle \, dS.$$

By averaging this over a small interval $(t, 0)$ and using the fact that the function $\nu\,\Phi \in C^\infty(\bar{M})$ and vanishes on bM, we find that the right side of (8.6.38) $\to 0$ as $t \to 0^-$. Thus we must have

$$(8.6.39) \qquad \mathfrak{b}\Phi = \omega, \quad \bar{\partial}\Phi = 0.$$

Since $\Phi\,(= N\,\bar{\partial}\omega) \in \mathfrak{L}^{p\,1} \ominus \mathfrak{H}^{p\,1}$, we may define

$$(8.6.40) \qquad \Omega = N\Phi, \quad \varphi = \mathfrak{b}\Omega.$$

Since $\bar{\partial}\Phi = 0$, we again have $\bar{\partial}\Omega = 0$, so that

$$L^{p\,0}\varphi = L^{p\,0}\,\mathfrak{b}\Omega = \mathfrak{b}L^{p\,1}\Omega = \mathfrak{b}\Phi = \omega.$$

As in (8.6.38), we see also that $\varphi \in \mathfrak{L}^{p\,0} \ominus \mathfrak{H}^{p\,0}$.

If there were another solution in $\mathfrak{D}(L^{p\,0})$, the difference φ_1 would be in $\mathfrak{D}(L^{p\,0})$ and we would have

$$\varphi_1 \in \mathfrak{D}^{p\,0}, \quad d(\varphi_1, \psi) \equiv (\bar{\partial}\varphi_1, \bar{\partial}\psi) = 0, \quad \psi \in \mathfrak{D}^{p\,0}$$

so that $\varphi_1 \in \mathfrak{H}^{p\,0}$, since we could take $\psi = \varphi_1$. From (8.6.39) and (8.6.40) we see that $N^{p\,0}$ is bounded. Finally, if $\omega_n \rightharpoonup \omega$, it would follow that $\mathfrak{b}\Phi_n \rightharpoonup \omega$, $\bar{\partial}\Phi_n = 0$, and $\Phi_n \rightharpoonup \Phi$, at least in a subsequence, since, by Theorem 8.4.5, $D(\Phi_n, \Phi_n)$ is uniformly bounded. Thus, by that theorem $\Phi_n \to \Phi$ and so $\varphi_n \to \varphi$, since

$$(\varphi, \varphi) = (\mathfrak{b}\Omega, \mathfrak{b}\Omega) = d(\Omega, \Omega) \le C_3 \|L\Omega\|^2 = C_3(\Phi, \Phi) \le C_3^2\, d(\Phi, \Phi)$$
$$= C_3^2 \|\omega\|^2.$$

Chapter 9

Introduction to parametric Integrals; two dimensional problems

9.1. Introduction. Parametric integrals

An integral (1.1.1) is said to be *in parametric form*, and we say f *is the integrand of a parametric problem*, if and only if the value of the integral is unchanged by an arbitrary diffeomorphism *with positive Jacobian* from G onto another domain G'; if we assume that $f(x, z, p)$ is defined everywhere, we see that we must have

$$\int_{G'} f[x', z'(x'), \nabla z'(x')]\, dx' = \int_G f[x, z(x), \nabla z(x)]\, dx, \tag{9.1.1}$$
$$z(x) = z'[x'(x)], \quad z^i_\beta(x) = \frac{\partial' z^i}{\partial' x^\alpha} \cdot \frac{\partial' x^\alpha}{\partial x^\beta}$$

for any G', and vector $z' \in C^1(\bar{G}')$, and any positive diffeomorphism $x' = x'(x)$ from $\bar{G}$ onto $\bar{G}'$. By taking x_0, z_0, $'p$, and x'_0 as arbitrary constants and defining

$$'z^i = z^i_0 + 'p^i_\alpha('x^\alpha - 'x^\alpha_0), \quad 'x^\alpha = 'x^\alpha_0 + a^\alpha_\beta(x^\beta - x^\beta_0), \tag{9.1.2}$$
$$\det \|a^\alpha_\beta\| > 0$$

and taking $G = B(x_0, \varrho)$ and then letting $\varrho \to 0$, we obtain

$$\det \|a^\alpha_\beta\| \cdot f[x'_0, z_0, p'_1, \ldots, p'_\nu] = f[x_0, z_0, a^\alpha_1 p'_\alpha, \ldots, a^\alpha_\nu p'_\alpha] \tag{9.1.3}$$

for all sets of constants as indicated. Thus, f must be independent of x and we must have $N \geq \nu$.

In case $N = \nu + 1$, we now show that f must have the form

$$f(z, p) = F(z, D_1, \ldots, D_{\nu+1}),$$
$$D_i = \begin{vmatrix} p^1_1 \cdots p^{i-1}_1 p^{i+1}_1 \cdots p^{\nu+1}_1 \\ \cdots\cdots \quad \cdot \quad \cdots\cdots \\ p^1_\nu \cdots p^{i-1}_\nu p^{i+1}_\nu \cdots p^{\nu+1}_\nu \end{vmatrix} \tag{9.1.4}$$

where F is positively homogeneous of the first degree in the D_i. This is easily seen as follows: (9.1.3) implies that if the p matrix is multiplied on the left by any $\nu \times \nu$ matrix with positive determinant, then f is multiplied by that determinant. Thus, if $D_{\nu+1} > 0$, we may multiply the p matrix on the left by the inverse of the matrix of $D_{\nu+1}$. The result is that the new p-matrix has the identity for its first ν columns and the elements in the last column are just $\pm D_i/D_{\nu+1}$. Thus, for some F,

$$f(z, p) = D_{\nu+1} \cdot F\left(z, \frac{D_1}{D_{\nu+1}}, \ldots, \frac{D_\nu}{D_{\nu+1}}, 1\right), \quad D_{\nu+1} > 0. \tag{9.1.5}$$

If $D_{\nu+1} < 0$, we begin by changing the sign of 1 row and obtain

$$f(z, p) = -D_{\nu+1} F\left(z, -\frac{D_1}{D_{\nu+1}}, \ldots, -\frac{D_\nu}{D_{\nu+1}}, -1\right), \quad D_{\nu+1} < 0 \tag{9.1.5'}$$

where the two F's join up continuously along $D_{\nu+1} = 0$. It is not necessarily (and usually not) true that

$$F(z, -D_1, \ldots, -D_{\nu+1}) = -F(z, D_1, \ldots, D_{\nu+1}). \tag{9.1.6}$$

It is clear from this discussion that an integrand of a parametric problem is ordinarily not of class C^1 on the set of p for which the p-matrix has rank $< \nu$ and, moreover, at points where f is of class C^2, the quadratic form in (1.10.7) and (1.10.8) necessarily has ν 0 eigenvalues and the biquadratic form in (1.5.5) degenerates.

We have already seen that the area integral (1.1.4) is in parametric form. Using our current notation

$$f(z, p) = \sqrt{D_1^2 + D_2^2 + D_3^2}. \tag{9.1.7}$$

Until H. A. Schwarz showed how to inscribe a polyhedron of arbitrarily large area in a portion of a cylindrical surface, it was believed that the area of a surface could be defined as the sup. of the areas of inscribed polyhedra, by analogy with the definition of the length of a curve. This upsetting discovery was the motivation for a lengthy study of surfaces and their areas. The principal surviving definitions are those of the Hausdorff two-dimensional measure of a point set in R^N, $N \geq 3$, and of the Lebesgue area of a Fréchet surface. The Hausdorff measure Λ^2 is generated in the usual sense of measure theory by the outer measure Λ^{2*} defined by

$$\Lambda^{2*}(S) = \lim_{\varrho \to 0} \Lambda_\varrho^{2*}(S), \quad \Lambda_\varrho^{2*}(S) = \inf_{r_i \leq \varrho} \sum \pi r_i^2 \tag{9.1.8}$$

for all coverings of S by a countable number of balls $B(x_i, r_i)$. To define the Lebesgue area, we must define a Fréchet surface.

The idea is to invent a notion of surface which is independent of its parametric representation. In case we have a point set S which is the homeomorphic image of a closed circular disc, we would naturally call S a surface and any homeomorphism from a closed Jordan region onto S a parametric representation. But there are times when it is desirable to consider more general surfaces. We therefore proceed formally as follows:

Definition 9.1.1. Let z_1 and z_2 be continuous vector functions with diffeomorphic[1] compact domains G_1 and G_2 which must be finite unions of disjoint manifolds, with or without boundary, of class C^1 at least. We define the *distance*

$$D(z_1, z_2) = \inf_\tau \left\{ \max_{x_1 \in G_1} | z_1(x_1) - z_2[\tau(x_1)] | \right\} \tag{9.1.9}$$

[1] Historically, *homeomorphic* domains and *homeomorphisms* were allowed instead of *diffeomorphic* domains and *diffeomorphisms*. In the one and two dimensional cases, the restriction which we are making results only in the restriction to domains of class C^1; in higher dimensions, Milnor has shown that two manifolds of class C^1 may be homeomorphic without being diffeomorphic. Our restriction results in a great simplification of the theory of integrals over Fréchet varieties.

for all diffeomorphisms τ from G_1 onto G_2. If $D(z_1, z_2) = 0$, we say that z_1 is *equivalent* to z_2 and write $z_1 \approx z_2$.

Theorem 9.1.1. (a) *The distance D so defined is non-negative, symmetric and satisfies the triangle inequality.*

(b) *The equivalence relation defined above is symmetric and transitive.*

Definition 9.1.2. We define a *Fréchet variety* S as a class of continuous vector functions which consists of all such which are equivalent to some given one. Any vector function in S is called a *parametric representation* of S. The range $\mathfrak{R}(S)$ is the common range of the parametric representations of S. The *topological type of* S is the class of domains of the vectors in S and the *dimension of* S is the common dimension of these domains.

Definition 9.1.3. If S_1 and S_2 are Fréchet varieties of the same topological type, we define the *distance*

$$D(S_1, S_2) = D(z_1, z_2)$$

where z_k is a parametric representation of S_k, $k = 1, 2$.

Theorem 9.1.2. (a) *With this definition of distance, the Fréchet varieties of a given topological type form a metric space.*

(b) *If S and S_n are all of the same topological type, z is a representation of S with domain G, and $D(S, S_n) \to 0$, then $\exists$ a representation z_n of S_n, with domain G, such that $z_n \to z$ uniformly on G.*

(c) *If z and z_n are representations of S and S_n, respectively, all having domain G, and if $z_n \to z$ uniformly on G, then $D(S, S_n) \to 0$.*

Proof. (a) and (c) are evident. To prove (b), let Z_n be a representation of S_n having domain G_n. From the definition of distance it follows that there exists, for each n, a diffeomorphism τ_n from G onto G_n such that

$$|z_n(x) - z(x)| \leq D(S, S_n) + 1/n, \quad z_n(x) = Z_n[\tau_n(x)], \quad x \in G.$$

The result follows.

Now LEBESGUE [1] was familiar with the following fact: Suppose z_n and z are each piecewise linear homeomorphisms on a polyhedral domain G and suppose z_n converges uniformly to z (nothing being said about derivatives) on G. Then

$$A(z) \leq \liminf_{n\to\infty} A(z_n). \tag{9.1.10}$$

$A(z)$ and $A(z_n)$ being given by the integral (1.1.4) and being the elementary areas of the corresponding polyhedra. In the next section, we shall prove a general lower-semicontinuity theorem from which we can conclude that (9.1.10) holds if G is of class C^1, z is Lipschitz, and each $z_n \in C^1(G)$; in fact z may merely $\in H_2^1(G)$ in the two dimensional case. Assuming these results, the theory of Lebesgue area proceeds as follows:

Definition 9.1.4. Given G of class C^1 and $z \in C^0(G)$, we define

$$L(z, G) = \inf \liminf_{n\to\infty} A(z_n, G) \tag{9.1.11}$$

for all sequences $\{z_n\}$ in which each $z_n \in C^1(G)$ and $z_n \to z$ uniformly on G.

The following theorem follows immediately:

Theorem 9.1.3. *We suppose that* $\nu = 2$.

(a) $L(z, G)$ *is given by* (1.1.4) *if* $z \in H_2^1(G) \cap C^0(G)$.

(b) $L(z, G) \leq \liminf_{n\to\infty} L(z_n, G)$ *if* $z_n \to z$ *uniformly on* G.

(c) *Given* $z \in C^0(G)$, *there exists a sequence* $\{z_n\}$ *in which each* $z_n \in C^1(G)$, $z_n \to z$ *uniformly on* G, *and* $L(z_n, G) \to L(z, G)$.

(d) $L(z_1) = L(z_2)$ *if* $D(z_1, z_2) = 0$.

Proof. (b) and (c) are evident. (a) follows since the integral is lower-semicontinuous according to Theorem 9.2.1 so that $L(z, G) \leq$ the integral $A(z, G)$. On the other hand if $z_n \to z$ uniformly on G and strongly in $H_2^1(G)$, then $A(z_n, G) \to A(z, G)$. To prove (d), let $Z_n \to z_2$ uniformly on G_2 so that $A(Z_n, G_2) \to L(z_2, G_2)$ and $Z_n \in C^1(G_2)$. There exists a sequence $\{\tau_n\}$ of diffeomorphisms from G_1 onto G_2 such that $z_2[\tau_n(x_1)] \to z_1(x_1)$ uniformly on G_1. Thus, if $\zeta_n(x_1) = Z_n[\tau_n(x_1)]$, then $\zeta_n \to z_1$ uniformly on G_1 and each $\zeta_n \in C^1(G_1)$ so that

$$L(z_1, G_1) \leq \liminf A(\zeta_n, G_1) = \lim A(Z_n, G_2) = L(z_2, G_2).$$

The argument is symmetric.

Remark. The foregoing definition and theorem can be carried over to the case where G is a manifold by using coordinate patches and partitions of unity.

Definition 9.1.5. We define the *Lebesgue area* $L(S)$ of the surface S as the common value of $L(z, G)$ for its representations z.

Theorem 9.1.4. (a) *Parts* (*b*) *and* (*c*) *of Theorem* 9.1.3 *hold with* (z, G) *and* (z_n, G) *replaced respectively by* S *and* S_n.

(b) *If* S *possesses the representation* $z \in H_2^1(G) \cap C^0(G)$, *then* $L(S) = A(z, G)$.

Now, there are some anomalies. If we define $z = z(x^1)$ for (x^1, x^2) on $G = \overline{B(0,1)}$, then it is obvious that $L(z, G) = 0$ even if the range of z fills a cube. But one might hope that if z is a homeomorphism with range S, then $L(z, G)$ would be equal to the Hausdorff measure $\Lambda^2(S)$. However, this is not the case. In fact, Besicovitch has given an example of such a z and S for which $L(z, G)$ is finite and the *three dimensional measure* of $S > 0$! On the other hand from an old theorem of the author [3], it follows that this S possess a "generalized conformal" representation $z = z^*(x^*)$ on $\overline{B(0,1)} = G$ in which z^* is continuous and in $H_2^1(G)$

and, moreover,

$$E^* = G^*, \quad F^* = 0 \quad \text{a.e.} \quad L(z, G) = L(z^*, G) = \int_G \frac{E^* + G^*}{2} dx^*$$
(9.1.12)
$$E^* = |z^*_{x^1*}|^2, \quad F^* = z^*_{x^1*} \cdot z^*_{x^2*}, \quad G^* = |z^*_{x^2*}|^2.$$

We notice that the area integrand

(9.1.13) $$\sqrt{E^* G^* - (F^*)^2} = \frac{E^* + G^*}{2} \quad \text{whenever } E^* = G^*, \quad F^* = 0$$

since

(9.1.14) $$AC - B^2 + \left(\frac{A - C}{2}\right)^2 + B^2 = \left(\frac{A + C}{2}\right)^2 (A = E^*, \quad B = F^*$$
$$C = G^*).$$

However, the reader can easily prove that if z is a diffeomorphism of G onto S, then $L(z, G) = \Lambda^2(S)$.

In dealing with more general integrals

(9.1.15) $$I_f(z, G) = \int_G f(z, p)\, dx$$

in which f satisfies (9.1.3), it is desirable to introduce the notion of *oriented Fréchet variety*. This is done by defining the distance $D(z_1, z_2)$ in Definition 9.1.1 allowing only *positive* (i.e. with positive Jacobian) diffeomorphisms τ and requiring the domains G_1 and G_2 to be oriented. Then Theorem 9.1.1 holds and we may define an oriented Fréchet variety as in Definition 9.1.2 using the "oriented distance" above. We may then define $D(S_1, S_2)$ for oriented varieties as in Definition 9.1.3 and then note that Theorem 9.1.2 holds.

For the special class of integrand functions $f(z, p)$ of (9.1.15) which is described in the next section, we can extend the notion of Lebesgue area to that of an integral over a Fréchet variety.

Definition 9.1.4'. Suppose f satisfies (9.1.3) and the conditions decsribed in § 9.2 below. Given G of class C^1 and $z \in C^0(G)$ we define

$$\mathfrak{J}_f(z, G) = \inf \{\liminf_{u \to \infty} I_f(z_n, G)\}$$

for all sequences $\{z_n\}$ as in Definition 9.1.4.

The following theorem follows as before using the theorem of the next section:

Theorem 9.1.3'. *If f satisfies the conditions above (with $\nu \geq 2$) then Theorem 9.1.3 holds with L replaced by $\mathfrak{J}_f$ and $A(z, G)$ by $I_f(z, G)$*, except *that we must require $z \in C^0_1(G)$ if $\nu > 2$ in part* (a).

Many people have made important contributions to the theory outlined above. We mention only E. J. McShane ([1], [2], [3], [4]), T. Radó ([3], [4]), L. Cesari ([1], [2], [3]) and the author ([1], [2]). L.C. Young ([2], [3], [4], [5]) introduced the idea of a "generalized surface" analogous to McShane's idea of a generalized curve; this idea was developed at some length by YOUNG, W. H. FLEMING and others ([1], [2], [3]). The idea is suggested by the following: Suppose $z \in H_\nu^1(G)$. Then the integral

$$\int_G F[z(x), D_1(x), \ldots, D_{\nu+1}(x)]\, dx, \tag{9.1.16}$$

where F is homogeneous of degree 1 in the D_i which are defined in (9.1.4), is a *linear functional* of F. Any linear functional of F, which is ≥ 0 whenever F is, is called a *generalized surface.* Unfortunately, an arbitrary generalized surface does not arise from a vector function z by means of a formula (9.1.16) even if fairly general vectors z and "generalized Jacobians" are allowed. The "integral currents" introduced by FEDERER and FLEMING seem like a more promising tool in the study of the existence and differentiability properties of solutions of variational problems involving such integrals.

9.2. A lower semi-continuity theorem

In § 4.4, we saw that if f satisfies (9.1.3) and (9.1.4) with $N = \nu + 1$, then f is weakly quasi-convex in p if and only if $F(z, X)$ is convex in the X_k (we shall use X_i instead of D_i from now on) which, in turn, implies that f is quasi-convex in p. In that case, if $F \in C^1$ for all (z, X) with $X \neq 0$, then f satisfies the conditions of Theorem 4.4.5 and hence the integral (9.1.15) is lower-semicontinuous if $z_n \to z$ *uniformly on* G *and weakly in* $H_\nu^1(G)$. This result, however, is not sufficient for the theorems of the preceding section. So we shall have to prove a special lower-semicontinuity theorem for such integrals. We shall use the method of MCSHANE [3]. This *method* can be extended to *certain* integrals (9.1.15) in which $N > \nu + 1$ as well as to *all* such integrals if $N = \nu + 1$.

Lemma 9.2.1. *Suppose* $f(z, p) \in C^1$ *for all* (z, p) *for which the* p*-matrix has rank* ν *and suppose* f *satisfies* (9.1.3). *Then if* (z_0, p_0) *is such that the rank of* p_0 *is* ν, *there are unique constants* a_I $(I = i_1, \ldots, i_\nu$ *with* $i_1 < i_2 < \cdots < i_\nu)$ *such that*

$$f(z_0, z_0) = \sum_I a_I X_0^I; \quad X_0^I = \det \begin{Vmatrix} p_{10}^{i_1} \cdots p_{10}^{i_\nu} \\ \cdots\cdots \\ p_{\nu 0}^{i_1} \cdots p_{\nu 0}^{i_\nu} \end{Vmatrix}$$

$$\frac{\partial}{\partial p_\alpha^i}\Big[f(z_0, p) - \sum_I a_I X^I\Big]_{p=p_0} = 0 \quad \text{for all } (i, \alpha) \tag{9.2.1}$$

$$\sum_I a_I\, dz^I = A\, d\zeta^1 \wedge \cdots \wedge d\zeta^\nu \quad (dz^I = dz^{i_1} \wedge \cdots \wedge dz^{i_\nu})$$

for some linearly independent linear functions $\zeta^1(z), \ldots, \zeta^\nu(z)$.

Proof. We first note that if we make a rotation of axes in the z-space and make the corresponding change in the p-matrix:

$$z^i = c_j^{i\prime} z^j, \; p_\alpha^i = c_j^{i\prime} p_\alpha^j, \quad f(z, p) = f'(z', p')$$

then the X^I transform in the corresponding way,

(9.2.2) $$X^I = \sum_J c_J^I X'^J, \quad C_J^I = \det c_{j\beta}^{i\alpha}$$

and the conditions (9.2.1) are preserved in the new variables. We may therefore assume that

(9.2.3) $$p_0 = \begin{pmatrix} p_{01}^1 \cdots p_{01}^\nu \, 0 \ldots 0 \\ \cdot \;\; \cdots \;\; \cdot \\ p_{0\nu}^1 \cdots p_{0\nu}^\nu \, 0 \ldots 0 \end{pmatrix}.$$

For p near p_0, we use (9.1.3) to conclude that

(9.2.4) $$f(z_0, p) = X^{1\ldots\nu} f(z_0, \bar{p}), \quad \bar{p} = \begin{pmatrix} 1 \ldots 0 \, \bar{p}_1^{\nu+1} \ldots \bar{p}_1^N \\ \cdots \;\; \cdot \;\; \cdot \;\; \cdots \\ 0 \ldots 0 \, \bar{p}_\nu^{\nu+1} \ldots \bar{p}_\nu^N \end{pmatrix}$$

$$X^I = X^{1\ldots\nu} \bar{X}^I, \quad \bar{p}_\alpha^i = (X^{1\ldots\nu})^{-1} \det \begin{pmatrix} p_1^1 \cdots p_1^{\alpha-1} \, p_1^i \, p_1^{\alpha+1} \cdots p_1^\nu \\ p_\nu^1 \cdots p_\nu^{\alpha-1} \, p_\nu^i \, p_\nu^{\alpha+1} \cdots p_\nu^\nu \end{pmatrix},$$

the $\bar{X}^I$ being formed from the $\bar{p}$ matrix.

From (9.2.4), we conclude that

$$f(z_0, p_0) = X_0^{1\ldots\nu} f(z_0, \bar{p}_0)$$

(9.2.5) $$f_{p_\alpha^i}(z_0, p_0) = \begin{cases} f(z_0, \bar{p}_0) \left(\dfrac{\partial X^{1\ldots\nu}}{\partial p_\alpha^i}\right)_{p=p_0}, & 1 \le i \le \nu \\[2ex] f_{p_\delta^i}(z_0, \bar{p}_0) \left(\dfrac{\partial X^{1\ldots\nu}}{\partial p_\alpha^\delta}\right)_{p=p_0}, & \nu + 1 \le i \le N \end{cases}$$

where $\bar{p}_0$ is, of course, the matrix obtained from $\bar{p}$ by setting all the $\bar{p}_\alpha^i$, with $i > \nu$, $= 0$. If we define

(9.2.6) $$f_1(z_0, p) = \sum_I a_I X^I = X^{1\ldots\nu} \sum_I a_I \bar{X}^I,$$

we obtain, using the formulas (9.2.5) for f_1, that

$$f_1(z_0, p_0) = a_{1\ldots\nu} X_0^{1\ldots\nu}, \quad f_1(z_0, \bar{p}_0) = a_{1\ldots\nu}$$

(9.2.7) $$f_{1\,p_\alpha^i}(z_0, p_0) = \begin{cases} a_{1\ldots\nu} \left(\dfrac{\partial X^{1\ldots\nu}}{\partial p_\alpha^i}\right)_{p=p_0}, & 1 \le i \le \nu \\[2ex] \sum_\delta (-1)^{\nu-\delta} a_{1\ldots\delta-1,\delta+1,\ldots,\nu i} \left(\dfrac{\partial X^{1\ldots\nu}}{\partial p_\alpha^\delta}\right)_{p=p_0}, & i > \nu. \end{cases}$$

So, from (9.2.5), (9.2.7) and the first two lines of (9.2.1), we conclude that we must have

(9.2.8) $$a_{1\ldots\nu} = f(z_0, \bar{p}_0), \quad a_{1\ldots\delta-1,\delta+1,\ldots,\nu i} = (-1)^{\nu-\delta} f_{p_\delta^i}(z_0, \bar{p}_0).$$

Now, let us introduce linearly independent ζ^i by

(9.2.9) $$\zeta^i = z^i + \sum_{j=\nu+1}^N c_j^i z^j, \quad i = 1, \ldots, \nu.$$

Then, if we define $c_j^i = \delta_j^i$ for $1 \leq j \leq \nu$, the c-matrix has the same form as the $\bar{p}$ matrix. The third line of (9.2.1), then requires that

$$\sum_I a_I dz^I = A \sum_I \det \begin{pmatrix} c_{i_1}^1 \dots c_{i_\nu}^1 \\ \cdot \cdot \cdot \cdot \cdot \\ c_{i_1}^\nu \dots c_{i_\nu}^\nu \end{pmatrix} dz^I. \tag{9.2.10}$$

Equating coefficients in (9.2.10) for $I = (1, \dots, \nu)$ or $(1, \dots, \delta - 1, \delta + 1, \dots, \nu\, i)$ yields

$$a_{1 \dots \nu} = A = f(z_0, \bar{p}_0), \quad c_i^\delta = [f(z_0, p_0]^{-1} f_{p_\delta^i}(z_0, p_0) \tag{9.2.11}$$

so that the remaining a_I are uniquely determined. The ζ^i are, of course, determined only up to a linear transformation among themselves which affects only the factor A.

In order to cary over McShane's proof, we impose the following requirements on f:

General assumptions on f. *We assume that*

(i) f *is continuous for all* (z, p) *and* $\in C^1$ *for all* (z, p) *for which the rank of the* p-*matrix is* ν;

(ii) f *satisfies* (9.1.3) *for all* (z, p), *etc.*,

(iii) $f(z, p) \geq m\,|X|, m > 0, |X|^2 = \sum_I (X^I)^2$,

(iv) *if the rank of* p_0 *is* ν *and the* a_I *are the constants of Lemma* 9.2.1 *for* (z_0, p_0), *then*

$$f(z_0, p) \geq \sum_I a_I X^I \equiv f_1(z_0, p) = f(z_0, p_0) + \sum_I a_I (X^I - X_0^I).$$

Remark 1. If $N = \nu + 1$, we have seen that these conditions (except for (iii)) are necessary and sufficient for the parametric integrand to be quasi-convex in p. For $N > \nu + 1$, it is not known that the conditions above are necessary for quasi-convexity.

Remark 2. We note that if f merely satisfies (9.1.3) then

$$f[z(x), p(x)]\, dx = \varphi[z(x), \Pi_\nu(x)] \cdot dA(z) \tag{9.2.12}$$

for any C^1 mapping $z = z(x)$, Π_ν being the oriented tangent ν-plane. This fact makes it easier to define $I(z, G)$ where G is a C^1 ν-manifold.

We must now introduce the notion of the *order of a point in* ν-*space with respect to an oriented* $(\nu - 1)$-*manifold*. We shall be concerned principally with such manifolds of the topological type of ∂G for some domain G in R_ν of class C^1. However, the generalization to other orientable manifolds will be evident. We allow ∂G to consist of several disjoint manifolds, each oriented as a part of ∂G in the usual way. If $\nu = 2$ and ∂G is just a Jordan curve, then this order is just the *winding number* of the point with respect to the curve.

Definition 9.2.1. Suppose G is of class C^1 and $\varphi \in C^1(G)$, $\varphi = (\varphi^1, \dots, \varphi^\nu)$ and let $\tau : z^\alpha = \varphi^\alpha(x)$, $\alpha = 1, \dots, \nu$. Suppose $z \notin \tau(\partial G)$. Then we define the order $0[z, \tau(\partial G)]$ by

$$(9.2.13) \quad 0[z, \tau(\partial G)] = \int_{\partial G} \sum K_{0,\alpha}[\varphi(x) - z] \cdot (-1)^{\alpha+\beta} \frac{\partial \varphi'_\alpha}{\partial x'_\beta} d x'_\beta$$

where K_0 is the elementary function for LAPLACE's equation, as defined in (2.4.2), and

$$(9.2.14) \quad \frac{\partial \varphi'_\alpha}{\partial x'_\beta} = \frac{\partial(\varphi^1, \ldots, \varphi^{\alpha-1}, \varphi^{\alpha+1}, \ldots, \varphi^\nu)}{\partial(x^1, \ldots, x^{\beta-1}, x^{\beta+1}, \ldots, x^\nu)}.$$

Lemma 9.2.2: *Suppose $\tau : z^\alpha = \varphi^\alpha(x)$ is of class C^1 on the open set Ω in R_ν. Let W be the set of x for which $\partial\varphi/\partial x = 0$. Then*

$$|\tau(W)| = 0.$$

Proof. Suppose $x_0 \in W$. Then we may write

$$\tau : z^\alpha - z_0^\alpha = [c_\beta^\alpha + \varepsilon_\beta^\alpha(x, x_0)](x^\beta - x_0^\beta),$$

$$z_0^\alpha = \varphi^\alpha(x_0), \quad c_\beta^\alpha = \varphi^\alpha_{,\beta}(x_0), \quad |\varepsilon(x, x_0)| \leq \varepsilon \text{ if } |x - x_0| \leq \delta.$$

Thus

$$(9.2.15) \quad \begin{gathered} z^\alpha - z_0^\alpha = c_\beta^\alpha(x^\beta - x_0^\beta) + r^\alpha(x, x_0), \quad |r(x, x_0)| \leq \varepsilon \cdot |x - x_0|, \\ \text{if } |x - x_0| \leq \delta. \end{gathered}$$

The image of $B(x_0, r)$ under the linear part of the mapping is an ellipsoid s in some plane of lower dimension. From (9.2.15), it follows that the total image of $B(x_0, r) \subset (s, \varepsilon r)$ if $r \leq \delta$. Since the derivative of the φ^α are uniformaly bounded on any compact subset of Ω, the result follows easily.

Lemma 9.2.3. (a) *$0[z, \tau(\partial G)]$ depends only on the values of the φ^α on ∂G and the values of their directional derivatives along ∂G. Moreover, it is independent of the parametric representation of ∂G. It is assumed in this lemma that $\varphi \in C^1(\bar{G})$ and that G is of class C^1.*

(b) *$0[z, \tau(\partial G)]$ is continuous in z if $z \notin \tau(\partial G)$.*

(c) *If $\varphi(x) \neq z$ for any $x \in G$* (i.e. $\bar{G}$), *then*

$$(9.2.16) \quad 0[z, \tau(\partial G)] = 0.$$

(d) *If the closed regions $G_1, \ldots, G_K$ are disjoint and $\subset G$, all being of class C^1, and if $\Gamma = G - \bigcup_{k=1}^K G_k^0$ and $z \notin \tau(\partial G) \cup \bigcup_{k=1}^K \tau(\partial G_k)$, then*

$$(9.2.17) \quad 0[z, \tau(\partial G)] = 0[z, \tau(\partial \Gamma)] + \sum_{k=1}^K 0[z, \tau(\partial G_k)].$$

(e) *If $z \notin \tau(\partial G)$, then $0[z, \tau(\partial G)]$ is an integer and*

$$(9.2.18) \quad N[z, \tau(G)] \geq |0[z, \tau(\partial G)]| \quad \text{(almost everywhere)},$$

$N[z, \tau(G)]$ being the number of x in G such that $\varphi(x) = z$.

(f) *If $\varphi(x, t) \in C^1(G)$ in x and is continuous on $G \times [0, 1]$ and if $(z \notin \tau_t(\partial G)$ for any t on $[0,1]$, then $0[z, \tau_t(\partial G)]$ is constant with respect to t, where $\tau_t : z = \varphi(x, t)$.*

(g) *The functions* $N[z, \tau(G)]$ *and* $O[z, \tau(\partial G)] \in L_1(R_\nu)$ *and*

$$(9.2.19)\qquad \int_{R_\nu} O[z, \tau(\partial G)]\, dx = \int_G \frac{\partial \varphi}{\partial x} dx, \quad \int_{R_\nu} N[z, \tau(G)]\, dz = \int_G \left|\frac{\partial z}{\partial x}\right| dx.$$

Proof. (a), (b), and (d) are evident. To prove (c), we note that

$$\int_{R_\nu} K_{0,\alpha}[\varphi(x) - z](-1)^{\alpha+\beta} \frac{\partial \varphi'_\alpha}{\partial x'_\beta} dx'_\beta$$

$$= \int_G \left\{K_{0,\alpha} \cdot (-1)^\alpha \cdot \sum_\beta (-1)^\beta \frac{\partial}{\partial x^\beta} \frac{\partial \varphi'_\alpha}{\partial x'_\beta} + (-1)^\alpha K_{0,\alpha\gamma} \cdot (-1)^\beta \varphi^\gamma_{,\beta} \frac{\partial \varphi'_\alpha}{\partial x'_\beta}\right\} dx$$

$$= \int_G \sum_\alpha K_{0,\alpha\alpha}[\varphi(x) - z] \frac{\partial \varphi}{\partial x} dx = 0$$

(9.2.20)

since the first term on the right vanishes by Lemma 4.4.6.

To prove (e), we begin by letting W be the set of x in G for which $\partial\varphi/\partial x = 0$. Since the φ^α can be extended, it follows from Lemma 9.2.2 that $|\tau(W)| = |\tau(\partial G)| = 0$ and $\tau(W)$ and $\tau(\partial G)$ are compact. Now if $z \in \tau(G) - [\tau(W) \cup \tau(\partial G)]$ then $N[z, \tau(R)]$ is finite ($R = G^{(0)} - W$) and $\partial\varphi/\partial x \neq 0$ at each x for which $\varphi(x) = z$. In fact, if ϱ is a small domain of class C^1, its counter image consists of $K = N[z, \tau(R)]$, small domain ϱ $G_1, \ldots, G_K$ and the restriction of φ to G_k is a diffeomorphism of G_k onto ϱ for each k. If z is interior to ϱ, all of its counter images are in the G_k, one in each, so that if Γ is the domain in (d), then

$$(9.2.21)\qquad O[z, \tau(\partial\Gamma)) = 0, \quad O[z, \tau(\partial G)] = \sum_{k=1}^K O[z, \tau(\partial G_k)].$$

To evaluate one of the terms on the right in (9.2.21) we begin by replacing K_0 in (9.2.20) ($G = G_k$) by $K_{0\sigma}$ where

$$(9.2.22)\qquad K_{0\sigma}(y) = \int_{B(0,\sigma)} K_0(y - \eta)\, \psi^*_\sigma(\eta)\, d\eta, \quad \psi^*_\varrho(\eta) = \varrho^{-\nu}\psi(\varrho^{-1}\eta),$$

where ψ is a mollifier. Thus, for $\sigma > 0$ but small,

$$\int_{\partial G_k} K_{0\sigma,\alpha}[\varphi(x) - z](-1)^{\alpha+\beta} \frac{\partial \varphi'_\alpha}{\partial x'_\beta} dx'_\beta = \int_{G_k} \psi^*_\sigma[\varphi(x) - z] \frac{\partial \varphi}{\partial x} dx$$

$$= sgn\left(\frac{\partial \varphi}{\partial x}\right) \int_\varrho \psi^*_\sigma(\zeta - z)\, d\zeta = sgn\, \frac{\partial \varphi}{\partial x}.$$

The result (e) follows. Moreover, we evidently have

$$(9.2.23)\qquad \begin{aligned} &\int_\varrho O[z, \tau(\partial G)]\, dz = \sum_{k=1}^K \int_{\tau_k} \frac{\partial \varphi}{\partial x} dx \\ &\int_\varrho N[z, \tau(G)]\, dz \sum_{k=1}^K \int_{\tau_k} \left|\frac{\partial \varphi}{\partial x}\right| dx. \end{aligned}$$

The results in (g) follow from (9.2.23) by covering almost all of $\tau(G) - [\tau(W) \cup \tau(\partial G)]$ by disjoint small regions ϱ. The result (f) is now obvious.

Definition 9.2.2. If G is of class C^1, $\varphi \in C^0(G)$, $\tau : z^\alpha = \varphi^\alpha(x)$ for $\alpha = 1, \ldots, \nu$, and $z \notin \tau(\partial G)$, we define

$$0[z, \tau(\partial G)] = \lim_{n\to\infty} 0[z, \tau_n(\partial G)], \quad \tau_n : z^\alpha = \varphi_n^\alpha(x)$$

for any sequence $\{\varphi_n\} \in C^1(\partial G)$ such that φ_n converges uniformly to φ on ∂G.

Lemma 9.2.3'. If *G is of class C^1 and $\varphi \in C^0(G)$, then all the results of Lemma* 9.2.3 *hold, except possibly those in* (g); *$\varphi(x, t)$ is merely required to* $\in C^0(G \times [0,1))$ *in* (f).

Lemma 9.2.4. *If $\varphi \in C_1^0(G)$, G arbitrary, there exist sets Z_α, $\alpha = 1, \ldots, \nu$, each of 1-dimensional measure 0 such that if R is a cell in G, none of whose faces lie along planes $x^\alpha = c^\alpha$ where $c^\alpha \in Z_\alpha$, then*

$$\int_R \frac{\partial \varphi}{\partial x}\, dx = \int_R 0[z, \tau(\partial R)]\, dz. \tag{9.2.24}$$

If $\nu = 2$, this holds if $\varphi \in C^0(G) \cap H_2^1(G)$.

Proof. We first assume $\varphi \in C_1^0(G)$. We begin by extending φ to be Lipschitz everywhere and then, by using mollifiers we may approximate to φ on G uniformly by vectors φ_n whose gradients $\nabla \varphi_n$ converge almost everywhere and boundedly to $\nabla \varphi$. Clearly this convergence holds almost everywhere on each plane $x^\alpha = c^\alpha$ for c^α not in a certain set Z_α of measure zero. For cells R as described, we have

$$\lim_{n\to\infty} \int_R \frac{\partial \varphi_n}{\partial x}\, dx = \int_R \frac{\partial \varphi}{\partial x}\, dx \tag{9.2.25}$$

and moreover, if z is not on $\tau(\partial R)$, we see also that $0[z, \tau(\partial R)]$ is given by (9.2.13). Since φ is Lipschitz, it follows that $|\tau(\partial R)| = 0$. Thus (9.2.13) holds for almost all z and this is also true for each n. Thus, for almost all z, $0[z, \tau_n(\partial R)] \to 0[z, \tau(\partial R)]$ and

$$\int_e |0[z, \tau_n(\partial R)]|\, dz \le \left(\frac{|e|}{\gamma_\nu}\right)^{1/\nu} \cdot M^{\nu-1} \Lambda^{\nu-1}(\partial R), \quad \text{(e measurable)}, \tag{9.2.26}$$

M being the common Lipschitz constant for the φ_n. Thus

$$\lim_{n\to\infty} \int_{R_\nu} 0[z, \tau_n(\partial R)]\, dz = \int_{R_\nu} 0[z, \tau(\partial R)]\, dz \tag{9.2.27}$$

and the first result follows.

In case $\nu = 2$ and $\varphi \in C^0(G) \cap H_2^1(G)$, we may approximate to φ uniformly and strongly in H_2^1 on any domain $D \subset\subset G$ by $\varphi_n \in C^\infty(D)$ so that $\varphi_n \to \varphi$ on ∂R strongly in $H_2^1(\partial R)$ for all R on a grating as indicated. Clearly (9.2.25) holds. For any such R, φ and φ_n are A.C. and the lengths $l[\tau_n(\partial R)]$ and $l[\tau(\partial R)]$ are uniformly bounded so that $|\tau_n(\partial R)| = |\tau(\partial R)| = 0$. Then, again $0[z, \tau_n(\partial R)] \to 0[z, \tau(\partial R)]$ for almost all z and

$$\int_e |0[z, \tau_n(\partial R)]| \, dz \leq \left(\frac{|e|}{\pi}\right)^{1/\nu} \cdot l[\tau_n(\partial R)]. \tag{9.2.26'}$$

Thus the result follows.

Remark. The only thing that prevents us from allowing $\varphi \in H^1_\nu(G)$ for any ν is that we do not know that $|\tau(\partial R)| = 0$ even if $\varphi \in H^1_\nu(\partial R)$ and is the strong limit in $H^1_\nu(\partial R)$ of φ_n in C^1. This is on account of the example of BESICOVITCH.

We can now prove MCSHANE's principal lemma:

Lemma 9.2.5. *Suppose $\varphi \in C^0_1(G)$, R is a cell for which* (9.2.24) *holds, $\varphi_n \in C^1(D)$ for some D with $R \subset D \subset\subset G$, and φ_n converges uniformly to φ on R. Then there are measurable subsets $V_n \subset R$ such that*

$$\lim_{n \to \infty} \int_{V_n} \frac{\partial \varphi_n}{\partial x} \, dx = \int_R \frac{\partial \varphi}{\partial x} \, dx.$$

If $\nu = 2$, the same result holds if $\varphi \in C^0(G) \cap H^1_2(G)$.

Proof. Let $\varepsilon_n = \max |\varphi_n(x) - \varphi(x)|$ for $x \in \partial R$. For each n, let W_n be the subset of R where $\partial \varphi_n / \partial x = 0$. Cover almost all of the set of z where $0[z, \tau(\partial R)] \neq 0$, $z \notin \tau_n(W_n)$, and z is at a distance $> \varepsilon_n$ from $\tau(\partial R)$ by small disjoint domains ϱ_{ni}, for each of which, $\tau_n^{-1}(\varrho_{ni}) = \bigcup_k G_{nik}$ and $\tau_n \mid G_{nik}$ is a diffeomorphism from G_{nik} onto ϱ_{ni}. Then

$$\sum_{i,k} \int_{G_{nik}} \frac{\partial \varphi_n}{\partial x} \, dx = \sum_i \int_{\varrho_{ni}} 0[z, \tau(\partial R)] \, dz \tag{9.2.28}$$

since $0[z, \tau_n(\partial R)] = 0[z, \tau(\partial R)]$ for $z \in \bigcup \varrho^{(0)}_{ni}$. Since the right side of (9.2.28) tends to $\int_R 0[z, \tau(\partial R)] \, dz$, we may take $V_n = \bigcup_{i,k} G_{nik}$ to obtain the result.

Theorem 9.2.1. *Suppose f satisfies the general conditions, G is of class C^1, $z = (z^1, \ldots, z^N) \in C^0_1(G)$, $z_n \in C^1(G)$, and $z_n \to z$ uniformly on G. Then*

$$I_f(z, G) \leq \liminf_{n \to \infty} I_f(z_n, G). \tag{9.2.29}$$

If $\nu = 2$, the result holds if z merely $\in C^0(G) \cap H^1_2(G)$.

Proof. We shall prove the first statement; the proof of the second is similar. If the right side of (9.2.29) is $+\infty$, the result follows. Otherwise, by taking a subsequence if necessary, we may assume that $I_f(z_n, G) \leq m M$ so that

$$A(z_n, G) = \int_G |X_n(x)| \, dx \leq M, \quad n = 1, 2, \ldots \tag{9.2.30}$$

From the representation (9.2.12), the differentiability condition on f, and the compactness of the totality of n.o. sets in R_N, it follows that there exists a number L and a function $\omega(\delta)$, depending only on f and A, such that

$$\begin{aligned} &|a| = \Big[\sum_I a_I^2\Big]^{1/2} \leq L, \quad |f(z_1, p) - f(z_2, p)| \leq \omega(\delta)\,|X|, \\ &z_k \in \overline{B(0, A)}, \quad |z_1 - z_2| \leq \delta, \quad \lim_{\delta \to 0^+} \omega(\delta) = 0 \end{aligned} \tag{9.2.31}$$

whenever a_I is the set of constants of Lemma 9.2.1 for f at some point (z_0, p_0) where $z_0 \in \overline{B(0, A)}$ and the rank of p_0 is ν, A being a common bound for all the $z_n(x)$.

Next we notice that there are sets Z_α of measure 0 such that if no face of the cell R lies along a plane $x^\alpha = c^\alpha \in Z_\alpha$, then (9.2.24) holds for every set Z^I, $I = i_1 < \cdots < i_\nu$. For almost all x_0

$$\begin{aligned} &\lim_{R \to x_0} |R|^{-1} \int_R f[z(x), p(x)]\, dx = f[z(x_0), p(x_0)] \\ &\lim_{R \to x_0} |R|^{-1} \int_R |X(x) - X(x_0)|\, dx = 0. \end{aligned} \tag{9.2.32}$$

Let $\varepsilon > 0$. Then we may cover almost all of G by a countable number of cells R_j, avoiding the Z_α, having maximum edge $\leq$ twice minimum edge, so small that

$$|z(x) - z(x_j)| \leq \omega(\delta) \leq \varepsilon/4M, \quad x, x_j \in R_j \tag{9.2.33}$$

and so that

$$\begin{aligned} &\sum_{j=1}^{\infty} \int_{R_j} |f[z(x), p(x)] - f[z(x_j), p(x_j)]|\, dx < \varepsilon/4 \\ &\sum_{j=1}^{\infty} \int_{R_j} |X(x) - X(x_j)|\, dx < \varepsilon/4L. \end{aligned} \tag{9.2.34}$$

We may choose a finite number of these R_j so that

$$\sum_j \int_{R_j} f[z(x_j), p(x_j)]\, dx > I_f(z, G) - 2\varepsilon/4. \tag{9.2.34}$$

Now, if for some j, the rank of $p(x_j) < \nu$, then

$$\begin{aligned} \liminf \int_{R_j} f[z_n(x), p_n(x)]\, dx &\geq 0 \\ &= \int_{R_j} f[z(x_j), p(x_j)]\, dx \quad (X(x_j) = 0). \end{aligned} \tag{9.2.35}$$

Otherwise, if $\eta_n = \max |z_n(x) - z(x)|$,

$$\begin{aligned} \int_{R_j} f[z_n(x), p_n(x)]\, dx &\geq \int_{R_j} f[z(x), p_n(x)]\, dx - \omega(\eta_n)\, A(z_n, R_j) \\ &\geq \int_{R_j} f[z(x_j), p_n(x)]\, dx - [\omega(\eta_n) + \varepsilon/4M]\, A(z_n, R_j). \end{aligned} \tag{9.2.36}$$

Now, for each fixed j for which $X(x_j) \neq 0$, let a_I be the constants of Lemma 9.2.1 and let $\zeta^1, \ldots, \zeta^\nu$ be the linear functions of z determined as in that lemma. Then, from our general assumptions and Lemma 9.2.5, it follows that there are sets $V_n \subset R_j$ such that

$$\begin{aligned} \int_{R_j} f[z(x_j), p_n(x)]\, dx &\geq \int_{V_n} f[z(x_j), p_n(x)]\, dx \\ &\geq \int_{V_n} \sum_I a_I X_n^I(x)\, dx = \int_{V_n} A \frac{\partial \zeta_n}{\partial x}\, dx \to \int_{R_j} A \frac{\partial \zeta}{\partial x}\, dx \\ &= \int_{R_j} \sum_I a_I X^I(x)\, dx \geq \int_{R_j} f[z(x_j), p(x_j)]\, dx \\ &\quad - L \int_{R_j} |X(x) - X(x_j)|\, dx. \end{aligned} \tag{9.2.37}$$

From (9.2.30) and (9.3.33)—(9.2.37), we obtain

$$\begin{aligned} \liminf_{n\to\infty} I_f(z_n, G) &\geq \liminf_{n\to\infty} \sum_j \int_{R_j} f[z_n(x), p_n(x)]\, dx \\ &\geq \sum_j \int_{R_j} f[z(x_j), p(x_j)]\, dx - 2\varepsilon/4 \geq I_f(z, G) - \varepsilon. \end{aligned}$$

The theorem follows.

9.3. Two dimensional problems; introduction; the conformal mapping of surfaces

The simplest possible multidimensional parametric problem is the problem of Plateau, i.e. the problem of proving the existence (and differentiability) of a surface of least area having a given boundary. This was solved in the one contour case by J. Douglas and T. Rado [3] in 1930—31. The next decade saw the solution (see Morrey [8] for more literature of this period) by Douglas, R. Courant, McShane, and others of this problem among surfaces of higher topological type bounded by one or more contours. The solution was greatly facilitated by the use of the theory of conformal mapping which enabled one to replace the area integral by the Dirichlet integral; as we have seen in (9.1.13) and (9.1.14) we see that

$$A(z, G) \leq \frac{1}{2} D(z, G), \tag{9.3.1}$$

the equality holding if and only if z is a conformal representation of the given surface. Thus the minimizing vectors are harmonic. This fact enabled M. Morse, ([4]) C. B. Tompkins (Morse and Tompkins [1]—[4]), Courant ([1]), M. Shiffman, and others (see Morrey [8]) to apply the Morse critical point theory to obtain interesting results concerning the existence of "unstable" minimal surfaces. Finally the author (MORREY [8]) generalized COURANT's solution in the case of k contours, the surface being of the type of a domain bounded by k circles, to the case of surfaces lying in a RIEMANNIAN manifold. The conformal mapping theorem still holds for such surfaces but the work of BOCHNER ([2]) on "harmonic" vectors in RIEMANNIAN spaces was not sufficient to enable the author to carry over COURANT's work. It was necessary to carry over the theory of functions of class H_2^1 to vectors in a RIEMANNIAN manifold in order to obtain the "in the large" solution of the minimum problem. By assuming that the manifold $\mathfrak{M}$ (of class C^1 at least) is "homogeneously regular", i.e. that there exist *positive* numbers m and M, independent of P_0, such that each point P_0 of $\mathfrak{M}$ is in the range of a coordinate patch with domain $B(0, 1)$ for which

$$m\,|\xi|^2 \leq g_{ij}(x)\,\xi^i\,\xi^j \leq M\,|\xi|^2, \quad x \in B(0,1), \quad \xi \text{ arbitrary},$$

the author could show that the minimizing surface is continuous. If $\mathfrak{M}$ is of class C^3 ($g_{ij} \in C^2$), the differentiability results for $\nu = 2$, N arbitrary, as stated in § 1.10 imply that the solution harmonic vectors $\in C_\mu^2$ for any μ, $0 < \mu < 1$; if $\mathfrak{M}$ is of class C_μ^n, $0 < \mu < 1$, for some $n \geq 3$, then the harmonic vectors $\in C_\mu^n$; if $\mathfrak{M}$ is C^∞ or analytic, so are the harmonic vectors. This work is presented in § 9.4 below.

In § 9.2, we extended MCSHANE's proof ([3]) of lower semicontinuity for more general parametric integrals. Shortly after the war, CESARI ([1], [2], [3]) extended MCSHANE's theorem to his more general integrals (we do not discuss these) and also showed that the convexity condition on $F(z, X)$ was a necessary, as well as a sufficient condition for lower-semicontinuity. But it was not until the concurrent results of CESARI [4], DANSKIN and SIGALOV [2] that an existence theorem for the integral I_f of § 9.2, with $\nu = 2$ and $N = 3$, was proved. The difficulty was that the lower semi-continuity theorem requires uniform convergence (at least on interior domains) and even though conformal mapping kept the DIRICHLET integral bounded, this is not enough to ensure equicontinuity. But the implied condition

$$m\,|X| \leq f(z, p) \equiv F(z, X) \leq M\,|X| \tag{9.3.2}$$

implies an a priori HÖLDER condition, on interior domains, for the solution vector. The author ([17]) was able to give a short proof of this theorem by minimizing a dominating integral I for $\mathfrak{J}_f$ among all z for which another integral $J(z, G) \leq K$, obtaining bounds independent of K, and then letting $K \to \infty$. We present this in § 9.5.

The author is not aware of any general differentiability theorems for the solutions which have been shown to exist. However, if the ratio M/m above is sufficiently small and if

$$|X|\,F_{X^\gamma X^\delta}(z, X)\,\xi^\gamma\,\xi^\delta \geq m_1[|\xi|^2 - |X|^{-2}(X \cdot \xi)^2] \tag{9.3.3}$$

where m_1 is sufficiently large, then the dominating function f_0 defined by

$$[f_0(z,p)]^2 = [f(z,p)]^2 + \left(\frac{E-G}{2}\right)^2 + F^2 \tag{9.3.4}$$

(E, F, G defined in (9.1.12)) is convex in all the p^i_α ($\alpha = 1, 2$). To be sure, f_0 does not $\in C^2$ for any p for which $X = 0$ but the second derivatives are bounded and the analysis of § 1.11 and §§ 5.2—5.5 progresses far enough to show that the solution vector $z \in C^1_\mu$ for any μ, $0 < \mu < 1$. Since the solution vector is obviously conformal, the higher differentiability follows in the usual way near points x where not all the derivatives $z^i_{,\alpha}(x)$ are zero; at such points $f_0 \notin C^2$ and so it is not likely, in general, that derivatives of order > 1 will be continuous at such points. For the case of minimal surfaces, we shall see that they may have branch point singularities.

For the general parametric integral with $\nu = 2$, the author has constructed a dominating function f^{**}, coinciding with f when $E = G$ and $F = 0$, which is regular only if the general sense (see § 9.5 below). However, by making certain a priori assumptions about the solution (i.e. that it has the form $z = f(x, y) \in C^2$, etc.) JENKINS and SERRIN have obtained interesting results concerning these solutions.

Moreover, in 1957, a student T. C. KIPPS, in his thesis, derived some a priori bounds for the HÖLDER continuity of the first derivatives of a conformal representation of a solution surface, assuming that these derivatives were continuous on an interior domain and assuming that the problem was regular. We present his results in § 9.5. Strangly enough, he was not able to use these results to demonstrate the continuity of the derivatives, in spite of the fact that his bounds do not depend on the moduli of continuity of the derivatives.

We shall now sketch preliminary results, including LICHTENSTEIN's conformal mapping theorem for surfaces, leading to the solution of the problem of Plateau which will be presented in § 9.4. We shall confine ourselves here to surfaces of *type k*, i.e. those having the topological type of a plane domain bounded by k disjoint JORDAN curves; surfaces of higher topological types are treated in the book [3] by COURANT. We shall not assume any theorems on conformal mapping; these will be proved; we shall, however, use the MÖBIUS (linear, fractional) transformations. We shall assume, also, if G is a domain of type k and class C^1_μ, there is a diffeomorphism of class C^1_μ of $\bar{G}$ onto a *circular domain* B, i.e. one bounded by k disjount circles, in such a way that the outer boundaries correspond.

Lemma 9.3.1. *Suppose* $\{z_n\}$ *is a sequence of vectors such that each* $z_n \in C^1_\mu(B_n)$, *where* B_n *is a circular domain of type* k. *We suppose that* $C_{n1} = \partial B(0,1)$ *is the outer boundary of* B_n *for each* n *and* $C_{n2}, \ldots, C_{nk}$ *are its other boundaries, that* $\Gamma_1, \ldots, \Gamma_k$ *are oriented* FRÉCHET *curves* *whose ranges are disjoint* JORDAN *curves, and that the* FRÉCHET *curves* Γ_{ni}, *each of which is defined by* z_n *restricted to* C_{ni}, *converge respectively to* Γ_i *as* $n \to \infty$. *We suppose also that there is an* $L\,(< +\infty)$ *such that*

$$D(z_n, B_n) \leq L, \quad n = 1, 2, \ldots \tag{9.3.5}$$

Finally, we suppose that the C_{ni} *converge to* C_i, *where each* C_i *is a circle or a point, and that* $C_1, \ldots, C_l$, $l \leq k$, *are circles and the* C_i *with* $l < i \leq k$ *(if any) are points.*

Then the $C_1, \ldots, C_l$ *are disjoint.*

Proof. If this were not so, two of the limiting *circles* C_i and $C_j\,(1 \leq i < j \leq l)$ would be externally tangent at some point x_0. Let $d\,(> 0)$ be the minimum distance from a point on any Γ_p to one on a different Γ_q and choose $R > 0$ but small enough so that every circle $\partial B(x_0, r)$ with $0 < r \leq R$ intersects C_i and C_j. Now, choose δ, $0 < \delta < R$. Then, there is an N such that if $n > N$ then every circle $\partial B(x_0, r)$ with $\delta \leq r \leq R$ contains an arc γ_{nr} which is in B_n and has its endpoints on two distinct C_{np} and C_{nq} ("usually" C_{ni} and C_{nj}, but there may be circles C_{ns} with $l < s \leq k$ close to x_0). Thus for such r

$$\int_{\gamma_{nr}} |z_{n\vartheta}(r, \vartheta)|^2\, d\vartheta \geq (2\pi)^{-1} \left[\int_{\gamma_{nr}} |z_{n\vartheta}|\, d\vartheta\right]^2 \geq \frac{d^2}{8\pi}$$

so that

$$L \geq \int_{\delta}^{R} r^{-1} \int_{\gamma_{nr}} |z_{n\vartheta}(r, \vartheta)|^2\, d\vartheta\, dr \geq \frac{d^2}{8\pi} \log\left(\frac{R}{\delta}\right), \quad n > N.$$

Since δ is arbitrarily small, we arrive at a contradiction.

As a result of this lemma, we easily conclude the following corollary:

Corollary 1. *Suppose that the sequences* $\{z_n\}$, $\{B_n\}$, $\{C_{ni}\}$, $\{\Gamma_{ni}\}$, C_i, *and* Γ_i *satisfy the conditions of Lemma* 9.3.1. *Then it may also be assumed that* $C_{ni} = C_i$ *for each* $i \leq l$ *and, moreover, this assumption may be made without altering the values of* $\liminf D(z_n, B_n)$.

Now suppose that all these conditions are satisfied. Since the Γ_i are Jordan curves and the Fréchet curves $\Gamma_{ni} \to \Gamma_i$ for each i, it is easy to see that there exist continuous vector functions $\zeta_{ni}(\varphi)$ which are periodic of period 2π, are not constant on any segment, and which converge uniformly to $\zeta^i(\varphi)$ where $z = \zeta^i(\varphi)$, $0 \leq \varphi \leq 2\pi$ is a topological representation of Γ_i, except that $\zeta^i(0) = \zeta^i(2\pi)$ (Theorem 9.1.2(b)). It follows easily that there are non-decreasing continuous functions φ_{ni} such that

$$\begin{aligned}
&\text{(i) } \zeta_{ni}[\varphi_{ni}(\theta)] = \xi_{ni}(\theta) = z_n[x_i + r_i\cos\theta,\, y_i + r_i\sin\theta], \\
&\qquad C_i = \partial B(x_i, y_i, r_i), \quad i = 1, \ldots, l; \\
&\text{(ii) } \varphi_{ni}(\theta) - \theta \text{ is periodic of period } 2\pi; \\
&\text{(iii) } -\pi < \varphi_{ni}(0) \leq \pi.
\end{aligned} \tag{9.3.6}$$

We may immediately conclude the following corollary:

Corollary 2. *Suppose the sequences* $\{z_n\}$, etc., *of Corollary* 1 *satisfy the conditions of that corollary and that the* ζ_{ni}, ζ_i, ξ_{ni}, *and* φ_{ni} *have been chosen as above. Then a subsequence, still called* $\{n\}$, *may be chosen so that the functions* $\varphi_{ni}(\theta)$ *converge to non-decreasing limit functions* $\varphi_i(\theta)$.

Remark. These limit functions need not be continuous but we prove the following lemma:

Lemma 9.3.2. *Suppose the sequences* $\{z_n\}$, $\{\varphi_{ni}\}$, etc., *satisfy the conditions above. Then, for a given* j, $1 \leq j \leq l$, *and* θ_0, *either* φ_j *is continuous at* θ_0 *or else*

$$\varphi_j(\theta_0+) - \varphi_j(\theta_0-) = 2\pi. \tag{9.3.7}$$

Proof. Suppose for some i and θ_0, we have

$$0 < \varphi_i(\theta_0^+) - \varphi_i(\theta_0^-) < 2\pi.$$

Then $\zeta_i[\varphi_i(\theta_0^+)]$ and $\zeta_i[\varphi_i(\theta_0^-)]$ are distinct points; let

$$|\zeta_i[\varphi_i(\theta_0^+)] - \zeta_i[\varphi_i(\theta_0^-)] = d > 0.$$

Let p_0 be the point of C_i corresponding to θ_0. From the uniform convergence of $\zeta_{ni}(\varphi)$ to $\zeta_i(\varphi)$ and the pointwise convergence of $\varphi_{ni}(\theta)$ to $\varphi_i(\theta)$, all being monotone, it follows that $\exists$ a $\delta > 0$ and an N such that C_i divides the disc $B(p_0, \delta)$, $B(p_0, \delta)$ intersects no other C_j with $j \leq l$, and if $\gamma(r)$ is the arc of $\delta B(p_0, r)$ which contains points of B_n then

$$z_n[\gamma(r)] \text{ is of length } \geq d/2, \quad 0 < r < \delta. \tag{9.3.8}$$

It is clear that (9.3.8) certainly holds even if that arc contains points of C_{ni} with $i > l$ if we interpret $\gamma(r) = \gamma_n(r)$ as the part of that arc in B_n. Thus, for each $\varepsilon > 0$

$$L \geq \int_\varepsilon^\delta r^{-1} \int_{\gamma_n(r)} |z_n(\psi)|^2 \, d\psi \geq \frac{d^2}{8\pi} \log\left(\frac{\delta}{\varepsilon}\right), \quad n > N. \tag{9.3.9}$$

This is impossible.

We wish now to prove the following theorem of LICHTENSTEIN [2]:

Theorem 9.3.1. *Any locally regular*[1] *surfaces of type k and of class* C_μ^1 *possesses a representation z which is conformal and of class* $C_\mu^1(\bar{B})$ *where B is a domain of type k bounded by circles. Three given points on one of the boundary curves of S can be made to correspond, respectively, to three given points on the outer boundary of B which may be taken as* $\partial B(0,1)$. *In case the surface is of class* C_μ^n *for* $n \geq 1$, C^∞, *or analytic near a point, the representation z has the corresponding class near the corresponding point. If the point is on the boundary of S and* ∂S *has the same class near that point, then z has the same class near the corresponding point on* ∂B.

Preparatory remarks. As was pointed out by COURANT in [3], this theorem is not necessary if one wishes only to prove the existence of a

[1] See (9.3.10) et seq. below.

minimal surface (in the sense of differential geometry) with given boundary. But the theorem is interesting in itself and is helpful in proving the existence of a *surface of least area having a given boundary which surface is simultaneously a minimal surface.* We shall prove this for the general k; LICHTENSTEIN [2] proved it for $k = 1$. Suppose that

$$S : z = z(u, v), \quad (u, v) \in \bar{G}, \quad z_u \times z_v \neq 0, \tag{9.3.10}$$

where G is of type k and of class C^1_μ and $z \in C^1_\mu(\bar{G})$. Using inversions and simple transformations, we may assume that the outer boundary of G corresponds to that bounding curve of S containing the three given points. We first prove several lemmas.

Lemma 9.3.3. *There exists a mapping*

$$\tau : u = u(x, y), \quad v = v(x, y), \quad (x, y) \in \bar{B} \tag{9.3.11}$$

from a circular domain $\bar{B}$ *of type* k *onto* $\bar{G}$ *in which* u *and* $v \in H^1_2(B) \cap C^0(B)$, *and which minimizes*

$$\begin{aligned} D(Z, B) &= \iint_B (|Z_x|^2 + |Z_y|^2)\, dx\, dy, \\ Z(x, y) &= z[u(x, y), v(x, y)], \end{aligned} \tag{9.3.12}$$

among all such mappings which carry the three given (distinct) points on the outer boundary $C_1 = \partial B(0,1)$ *of* B *into the three given points on the outer boundary* Γ_1 *of* G, *where* $\partial G = \Gamma_1 \cup \ldots \cup \Gamma_k$.

Proof. First of all, it is clear that there is a diffeomorphism from some circular domain $\bar{B}'$ with outer boundary C_1 onto $\bar{G}$. A map of the desired type is then obtained by preceding this with a Möbius transformation of $B(0,1)$ into itself to establish the three point correspondence; such a transformation carries circular domains into circular domains. We note that

$$\begin{aligned} D(Z, B) = \iint_B [&E(u, v)(u_x^2 + u_y^2) + 2F(u_x v_x + u_y v_y) + \\ &+ G(v_x^2 + v_y^2)]\, dx\, dy \\ E(u, v) = z_u \cdot z_u, \quad &F = z_u \cdot z_v, \quad G = z_v \cdot z_v \end{aligned} \tag{9.3.13}$$

where E, F, and $G \in C^0_\mu(\bar{G})$ and there are numbers m and M such that

$$m(\lambda^2 + \mu_1^2) \le E\lambda^2 + 2F\lambda\mu_1 + G\mu_1^2 \le M(\lambda^2 + \mu_1^2), \quad 0 < m \le M. \tag{9.3.14}$$

To show the existence of the minimum, let $\{\tau_n\}$ be a minimizing sequence. From (9.3.11)—(9.3.14), it follows that $\{\tau_n\}$ satisfies the conditions of Lemma 9.3.1 and so, by choosing a subsequence, we may assume that our minimizing sequence satisfies all the conditions of Corollary 2. From the three point condition and Lemma 9.3.2, it follows that the restriction of τ_n to C_1 converges uniformly to a representation of Γ_1

which satisfies the three point condition. There are now three types of degeneracy which can occur with respect to the boundary conditions:

case *i*: $k > 1$, $1 \leq l \leq k$, one or more C_i with $i > l$

are points interior to the limiting domain B bounded by $C_1, \ldots, C_l$;

case *ii*: some such C_i is a point on some C_j with $j \leq l$;

case iii: some φ_i with $i \leq l$ satisfies (9.3.7).

We first consider case i. Let $R > 0$ be chosen small enough so that $\overline{B(p_0, R)} \subset B$ and contains no other C_p than those which reduce to $C_i = p_0$. Let $0 < \delta < R$. Then there is an N so large that all the C_{np} which are approaching p_0 lie inside $B(p_0, \delta)$. Thus, for each r, $\delta \leq r \leq R$, $\tau_n[\partial B(p_0, r)]$ encloses all the corresponding Γ_p and so has length $\geq$ some $d > 0$ so that

$$(9.3.15) \qquad D(Z_n, B_n) \geq m \int_{\delta}^{R} r^{-1}(d^2/2\pi)\, dr = (m\, d^2/2\pi) \log(R/\delta)$$

which is impossible since δ is arbitrary. The reasoning to eliminate case ii is the same. So we may assume that $l = k$ and all the $B_n = B$.

We now suppose that case iii holds. Let p_0 be the point of C_i having coordinate θ_0. Choose $R > 0$ so that $B(p_0, r) \cap B$ is bounded by an arc γ' of C_i and an arc γ'' of $\partial B(p_0, r)$ for each $r < R$, and suppose $0 < \delta \leq r \leq R$. Let d denote the diameter of Γ_i. Since $\tau_n(\gamma'') \cup \tau_n(\gamma''')$ encloses the domain bounded by Γ_i, we see that

$$\operatorname{diam} \tau_n(\gamma'') + \operatorname{diam} \tau_n(\gamma''') \geq d. \quad (\gamma' \cup \gamma''' = C_i)$$

But, from Lemma 9.3.2 and the convergence of φ_{ni} to φ_i, it follows that

$$\operatorname{diam} \tau_n(\gamma''') \leq d/2 \text{ so } l[\tau_n(\gamma'')] \geq d/2, \quad \delta \leq r \leq R, \quad n > N.$$

As in (9.3.9) we arrive at a contradiction. Thus for each i, the φ_i are continuous so the convergence of τ_n is uniform on each C_i to a representation of Γ_i.

Finally, we note that there is an $R > 0$, independent of p, such that if $p \in \bar{B}$, then either $B(p, r) \subset B$ or $B(P, r) \cap B$ is bounded by an arc γ' of some one C_i and an arc γ'' of $\partial B(P, r)$ whenever $0 < r < R$; in the former case, we set $\gamma'' = \partial B(P, r)$. Now, let $\varepsilon > 0$ and choose $\delta_1 > 0$ so small that $\operatorname{diam} \tau_n(\gamma') < \varepsilon/2$ whenever $r < \delta_1$; this is possible on account of the equicontinuity. Now, suppose for some δ, $0 < \delta < \delta_1$, that osc τ_n on $\overline{B(p, \delta)} \cap \bar{B} \geq \varepsilon$. Then, for $\delta \leq r \leq \delta_1$, the length of $\tau_n(\gamma'') \geq \varepsilon/2$ so that (see (9.3.15))

$$(9.3.16) \qquad D(Z_n, B_n) \geq m \int_{\delta}^{\delta_1} r^{-1}(\varepsilon^2/8\pi)\, dr = (m\, \varepsilon^2/8\pi) \log(\delta_1/\delta).$$

The equicontinuity of the τ_n now follows from (9.3.16) and a subsequence converges uniformly to a minimizing map τ.

Our aim now is to discuss the minimizing map τ. Since a similarity transformation or, more generally, a MÖBIUS transformation in the (x, y) plane carries one circular domain into another and leaves the Dirichlet integral invariant, we see that our τ minimizes $D(Z, B)$ among a larger class of mappings from circular domains in which the outer boundary need not be $\partial B(0,1)$ and three points do not need to be fixed. So, suppose that

$$\begin{aligned} &\sigma_\lambda : x' = \xi(x, y; \lambda), \quad y' = \eta(x, y; \lambda), \quad |\lambda| < \lambda_0, \\ (9.3.17) \qquad &\xi(x, y; 0) = x, \quad \eta(x, y; 0) = y, \\ &\xi_\lambda(x, y; 0) = \nu(x, y), \quad \eta_\lambda(x, y; 0) = \omega(x, y) \end{aligned}$$

is a family of diffeomorphisms, of class C^∞ in (x, y, λ), of B into a circular domain B_λ for each λ. Then ν and ω must satisfy the following boundary conditions on each C_j, $j = 1, \ldots, k$:

$$\begin{aligned} &(\nu, \omega) = (\nu_1, \omega_1) + (\nu_2, \omega_2) + (\nu_3, \omega_3), \\ &\nu_1 \cos\theta + \omega_1 \sin\theta = 0, \quad (\nu_2, \omega_2) = (c_j, d_j) = \text{const.} \\ (9.3.18) \qquad &\nu_3 + i\,\omega_3 = (e_j + i f_j)(\cos\theta + i \sin\theta), \quad (e_j + i f_j) = \text{const.} \\ &C_j : x = x_j + r_j \cos\theta, \quad y = y_j + r_j \sin\theta. \end{aligned}$$

Moreover, if we set

$$\begin{aligned} (9.3.19) \qquad &\varphi(\lambda) = D(Z', B_\lambda), \quad Z'(x', y') = Z[\xi'(x', y'; \lambda), \ \eta'(x', y'; \lambda)], \\ &\sigma_\lambda^{-1} : x = \xi'(x', y'; \lambda), \quad y = \eta'(x', y'; \lambda), \end{aligned}$$

then $\varphi'(0) = 0$ since $\varphi(\lambda)$ takes on its minimum for $\lambda = 0$, since $\tau\sigma_\lambda^{-1}$ is an admissible map of B_λ onto G. By expressing $D(Z', B_\lambda)$ as an integral over B, we obtain

$$\begin{aligned} (9.3.20) \qquad &Z'_{x'} = J^{-1}(Z_x \eta_y - Z_y \eta_x), \quad Z'_{y'} = J^{-1}(-Z_x \xi_y + Z_y \xi_x), \\ &J = \xi_x \eta_y - \xi_y \eta_x. \end{aligned}$$

$$\begin{aligned} &\varphi(\lambda) = \int\limits_B\int [\mathfrak{E}(\xi_y^2 + \eta_y^2) - 2\mathfrak{F}(\xi_x \xi_y + \eta_x \eta_y) + \mathfrak{G}(\xi_x^2 + \eta_x^2)]\, J^{-1}\, dx\, dy\,, \\ (9.3.21) \qquad &\mathfrak{E} = |Z_x|^2, \quad \mathfrak{F} = Z_x \cdot Z_y, \quad \mathfrak{G} = |Z_y|^2 \end{aligned}$$

$\mathfrak{E}$, $\mathfrak{F}$, and $\mathfrak{G}$ being independent of λ and in $L_1(B)$. Differentiating with respect to λ and setting λ (and ξ_y and η_x) $= 0$, we obtain

$$\begin{aligned} (9.3.22) \qquad &-\varphi'(0) = \int\limits_B\int [U(\nu_x - \omega_y) - V(\nu_y + \omega_x)]\, dx\, dy = 0, \\ &U = \mathfrak{E} - \mathfrak{G}, \quad V = -2\mathfrak{F}. \end{aligned}$$

Lemma 9.3.4. *Suppose ν and $\omega \in C^\infty(\bar{B})$ and satisfy the boundary conditions* (9.3.18). *Then there exists a family σ_λ of maps satisfying the conditions in and following* (9.3.17).

Proof. By using a partition of unity of class C^∞, we may write any such (ν, ω) as a sum of one which $\in C_c^\infty(B)$ and three for each C_j with

$j \geq 1$, each of which vanishes except near C_j. Corresponding families σ_λ in (9.3.17) can be found simply by setting

$$\xi(x, y; \lambda) = x + \lambda \nu(x, y), \quad \eta(x, y; \lambda) = y + \lambda \omega(x, y)$$

for the $(\nu, \omega) \in C_c^\infty(B)$ or of the types (ν_2, ω_2) and (ν_3, ω_3) for each C_j. The remaining ones are handled by polar coordinates and these latter each carry B into itself.

From this lemma and the duscussion preceding it, we obtain the following corollary:

Corollary. *If $\mathfrak{E}$, $\mathfrak{F}$ and $\mathfrak{G}$, as defined in (9.3.21), correspond via (9.3.13) to our minimizing map τ, then $\varphi'(0) = 0$ for all $(\nu, \omega) \in C^\infty(\bar{B})$ and satisfying (9.3.18).*

Lemma 9.3.5. *If the conclusion of the corollary holds, then*

$$(9.3.23) \qquad\qquad U = V = 0.$$

Remark. The essential idea of the last part of this proof is due to H. LEWY; the proof given here follows in a general way the argument given in COURANT's book [3], pp. 169—178.

Proof. Suppose, first, that ν and $\omega \in C_c^\infty(B)$, let ψ be a mollifier, and let ν_ϱ and ω_ϱ be the ψ-mollified functions. Then, defining $U = V = \nu = \omega = 0$ outside B and keeping ϱ small, we obtain

$$\iint_B [U(\nu_{\varrho x} - \omega_{\varrho y}) - V(\nu_{\varrho y} + \omega_{\varrho x})]\, dx\, dy$$

$$= \iint_{R_2} \iint_{R_2} \psi_\varrho^*(\xi - x)\{U(x, y)(\nu_\sigma - \omega_\tau) - V(x, y)(\nu_\tau + \omega_\sigma)\}\, dx\, dy\, d\sigma\, d\tau$$

$$= \iint_B [U_\varrho(\nu_x - \omega_y) - V_\varrho(\nu_y + \omega_x)]\, dx\, dy$$

$$= \iint_B [\omega(U_{\varrho y} + V_{\varrho x}) - \nu(U_{\varrho x} - V_{\varrho y})]\, dx\, dy.$$

(9.3.24)

Thus if this $= 0$ for all such (ν, ω), we see that $U_\varrho + i V_\varrho$ is analytic for each $\varrho > 0$ and hence we have

$$(9.3.25) \qquad\qquad \Phi(z) = U + i V \quad \text{analytic on } B.$$

Now, since U and $V \in L_1(B)$, we see that

$$(9.3.26) \qquad \lim_{\varrho \to 0} \int_{\partial B_\varrho} [U(\nu\, dy + \omega\, dx) - V(\omega\, dy - \nu\, dx)] = 0$$

for any $(\nu, \omega) \in C^\infty(\bar{B})$ and satisfying (9.3.18) (B_ϱ is the set of (x, y) in B such that $B(x, y; \varrho) \subset B$). Setting

$$(9.3.27) \quad \nu + i\omega + P(r)[c_1 + i d_1 + (e_1 + i f_1) e^{i\theta}], \quad 1 - h \leq r \leq 1$$

where $P(r) = 1$ near $r = 1$ and $P(r) = 0$ near $r = 1 - h$, $\nu + i\omega$ being 0 elsewhere, (9.3.26) for all c_1, d_1, e_1, f_1, implies that

$$(9.3.28) \qquad \int_{|z|=1-\varrho} \Phi(z)\, dz = \int_{|z|=1-\varrho} z\, \Phi(z)\, dz = 0, \quad 0 < \varrho < \varrho_0.$$

By choosing the corresponding variations for each C_j, we obtain

$$\int_{C_{j\varrho}} \Phi(z)\,dz = \int_{C_{j\varrho}} (z - z_j)\,\Phi(z)\,dz = 0, \quad j = 2, \ldots, k, \quad 0 < \varrho < \varrho_0, \tag{9.3.29}$$

$C_{j\varrho}$ being the circle in B concentric with and at a distance ϱ from C_j. From (9.3.29), we conclude that there is an analytic function $\Psi(z)$ on B such that

$$\Psi''(z) = \Phi(z).$$

Since $\partial\Psi'(z)/\partial x = \Phi(z)$ and $\partial\Psi'(z)/\partial y = i\,\Phi(z)$, it follows that $\Psi' \in H_1^1(B)$ and so has strong boundary values in $L_1(\partial B)$. Thus Ψ is continuous on $\bar{B}$.

Now, let us choose

$$v = -P(r)\,\Lambda(\theta)\sin\theta, \quad \omega = P(r)\,\Lambda(\theta)\cos\theta \tag{9.3.30}$$

where $P(r)$ was defined above (in (9.3.27)). Then (9.3.26) yields

$$-\lim_{\varrho\to 0} \int_{|z|=1-\varrho} \Lambda(\theta)\,Im\{\Psi''[(1-\varrho)\,e^{i\theta}]\,e^{2i\theta}\}\,d\theta = 0. \tag{9.3.31}$$

Integrating by parts once, we may then let $\varrho \to 0$ in (9.3.31) to obtain

$$Im \int_{|z|=1} \Psi'(e^{i\theta})\,[i\,e^{i\theta}\,\Lambda'(\theta) - e^{i\theta}\,\Lambda(\theta)]\,d\theta = 0. \tag{9.3.32}$$

If we break this into two integrals, change the variable to φ in the second integral, write $\Lambda(\varphi) = \Lambda(0) + \int_0^{\varphi} \Lambda'(\theta)\,d\theta$, and then interchange the order of integrations, (9.3.32) becomes

$$Im\left\{\int_0^{2\pi} \Lambda'(\theta)\left[i\,e^{i\theta}\,\Psi'(e^{i\theta}) - \int_\theta^{2\pi} \Psi'(e^{i\varphi})\cdot e^{i\varphi}\,d\varphi\right]d\theta + i\,\Lambda(0)\int_0^{2\pi} \frac{d}{d\varphi}\,\Psi(e^{i\varphi})\,d\varphi\right\} \tag{9.3.33}$$

in which the coefficient of $\Lambda(0)$ vanishes. Since Λ' is an arbitrary real function $\in C^\infty$ except that

$$\int_0^{2\pi} \Lambda'(\theta)\,d\theta = 0, \tag{9.3.34}$$

it follows from the Lemma of DU BOIS RAYMOND (Lemma 2.3.1), that the imaginary part of the quantity in the bracket in (9.3.33) is a constant. Carrying out the integration, we see that

$$Im[i\,z\,\Psi'(z) - i\,\Psi(z)] = \text{const. on } C_1, \tag{9.3.35}$$

so that $z\,\Psi'(z) - \Psi(z) = \chi(z)$ where χ is analytic across C_1. Thus we see that $\Psi \in C^\infty$ along C_1 (since Ψ is continuous). Differentiating (9.3.35) with respect to θ we obtain

$$Im\left\{i\,z\,\frac{d}{dz}[i\,z\,\Psi' - i\,\Psi]\right\} = -Im\,z^2\,\Phi(z) = 0. \tag{9.3.36}$$

By repeating the argument for each C_j, we obtain

$$(9.3.37)\qquad Im(z-z_j)^2\Phi(z)=0,\quad z\in C_j,\quad j=2,\ldots,k.$$

Accordingly we see that Φ can be extended analytically accross each boundary circle of B.

Now, on C_1, let $f(\theta)=z^2\Phi(z)$. Then $f(\theta)$ is real and it follows from (9.3.28) that

$$(9.3.38)\qquad \int_0^{2\pi} f(\theta)\,d\theta=\int_0^{2\pi} z\Phi(z)\cdot z\,d\theta=-i\int_0^{2\pi} z\,\Phi(z)\,dz=0.$$

Since f is analytic and periodic, it follows that f, and hence Φ has at least two zeros on C_1. By using (9.3.29), we conclude also that Φ has at least two zeros on each C_j. Now, either $\Phi(z)\equiv 0$ on $\bar{B}$ or else it has a finite and non-negative number of zeros in B and no poles there. The number n of zeros of Φ interior to B is given by

$$(9.3.39)\qquad \begin{aligned} n&=\frac{1}{2\pi i}\lim_{\varrho\to 0^+}\int_{\partial B'_\varrho}\frac{\Phi'(z)}{\Phi(z)}\,dz,\\ B'_\varrho&=B-\bigcup_s\overline{[B\cap B(\zeta_s,\varrho)]},\end{aligned}$$

the ζ_s being the zeros of Φ on ∂B and ϱ being suitably small. But the value of the integral in (9.3.39) is easily seen to yield

$$n\le -k$$

since there are at least two ζ_s on each C_j, $j=1,\ldots,k$. Thus we must have $\Phi(z)\equiv U+iV\equiv 0$ and the lemma is proved.

Lemma 9.3.6. *Suppose $z=z(w)$, $w=(u,v)\in\bar{G}$, is a diffeomorphism of class C^1, G being of class C^1 $(z=(z^1,\ldots,z^N))$. Suppose S is the (unorientedted) Fréchet surface having the representation z and suppose $z=Z(X)$, $X=(x,y)\in\bar{B}$, is a representation of S in which $Z\in C^0(\bar{B})\cap H_2^1(B)$ where $B\in C^1$. Then $\exists$ a unique mapping $\tau: w=w(X)$, $X\in\bar{B}$, such that*

$$(9.3.40)\qquad Z(X)=z[w(X)],\quad X\in\bar{B},$$

$\tau\in C^0(\bar{B})\cap H_2^1(B)$, and $\exists$ a sequence $\tau_n: w=w_n(X)$ of diffeomorphisms from $\bar{B}$ onto $\bar{G}$ which converge uniformly to τ.

Proof. Let $|S|$ denote the range of S, i. e. the ranges of z and Z. Since z is $1-1$, z^{-1} is a homoemorphism from $|S|$ onto $\bar{G}$. Since Z is a representation of S, $\exists$ a sequence $\{\tau_n\}$ as above such that $z[\tau_n(X)]\equiv Z_n(X)$ converges uniformly to $Z(X)$. If we define $\tau(X)=z^{-1}[Z(X)]$, we see that (9.3.40) holds and also that

$$\tau_n(X)=z^{-1}[Z_n(X)]$$

so that τ_n converges uniformly to τ on $\bar{B}$. Clearly τ is uniquely determined by (9.3.40). To show that $\tau\in H_2^1(B)$, let $X_0\in B$, let $w_0=\tau(X_0)$, and $z_0=z(w_0)=Z(X_0)$. It is clear that $z(u^1,u^2)$ can be extended to a diffeomorphism $\zeta(u^1, \ldots, u^N)$ from a neighborhood in R_N of $(u_0^1, u_0^2, 0, \ldots, 0)$ where $w_0 = (u_0^1, u_0^2)$ onto a neighborhood of z_0 in R_N. Then

$$\tau(X) = \zeta^{-1}[Z(X)]$$

for X in B near X_0. That $\tau \in H_2^1(B)$ follows from Theorem 3.1.9.

Lemma 9.3.7. *Suppose* $\tau : w = w(X)$ *satisfies the conclusions of Lemma* 9.3.6. *Then the Jacobian* $u_x v_y - u_y v_x$ *has the same sign almost everywhere in* B.

Proof. From Lemma 9.2.4, it follows that there are sets J_1 and J_2 of 1-dimensional measure 0 such that if none of the sides of the cell R lie along lines $x^\alpha = c^\alpha$ where $c^\alpha \in J_\alpha$, $\alpha = 1, 2$, then (9.2.24) holds. From Lemma 9.2.5, it follows that if R is a cell for which (9.2.24) holds, there exists a sequence of measurable sets $V_n \subset R$ such that

$$\lim_{n\to\infty} \iint\limits_{V_n} \frac{\partial(u_n, v_n)}{\partial(x, y)}\, dx\, dy = \iint\limits_{R} \frac{\partial(u, v)}{\partial(x, y)}\, dx\, dy. \tag{9.3.41}$$

It is clear, since we may confine ourselves to a subsequence that we may assume that all the Jacobians $u_{nx} v_{ny} - u_{ny} v_{nx}$ have the same sign throughout B. The result then follows using (9.3.41).

We can now prove the conformal mapping Theorem 9.3.1.

Proof of Theorem 9.3.1. We let τ be the mapping of Lemma 9.3.3. From Lemma 9.3.5, it follows that

$$\begin{aligned} \mathfrak{E} - \mathfrak{G} + 2i\,\mathfrak{F} = 0 = E(u_x + i\,u_y)^2 + 2F(u_x + i\,u_y)\, \times \\ \times\, (v_x + i\,v_y) + G(v_x + i\,v_y)^2; \end{aligned} \tag{9.3.42}$$

here $\mathfrak{E}$, $\mathfrak{F}$, and $\mathfrak{G}$ are defined in (9.3.21) and E, F, G in (9.3.13). It follows from (9.3.42) that

$$v_x = -G^{-1}(F\,u_x \pm H\,u_y), \quad v_y = G^{-1}(\pm H\,u_x - F\,u_y),$$
$$H = \sqrt{E\,G - F^2}.$$

From Lemma 9.3.6 and 9.3.7, it follows that the $\pm$ sign may be replaced by k, where $k = \pm 1$ *and is constant on* $\bar{B}$. Thus

$$v_x = -G^{-1}(F\,u_x + k\,H\,u_y), \quad v_y = G^{-1}(k\,H\,u_x - F\,u_y), \text{ or} \tag{9.3.43}$$

$$\begin{aligned} u_y = -k(b\,u_x + c\,v_x), \quad v_y = k(a\,u_x + b\,v_x), \\ a = H^{-1}E, \quad b = H^{-1}F, \quad c = H^{-1}G, \quad a\,c - b^2 = 1. \end{aligned} \tag{9.3.44}$$

Since τ is continuous, the coefficients in (9.3.43) are continuous and (9.3.43) implies (since $\tau \in H_2^1(B)$ also) that

$$\iint\limits_B G^{-1}[\zeta_x(k\,H\,u_x - F\,u_y) + \zeta_y(F\,u_x + k\,H\,u_y)]\,dx\,dy = 0, \quad \zeta \in C_c^1(B). \tag{9.3.45}$$

From the theorem of § 5.5, it follows first that $\tau \in H_p^1(D)$ for each $p > 2$ and each $D \subset\subset B$. This implies that $\tau \in C_{\mu'}^0(D)$, that the coefficients

$\in C_\mu^0(D)$, and hence that u and $v \in C_\mu^1(D)$ for $D \subset\subset B$. The higher differentiability results on the interior follow as usual.

Let $(x_0, y_0) \in \partial B$ and $(u_0, v_0) = \tau(x_0, y_0)$. A neighborhood $\mathfrak{N}$ of (u_0, v_0) may be mapped by a diffeomorphism of class C_μ^1:

$$U = p(u, v), \quad V = q(u, v), \quad u = P(U, V), \quad v = Q(U, V)$$

onto a disc $B(0, 0, R)$ so (u_0, v_0) corresponds to $(0, 0)$ and $\mathfrak{N} \cap \partial G$ corresponds to $U = 0$. Boundary coordinates (t, θ), where $t = \log r$ and (r, θ) are polar coordinates with pole at the center of the circle of ∂B containing (x_0, y_0), may be chosen and the equations (9.3.43), (9.3.44), and (9.3.45) go over into equations of the same types. Since $U = 0$ when $t =$ const., we conclude from the theory of § 5.5, first that U and hence $V \in$ every $H_p^1(D)$, U and $V \in$ every $C_{\mu'}^0$, the coefficients $\in C_\mu^0$, and hence U and $V \in C_\mu^1(D)$ for D a neighborhood of (x_0, y_0) on $\bar{B}$. Thus $\tau \in C_\mu^1(\bar{B})$. Again the higher differentiability results follow as usual.

9.4. The problem of Plateau

We now resume our discussion of the problem of PLATEAU. Let Γ denote the system $(\Gamma_1, \ldots, \Gamma_k)$, each Γ_i being an oriented closed FRÉCHET curve with range $|\Gamma_i|$.

Definition 9.4.1. We define

$$(9.4.1) \qquad \begin{aligned} l(\Gamma) &= \inf\{\liminf L(z_n, G_n)\} \\ d(\Gamma) &= \inf\{\liminf D(z_n, G_n)\} \end{aligned}$$

for all sequences $\{z_n\}$ where $z_n \in C^0(\bar{G}_n)$, each G_n is of type k and of class C^1 with boundary curves $C_{n1}, \ldots, C_{nk}$, each oriented as a part of ∂G_n, and z_n restricted to C_{ni} is a representation of Γ_{ni} where $\Gamma_{ni} \to \Gamma_i$, $i = 1, \ldots, k$; in the case of $d(\Gamma)$ we require the z_n also $\in H_2^1(G_n)$. We also define

$$(9.4.2) \qquad \begin{aligned} &l^*(\Gamma) = d^*(\Gamma) = +\infty \quad \text{if} \quad k = 1 \\ &l^*(\Gamma) = \min \sum_{i=1}^{p} l(\Gamma^{(i)}), \quad d^*(\Gamma) = \min \sum_{i=1}^{p} d(\Gamma^{(i)}) \end{aligned}$$

for all possible systems $\Gamma^{(1)}, \ldots, \Gamma^{(p)}$ where each $\Gamma^{(i)}$ consists of the curves Γ_j for which $j \in T_i$ where $\bigcup_{i=1}^{p} T_i = \{1, 2, \ldots, k\}$.

Lemma 9.4.1. (a) *In the definitions of* $l[\Gamma)$, $d(\Gamma)$, $l^*(\Gamma)$, *and* $d^*(\Gamma)$ *we may assume that each* G_n *is of class* C_μ^1 *and that each* $z_n \in C_\mu^1(\bar{G}_n)$ *and is locally regular.*

(b) $d(\Gamma) = 2l(\Gamma)$, $d^*(\Gamma) = 2l^*(\Gamma)$.

(c) $d(\Gamma) \le d^*(\Gamma)$.

Proof. All the statements in (a) are evident from the definitions except possibly the one concerning local regularity. To see this, let

$\{z_n\} \in C^1_\mu(\bar{G}_n)$. We may define Z_n by

$$\begin{aligned} Z^i_n(x) &= z^i_n(x), \quad i = 1, \ldots, N, \\ Z^{N+1}_n(x) &= \lambda_n x^1, \quad Z^{N+2}_n = \lambda_n x^2, \end{aligned} \qquad x = (x^1, x^2) \in \bar{G}_n$$

where λ_n is chosen sufficiently small. (b) follows from (a) and Theorem 9.3.1.

If $k = 1$, (c) is evident. If $k > 1$, let $\Gamma = \Gamma^{(1)} \cup \Gamma^{(2)}$. By reordering the Γ_i, we may assume $\Gamma^{(1)} = \{\Gamma_1, \ldots, \Gamma_m\}$, $\Gamma^{(2)} = \{\Gamma_{m+1}, \ldots, \Gamma_k\}$. Let B_1 and B_2 be fixed circular regions of types m and $k - m$, respectively and let $z_{1n} \in C^1_\mu(B_1)$ and $z_{2n} \in C^1_\mu(B_2)$ for each n and suppose that

$$L(z_{1n}, B_1) \to l(\Gamma^{(1)}), \quad L(z_{2n}, B_2) \to l(\Gamma^{(2)}).$$

Choose x_1, x_2, R_1 and R_2 so that $\overline{B(x_i, 3R_i)} \subset B_i$, $i = 1, 2$. By altering z_{in} only in $B(x_i, 3R_i)$, we may obtain $z'_{ni} \in C^1_\mu(B_i)$ but so that $z'_{in}(x) = z_{in}(x_i)$ for $x \in \overline{B(x_i, 2R_i)}$, $i = 1, 2$. It is clear that we may assume $x_1 = x_2 = 0$ and $R_1 = R_2 = R$. Next, let $z_0 \in R_N$. We may alter z'_{in} on $B(0, 2R)$ only to obtain z''_{in}, where z''_{in} is constant on each circle $\partial B(0, r)$ with $R \leq r \leq 2R$, $z''_{in}(x) = z_0$ for $x \in \overline{B(0, R)}$ and $z''_{in} \in C^1_\mu(B_i)$, $i = 1, 2$ (of course the local regularity is sacrificed). Now, define $\bar{B} = [\bar{B}_1 - B(0, R)] \cup \sigma[\bar{B}_2 - B(0, R)]$ where σ is the inversion with respect to $\partial B(0, R)$ followed by a reflection in the x^1 axis. On $\bar{B}$, we define

$$z_n(x) = \begin{cases} z''_{1n}(x), & x \in \bar{B}_1 - B(0, R) \\ z''_{2n}[\sigma^{-1}(x)], & x \in \sigma[\bar{B}_2 - B(0, R)]. \end{cases}$$

Then $z_n \in C^1_\mu(\bar{B})$ and we have

$$L(z_n, B) = L(z''_{1n}, B_1) + L(z''_{2n}, B_2) = L(z_{1n}, B_1) + L(z_{2n}, B_2).$$

Thus

$$l(\Gamma) \leq \liminf_{n \to \infty} L(z_n, B) \leq l(\Gamma^{(1)}) + l(\Gamma^{(2)}).$$

By induction, we may extend this to p systems for any $p \leq k$ and thus show that $l(\Gamma) \leq l^*(\Gamma)$ from which the result follows.

Lemma 9.4.2. *Suppose $\zeta \in H^1_2[\partial B(x_0, R)]$ and H is the harmonic function on $B(x_0, R)$ coinciding with ζ on $\partial B(x_0, R)$. Then*

$$D[H, B(x_0, R)] \leq \int_0^{2\pi} |\zeta'(\theta)|^2 \, d\theta.$$

Proof. This follows by expanding in Fourier series.

Theorem 9.4.1. *Suppose $\Gamma = \{\Gamma_1, \ldots, \Gamma_k\}$, $k \geq 1$, is a set of oriented* FRÉCHET *curves whose ranges $|\Gamma_i|$ are disjoint Jordan curves. Suppose also that*

$$d(\Gamma) < d^*(\Gamma) \quad (d(\Gamma) \text{ finite if } d^*(\Gamma) = +\infty).$$

Then there exists a vector $z \in C^0(\bar{B})$, B being a circular region of type k, where z is harmonic and conformal on B and z restricted to each properly *oriented boundary circle C_i of B is a representation of Γ_i. Moreover z is a representation of a surface of least area; i.e. $L(z, B) = l(\Gamma)$.*

Proof. Let $\{z_n\}$ be a sequence $\ni L(z_n, G_n) \to l(\Gamma)$. We may first assume each $G_n \in C^1_\mu$ and each $z_n \in C^1_\mu(\bar{G}_n)$ and locally regular. On account of Theorem 9.3.1, we may assume that each $z_n \in C^1_\mu(\bar{B}_n)$, each B_n being a circular region, in which we have

$$D(z_n, B_n) \to d(\Gamma). \tag{9.4.3}$$

Finally we may assume that the z_n and the B_n satisfy all the conditions in Lemma 9.3.1 and its two corollaries as well as (9.4.3). Then there are three possible types of degeneracy as in the proof of Lemma 9.3.3. We shall show that each of the cases considered there implies that $d(\Gamma) = d^*(\Gamma)$ and is therefore impossible. Let $M = \sup D(z_n, B_n)$.

Suppose case i holds and let $x_0 = |C_i|$. Choose $\delta > 0$ but so small that $\overline{B(x_0, \delta^{1/2})} \subset B$ and contains no $|C_j|$ other than those for which $|C_j| = x_0$. For $n > N$, all the $|C_{nj}|$ with $|C_j| = x_0$ are inside $B(x_0, \delta)$. For such n, since

$$M \geq \int_\delta^{\delta^{1/2}} \int_0^{2\pi} r^{-1} |z_{n\theta}(r, \theta)|^2 \, dr \, d\theta,$$

it follows that there is an $\bar{r}_n$, $\delta \leq \bar{r}_n \leq \delta^{1/2}$ such that

$$\int_0^{2\pi} |z_{n\theta}(\bar{r}_n, \theta)|^2 \, d\theta \leq 2M/|\log\delta|$$

and $z_n(\bar{r}_n, \theta)$ is A.C. in θ. Let $B_{n1} = B_n \cup B(x_0, \bar{r}_n)$, B'_{n2} be the domain outside the C_{nj} interior to $B(x_0, \bar{r}_n)$, and B_{n2} be $\sigma_n(B'_{n2})$, where σ_n is the inversion in ∂C_{ni} followed by the reflection in the x^1 axis. Let H_{n1} be the harmonic function in $B(x_0, \bar{r}_n)$ coinciding on $\partial B(x_0, \bar{r}_n)$ with z_n and let H_{n2} be the transform of H_{n1} under the inversion in $\partial B(x_0, \bar{r}_n)$. Define

$$z_{n1}(x) = \begin{cases} z_n(x), & x \in B_n - B(x_0, \bar{r}_n) \\ H_{n1}(x), & x \in B(x_0, \bar{r}_n), \end{cases}$$

$$z'_{n2}(x) = \begin{cases} z_n(x), & x \in B'_{n2} \cap B(x_0, \bar{r}_n) \\ H_{n2}(x), & x \in B'_{n2} - B(x_0, \bar{r}_n) \end{cases}$$

$$z_{n2}(x) = z'_{n2}[\sigma_n^{-1}(x)], \quad x \in B_{n2}.$$

Then we see from Lemma 9.4.2 that

$$D(z_{n1}, B_{n1}) + D(z_{n2}, B_{n2}) \leq D(z_n, B_n) + 4M/|\log\delta|.$$

Since δ can be arbitrarily small, it follows that $d^*(\Gamma) \leq d(\Gamma)$.

Suppose case ii holds and let $x_0 = C_i$, $x_0 \in C_j$, and let z_0 be the image of x_0 under the limiting continuous map of C_j onto Γ_j defined by the function $\Phi_j(\theta)$ of Lemma 9.3.2 and equation (9.3.6) (recall that $1 \leq j \leq l < i$). We may choose a sequence $\{\delta_n\}$ in which each $\delta_n > 0$ and

$\delta_n \to 0$, and a sequence $\bar{r}_n$ with $\delta_n \leq \bar{r}_n \leq \delta_n^{1/2}$ such that $\partial B(x_0, \delta_n^{1/2})$ intersects only C_j and $B(x_0, \delta_n^{1/2})$ contains only those C_{qn} for which $C_q = x_0$, all of these being inside $B(x_0, \delta_n)$ and such that

$$\int_{B_n \cap \partial B(x_0, \bar{r}_n)} |z_{n\theta}(\bar{r}_n, \theta)|^2 \, d\theta \leq 2M/|\log \delta_n|. \tag{9.4.4}$$

Since the z_n converge uniformly along C_j we see that

$$\max|z_n(x) - z_0| \text{ for } x \in \gamma_n = \left[C_j \cap \overline{B(x_0, r_n)}\right] \cup [\partial B(x_0, \bar{r}_n) \cap B] < \varepsilon_n,$$

$$\lim \varepsilon_n = 0. \tag{9.4.5}$$

Let $R_n \to 0$ with $\varepsilon_n/R_n \to 0$ and define the mapping

$$\omega_n(z) = z_0 + \Phi_n(|z - z_0|) \cdot |z - z_0|^{-1} \cdot (z - z_0) \tag{9.4.6}$$

where $\Phi_n \in C^1(R_1)$, Φ_n is non-decreasing, and

$$\begin{aligned} &\Phi_n(s) = 0 \text{ for } s \leq \varepsilon_n \text{ and } \Phi(s) = s \text{ for } s \geq R_n, \\ &s^{-1}\Phi_n(s) \text{ and } \Phi_n'(s) \leq R_n/(R_n - 2\varepsilon_n) \text{ for all s}. \end{aligned} \tag{9.4.7}$$

If we then define

$$z_n'(x) = \omega_n[z_n(x)] \tag{9.4.8}$$

we see that (9.4.3) still holds and $z_n'(x) = z_0$ in an open set $\supset \gamma_n$. Consequently, for each n, there exists a simple closed curve $C_{k+1,n}$ of class C_μ^1 which is in this open set, is in B, and encloses the C_{pn} for which $C_p = x_0$. Let us define $B_{n1} = B_n \cup \underset{p}{U} C_{pn} \cup \underset{p}{U} B_{pn}$ where B_{pn} is the interior of C_{pn}, and let B_{n2} be the exterior of the C_{pn} and define

$$\begin{aligned} z_{n1}(x) &= \begin{cases} z_n'(x), & x \in B_{n1} \text{ and outside } C_{k+1,n} \\ z_0, & x \textit{ on and inside } C_{k+1,n}. \end{cases} \\ z_{n2}(x) &= \begin{cases} z_n'(x), & x \text{ inside } C_{k+1,n} \text{ and outside and on the } C_{pn}, \\ z_0, & x \text{ outside and on } C_{k+1,n}. \end{cases} \end{aligned} \tag{9.4.9}$$

Then, again

$$D(z_{n1}, B_{n1}) + D(z_{n2}, B_{n2}) = D(z_n', B_n) \tag{9.4.10}$$

and we see that $d^*(\Gamma) = d(\Gamma)$.

Finally, we suppose case iii holds; we assume that $1 \leq j \leq l$ and that the function Φ_j of Lemma 9.3.2 satisfies (9.3.7). Let x_0 be the point on C_j corresponding to θ_0 and let $z_0 = \Phi_j(\theta)$ for $\theta_0 < \theta < \theta_0 + 2\pi$. Then, as above, we may choose δ_n and $\bar{r}_n$ so that (9.4.4) holds, $\partial B(x_0, \delta_n^{1/2})$ intersects C_j only, and the C_{np} for which $|C_p| = x_0$ are all inside $B(x_0, \delta_n)$. This time (9.4.5) holds for all x on $\gamma_n = [C_j - B(x_0, \bar{r}_n)] \cup [\partial B(x_0, r_n) \cap B]$. We again choose R_n and define $\omega_n(z)$ by (9.4.6) and (9.4.7) and z_n' by (9.4.8) and see that (9.4.3) still holds and $z_n'(x) = z_0$ on an open set $\supset \gamma_n$. Consequently for each n, there exists a JORDAN curve $C_{k+1,n}$ of class C_μ^1 in this open set and in B_n which encloses C_j and the C_{pn} for which $C_p = x_0$ (if any). Then we may define $B_{n1} = B_n \cup C_j \cup \bigcup\limits_p C_{pn} \cup$ (their interiors) and B_{n2} as the exterior of C_j and C_{pn} and can define z_{n1} and z_{n2} as in (9.4.9) and find again that $d^*(\Gamma) = d(\Gamma)$.

Thus none of the degeneracies occur, $B_n = B$ for every n, and the z_n converge uniformly on ∂B. The result follows by replacing each z_n by the harmonic function with the same boundary values and using the lower-semicontinuity. The conformality follows from Lemma 9.4.1.

As was mentioned above, the author solved the PLATEAU problem, in the k-contour case, for surfaces situated in a "homogeneously regular" Riemannian manifold (MORREY [8]); these were defined in the first paragraph of § 9.3. We now present a simplified proof of this result. We shall assume that this manifold $\mathfrak{M}$ is of class C^1_μ, $0 < \mu < 1$, at least and that $\mathfrak{M}$ *is connected*. It follows that $\mathfrak{M}$ is separable. We shall consider vector functions z from various sets having as values points of $\mathfrak{M}$.

Notation. If z_1 and z_2 are two points in $\mathfrak{M}$, we let $|z_1 - z_2|$ be the *geodesic distance* between them, i.e. the inf of the lengths of all paths in $\mathfrak{M}$ joining them.

Definition 9.4.2. A vector function z defined on a measure space (E, μ) is said to be *measurable-μ* iff for every open set $\mathfrak{O} \subset \mathfrak{M}$, the subset of x in E where $z(x) \in \mathfrak{O}$ is measurable-μ.

Lemma 9.4.3. *If z_1 and z_2 are μ-measurable on E, then $|z_1(x) - z_2(x)|$ is measurable on E.*

Proof. If z is measurable-μ on E and S is a BOREL subset of $\mathfrak{M}$, the subset of x where $z(x) \in S$ is measurable. For each n, divide $\mathfrak{M}$ into at most a countable number of disjoint BOREL sets S_{nj}, each of diameter $< n^{-1}$; it is easy to see how to do this using coordinate patches. In each S_{nj}, choose a point z_{nj}. Let E_{njk} be the subset of x in E for which $z_1(x) \in S_{nj}$ and $z_2(x) \in S_{nk}$. For each n, the E_{njk} are disjoint and measurable and $E = \bigcup E_{njk}$ (n fixed). Define $\Phi_n(x) = |z_{nj} - z_{nk}|$ for $x \in E_{nkj}$. Then each Φ_n is measurable-μ on E and $\Phi_n(x) \to |z_1(x) - z_2(x)|$ for each $x \in E$.

Definition 9.4.3. A function z (into $\mathfrak{M}$) defined on a measure space (E, μ) is said to be *of class $L_p(E, \mu)$* $\Leftrightarrow$ z is measurable-μ on E and $|z(x) - z_0| \in L_p(E, \mu)$ $(p \geq 1)$ for each z_0 in $\mathfrak{M}$.

The following lemma is elementary and we omit its proof.

Lemma 9.4.4. *If z_1 and $z_2 \in L_p(E, \mu)$ then $|z_1(x) - z_2(x)| \in L_p(E, \mu)$. If we identify equivalent functions and define the distance*

$$l_p(z_1, z_2) = \left\{ \int_E |z_1(x) - z_2(x)|^p \, d\mu \right\}^{1/p},$$

the resulting space is a complete metric space ($p \geq 1$).

Lemma 9.4.5. *Suppose z is of class $L_p(E)$ for some measurable set E in R_ν, $\nu \geq 1$. Then there exists a function $\bar{z}$ defined almost everywhere on E and coinciding with z almost everywhere on E such that*

$$\lim_{e \to x_0} [m(e)]^{-1} \int_e |z(x) - \bar{z}(x_0)|\, dx = 0$$

for every x_0 where $\bar{z}(x_0)$ is defined and every regular family of sets $e \subset E$ about x_0.

Proof. Let $\{z_n\}$ be a dense subset of $\mathfrak{M}$. Since $|z(x) - z_n| \in L_1(E)$ for each n, $\exists$ a set Z with $m(Z) = 0 \ni$ if $x_0 \in E - Z$, then

$$\begin{aligned} &\lim_{h \to 0} (2h)^{-\nu} \int_{E \cap R_h} |z(x) - z_n|\, dx = |z(x_0) - z_n|. \\ &\lim_{e \to x_0} [m(e \cap E)/m(e)] = 1 \end{aligned} \tag{9.4.11}$$

for every n and every regular family of sets about x_0. It follows easily that if $x_0 \in E - Z$, then (9.4.11) holds with z_n replaced by any z_0.

Definition 9.4.4. A vector $z \in C^1_\varrho(\bar{G})$ $(G \subset R_\nu,\ 0 \leq \varrho \leq \mu) \Leftrightarrow$ for each x_0 in $\bar{G}$, the components z^i with respect to a coordinate patch having range $\supset z(x_0)$ are of class C^1_ϱ in some neighborhood of x_0 on $\bar{G}$, A vector z is *absolutely continuous* (A.C.) on $[a, b] \Leftrightarrow$ for each $\varepsilon > 0$, $\exists$ a $\delta > 0$ such that

$$\sum_j |z(x_j'') - z(x_j')| \leq \varepsilon \quad \text{whenever} \quad \sum_j (x_j'' - x_j') \leq \delta, \tag{9.4.12}$$

and the intervals (x_j', x_j'') are disjoint. We say that *z is of class $\bar{H}^1_p(\bar{G})$* $\Leftrightarrow$ z is continuous on $\bar{G}$, z is A.C. in each x^α for almost all values of the other variables, and the partial derivatives $|z_{,\alpha}|$ (i.e. the derivatives of the arc-length along $x'_\alpha =$ const. with respect to x^α) $\in L_p(G)$.

Lemma 9.4.6. z is A. C. *on $[a, b]$ (of class $\bar{H}^1_p(\bar{G})$) $\Leftrightarrow$ for each x_0 on $[a, b]$ (x_0 on $\bar{G}$), the components z^i with respect to a coordinate patch with range $\supset z(x_0)$ are* A.C. *(of class $\bar{H}^1_p$) in some neighborhood of x_0 on $[a, b]$ $(\bar{G})$. If z is of class $\bar{H}^1_p(\bar{G})$ the quantities*

$$G_{\alpha\beta}(x) = g_{ij}[z(x)]\, z^i_{,\alpha}(x)\, z^j_{,\beta}(x), \quad |z_{,\alpha}(x)| = \sqrt{G_{\alpha\alpha}(x)} \tag{9.4.13}$$

*$\in L_{p/2}(G)$ and are independent of the local (z)-coordinates used. If z is of class $\bar{H}^1_p(\bar{G})$ and $x = x(y)$ is a $1-1$ bi-*LIPSCHITZ *map of Γ onto $\bar{G}$, then $w(y) = z[x(y)]$* is *of class $\bar{H}^1_p(\Gamma)$ and the derivatives transform in the ususal way.*

Notation. For vectors defined on domains in R_2, we use the notations E, F, and G instead of G_{11}, G_{12}, and G_{22}, or perhaps,

$$|z_x|^2 = E, \quad z_x \cdot z_y = F, \quad |z_y|^2 = G. \tag{9.4.14}$$

Definition 9.4.5. We carry over the definitions of $D(z_1, z_2)$, oriented and non-oriented FRÉCHET variety, parametric representation, etc. as given in Definitions 9.1.1, 9.1.2, and 9.1.3. For a vector $z \in C^1(\bar{G})$, G of class C^1, we define

$$(9.4.15)\quad A(z, G) = \iint_G \sqrt{EG - F^2}\, dx\, dy, \quad D(z, G) = \iint_G (E + G)\, dx\, dy,$$

E, F, and G being defined locally by (9.4.13) and (9.4.14). Then, for $z \in C^0(\bar{G})$, we define $L(z, G)$ as in (9.1.11) and $L(S)$ as in Definition 9.1.5.

Lemma 9.4.7. *The theorems of § 9.1 generalize in the obvious way to vectors z and surfaces S with ranges in $\mathfrak{M}$. We interpret the space $H_p^1(G) \cap C^0(\bar{G})$ to be our new space $\bar{H}_p^1(\bar{G})$. Lemmas 9.3.1—9.3.7 and their corollaries and Theorem 9.3.1 all hold for vectors z with ranges in $\mathfrak{M}$.*

Definition 9.4.6. We define $l(\Gamma)$, $d(\Gamma)$, $l^*(\Gamma)$, and $d^*(\Gamma)$ as in Definition 9.4.1; in the definitions of $d(\Gamma)$ and $d^*(\Gamma)$, we require that the $z_n \in \bar{H}_2^1(\bar{G}_n)$.

Lemma 9.4.8. (a) *Lemma 9.4.1 holds.*

(b) *Instead of Lemma 9.4.2, the following is true: Suppose ζ is A.C. on $\partial B(x_0, R)$ with*

$$(9.4.16)\qquad \int_0^{2\pi} |\zeta'(\theta)|^2\, d\theta \leq m/\pi.$$

Then $\exists$ a vector H of class $\bar{H}_2^1[\overline{B(x_0, R)}]$ which coincides with ζ on $\partial B(x_0, R)$ for which

$$(9.4.17)\qquad D[H, B(x_0, R)] \leq K \cdot \int_0^{2\pi} |\zeta'(\theta)|^2\, d\theta, \quad K = M/m.$$

(c) *Instead of Theorem 9.4.1, the following is true: Suppose that Γ satisfies the hypotheses of Theorem 9.4.1. Then $\exists$ a circular region B of type k and a sequence $\{z_n\}$ of vectors of class $\bar{H}_2^1(\bar{B})$ such that $D(z_n, B) \to d(\Gamma)$ and the restrictions of z_n to the properly oriented C_i converge uniformly along C_i to continuous representatives of Γ_i, $i = 1, \ldots, k$.*

Proof. (a) is evident except for the local regularity. But here again we may embed $\mathfrak{M}$ in the manifold $\mathfrak{M} \times R_2$, the metric in any local coordinate system being given by

$$ds^2 = \sum_{i,j=1}^{N} g_{ij}(z^1, \ldots, z^N)\, dz^i\, dz^j + \bar{m}[(dz^{N+1})^2 + (dz^{N+2})^2],$$
$$\bar{m} = (m + M)/2,$$

the relation between any two coordinate systems being

$$'z^i = f^i(z^1, \ldots, z^N), \quad i = 1, \ldots, N, \quad 'z^i = z^i, \quad i = N + 1, N + 2.$$

the f^i being defined as for $\mathfrak{M}$. Clearly $\mathfrak{M}'$ is homogeneously regular with the same numbers m and M.

(b) Extend ζ to be periodic in θ and choose θ_0. There is a θ_1 with $|\theta_1 - \theta_0| \leq \pi$ such that $|\zeta(\theta) - \zeta(\theta_0)| \leq |\zeta(\theta_1) - \zeta(\theta_0)|$ for all θ.

From (9.4.16), it follows that

$$|\zeta(\theta_1) - \zeta(\theta_0)|^2 < \pi \int_0^{2\pi} |\zeta'(\theta)|^2 \, d\theta \leq m. \tag{9.4.18}$$

Now, since $\mathfrak{M}$ is homogeneously regular, we may choose a special coordinate patch in which $\zeta(\theta_0)$ is the origin and

$$m\,|\xi|^2 \leq g_{ij}(z)\,\xi^i \xi^j \leq M\,|\xi|^2, \quad z \in B(0,1). \tag{9.4.19}$$

From (9.4.19), it follows that for any path $z^i = z^i(s)$ (s arc length),

$$\int_0^l \sqrt{g_{ij}[z(s)]\, z_s^i z_s^j}\, ds \geq \sqrt{m} \int_0^l |z_s|\, ds \geq l\sqrt{m}$$

so that the range of this patch contains all points within a distance $\sqrt{m}$ of $\zeta(\theta_0)$ and hence all the $\zeta(\theta)$. So, if we define H as that vector on $\overline{B(x_0, R)}$ whose components H^i with respect to this patch are the harmonic functions coinciding with $\zeta^i(\theta)$ on $\partial B(x_0, R)$, we see, using Lemma 9.4.2, that

$$D[H, B(x_0, R)] = \int\!\!\int_{B(x_0,R)} g_{ij}[H(x,y)]\,(H_x^i H_x^j + H_y^i H_y^j)\, dx\, dy$$

$$\leq M \sum_i \int\!\!\int_{B(x,R)} |\nabla H^i|^2\, dx\, dy \leq M \int_0^{2\pi} \sum_i |\zeta_\theta^i|^2\, d\theta \leq K \int_0^{2\pi} |\zeta_\theta|^2\, d\theta.$$

(c) The proof of Theorem 9.4.1 carries over right down to the last paragraph. The mappings ω_n are defined as follows: Since $\mathfrak{M}$ is of class C^1, there exists a coordinate patch (Z) in which z_0 corresponds to the origin in the Z-space and $g_{ij}(0) = \delta_{ij}$. Then, for n sufficiently large we define ω_n as the identity outside the range of that patch and define it as in (9.4.6) and (9. 4.7) in terms of the Z coordinates inside the patch.

In order to complete the proof of Theorem 9.4.1 for surfaces in $\mathfrak{M}$, the author found it expedient to introduce a theory of functions of class $H_p^1(G)$ with values in $\mathfrak{M}$. Since $\mathfrak{M}$ is not a linear space, some of the theorems of chapter 3 do not make sense and new proofs have to be found for most of the remaining theorems. A useful tool is the well-known device, used extensively by E. SILVERMAN, of mapping $\mathfrak{M}$ isometrically into the space m of all bounded sequences. Although this space is linear, it is not separable and so this device cannot be used to prove all theorems.

Remark. Many years ago, H. WHITNEY [1] proved that any manifold $\mathfrak{M}$ of class C^n could be mapped by a diffeomorphism Φ of class C^n onto an analytic manifold $\mathfrak{M}'$ in some R_P (actually P may be taken $= 2N+1$). If $\mathfrak{M}$ is a Riemannian manifold and this mapping could be done in such a way that there were numbers m_1 and M_1 with $0 < m_1 \leq M_1$ such that the *geodesic distances* along $\mathfrak{M}$ and $\mathfrak{M}'$ satisfied

$$m_1\,|p - q| \leq |\Phi(p) - \Phi(q)| \leq M_1\,|p - q|,$$

it would then follow that a vector $z(x)$ would be A.C. on $[a, b] \Leftrightarrow X(x) = \Phi[z(x)]$ is A.C. on $[a, b]$ and we would have

$$m_1 |z'(x)| \leq |X'(x)| \leq M_1 \cdot |z'(x)|.$$

In this case, it would be sufficient to define z to be of class $H_p^1(G) \Leftrightarrow \Phi(z) \in H_p^1(G)$; then the whole theory of these functions would be available. Actually, J. NASH ([1], [2]) has proved that Φ can be taken to be isometric (i.e. $m_1 = M_1 = 1$) if P is sufficiently large. But all these facts require proof and the last-mentioned theorem is somewhat difficult. Therefore, in order to keep the exposition relatively self-contained, we present an alternate development of the theory of H_p^1 spaces using SILVERMAN's idea and terminating with the DIRICHLET growth Lemma 9.4.18.

Definition 9.4.7. A vector function $X(x)$ $(=\{X^1(x), X^2(x), \ldots\})$ into m is A.C. $\Leftrightarrow$ (9.4.12) holds with z replaced by X; here $|X(\beta) - X(\alpha)| = \sup_i |X^i(\beta) - X^i(\alpha)|$. $X \in L_p(G)$, $G \subset R_\nu$, $\Leftrightarrow$ each X^i is measurable on G and $|X(x)| = \sup_i |X^i(x)| \in L_p(G)$.

Lemma 9.4.9. *Suppose X is* A. C. *on m and $\Phi(x)$ is the length of the arc $X = X(t)$, $a \leq t \leq x$. Then Φ is* A.C. *on $[a, b]$ and*

$$\Phi'(x) = \sup_i |X_x^i(x)| \quad \text{a.e.} \tag{9.4.20}$$

If the X^i are A.C. on $[a, b]$ and $\sup|X_x^i(x)| \in L_1([a, b])$, *then X is A.C. on $[a, b]$.*

Proof. (a) That Φ is A.C. is immediate. Then if $a \leq x < b$ and $0 < h < b - x$ and $\Phi'(x)$ and all the $X_a^i(x)$ exist, then

$$h^{-1}[\Phi(x + h) - \Phi(x)] \geq h^{-1} |X^i(x + h) - X^i(x)|, \quad i = 1, 2, \ldots$$

from which it follows that $\Phi'(x) \geq \sup_i |X_x^i(x)|$. On the other hand if the X^i are A.C. on $[a, b]$ and $\sup_i |X_x^i(x)| \in L_1([a, b])$, then

$$|X^i(\beta) - X^i(\alpha)| \leq \int_\alpha^\beta |X_x^i(x)|\, dx \leq \int_\alpha^\beta \sup_i |X_x^i(x)|\, dx$$

from which the absolute continuity follows easily.

Definition 9.4.8. *A vector function z from $G \subset R$ into $\mathfrak{M}$ is of class $H_p^1(G) \Leftrightarrow$ (i) $z \in L_p(G)$,* (ii) z is equivalent to a function z_0 which is A.C. in each variable x^α for almost all values of the others, and (iii) the partial derivatives $|z_{0,\alpha}(x)| \in L_p(G)$. A vector X from G into m is of class $H_p^1(G) \Leftrightarrow$ each $X^i \in H_p^1(G)$ and if $|X(x)| = \sup_i |X^i(x)|$ and $|X_{,\alpha}(x)| = \sup_i |X_{,\alpha}^i(x)| \in L_p(G)$. Let $\{z_n\}$ be a countable dense subset of $\mathfrak{M}$; we define the mapping τ from $\mathfrak{M}$ into m by $\tau(z) = \{|z - z_1|, |z - z_2|, \ldots\}$.

Lemma 9.4.10. (a) *If $z \in H_p^1(G)$, then $z \in H_p^1(D)$ for each $D \subset G$.*

(b) *If $z \in H_p^1(D)$ for each $D \subset\subset G$ and z and $|z_{0,\alpha}| \in L_p(G)$,* then $z \in H_p^1(G)$.

(c) *The mapping* τ *is isometric.*

(d) *A vector* z *into* $\mathfrak{M}$ *is A.C.* $\Leftrightarrow$ *the vector* $X = \tau z$ *into* m *is* A.C. *A vector* z *into* $\mathfrak{M} \in H_p^1(G) \Leftrightarrow X = \tau z \in H_p^1(G)$.

(e) *If* $z \in H_p^1(G)$, $x = x(y)$ *is a bi-Lipschitz map from* Γ *onto* G, *and* $w(y) = z[x(y)]$, *then* $w \in H_p^1(\Gamma)$.

(f) *If* $z \in H_p^1(G)$, *then* z *is equivalent to a function* $\bar{z}$ *which has the absolute continuity properties of the* z_0 *in definition* 9.4.8 *and, in addition, if* $x = x(y)$ *is a bi-Lipschitz map of* $\bar{\Gamma}$ *onto* $\bar{G}$ *and we define* $\bar{w}(y) = \bar{z}[x(y)]$, *then* $\bar{w}$ *also possesses those absolute continuity properties. In fact* $\bar{z}$ *may be chosen as the function of Lemma* 9.4.5.

Proof. (c) Clearly $\| z' - z_n| - |z'' - z_n\| = |X^n(z') - X^n(z'')| \leq |z' - z''|$ for every n. But, by taking z_k near z' and z_l near z'', we see that $|X(z') - X(z'')| = \sup_i |X^i(z') - X^i(z'')|$. (a) and (b) are evident. (d) follows from (c) and Theorems 3.1.2 (g) and Lemma 3.1.1.

To prove (e), it is sufficient to use (d) and then Theorem 3.1.7 for each component X^i. If we let $Y^i(y) = X^i[x(y)]$, we have

$$|Y_{,\alpha}(y)| = \sup_i |X^i_{,\beta}[x(y)] \cdot x^\beta_{,\alpha}(y)| \leq M \sum_\beta \{\sup_i |X^i_{,\beta}[x(y)]|\},$$
$$M = \sup_\beta |x^\beta_{,\alpha}(y)|.$$

Clearly each $|Y_{,\alpha}| \in L_p(\Gamma)$.

Part (f) follows from (e) and Theorem 3.1.8 for each X^i. The last statement in (f) must be proved by returning to the vector z in $\mathfrak{M}$ and repeating part of the proof of Theorem 3.1.8.

It is exceedingly important to obtain the usual formulas for the derivatives in the change of variable theorem. We prove what is necessary in the next lemma.

Lemma 9.4.11. *Let* $z(x, y) \in H_p^1(G)$, *let* $\bar{z}(x, y)$ *be the equivalent function of Lemma* 9.4.10(f), *let* E, F, *and* G *be defined by* (9.4.13) *and* (9.4.14) *in terms of the partial derivatives of the* $\bar{z}^i$, *let* $\sigma: x = x(s, t)$, $y = y(s, t)$ *be a bi-Lipschitz map of* Γ *onto* G, *and let* $\bar{w}(s, t) = \bar{z}[x(s, t), y(s, t)]$. *Then*

$$|\bar{w}_s(s, t)|^2 = E(x, y)\, x_s^2 + 2F\, x_s, y_s + G\, y_s^2$$
$$|\bar{w}_t(s, t)|^2 = E(x, y)\, x_t^2 + 2F\, x_t\, y_t + G\, y_t^2 \quad \text{(a.e.)},$$

$|\bar{w}_s|^2$ *and* $|\bar{w}_t|^2$ *being defined similarly in terms of the partial derivatives of the* $\bar{w}^i$.

Proof. Since the expressions for E, F, G, $|\bar{w}_s|^2$, and $|\bar{w}_t|^2$ are independent of the coordinate systems (z) on $\mathfrak{M}$, we confine ourselves to the range $\mathfrak{R}$ of an admissible coordinate patch. Let S and S' be the respective sets of (x, y) and (s, t) where $\bar{z}(x, y)$ and $\bar{w}(s, t) \in \mathfrak{R}$; clearly $S = \sigma(S')$. If $\bar{z}(x_0, y)$ and $\bar{z}(x, y_0)$ are A.C., and $(x_0, y_0) \in S$, then $\exists$ an $\alpha(x_0, y_0) \ni (x_0, y)$ and $(x, y_0) \in S$ for all x and $y \ni |x - x_0| < \alpha$ and $|y - y_0| < \alpha$. Let $\varepsilon > 0$. There exists a compact subset $\sum$ of S such that all the functions z^i, z^i_x, and z^i_y are defined and continuous on $\sum$, w^i, w^i_s,

w_t^i, x_s, y_s, x_t, and y_t are all continuous on $\sum' = \sigma^{-1}(\sum)$, $m(S - \sum) < \varepsilon$, and

$$\bar{z}^i(x+h, y) - \bar{z}^i(x, y) = [\bar{z}_x^i(x, y) + \varepsilon_1^i(x, y, h)]\, h$$

$$\bar{z}^i(x, y+h) - \bar{z}^i(x, y) = [\bar{z}_y^i(x, y) + \varepsilon_2^i(x, y, h)]\, h$$

whenever the points involved are in $\sum$; we also have

$$|\varepsilon_1^i(x, y, h)|, \quad |\varepsilon_2^i(x, y; h)| \leq \varepsilon(h), \quad \lim_{h\to 0} \varepsilon(h) = 0, \quad (x, y) \in \textstyle\sum.$$

Now, further, if (x_0, y_0) is not in a subset Z of $\sum$ of measure 0, the linear metric density at (x_0, y_0) of the part of $\sum$ on the lines $x = x_0$ and $y = y_0$ is 1 and the linear metric density at $(s_0, t_0) = \sigma^{-1}(x_0, y_0)$ of the part of $\sum'$ on the lines $s = s_0$ and $t = t_0$ is 1. Consequently the projection of $\sigma[\sum'(t_0)]$ on at least one of the lines $x = x_0$ of $y = y_0$ has metric density 1 at (x_0, y_0); here $\sum'(t_0)$ denotes the part of $\sum'$ on the line $t = t_0$. If that on $y = y_0$ has that property, the intersection of $\sum(y_0)$ and the projection $P_{y_0}[\sigma\{\sum'(t_0)\}]$ still has metric density unity at (x_0, y_0) and so the subset Δ of $\sum'(t_0)$ such that $P_{y_0}[\sigma(\Delta)]$ lies in the intersection above still has metric density 1 at (s_0, t_0). The usual relations

$$\bar{w}_s^i(s_0, t_0) = \bar{z}_x^i(x_0, y_0)\, x_s(s_0, t_0) + \bar{z}_y^i(x_0, y_0)\, y_s(s_0, t_0)$$

follow by letting $s \to s_0$ on Δ. The formula for $\bar{w}_t$ and hence those in the theorem follow similarly. The lemma follows since ε is arbitrary.

Definition 9.4.9. A domain $G \subset R_\nu$ is said to be *of class D'* $\Leftrightarrow$ it is the union of a finite number of diffeomorphic images of ν-simplices which are joined together as are the simplices in a simplicial manifold.

Lemma 9.4.12. *Suppose $z \in H_p^1(G)$, G being of class D'. Then the function $\bar{z}$ of Lemmas 9.4.5 and 9.4.10(f) is defined almost everywhere on ∂G and $\in L_p(\partial G)$. If $\bar{G}$ is the cell $[a, b]$, then*

$$\lim_{x^\alpha \to a^\alpha +} \int_{a'_\alpha}^{b'_\alpha} |\bar{z}(x^\alpha, x'_\alpha) - \bar{z}(a^\alpha, x'_\alpha)|^p\, dx'_\alpha = 0, \quad \alpha = 1, \ldots, \nu. \tag{9.4.21}$$

If $x'_{0\alpha} \notin Z_\alpha$, $m(Z_\alpha) = 0$, then the line $x'_\alpha = x'_{0\alpha}$ intersects ∂G in a finite number of points, at each of which ∂G has a tangent plane which does not contain the line, and $\bar{z}(x^\alpha, x'_{0\alpha})$ is A.C. *in x^α on each closed segment of the line in $\bar{G}$.*

Proof. In order to prove the first two statements, it is sufficient to prove them for a cell $[a, b]$, on account of Lemmas 9.4.10(f) and (9.4.11). In this case, choose α and extend $\bar{z}$ across the face $x^\alpha = a^\alpha$ by reflection obtaining z_0 with the usual A.C. properties. Then $\bar{z}_0$ is defined almost everywhere on $x^\alpha = a^\alpha$ and $\bar{z}_0(a^\alpha, x'_\alpha) = \lim_{x^\alpha \to a^\alpha +} \bar{z}(x^\alpha, x'_\alpha)$ a.e. Then (9.4.21) follows from equation (3.4.10) which reduces in our case to

$$\int_{a'_\alpha}^{b'_\alpha} |\bar{z}(x^\alpha, x'_\alpha) - \bar{z}(a^\alpha, x'_\alpha)|^p\, dx'_\alpha \leq |x^\alpha - a^\alpha|^{p-1} \int_{a^\alpha}^{x^\alpha} \int_{a'_\alpha}^{b'_\alpha} |\bar{z}_{,\alpha}(x)|^p\, dx. \tag{9.4.22}$$

The first part of the last statement (about ∂G) is well known (see Lemma 9.2.2). The rest follows, since if $x'_{0\alpha}$ is not in Z_α, a neighborhood $\mathfrak{N}$ of a point $x_0 = (x_0^\alpha, x'_{0\alpha})$ on ∂G can be mapped by a diffeomorphism of the form

$$y'_\alpha = x'_\alpha, \quad y^\alpha = f_\alpha(x^\alpha, x'_\alpha), \quad x'_\alpha \in [a'_\alpha, b'_\alpha]$$

onto a cell in such a way that $\partial G \cap \mathfrak{N}$ corresponds to a plane $y^\alpha = \text{const}$.

The following substitution lemma is an immediate consequence of Lemma 9.4.12:

Lemma 9.4.13. *Suppose G and Δ are of class D' with $\Delta \subset G$, suppose $z \in H_p^1(G)$, $w \in H_p^1(\Delta)$, $\overline{w}(x) = \overline{z}(x)$ a.e. on $\partial\Delta$, and $Z(x) = \overline{w}(x)$ on Δ and $Z(x) = \overline{z}(x)$ on $\bar{G} - \Delta$. Then $Z \in H_p^1(G)$ and $\bar{Z}(x) = \bar{z}(x)$ a.e. on $\partial\Delta \cup \partial G$.*

Lemma 9.4.14. *Suppose $\{z\}$ is a family of functions in $H_p^1(G)$, G Lipschitz, such that $D_p(z, G)$ is uniformly bounded. Then*

(i) *if $\int_G |z(x) - z_0|^p\,dx$ is uniformly bounded, so is $\int_{\partial G} |z(x) - z_0|^p\,dS$;*

(ii) *if $\int_\tau |z(x) - z_0|^p\,dx$ or $\int_\sigma |z(x) - z_0|^p\,dS$ is uniformly bounded for some cell $\tau \subset G$ or some open set σ on ∂G, then $\int_G |z(x) - z_0|^p\,dx$ is uniformly bounded.*

Proof. Since $\bar{G}$ is the union of a finite number of bi-Lipschitz images of cells, it is sufficient to prove this when G and σ are cells. Then (i) follows immediately from (9.4.22). Moreover, if the integral over σ (part of a face of ∂G) is uniformly bounded, (9.4.22) shows that the integral over a cell τ adjacent to σ is also. If $\tau = [\gamma, \delta]$, we see by applying (9.4.22) with $\alpha = 1$ that the integral over $\tau_1 : a' \le x' \le b'$, $x'_1 \in [\gamma'_1, \delta'_1]$ is uniformly bounded. The result (ii) follows by applying this procedure for $\alpha = 2, 3, \ldots, \nu$ in turn.

Definition 9.4.9. We say that z_n *tends weakly to* $z (z_n \rightharpoonup z)$ in $H_p^1(G)$ $\Leftrightarrow$ each z_n and $z \in H_p^1(G)$, $D_p(z_n, G)$ is uniformly bounded and $z_n \to z$ in $L_p(G)$; here we assume G of class C_1^0. In case $p = 1$, we assume also that the set functions $D_1(z_n, e)$ are uniformly absolutely continuous.

Lemma 9.4.15. *If G is of class C_1^0 and $z_n \rightharpoonup z$ in $H_p^1(G)$, then $z_n \to z$ in $L_p(\partial G)$. Moreover $D_p(z, G) \le \liminf D_p(z_n, G)$.*

Proof. It is clearly sufficient to prove this for G a cell. To prove the convergence in L_p of $z_n(a^\alpha, x'_\alpha)$ to $z(a^\alpha, x'_\alpha)$, we let $\{r\}$ be a subsequence of $\{n\}$ and let $\varepsilon > 0$. We may choose a subsequence $\{z_s\}$ such that $\bar{z}_s(x^\alpha, x'_\alpha) \to \bar{z}(x^\alpha, x'_\alpha)$ in $L_p[a'_\alpha, b'_\alpha]$ for almost all x^α. From (9.4.22) and the weak convergence (and the uniform absolute continuity in the case $p = 1$), it follows that we may choose an $\bar{x}^\alpha$ such that $\bar{z}_s(\bar{x}^\alpha, x'_\alpha) \to \bar{z}(\bar{x}^\alpha, x'_\alpha)$ in L_p and the integrals on the left in (9.4.22) (with $\bar{z} = \bar{z}$ or

$\bar{z}_s$ and $x^\alpha = \bar{x}^\alpha$) are all $< (\varepsilon/3)^p$. It follows easily that

$$\left\{ \int_{a'_\alpha}^{b'_\alpha} |\bar{z}_s(a^\alpha, x'_\alpha) - \bar{z}(a^\alpha, x'_\alpha)|^p \, dx \right\}^{1/p} < \varepsilon, \quad s > S(\varepsilon).$$

Thus $\bar{z}_s(a^\alpha, x'_\alpha) \to \bar{z}(a^\alpha, x'_\alpha)$ and, since $\{r\}$ was any subsequence, it follows that $\bar{z}_n(a^\alpha, x'_\alpha) \to \bar{z}(a^\alpha, x'_\alpha)$ in L_p.

To prove the lower semicontinuity, choose a subsequence $\{z_r\}$ so $D_p(z_n, G) \to \liminf_n D_p(z_n, G)$, and so that $\bar{z}_r(x^\alpha, x'_{0\alpha}) \to \bar{z}(x^\alpha, x'_{0\alpha})$ in $L_p[a^\alpha, b^\alpha]$ for almost all $x'_{0\alpha}$, $\alpha = 1, \ldots, \nu$. In case

$$\int_{a^\alpha}^{b^\alpha} |\bar{z}_{r,\alpha}(x^\alpha, x'_{0\alpha})|^p \, dx^\alpha$$

is uniformly bounded for some subsequence, the convergence of $\bar{z}_r$ to $\bar{z}$ along the line is uniform, so that

$$\int_{a^\alpha}^{b^\alpha} |\bar{z}_{,\alpha}(x^\alpha, x'_{0\alpha})|^p \, dx^\alpha \leq \liminf_r \int_{a^\alpha}^{b^\alpha} |\bar{z}_{r,\alpha}(x^\alpha, x'_{0\alpha})|^p \, dx^\alpha; \tag{9.4.23}$$

of course this holds if the right side $= +\infty$. The result follows by integration with respect to $x'_{0\alpha}$ and summing on α.

Lemma 9.4.16. *Suppose $z_n \in H_p^1(G)$ for each $n, p > 1$, G is of class C_1^0, and $D_p(z_n, G)$ and $\iint_G |z_n(x, y) - z_0|^p \, dx \, dy$ are uniformly bounded. Then a subsequence converges weakly on G to some z in $H_p^1(G)$.*

Proof. It is sufficient to prove this for the cell G: $a \leq x \leq b$, $c \leq y \leq d$. Let U_{nk} be the set of y such that $\bar{z}_n(x, y)$ is A.C. and

$$\int_a^b [|\bar{z}_n(x, y) - z_0|^p + |\bar{z}_{nx}(x, y)|^p] \, dx \leq k. \tag{9.4.24}$$

Then $m(U_{nk}) \geq (d - c) - M/k$, M being the obvious bound. Let

$$U_k = \bigcap_{N=1}^{\infty} \bigcup_{n=N}^{\infty} U_{nk}, \qquad U = \bigcup_{k=1}^{\infty} U_k.$$

Then U_k consists of all y such that (9.4.24) holds for infinitely many n and U consists of all y for which (9.4.24) holds for infinitely many n and some k. Clearly $m(U) = d - c$ and if $\bar{y} \in U$, then the $\bar{z}_n(x, \bar{y})$ are equicontinuous with $|\bar{z}_n(x, \bar{y}) - z_0|$ uniformly bounded (for some subsequence). Thus we may find a subsequence $\bar{z}_{1s}$ converging uniformly on $[a, b]$ to $\zeta_1(x)$ for $y = y_1$, y_1 being in the middle third of $[c, d]$. By repeating the whole argument on the subsequence $\bar{z}_{1s}$, we can find a second subsequence $\bar{z}_{2s}$ converging uniformly to $\zeta_2(x)$ on $y = y_2$ in the middle third of $[c, y_1]$. By continuing this and then taking the diagonal sequence, we obtain a sequence $\bar{z}_s$ such that $\bar{z}_s(x, y_q)$ converges uniformly to some $\zeta_q(x)$ for each y_q, y_q being dense in $[c, d]$.

We now show that $\{\bar{z}_s\}$ is a Cauchy sequence. Let $\varepsilon > 0$. From (9.4.22), it follows that there is a finite subset $T(\varepsilon)$ of the y_q such that if $c \leq y \leq d$, there is a y_q in $T(\varepsilon)$ such that

$$\int_a^b |\bar{z}_s(x, y) - \bar{z}_s(x, y_q)|^p \, dx < (\varepsilon/4)^p, \quad s = 1, 2, \ldots \tag{9.4.25}$$

Moreover, $\exists$ an $S(\varepsilon) \ni$

$$\int_a^b |\bar{z}_s(x, y_q) - \xi_q(x)|^p \, dx < (\varepsilon/4)^p, \quad s > S(\varepsilon), \quad q \in T(\varepsilon). \tag{9.4.26}$$

Combining (9.4.25) and (9.4.26) for s and t, we find that

$$\int_a^b |\bar{z}_s(x, y) - \bar{z}_t(x, y)|^p \, dx < \varepsilon^p, \quad c \leq y \leq d$$

from which the result follows.

Lemma 9.4.17. *Suppose $z \in H_2^1[B(p_0, a)]$, $p_0 = (x_0, y_0)$, with*

$$D[z, B(p_0, r)] \leq k^2(r), \quad \int_0^r \varrho^{-1} k(\varrho) \, d\varrho = K(r) < \infty, \quad 0 \leq r \leq a.$$

Then $\bar{z}(x_0, y_0)$ is defined and if $\overline{w}(r, \theta) = \bar{z}(x_0 + r\cos\theta, y_0 + r\sin\theta)$, then $\overline{w}$ is A.C. *in r on $[0, a]$ for almost all θ and*

$$|\bar{z}(x, y) - \bar{z}(x_0, y_0)| \leq r \int_0^1 l[(1-t)\, x_0 + t\, x, (1-t)\, y_0 + t\, y] \, dt,$$

$$l(x, y) = [E(x, y) + G(x, y)]^{1/2}, \quad r^2 = (x - x_0)^2 + (y - y_0)^2$$

for almost all (x, y).

Proof. From the change of variable theorem, we conclude that if we let $L(r, \theta) = l(x_0 + r\cos\theta, y_0 + r\sin\theta)$, then

$$L(r, \theta) \geq 0, \quad L^2(r, \theta) = |\overline{w}_r(r, \theta)|^2 + r^{-2} |\overline{w}_\theta(r, \theta)|^2.$$

Define

$$h(r) = \int_0^r \int_0^{2\pi} \varrho^{1/2} L(\varrho, \theta) \, d\varrho \, d\theta.$$

Then

$$h(r) \leq (2\pi)^{1/2} r^{1/2} k(r)$$

$$\int_0^r \int_0^{2\pi} L(\varrho, \theta) \, d\varrho \, d\theta = \int_0^r \varrho^{-1/2} h'(\varrho) \, d\varrho$$

$$= r^{-1/2} h(r) + \frac{1}{2} \int_0^r \varrho^{-3/2} h(\varrho) \, d\varrho < (2\pi)^{1/2} [k(r) + K(r)] \to 0$$

as $r \to 0$. Thus, since $L(r, \theta) \geq |\overline{w}_r(r, \theta)|$, $r^{-1} |\overline{w}_\theta(r, \theta)|$, it follows that $\overline{w}$ is A.C. in r on $[0, a]$ for almost all θ with $\overline{w}(0, \theta)$ in L_1 and there exists a sequence $r_n \to 0 \ni \overline{w}_\theta(r_n, \theta) \to 0$ in L_1 so that $\overline{w}(0, \theta)$ is a constant. The result follows easily.

Lemma 9.4.18 (DIRICHLET growth lemma). *Theorem 3.5.2 holds in the case $p = \nu = 2$ and u replaced by z.*

Sketch of proof. For if (x_1, y_1) and (x_2, y_2) are any points in $B(x_0, y_0; R)$, $\bar{z}(x_1, y_1)$ and $\bar{z}(x_2, y_2)$ are both defined and the result of Lemma 9.4.17 may be used for $(x_0, y_0) = (x_1, y_1)$ and (x_2, y_2) in turn and almost all (x, y). Then the proof of Theorem 3.5.2 carries over with $|\nabla u|$ replaced by L.

We can now prove our version of the DIRICHLET problem:

Theorem 9.4.2. *Suppose $z^* \in H_2^1(G)$ and G is of class C_1^0. Then there exists a $z \in H_2^1(G)$ such that $\bar{z} = \bar{z}^*$ a.e. on ∂G and minimizes $D(Z, G)$ among all such vectors. If $\mathfrak{M}$ is of class C^3, then $z \in C_{\mu'}^2(\bar{D})$ for every $D \subset\subset G$ and every μ', $0 < \mu' < 1$. If $\mathfrak{M}$ is of class C_μ^n for some $n \geq 3$, then $z \in C_\mu^n(\bar{D})$ for such D. If $\mathfrak{M}$ is of class C^∞ or analytic, so is z on each such D. If G is a circular region and z^* is continuous along ∂G, then z is continuous on $\bar{G}$.*

Proof. The existence of a minimizing z is immediate, using a minimizing sequence and Lemmas 9.4.14, 9.4.16, and 9.4.15. Now, suppose $B(x_0, y_0; a) \subset B$ and let $\bar{w}(r, \theta) = \bar{z}(x_0 + r\cos\theta, y_0 + r\sin\theta)$. For almost all r, $\bar{w}$ is A.C. in θ with $|\bar{w}_\theta|$ in L_2. Let

$$\Phi(r) = D[z, B(x_0, y_0; r)], \quad L = \max(M/m, \pi\, d(z^*, B)/m).$$

Then, from Lemmas 9.4.8(b) and 9.4.13, it follows that

$$\Phi(r) \leq L\, r\, \Phi'(r), \quad \Phi(a) \leq d(z^*, B) = D(z, B)$$

for almost all r. Hence z satisfies the hypotheses of Lemma 9.4.18 and so $z \in C_\mu^0(\bar{D})$ for any $D \subset\subset B$. The differentiability results then follow from Theorem 1.10.4 since the continuity allows us to conclude that z is an extremal for the integral $I(z, D)$ whenever D is small enough for $z(x, y)$ to $\in$ the range of a coordinate patch of $\mathfrak{M}$ for $(x, y) \in \bar{D}$, where

$$f(x, y, z, p, q) = g_{ij}(z^1, \ldots, z^N)(p^i p^j + q^i q^j).$$

Now, let B be a circular domain and suppose $(x_0, y_0) \in \partial B$. Suppose that R is small enough so that $\overline{B(x_0, y_0; R)}$ intersects no part of ∂B except the circle C containing (x_0, y_0) and define

$$D[z, B(x_0, y_0; R)] = \varepsilon^2(R)/2\pi, \quad \bar{w}(r, \theta) = \bar{z}(x_0 + r\cos\theta, y_0 + r\sin\theta).$$

It follows that there is an $\bar{r}$ such that

$$\int_{B \cap \partial B(p_0, \bar{r})} |\bar{w}_\theta(\bar{r}, \theta)|^2\, d\theta \leq D[z, B(p_0, R)], \quad R/e \leq \bar{r} \leq R, \quad p_0 = (x_0, y_0).$$

and $\bar{z}$ is A.C. along $\bar{B} \cap \partial B(p_0, \bar{r})$ (so $\bar{w}$ is A.C. in θ up to ∂B). From the SCHWARZ inequality and the continuity of z^* along ∂B, it follows that if $\zeta^* = \bar{z}$ on $\bar{B} \cap \partial B(p_0, \bar{r})$ and $= z^*$ on $\overline{B(p_0, \bar{r})} \cap \partial B$, then

$$\text{osc}\ \zeta^* \ \text{on}\ \partial[B \cap B(p_0, \bar{r})] \leq \varepsilon(R) + \omega(R) \tag{9.4.27}$$

where $\omega(R)$ is the oscillation of z^* on $\overline{B(p_0, \bar{r})} \cap \partial B$. Now, $\bar{B} \cap \overline{B(p_0, \bar{r})}$ can be mapped by a conformal map on the closed unit disc $\overline{B(0, 1)}$ so that a given point p_1 of $B \cap B(p_0, \bar{r})$ is carried into the origin. By using maps of the form $w - w_0 = (z - z_0)^\alpha$ to get rid of the corners and then using Theorem 9.3.1 we see that the resulting map is analytic on the closed domain except at the corners. Let ζ be the transform of ζ^* and Z be that of z. Then Z is continuous on $B(0,1)$, $Z \in H_2^1[B(0,1)]$, $\bar{Z} = \zeta$ a.e. on $\partial B(0,1)$, and Z minimizes $D[Z', B(0,1)]$ among all such functions. If R is small enough, $D[z, B(p_0, R)] \leq m/\pi$ so that

$$\Phi(r) \leq L\, r\, \Phi'(r), \quad 0 \leq r \leq 1, \quad \Phi(1) \leq \varepsilon^2(R)/2\pi,$$
$$\Phi(r) = D[Z, B(0, r)].$$

By the method of proof of Lemma 9.4.17, we find that

$$\Phi(r) \leq (2\pi)^{-1}\, \varepsilon^2(R) \cdot (r/R)^{2\mu} = k^2(r), \quad 2\mu = 1/L,$$

(9.4.28)
$$\int_0^{2\pi} |\zeta(\theta) - \bar{Z}(0, 0)|\, d\theta \leq \int_0^1 \int_0^{2\pi} L(r, \theta)\, dr\, d\theta \leq \varepsilon(R)\,(1 + \mu^{-1}).$$

Since $\operatorname{osc}\zeta = \operatorname{osc}\zeta^*$ and (9.4.27) and (9.4.28) hold we see that

$$\max_\theta |Z(0, 0) - \zeta(\theta)| \leq \varepsilon(R)\,(1 + \mu^{-1})/2\pi + \varepsilon(R) + \omega(R).$$

Since $Z(0, 0) = z(p_1)$ and p_1 was arbitrary, the *two dimensional* continuity of $\bar{z}$ at (x_0, y_0) follows.

We can now prove our main theorem.

Theorem 9.4.3. *Under the hypotheses of Theorem* 9.4.1, $\exists$ *a vector* z *of class* $\bar{H}_2^1(\bar{B})$, B *being a circular region of type* k *such that*

$$D(z, B) = d(\Gamma) \quad \text{and} \quad L(z, B) = l(\Gamma);$$

z *is (generalized) conformal and satisfies the boundary conditions as in Theorem* 9.4.1. *The differentiability results for* z *are those of Theorem* 9.4.2.

Proof. Let $\{z_n\}$ be the minimizing sequence of Lemma 9.4.8 (c). Then a subsequence converges weakly in $H_2^1(B)$ to a vector Z which satisfies the boundary conditions and is continuous along ∂B. From the lower-semicontinuity, it follows that $D(Z, B) \leq d(\Gamma)$. From Theorem 9.4.2, it follows that $\exists$ a $z \in H_2^1(B)$, $\bar{z} = \bar{Z}$ a.e. on ∂B, and z minimizes $D(z', B)$ among all such z'. Moreover z *is continuous* on $\bar{B}$ and so $\in \bar{H}_2^1(\bar{B})$. Consequently

$$d(\Gamma) \leq D(z, B) \leq D(Z, B) \leq d(\Gamma)$$

so that z is our desired solution. Since $d(\Gamma) \leq 2\, l(\Gamma)$ and $D(z', B) \geq 2\, L(z', B)$ for every $z' \in \bar{H}_2^1(\bar{B})$, it follows that z is conformal.

Remark. Z is also a minimizing function and so has the properties of z.

9.5. The general two-dimensional parametric problem

We first present the author's simplification (MORREY [17]) of the existence proofs of CESARI, DANSKIN, and SIGALOV referred to above. We shall assume that the integrand function satisfies the general assumptions on f given in § 9.2 (just before equation (9.2.12)) and, in addition the condition

$$M \cdot |X| \geq f(z, p) \; (\geq m\,|X| \text{ by (iii)}, \quad 0 < m \leq M). \tag{9.5.1}$$

We recall that if $N = 3$ (and $\nu = 2$) then $f(z, p) = \Phi(z, X)$ and the conditions hold if $\Phi \in C^1$ for $X \neq 0$, Φ convex in X, and Φ satisfies

$$m\,|X| \leq \Phi(z, X) \leq M\,|X|, \quad X^1 = (p^2 q^3 - p^3 q^2), \tag{9.5.2}$$
$$X^2 = (p^3 q^1 - p^1 q^3), \quad X^3 = (p^1 q^2 - p^2 q^1).$$

We shall confine ourselves to the consideration of surfaces of type k with $k = 1$, i.e. of the type of the disc.

Definition 9.5.1. If Γ is an oriented closed FRÉCHET curve, we define

$$d(\Gamma) = \inf \left\{\liminf_{n\to\infty} \mathfrak{J}_f(z_n, B_1)\right\}^1, \quad B_1 = B(0, 1) \tag{9.5.3}$$

for all sequences $\{z_n\}$ which converge uniformly on ∂B_1 to a representation of Γ.

The following lemma is immediate:

Lemma 9.5.1. (a) *If $\Gamma_n \to \Gamma$ in the sense of Fréchet, then*

$$d(\Gamma) \leq \liminf d(\Gamma_n).$$

(b) *If $d(\Gamma) < \infty$, $\exists$ a sequence $\{\Gamma_n\} \ni$ each Γ_n is regular and of class C^∞, $\Gamma_n \to \Gamma$, and $d(\Gamma_n) \to d(\Gamma)$. If Γ is a Jordan curve, the Γ_n may be chosen to be regular Jordan curves of class C^∞.*

(c) *If Γ is a regular Jordan curve of class C^∞,*

$$d(\Gamma) = \inf \mathfrak{J}_f(z, B_1)$$

for all locally regular $z \in C^\infty(\bar{B}_1)$ for which the restriction of z to ∂B_1 furnishes a representation of Γ.

Remark. In order to prove the local regularity in (c) above without increasing the number of dimensions N, which would require the extension of f, one can begin by approximating by polyhedra Π_n with boundaries Γ_n, spanning Γ_n and Γ by a piecewise regular C^∞ band, and then rounding off the vertices and edges.

It is convenient to have at our disposal the "dominating function" Φ defined in the following lemma. It is not necessary for the existence theorem; the rough function Ψ defined by

$$\Psi^2 = f^2 + \left(\frac{E - G}{2}\right)^2 + F^2$$

---

[1] See Equation (9.1.15) and Definition 9.1.4′.

is sufficient for that purpose. However, it yields a simple proof that the solution vector is (generalized) conformal and is also helpful in proving the higher differentiability results once the solution vector is shown to $\in C^1_\mu(\bar{D})$ for $D \subset\subset G$.

Let us suppose that $f(p, q)$ is of class C^2 when the (p, q) matrix has rank 2 and suppose we perform a rotation of axes in R_N and define

$$'p^i = c_{ij} p^j, \quad 'q^i = c_{ij} q^j, \quad 'f(p', q') = f(p, q).$$

Then we observe that

$$\begin{aligned} &\sum_{i,j} [\lambda^2 f_{p^i p^j} + \lambda\mu (f_{p^i q^j} + f_{q^i p^j}) + \mu^2 f_{q^i q^j}] \xi^i \xi^j \\ (9.5.4) \qquad &= \sum_{i,j}' [\lambda^2 \, 'f_{'p^i 'p^j} + \lambda\mu ('f_{'p^i 'q^j} + 'f_{'q^i 'p^j}) + \mu^2 \, 'f_{'q^i 'q^j}] \, '\xi^i \, '\xi^j, \\ &'\xi^i = c_{ik} \xi^k. \end{aligned}$$

Thus if f is weakly quasi-convex in (p, q), $'f$ is also in $('p, 'q)$. If (p_0, q_0) is any point where the (p, q) matrix has rank 2, we may perform a rotation of axes so that

$$(9.5.5) \qquad 'p_0^i = 'q_0^i = 0 \quad \text{for} \quad i \geq 3, \quad 'p_0^1 \, 'q_0^2 - 'p_0^2 \, 'q_0^1 > 0.$$

Let us consider the set of all $('p, 'q)$ satisfying (9.5.5) and let us drop the primes. The relation (9.1.3) becomes

$$(9.5.6) \quad f(\alpha p + \beta q, \gamma p + \delta q) = (\alpha\delta - \beta\gamma) f(p, q), \quad \alpha\delta - \beta\gamma > 0.$$

Differentiating (9.5.6) with respect to p^i and q^i and solving, we obtain

$$\begin{aligned} &f_{p^i}(P, Q) = \delta f_{p^i} - \gamma f_{q^i}, \quad f_{q^i}(P, Q) = -\beta f_{p^i} + \alpha f_{q^i} \\ &f_{p^i p^j}(P, Q) = \Delta^{-1} [\delta^2 f_{p^i p^j} - \gamma\delta (f_{p^i q^j} + f_{q^i p^j}) + \gamma^2 f_{q^i q^j}] \\ (9.5.7) \quad &f_{p^i q^j}(P, Q) = \Delta^{-1} [-\beta\delta f_{p^i p^j} + \alpha\delta f_{p^i q^j} + \beta\gamma f_{q^i p^j} - \alpha\gamma f_{q^i q^j}] \\ &f_{q^i q^j}(P, Q) = \Delta^{-1} [\beta^2 f_{p^i p^j} - \alpha\beta (f_{p^i q^j} + f_{q^i p^j}) + \alpha^2 f_{q^i q^j}] \end{aligned}$$

$$P = \alpha p + \beta q, \quad Q = \gamma p + \delta q, \quad \Delta = \alpha\delta - \beta\gamma, \quad f_{p^i} = f_{p^i}(p, q), \quad \text{etc.}$$

Thus, if we let $e_1 = (1, 0, \ldots, 0)$, $e_2 = (0, 1, \ldots, 0)$ and define

$$(9.5.8) \quad \begin{aligned} &f_{p^i p^j}(e_1, e_2) = a_{ij}, \quad f_{p^i q^j}(e_1, e_2) = b_{ij}, \quad f_{q^i q^j}(e_1, e_2) = c_{ij}, \\ &f_{p^i}(e_1, e_2) = d_i, \quad f_{q^i}(e_1, e_2) = e_i', \quad f(e_1, e_2) = h_0 \end{aligned}$$

we find that, on the space (9.5.5)

$$f(p, q) = h_0(p^1 q^2 - p^2 q^1), \quad f_{p^i} = d_i q^2 - e_i' q^1, \quad f_{q^i} = -d_i p^2 + e_i' p^1, \quad i \geq 3,$$

$$\begin{aligned} &Q(f) = K^{-1} \sum_{i,j=3}^{N} [a_{ij} (q^2 \lambda - p^2 \mu)^2 - (b_{ij} + b_{ji}) (q^2 \lambda - p^2 \mu) \times \\ (9.5.9) \qquad &\times (q^1 \lambda - p^1 \mu) + c_{ij} (q^1 \lambda - p^1 \mu)^2] \xi^i \xi^j \\ &K = \sqrt{EG - F^2} = p^1 q^2 - p^2 q^1. \end{aligned}$$

$Q(f)$ being the form in (9.5.4). Thus the form Q necessarily degenerates and has rank $\leq N - 2$ in ξ for all (λ, μ) with $\lambda^2 + \mu^2 = 1$.

Definition 9.5.2. We say that an f of the type discussed is *the integrand of a regular parametric problem* $\Leftrightarrow$ the form $Q(f)$ is non-negative definite with rank $= N - 2$ for all (λ, μ) with $\lambda^2 + \mu^2 = 1$ and all (p, q) of rank 2.

Remarks. In the case $N = 3$, this corresponds to the existence of numbers m_1 and M_1 such that

$$m_1 [|\xi|^2 - |X|^{-2}(X \cdot \xi)^2] \leq |X| \Phi_{X^i X^j} \xi^i \xi^j \leq M_1 [|\xi|^2 - |X|^{-2}(X \cdot \xi)^2],$$
$$0 < m_1 \leq M_1.$$

Since the set of all normal orthogonal pairs of vectors (e_1, e_2) is compact, the condition is equivalent to the existence of m_1 and M_1 such that

$$m_1(\lambda^2 + \mu^2) |\hat{\xi}|^2 < \sum_{i,j=3}^{N} [a_{ij} \lambda^2 + (b_{ij} + b_{ji}) \lambda \mu + c_{ij} \mu^2] \xi^i \xi^j$$
$$(9.5.10) \qquad < M_1(\lambda^2 + \mu^2) |\hat{\xi}|^2, \ |\hat{\xi}|^2 = \sum_{i=3}^{N} (\xi^i)^2$$

or, at the general point in the space (9.5.5)

$$(9.5.11) \qquad \begin{aligned} &m_1 K^{-1}(G\lambda^2 - 2F\lambda\mu + E\mu^2) |\hat{\xi}|^2 \leq Q(f) \\ &\leq M_1 K^{-1}(G\lambda^2 - 2F\lambda\mu + E\mu^2) |\hat{\xi}|^2. \end{aligned}$$

By multiplying by a constant we may assume that f satisfies the $m - M$ condition with $2 = m \leq M$.

Lemma 9.5.2. *If f is the integrand of a regular parametric problem with $\nu = 2$ and $N \geq 3$ and if $f \in C_\mu^n$ (or C^∞) whenever the (p, q) matrix has rank 2, then there is a function Φ which is homogeneous of degree 2 and of class C_μ^n (or C^∞) for all $(p, q) \neq (0, 0)$, which is the integrand of a regular non-parametric problem* (i.e. $Q(\Phi)$ *has rank* N), *and which satisfies*

$$(9.5.12) \qquad \begin{aligned} &D \leq \Phi(z, p, q) \leq M D, \quad D = \frac{E + G}{2}, \\ &\Phi(z, p, q) \geq f(z, p, q), \end{aligned}$$

the equality holding $\Leftrightarrow E = G$ *and* $F = 0$, *M being the bound for f.*

Proof. Since z will act as a parameter in the following construction, we shall suppress it. Since f satisfies the $m - M$-condition of (9.5.1), we may suppose (by multiplying by a constant) that $m = 2$. We define Φ to have the form

$$(9.5.13) \qquad \begin{aligned} &\Phi(p, q) = D[1 + \omega(\tau, h)], \quad D = \frac{E + G}{2}, \quad E = |p|^2, \quad \text{etc.} \\ &\tau(p, q) = D^{-1} K, \quad h(p, q) = K^{-1} f(p, q), \quad K = \sqrt{EG - F^2}. \end{aligned}$$

Clearly we may apply the rotation procedure to D, K, τ, and h as well as to f. Thus it is sufficient to compute all the derivatives on the space (9.5.5). All the requisite derivatives of f can be computed from (9.5.7), (9.5.8), and (9.5.9). The derivatives of K are easily found on our space from the equation $K^2 = EG - F^2$; we obtain

$$K = p^1 q^2 - p^2 q^1, \quad K_{p^i} = K_{q^i} = 0, \quad i \geq 3.$$

A straightforward computation leads tothe result that

$$Q(\Phi) = (1 + \omega - \tau\,\omega_\tau)(\lambda^2 + \mu^2)\,|\xi|^2 + \tau^{-2}(\tau\,\omega_\tau - h\,\omega_h) \times \\ \times D^{-1}(G\lambda^2 - 2F\lambda\mu + E\mu^2)\,|\hat{\xi}|^2 + \omega_{\tau\tau} X^2 + \\ + 2\tau^{-1}(\omega_{\tau h} - \tau^{-1}\omega_h)\,X\,Y + \tau^{-2}\omega_{hh} Y^2 + \tau^{-2}\omega_h D^{-1} \times \\ \times \sum_{i,j=3}^{N} [a_{ij}\varrho^2 - (b_{ij} + b_{ji})\,\varrho\,\sigma + c_{ij}\sigma^2]\,\xi^i\xi^j, \tag{9.5.14}$$

$$X = D^{1/2}(\lambda\,\tau_p\cdot\xi + \mu\,\tau_q\cdot\xi), \quad Y = D^{1/2}\tau(\lambda\,h_p\cdot\xi + \mu\,h_q\cdot\xi),$$

$$\varrho = \lambda\,q^2 - \mu\,p^2, \quad \sigma = \lambda\,q^1 - \mu\,p^1.$$

X and Y are seen to be bounded and defined if $D \neq 0$, $\tau \neq 0$, and the last term in the expansion for $Q(\Phi)$ is between

$$m_1\tau^{-2}\omega_h W \text{ and } M_1\tau^{-2}\omega_h W, \quad W = D^{-1}(G\lambda^2 - 2F\lambda\mu + \\ + E\mu^2)\,|\hat{\xi}|^2. \tag{9.5.15}$$

Clearly we may assume

$$0 < m_1 \leq 1. \tag{9.5.16}$$

Accordingly, we want there to exist an $m_2 > 0$ such that

$$1 + \omega - \tau\,\omega_\tau \geq m_2 > 0, \quad \tau\,\omega_\tau - h\,\omega_h \geq -m_1\,\omega_h. \tag{9.5.17}$$

We must also have

$$\Phi = D(1 + \omega) \geq f = \tau\,h\,D, \quad \omega(1, h) = h - 1, \tag{9.5.18}$$

the equality holding only if $\tau = 1$ (corresponding to $E = G$, $F = 0$). We notice also that

$$\left(1 - \sqrt{1 - \tau^2}\right)|\hat{\xi}|^2 \leq W \leq \left(1 + \sqrt{1 - \tau^2}\right)|\xi|^2. \tag{9.5.19}$$

We define

$$\omega(\tau, h) = [1 + (h - 2)\,\tau]\,\varphi(u), \quad u = (h - 1)\,\tau/[1 + (h - 2)\,\tau], \\ 0 \leq \tau \leq 1, \quad 2 \leq h, \quad \tau = u/[(h - 1) - (h - 2)\,u], \tag{9.5.20}$$

Then it follows that u increases from 0 to 1 as τ does, for each h, and

$$\omega_\tau = [(h - 1) - (h - 2)\,u]\,\varphi'(u) + (h - 2)\,\varphi(u)$$

$$\tau\,\omega_\tau - \omega = u\,\varphi'(u) - \varphi(u), \quad \omega_h = \tau\,[\varphi(u) + (1 - u)\,\varphi'(u)]$$

$$\tau\,\omega_\tau - (h - m_1)\,\omega_h = \tau\{[(2 - m_1)\,u - (1 - m_1)]\,\varphi'(u) - \\ - (2 - m_1)\,\varphi(u)\} \tag{9.5.21}$$

$$\omega_{\tau\tau}\,\omega_{hh} - (\omega_{\tau h} - \tau^{-1}\omega_h)^2 \equiv 0$$

$$\omega_{\tau\tau} = [(h - 1) - (h - 2)\,u]\,\varphi''(u)\,(\partial u/\partial\tau) \geq 0 \Leftrightarrow \varphi''(u) \geq 0.$$

For $\tau = 1$, $(1 + \omega) - \tau\,h = 0$ and

$$\frac{\partial}{\partial\tau}[(1 + \omega) - \tau\,h] = \omega_\tau - h \leq 0 \;\; \textit{if} \;\; \varphi''(u) \geq 0$$

$$\frac{\partial}{\partial u}\omega_\tau = [(h - 1) - (h - 2)\,u]\,\varphi''(u) \geq 0$$

so ω_τ takes its max. for $\tau = 1$ where its value is

$$\omega_\tau(1, h) = \varphi'(1) + h - 2.$$

But for $\tau = 1$, we must have ((9.5.17))

$$(9.5.22)\qquad \begin{aligned} \omega_\tau(1, h) - \omega(1, h) \leq 1 - m_2 &\Leftrightarrow \varphi'(1) + h - 2 \leq h - m_2 \\ &\Leftrightarrow \varphi'(1) \leq 2 - m_2. \end{aligned}$$

Thus $\omega_\tau \leq h$ and so (9.5.18) holds. The first inequality in (9.5.17) holds if the last one in (9.5.22) does. If $\varphi''(u) = 0$ for $u \leq (1 - m_1)/(2 - m_1)$ after which $\varphi''(u) \geq 0$, all the above hold and $\tau\,\omega_\tau - (h - m_1)\,\omega_h \geq 0$ so we see that

$$(9.5.23)\qquad \varphi'(1) \geq 2 - m_1.$$

So we take $\varphi = 0$ for $0 \leq u \leq (1 - m_1)/(2 - m_1)$, choose $\varphi \in C^\infty[0, 1]$ and analytic with $\varphi''(u) \geq 0$ for $(1 - m_1)/(2 - m_1) < u \leq 1$ and so that $\varphi(1) = 1$, $2 - m_1 < \varphi'(1) \leq 2 - m_2$ where $0 < m_2 < m_1$.

Definition 9.5.3. We define the integral

$$I(z, G) = \iint_G \Phi(z, p, q)\, dx\, dy.$$

$$J(z, B_1) = \iint_{B_1} (|z_{xx}|^2 + 2\,|z_{xy}|^2 + |z_{yy}|^2)\, dx\, dy.$$

Remark. We note that

$$I(z, G) = \mathfrak{J}_f(z, G) \Leftrightarrow z \in \bar{H}_2^1(G) \text{ and } z \text{ is conformal.}$$

Lemma 9.5.3. *Suppose Γ is a regular Jordan curve of class C^∞ and $z = \zeta(p)$, $p \in \partial B_1$ is a regular representation of Γ of class C^∞. Then, for K sufficiently large, the class $\mathfrak{J}(K)$ of vectors $z \in \bar{H}_2^1(B_1) \cap H_2^2(B_1)$ for which*

$$(9.5.24)\qquad J(z, B_1) \leq K, \quad z(p) = \zeta(p) \text{ for } p = (1,0), \left(-\tfrac{1}{2}, \pm\tfrac{\sqrt{3}}{2}\right),$$

is not empty. For each such K, there is a vector z_K in $\mathfrak{J}(K)$ which minimizes $I(z, B_1)$ among all z in $\mathfrak{J}(K)$. Moreover

$$(9.5.25)\qquad d(\Gamma) = \lim_{K \to \infty} I(z_K, B_1).$$

Proof. The first statement is evident since there is a locally regular $Z \in C^\infty(\bar{B}_1)$ such that $z = Z(p)$, $p \in \partial B_1$, gives a representation of Γ; the three point condition may be secured by performing a Möbius transformation, the transform of z being in $C^\infty(\bar{B}_1)$. Now let $\{z_p\}$ be a minimizing sequence for I in $\mathfrak{J}(K)$. Then the norms of the z_p in $H_2^2(B_1)$ are uniformly bounded so that the z_p are equicontinuous on $\bar{B}_1$ by Sobolev's lemmas (see § 3.5). Thus, a subsequence converges weakly in $H_2^2(B_1)$ and uniformly on $\bar{B}_1$ to some function z_K in $H_2^2(B_1)$. Since this convergence implies weak convergence of the second derivatives and strong convergence of the first derivatives in $L_2(B_1)$, we see that z_K is admissible. Since I and J are both lower-semicontinuous (I involves only first derivatives), z_K is a minimizing function. The last statement follows from the fact that we may select a sequence $\{z_p\}$ of regular surfaces of class C^∞, *each bounded by* Γ, *such that* $\mathfrak{J}_f(z_p, B_1) \to d(\Gamma)$; each of these may be represented conformally by some admissible vector z_p^*.

Lemma 9.5.4. *For each* K, z_K *satisfies the condition*

$$D[z_K, B(P_0, R)] \leq D[z_K, B(P_0, a)] \cdot \left(\frac{R}{a}\right)^\lambda, \quad \lambda = \frac{m}{2M}, \quad 0 \leq R \leq a, \tag{9.5.26}$$

for every circle $B(P_0, a) \subset B_1$. *Thus the* z_K *satisfy (if* K *is large enough) a uniform Hölder condition on* $\bar{B}_A$, *which depends only on* A, m, M, *and* $d(\Gamma)$, *for each* $A < 1$ *and are equicontinuous along* ∂B_1. *There exists a vector* $z \in \bar{H}_2^1(\bar{B}_1)$ *such that* $z = z(p)$, $p \in \partial B_1$, *gives a representation of* Γ *which satisfies the three point condition in* (9.5.24) *and for which* $\mathfrak{J}_f(z, B_1) = d(\Gamma)$; *the vector* z *satisfies the Hölder condition satisfied by the* z_K *on each* B_A. *Moreover* z *is (generalized) conformal.*

Proof. The equicontinuity on each $\bar{B}_A$ follows from the first result and the writer's "Dirichlet growth theorem" (Theorem 3.5.2). The equicontinuity along ∂B_1 follows from a well known lemma of Courant (Lemma 9.3.2) since we have $D(z_K, B_1)$ uniformly bounded and have a three point condition.

To prove (9.5.26), we use (9.5.12) and the minimizing property of z_K to show that

$$\begin{aligned} \frac{1}{4} m D[z_K, B(P_0, R)] &\leq I(z_K, B(P_0, R)] \leq I[Z, B(P_0, R)] \\ &\leq \frac{1}{2} M D[Z, B(P_0, R)], \quad (m = 2) \end{aligned} \tag{9.5.27}$$

where Z is the biharmonic function such that $Z - z_k \in H_{20}^2[B(P_0, R)]$. We shall omit the routine justification of the following formal calculations.

Let (r, θ) be polar coordinates with pole at P_0 and let

$$z_K(r, \theta) = \frac{a_0(r)}{2} + \sum_{n=1}^{\infty} [a_n(r) \cos n\theta + b_n(r) \sin n\theta], \quad 0 \leq r \leq a,$$

$$Z(r, \theta) = \frac{A_0(r)}{2} + \sum_{n=1}^{\infty} [A_n(r) \cos n\theta + B_n(r) \sin n\theta], \quad 0 \leq r \leq R.$$

Since Z is biharmonic and has the same Dirichlet data as z on $\partial B(P_0, R)$, we obtain

$$A_n(r) = c_n \left(\frac{r}{R}\right)^n + d_n \left(\frac{r}{R}\right)^{n+2}, \quad B_n(r) = e_n \left(\frac{r}{R}\right)^n + f_n \left(\frac{r}{R}\right)^{n+2}$$

where c_n, d_n, e_n, and f_n are constants defined by

$$2c_n = (n+2)\,\alpha_n - \beta_n, \quad 2d_n = \beta_n - n\,\alpha_n,$$
$$\alpha_n = a_n(R), \quad \beta_n = R\,a_n'(R), \quad n > 0,$$

and similar formulas hold for e_n and f_n for $n > 0$. Thus if we set

$$\Psi(R) = D[z, B(P_0, R)],$$

equation (9.5.27) and a computation of $D[Z, B(P_0, R)]$ shows that

$$\begin{aligned}\Psi(R) &\le L\,D[Z, B(P_0, R)]\\ &= L\int_0^R r\left\{\frac{A_0'^2}{2} + \sum_{n-1}^{\infty}[A_n'^2 + B_n'^2 + r^{-2}n^2(A_n^2 + B_n^2)]\right\}dr\\ &\le 2LR\Psi'(R), \quad L = \frac{2M}{m}.\end{aligned}$$

The result (9.5.26) follows.

Thus from the z_K, we may extract a subsequence $\{z_p\}$ which converges weakly in $H_2^1(B_1)$ and uniformly along ∂B_1 and on each $\bar{B}_A$ with $A < 1$ to some $z \in \bar{H}_2^1(\bar{B}_A)$ for each $A < 1$. Since both I and $\mathfrak{J}_f$ are lower semicontinuous with respect to this type of convergence, (Theorem 4.4.5) we conclude that

$$\mathfrak{J}_f(z, B_1) \le I(z, B_1) \le d(\Gamma).$$

But suppose we define $Z_R = z$ on $\bar{B}_R$ and Z_R to be the harmonic function on $B_1 - \bar{B}_R$ which coincides with z on $\partial B_1 \cup \partial B_R$. Since each Z_R is continuous and $\mathfrak{J}_f(Z_R, B_1) \to \mathfrak{J}_f(z, B_1)$, etc., we find that

$$d(\Gamma) \le \mathfrak{J}_f(z, B_1) \le I(z, B_1) \le d(\Gamma).$$

From this we conclude that z is conformal (on the interior). From the lower-semicontinuity of I on each domain D of type k, we obtain

$$I(z, G) = \lim_{p\to\infty} I(z_p, G).$$

But now, from the lower semicontinuity of D, we have

$$\begin{aligned}\frac{1}{4}m\,D(z,G) &\le \frac{1}{4}m \liminf_{p\to\infty} D(z_p, G) \le \lim_{p\to\infty} I(z_p, G)\\ &= I(z, G) \le I(H, G) \le \frac{1}{2}M\,D(H, G),\end{aligned} \tag{9.5.28}$$

H being the usual harmonic function. The continuity of z at points of ∂B_1 now follows as in the proof of Theorem 9.4.2. The result (9.5.28) gives the Hölder continuity of z on interior domains.

We can now prove the existence theorem:

Theorem 9.5.1. *Suppose Γ is a Jordan curve for which $d(\Gamma) < \infty$. Then $\exists$ a continuous vector $z \in \bar{H}_2^1(\bar{B}_1)$ such that*

$$\mathfrak{J}_f(z, B_1) = d(\Gamma). \tag{9.5.29}$$

Moreover z is conformal.

Proof. Choose $\{\Gamma_n\}$ so that each Γ_n is regular and of class C^∞ and so that $\Gamma_n \to \Gamma$ and $d(\Gamma_n) \to d(\Gamma)$ (Lemma 9.5.1). Choose regular representations $z = \zeta_n(p)$, $p \in \partial B_1$, of class C^∞ of the Γ_n so that the $\zeta_n(p)$ converge uniformly on ∂B_1 to $\zeta(p)$ where $z = \zeta(p)$ is a topological representation of Γ. For each n, let z_n be the minimizing vector of Lemma 9.5.4. From Lemmas 9.4.2 and 9.5.4, we conclude that a subsequence, still called $\{z_n\}$, converges uniformly along ∂B_1 and on each $\bar{B}_A$ as in the proof of Lemma 9.5.4. By repeating the argument in the last part of that proof, we find that $z (= \lim z_n) \in \bar{H}_2^1(\bar{B}_1)$, that (9.5.29) holds, and the z is conformal.

We conclude this section with the results of KIPPS mentioned above. In fact, we generalize his results somewhat.

Lemma 9.5.5. *Suppose p and $q \in H_2^1[B(P_0, a)]$.*

(a) *Then, for almost all r, $0 < r < a$, $\bar{p}$ and $\bar{q}$ (Lemma 9.4.5) are* A.C. *on $\partial B(P_0, r)$ with p_θ and $q_\theta \in L_2$, and*

$$2 \int\int_{B(P_0,r)} (p_x q_y - p_y q_x)\, dx\, dy = \int_{\partial B(P_0,r)} (\bar{p}\, d\bar{q} - \bar{q}\, d\bar{p}).$$

(b) *If, also, π and $\varkappa \in H_2^1[B(P_0, r)]$ for such an r with $\pi - p$ and $\varkappa - q \in H_{20}^1[B(P_0, r)]$, then*

$$\int\int_{B(P_0,r)} (\pi_x \varkappa_y - \pi_y \varkappa_x)\, dx\, dy = \int\int_{B(P_0,r)} (p_x q_y - p_y q_x)\, dx\, dy.$$

(c) *If $p = z_x$ and $q = z_y$ for some $z \in H_2^2[B(P_0, a)]$, then $q_x = p_y$ a.e. on $B(P_0, a)$.*

Proof. (a) and (c) are proved by approximating by the mollified functions. To prove (b) write $\pi = p + \pi_0$, $\varkappa = q + \varkappa_0$; then π_0 and $\varkappa_0$ may be approximated strongly in $H_2^1[B(P_0, r)]$ by functions $\in C_c^\infty[B(P_0, r)]$ and p and q may be approximated similarly by functions $\in C^\infty\,[\overline{B(P, r)}]$.

Lemma 9.5.6. (a) *The conformal minimizing vector z of Theorem 9.5.1 satisfies*

$$(9.5.30)\qquad \int\int_{B_1} (\zeta_x^i f_{p^i} + \zeta_y^i f_{q^i} + \zeta^i f_{z^i})\, dx\, dy = 0, \quad \zeta \in H_{20}^1(B_1).$$

(b) *If also, p and $q \in \bar{H}_2^1[B(P_0, a)]$ for some $B(P_0, a) \subset\subset B_1$, then f_{p^i}, f_{q^i}, and $f_{z^i} \in \bar{H}_2^1[B(P_0, a)]$ and z satisfies*

$$(9.5.31)\qquad L_r \equiv \frac{\partial}{\partial x} f_{p^r} + \frac{\partial}{\partial y} f_{q^r} - f_{z^r} = 0 \quad \text{a.e. on } B(P_0, a).$$

(c) *For almost all (x, y) where $p = q = 0$, we have $f_p = f_q = f_z = 0$ and the derivatives of p, q, f_p, and f_q vanish a.e. on that set. Moreover p and q satisfy the relations*

$$(9.5.32)\qquad \begin{aligned} p \cdot p_x - q \cdot q_x = q \cdot p_x + p \cdot q_x = p \cdot p_y - q \cdot q_y \\ = q \cdot p_y + p \cdot q_y = 0 \quad \text{a.e.} \end{aligned}$$

(d) *If* $N = 3$ *and we define* $X = p \times q$ *and* $f(z, p, q) = \Phi(z, X)$, *then*

$$\frac{\partial}{\partial x} f_{p^r} + \frac{\partial}{\partial y} f_{q^r} + f_{z^r} = k^r [(A p_x + B p_y + B q_x + C q_y) \cdot k - H D],$$

$$(9.5.33) \quad k = |X|^{-1} X, \quad A = \Phi_{X^\varrho X^\sigma} p^\varrho p^\sigma, \quad B = \Phi_{X^\varrho X^\sigma} p^\sigma q^\varrho,$$
$$C = \Phi_{X^\varrho X^\sigma} q^\varrho q^\sigma, \quad H = \Phi_{X^\varrho z^\varrho}$$

for almost all (x, y) *where* $X \neq 0$, i.e. $(p, q) \neq (0, 0)$.

Proof. (a) follows since z minimizes the non-parametric integral I and so satisfies (9.5.30) with f replaced by Φ. Since $\Phi \geq f$ everywhere with the equality holding along our solution,

$$(9.5.34) \qquad \Phi_{p^i} = f_{p^i}, \quad \Phi_{q^i} = f_{q^i}, \quad \Phi_{z^i} = f_{z^i} \quad \text{a.e.}$$

along our solution. Since Φ is homogeneous of degree 2 and the derivatives Φ_{pp}, Φ_{pq}, and Φ_{qq} are continuous and bounded if $(p, q) \neq (0, 0)$ and the derivatives Φ_{pz} and Φ_{qz} are continuous everywhere, we easily see that $f_p = \Phi_p$ and $f_q = \Phi_q \in \bar{H}_2^1[\overline{B(P_0, a)}]$. We have seen elsewhere (§. 3.1) that the derivatives of p, q, f_p, f_q, and $f_z = 0$ almost everywhere on the set where $p = q = 0$. The relations (9.5.32) follow by differentiating the conformality relations.

The relations in (9.5.33) follow by simply carrying out the differentiations in (9.5.31) and using the form of f: Since $q_x = p_y$, we see, for almost all (x, y) where $X \neq 0$, that

$$L_r = \mathfrak{A}_{rs} p_x^s + \mathfrak{B}_{rs}(p_y^s + q_x^s) + \mathfrak{C}_{rs} q_y^s + \mathfrak{D}_{rs} p^s + \mathfrak{E}_{rs} q^s - \Phi_{z^r}$$

$$\mathfrak{A}_{rs} = \varphi_{X^\varrho X^\sigma} X^\varrho_{p^r} X^\sigma_{p^s}, \quad \mathfrak{B}_{rs} = \varphi_{X^\varrho X^\sigma} X^\varrho_{p^r} X^\sigma_{q^s}, \quad \mathfrak{C}_{rs} = \varphi_{X^\varrho X^\sigma} X^\varrho_{q^r} X^\sigma_{q^s},$$

$$\mathfrak{D}_{rs} = \varphi_{X^\varrho z^s} X^\varrho_{p^r}, \quad \mathfrak{E}_{rs} = \varphi_{X^\varrho z^s} X^\varrho_{q^r}, \quad (f(z, p, q) = \varphi(z, X)).$$

Using the conformality relations, one easily sees that

$$k^r X^\varrho_{p^r} = -p^\varrho, \quad k^r X^\varrho_{q^r} = -q^\varrho.$$

Since

$$p^r X^\varrho_{p^r} = q^r X^\varrho_{q^r} = X^\varrho, \quad p^r X^\varrho_{q^r} = q^r X^\varrho_{p^r} = 0,$$

the results in (9.5.33) now follow easily from the homogeneity of φ as a function of X.

Our method is to set up a non-parametric integral of the type discussed in § 5.4 with unknown 3-vector functions π and $\varkappa$ which is minimized by taking $(\pi, \varkappa) = (p, q)$. The Hölder continuity then follows from the results of that section.

Definition 9.5.4. Given a conformal vector $z \in \bar{H}_2^1(\bar{B}_1)$ with p and q also $\in \bar{H}_2^1[\overline{B(P_0, a)}]$, $\overline{B(P_0, a)} \subset B_1$. We define

$$(9.5.35) \quad I^*[\pi, \varkappa, B(P_0, a)] = \int_{B(P_0, a)} \int \Psi(x, y, \pi, \varkappa, \pi_x, \varkappa_x, \pi_y, \varkappa_y)\, dx\, dy$$

where, for $X \neq 0$, i.e. $(p, q) \neq 0$,

$$\begin{aligned}\Psi = & |\pi_y - \varkappa_x|^2 + |X|^{-1}[(p \cdot \pi_x - q \cdot \varkappa_x)^2 + (q \cdot \pi_x + p \cdot \varkappa_x)^2 + \\ & + (p \cdot \pi_y - q \cdot \varkappa_y)^2 + (q \cdot \pi_y + p \cdot \varkappa_y)^2] + \frac{l}{AC}[(A \cdot \pi_x + B \cdot \pi_y + \\ & + B \cdot \varkappa_x + C \cdot \varkappa_y) \cdot k - H(p \cdot \pi + q \cdot \varkappa)/2]^2 - 2l(\pi_x \cdot \varkappa_y - \pi_y \cdot \varkappa_x) \\ & + |\pi - p|^2 + |\varkappa - q|^2, \quad k = |X|^{-1} X, \quad l = (2 - \sqrt{2})/(1 + m_1/M_1)\end{aligned}$$

(9.5.36)

where A, B, C, and H are defined in (9.5.33) and m_1 and M_1 refer to (9.5.9). If $p = q = 0$, we define

$$\begin{aligned}\Psi &= \overline{m}_2(|\pi_x|^2 + |\pi_y|^2 + |\varkappa_x|^2 + |\varkappa_y|^2) + |\pi - p|^2 + |\varkappa - q|^2, \\ \overline{m}_2 &= (m_2 + M_2)/2, \quad m_2 = l\, m_1/M_1, \\ M_2 &= \text{ larger of } 2 + \sqrt{2} - l \text{ and } l\, M_1/m_1.\end{aligned} \tag{9.5.37}$$

Theorem 9.5.2. *If Ψ_0 denotes the sum of all the terms in Ψ which are quadratic in π_x, π_y, $\varkappa_x$, and $\varkappa_y$, then*

$$\begin{aligned}m_2(|\pi_x|^2 + |\pi_y|^2 + |\varkappa_x|^2 + |\varkappa_y|^2) \leq \Psi_0 \leq M_2(|\pi_x|^2 + \\ + |\pi_y|^2 + |\varkappa_x|^2 + |\varkappa_y|^2), \quad m_2 + M_2 \geq 2.\end{aligned} \tag{9.5.38}$$

Suppose that z is the minimizing vector of Theorem 9.5.1 *and that p and $q \in \overline{H}_2^1[B(P_0, a)]$, $[\overline{B(P_0, a)}] \subset B_1$. Moreover the pair of vectors $(\pi, \varkappa) = (p, q)$ minimizes $I^*[\pi, \varkappa, B(P_0, a)]$ among all such pairs $(\pi, \varkappa) \in H_2^1[B(P_0, a)] \ni \pi - p$ and $\varkappa - q \in H_{20}^1[B(P_0, a)]$. Thus, p and $q \in C_\mu^0[\overline{B(P_0, R)}]$ for each $R < a$. Moreover the Hölder condition depends only on the bounds m, M, m_1, and M_1 and not on any bounds or moduli of continuity for z, p, and q other than those holding by virtue of the minimizing property of z. If Φ is of class C_μ^n (C^∞, or analytic) for $|X| \neq 0$; then so is z away from the locally compact subset where $p = q = 0$.*

Proof. At points where $(p, q) \neq 0$, let us define the n.o. set (i, j, k) by

$$i = |p|^{-1} p, \quad j = |q|^{-1} q, \quad k = |X|^{-1} X$$

and let us set

$$\begin{aligned}\pi_x &= \alpha_1 i + \beta_1 j + \gamma_1 k & \varkappa_x &= \alpha_3 i + \beta_3 j + \gamma_3 k \\ \pi_y &= \alpha_2 i + \beta_2 j + \gamma_2 k & \varkappa_y &= \alpha_4 i + \beta_4 j + \gamma_4 k.\end{aligned}$$

Then, for such points,

$$\begin{aligned}\Psi_0 = & (\alpha_2 - \alpha_3)^2 + (\beta_2 - \beta_3)^2 + (\beta_2 - \gamma_3)^2 + (\alpha_1 - \beta_3)^2 + \\ & + (\alpha_3 + \beta_1)^2 + (\alpha_2 - \beta_4)^2 + (\alpha_4 + \beta_2)^2 + \frac{l}{AC}(A\gamma_1 + \\ & + B\gamma_2 + B\gamma_3 + C\gamma_4)^2 + l[(\alpha_2\alpha_3 - \alpha_1\alpha_4) + (\beta_2\beta_3 - \beta_1\beta_4) + \\ & + (\gamma_2\gamma_3 - \gamma_1\gamma_4)].\end{aligned} \tag{9.5.39}$$

A surprisingly simple computation shows that the characteristic roots are (some are multiple roots)

$$l, l + 2, \quad -l + 2 \pm \sqrt{2}, \quad 2 - l, \quad lA/C, \quad \text{and} \quad l\,C/A.$$

Since the latter two are between $l\, m_1/M_1$ and $l\, M_1/m_1$, (9.5.38) follows.

In case $\pi = p$ and $\varkappa = q$, we conclude from Lemmas 9.5.5 and 9.5.6 that

$$\Psi = -2\,l(p_x \cdot q_y - p_y \cdot q_x) \quad (p, q) \neq (0, 0). \tag{9.5.40}$$

For other $(\pi, \varkappa)$ we see that

$$\begin{gathered} \Psi \geq -2l(\pi_x \cdot \varkappa_y - \pi_y \cdot \varkappa_x), \\ I^*[\pi, \varkappa, B(P_0, a)] \geq -2l \int_{B(P_0,a)}\!\!\int (p_x \cdot q_y - p_y \cdot q_x)\, dx\, dy \end{gathered} \tag{9.5.41}$$

since (9.5.40) and the first inequality in (9.5.41) hold almost everywhere where $(p, q) = (0, 0)$ or, respectively, $(\pi, \varkappa) = (0, 0)$. Hence the minimizing statement follows.

Finally, if z is the minimizing vector, we know that $(p, q) \in L_2(B_1)$ and also satisfy a Dirichlet growth condition of the form

$$\int_{B(P_1,r)}\!\!\int (|p|^2 + |q|^2)\, dx\, dy \leq D(z, B_1) \cdot (r/b)^{m_1/M_1}, \quad b = 1 - |O\,P_1|.$$

Thus the integrand Ψ satisfies the conditions in § 5.4 whether p and q are continuous or not. The remaining results follow.

Chapter 10

The higher dimensional plateau problems

10.1. Introduction

Until recently, no general results had been obtained concerning the existence and/or differentiability of the solutions of parametric problems in more than two variables. The greatest single stumbling block was the non-existence of a useful generalization of a conformal map to higher dimensions. Now, by imitating the proof of the author's old conformal mapping theorem (MORREY [3]), one can prove that a "non-degenerate" Fréchet variety of the topological type of the ν-ball (i.e. a Fréchet variety which possesses a representation on $\overline{B(0,1)}$ in which no continuum is carried into a point) which possesses a representation of class $H_\nu^1[B(0,1)]$ possesses such a representation which minimizes $\int_{B(0,1)} |\nabla z|^\nu\, dx$ among all such. However, one can not conclude that the value of this integral $\leq C \cdot L[z, B(0,1)]$ or even that $L[z, B(0,1)]$ is given by the area integral for such a representation. So the methods which had been successful in the two dimensional problems did not lead to results in the higher dimensional cases.

Almost simultaneously, results on the ν-dimensional PLATEAU problem with $\nu > 2$ were obtained by DE GIORGI [2], REIFENBERG [1], FEDERER and FLEMING and FLEMING [2]. DE GIORGI proved that a portion of minimum area of a part of the ν-dimensional boundary of an open set in $(\nu + 1)$-space is a regular analytic manifold. REIFENBERG proved that if A is any *compact point set* in R_N, there is a *compact point* set X in R_N which is bounded by A in a certain sense (see Definition 10.2.6) which minimizes $\Lambda^\nu(X - A)$ among all such X and which has the additional property that each point p of $X - A$, not in a relatively compact subset Z of $X - A$ with $\Lambda^\nu(Z) = 0$, is in a neighborhood on $X - A$ which is a topological ν-disc. Recently (REIFENBERG [2] and [3]) he has proved that these topological ν-discs are in fact analytic. Very recently, since delivering the Colloquium Lectures before the American Mathematical Society in August, 1964, the author (MORREY [21]) has found that these results may be carried over to sets X and A in a Riemannian manifold of considerable generality (see below). Almost concurrently, FEDERER and FLEMING (FEDERER [2], FLEMING [2], FEDERER and FLEMING) have approached this problem using their integral currents and have obtained results more or less comparable with those of REIFENBERG. Since FEDERER is writing up their results in book form, we shall present the author's extension to Riemannian manifolds of his simplification of the work of REIFENBERG.

Suppose that $\mathfrak{M}$ is a Riemannian manifold without boundary of class C^n, $n \geq 2$. Let $P_0 \in \mathfrak{M}$. It is well known that there is a (non-unique) coordinate patch τ of class C^n having a domain containing the origin such that

$$h_{ij}(0) = \delta_{ij}, \quad h_{ijw^k}(0) = 0, \tag{10.1.1}$$

the h_{ij} being the components of the metric tensor with respect to τ. To obtain such a mapping, let ω be any coordinate patch of class C^n with range containing P_0; we may clearly assume that its domain D in the y-space contains the origin. By letting T be a properly chosen linear transformation from the y-space to the z-space, we may arrange that $g_{ij}(0) = \delta_{ij}$, the g_{ij} being the components of the metric tensor in the z coordinate system. We obtain τ by letting

$$w^i = z^i + a^i_{jk} z^j z^k \tag{10.1.2}$$

where the a^i_{jk} are constants to be determined. Differentiating the relation

$$g_{ij}(z) = h_{lm}(w) \frac{\partial w^l}{\partial z^i} \frac{\partial w^m}{\partial z^j}$$

with respect to z^k, setting $z = 0$, and using (10.1.1) and (10.1.2), we see that the a^i_{jk} must satisfy

$$a^i_{jk} + a^j_{ik} = g_{ijz^k}(0). \tag{10.1.3}$$

The reader may easily verify that we may take

$$a^i_{jk} = \frac{1}{2}[g_{ijz^k}(0) + g_{ikz^j}(0) - g_{jkz^i}(0)].$$

Definition 10.1.1. A coordinate system ω of class C^2 (at least) is a *normal coordinate system centered at* a point q on $\mathfrak{M} \Leftrightarrow$ the domain G of ω contains the origin, $\omega(0) = q$, $g_{ij}(0) = \delta_{ij}$, the g_{ij} being the components of the metric tensor with respect to ω, the arcs on $\mathfrak{M}$ which correspond to segments in G through 0 are arcs of geodesics, and the distance along such segments equals the corresponding distance along the geodesic arcs.

Lemma 10.1.1. *Suppose $\mathfrak{M}$ is of class C^4. Then*

(a) *Each point q of $\mathfrak{M}$ is the range of a normal coordinate system of class C^2 centered at q.*

(b) *If ω_1 and ω_2 are two such systems both centered at the same point q and both having the domain $B(0, R)$, then $\omega_2^{-1}\omega_1$ is the restriction to $B(0, R)$ of an orthogonal transformation.*

(c) *If ω is such a system and $g_{ij}(z)$ are the components of the metric tensor with respect to ω, then*

$$g_{ij}(t\lambda)\lambda^i\lambda^j = 1 \quad \text{if} \quad \sum_{i=1}^N (\lambda^i)^2 = 1 \quad \textit{and} \quad 0 \leq t \leq R; \tag{10.1.4}$$
$$g_{ijz^k}(0) = 0 \quad \textit{for all} \quad i, j, k.$$

(d) *Given a point P_0 and a mapping τ of class C^4 satisfying* (10.1.1), *there exists a number $R_0 > 0$ and a family ω_p of normal coordinate systems, one ω_p being centered at each point p in $\overline{B(P_0, R_0/3)}$ (on $\mathfrak{M}$) and having the domain $\overline{B(0, R_0)}$, such that the vectors in the tangent space to at p which correspond under ω_p to the unit vectors $e_1, \ldots, e_N$ in R_N are obtained from those corresponding to $e_1, \ldots, e_N$ under τ by the Gram-Schmidt process. If p and $q \in \overline{B(P_0, R_0/3)}$, the mapping*

$$w = U(y; p, q) \equiv \omega_q^{-1}[\omega_p(y)] \tag{10.1.5}$$

is of class C^2 for $(y, p, q) \in \bar{G} \times \overline{\mathfrak{N}} \times \overline{\mathfrak{N}}$ where $\bar{G} = \overline{B(0, R_0/3)} \subset R_N$ and $\overline{\mathfrak{N}} = \overline{B(P_0, R_0/3)}$.

Proof. Clearly part (a) follows from part (d). Part (b) is evident since there is a unique geodesic passing through a given point and having a given direction. The first equality in (10.1.4) follows since the lines $z^i = \lambda^i t$ correspond to geodesics and t is the arc length along the geodesic and

$$(ds/dt)^2 = g_{ij}(t\lambda)\lambda^i\lambda^j \equiv 1. \tag{10.1.6}$$

By differentiating this with respect to t and λ^k we obtain

$$g_{ijz^k}(0)\lambda^i\lambda^j\lambda^k = 0 \quad (\text{all } \lambda) \tag{10.1.7}$$
$$g_{ijz^k}(0)\lambda^i\lambda^j + 2\lim_{t\to 0} t^{-1}[g_{kj}(t\lambda) - \delta_{kj}]\lambda^j = 0$$

for all λ. The second result in (10.1.4) follows from (10.1.7).

To prove (d), we first note that the EULER equations for minimizing the length integral become, *if t is a parameter proportional to arc length*

$$(10.1.8)\qquad \begin{aligned} &\frac{d^2 w^i}{dt^2} + H^i_{jk}(w)\frac{d w^j}{dt}\frac{d w^k}{dt} = 0 \\ &H^i_{jk} = \frac{1}{2} h^{il}[h_{lj\,w^k} + h_{lk\,w^j} - h_{jk\,w^l}]. \end{aligned}$$

So, let P_0 and τ be given as stated. From the existence theorem for systems of ordinary differential equations as applied to the system (10.1.8), we see that the solution function $\mathfrak{U}(t; y, z)$ of the system

$$\begin{aligned} &\mathfrak{U}^i_{tt}(t; y, z) + H^i_{jk}[\mathfrak{U}(t; y, z)]\,\mathfrak{U}^j_t\,\mathfrak{U}^k_t = 0 \\ &\mathfrak{U}^i(0; y, z) = z^i, \quad \mathfrak{U}^i_t(0; y, z) = a^i_j(z)\, y^j \end{aligned}$$

where the vectors $a^i_j(z)\, e_i$ are the vectors in R_N corresponding under τ to the vectors in the tangent space at $p = \tau(z)$ which are obtained from those corresponding under τ to $e_1, \ldots, e_N$ by the Gram-Schmidt process. That is, the $a^i_j(z)$ are uniquely determined by the conditions that

$$(10.1.9)\qquad h_{ij}(z)\, a^i_k(z)\, a^j_l(z) = \delta_{kl}, \quad a^i_k(z) = 0 \quad \text{for} \quad i > k.$$

It is easy to see that $\mathfrak{U} \in C^2(\Omega)$ for some open set Ω containing $(0; 0, 0)$ and that

$$(10.1.10)\qquad \mathfrak{U}(t; \tau y, z) = \mathfrak{U}(t\tau; y, z) \ \text{so}\ \mathfrak{U}(t; y, z) = \mathfrak{U}(1; t y, z).$$

If we define

$$(10.1.11)\qquad U(y, z) = \mathfrak{U}(1; y, z), \quad \omega_p(y) = \tau\{U[y, \tau^{-1}(p)]\}$$

we see that $U \in C^2$ in a domain of the type desired and that the ω_p satisfy the conditions stated.

Remark. Of course $\mathfrak{U}$ is of class C^4 in t and, since the $\Gamma^i_{jk} \in C^2$ (and, in general, nothing more can be said if $\mathfrak{M}$ is only of class C^4), we see that the first and second partial derivatives of $\mathfrak{U}$, $\mathfrak{U}_t$, and $\mathfrak{U}_{tt}$ with respect to y and z are of class C^0 in (t, y, z). Since the h_{ij} are of class C^3, so are the $a^i_j(z)$.

We now state our general assumptions on $\mathfrak{M}$:

General assumptions on $\mathfrak{M}$. *We assume that $\mathfrak{M}$ is a separable Riemannian manifold without boundary of class C^4 and that there are positive numbers R_0, C_0, η_0, K_0, and K_1, which are independent of P_0, such that each point P_0 of $\mathfrak{M}$ is in the range of a coordinate system τ of class C^4 which satisfies* (10.1.1), *has domain $B(0, 4R_0)$, is such that $\tau(0) = P_0$, and for which*

$$\sum_{i,j} (|\nabla^2 h_{ij}(w)| + |\nabla^3 h_{ij}(w)|) \leq K_0.$$

We suppose also that R_0, C_0, and K_0 are so related that

$$(10.1.12)\qquad \begin{aligned} &(1 + C_0 r^2)^{-2} \leq h_{ij}(w)\,\lambda^i \lambda^j \leq (1 + C_0 r^2)^2, \\ &r = |w|, \quad \sum_{i=1}^{N} (\lambda^i)^2 = 1, \quad (1 + 16\, C_0 R_0^2)^2 \leq 5/4. \end{aligned}$$

With each P_0, we associate a definite τ as above, and we associate a family $\{\omega_p\}$ of normal coordinate systems related to $\tau(P_0)$ as in Lemma 10.1.1; *we assume that the number R_0 mentioned there is the same as that mentioned above. Instead of* (10.1.12) *we assume that, for each fixed p, ω_p satisfies*

$$(10.1.13)\quad \begin{aligned} &(1+\eta_0 r)^{-2} \le g_{ij}(y;p)\,\lambda^i\lambda^j \le (1+\eta_0 r)^2, \quad \sum_i (\lambda^i)^2 = 1 \\ &|\Gamma^i_{jk}(y;p)\,\lambda^i\mu^j\mu^k| \le \eta_0\,|\lambda|\cdot|\mu|^2, \quad 0 \le |y| = r \le R_0, \end{aligned}$$

the $g_{ij}(y;p)$ being the ω_p components of the metric tensor and the $\Gamma^i_{jk}(y;p)$ being obtained from the g_{ij} by the corresponding formulas (10.1.8). *Finally, we assume that the second gradients of the function $U(y,p,q)$ in* (10.1.5) *are uniformly bounded by K_1 for each P_0 and all (y,p,q) in $\bar{G} \times \mathfrak{M} \times \mathfrak{M}$.*

Remarks 1. It is probable that many of the results can be carried over to manifolds of class C^n_μ with $n < 4$ $(0 \le \mu \le 1)$. However, a relaxation of the differentiability requirements appears to necessitate a different method of proof. It is shown in § 10.7 that the topological ν-discs (on the minimizing set) are of class C^3_μ for any μ, $0 < \mu < 1$, if $\mathfrak{M}$ is of class C^4, are of class C^n_μ if $\mathfrak{M}$ is of class C^n_μ with $n \ge 4$, and are of class C^∞ or analytic if $\mathfrak{M}$ is.

Remarks 2. It is seen that any compact manifold of class C^4 satisfies all the conditions.

Lemma 10.1.2. *Let ω be a normal coordinate system with domain $\overline{B(0,R_0)}$ and suppose that the $g_{ij}(z)$ are the components of the metric tensor. Then, writing $z = (z^N, z'_N)$, we obtain*

(a) $g_{Nm}(z^N, 0) = \delta_{Nm}$, $|z^N| \le R_0$;

(b) *the angle on $\mathfrak{M}$ at p between an arc γ through p and the geodesic from $q = \omega(0)$ to p is the same as the angle at $y = \omega^{-1}(p)$ between $\omega^{-1}(\gamma)$ and the line $0\,y$.*

Proof. Setting $z^i = t\lambda^i$ in (10.1.6) and using the fact that lines through the origin are solutions of (10.1.8), we obtain

$$(10.1.14)\quad g_{ij}(z)\,z^i z^j \equiv |z|^2, \quad \Gamma^i_{jk}(z)\,z^j z^k \equiv 0, \quad i = 1, \ldots, N.$$

Setting $z^i = 0$ for $i < N$ and using the formulas for Γ^i_{jk}, we obtain

$$(10.1.15)\quad g_{NN}(z^N, 0) = 1, \quad 2g_{Ni\,z^N}(z^N, 0) - g_{NN\,z^i}(z^N, 0) = 0.$$

Differentiating (10.1.14) with respect to z^m, setting $z^i = 0$ for $i < N$, and then using (10.1.15), we arrive at the equation

$$z^N \frac{\partial}{\partial z^N}\left[z^N g_{Nm}(z^N, 0)\right] = 0$$

which leads to (a). (b) along the z^N axis follows from (a) and along any line through the origin follows by a rotation of axes.

Lemma 10.1.3. *For each $\eta > 0$, there is an R'_0 with $0 < R'_0 < R_0/3$, which depends only on $\mathfrak{M}$* (i.e. *R_0, C_0, η_0, K_0, and K_1*) *and η and which has* *the following property: Suppose* $0 < R \leq R_0'$, $B(p, r) \subset B(q, R)$, ω *is a normal coordinate system centered at* q, *and* y' *is any point on* $\omega^{-1}[\partial B(p, r)]$, *then*

$$B_1 \subset \omega^{-1}[B(p, r)] \subset B_2$$

where B_1 *and* B_2 *are the balls in* R_N *of respective radii* $r(1+\eta)^{-1}$ *and* $r(1+\eta)$ *whose boundaries are tangent to* $\omega^{-1}[\partial B(p, r)]$ *at* y'.

Proof. Let $p' = \omega(y')$. Since our results are unaltered by orthogonal transformations in R_N, we may suppose that $\omega = \omega_q$, and that ω_q, ω_p, and $\omega_{p'}$ are all related to a single τ as in Lemma 10.1.1 and our general assumptions. Let σ be the non-homogeneous linear transformation on R_N which carries y' into the origin and which osculates $\omega_{p'}^{-1}\omega_q$ at y'; we write σ in the form

$$z = \sigma(y), \quad \sigma(y') = 0.$$

The angle at the origin in the z-space between two arcs which intersect there is the same as the angle at p' between the arcs which correspond under the transformation $\omega_q \sigma^{-1}$. Also lines in the y-space correspond to those in the z-space.

Let γ be the directed arc in $\overline{B(0, r)}$ which starts at the point $w_0 = \omega_p^{-1}(p') \in \partial B(0, r)$ and is such that $\sigma \omega_q^{-1} \omega_p(\gamma)$ is a segment starting at 0 and making an angle $\theta < \pi/2$ with the inner normal to $\sigma \omega_q^{-1} \omega_p[\partial B(0, r)]$. From Lemma 10.1.2, it follows that γ makes the angle θ with the radius vector from w_0 to 0. Since $U(y; q, q) = y$ and the second gradients of U (including mixed ones) are bounded by K_1, we see that

$$(1+K_1 R)^{-1} \leq \frac{ds(z)}{ds(y)} \leq (1+K_1 R), \quad (1+K_1 R)^{-1} \leq \frac{ds(y)}{ds(w)} \leq (1+K R),$$

$$(1+K_1 R)^{-2} \leq \frac{ds(z)}{ds(w)} \leq (1+K_1 R)^2, \tag{10.1.16}$$

w being the coordinates in the ω_p system.

Now, let us introduce the Euclidean metric in the z-space and let

$$g_{0ij}(w) = \sum_{k=1}^{N} \frac{\partial z^k}{\partial w^i} \frac{\partial z^k}{\partial w^j}.$$

Then the arc γ is a solution of the equations

$$\frac{d^2 w^i}{dt^2} + \Gamma^i_{0jk}(w) \frac{dw^j}{dt} \frac{dw^k}{dt} = 0 \tag{10.1.17}$$

where the Γ^i_{0jk} are defined as usual in terms of the g_{0kj} and t is the distance in the z-space. From our assumptions it follows that

$$|\Gamma^i_{0jk}(w) \lambda^i \mu^j \mu^k| \leq Z_1(\mathfrak{M}) |\lambda| \cdot |\mu|^2. \tag{10.1.18}$$

Now define

$$v(t) = [d(p_t, p)]^2 = \sum_{i=1}^{N} (w^i)^2.$$

Letting dots denote differentiation with respect to t and using (10.1.17), we obtain

$$(10.1.19)\quad \begin{aligned} v(0) &= r^2, \quad \dot v(0) = 2\sum_{i=1}^{N} w^i(0)\, \dot w^i(0) = -2r\,|\dot w(0)|\cos\theta \\ \ddot v(t) &= 2\sum_{i=1}^{N} (\dot w^i)^2 - 2\sum_{i,j,k=1}^{N} \Gamma^i_{0jk}(w)\, w^i\, \dot w^j\, \dot w^k . \end{aligned}$$

Since t is distance in the z-space, (10.1.16), (10.1.18), and (10.1.19) yield

$$(10.1.20)\quad \begin{aligned} & r^2 - 2(1+K_1 R)^2\, t\, r\cos\theta + (1+K_1 R)^{-4}(1-Z_1 R)\, t^2 \le v(t) \\ & \le r^2 - 2(1+K_1 R)^{-2}\, t r\cos\theta + (1+K_1 R)^4 (1+Z_1 R)\, t^2, \\ & 1 - Z_1 R > 0 . \end{aligned}$$

By letting t^* be the first positive value of t for which $v(t) = r^2$, we find from (10.1.20) that

$$(10.1.21)\quad \begin{aligned} & 2(1+K_1 R)^{-6}(1+Z_1 R)^{-1} r\cos\theta \le t^* \\ & \le 2(1+K_1 R)^6 (1-Z_1 R)^{-1} r\cos\theta . \end{aligned}$$

This proves the theorem with $\omega^{-1}[B(p, r)]$ replaced by $\sigma\,\omega^{-1}[B(p, r)]$.

Call the balls in the z-space B_1' and B_2'. Their respective radii are

$$(10.1.22)\quad (1+K_1 R)^{-6}(1+Z_1 R)^{-1}\cdot r \text{ and } (1+K_1 R)^6 (1-Z_1 R)^{-1}\cdot r,$$

and we have

$$\sigma^{-1}(B_1') \subset \omega^{-1}[B(p, r)] \subset \sigma^{-1}(B_2')$$

where $\sigma^{-1}(B_1')$ and $\sigma^{-1}(B_2')$ are ellipsoids with min. and max. radii

$$\sigma^{-1}(B_1'):\ (1+K_1 R)^{-8}(1+Z_1 R)^{-1} r \text{ and } (1+K_1 R)^{-4}(1+Z_1 R)^{-1} r,$$

$$\sigma^{-1}(B_2'):\ (1+K_1 R)^{4}(1-Z_1 R)^{-1} r \text{ and } (1+K_1 R)^{8}(1-Z_1 R)^{-1} r,$$

respectively; these ellipsoids are tangent to $\omega_q^{-1}[\partial B(p, r)]$ at $\omega_q^{-1}(p') = y'$. If R is sufficiently small, the ellipsoids B_1' and B_2' are nearly balls and balls B_1 and B_2 can be chosen with $B_1 \subset B_1'$ and $B_2 \supset B_2'$, which have respective radii $(1+\eta)^{-1} r$ and $(1+\eta)\, r$ and which are tangent to $\omega_q^{-1}[\partial B(p, r)]$ at y'. The last statement is proved similarly.

Definition 10.1.2. Given a set S, we define (S, ϱ) as the set of all points within a distance ϱ of S; i.e.

$$(S, \varrho) = \bigcup_{P\in S} B(P, \varrho) .$$

If S_1 and S_2 are compact sets, we define their *point set distance* $D(S_1, S_2)$ as the smallest number ϱ such that $S_1 \subset \overline{(S_2, \varrho)}$ and $S_2 \subset \overline{(S_1, \varrho)}$.

Definition 10.1.3. *A geodesic k-plane centered at a point P of $\mathfrak{M}$* is a locus of the form $\omega[\Pi \cap B(0, R_0)]$ where Π is a k-plane in R_N through 0 and ω is a normal coordinate system with domain $B(0, R_0)$ for which $\omega(0) = P$.

Lemma 10.1.4. *Suppose ω is a normal coordinate system with domain $B(0, R_0)$ and range $B(P_0, R_0)$ for which $\omega(0) = P_0$. Let P be a point in $B(0, R)$, let S_1 and S_2 be compact subsets of $B(0, R)$, and let $\omega(P) = Q$ and $\omega(S_k) = T_k$, $k = 1, 2$, $0 < R \leq R_0$. Then*

$$\begin{aligned} &(1+\eta_0 R)^{-1} d(P, S_1) \leq d(Q, T_1) \leq (1+\eta_0 R)\, d(P, S_1) \\ (10.1.23)\quad &(1+\eta_0 R)^{-1} D(S_1, S_2) \leq D(T_1, T_2) \leq (1+\eta_0 R)\, D(S_1, S_2) \\ &(1+\eta_0 R)^{-1} \varrho(S_1) \leq \varrho(T_1) \leq (1+\eta_0 R)\, \varrho(S_1) \end{aligned}$$

where $\varrho(S)$ denotes the radius of the smallest sphere containing S_1 (compact).

Proof. If S_1 consists of a single point, the first line of (10.1.23) follows from our assumption (10.1.13). There is a point P_1 in S_1 such that $d(P, P_1) = d(P, S_1)$. If we let $Q_1 = \omega(P_1)$, then

$$d(Q, T_1) \leq d(Q, Q_1) \leq (1+\eta_0 R)\, d(P, P_1) = (1+\eta_0 R)\, d(P, S_1).$$

The other inequalities are proved similarly.

We conclude this section with the following lemma:

Lemma 10.1.5. *There are constants $R_0'' = R_0''(\mathfrak{M}) \leq R_0/3$ and $K_2 = K_2(\mathfrak{M})$ with the following property: Suppose that $\sum$ is a geodesic k-plane centered at a point $p \in B(q, R)$ where $0 < R \leq R_0''$. Suppose ω is a normal coordinate system with domain $B(0, R_0)$ for which $\omega(0) = q$ and suppose $\sum_0$ is the k-plane in R_N which is tangent at $\omega^{-1}(p)$ to $\omega^{-1}(\sum)$. Then*

$$\begin{aligned} &\omega^{-1}[\textstyle\sum \cap B(q, R)] \subset (\textstyle\sum_0, K_2 R^2) \\ &\omega^{-1}[\textstyle\sum \cap B(p, r)] \subset (\textstyle\sum_0, K_2 r^2), \quad r \leq R_0''. \end{aligned}$$

Proof. If the coordinates in the system ω are w and the Γ^i_{jk} are defined as usual, the arcs in $B(0, R_0)$ corresponding to the geodesics are solutions of the system (10.1.17) with Γ^i_{0jk} replaced by Γ^i_{jk}. Since the geodesics making up $\sum \cap B(q, R)$ correspond to arcs tangent to $\sum_0$ at $\omega^{-1}(p)$ and since $\sum \cap B(q, R) \subset \sum \cap B(p, 2R)$ so that no such arc is of length $> 2R$, the result follows from the assumptions (10.1.13). The last statement follows since the second derivatives of the transformations $\omega_{0p}^{-1}\, \omega_{0q}$, etc., are uniformly bounded.

10.2. ν surfaces, their boundaries, and their Hausdorff measures

In this section, we define the ν-dimensional Hausdorff measures of sets in a metric space and prove certain theorems about these measures. Most of the less standard theorems presented here are due to REIFENBERG [1]. We also introduce our notions of ν-surfaces and their "algebraic boundaries". Most of the topological results needed for the remainder of this chapter are presented in the following section. The presentation there given makes extensive use of the general theory as presented in the book "Foundations of Algebraic Topology" by EILENBERG and STEENROD, especially Chapter 1, §§ 1—14, Chapter 9, § 7, and Chapter 10, §§ 2 and 5.

Definition 10.2.1. Suppose S is a set in a metric space and that $\nu \geq 1$ and $\delta > 0$. If S is empty, we define ${}^*\Lambda^\nu_\delta(S) = 0$; otherwise, we define ${}^*\Lambda^\nu_\delta(S)$ as the inf. of $\sum_i \gamma_\nu r_i^\nu$ for all coverings of S by finite or countable families of balls $\{B(P_i, r_i)\}$ in which each $r_i < \delta$. We then define the *Hausdorff outer measure* ${}^*\Lambda^\nu$ by

$$(10.2.1) \qquad {}^*\Lambda^\nu(S) = \lim_{\delta\to 0^+} {}^*\Lambda^\nu_\delta(S).$$

Λ^ν-measurable sets are then defined as usual. If $\nu = 0$, we define ${}^*\Lambda^0_\delta(S)$ as the inf. of the number of balls of radii $<\delta$ required to cover S and then define ${}^*\Lambda^0(S)$ by (10.2.1).

The following lemma is well-known and we omit the proof:

Lemma 10.2.1. (a) *Borel sets are measurable Λ^ν for each ν.*

(b) *Any set is measurable Λ^0 and $\Lambda^0(S)$ is the number of points in S.*

(c) *If S is a set in a metric space $\mathfrak{M}$ and $\tau : S \to \mathfrak{M}'$ is a mapping in which $\mathfrak{M}'$ is a metric space and $d[\tau(p), \tau(q)] \leq \lambda\, d(p, q)$ for p and q on S, then*

$${}^*\Lambda^\nu[\tau(S)] \leq \lambda^\nu\, {}^*\Lambda^\nu(S).$$

Definition 10.2.2. Suppose G is a domain on a manifold X of class C^1 and suppose $\tau \in C^1(\bar{G})$ where τ is a mapping into the Riemannian manifold $\mathfrak{M}$ of class C^1. We define

$$L_\nu(\tau) = \int_G F(p; \tau)\, dS(p)$$

where if $\sigma : \Gamma \to G$ and $\omega : \Omega \to \mathfrak{M}$ are coordinate patches with domains $\Gamma \subset R_\nu$ and $\Omega \subset R_N$ and ranges containing p and $\tau(p)$, respectively, we have

$$F(p; \tau)\, dS(p) = [\gamma(x)]^{1/2}\, dx, \quad \gamma(x) = \det(\gamma_{\alpha\beta}(x))$$

$$\gamma_{\alpha\beta}(x) = \sum_{i,j=1}^{N} G_{ij}[z(x)] \frac{\partial z^i}{\partial x^\alpha} \frac{\partial z^j}{\partial x^\beta}, \quad (p = \sigma(x))$$

$$dS(p) = [g(x)]^{1/2}\, dx, \quad g(x) = \det(g_{\alpha\beta}(x))$$

where $G_{ij}(z)$ and $g_{\alpha\beta}(x)$ are the components of the metric tensors on $\mathfrak{M}$ and X with respect to the coordinate patches ω and σ, respectively, and $z(x) = \omega^{-1}\tau\sigma(x)$.

Lemma 10.2.2. *Suppose that G, X, $\mathfrak{M}$, and τ have their significance as in Definition* 10.2.2 *and suppose that τ is a diffeomorphism. Then*

$$L_\nu(\tau) = \Lambda^\nu[\tau(G)].$$

This is proved by first proving the formula

$$\int_G f[\tau(p)] \cdot F[p; \tau]\, dS(p) = \int_{\tau(G)} f(q)\, d\Lambda^\nu(q)$$

for $f \in C^1[\tau(G)]$. This is proved first for functions with support in a small neighborhood of a given point. The proof in that case is elementary.

Definition 10.2.3. A family $\mathfrak{F}$ of sets T is said to *cover a set S in the sense of Vitali* iff each point P of S is in a set T of $\mathfrak{F}$ of arbitrarily small diameter.

Definition 10.2.4. Let $\mathfrak{F}$ be a family of sets T. By $\sigma(\mathfrak{F})$, we mean the union of all the sets T for $T \in \mathfrak{F}$. In case $\mathfrak{F}$ is countable and $\mathfrak{F} = \{B(P_i, r_i)\}$ we still use the notation

$$\sigma(\mathfrak{F}) = \bigcup_{i=1}^{\infty} B(P_i, r_i).$$

Lemma 10.2.3. (cf. MORSE, A. P. [2]) *Suppose that $\mathfrak{M}$ satisfies the conditions of* § 10.1. *Suppose that S is a bounded set and that $\mathfrak{F}$ is a family of closed balls in $\mathfrak{M}$ which covers S in the sense of Vitali. Then there is a countable disjoint sub-family $\{\overline{B(P_i, r_i)}, i = 1, 2, \ldots\}$ such that*

$$S - \bigcup_{i=1}^{\infty} \overline{B(P_i, r_i)} \subset \bigcup_{i=k}^{\infty} \overline{B(P_i, 5r_i)}, \quad k = 1, 2, \ldots$$

Proof. We suppose that $S \subset B(P_0, r_0)$ and let $\mathfrak{F}'$ be the family of all balls $B(P, r) \in \mathfrak{F}$ such that $B(P, r) \subset B(P_0, r_0)$ and $B(P, r) \cap S$ is not empty. Then $\mathfrak{F}'$ still covers S in the sense of Vitali.

We now define the radii R_i, the balls $B(P_i, r_i)$, and the families $\mathfrak{F}_i$ by induction as follows: $R_1 = \sup r$ for $B(P, r) \in \mathfrak{F}'$, $\overline{B(P_1, r_1)}$ is a ball of $\mathfrak{F}'$ in which $r_1 > R_1/2$, and $\mathfrak{F}_1$ consists of those balls of $\mathfrak{F}'$ which do not intersect $B(P_1, r_1)$. Having defined $R_1, \ldots, R_k$, $\overline{B(P_i, r_i)}$ for $i = 1, \ldots, k$, and $\mathfrak{F}_1, \ldots, \mathfrak{F}_k$, we define $R_{k+1} = \sup r$ for $B(P, r) \in \mathfrak{F}_k$, $\overline{B(P_{k+1}, r_{k+1})}$ as a ball in $\mathfrak{F}_k$ for which $r_{k+1} > R_{k+1}/2$, and $\mathfrak{F}_{k+1}$ as those balls in $\mathfrak{F}_k$ which do not intersect $\overline{B(P_{k+1}, r_{k+1})}$. We see by induction that any ball in $\mathfrak{F}' - \mathfrak{F}_1$ must lie in $\overline{B(P_1, 5r_1)}$ and, in general, any ball in $\mathfrak{F}_k - \mathfrak{F}_{k+1}$ must lie in $\overline{B(P_{k+1}, 5r_{k+1})}$. By induction, we conclude that the $\overline{B(P_i, r_i)}$ are disjoint and that

$$\sigma(\mathfrak{F}_k) \subset \bigcup_{i=k+1}^{\infty} \overline{B(P_i, 5r_i)} \subset \bigcup_{i=j}^{\infty} \overline{B(P_i, 5r_i)} \quad \text{if} \quad j \leq k+1.$$

Now, suppose $P \in S - \bigcup_{i=1}^{k} \overline{B(P_i, r_i)}$. Then, since the finite union is closed $P \in$ some ball $\overline{B(P', r)}$ which does not intersect any $\overline{B(P_i, r_i)}$ with $i \leq k$. Thus $B(P', r) \in \mathfrak{F}_k$ and hence $P \in \sigma(\mathfrak{F}_k)$. The result follows since

$$S - \bigcup_{i=1}^{\infty} \overline{B(P_i, r_i)} = \bigcap_{j=1}^{\infty} \left[S - \bigcup_{i=1}^{j} \overline{B(P_i, r_i)} \right].$$

Definition 10.2.5. Suppose S is a set, P is a point, and Π is a k-plane, all being in R_N. We define the *cone* $C(P, S)$ to consist of all the segments PQ for Q in S. We define $C(\Pi, S)$ to consist of all segments PQ where $P \in S$ and Q is its projection on Π. If $\mathfrak{M}$ satisfies the conditions of § 10.1, ω is a normal coordinate system with origin at P and domain containing

$C(P, S)$, we define the *geodesic cone* $C(P', S') = \omega[C(P, S)]$, where $P' = \omega(P)$ and $S' = \omega(S)$.

Theorem 10.2.1. *Suppose $S \subset R_N$, S is a Borel set, and $\Lambda^{\nu-1}(S) < \infty$. Then, for any P,*

$$\Lambda^{\nu}[C(P, S)] \leq \nu^{-1} r \Lambda^{\nu-1}(S) \quad \text{if} \quad S \subset \overline{B(P, r)}.$$

The proof is similar to but simpler than that of the following theorem and is left to the reader.

Theorem 10.2.2. *Suppose $S \subset R_N$, $\Lambda^{\nu-1}(S) < \infty$, and Π is a p-plane with $p < N$. Then*

$$\Lambda^{\nu}[C(\Pi, S)] \leq C_1(\nu) \cdot r \cdot \Lambda^{\nu-1}(S) \quad \text{if} \quad S \subset (\Pi, r) \equiv \bigcup_{P \in \Pi} B(P, r).$$

Proof. The theorem is evident if $\nu = 1$. So we suppose $\nu > 1$. We cover S with a countable family of balls $\{B(P_i, r_i)\}$ such that

$$(10.2.2) \qquad \sum_i \gamma_{\nu-1} r_i^{\nu-1} < \Lambda^{\nu-1}(S) + \varepsilon, \quad r_i < \delta, \quad i = 1, 2, \ldots,$$

ε and δ being arbitrary positive numbers. Let P_{ij}, $0 \leq j \leq (r - r_i)/r_i$ be the point on the segment $I(P_i)$, joining P_i to its projection Q_i on Π, which is at a distance $j r_i$ from P_i. If $P \in B(P_i, r_i)$, we see that

$$I(P) \subset \bigcup_j B(P_{ij}, 2r_i).$$

Accordingly, since $j \leq r/r_i$, we obtain

$$\Lambda^{\nu}_{2\delta}[C(\Pi, S)] \leq \sum_i \frac{r}{r_i} \gamma_\nu (2r_i)^\nu \leq r(2^\nu \gamma_\nu/\gamma_{\nu-1}) \cdot \sum_i \gamma_{\nu-1} r_i^{\nu-1}$$
$$\leq C_1(\nu) \cdot r \cdot [\Lambda^{\nu-1}_\delta(S) + \varepsilon].$$

The result follows from the arbitrariness of δ and ε.

Theorem 10.2.3. *Suppose $\mathfrak{M}$ satisfies the conditions of § 10.1, S is Λ^ν-measurable with $\Lambda^\nu(S) < \infty$, $S \subset G \subset \mathfrak{M}$ where G is open, and $U \in C^1(G)$, $\nabla U \neq 0$ on G and U satisfies a Lipschitz condition with Lipschitz constant M. Then*

$$\int_{-\infty}^{\infty} \Lambda^{\nu-1}(S \cap C_h)\, dh \leq M \Lambda^\nu(S),$$

C_h being the subset of G where $U(P) = h$.

Proof. It is sufficient to prove this for S compact; then $|\nabla U(P)| \geq c > 0$ for P on S. Let ε and δ be small positive numbers and cover S with a family $\{B(P_i, r_i)\}$ of balls such that

$$\sum_i \gamma_\nu r_i^\nu < \Lambda^\nu_\delta(S) + \varepsilon, \quad B(P_i, r_i) \subset G. \quad (r_i < \delta).$$

Now, for each i,

$$\Lambda^{\nu-1}_\delta[S_h \cap B(P_i, r_i)] \leq \gamma_{\nu-1}[\varrho_i(h)]^{\nu-1}, \quad S_h = S \cap C_h,$$

where $\varrho_i(h)$ is the radius of the smallest closed ball containing $C_h \cap B(P_i, r_i)$. This intersection is non empty only for

$$U(P_i) - [r_i \cdot |\nabla U(P_i)| + 0(\delta)] \leq h \leq U(P_i) + [r_i |\nabla U(P_i)| + 0(\delta)].$$

For such h,

$$\varrho_i(h) = \sqrt{r_i^2 - h'^2}, \quad h' = [|\nabla U(P_i)| + 0(\delta)]^{-1} \cdot [h - U(P_i)].$$

Thus

$$\int \Lambda_\delta^{\nu-1}(S_h)\, dh \leq \sum_i \int_{-r_i}^{r_i} |\nabla U(P_i)|\, \gamma_{\nu-1} [\varrho_i(h)]^{\nu-1}\, dh' + 0(\delta)$$

$$\leq M \sum_i \gamma_\nu r_i^\nu + 0(\delta) \leq M \Lambda_\delta^\nu(S) + 0(\delta) + \varepsilon.$$

The result follows.

Definition 10.2.6. A ν-surface is merely a compact set (in $\mathfrak{M}$) X. In case A is a compact subset of X and $\nu > 1$, we define the *algebraic boundary* $b(X, A)$ of X *with respect to* A (more properly $b(X, A, G)$, G being a group of coefficients; but we shall suppress G) as the kernel of the homomorphism $i_*: H_{\nu-1}(A) \to H_{\nu-1}(X)$ (i.e. $H_{\nu-1}(A, G) \to H_{\nu-1}(X, G)$); here i is the inclusion mapping from A into X and i_* is the corresponding homomorphism. The Čech homology theory is used.

Remarks. In case X is a compact orientable ν-manifold of class C^2 with ordinary boundary the connected $(\nu - 1)$-manifold A of class C^2, then $b(X, A) = H_{\nu-1}(A)$ (for any G). Of course our definition allows X and A to be arbitrary compact sets with $A \subset X$. If $A = X$, then $b(X, A) = 0$, an uninteresting case. But if L is a non-zero subgroup of $H_{\nu-1}(A)$, the class $\mathfrak{C}(A, L)$ of sets X for which $b((X, A) \supset L$ is an interesting class and, as we shall see, large enough to contain an X such that $X - A$ has minimum HAUSDORFF Λ^ν-measure. The topological theorems which ensure this are the following:

(i) *Each class* $\mathfrak{C}((A, L)$ *is closed under point set convergence* (see Definition 10.1.2) of X_n to X (i.e. if each $X_n \in \mathfrak{C}((A, L)$, then $X \in \mathfrak{C}((A, L))$.

(ii) *Suppose* $X \in \mathfrak{C}(A, L)$, G *is open* (in $\mathfrak{M}$), $\bar{G} \cap A$ *is empty*, $X \cap \partial G = B$, U *is compact*, $U \subset \bar{G}$, $U \cap \partial G = B$, *and* $b(U, B) \supset B(X_1, B)$ *where* $X_1 = \overline{(X \cap G)}$. *Then the surface* X', *obtained from* X *by replacing* $X \cap G$ *by* $U \cap G$, *also* $\in \mathfrak{C}(A, L)$.

(i) is just Theorem 10.3.16 and (ii) is Theorem 10.2.5 below. The theorems in § 10.3 and many others were proved by J. F. ADAMS in the Appendix to the paper [1] of REIFENBERG mentioned above. That Appendix contains many further examples illustrating the notion of algebraic boundary.

Lemma 10.2.4. *Suppose* A *is a compact set in* R_N, P *is a point in* R_N, *and* $X = C(P, A)$. *Then* $b(X, A) = H_{\nu-1}(A)$ *if* $\nu > 1$.

This follows from Theorem 10.3.2 since X is contractible.

The following isoperimetric inequality of REIFENBERG is important:

Theorem 10.2.4. *Suppose* $\mathfrak{M}$ *and* R_0 *satisfy the conditions of* § 10.1. *Then there are constants* $C_2(\nu, \mathfrak{M})$ *and* $C_3(\nu, \mathfrak{M})$, $2 \leq \nu \leq N$, *with the following property: If* A *is compact*, $A \subset$ *some ball* $B(P_0, R_0)$, *and* $\Lambda^{\nu-1}(A) = l^{\nu-1} < +\infty$, *there exists a surface* X *such that* $b(X, A)$

$\supset H_{\nu-1}(A)$, *X is in the geodesically convex hull of A and within a distance* $\leq C_2\, l$ *of A, and*

$$\Lambda^\nu(X) \leq C_3 \cdot l^\nu. \tag{10.2.3}$$

Proof. By virtue of the results of § 10.1 and Lemma 10.2.1 (c), it is sufficient to prove this for $\mathfrak{M} = R_N$ in which case R_0 may be arbitrary. Accordingly we assume that $A \subset R_N$ and carry out the construction in R_N.

We prove this by induction on ν. We define

$$\varphi(a) = \sum_{s=-\infty}^{\infty} \Lambda^0(A \cap \Pi^1_{a+1.1ls}); \quad \Pi^1_a : x^1 = a.$$

From Theorem 10.2.3, it follows that

$$\int_0^{1.1l} \varphi(a)\, da \leq \Lambda^1(A) = l.$$

Accordingly there is a value of a for which $\varphi(a) < 1$ and hence 0. By repeating this construction along each axis, we see that we may divide R_N into cubes of side 1.1 l, no one of which contains a point of A on its boundary. Of course only a finite number contain points of A; call them $R_1, \ldots, R_S$. In each R_s, select a P_s interior to the convex hull of $A \cap R_s$ and define

$$X = \bigcup_{s=1}^{S} C(P_s, A_s), \quad A_s = A \cap R_s.$$

That $b(X, A) = H_{\nu-1}(A)$ follows from Theorem 10.3.3. The other properties are evident.

Suppose, now that $\nu > 2$, $N \geq \nu$, and that the theorem has been proved for all (ν', N') in which $2 \leq \nu' < \nu$ and $N' \geq \nu'$. Let A be given and define

$$\varphi^1(a^1) = \sum_{s=-\infty}^{\infty} \Lambda^{\nu-2}(A \cap \Pi^1_{a^1+ls}).$$

Then there is a value $\bar{a}^1$, $0 \leq \bar{a}^1 < l$, such that $\varphi^1(\bar{a}^1) \leq l^{\nu-2}$. Let $\Pi^1_{s^1} = \Pi^1_{\bar{a}^1+ls^1}$ and $D^1_{s^1} = A \cap \Pi^1_{s^1}$. Then

$$\sum_{s^1=-\infty}^{\infty} \Lambda^{\nu-2}(D^1_{s^1}) \leq l^{\nu-2}.$$

From our induction hypothesis it follows that we can find $(\nu-1)$-surfaces $B^1_{s^1} \subset \Pi^1_{s^1}$ such that $b(B^1_{s^1}, D^1_{s^1}) = H_{\nu-2}(D^1_{s^1})$, $B^1_{s^1}$ is in the convex cover of $D^1_{s^1}$ and within a distance $C\, l^1_{s^1} \leq C\, l$ of $D^1_{s^1}$ and satisfies (10.2.3) for $\nu - 1$. If we let $A^1_{s^1}$ be the part of A for which $\bar{a}^1 + l(s^1 - 1) \leq x^1 \leq \bar{a}^1 + l\, s^1$ and define

$$C^1 = A \cup \bigcup_{s^1} B^1_{s^1}, \quad C^1_{s^1} = B^1_{s^1-1} \cup A^1_{s^1} \cup B^1_{s^1}, \quad (l^1_{s^1})^{\nu-2} = \Lambda^{\nu-2}(D^1_{s^1}),$$

then the hypotheses of Theorem 10.3.11 are satisfied (with $r = s^1 +$ $+$ some integer) and also

$$\sum_{s^1} \Lambda^{\nu-1}(B^1_{s^1}) \le C \cdot \sum_{s^1} [\Lambda^{\nu-2}(D^1_{s^1})]^{(\nu-1)/(\nu-2)}$$
$$\le C \left\{\sum_{s^1} \Lambda^{\nu-2}(D^1_{s^1})\right\}^{(\nu-1)/(\nu-2)} \le C\, l^{\nu-1}.$$

For each s^1, we replace A by $C^1_{s^1}$ and repeat the construction above along the x^2 axis obtaining the surfaces $D^{12}_{s^1 s^2}$ spanned as above by $B^{12}_{s^1 s^2}$ satisfying (10.2.3). We let $A^{12}_{s^1 s^2}$ be the part of $C^1_{s^1}$ for which $\bar{a}^{12}_{s^1} +$ $+ l(s^2 - 1) \le x^2 \le \bar{a}^{12}_{s^1} + l\, s^2$ and define

$$C^{12}_{s^1} = C^1_{s^1} \cup \bigcup_{s^2} B^{12}_{s^1 s^2}, \quad C^{12}_{s^1 s^2} = B^{12}_{s^1, s^2-1} \cup A^{12}_{s^1 s^2} \cup B^{12}_{s^1 s^2}.$$

Again, the hypotheses of Theorem 10.3.11 are satisfied (with $r = s^2 +$ $+$ some integer) and also

$$\sum_{s^1, s^2} \Lambda^{\nu-1}(B^{12}_{s^1 s^2}) \le C\, l^{\nu-1}.$$

Then, for each (s^1, s^2), we perform the same construction along the x^3 axis with A replaced by $C^{12}_{s^1 s^2}$. This process is continued until all the axes are exhausted.

Then each $C^{1\ldots N}_{s^1 \ldots s^N}$ is in the closed cube

$$\bar{a}^1 + l(s^1 - 1) \le x^1 \le \bar{a}^1 + l\, s^1, \quad \bar{a}^{12}_{s^1} + l(s_2 - 1) \le x^2 \le \bar{a}^{12}_{s^2;} + l\, s_2, \ldots,$$

of side l. For each $C^{1\ldots N}_{s^1 \ldots s^N}$, select a point $P^{1\ldots N}_{s^1 \ldots s^N}$ interior to the convex cover of $C^{1\ldots N}_{s^1 \ldots s^N}$ and define $X^{1\ldots N}_{s^1 \ldots s^N} = C(P^{1\ldots N}_{s^1 \ldots s^N}, C^{1\ldots N}_{s^1 \ldots s^N})$

$$X^{1\ldots N-1}_{s^1 \ldots s^{N-1}} = \bigcup_{s^N} X^{1\ldots N}_{s^1 \ldots s^N}, \ldots, X = \bigcup_{s^1} X^1_{s^1}.$$

Then, by using Lemma 10.2.4 first and then Theorems 10.3.9, 10.3.10, and 10.3.11 repeatedly, with $A = C^{1\ldots k}_{s^1 \ldots s^k}$ and $A_{s^{k+1}} = C^{1\ldots k+1}_{s^1 \ldots s^{k+1}}$, X_{s+1} $= X^{1\ldots k+1}_{s^1 \ldots s^{k+1}}$, we see in turn that

$$b(X^{1\ldots N-1}_{s^1 \ldots s^{N-1}}, C^{1\ldots N-1}_{s^1 \ldots s^{N-1}}) = H_{\nu-1}(C^{1\ldots N-1}_{s^1 \ldots s^{N-1}}), \ldots,$$
$$b(X, A) = H_{\nu-1}(A)$$

and that all the other conditions are satisfied.

Theorem 10.2.5. *Suppose L is a subgroup of $H_{\nu-1}(A)$, $b(X, A) \supset L$, and G is an open set such that $\bar{G} \cap A$ is empty. Define $X_1 = \overline{(X \cap G)}$, $A_1 = X_1 - G$, and $Y = Y_1 \cup Y_2$, where $Y_2 = X - G$ and Y_1 is any surface such that $b(Y_1, A_1) \supset b(X_1, A_1)$ and $Y_1 \cap Y_2 \subset A_1$. Then $b(Y, A) \supset L$.*

Proof. Define $X_2 = Y_2$, $A_2 = A \cup A_1 = B$, $L_1 = b(X_1, A_1)$, L_2 $= b(X_2, A_2)$, $L_1' = b(Y_1, A_1)$, and $L_2' = L_2 = b(Y_2, A_2)$. Then the hypotheses of both Theorems 10.3.9 and 10.3.10 are satisfied by $(X, X_1,$

X_2) and (Y, Y_1, Y_2). Thus

$$b(X, A) = i(B, A)_*^{-1}[i(B, A_1)_* L_1 + i(B, A_2)_* L_2]$$

$$b(Y, A) = i(B, A)_*^{-1}[i(B, A_1)_* L_1' + i(B, A_2)_* L_2'].$$

Since $L_1' \supset L_1$ and $L_2' = L_2$, it follows that $b(Y, A) \supset b(X, A)$.

10.3. The topological results of Adams (see REIFENBERG [1])

We shall assume that all sets are subsets of a HAUSDORFF space. (X, A) is a pair (of compact sets in which $A \subset X$). $b_A(X)$ or $b(X; A) = \text{Ker}\, i_*$: $: H_{\nu-1}(A) \to H_{\nu-1}(X)$ where $i: A \to X$ is the inclusion map. We shall denote the inclusion map of A into X by $i(X, A) : i(X, A)(a) = a$ if $a \in A$. Let P_0 denote a "base point" and also the set $\{P_0\}$. Given X, we let $f_X : X \to P_0$ be defined by $f_X(x) = P_0$ for $x \in X$. We let ε_X (or $\varepsilon(X)$) : $H_0(X) \to G$ be the map γf_{X*} where γ is the isomorphism from $H_0(P_0)$ onto G (Actually, in the book by EILENBERG-STEENROD, G is defined as $H_0(P_0)$, in which case γ is just the identity). ε_X is called the "augmentation homomorphism" by ADAMS (REIFENBERG [1]). Theorems 1—8 below are just the respective Lemmas 1A—8A of the paper REIFENBERG [1].

Theorem 10.3.1. *If $X = A$, then $b(X, A) = 0$.*

Proof. For $i(X, A)_*: H_{\nu-1}(A) \to H_{\nu-1}(X)$ is the identity so no elements $a \neq 0$ in A are carried into 0.

Definition 10.3.1. A set X is said to be *contractible* (on itself to a point) iff $\exists$ a continuous map $h : X \times I \to X$ such that $h(x, 0) = x$ and $h(x, 1) = x_0$ for all x on X, x_0 being a fixed point.

Theorem 10.3.2. *If X is contractible, then $b(X, A) = H_{\nu-1}(A)$ if $\nu > 1$, and $b(X, A) = Ker\, \varepsilon_A$ if $\nu = 1$.*

Proof. From EILENBERG-STEENROD, Chapter 1, Theorem 11.5, it follows that $H_q(X) = 0$ for every $q \neq 0$ and $\tilde{H}_0(X) = 0$ (see Chap. 1, § 7) and $H_0(X)$ is isomorphic with G. By definition

$$b(X, A) = Ker\, i(X, A)_* = H_{\nu-1}(A) \text{ if } \nu > 1$$

since $H_{\nu-1}(X) = 0$ if $\nu > 1$. In case $\nu = 1$

$$b(X, A) = Ker\, i_*(X, A) = Ker\, \varepsilon_A$$

since $\varepsilon_A = \gamma f_{X*} i_*(X, A)$ and γf_{X*} is an isomorphism.

Theorem 10.3.3. *Suppose that $X = \bigcup_{r=1}^{n} X_r$ where the X_r are disjoint and contractible. Define $A_r = A \cap X_r$, $\varepsilon_r = \varepsilon(A_r)$, and*

$$K_0 = \sum_r i(A, A_r)_*\, Ker\, \varepsilon_r \subset H_0(A).$$

Then $b(X, A) = H_{\nu-1}(A)$ if $\nu > 1$ and $b(X, A) = K_0$ if $\nu = 1$.

Proof. From $E-S$[1], Chap. 1, Theorem 13.2, it follows, by replacing the general pair (X, A) by $(A, 0)$ and $(X, 0)$, that

$$H_{\nu-1}(X) \simeq \sum_{r=1}^{n} H_{\nu-1}(X_r), \quad H_0(A) \simeq \sum_{r} H_0(A_r)$$

and, in fact, each a in $H_0(A)$ is uniquely representable in the form

$$a = \sum_{r=1}^{n} i(A, A_r)_* \, a_r, \quad a_r \in H_0(A_r) \tag{10.3.1}$$

and the correspoinding result holds for X and X_r. If $\nu > 1$, it follows as in Theorem 10.3.2 that each $H_{\nu-1}(X_r) = 0$ so that $H_{\nu-1}(X) = 0$ and the theorem follows in that case as before.

Now, let δ and θ be the isomorphisms above from $H_0(A)$ onto $\sum_r H_0(A_r)$ and from $H_0(X)$ onto $\sum_r H_0(X_r)$ and let I be the homomorphism from $\sum_r H_0(A_r)$ to $\sum_r H_0(X_r)$ in which $x_r = i(X_r, A_r)_* (a_r)$. Then, it follows that $i(X, A)_* = \theta^{-1} I \delta$. Thus $a \in b(X, A) \Leftrightarrow i(X, A)_* a = 0 \Leftrightarrow i(X_r, A_r)_* a_r = 0$ for $r = 1, \ldots, n$. But since $\varepsilon_r = \gamma f_{X_r *} i(X_r, A_r)_*$ and $\gamma f_{X_r *}$ is an isomorphism, it follows that $a \in b(X, A) \Leftrightarrow \varepsilon_r(a_r) = 0$, $r = 1, \ldots, n$. But this holds

$$\Leftrightarrow a \in \sum_r i(A, A_r)_* \, Ker\, \varepsilon_r = K_0.$$

Theorem 10.3.4. *Suppose that X is an ν-disc with boundary A. Let L denote $H_{\nu-1}(A)$ if $\nu > 1$ and $Ker\, \varepsilon_A$ if $\nu = 1$. Then $b(X, A) \supset L$.*

Proof. This follows from Theorem 10.3.2 since X is contractible.

Theorem 10.3.5. *If A is a Jordan arc, then $H_1(A) = 0$.*

Proof. This follows since A is contractible.

Theorem 10.3.6. *Suppose that $f: (X, A) \to (Y, B)$ and L_A is a subgroup of $H_{\nu-1}(A)$. Let $L_B = (f|A)_* L_A$ and suppose $b(X, A) \supset L_A$. Then $b(Y, B) \supset L_B$.*

Proof. From Theorem 4.1, Chapter I of $E-S$ and its proof, it follows that the following diagram is commutative:

$$\begin{array}{ccc} H_{\nu-1}(A) & \xleftarrow{\;i_*\;} & H_{\nu-1}(A) \\ {\scriptstyle f_{1*}} \downarrow & & \downarrow {\scriptstyle f_{2*}} \\ H_{\nu-1}(Y) & \xleftarrow{\;i'_*\;} & H_{\nu-1}(B) \end{array}$$

where we have used the notation of that theorem. In particular

$$f_{1*} \, i_* = i'_* \, f_{2*}.$$

We are given that $i_*(L_A) = 0$ and that $L_B = f_{2*}(L_A)$. It follows that $i'_*(L_B) = f_{1*}(0) = 0$ which is to say that $L_B \subset Ker\, i(Y, B)_* = b(Y, B)$.

Theorem 10.3.7. *Suppose L is a subgroup of $H_{\nu-1}(A)$, that $L \subset b(X, A)$, and that $Y \supset X$. Then $L \subset b(Y, A)$.*

---

[1] i. e. EILENBERG-STEENROD.

Proof. This follows from Theorem 10.3.6 by taking f as the inclusion map.

Theorem 10.3.8. *Suppose $N = \nu$ and A is the unit $(\nu - 1)$-sphere in R_ν $(A = \partial B(0, 1))$. Then*

(a) *If $X \supset \overline{B(0,1)}$, $b(X, A) = H_{\nu-1}(A)$ if $\nu > 1$ or $b(X, A) \supset Ker\, \varepsilon_A$ if $\nu = 1$.*

(b) *If X does not contain $\overline{B(0,1)}$, $b(X, A) = 0$.*

Proof. Part (a) follows from Theorems 10.3.4 and 10.3.7. In part (b), there is a point in $B(0,1) - X$, since X is compact. Let x_0 be such a point and let Y consist of all points $(1 - t)\,x + t\,\xi$, $0 \le t \le 1$, $x \in X$, where ξ is the intersection of the ray $(x_0\, x)$ with A. Then (A, A) is a strong deformation retract of (Y, A). By Theorem 11.8, Chapter 1 of E—S, the homology sequences for (Y, A) and (A, A) are isomorphic. Accordingly, in considering the i_*-parts of these sequences which are, respectively,

$$H_{m-1}(Y) \xleftarrow{i_*} H_{m-1}(A) \text{ and } H_{m-1}(A) \xleftarrow{i_*} H_{m-1}(A),$$

we conclude that $H_{m-1}(A) \simeq H_{m-1}(Y)$ so $b(Y, A) = 0$. Hence $b(X, A) = 0$ by Theorem 10.3.7.

Theorem 10.3.9. (R 1[1], Lemma 11 A). *Suppose that $X = U_1^N X_r$. Suppose $A_r \subset X_r$ and $A \subset X$ and we define $B = A \cup \underset{r}{U} A_r$. Suppose L and L_r are subgroups of $H_{\nu-1}(A)$ and $H_{\nu-1}(A_r)$, respectively, and suppose $b(X_r, A_r) \supset L_r$. Suppose also that*

$$i(B, A)_* \, L \subset \sum_r i(B, A_r)_* \, L_r. \tag{10.3.2}$$

Then $b(X, A) \supset L$.

Proof. Suppose $h \in L$. Then

$$\begin{aligned} i(X, A)_* \, h &= i(X, B)_* \, i(B, A)_* \, h \in i(X, B)_* \sum_r i(B, A_r)_* \, L_r \\ &= \sum_r i(X, A_r)_* \, L_r = \sum_r i(X, X_r)_* \, i(X_r, A_r)_* \, L_r \\ &= \sum_r i(X, X_r)_* \, 0 = 0 \end{aligned}$$

since $b(X_r, A_r) \supset L_r$. Thus $h \in Ker\, i(X, A)_* = b(X, A)$.

Lemma 10.3.1. *Suppose that $X = U_1^n X_r$, $B = U_1^n A_r$, $A_r \subset X_r$ for each r and $X_r \cap X_s = A_r \cap A_s$ whenever $r \neq s$. Then, for each q, each $u \in H_q(X, B)$ can be represented uniquely in the form*

$$u = \sum_r i_{r*} \, u_r, \quad u_r \in H_q(X_r, A_r), \quad i_r : (X_r, A_r) \subset (X, B). \tag{10.3.3}$$

Proof. We shall prove this for $n = 2$; the theorem can be proved for any n by induction.

To do this we consider the triad $(Y_1 \cup Y_2;\, Y_1, Y_2)$ and the inclusion maps $k_2\colon (Y_1, Y_1 \cap Y_2) \subset (Y_1 \cup Y_2, Y_2)$ and $k_1\colon (Y_2, Y_1 \cap Y_2) \subset (Y_1$

[1] i. e. REIFENBERG [1].

$\cup Y_2, Y_1)$, where $Y_1 = X_1 \cup B$ and $Y_2 = X_2 \cup B$ (see $E-S$, Chapter 1, § 14). Since each k_r is an identity map, $Y_1 - (Y_1 \cap Y_2) = (Y_1 \cup Y_2) - Y_2 = Y_1 - Y_2$ and is open in $Y_1 \cup Y_2$, and k_2 maps $Y_1 - (Y_1 \cap Y_2)$ in a $1-1$ way onto $(Y_1 \cup Y_2) - Y_2$, it follows that k_2 is a relative homeomorphism; the same is true of k_1 (see $E-S$, Chapter 10, § 5, p. 266). From Theorem 5.4 on that page (generalized excision), it follows that the maps k_1 and k_2 induce isomorphisms of the homology groups in all dimensions, so that the triad above is proper. Thus it follows from $E-S$, Chapter 1, Theorem 14—2 that each $u \in H_q(Y_1 \cup Y_2, Y_1 \cap Y_2)$ is uniquely expressible in the form $u = \tilde{i}_{1*} \tilde{u}_1 + \tilde{i}_{2*} \tilde{u}_2$, where $\tilde{u}_r \in H_q(Y_r, Y_1 \cap Y_2)$ and $\tilde{i}_r : (Y_r, Y_1 \cap Y_2) \subset (Y_1 \cup Y_2, Y_1 \cap Y_2)$. From the hypotheses and the definitions of Y_1 and Y_2, it follows that

$$Y_1 \cup Y_2 = X, \quad Y_1 \cap Y_2 = B, \quad H_q(Y_1 \cup Y_2, Y_1 \cap Y_2) = H_q(X, B).$$
$$Y_r - (Y_r - X_r) = X_r, \quad B - (Y_r - X_r) = A_r,$$

and $Y_r - X_r$ is open (in X) and $\subset Y_r \cap B$. Thus, from the generalized excision ($E-S$, Chapter 10, Theorem 5.4), it follows that the inclusion maps $\tau_r : (X_r, A_r) \subset (Y_r, Y_1 \cap Y_2)$ induce isomorphisms of the homology groups in all dimensions. The result follows ($i_{r*} = \tilde{i}_{r*} \tau_{r*}$, $u_r = \tau_{r*} \tilde{u}_r$).

Theorem 10.3.10. (R 1. Lemma 12 A) *Using the notation of Theorem 10.3.9, suppose further that $A \cap X_r \subset A_r$ and that $X_r \cap X_s \subset A_r \cap A_s$ whenever $r \neq s$. Let $K_r = b(X_r, A_r) \subset H_{\nu-1}(A_r)$. Then*

$$b(X, A) = i(B, A)_*^{-1} \left[\sum_r i(B, A_r)_* K_r\right]$$

Proof. If we let L denote the right side, then (10.3.2) holds, so that $b(X, A) \supset L$. To show that $L \supset b(X, A)$ we must show that if $h \in H_{\nu-1}(A)$ and $i(X, A)_* h = 0$, then

$$i(B, A)_* h \in \sum_r i(B, A_r)_* K_r,$$

i.e. there are elements $h_r \in H_{\nu-1}(A_r)$ such that

$$(10.3.4) \qquad i(X_r, A_r)_* h_r = 0 \text{ and } i(B, A)_* h = \sum_r i(B, A_r)_* h_r.$$

Since $i(X, A)_* h = i(X, B)_* \cdot i(B, A)_* h$, this will follow if we can show that any $k \in H_{\nu-1}(B)$ with $i(X, B)_* k = 0$ can be represented by the sum on the right in (10.3.4) in which each h_r satisfies $i(X_r, A_r)_* h_r = 0$. This is proved by diagram chasing in the following diagram:

$$\begin{array}{ccccc} H_\nu(X, B) & \xrightarrow{\partial} & H_{\nu-1}(B) & \xrightarrow{i_*} & H_{\nu-1}(X) \\ \downarrow \simeq & & \downarrow & & \\ \sum_r H_\nu(X_r, A_r) & \xrightarrow{\sum_r \partial} & \sum_r H_{\nu-1}(A_r) & \xrightarrow{\sum i_*} & \sum_r H_{\nu-1}(X_r). \end{array}$$

The marked isomorphism follows from Lemma 10.3.1 and $\sum_r \partial$ denotes the transformation defined by

$$\Big(\sum_r \partial\Big)(u_1, \ldots, u_n) = (\partial u_1, \ldots, \partial u_n), \quad \text{etc.}$$

If $k \in H_{\nu-1}(B)$ and $i(X, B)_* \, k = 0$, then $k \in \mathit{Ker}\, i(X, B)_* = \partial H_\nu(X, B)$.

This follows from the exactness of the Čech homology theory over the category $\mathfrak{G}_C$ of compact pairs and maps of such ($E-S$, Chapter 9, Theorem 7.6). Thus $\exists$ a u in $H_\nu(X, B) \ni \partial u = k$. From the marked isomorphism, it follows that u is uniquely representable in the form (10.3.3) with q replaced by ν. Then

$$u = \sum_r i_{r*}\, u_r, \quad u_r \in H_\nu(X_r, A_r), \quad i_r : (X_r, A_r) \to (X, B).$$

Then

$$\partial u = \sum_r \partial i_{r*}\, u_r = \sum_r i(B, A_r)_*\, h_r, \quad h_r = \partial u_r$$

as desired and since each of the rows

$$\longrightarrow H_\nu(X_r, A_r) \xrightarrow{\ \partial\ } H_{\nu-1}(A_r) \xrightarrow{\ i_*\ } H_{\nu-1}(X_r) \longrightarrow$$

is exact, it follows that $i(X_r, A_r)_*\, h_r = 0$ as required.

Theorem 10.3.11. *Suppose that*

$$A = \bigcup_{r=1}^n A_r, \quad A_r \cap A_{r+1} = D_r \quad \text{for} \quad r = 1, \ldots, n-1,$$
$$A_r \cap A_s = 0 \quad \text{if} \quad |r - s| > 1.$$

Let K_r denote $H_{\nu-2}(D_r)$ if $\nu > 2$ and $K_r = \mathit{Ker}\, \varepsilon : H_0(D_r) \to G$ if $\nu = 2$. Suppose that

$$b(B_r, D_r) \supset K_r, \quad r = 1, \ldots, n; \quad B_0 = B_{n+1} = 0$$
$$C = A \cup \bigcup_{r=1}^n B_r, \quad C_r = B_{r-1} \cup A_r \cup B_r, \quad r = 1, \ldots, n+1,$$
$$A \cap B_r = D_r, \quad r = 0, \ldots, n+1.$$

Then

$$\sum_{1 \le r \le n} i(C, C_r)_*\, H_{\nu-1}(C_r) \supset i(C, A)_*\, H_{\nu-1}(A).$$

Proof. We prove this for $n = 2$; the proof for the general n is by induction. In this case we change our notation as follows:

$$A = A_1 \cup A_2, \quad A_1 \cap A_2 = D, \quad C_1 = A_1 \cup B, \quad C_2 = A_2 \cup B,$$
$$C = C_1 \cup C_2 = A \cup B, \quad A \cap B = D, \quad b(B, D) \supset K$$

where $K = H_{\nu-2}(D)$ if $\nu > 2$ and $K = \mathit{Ker}\, \varepsilon : H_0(D) \to G$ if $\nu = 2$. The result is proved by diagram chasing in the following diagram:

$$\begin{array}{ccccccccc}
H_{\nu-1}(A_1) & \xrightarrow[i_{1*}]{} & H_{\nu-1}(C_1) & \xrightarrow[j_{1*}]{} & H_{\nu-1}(C_1, B) & \xleftarrow[\tau_1]{} & H_{\nu-1}(A_1, D) & \longrightarrow & \\
 & & \downarrow \varrho_1 & & \downarrow & & \downarrow \sigma_1 & & \downarrow \partial \\
H_{\nu-1}(A) & \xrightarrow[i_*]{} & H_{\nu-1}(C) & \xrightarrow[j_*]{} & H_{\nu-1}(C, B) & \xleftarrow[\tau]{} & H_{\nu-1}(A, D) & \xrightarrow{\ \partial\ } & H_{\nu-2}(D) \\
 & & \uparrow \varrho_2 & & \uparrow & & \uparrow \sigma_2 & & \uparrow \partial \\
H_{\nu-1}(A_2) & \xrightarrow[i_{2*}]{} & H_{\nu-1}(C_2) & \xrightarrow[j_{2*}]{} & H_{\nu-1}(C_2, B) & \xleftarrow[\tau_2]{} & H_{\nu-1}(A_2, D) & \longrightarrow &
\end{array}$$

Since $C - A = B - D = C_1 - A_1 = C_2 - A_2$ and is open on C, it follows from the strong excision theorem ($E-S$, Theorem 5.4, p. 266) that the maps τ, τ_1, and τ_2 are isomorphisms (onto). Clearly the images in $H_{\nu-1}(A, D)$ of the $H_{\nu-1}(A_r)$ are included in those of the $H_{\nu-1}(C_r)$, i.e.

$$\sigma_r \tau_r^{-1} j_{r*} i_{r*} [H_{\nu-1}(A_r)]$$
$$= \tau^{-1} j_* \varrho_r i_{r*} [H_{\nu-1}(A_r)] \subset \tau^{-1} j_* \{\varrho_r [H_{\nu-1}(C_r)]\}.$$

Next, the sequences

$$H_{\nu-1}(A_r) \xrightarrow{\tau_r^{-1} j_{r*} i_{r*}} H_{\nu-1}(A_r, D) \xrightarrow{\partial} H_{\nu-2}(D), \quad r = 1,2,$$
$$H_{\nu-1}(A) \xrightarrow{\tau^{-1} j_* i_*} H_{\nu-1}(A, D) \xrightarrow{\partial} H_{\nu-2}(D),$$

are exact, so the image of $H_{\nu-1}(A_r)$ in $H_{\nu-1}(A_r, D)$ consists of exactly those elements u_r of the latter group for which $\partial u_r = 0$, $r = 1, 2$; and the corresponding statement holds for the image of $H_{\nu-1}(A)$ in $H_{\nu-1}(A, D)$. Finally, from Lemma 10.3.1, it follows that each element u of $H_{\nu-1}(A, D)$ is uniquely representable in the form

$$u = \sigma_1 u_1 + \sigma_2 u_2, \quad u_r \in H_{\nu-1}(A_r, D).$$

Taking inverse images under $j_*^{-1} \tau$, it follows that

$$i(C, A)_* H_{\nu-1}(A) \subset i(C, C_1)_* H_{\nu-1}(G) + i(C, C_2)_* H_{\nu-1}(C_2)$$

which is the desired result.

Theorem 10.3.12. *Suppose that* $X = I \times Y$ *and* $A = (0 \times Y) \cup (1 \times Y)$. *Let* I_0 *and* I_1 *be the natural embeddings of* Y *in* A *as* $(0 \times Y)$ *and* $(1 \times Y)$, *respectively. Then* $b(X, A) \supset$ *the totality of elements in* $H_{\nu-1}(A)$ *of the form* $I_1^* h - I_0^* h$ *for* $h \in H_{\nu-1}(Y)$.

Proof. Let $J_0 = i(X, A) I_0$ and $J_1 = i(X, A) I_1$. Then J_0 and J_1 are homotopically equivalent in X. Thus $J_{1*} = J_{0*}$ ($E-S$, Chapter 1, Axiom 5), so that

$$i(X, A)_* (I_{1*} h - I_{0*} h) = J_{1*} h - J_{0*} h = 0 \text{ for all } h \in H_{\nu-1}(Y).$$

This is the result.

Theorem 10.3.13. *Suppose that* $X' = I \times Y$, $A'_0 = 0 \times Y$, $A'_1 = 1 \times Y$, $A' = A'_0 \cup A'_1$ *and suppose that* $f: X' \to X$ *is continuous and that* $f(A'_0) = A_0$, $f(A'_1) = A_1$. *Suppose that* $f_0 = f \mid A'_0$, *that* $f_1 = f \mid A'_1$, *and that* f_0 *is a homeomorphism. Suppose* $K = b(X, A)$ *and* L_0 *is a subgroup of* $H_{\nu-1}(A_0)$. *Then there is a subgroup of* L_1 *of* $H_{\nu-1}(A_1)$ *such that*

$$K + i(A, A_0)_* L_0 = K + i(A, A_1)_* L_1. \tag{10.3.5}$$

Proof. Let $i_0: Y \to A'_0$ and $i_1: Y \to A'_1$ be the homeomorphisms defined by $i_0(y) = (0, y)$ and $i_1(y) = (1, y)$, respectively. From $E-S$, Chapter 1, § 5, it follows that f_{0*}, i_{0*}, and i_{1*} are isomorphisms. Thus we define L_1 as the totality of elements of the form

$$h_1 = f_{1*} i_{1*} h \text{ where } h = i_{0*}^{-1} f_{0*}^{-1} y_0 \text{ and } h_0 \in L_0. \tag{10.3.6}$$

Now, suppose $k \in K$ and $h_0 \in L_0$. There exists a unique h in $H_{\nu-1}(Y)$ defined by (10.3.6); let h_1 be defined by (10.3.6). Then

$$i(A, A_0)_* h_0 = i(A, A_0)_* f_{0*} i_{0*} h = F_{0*} h,$$
$$i(A, A_1)_* h_1 = i(A, A_1)_* f_{1*} i_{1*} h = F_{1*} h,$$
$$F_0 : Y \to A, \quad F_0 = i(A, A_0) f_0 i_0,$$
$$F_1 : Y \to A, \quad F_1 = i(A, A_1) f_1 i_1.$$

But now the maps $G_0 = i(X, A) F_0$ and $G_1 = i(X, A) F_1$ are seen to be homotopically equivalent in X since we may define

$$G_t : Y \to X \quad \text{by} \quad G_t(y) = f(t, y), \quad 0 \leq t \leq 1.$$

Thus from $E-S$, Chapter 1, Axiom 5, it follows that

$$i(X, A)_* [i(A, A_0)_* h_0 - i(A, A_1)_* h_1] = 0, \quad \text{or}$$

$$(10.3.7) \quad i(A, A_0)_* h_0 = k' + i(A, A_1)_* h_1, \quad k' \in K = b(X, A), \quad \text{or}$$

$$k + i(A, A_0)_* h_0 = k + k' + i(A, A_1)_* h_1 \in K + i(A, A_1) L_1.$$

In like manner if $k \in K$ and $h_1 \in L_1$, there is at least one h and hence one h_0 satisfying (10.3.6) and hence (10.3.7). Thus

$$k + i(A, A_1)_* h_1 = k - k' + i(A, A_0)_* h_0 \in K + i(A, A_0)_* L_0.$$

Theorem 10.3.14. *Suppose that Π is a ν-plane (in R_N) through P, that $A = \Pi \cap \partial B(P, r)$, and that $D = \Pi \cap \overline{B(P, r)}$. Then $b(D, A) = H_{\nu-1}(A)$ and if X is any surface $\supset A$ for which $b(X, A) = L$ where L is a non-zero subgroup of $H_{\nu-1}(A)$, then*

$$\Lambda^\nu(X) \geq \Lambda^\nu(D) = \gamma_\nu r^\nu.$$

Proof. That $b(D, A) = H_{\nu-1}(A)$ follows from Theorem 10.3.8. Let $f = p_\Pi$ and $Y = f(X)$. Then $f : (X, A) \to (Y, A)$ and $f_2 = f|_A$ is the identity. From $E-S$, Chapter 1, § 4, the diagram

$$\begin{array}{ccc} H_{\nu-1}(A) & \xrightarrow{i_*} & H_{\nu-1}(X) \\ \downarrow{f_{2*}} & & \downarrow{f_{1*}} \\ H_{\nu-1}(A) & \xrightarrow{i'_*} & H_{\nu-1}(Y) \end{array}$$

is commutative and f_{2*} is just the identity (Axiom 1). Now, suppose $y \in b(X, A)$. Then $y \in H_{\nu-1}(A)$ and $i_*(X, A)\, y = 0$. So

$$i'_*(y) = f_{1*}\, i_* f_{2*}^{-1} y = 0.$$

Thus $b(Y, A) \supset b(X, A) \neq 0$. The theorem then follows from Theorem 10.3.8 and the fact that $\Lambda^\nu(X) \geq \Lambda^\nu(Y)$.

Theorem 10.3.15 (R 1, Lemma 21 A). *Suppose $X_r \supset X_{r+1}$ and $b(X_r, A) \supset L$ for $r = 1, 2, \ldots$, L being a subgroup of $H_{\nu-1}(A)$. Let $X = \bigcap X_r$. Then $b(X, A) \supset L$.*

Proof. Suppose $h \in L$. Then $i(X_r, A)_* h = 0$. By the continuity of the Čech homology ($E-S$, pp. 260—261) the injections $i(X_r, X)$ yield an isomorphism of $H_{\nu-1}(X)$ onto the inverse limit group $\operatorname{Lim} H_{\nu-1}(X_r)$. This isomorphism maps the element $i(X, A)^* h$ into the sequence $\{i(X_r, A)_* h\}$ which is a sequence of zeros and is the zero element of the inverse limit group. Thus $i(X, A)_* h = 0$. Since h is arbitrary $b(B, A) \supset L$.

Theorem 10.3.16. *Suppose $b(X_r, A) \supset L$ for $r = 1, 2, \ldots, L$ being a subgroup of $H_{\nu-1}(A)$, and suppose $X_r \to X$ in the point set sense. Then $b(X, A) \supset L$.*

Proof. From Theorem 10.3.7, it follows that $b(Y_r, A) \supset L$ for each r, where $Y_r = \bigcup_{s \geq r} X_s$. Clearly $Y_r \supset Y_{r+1}$ and $\bigcap Y_r = X$ so the result follows from Theorem 10.3.15.

10.4. The minimizing sequence; the minimizing set

In this section we construct a minimizing sequence and a minimizing set and prove a first smoothness peoperty of the minimizing set. We consider compact sets X and A on $\mathfrak{M}$ in which

$$A \subset X, \quad b(X, A) \supset L, \quad L \text{ a subgroup of } H_{\nu-1}(A). \tag{10.4.1}$$

We define, for a given compact A and subgroup L,

$$d(A, L) = \inf \Lambda^\nu(S) \text{ for all } X \text{ with } b(X, A) \supset L, \; S = X - A. \tag{10.4.2}$$

We shall assume that $\mathfrak{M}$ satisfies the conditions of § 10.1 and

$$d(A, L) < +\infty. \tag{10.4.3}$$

We now define $R(P) = \min[d(P, A), R_0]$. If $b(X, A) \supset L$, we define

$$\left.\begin{aligned} \varphi(r, P; X) &= \int_0^r \Lambda^{\nu-1}[X \cap \partial B(P, t)]\, dt \\ \psi(r, P; X) &= \Lambda^\nu[X \cap B(P, r)] \end{aligned}\right\}, R(P) > 0, \quad 0 < r < R(P). \tag{10.4.4}$$

Lemma 10.4.1. *Suppose $b(X, A) \supset L$. Then*

$$\begin{aligned} &\varphi(r_2, P; X) - \varphi(r_1, P; X) \leq \psi(r_2, P; X) - \psi(r_1, P; X), \\ &\qquad 0 \leq r_1 < r_2 < R(P), \end{aligned} \tag{10.4.5}$$

$$\begin{aligned} &\psi(r', P'; X) \leq \psi(r' + |P P'|, P; X), \quad r' < R(P'), \\ &\qquad r' + |P P'| < R(P), \end{aligned} \tag{10.4.6}$$

and φ and ψ are non-decreasing on r for each $P \notin A$.

Proof. (10.4.5) follows from Theorem 10.2.3. with $U(Q) = d(P, Q)$ taking $S = X \cap [B(P, r_2) - B(P, r_1)]$. (10.4.6) follows since $B(P', r') \subset B(P, r') + |P P'|)$.

Lemma 10.4.2. *Suppose $b(X, A) \supset L$ and*

$$\Lambda^\nu(S) = d(A, L) + \varepsilon. \tag{10.4.7}$$

For $0 \leq r < R(P)$, *define*

$$\begin{aligned}
&\hat{\varphi}(r, P; X) = \max[0, \varphi(r, P; X) - \varepsilon],\\
&\hat{\psi}(r, P; X) = \max[0, \psi(r, P; X) - \varepsilon],\\
&\varrho(P, X) = \sup r \text{ for those } r \text{ for which } \hat{\varphi}(r, P; X) = 0\\
&\varrho^*(P, X) = \sup r \text{ for those } r \text{ for which } \hat{\psi}(r, P; X) = 0.
\end{aligned} \tag{10.4.8}$$

Then, for almost all r, $0 \leq r < R(P)$,

$$\psi(r, P; X) \leq \begin{cases} C_3 [\varphi_r(r, P; X)]^{\nu/(\nu-1)} + \varepsilon, \\ \nu^{-1} r(1 + h r)\, \varphi_r(r, P; X) + \varepsilon, \end{cases} \tag{10.4.9}$$

where C_3 *denotes the constant of Theorem* 10.2.4 *and* h *is chosen so that* $(1 + \eta_0 r)^{2\nu-1} \leq (1 + h r)$ for $0 \leq r \leq R_0$. *Also*

$$\varrho^*(P) \leq \varrho(P), \quad \hat{\varphi}_r(r, P; X) \leq \hat{\psi}_r(r, P; X) \quad (r \notin Z(P));$$

$$\hat{\varphi}(r_2, P; X) - \hat{\varphi}(r_1, P; X) \leq \hat{\psi}(r_2, P; X) - \hat{\psi}(r_1, P; X), \quad 0 \leq r_1 < r_2 < R(P); \tag{10.4.10}$$

$$r^{-\nu}(1 + h r)^{\nu}\, \hat{\varphi}(r, P; X) \text{ and } r^{-\nu}(1 + h r)^{\nu}\, \hat{\psi}(r, P; X) \text{ are non-decreasing;} \tag{10.4.11}$$

$$\hat{\varphi}(r, P; X) \geq \varkappa[r - \varrho(P, X)]^{\nu}, \quad \varrho(P, X) \leq r < R(P), \quad \varkappa = \nu^{-\nu} C_3^{1-\nu} > 0; \tag{10.4.12}$$

$$\psi(r, P; X) \leq \varepsilon + \nu^{-1}(1 - k)^{-1}[1 + \varrho(P, X)]\, \varepsilon, \quad 0 \leq r \leq k\, \varrho(P, X), \quad 0 < k < 1. \tag{10.4.13}$$

Proof. For almost all $r < R(P)$

$$\varphi_r(r, P; X) = \Lambda^{\nu-1}[X \cap \partial B(P, r)].$$

From Theorem 10.2.5 with $G = B(P, r)$, $A_1 = X \cap \partial B(P, r)$, and Y_1 any surface with $b(Y_1, A_1) \supset H_{\nu-1}(A_1)$. It follows that $[\overline{X - B(P, r)}] \cup Y_1 = Y$ is a surface with $b(Y, A) \supset L$. By Theorem 10.2.4 we may take for Y_1 a surface with $\Lambda^{\nu}(Y_1) \leq C_3[\varphi_r(r, P; X)]^{\nu/(\nu-1)}$. Thus if the first inequality in (10.4.9) did not hold, we would have

$$\Lambda^{\nu}(S) > \Lambda^{\nu}(Y_1) + \Lambda^{\nu}[S - \overline{B(P, r)}] + \varepsilon \geq d(A, L) + \varepsilon \quad (S = X - A).$$

The other inequality follows similarly since we may also take $Y_1 = \omega[C(0, A_{10})]$ where ω is a normal coordinate system with domain $B(0, R_0)$ such that $\omega(0) = P$, $A_{10} = \omega^{-1}(A_1)$. Then we would arrive at the same contradiction, since in this case

$$\begin{aligned}
\Lambda^{\nu}(Y_1) &\leq (1 + \eta_0 r)^{\nu} \Lambda^{\nu}[C(0, A_{10})] = (1 + \eta_0 r)^{\nu} \cdot \nu^{-1} r\, \Lambda^{\nu-1}(A_{10})\\
&\leq (1 + \eta_0 r)^{2\nu-1} \nu^{-1} r\, \varphi_r(r, P; X),
\end{aligned}$$

as is seen from our assumptions (10.1.13), Theorem 10.2.1, and Lemma 10.2.1 (c).

The results in (10.4.10) follow from (10.4.5) and (10.4.8).

From (10.4.5) and (10.4.9), it follows that

$$\hat{\varphi}(r, P; X) \le \begin{cases} \nu^{-1} r(1 + h r)\, \hat{\varphi}_r(r, P; X), \\ \\ C_3 [\hat{\varphi}_r(r, P; X)]^{\nu/(\nu-1)}, \end{cases} \qquad (r \notin Z(P)) \tag{10.4.14}$$

from which (10.4.12) and the first inequality in (10.4.11) follow. To prove (10.4.13), we assume $0 < k < 1$, $0 \le r \le k s$, $s < \varrho(P, X)$ and use (10.4.9) to obtain

$$\begin{aligned} \psi(r, P; X) &\le \psi(k s, P; X) \le \frac{1}{(1-k)\, s} \int_{ks}^{s} \nu^{-1} t(1 + h t)\, \varphi_r(t, P; X)\, dt + \varepsilon \\ &\le [1 + h \varrho(P, X)]\, \nu^{-1}(1-k)^{-1} \varphi(s, P; X) + \varepsilon \\ &\le \varepsilon + \nu^{-1}(1-k)^{-1} [1 + \varrho(P, X)]\, \varepsilon \end{aligned}$$

from which (10.4.13) follows.

To prove the second inequality in (10.4.11), we define

$$\hat{\chi}(r, P; X) = \int_0^r \hat{\psi}_r(t, P; X)\, dt, \quad \omega(r, P; X) = \hat{\psi}(r, P; X) - \hat{\chi}(r, P; X).$$

From (10.4.10) and the definitions it follows that

$$\omega(r_2) - \omega(r_1) \ge \int_{r_1}^{r_2} [\hat{\varphi}_r(r, P, X) - \hat{\psi}_r(r, P, X)]\, dr, \quad 0 < r_1 < r_2 < R(P).$$

But since ω is singular, i.e. $\omega'(r) = 0$ a.e., it follows that ω is non-decreasing. From (10.4.9) and (10.4.10) and the definitions, we obtain

$$\frac{\partial}{\partial r} [h(r)]^{-1} \hat{\psi}(r, P; X) \ge 0, \quad h(r) = \gamma_\nu r^\nu (1 + h r)^{-\nu}, \quad \text{or}$$

$$\begin{aligned} \frac{\partial}{\partial r} [h(r)]^{-1} \hat{\chi}(r, P; X) &\ge \frac{\partial}{\partial r} [h(r)]^{-1} \omega(r, P; X) \\ &= -\{[h(r)]^{-1}\}' \omega(r, P; X) \quad (r \notin Z_2(P)) \end{aligned}$$

$$\begin{aligned} [h(r_2)]^{-1} \hat{\chi}(r_2) - [h(r_1)]^{-1} \hat{\chi}(r_1) &\ge -\int_{r_1}^{r_2} \omega(r, P; X)\, d[h(r)]^{-1} \\ = -\left[h^{-1}(r_2)\, \omega(r_2) - h^{-1}(r_1)\, \omega(r_1) - \int_{r_1}^{r_2} h^{-1}(r)\, d\omega(r) \right] \\ (\omega(r) = \omega(r, P; X)) \end{aligned}$$

from which the desired inequality follows, since ω is non-decreasing.

Now, suppose that $A \subset \mathfrak{M}$ and that $\{X_n\}$ is a minimizing sequence; i.e. $b(X_n, A) \supset L$ and $\Lambda^\nu(X_n - A) = d(A, L) + \varepsilon_n$ for each n, where $\varepsilon_n \to 0$. Let $\{Q_i\}$ be a countable dense subset of $\mathfrak{M}$. There is a subsequence, still called $\{X_n\}$, such that the functions $\psi(r, Q_i; X_n)$ converge for each i

and each r, $0 \leq r \leq R(Q_i)$ to functions $\tilde{\psi}(r, Q_i)$. Next, we define (for that subsequence $\{X_n\}$)

$$\psi^+(r, Q) = \limsup_{(r', Q_i)\to(r,Q)} \tilde{\psi}(r, Q_i), \quad \psi^-(r, Q) = \liminf_{(r', Q_i)\to(r,Q)} \tilde{\psi}(r, Q_i)$$

$$\varphi^+(r, Q) = \limsup_{n\to\infty} \varphi(r, Q; X_n), \quad \varphi^-(r, Q) = \liminf_{n\to\infty} \varphi(r, Q; X_n).$$

Then we have the following results:

Theorem 10.4.1. (a) *$\varphi^\pm$ and $\psi^\pm$ are all non-decreasing in r and $\psi^+(r, P)$ is upper-semicontinuous in (r, P).*

(b) $\psi^-(r, P) \leq \liminf_{n\to\infty} \psi(r, P; X_n) \leq \limsup_{n\to\infty} \psi(r, P; X_n) \leq \psi^+(r, P)$.

(c) $\psi^-(r_2, P) \geq \psi^+(r_1, P)$ *whenever* $0 \leq r_1 < r_2 < R(P)$. *Thus* $\psi^+(r, P) = \psi^-(r, P)$ *whenever either ψ^+ or ψ^- is continuous in r.*

(d) $\varphi^-(r, P) \leq \varphi^+(r, P) \leq \psi^+(r, P)$, $0 \leq r < R(P)$.

(e) $\psi^+(r', P') \leq \psi^-(r, P)$ *if* $r' < R(P')$ *and* $r' + d(P, P') < r < R(P)$.

(f) For each P in $\mathfrak{M} - A$, $[h(r)]^{-1}\psi^\pm(r,)$ *and* $[h(r)]^{-1}\varphi^\pm(r, P)$ are all non-decresing in r and

$$\psi(P) \equiv \lim_{r\to 0^+} [h(r)]^{-1} \psi^+(r, P) \quad (h(r) = \gamma_\nu r^\nu (1 + h\, r)^{-\nu})$$

is upper-semicontinuous.

(g) *There is a positive number β such that $\psi(P) \geq \beta$ whenever $\psi(P) > 0$, so this latter set is closed in $\mathfrak{M} - A$.*

(h) $\varphi^-(r, P) \geq \psi(P) \cdot h(r)$, $0 \leq r < R(P)$.

(i) *If* $\psi(P) = 0$, *then* $\psi^+(r, P) = 0$ for $0 \leq r < \varrho(P)$ for some positive number $\varrho(P)$.

(j) *If we define S as the set of P in $\mathfrak{M} - A$ for which $\psi(P) > 0$, then S is bounded and $X_0 = S \cup A$ is compact.*

(k) *An altered sequence, still called $\{X_n\}$, exists, which has the same functions $\psi^\pm$, for which the new functions '$\varphi^\pm$ are still non-decreasing in r and satisfy the conditions in* (a), (f), *and* (h), *and for which $D(X_n, X_0) \to 0$.*

(l) *If $P \in S$, $0 < r < R(P)$ and $\{B(P_i, r_i)\}$ is a finite or countable disjoint family in which each $P_i \in S$ and each $B(P_i, r_i) \subset B(P, r)$, then*

$$\psi^\pm(r, P) \geq \sum_i \psi^-(r_i, P_i).$$

Proof. (a) If f_1 and f_2 are continuous and non-decreasing on an interval I, then the functions $\max[f_1(x), f_2(x)]$ and $\min[f_1(x), f_2(x)]$ are easily seen to be non-decreasing. This result is easily extended to imply that the sup and inf of a countable family of non-decreasing functions (continuous or not) are non-decreasing. Thus (a) follows easily.

The proof of (b) uses the same ideas as that of (c). So we prove (c). Choose $\varrho > 0$ so $4\varrho < r_2 - r_1$. We have

$$\psi^-(r_2, P) \geq \inf_{\substack{r' > r_2 - \varrho \\ d(P, Q_i) < \varrho}} \tilde{\psi}(r', Q_i), \quad \psi^+(r_1, P) \leq \sup_{\substack{r'' < r_1 + \varrho \\ d(P, Q_j) < \varrho}} \tilde{\psi}(r'', Q_j).$$

But now, if $r' > r_2 - \varrho$, $d(P, Q_i) < \varrho$, $r'' < r_1 + \varrho$, $d(P, Q_j) < \varrho$, then $d(Q_i, Q_j) < 2\varrho$ and

$$\tilde{\psi}(r', Q_i) \geq \tilde{\psi}[r' - d(Q_i, Q_j), Q_j] \geq \tilde{\psi}(r_2 - 3\varrho, Q_j) \geq \tilde{\psi}(r'', Q_j),$$

using (10.4.6) and a limit process. Parts (d) and (e) follow from (10.4.5) and (10.4.6) and limit processes. That $h^{-1}\varphi^{\pm}$ and $h^{-1}\psi^{\pm}$ are non-decreasing follows from (10.4.11) and the argument in (a).

To prove (h), we first note that

$$\varphi_r(r, P; X_n) \geq \nu r^{-1}(1 + h r)^{-1}[\psi(r, P; X_n) - \varepsilon_n].$$

By integrating with respect to r between r_0 and r, $0 < r_0 < r$, and then letting $n \to \infty$ first and then $r_0 \to 0$, we obtain

$$\varphi^-(r, P) \geq \int_0^r \nu t^{-1}(1 + h t)^{-1} \psi^-(t, P)\, dt$$

from which the result follows, using parts (f), (b), and (c).

To prove (g) and (i), we consider two cases:

Case 1 — — $\lim_{n\to\infty} \varrho(P, X_n) = 0$. Then it follows from (10.4.12) that

$$\varphi^-(r, P) \geq \beta \gamma_\nu r^\nu, \quad 0 \leq r < R(P), \quad \beta = \varkappa \gamma_\nu^{-1}.$$

Case 2 — — $\lim_{n\to\infty} \varrho(P, X_n) = \varrho(P) > 0$ as n runs through a subsequence. In that case we see from (10.4.13) by letting $n \to \infty$ that $\psi^-(r, P) = 0$ for $0 \leq r < \varrho(P)$ so that $\psi(P) = 0$.

To prove (j), let us suppose that S is not bounded. Then there exists a disjoint countable family of balls $B(P_i, R_0)$ with $P_i \in S$ and $\overline{B(P_i, R_0)} \cap A$ empty for each i. But then

$$d(A, L) = \lim_{n\to\infty} \Lambda^\nu(S_n) \geq \liminf_{n\to\infty} \sum_{i=1}^{\infty} \Lambda^\nu[X_n \cap B(P_i, R_0)]$$

$$\geq \liminf_{n\to\infty} \sum_{i=1} \psi(R_0, P_i; X_n) \geq \sum_{i=1}^{N} \varphi^-(R_0, P_i) \geq \beta N h(R_0)$$

for every N. This is impossible.

To prove (k), choose for each p a finite number of regular domains of class C^2 whose union G_p contains $(X_0, 1/2p)$ and is contained in $(X_0, 1/p)$. Let G_p' be the set (G_p, ϱ_p) where ϱ_p is so small that G_p' and $\Gamma_p = G_p' - \bar{G}_p$ are of class C^1, and define U_p on Γ_p by defining $U_p(x) = t$ for $x \in \partial G_{pt}$ the parallel surface "t of the way" from ∂G_p to $\partial G_p'$. Then $U_p \in C^1(\bar{\Gamma}_p)$ and $\nabla U_p \neq 0$ there. Clearly it follows from part (i) and the compactness of $\bar{\Gamma}_p$ that $\Lambda^\nu(\bar{\Gamma}_p \cap X_n) \to 0$. Thus, for each p, we may choose an n_p and a t_p, $0 \leq t_p \leq 1$, so that $\Lambda^{\nu-1}[\partial G_{p t_p} \cap X_{n_p}] \leq 1/p^{\nu-1}$. From Theorem 10.2.4 it follows that the part of X_{n_p} outside $G_{p t_p}$ can be replaced by a set Y_p such that $b(Y_p, A_p) \supset H_{\nu-1}(A_p)$, $A_p = \partial G_{p t_p} \cap X_{n_p}$, $Y_p \subset (A_p, C_1 p^{-1})$, and $\Lambda^\nu(Y_p) \leq C_2 p^{-\nu}$. Clearly also, if $P \in S$, there are points of the new (and of the previous) X_{n_p} converging to P as $p \to \infty$. The result follows.

To prove (l), we note that

$$\psi^+(r, P) \geq \limsup_{n\to\infty} \psi(r, P; X_n) \geq \liminf_{n\to\infty} \Lambda^\nu[X_n \cap B(P, r)]$$
$$\geq \liminf_{n\to\infty} \sum_i \Lambda^\nu[X_n \cap B(P_i, r_i)] \geq \sum_i \left\{\liminf_{n\to\infty} \Lambda^\nu[X_n \cap B(P_i, r_i)]\right\}$$

from which the result follows.

Lemma 10.4.3 (cf. REIFENBERG [1], Lemma 5*). *Suppose X_n, X_0, S, and all the various other functions have their significance above. Suppose $p > \nu \geq 3$. For each $\eta > 0$, there are positive numbers $\varepsilon'(\eta, \beta, p, \nu, \mathfrak{M})$, $\eta'(\eta, \beta, p, \nu, \mathfrak{M})$, and $\lambda(\eta, \beta, p, \nu, \mathfrak{M})$ with the following property. If $P' \in S$, $0 < r' < R(P')$, and $K(P', r') \equiv S \cap B(P', r') \subset (\Pi', \eta' r')$ for some geodesic p-plane Π' centered at P' and if*

$$\beta \leq [h(r)]^{-1} \varphi^-(r, P) \leq [h(r)]^{-1} \psi^+(r, P) \leq \beta + \varepsilon' \qquad (10.4.15)$$
$$(h(r) = \gamma_\nu r^\nu (1 + h r)^{-\nu})$$

for every $B(P, r) \subset B(P', r')$ for which $P \in S$ and $0 < r < R(P)$, then there exists a P^* in S such that $K(P^*, \lambda r') \subset (\Pi, \eta \lambda r')$ for some $p - 1$ dimensional geodesic plane Π centered at P^* and $B(P^*, \lambda r') \subset B(P', r')$. In fact we may take ε', η', and λ all of the form $C(\beta, p, \nu, \mathfrak{M}) \eta^h$, $h = h(\beta, p, \nu, \mathfrak{M})$.

Proof. We first show that there is an $R' = c(p, \nu, \mathfrak{M}) r'$, $c > 0$, such that there is a Q in Π' such that $B(Q, R') \subset B(P', r'/2)$ and $K(Q, R')$ is empty. If this were not so, every sphere $B(Q, R')$ with $Q \in \Pi'$ would contain a point Q' of S and so every sphere $B(Q, 2R')$ would contain a sphere $B(Q', R')$. But there are at most $Z_1(p, \mathfrak{M}) \cdot (r'/R')^p$ disjoint spheres $B(Q_i, 2R') \subset B(P', r'/2)$ $(Q_i \in \Pi')$, so that (recall Theorem 10.4.1(l) and the fact that $R(P) \leq R_0$ for every $P \in \mathfrak{M} - A$).

$$(\beta + \varepsilon') h(r'/2) \geq \psi^+\left(\frac{r'}{2}, P'\right) \geq \sum_i \varphi^-(2R', Q_i) \geq Z_1 \cdot \left(\frac{r'}{R'}\right)^p \cdot \beta \cdot h(2R').$$

Thus, if we take $\varepsilon' \leq \beta$, we conclude that

$$(r'/R')^{p-\nu} \leq 2^{1-2\nu} \cdot Z_1^{-1} \cdot (1 + 2h R')^\nu. \qquad (10.4.16)$$

Now, let $B(Q, R)$ be the largest sphere in $B(P', r'/2)$ with Q on Π' such that $K(Q, R)$ is empty. Let $P \in S \cap \partial B(Q, R)$ and let P_0 be a point on $Q P$ produced beyond P and let $x = |P P_0|$. Let $l_n \equiv l_n(P, r) \equiv X_n \cap \partial B(P, r)$, let C_θ be the geodesic half-cone consisting of the geodesic rays through P which make an angle $\leq \theta$ at P with the geodesic $Q P P_0$, and let $l_{n\theta} \equiv l_{n\theta}(P, r) \equiv l_n(P, r) \cap C_\theta$. For almost all r and θ

$$\Lambda^\nu[C(P_0, l_n)] = \Lambda^\nu[C(P_0, l_{n\theta})] + \Lambda^\nu[C(P_0, l_n - l_{n\theta})]. \qquad (10.4.17)$$

Let τ be a normal coordinate patch centered at P. Then

$$\begin{aligned} &l_{n\theta} \subset B(P_0, \varrho), \quad \tau^{-1}(l_{n\theta}) \subset B(P_0, \varrho_0), \quad l_n - l_{n\theta} \subset B(P_0, \sigma) \\ &l_n - l_{n\theta} \subset \mathfrak{M} - B(Q, R), \quad \varrho_0^2 = r^2 + x^2 - 2r x \cos\theta. \end{aligned} \qquad (10.4.18)$$

From Lemma 10.1.3, it follows that $B[\tau^{-1}(Q'), R/2] \subset \tau^{-1}[B(Q, R)] \subset B[\tau^{-1}(Q''), 2R]$ where Q' and Q'' are on the ray PQ at respective distances $R/2$ and $2R$ from P (this assumes that R_0 is small enough). Accordingly, if we let ϱ and σ be the smallest radii satisfying (10.4.18), we must have

$$\begin{aligned} &\varrho^2 \leq (1 + Z_1 r)^2 (r^2 + x^2 - 2r\,x \cos\theta), \\ (10.4.19) \qquad &\sigma^2 \leq (1 + z_1 r)^2 (r^2 + x^2 + 2r^2\, x/R) \\ &\sigma^2 \geq (1 + z_1 r)^{-2} (r^2 + x^2), \quad Z_1 \leq 2\eta_0 \quad \text{if} \quad x \leq r(\sqrt{2} - 1) \end{aligned}$$

on account of Lemma 10.1.3. From the minimizing property of the X_n, we conclude, as in the proof of Lemma 10.4.2, that

$$\Lambda^\nu(K_n) \leq \Lambda^\nu[C(P_0, l_n)] + \varepsilon_n, \quad K_n \equiv K_n(P, r) \equiv X^n \cap B(P, r),$$
$$\Lambda^\nu(K_n) \leq \nu^{-1}[\varrho(1 + h\varrho)\, \Lambda^{\nu-1}(l_{n\theta}) + \sigma(1 + h\sigma)\, \Lambda^{\nu-1}(l_n - l_{n\theta})] + \varepsilon_n.$$

Solving for $\Lambda^{\nu-1}(l_{n\theta})$, we obtain

$$(10.4.20) \quad \Lambda^{\nu-1}(l_{n\theta}) \leq (\sigma - \varrho)^{-1}[\sigma(1 + h\sigma)\, \Lambda^{\nu-1}(l_n) - \nu\, \Lambda^\nu(K_n) + \nu\, \varepsilon_n].$$

Now, let us choose r_0, r_1, x, and $\cos\theta$ as follows:

$$(10.4.21) \qquad \begin{aligned} &0 < r_0 < R/3, \quad r_1 = r_0 (r_0/R)^{1/3}, \quad x = r_0 (r_0/R)^{5/6}, \\ &\cos\theta = (r_0/R)^{1/12}; \quad \varepsilon' = \min[\beta, (r_0/R)]. \end{aligned}$$

We shall consider only values of r with

$$r_1 \leq r \leq r_0.$$

Then, using (10.4.19) and (10.4.21), we obtain (using $\sqrt{1 \pm y} \leq 1 \pm y/2$, etc.)

$$\begin{aligned} \sigma^2 - \varrho^2 \geq &-[(1 + Z_1 r)^2 - (1 + Z_1 r)^{-2}]\,(r^2 + x^2) + \\ &+ 2r\,x(1 + Z_1 r)^2 \cos\theta \geq (1 + Z_1 r)\,[-4 Z_1 r (r^2 + x^2) + \\ &+ 2r\,x \cos\theta] \geq (1 + Z_1 r) \cdot 2r\, r_0 \{(r_0/R)^{11/12} - \\ &- 2 Z_1 r_0 [1 + (r_0/R)]\} \quad (Z_1 \leq 2\eta_0) \end{aligned}$$

$$\begin{aligned} \sigma + \varrho &\leq (1 + Z_1 r)\, r \left[2 + \frac{x}{R} - \frac{x}{r}\cos\theta + \frac{x^2}{r^2}\right] \left(\frac{x^2}{r^2} \leq \frac{x}{r}\cos\theta\right) \\ &\leq (1 + Z_1 r) \cdot 2r \left[1 + \frac{1}{2}(r_0/R)^{11/6}\right]. \end{aligned}$$

$$\therefore \sigma - \varrho \geq \frac{1}{2} r_0 (r_0/R)^{11/12} \quad \text{if} \quad 2\eta_0 R \leq \lambda_0 \quad \text{and} \quad r_0/R \leq \lambda_0.$$

Now, since $\Lambda^\nu(K_n) - \varepsilon_n \leq \nu^{-1} r(1 + h r)\, \Lambda^{\nu-1}(l_n)$ and $\Lambda^\nu(K_n) \geq \beta\, h(r)$ by Lemma 10.4.2, we may use the inequalities above to conclude that (if $r_0/R \leq \lambda_0$, etc.)

$$(10.4.22) \qquad \begin{aligned} 0 &\leq \sigma(1 + h\sigma)\, \Lambda^{\nu-1}(l_n) - \nu\, \Lambda^\nu(K_n) + \\ &+ \nu\, \varepsilon_n \,(\sigma \leq r(1 + Z_1 r)\,(1 + 2r_0/R) \leq 2r_0) \\ &\leq r(1 + Z_1 r_0)\,(1 + 2r_0/R)\,(1 + 2h\, r_0)\, \Lambda^{\nu-1}(l_n) - \\ &- \nu\, \beta\, \gamma_\nu\, r^\nu (1 + h r)^{-\nu} + \nu\, \varepsilon_n \\ &\leq r_0 [(1 + 3r_0/R)\, \Lambda^{\nu-1}(l_n) - \beta\, h'(r)] + \nu\, \varepsilon_n, \\ &(h(r) = \gamma_\nu\, r^\nu (1 + h r)^{-\nu}). \end{aligned}$$

Using (10.4.22), integrating (10.4.20), and letting $n \to \infty$, we obtain

$$\limsup_{n\to\infty} \int_{r_1}^{r_0} \Lambda^{\nu-1}(l_{n\theta})\,dr \leq (\sigma - \varrho)^{-1} r_0\, h(r_0)\,[(1 + 3r_0/R)(\beta + \varepsilon') -$$

$$- \beta + \beta\, h(r_1)/h(r_0)] \leq Z_2 (r_0/R)^{1/12} h(r_0)$$

since $\nu \geq 3$ so $h(r_1)/h(r_0) \leq Z_3 \cdot (r_0/R)$. Using a similar procedure we obtain

$$\limsup_{n\to\infty} \Lambda^{\nu}[K_n(P, r_0) \cap C_\theta] \leq \limsup \Lambda^{\nu}[K_n(P, r_0)] -$$

$$- \liminf \Lambda^{\nu}[K_n(P, r_0) \cap (\mathfrak{M} - C_\theta)]$$

$$\text{(10.4.23)} \qquad \leq (\beta + \varepsilon')\, h(r_0) - \liminf \int_0^{r_0} \Lambda^{\nu-1}(l_n)\,dr +$$

$$+ \limsup \int_{r_1}^{r_0} \Lambda^{\nu-1}(l_{n\theta})\,dr + \limsup \int_0^{r_1} \Lambda^{\nu-1}(l_n)\,dr$$

$$\leq Z_4 (r_0/R)^{1/12} h(r_0)\,.$$

Let Π_1 be the geodesic $(N-1)$-plane through $P \perp QP$. Let $\lambda = r_0/2r'$. Then

$$\text{(10.4.24)} \qquad r_0/r' = 2\lambda \leq r_0/R \leq (r_0/r')\,(r'/R') \leq Z_5(p, \nu, \mathfrak{M}) \cdot \lambda.$$

We now show that if λ, or r_0/R, is chosen small enough, then $K(P, r_0/2) \equiv K(P, \lambda r') \equiv S \cap B(P, \lambda r') \subset (\Pi_1, \eta\lambda r'/2)$. For suppose $P^* \in K(P, r_0/2)$ and is at a distance $\geq \eta r_0/4 = \eta \lambda r'/2$ from Π_1. Let $P_0^* = \tau^{-1}(P^*)$, $Q_0' = \tau^{-1}(Q')$, $\Pi_{10} = \tau^{-1}(\Pi_1)$, and $C_{\theta 0} = \tau^{-1}(C_\theta)$. Since τ is centered at P, $\tau^{-1}(P) = 0$, $\tau^{-1}[B(P, r)] = B(0, r)$ if $r \leq R_0$, Π_{10} is an actual $(N-1)$-plane through 0, and $C_{\theta 0}$ is an actual half-cone whose generators make an angle θ with the normal $0\,P_{00}$ to Π_1, $P_{00} = \tau^{-1}(P_0)$. From the definition of Q' given just below equation (10.4.18) it follows that $B(Q_0', R/2)$ is tangent to Π_{10} at the origin. We have remarked above and can also conclude from Lemma 10.1.3 that $B(Q_0', R/2) \subset \tau^{-1}[B(Q, R)]$. Finally P_0^* is at a distance ϱ from Π_{10} where (Lemma 10.1.4)

$$(1 + \eta_0 r_0/2)^{-1} \eta\, r_0/4 \leq \varrho \leq (1 + \eta_0 r_0/2)\, \eta\, r_0/4\,.$$

Since P^* is not in $B(Q, R)$, P_0^* is not in $B(Q_0', R/2)$ and so, since it is in $B(0, r_0/2)$, it must be on the same side of Π_{10} as is $C_{\theta 0}$ if r_0/R is small enough $(<\eta/(1 + \eta_0 R/6))$. By passing a 2-plane through 0, P_{00}, and P_0^*, we see that the distance from P_0^* to $C_{\theta 0}$ is

$$d = \varrho \sec\varphi \sin(\theta - \varphi)\,, \quad \varphi = \text{angle } P_{00}\, 0\, P_0^*\,.$$

It is easy to see that d takes on its minimum with ϱ given when $P_0^* \in \partial B(0, r_0/2)$. For that value of φ.

$$d = \varrho \sin\theta - \cos\theta \sqrt{(r_0^2/4) - \varrho^2}\,.$$

Using Lemma 10.1.4 again, we see that $B[P^*, (\eta r_0 - 3r_0 \cos\theta)/6] \subset B(P, r_0) \cap C_\theta$ which implies (see (10.4.23)) that

$$\beta h[(\eta r_0 - 3r_0 \cos\theta)/6] \leq Z_4 h(r_0) \cdot (r_0/R)^{1/12}$$

which is false if r_0/R is small enough.

Now, suppose $K(P', r') \subset (\Pi', \eta' r')$, let $\tilde{P}$ be the (geodesic) projection of P on Π', and let Π^* be the geodesic p-plane through P such that $\tau^{-1}(\Pi^*)$ is parallel to Π'_0, the tangent plane at $\tilde{P}_0 = \tau^{-1}(\tilde{P})$ to $\tau^{-1}(\Pi')$. Since $B(P, \lambda r') \subset B(P', r')$ it follows that $K(P, \lambda r') \subset (\Pi', \eta' r')$ so that $d(P, \tilde{P}) \leq \eta' r'$. From Lemmas 10.1.4 and 10.1.5 we conclude that

$$\begin{aligned}\tau^{-1}[K(P, \lambda r')] &\subset (\tau^{-1}(\Pi'), (1 + \eta_0 \lambda r') \eta' r') \\ &\subset (\Pi'_0, (1 + \eta_0 \lambda r') \eta' r' + K_2 \lambda^2 r'^2) \\ &\subset (\Pi_0^*, 2(1 + \eta_0 \lambda r') \eta' r' + K_2 \lambda^2 r'^2).\end{aligned}$$

Consequently $K(P, \lambda r') \subset (\Pi^*, 3\eta' r')$ if λ is small enough. Since $d(P, \tilde{P})/d(Q P) \leq \eta' r'/R' \leq Z_6 \eta' \leq 1/2$, it follows that Π^* and Π_1 are almost orthogonal so that $K(P, \lambda r') \subset (\Pi, \eta \lambda r')$ if

$$\eta' \leq \eta \lambda/4 \quad \text{and} \quad \eta' \leq 1/2 Z_6, \quad \text{and} \quad \Pi = \Pi^* \cap \Pi_1.$$

The fact, that ε' η', and λ may be taken in the form $C(\beta, p, \nu, \mathfrak{M}) \eta^h$ follows from an inspection of the proof.

Theorem 10.4.2. (cf. REIFENBERG [1], Lemma 6*). *For each $\xi > 0$, there exists an $\varepsilon_0(\xi, \beta, \nu, \mathfrak{M}) > 0$ and a $v(\xi, \beta, \nu, \mathfrak{M}) > 0$ such that if $P' \in S$ and $r' < R(P')$ and*

$$(10.4.25) \quad \beta \leq [h(r)]^{-1} \varphi^-(r, P) \leq [h(r)]^{-1} \psi^+(r, P) \leq \beta + \varepsilon_0$$

for every (r, P) with $P \in S$ and $B(P, r) \subset B(P', r')$, then to each such $B(P, r)$ will correspond a $P^* \in S$ and a geodesic ν-plane Π through P^* such that

$$(10.4.26) \quad B(P^*, v r) \subset B(P, r) \quad \text{and} \quad K(P^*, v r) \subset (\Pi, \xi v r).$$

Moreover ε_0 and v may be taken to have the form $C(\beta, \nu, \mathfrak{M}) \cdot \xi^h$, $h = h(\beta, \nu, \mathfrak{M})$.

Proof. $K(P, r)$ lies at a distance 0 from the whole of $\mathfrak{M}$. Then apply Lemma 10.4.3.

Lemma 10.4.4. *Suppose G is open and does not intersect A and suppose $S \cap G$ is not empty and $\gamma =$ the inf $\psi(P)$ for $P \in S \cap G$. Then if $\varepsilon_0 > 0$, there exists a sphere $B(P_1, r_1)$ such that*

$$\psi^+(r, P) \leq (\gamma + \varepsilon_0) h(r), \quad P \in S, \quad r < R(P), \quad B(P, r) \subset B(P_1, r_1).$$
(10.4.27)

Moreover, for all $B(P, r) \ni B(P, r) \subset G$, $P \in S$, and $r < R(P)$, it follows that

$$\varphi^-(r, P) \geq \gamma \cdot h(r).$$

Proof. Choose P_1 in $S \cap G$ so that $\psi(P_1) < \gamma + \varepsilon_0/4$, choose $r_2 > 0$ so that $B(P_1, r_2) \subset G$ and $\psi^+(r_2, P_1) < (\gamma + \varepsilon_0/2)\, h(r_2)$, and choose r_1, $0 < r_1 < r_2$, so that

$$(\gamma + \varepsilon_0)\, h(r_2 - r_1) > (\gamma + \varepsilon_0/2)\, h(r_2).$$

Then, if for some $B(P, r)$ in $B(P_1, r_1)$, with $P \in S$ and $r < R(P)$, (10.4.27) does not hold it follows from Theorem 10.4.1 that (since $r \leq r_1 - |P_1 P|$)

$$(\gamma + \varepsilon_0/2)\, h(r_2) > \psi^+(r_2, P_1) \geq \psi^-(r + r_2 - r_1, P) \geq \psi^+(P, r_2 - r_1)$$
$$> (\gamma + \varepsilon_0)\, h(r_2 - r_1) > (\gamma + \varepsilon_0/2)\, h(r_2)$$

which is impossible. The second statement follows from Theorem 10.4.1(h).

Theorem 10.4.3. $\psi(P) \geq 1$ *for all* $P \in S$.

Proof. Let $\beta = \inf \psi(P)$ for $P \in S$. Choose $\xi > 0$ and define ε and v as in Theorem 10.4.2. There is a sphere $B(P_1, r_1)$ with $P_1 \in S$ and $r_1 < R(P_1)$ such that (10.4.25) holds. We may apply Theorem 10.4.2 to conclude that each $B(P, r) \subset B(P_1, r_1)$ with $P \in S$ contains a point $P^* \in S$ and that $\exists$ a (geodesic) ν-plane Π through P^* so that (10.4.26) holds. We shall assume that the minimizing sequence, called $\{X_n\}$, satisfies the additional condition in Theorem 10.4.1(k). Then, for large n and fixed P and r

$$\begin{aligned} &K_n(P^*, v\,r) \subset (\Pi, 2\xi\, v\,r), \quad \Lambda^\nu[K_n(P^*, v\,r/2)] > \frac{\beta}{2} \cdot h(v\,r/2) \\ &\varphi(v\,r, P^*, X_n) < (\beta + 2\varepsilon_0) \cdot h(v\,r), \\ &(\Lambda^\nu[K_n(P^*, r)] = \psi(r, P^*; X_n), \ \text{etc.}). \end{aligned} \tag{10.4.28}$$

Hence, for each n, there is a ϱ_n, $v\,r/2 \leq \varrho_n \leq v\,r$ such that

$$\Lambda^{\nu-1}[X_n \cap \partial B(P^*, \varrho_n)] \leq 2(\beta + 2\varepsilon_0)\,(v\,r)^{-1}\, h(v\,r). \tag{10.4.29}$$

Let us choose a normal coordinate system τ centered at P^* and define

$$\begin{aligned} Y_{n0} &= \tau^{-1}[X_n \cap \overline{B(P^*, \varrho_n)}], \quad A_{n0} = \tau^{-1}[X_n \cap \partial B(P^*, \varrho_n)], \\ \Pi_0 &= \tau^{-1}(\Pi), \quad \Gamma_{n0} = f_n[C(\Pi_0, A_{n0})], \quad A^*_{n0} = f_n\, \pi\, A_{n0} = \Gamma_{n0} \cap \Pi_0, \\ A'_{n0} &= A_{n0} \cup A^*_{n0}, \quad Y_n = \tau\, Y_{n0}, \quad A_n = \tau\, A_{n0}, \quad \Gamma_n = \tau\, \Gamma_{n0}, \\ A^*_n &= \tau\, A^*_{n0}, \quad A'_n = \tau(A'_{n0}) \end{aligned}$$

where π denotes the projection (in R_N) into Π_0, and f_n denotes the radial projection onto $\partial B(0, \varrho_n)$ outside $B\left(0, \varrho_n \sqrt{1 - 25\xi^2}\right)$ and the expansion from $B\left(0, \varrho_n \sqrt{1 - 25\xi^2}\right)$ onto $B(0, \varrho_n)$ otherwise. Since $K_n(P^*, v\,r) \subset (\Pi, 2\xi\, v\,r)$, it follows that

$$Y_{n0} \subset (\Pi_0, 2(1 + \eta_0\, v\,r)\, \xi\, v\,r) \subset (\Pi_0, 5\xi\, \varrho_n).$$

From Lemma 10.2.2 and Theorem 10.2.2, it follows that

$$\begin{aligned} &\Lambda^\nu(\Gamma_n) \leq Z_1(\beta + 2\varepsilon_0) \cdot h(v\,r) \cdot \xi, \\ &Z_1 = 8\, C_1\, \varrho_n\, (v\,r)^{-1}(1 + \eta_0\, \varrho_n)^\nu (1 - 25\xi^2)^{-\nu/2}, \end{aligned} \tag{10.4.30}$$

if we assume that $1 + \eta_0\, \varrho_n \leq 5/4$.

Now, suppose that $\pi(Y_{n0})$ doesn't contain the point $Q_n \in \Pi_0 \cap B(0, \varrho_n \sqrt{1-25\xi^2})$, let $Q'_n = f_n(Q_n) \in \Pi_0 \cap B(0, \varrho_n)$, and let $g_n(Q)$ for $Q \neq Q'_n$ be the radial projection from Q'_n onto $\partial B(0, \varrho_n)$. Let $\varphi_n = g_n f_n \pi$ and let $Z_{n0} = \varphi_n(Y_{n0})$. We note also that $A^*_{n0} = \varphi_n(A_{n0})$. In the next paragraph, we shall apply Theorems 10.3.9 and 10.3.10 with

$$
\begin{gathered}
X_1 = Z_{n0}, \quad X_2 = \Gamma_{n0}, \quad X = Z_{n0} \cup \Gamma_{n0}, \quad A_1 = A^*_{n0}, \\
A_2 = A_{n0} \cup A^*_{n0}, \quad A = A_{n0}
\end{gathered} \tag{10.4.31}
$$

to show that $b(Z_{n0} \cup \Gamma_{n0}, A_{n0}) \supset b(Y_{n0}, A_{n0})$. From this, it follows from Theorem 10.3.6 that

$$
b(Z_n \cup \Gamma_n, A_n) \supset b(Y_n, A_n), \quad Z_n = \tau(Z_{n0}). \tag{10.4.32}
$$

From (10.4.32) it follows that we may replace Y_n by $Z_n \cup \Gamma_n$ and obtain another surface in $\mathfrak{C}(A, L)$. But, since $Z_{n0} \subset \Pi_0 \cap \partial B(0, \varrho_n)$, we see that $\Lambda^\nu(Z_{n0}) = \Lambda^\nu(Z_n) = 0$ and so

$$
\begin{aligned}
&\Lambda^\nu(Z_n \cup \Gamma_n) = \Lambda^\nu(\Gamma_n) \leq Z_2(\beta + 2\varepsilon_0) \cdot h(v\,r) \cdot \xi \\
&\Lambda^\nu[(X_n - Y_n) \cup Z_n \cup \Gamma_n] \leq d(A, L) + \varepsilon_n - 2^{-\nu-1} \cdot \beta \cdot h(v\,r) + \\
&\qquad\qquad + Z_2 \cdot (\beta + 2\varepsilon_0)\, h(v\,r) \cdot \xi < d(A, L)
\end{aligned}
$$

if ξ and ε_0 are sufficiently small and n is sufficiently large. Thus, for such ξ, ε_0, and n, $\pi(Y_{n0}) \supset \Pi_0 \cap B(0, \varrho_n \sqrt{1-25\xi^2})$, so

$$
\begin{aligned}
\Lambda^\nu(Y_{n0}) &\geq \gamma_\nu\, \varrho_n^\nu (1 - 25\xi^2)^{-\nu/2}, \\
\Lambda^\nu(Y_n) &\geq \gamma_\nu\, \varrho_n^\nu (1 - 25\xi^2)^{-\nu/2} (1 + \eta_0 v\,r)^{-\nu}.
\end{aligned}
$$

The theorem then follows from the arbitrariness of ξ and ε_0 by letting $n \to \infty$ through a subsequence so $\varrho_n \to \varrho$ and then using the arbitrariness of r.

We notice first, using Theorem 10.3.12 that

$$
\begin{gathered}
b(I \times X_{n0}, D_{n0}) \supset \{h' \mid h' = I_{n1*}(h) - I_{n0*}(h), \quad h \in H_{\nu-1}(A_{n0})\}, \\
D_{n0} = (0 \times A_{n0}) \cup (1 \times A_{n0})
\end{gathered}
$$

where I_{n0} and I_{n1} are defined in that theorem. Then, using Theorem 10.3.6 with $F_n : (I \times A_{n0}, D_{n0}) \to (\Gamma_{n0}, A_{n0} \cup A^*_{n0})$, where

$$
F_n(t, Q) = f_n[(1-t)\,Q + t\,\pi(Q)], \quad Q \in A_{n0},
$$

we see that

$$
\begin{aligned}
&b(\Gamma_{n0}, A_{n0} \cup A^*_{n0}) \supset \{k \mid k = J_{n1*}(h) - J_{n0*}(h),\ h \in H_{\nu-1}(A_{n0})\} \equiv L_2, \\
&J_{n1}(Q) = (\varphi_n \mid A_{n0})\,(Q), \quad Q \in A_{n0}, \quad J_{n1}(Q) \in A_{n0} \cup A^*_{n0}, \\
&J_{n0}(Q) = Q \qquad\qquad\quad , \quad Q \in A_{n0}, \quad J_{n0}(Q) \in A_{n0} \cup A^*_{n0}.
\end{aligned}
$$

Also, using the same theorem with F_n replaced by φ_n, we obtain

$$
b(X_1, A_1) \equiv b(Z_{n0}, A^*_{n0}) \supset (\varphi_n \mid A_{n0})_* \,[b(Y_n, A_{n0})] \equiv L_1.
$$

In order to prove (10.4.32), we conclude from Theorems 10.3.9 and 10.3.10 that it is sufficient to show that

$$\begin{aligned} &i(A_{n0} \cup A^*_{n0}, A_{n0})_* [b(Y_{n0}, A_{n0})] \\ (10.4.33) \qquad &\subset i(A_{n0} \cup A^*_{n0}, A^*_{n0})_* (\varphi_n \mid A_{n0}) [b(Y_{n0}, A_{n0})] + L_2, \\ &\text{or} \quad J_{n0*}[b(Y_{n0}, A_{n0})] \subset J_{n1*}[b(Y_{n0}, A_{n0})] + L_2. \end{aligned}$$

But if $h \in b(Y_{n0}, A_{n0})$ $(\subset H_{\nu-1}(A_{n0}))$, we see that

$$J_{n0*}(h) = J_{n1*}(h) + [J_{n1*}(-h) - J_{n0*}(-h)]$$

so that (10.4.33), and hence (10.4.32) holds.

Theorem 10.4.4. *The set X minimizes $\Lambda^\nu(X' - A)$ among all $X' \in \mathfrak{C}(A, L)$. Moreover $\psi(P) = 1$ almost everywhere on S and*

$$\psi^-(r, P) = \Lambda^\nu[S \cap B(P, r)]$$

for all $r < R(P)$ not in a countable set.

Proof. We have seen, using Theorems 10.4.1 (k) and Theorem 10.3.16, that $X_0 \in \mathfrak{C}(A, L)$. For each $\varrho > 0$, we can find a disjoint family $\{B(P_i, r_i)\}$, where $5r_i < \varrho$, $P_i \in S$, and $B(P_i, r_i) \cap A$ is empty for each i, such that

$$S \subset \bigcup_{i=1}^{\infty} B(P_i, r_i) \cup \bigcup_{i=k+1}^{\infty} B(P_i, 5r_i), \quad k = 0, 1, 2, \ldots \qquad (10.4.34)$$

(recall Lemma 10.2.3). From Theorem 10.4.3, we obtain

$$\begin{aligned} \sum_{i=1}^{\infty} \gamma_\nu r_i^\nu &\le (1 + h\varrho)^\nu \sum_{i=1}^{\infty} h(r_i) \le (1 + h\varrho)^\nu \sum_{i=1}^{\infty} \psi^-(r_i, P_i) \\ &\le (1 + h\varrho)^\nu \sum_{i=1}^{\infty} \liminf_n \psi(r_i, P_i; X_n) \\ &\le (1 + h\varrho)^\nu \liminf_n \sum_{i=1}^{\infty} \psi(r_i, P_i; X_n) \\ &\le (1 + h\varrho)^\nu \lim_n \Lambda^\nu(X_n) = (1 + h\varrho)^\nu d(A, L). \end{aligned} \qquad (10.4.35)$$

From (10.4.35), we see that the series $\gamma_\nu r_1^\nu + \ldots$ converges. From this and (10.4.34) we see that

$$\Lambda^\nu_\varrho(S) \le \sum_{i=1}^{\infty} \gamma_\nu r_i^\nu \le (1 + h\varrho)^\nu d(A, L). \qquad (10.4.36)$$

The first result follows by letting $\varrho \to 0$ in (10.4.36).

By replacing S by $S \cap B(P, r)$ and requiring the $B(P_i, r_i) \subset B(P, r)$, we conclude by the argument above that

$$\Lambda^\nu[S \cap B(P, r)] \le \psi^-(r, P) \le \liminf_n \Lambda^\nu[X_n \cap B(P, r)]. \qquad (10.4.37)$$

If, now, $P \in S$, $r < R(P)$, and $\Lambda^\nu[S \cap \partial B(P, r)] = 0$ and $\Lambda^\nu[S_n \cap \partial B(P, r)] = 0$ for every n, we find by repeating the argument for the set $S - \overline{B(P, r)}$ that

$$\Lambda^\nu[S - B(P, r)] \le \liminf_n \Lambda^\nu[S_n - B(P, r)]. \qquad (10.4.38)$$

Since $\Lambda^\nu(X_n - A) \to \Lambda^\nu(X_0 - A) = \Lambda^\nu(S)$, the equality must hold in (10.4.37) and (10.4.38). Consequently

$$\psi(P) = \lim_{r\to 0}(\gamma_\nu r^\nu)^{-1}\Lambda^\nu[S \cap B(p, r)], \quad P \in S.$$

Suppose $\Lambda^\nu(Z_k) > 0$ and choose ε with $0 < \varepsilon < \Lambda^\nu(Z_k)/2k$. Cover Z_k by an open set G so that $\Lambda^\nu[G \cap (S - Z_k)] < \varepsilon$. Again we may find a disjoint family $\{B(P_i, r_i)\}$ in which $P_i \in Z_k$, $r_i < R(P_i)$, $5r_i < \varrho$, and $B(P_i, r_i) \subset G$ such that (10.4.34) holds with S replaced by Z_k. Clearly the $B(P_i, r_i)$ cover all of Z_k except possibly a set ζ_k with $\Lambda^\nu(\zeta_k) = 0$. Then

$$\begin{aligned}\Lambda^\nu(Z_k) + \varepsilon &\geq \sum_{i=1}^{\infty}\Lambda^\nu[S \cap B(P_i, r_i)] \geq (1 + h\varrho)^{-\nu}(1 + k^{-1})\sum_{i=1}^{\infty}\gamma_\nu r_i^\nu \\ &\geq (1 + k^{-1})(1 + h\varrho)^{-\nu}\Lambda_\varrho^\nu(Z_k).\end{aligned}$$

Since this holds for each $\varepsilon > 0$ and $\varrho > 0$, it follows that $\Lambda^\nu(Z_k) \geq (1 + k^{-1})\Lambda^\nu(Z_k)$, so that $\Lambda^\nu(Z_k) = 0$. Since this is true for each k, the result follows from Theorem 10.4.3.

Our aim now is to show that $S = X_0 - A$ satisfies Reifenberg's (ε, R_1) condition at any point P_0 where $\psi(P_0) < 1 + \varepsilon_0$ where ε_0 depends only on ν, N, $\mathfrak{M}$, and ε (of course ε, $\varepsilon_0 > 0$).

Definition 10.4.1. A locally compact set S is said to satisfy Reifenberg's (ε, R_1) condition at $P_0 \Leftrightarrow$ there is a number $R_1 > 0 \ni$ to each point $P \in \overline{B(P_0, 2R_1)}$ and each $R \leq R_1$ there corresponds a geodesic ν-plane $\sum(P, R)$ centered at P and a geodesic ν-plane $\sum$ centered at P_0 such that

$$D[S \cap B(P, R), \textstyle\sum(P, R) \cap B(P, R)] \leq \varepsilon R,$$

$$D[S \cap \overline{B(P_0, 2R_1)}, \textstyle\sum \cap B(P_0, 2R_1)] \leq \varepsilon R_1.$$

Lemma 10.4.5. *With each $\varepsilon > 0$, there is associated a positive number R_0', with $R_0' \leq R_0$, which has the following property: If $0 < r \leq R_0'$, $0 < \varrho \leq \delta \leq \delta_0$, and $B(Q, \varrho) \subset B(P, r)$ and $P \notin B(Q, \varrho)$, then*

$$\Lambda_\delta^{\nu-1}(S) \leq (1 + \varepsilon)\Lambda_\delta^{\nu-1}(S'), \quad S = B(Q, \varrho) \cap \partial B(P, t), \quad S' = S \cap \Pi$$

Π being any geodesic ν-plane centered at P and containing Q. Also

$$\int_{r_1}^{r_2}\Lambda^{\nu-1}[\Pi \cap B(Q, \varrho) \cap \partial B(P, t)]\, dt \tag{10.4.39}$$
$$= \Lambda^\nu[\Pi \cap B(Q, \varrho) \cap B(P, r_2)] - \Lambda^\nu[\Pi \cap B(Q, \varrho) \cap B(P, r_1)].$$

Remark. If $\mathfrak{M} = R_N$, $\Lambda_\delta^{\nu-1}(S) = \Lambda_\delta^{\nu-1}(S')$ without restriction on r.

Proof. The last statement is evident since $\Pi \cap B(Q, \varrho) \cap \partial B(P, t)$ moves perpendicular to itself on $\mathfrak{M}$ as t varies. To prove the first statement, let ω be a normal coordinate system centered at P with z^N axis through $Q_0 = \omega^{-1}(Q)$. Let us write coordinates of points on the z^N axis in the form $(a, 0)$. Let $\eta > 0$ and choose R_0' to satisfy the condition of Lemma 10.1.3 for η instead of ε; η is to be chosen later. Let $S_0 = \omega^{-1}(S)$, $S'_0 = \omega^{-1}(S')$, and $\Pi_0 = \omega^{-1}(\Pi)$. Then Π_0 is a ν-plane through the z^N axis and $S'_0 = S_0 \cap \Pi_0$. Since $\varrho \leq \delta$, it is clear that

$$\Lambda_\delta^{\nu-1}(S) = \gamma_{\nu-1}[\varrho(S)]^{\nu-1}, \quad \Lambda_\delta^{\nu-1}(S') = \gamma_{\nu-1}[\varrho(S')]^{\nu-1} \tag{10.4.40}$$

where $\varrho(S)$ was defined in Lemma 10.1.4. From that lemma, we obtain

$$\varrho(S) \leq (1 + \eta_0 r)\,\varrho(S_0), \quad \varrho(S'_0) \leq (1 + \eta_0 r)\,\varrho(S') \tag{10.4.41}$$

If we write $Q_0 = (c, 0)$, $c \geq \varrho$, we see from Lemma 10.1.3 that

$$\begin{aligned} &S_0^- \subset S_0 \subset S_0^+, \quad S_0^- = \bigcup_{-\eta \leq k \leq \eta} B[c + k\varrho, 0; (1-\eta)\varrho] \cap \partial B(0, t) \\ &S_0^+ = \partial B(0, t) \cap B[c + \eta\varrho, 0; (1+\eta)\varrho] \cap B[c - \eta\varrho, 0; (1+\eta)\varrho]. \end{aligned} \tag{10.4.42}$$

From (10.4.41), it is possible to conclude, for η small enough, that

$$\begin{aligned} &\varrho(S_0^-) \leq \varrho(S_0) \leq \varrho(S_0^+), \quad \varrho(S_0^+) \leq (1 + Z\eta)\,\varrho(S_0^-) \\ &\varrho(S_0^- \cap \Pi_0) = \varrho(S_0^-) \leq \varrho(S') \leq \varrho(S_0^+ \cap \Pi_0) = \varrho(S_0^+) \\ &\varrho(S_0) \leq (1 + Z\eta)\,\varrho(S'_0). \end{aligned}$$

The desired result follows from this, (10.4.40), (10.4.41) and proper choices of η and R'_0.

Lemma 10.4.6. *Suppose $\Lambda^\nu(X) < \infty$ and $\varepsilon > 0$, suppose R'_0 satisfies the conditions of Lemma 10.4.5, and suppose that $r \leq R'_0$, $\theta > 0$, $l_\theta > 0$, and $0 < \xi(1 + \eta_0\delta_0) < 1 - \cos\theta$. Suppose that for arbitrarily small $\delta \leq \delta_0$ that there exists a disjoint family of balls $\{B(P_i, r_i)\}$ such that*

$$\begin{aligned} &B(P_i, r_i) \subset B(P, r), \quad r_i < \delta \quad \text{(for each } i) \\ &h(r_i) \leq \Lambda^\nu[K(P_i, r_i)] \leq M\,h(r_i), \quad \sum_i \gamma_\nu r_i^\nu \geq e_\theta \\ &P \in B(P_i, r_i) \text{ for some } i,\ P_i \text{ and } P \in X \text{ for each } i \\ &K(P_i, r_i) \subset (\Pi_i, \xi r_i) \quad (K(Q, \varrho) \equiv X \cap B(Q, \varrho)) \end{aligned}$$

where Π_i is a geodesic ν-plane centered at P_i and making an angle $> \theta$ there with the geodesic $P P_i$. Then

$$\begin{aligned} \varphi(r, P; X) \leq{} &(1 + \varepsilon)(1 + \eta_0 r)^\nu \psi(r, P; X) - \\ &- 2^{-\nu}(1 + \eta_0 r)^{-\nu}[1 - \cos\theta - \xi(1 + \eta_0\delta_0)]^\nu \cdot e_\theta. \end{aligned}$$

Proof. Let us cover $K(P, r) - \bigcup_i K(P_i, r_i)$ by a countable family $\{B(Q_j, \varrho_j)\}$ of balls $\subset B(P, r)$ and such that

$$\varrho_j < \delta, \quad \sum_j \gamma_\nu \varrho_j^\nu \leq \Lambda^\nu[K(P, r) - \bigcup_i K(P_i, r_i)] + \delta. \tag{10.4.43}$$

Now, using Lemma 10.4.5 with Π_j centered at P and passing through Q_j, we obtain

$$\begin{aligned} \int_0^r \Lambda_\delta^{\nu-1}[B(Q_j, \varrho_j) \cap l(P, t)]\,dt &\leq (1 + \varepsilon)\,\Lambda^\nu[\Pi_j \cap B(Q_j, \varrho_j)] \\ &\leq (1 + \varepsilon)(1 + \eta_0 r)^\nu \gamma_\nu \varrho_j^\nu \quad (l(P, t) = X \cap \partial B(P, t)). \end{aligned} \tag{10.4.44}$$

If $P \notin B(P_i, r_i)$, we note first that $B(P_i, r_i) \cap l(P, t)$ is empty for $t < s_i$, where

$$s_i = d(P\,P_i) - r_i \cos\theta - r_i\,\xi\,(1 + \eta_0\,\delta). \tag{10.4.45}$$

This expression for s_i is arrived at by choosing a geodesic coordinate system centered at P_i and using the result of Lemma 10.1.4 in this system. Using the same idea, we conclude that $B(\bar{P}_i, t_i) \subset B(P_i, r_i) \cap B(P, t)$ for a suitable point $\bar{P}_i$ on PP_i, where

$$\begin{aligned} 2t_i = s_i - (|P\,P_i| - r_i) &= r_i[1 - \cos\theta - \xi(1 + \eta_0\,\delta)] + \\ &+ 2r_i\,\sigma, \quad t \geq s_i. \end{aligned} \tag{10.4.46}$$

Thus, if Π_i^* is centered at P and passes through P_i, we conclude that

$$\begin{aligned} &\int_0^r \Lambda_\delta^{\nu-1}[B(P_i, r_i) \cap l(P, t)]\,dt \leq (1 + \varepsilon) \times \\ &\times \int_{s_i}^r \Lambda^{\nu-1}[\Pi_i^* \cap B(P_i\,r_i) \cap l(P, t)]\,dt \\ &= (1 + \varepsilon)\,\Lambda^\nu[\Pi_i^* \cap B(P_i, r_i)] - (1 + \varepsilon)\,\Lambda^\nu([\Pi_i^* \cap B(P_i, r_i) \\ &\cap B(P, s_i)] \leq (1 + \varepsilon)\,(1 + \eta_0\,r)^\nu\,\gamma_\nu\,r_i^\nu - \\ &- (1 + \varepsilon)\,(1 + \eta_0\,r)^{-\nu}\,\sigma^\nu\,\gamma_\nu\,r_i^\nu \end{aligned} \tag{10.4.47}$$

using (10.4.46) and Lemma 10.1.4.

Since the $B(P_i, r_i)$ are disjoint, at most one contains P and

$$\int_0^r \Lambda_\delta^{\nu-1}[B(P_{i0}, r_{i0}) \cap l(P, t)]\,dt \leq \Lambda^\nu[K(P_{i0}, r_{i0})] \leq M\,h(r_{i0}). \tag{10.4.48}$$

Adding the results in (10.4.44), (10.4.47), and (10.4.48) and using (10.4.43), we obtain (since $\gamma_\nu\,r_i^\nu \leq (1 + h\,\delta)^\nu\,h(r_i)$)

$$\begin{aligned} \int_0^r \Lambda_\delta^{\nu-1}[l(P, t)]\,dt &\leq (1 + \varepsilon)\,(1 + \eta_0\,r)^\nu\,\Lambda^\nu\Big[K(P, r) - \bigcup_i K(P_i, r_i)\Big] + \\ &+ (1 + \varepsilon)\,(1 + \eta_0\,r)^\nu (1 + h\,\delta)^\nu\,\Lambda^\nu\Big[\bigcup_i K(P_i, r_i)\Big] - \\ &- (1 + \varepsilon)\,(1 + \eta_0\,r)^{-\nu}\,\sigma^\nu\,e_\theta \end{aligned}$$

from which the lemma follows by letting $\delta \to 0$.

Definition 10.4.2. Suppose Π is a geodesic p-plane centered at P and suppose U is a set $\subset B(P, R_0)$. We define the *projection $p_\Pi(U)$ of U on Π* as follows: Let τ be a normal coordinate system centered at P. Then we define

$$p_\Pi(S) = \tau \circ p_{0\Pi} \circ \tau^{-1}(S)$$

where p_0 denotes the ordinary orthogonal projection into $\tau^{-1}(\Pi)$ (which is an actual p-plane in R_N). From Lemma 10.1.1(b), it follows that $p_\Pi(S)$ is independent of the particular τ chosen.

The method of proof of Theorem 10.4.3 leads to the following lemma:

Lemma 10.4.7. *Suppose that* $X \in \mathfrak{C}(A, L), \Lambda^\nu(X - A) < d(A, L) + \eta$, $\varrho \leq R_0$, $\overline{B(P, \varrho)} \cap A$ *is empty,* Π *is a geodesic* ν*-plane centered at* P, $K(P, \varrho) = X \cap B(P, \varrho)$, $l(P, \varrho) = X \cap \partial B(P, \varrho)$, *and*

$$K(P, \varrho) \subset (\Pi, \xi\varrho), \quad \Lambda^\nu[K(P, \varrho)] \geq \beta h(\varrho), \quad Z\xi \leq 1/2, \quad Z = 1 + \eta_0 \varrho,$$
$$\eta + C_1 \xi \varrho (1 + \eta_0 \varrho)^{2\nu} (1 - Z^2 \xi^2)^{-\nu/2} \Lambda^{\nu-1}[l(P, \varrho)] < \beta h(\varrho)$$

C_1 *being the constant of Theorem* 10.2.2. *Then*

$$p_\Pi[K(P, \varrho)] \supset \Pi \cap B\left(P, \varrho\sqrt{1 - Z^2\xi^2}\right).$$

Proof. For, otherwise $Y = K(P, \varrho) \cup l(P, \varrho)$ could be replaced by $\Gamma \cup Z$, constructed as in the proof of Theorem 10.4.3, in which $\Lambda^\nu(Z) = 0$ and

$$\Lambda^\nu(\Gamma \cup Z) + \Lambda^\nu(\Gamma) \leq C_1 \xi \varrho (1 + \eta_0 \varrho)^{2\nu} (1 - Z^2 \xi^2)^{-\nu/2} \times$$
$$\times \Lambda^{\nu-1}[l(P, \varrho)] < \Lambda^\nu(Y) - \eta.$$

Theorem 10.4.5. *There are positive constants* $c_1, c_2, h \geq 1$, *and* R_0'' *with the following property: Suppose* X_0 *is a minimizing set as in Theorem* 10.4.4 *and suppose* $P_0 \in S \equiv X_0 - A$. *Then, if*

$$0 < r_2 \leq R_0'', \quad r_2 < R(P_0), \quad r_2 \leq c_1 \varepsilon^*, \quad \varepsilon_2 \leq c_2 (\varepsilon^*)^h, \quad \textit{and}$$
$$h(r) \leq \psi^+(r, P) \leq h(r)(1 + 2\varepsilon_2) \quad \textit{for all} \tag{10.4.49}$$
$$B(P, r) \subset B(P_0, r_2),$$

S *satisfies an* (ε^*, R_1) *condition at* P_0 *for any* $R_1 \leq r_2/32$.

If $\psi(P_0) < 1 + \varepsilon_2$, *there is an* r_2 *such that* $r_2 \leq R_0''$ *and* $r_2 < R(P_0)$ *such that the second line holds in* (10.4.49).

Proof. The proof of the last statement is like that of Lemma 10.4.4. Suppose $\varepsilon_2 > 0$, $\xi > 0$, and R_0' is chosen so that Lemma 10.4.5 holds with $\varepsilon = 1/4$, say. We suppose that $r_2 \leq R_0'$. From Theorem 10.4.2, it follows that if $\varepsilon_2 \leq \varepsilon_2(\nu, \mathfrak{M})$, there is a $v > 0$ such that each $B(P, r)$ $\subset B(P_0, r_2)$ contains a point P^* in S through which passes a geodesic ν-plane $\sum(P^*)$, centered at P^*, such that

$$B(P^*, vr) \subset B(P, r), \quad K(P^*, vr) \equiv S \cap B(P^*, vr) \subset (\textstyle\sum(P^*), \xi vr).$$

Suppose, now, that

$$P' \in S, \quad B(P', r') \subset B(P_0, r_2), \quad Q_j \in K(P', r'/4), \quad j = 1, \ldots, \nu + 2.$$

We suppose that $0 < \delta \leq \delta_0$, where

$$\lambda \equiv (1 + \eta_0 \delta_0) \leq (16/15)^2.$$

From Lemma 10.2.3, it follows that we may find a disjoint family of balls $\{B(P_i, r_i)\}$ such that

$$K(P', r'/4) \subset \bigcup_{i=1}^\infty K(P_i, \varrho_i) \cup \bigcup_{i=k+1}^\infty K(P_i, 5\varrho_i), \quad k = 1, 2, \ldots$$
$$P_i \in S, \quad 5\varrho_i < \delta, \quad B(P_i, 5\varrho_i) \subset B(P', r'/4).$$

Since $\Lambda^\nu[K(P_i, \varrho_i)] \geq h(\varrho_i)$, it follows that the series of Λ^ν-measures converges. Hence we may choose a finite subfamily such that

$$\sum_i \Lambda^\nu K(P_i, \varrho_i) \geq \frac{1}{2} \Lambda^\nu K(P', r'/4)$$

$$\text{(10.4.50)} \qquad \sum_i h(\varrho_i) \geq 2^{-1}(1 + 2\varepsilon_2)^{-1} \Lambda^\nu K(P, r'/4) \geq 2^{-1}(1 + 2\varepsilon_2)^{-1} h(r'/4).$$

For each $j \leq \nu + 2$ and each i,

$$B(P_i, \varrho_i) \subset B(P', r'/4) \subset B(Q_j, r'/2) \subset B(P', r').$$

Thus we have a set of balls $\{B(P_i^*, v\varrho_i)\}$ such that $P_i^* \in S$ and there exists a geodesic ν-plane $\sum(P_i^*)$ centered at P_i^* such that

$$K(P_i^*, \varrho_i) \subset (\textstyle\sum(P_i^*), \xi v \varrho_i), \quad B(P_i^*, v\varrho_i) \subset B(P_i, \varrho_i).$$

Now suppose $\theta > 0$ and suppose

$$\text{(10.4.51)} \qquad (1 - \cos\theta) = 3\lambda\xi, \quad \lambda = 1 + \eta_0\delta_0.$$

Let σ_j be the set of $i \ni$ the geodesic $P_i^* Q_j$ makes an angle $> \theta$ at P_i^* with $\sum(P_i^*)$. Suppose for some j and some $\mu > 0$ that, for arbitrarily small δ, we would have

$$\text{(10.4.52)} \qquad \sum_{i \in \sigma_j} \gamma_\nu v^\nu \varrho_i^\nu \geq e_\theta = \mu v^\nu (r'/4)^\nu, \quad \mu > 0.$$

Then, from Lemma 10.4.6 and (10.4.51), it would follow that

$$\text{(10.4.53)} \quad \varphi(r'/2, Q_j) \leq \psi(r'/2, Q_j) - e_\theta \lambda^\nu \xi^\nu (1 + \varepsilon_1)(1 + \eta_0 r'/2)^{-\nu}.$$

But, from (10.4.49), we have

$$\psi(r'/2, Q_j) - \varphi(r'/2, Q_j) \leq 2\varepsilon_2(1 + h r'/2)^{-\nu} \gamma_\nu (r'/2)^\nu$$

which contradicts (10.4.53) if

$$\text{(10.4.54)} \quad \varepsilon_2 < 2^{-1-\nu} \mu (\lambda v \xi)^\nu (1 + \varepsilon_1)(1 + c_0 r'^2/4)^{-\nu}(1 + h r'/2)^\nu.$$

Thus the reverse inequality must hold in (10.4.52) for all sufficiently small δ.

Since there are only $(\nu + 2)$ Q_j, we may choose a δ so small that there is an i_0 which is not in any σ_j, provided merely that μ is chosen small enough so that

$$\text{(10.4.55)} \qquad (\nu + 2)\mu < 2^{-1}(1 + 2\varepsilon_2)^{-1}(1 + h r'/4)^{-\nu} \gamma_\nu;$$

this follows by comparing (10.4.52) (with the inequality reversed) and (10.4.50). Thus for some i_0 the geodesic $P_{i_0}^* Q_j$ makes an angle $\leq \theta$ at $P_{i_0}^*$ with $\sum(P_{i_0}^*)$ for each j. Thus, using this fact, (10.4.51), and the proof of Lemma 10.1.5 (the curvature of arcs corresponding to geodesics), we have shown that if $Q_1, \ldots, Q_{\nu+2}$ are any $(\nu + 2)$ points in $\overline{B(P', r'/4)}$, there is a geodesic ν-plane $\sum$ centered at a point of $\overline{K(P', r'/4)}$ such that each Q_j is within a distance

$$\text{(10.4.56)} \quad \leq \frac{r'}{2}\left[\sin\theta + \frac{1}{2} K_2 \left(\frac{r'}{2}\right)\right] \leq \frac{r'}{2}\left[\frac{5}{2}(\lambda\xi)^{1/2} + \frac{1}{2} K_2 \left(\frac{r'}{2}\right)\right] = \sigma r'/2$$

of $\sum (= \sum(P_{i_0}^*))$.

Let τ be a geodesic coordinate system with domain $\overline{B(0, R_0)}$ for which $\tau(0) = P'$. Choose $Q_1^*, \ldots, Q_{\nu+1}^*$ in $\overline{K(P', r'/4)}$ so that $\Lambda^\nu(\Delta^*)$ is a maximum, Δ^* being the ν-simplex in R_N having the $\tau^{-1}(Q_j^*)$ as vertices. Let Q be any other point of $\overline{K(P', r'/4)}$, let $\tilde{\sum}$ be the geodesic ν-plane of the preceding paragraph determined by Q and the Q_k^*, suppose $\tilde{\sum}$ is centered at $\tilde{P}(\in K(P', r'/4))$, and suppose $\tilde{\sum}_0$ is the tangent ν-plane in R_N to $\tau^{-1}(\tilde{\sum})$ at $\tau^{-1}(\tilde{P})$. Then it is easy to see, using Lemmas 10.1.4 and 10.1.5 that the distances from $\tilde{\sum}_0$ of $\tau^{-1}(Q)$ and the $\tau^{-1}(Q_j^*)$ are all

$$\begin{aligned} &\leq (1 + \eta_0 r'/4)\, \sigma r'/2 + K_2 r'^2/16 = \sigma' r'/4, \\ &\sigma' = 2(1 + \eta_0 r'/4)\, \sigma + K_2 r'/4. \end{aligned} \tag{10.4.57}$$

Let Δ_0 be the $(\nu + 1)$-simplex having Δ^* as one face and $\tau^{-1}(Q)$ as opposite vertex. From our construction it follows that the Λ^ν measure of each face of Δ_0 is $\leq D = \Lambda^\nu(\Delta^*)$. Consequently the Λ^ν measure of the projection of Δ_0 on $\tilde{\sum}_0$ is $\leq (\nu + 2)\, D$ and hence

$$\Lambda^{\nu+1}(\Delta_0) \leq 2D\, \sigma' r'/4.$$

Since Δ^* is one face of Δ_0, it follows that the distance of $\tau^{-1}(Q)$ from the ν-plane of Δ^* is $\leq (\nu + 2)\, \sigma' r'/2$. Since Q is any point of $K(P', r'/4)$, it follows that $\tau^{-1}(P') = O$ is at a distance $\leq (\nu + 2)\, \sigma' r'/2$ from that plane. Thus $\tau^{-1}(Q)$ is at a distance $\leq (\nu + 2)\, \sigma' r'$ from the plane through O parallel to that of Δ^*. The image of this plane under τ is a geodesic ν-plane $\sum$ through P' and we have shown that

$$K(P', r'/4) \subset (\textstyle\sum, \varepsilon' r'/4) \tag{10.4.58}$$

provided that

$$4(1 + \eta_0 r'/4)\, (\nu + 2)\, \sigma' \leq \varepsilon'. \tag{10.4.59}$$

Now, from (10.4.49) and (10.4.58), it follows that there exists an x between $r'/8$ and $r'/4$ such that

$$\frac{r'}{8} \Lambda^{\nu-1}[l(P', x)] \leq (1 + 2\varepsilon_2)\, h(r'/4) \quad \text{or} \quad \Lambda^{\nu-1}[l(P', x)] \leq Z_1 x^{-1} h(x),$$
$$\Lambda^\nu[K(P', x)] \geq h(x), \quad K(P', x) \subset (\textstyle\sum, 2\varepsilon' x).$$

Since X_0 is minimizing, it follows from Lemma 10.4.7 that the projection of $K(P', x)$ on $\sum$ contains $\sum \cap B(P', x\sqrt{1 - 4Z^2 \varepsilon'^2})$, Z being the number (near 1) mentioned there, provided merely that ξ is small.

We now show that S satisfies an (ε^*, R_1) condition at P_0 if

$$R_1 \leq r_2/32. \tag{10.4.60}$$

Let $P \in K(P_0, 2R_1)$ and $R \leq R_1$. Then $B(P, R) \subset B(P_0, r_2/8)$. If we take $P' = P$ and $r' = 8R$, then $B(P', r') \subset B(P_0, r_2)$ and we may take $\sum(P, R) = \sum$. Using (10.4.58) we find that

$$K(P, R) = K(P', r'/8) \subset (\textstyle\sum (P, R), 2\varepsilon' R).$$

From the preceding paragraph we conclude that the projection of $K(P, R)$ on $\sum = \sum(P, R) \supset \sum \cap B(P, R\sqrt{1 - 4Z^2 \varepsilon'^2})$, so that

$$\begin{aligned} \sum(P, R) \cap B(P, R) &\subset [K(P, R), 2\varepsilon' R + R(1 - \sqrt{1 - 4Z^2 \varepsilon'^2})] \\ &\subset [K(P, R), 3\varepsilon' R] \qquad (10.4.61) \\ (1 - \sqrt{1 - 4Z^2 \varepsilon'^2}) &\leq \varepsilon'. \end{aligned}$$

The cases where $P = P_0$ are taken care of in a similar way by setting $P' = P_0$, $r' = 16R_1$.

Now, by following back through equations (10.4.61), (10.4.59), (10.4.58.), (10.4.57), (10.4.56), and by choosing δ_0 and R_0'' so that

$$\lambda \equiv 1 + \eta_0 \delta_0 \leq (16/15)^2, \quad \eta_0 R_0'' \leq 1, \quad R_0'' \leq R_0',$$

we see that it is sufficient to have

$$(\nu + 2)(100\xi^{1/2} + 105 K_2 r_2/8) \leq \varepsilon^*,$$

and from Theorem 10.4.2, we may take ε_2 of the form $Z \cdot \xi^s$.

10.5. The local topological disc property

In this section, we prove what is probaly the most interesting theorem in REIFENBERG's first paper:

Theorem 10.5.1. *There are positive numbers $\varepsilon_0 = \varepsilon_0(N, \mathfrak{M})$ and $R_0' = R_0(N, \mathfrak{M}) \leq R_0$ such that if $0 < \varepsilon \leq \varepsilon_0$ and if the locally compact set S satisfies the (ε, R_1) condition at P_0 for some positive number $R_1 \leq R_0'$, then a neighborhood of P_0 on S is a topological disc.*

Definition 10.5.1. If $\sum$ and $\sum'$ are ν-planes in R_N we define

$$d_0(\sum, \sum') = D[\tilde{\sum} \cap \overline{B(0, 1)}, \quad \tilde{\sum}' \cap \overline{B(0, 1)}]$$

where $\tilde{\sum}$ and $\tilde{\sum}'$ are the respectively parallel ν-planes through 0. If c and c' are orthogonal matrices, we define

$$|c - c'| = \Big[\sum_{k,l} |c_k^l - {}'c_k^l|^2\Big]^{1/2} \quad (c = (c_k^l), \text{ etc.}).$$

We notice the following facts:

Lemma 10.5.1. (a) *If c is any orthogonal matrix, $|c| = \sqrt{N}$.*

(b) *If S_1 and S_2 are compact sets, P is a point (in $\mathfrak{M}$), $R, \eta, \eta' > 0$, and $S_1 \cap B(P, R) \subset (S_2, \eta') \cap B(P, R)$, then*

$$\overline{(S_1, \eta)} \cap \overline{B(P, R)} \subset \overline{(S_2, \eta + \eta')} \cap \overline{B(P, R)}.$$

(c) *If $\sum$ and $\sum'$ are ν-planes through P in R_N and if $D[\sum \cap B(P, R), \sum' \cap B(P, R)] = \eta R$, then $d_0(\sum, \sum') = \eta$.*

(d) *If c and c' are orthogonal matrices and $\sum_c$ and $\sum_{c'}$ are ν-planes in* R_N spanned respectively by *$c_1, \ldots, c_\nu$ and $'c_1, \ldots, 'c_\nu$, then*

$$d_0(\sum_c, \sum_{c'}) \leq |c - c'| \quad (c = (c_k^l), \quad c_k = (c_k^1, \ldots, c_k^N)).$$

Proof. Parts (a), (b), and (c) are easily verified by the reader. To prove (d), we first note that if $\sum_c$ and $\sum_{c'}$ pass through O, then $d_0(\sum_c, \sum_{c'})$ is just the maximum distance from $\sum_c$ of any point of $\sum_{c'} \cap \partial B(0, 1)$. Let us now write

$$c'_j = c_j + \sum_{k=1}^{N} e_j^k c_k. \tag{10.5.1}$$

A point P will be on $\sum_{c'} \cap \partial B(0, 1) \Leftrightarrow$

$$\overrightarrow{OP} = \sum_{j=1}^{\nu} {'x_p^j} c'_j = \sum_{j=1}^{\nu} \left[c_j + \sum_{p=1}^{N} e_j^p c_p\right] {'x_p^j}, \quad |'x_p| = 1. \tag{10.5.2}$$

If we let $d = d('x_p)$ denote the distance of the corresponding point P from $\sum$, we see that

$$d^2 = \sum_{k=\nu+1}^{N} (x_p^k)^2, \quad x_p^k = \sum_{j=1}^{\nu} e_j^k {'x_p^j}, \quad k \geq \nu + 1. \tag{10.5.3}$$

From (10.5.1) and (10.5.3), we obtain (by taking the max.)

$$[d(\textstyle\sum_c, \sum_{c'})]^2 \leq \displaystyle\sum_{j=1}^{\nu} \sum_{k=\nu+1}^{N} (e_j^k)^2, \quad |c - c'|^2 = \sum_{j,k=1}^{N} (e_j^k)^2,$$

from which the result follows.

Lemma 10.5.2 (Extension lemma). (a) *There is a constant $C_4(\mathfrak{M}) \geq 1$ with the following property: Suppose $\{p_i\}$ is a finite set in $\mathfrak{M}$, $r > 0$, $\eta > 0$, the f_i are vectors in some R_Q and we have*

$$d(p_i, p_{i'}) \geq r \quad \text{if} \quad i' \neq i$$

$$|f_i - f_{i'}| \leq \eta \quad \text{if} \quad i' \neq i \quad \text{and} \quad |p_i - p_{i'}| \leq 6r.$$

Then there is a vector function f (into R_Q) $\in C^2$ *on $\bigcup_i \overline{B(p_i, 5r/2)}$ such that*

$$f(p_i) = f_i \text{ for each } i$$

$$|\nabla f(p)| \leq C_4 \eta/r, \quad |\nabla^2 f(x)| \leq C_4 \cdot \eta/r^2, \quad p \in \bigcup \overline{B(p_i, 5r/2)}.$$

If $|f_i| \leq M$ for all i, we may take f so that $|f(p)| \leq M$.

(b) *There exist constants $C_5(M) \geq 1$ and $d_1(M) > 0$ such that the extension theorem of part* (a) *holds for orthogonal matrix functions provided that $\eta \leq d_1(M)$ and we replace C_4 by C_5.*

Proof. (a) Let $\varphi(s) \in C^\infty(R_1)$ with $\varphi'(s) \leq 0$, $\varphi(s) = 1$ for $s \leq 0$, and $\varphi(s) = 0$ for $s \geq 1$, and define

$$\hat{\lambda}_i(p) = \varphi\left[\frac{d(p, p_i) - r/4}{3r}\right], \tag{10.5.4}$$

$$\tilde{\lambda}_i(p) = \hat{\lambda}_i(p) \prod_{j \neq i} [1 - \hat{\lambda}_j(p)], \quad \lambda_i(p) = \tilde{\lambda}_i(p) / \sum_j \tilde{\lambda}_j(p).$$

It follows that

$$\lambda_i(p) = 1 \text{ if } d(p, p_i) \leq r/4, \quad \lambda_i(p) = 0 \text{ if } d(p, p_j) \leq r/4, \quad j \neq i,$$

$$\lambda_i(p) = 0 \text{ if } d(p, p_i) \geq 13r/4, \quad |\nabla^k \lambda_i(p)| \leq Z_1(\mathfrak{M}) \cdot r^{-k}, \tag{10.5.5}$$

$$\sum_i \lambda_i(p) = 1, \quad \sum_i \nabla^k \lambda_i(p) = 0, \quad p \in \bigcup \overline{B(p_i, 5r/2)}, \quad k = 1, 2.$$

The bound for $|\nabla \lambda_i(p)|$ does not depend on the number of points p_i since the number of non-zero terms in the expressions for $\nabla \tilde{\lambda}_i(p)$ and $\nabla \lambda_i(p)$ for a given p is $\leq$ the maximum number of p_j in the ball $B(p, 13r/4)$.

Then we define

$$(10.5.6) \qquad f(p) = \sum_i f_i \cdot \lambda_i(p), \quad p \in \tilde{S} = U \overline{B(p_i, 5r/2)}.$$

Now, suppose $p \in \tilde{S}$ and call p_0 the nearest p_i and $f_0 = f(p_0)$. Then since $\sum \nabla \lambda_i(p) \equiv 0$, we have

$$\nabla f(p) = \sum_i (f_i - f_0) \nabla \lambda_i(p) = \sum_{d(p,p_i) < 13r/4} (f_i - f_0) \nabla \lambda_i(p).$$

Since p_0 is the nearest p_i to p, we have $d(p, p_0) \leq 5r/2$ so that the number of terms in the latter sum is again $\leq$ maximum number of p_j in $B(p, 13r/4)$ and this depends only on $\mathfrak{M}$. Clearly, if $d(p, p_i) < 13r/4$, then $d(p_0, p_i) < 6r$ so that each $|f_i - f_0| \leq \eta$. The first result follows and the last statement follows from (10.5.6).

To prove (b), we merely extend each of the vectors c_j as in (a). If η is sufficiently small, the $c_1, \ldots, c_N$ as extended are linearly independent on $\tilde{S}$ and may be replaced by an orthonormal sequence by applying the Gram-Schmidt process.

Definition 10.5.2. Given an n.o. set $c = (c_1, \ldots, c_N)$ and a ν-plane $\sum$ in R_N. Let Π be an $(N - \nu)$-plane $\perp \sum$ and suppose that $c_1^*, \ldots, c_\nu^*$ are the respective projections of $c_1, \ldots, c_\nu$ on $\sum$ and $c_{\nu+1}^*, \ldots, c_N^*$ are those of $c_{\nu+1}, \ldots, c_N$ on Π. For those $(c, \sum)$ for which $c_1^*, \ldots, c_\nu^*$ span $\sum$ and $c_{\nu+1}^*, \ldots, c_N^*$ span Π, we define

$$\Lambda(c, \textstyle\sum) = c' = (c_1', \ldots, c_N')$$

where $c_1', \ldots, c_\nu'$ and $c_{\nu+1}', \ldots, c_N'$ are obtained from the respective sets $c_1^*, \ldots, c_\nu^*$ and $c_{\nu+1}^*, \ldots, c_N^*$ by the Gram-Schmidt process.

Definition 10.5.3. A finite set $\{p_i\}$ is an *r-set for a set S* $\Leftrightarrow d(p_i, p_{i'}) \geq r$ if $i' \neq i$ and $S \subset U B(p_i, r)$.

Lemma 10.5.3. *There are numbers $d_2(\mathfrak{M}) > 0$ and $C_6(\mathfrak{M}) \geq 1$ such that*

$$|\Lambda(c, \textstyle\sum) - \Lambda(c', \sum')| \leq C_6[|c - c'| + d_0(\sum, \sum')],$$

provided that

$$|c - c'| \leq d_2, \quad d_0(\textstyle\sum, \sum') \leq d_2, \quad d_0(\sum, \sum_c) \leq d_2,$$
$$d_0(\textstyle\sum', \sum_{c'}') \leq d_2$$

where $\sum_c$ and $\sum_{c'}'$ are the ν-planes through 0 spanned by $(c_1, \ldots, c_\nu)$ and $(c_1', \ldots, c_\nu')$. In fact, Λ is analytic in $(c, \sum)$ where it is defined.

Proof. This last statement means that if we suppose that $c'' = \Lambda(c, \sum)$, where $\sum$ is spanned by the vectors $c_1''', \ldots, c_\nu'''$, then the vectors c_j'' are analytic functions of the components of the vectors $(c_1, \ldots, c_N, c_1''', \ldots,$

c_ν'''). Since the domain of definition of these functions is open and contains the compact subset of vector sequence $(c_1, \ldots, c_N, c_1''', \ldots, c_\nu''')$ in which $c_1, \ldots, c_N$ is an n.o. set and $c_j''' = c_j$, $j = 1, \ldots, \nu$, the lemma follows.

Lemma 10.5.4. *There exists a positive number $R_0' \leq R_0$ with the following property: Suppose S is a locally compact set on $\mathfrak{M}$, $Y_0 \in \mathfrak{M}$, $S \subset B(Y_0, R_0')$, and τ is a normal coordinate system centered at Y_0 with domain $B(0, R_0')$ (and $\tau(0) = Y_0$). Then*

(i) *if S satisfies an (ε, R_1) condition at a point P and $B(P, 3R_1) \subset B(Y_0, R_0')$, then $\tau^{-1}(S)$ satisfies an $(\varepsilon', R_1/2)$ condition at $\tau^{-1}(P)$, where*

$$\varepsilon' = 2\varepsilon + 4KR;$$

(ii) *if $S_0 \equiv \tau^{-1}(S)$ satisfies an (ε, R_1) condition at a point P_0 and $B(P_0, 3R_1) \subset B(0, R_0')$, then S satisfies an $(\varepsilon', R_1/2)$ condition at $P = \tau(P_0)$.*

Proof. We choose $R_0' \leq R_0$ so that

$$\frac{1}{\sqrt{2}} d(p, q) \leq d[\tau(p), \tau(q)] \leq \sqrt{2}\, d(p, q)$$

whenever p and q are in $B(p, R_0')$ and τ is a normal coordinate system. Suppose that S_0 satisfies an (ε, R_1) condition at P_0 and that $B(P_0, 3R_1) \subset B(0, R_0')$. Then $\tau[B(P_0, 3R_1)] \subset B(Y_0, R_0')$. Choose R, $0 < R \leq R_1/2$, and $Q \in B(P, R_1)$ where $P = \tau(P_0)$. Then

$$\begin{aligned} S \cap B(Q, R) &\subset S \cap \tau\big[B(Q_0, R\sqrt{2})\big] \cap B(Q, R) \quad (Q_0 = \tau^{-1}(Q)) \\ &= \tau\big[S_0 \cap B(Q_0, R\sqrt{2})\big] \cap B(Q, R) \\ &\subset \tau\big[\textstyle\sum_0(Q_0, R\sqrt{2}), \varepsilon R\sqrt{2}\big] \cap B(Q, R) \\ &\subset \big\{\tau\big[\textstyle\sum_0(Q_0, R\sqrt{2})\big], 2\varepsilon R\big\} \cap B(Q, R) \\ &\subset [\textstyle\sum(Q, R),\ 2\varepsilon R + 4KR^2] \cap B(Q, R) \\ &= [\textstyle\sum(Q, R), \varepsilon' R] \end{aligned}$$

$\sum(Q, R)$ being the geodesic plane tangent at Q to $\tau\big[\sum_0(Q_0, R\sqrt{2})\big]$ (see Lemma 10.1.5). Likewise,

$$\begin{aligned} \textstyle\sum(Q, R) \cap B(Q, R) &\subset \tau\big\{\tau^{-1}[\textstyle\sum(Q, R)] \cap B(Q_0, R\sqrt{2})\big\} \cap B(Q, R) \\ &\subset \tau\big\{\textstyle\sum_0(Q_0, R\sqrt{2}) \cap B(Q_0 R\sqrt{2}), 2KR^2\big\} \cap B(Q, R) \\ &\subset \tau\big\{S_0 \cap B(Q_0, R\sqrt{2}), 2KR^2 + \varepsilon R\sqrt{2}\big\} \cap B(Q, R) \\ &\subset (S \cap \varepsilon' R) \cap B(Q, R), \quad 0 < R \leq R_1/2. \end{aligned}$$

The remainder of the proof of (ii) and that of (i) is similar.

We now recall the developments of § 10.1. We suppose that $P_0 \in \mathfrak{M}$, that τ is the associated coordinate system of class C^4 and satisfying (10.1.1), and that ω_{0p} is a normal coordinate system centered at p, the ω_{0p} and τ being related as in Lemma 10.1.1(d) and satisfying the bounds in our general conditions on $\mathfrak{M}$. The various Z_i will depend only on ν, N, and $\mathfrak{M}$, unless otherwise specified, and may denote different constants in different proofs.

Lemma 10.5.5. *There are positive numbers $\eta_0'(\mathfrak{M})$, $R_0''(\mathfrak{M})$, $K_3(\mathfrak{M})$, and $K_4(\mathfrak{M})$, with $R_0'' \leq R_0$ and $\eta_0' \leq 1/2$, with the following property: Suppose p and $q \in B(0, R_0/3)$, $d(p, q) = r_1 \leq R_0''$, $0 < r_2 \leq R_0''$, and $d(p, \sum_q) = r_3 \leq \eta_0' \cdot r_2$. Suppose also that $\sum_0$ is a ν-plane R_N and $\sum_p = \omega_{0p}(\sum_0)$, and $\sum_q = \omega_{0q}(\sum_0)$. Then*

$$(10.5.7) \quad D[\textstyle\sum_p \cap B(p, r_2), \quad \sum_q \cap B(p, r_2)] \leq K_3(r_1 + r_2)\, r_2 + K_4\, r_3.$$

Proof. Let $\sum_{0p}$ be the ν-plane in R_N tangent at $p_0 = \omega_{0q}^{-1}(p)$ to $\sum_{0p}^* = \omega_{0q}^{-1}(\sum_p)$. Clearly $\nabla\, \omega_{0q}^{-1} \circ \omega_{0p}(0)$ differs from the identity matrix by a matrix of norm $\leq Z_1\, d(p, q)$. Thus

$$(10.5.8) \qquad d(\textstyle\sum_{0p}, \sum_0) \leq Z_2\, r_1, \quad d(p_0, \sum_0) \leq (1 + \eta_0\, r_1)\, r_3.$$

Let us let $\Gamma = \omega_{0q}^{-1}[B(p, r_2)]$ and let $\sum_0'$ be the ν-plane $\|\sum_0$ through p_0. Then, we conclude from (10.5.8) and Lemma 10.1.5, etc., that

$$D[\textstyle\sum_{0p} \cap \Gamma, \sum_0' \cap \Gamma] \leq Z_3\, r_1\, r_2, \quad D[\sum' \cap \Gamma, \sum_0 \cap \Gamma] \leq Z_4\, r_3$$
$$D[\textstyle\sum_{0p} \cap \Gamma, \sum_{0p}^* \cap \Gamma] \leq K_2\, r_2^2.$$

It follows that

$$D[\textstyle\sum_{0p}^* \cap \Gamma, \sum_0 \cap \Gamma] \leq Z_3\, r_1\, r_2 + K_2\, r_2^2 + Z_4\, r_3.$$

The result follows since $\omega_{0q}(\sum_{0p}^* \cap \Gamma) = \sum_p \cap B(p, r_2)$, $\omega_{0q}(\sum_0 \cap \Gamma) = \sum_q \cap B(P, r_2)$.

Definition 10.5.4. The geodesic ν-planes $\sum_p$ and $\sum_q$ are said to be *parallel with respect to τ and the family $\{\omega_{0p}\}$.*

Lemma 10.5.6 (The c-lemma). *There are positive numbers $\varrho_0(\mathfrak{M}) < 1/2$ and $d_3(\mathfrak{M})$ such that to each pair of numbers (η_1, ϱ) for which $0 < \varrho \leq \varrho_0(\mathfrak{M})$ and $0 < \eta_1 \leq d_3(\mathfrak{M})\, \varrho$, there correspond positive numbers $\varepsilon_1(\eta_1, \varrho, \mathfrak{M})$ and $R_0'''(\eta_1, \varrho, \mathfrak{M})$ such that if the locally compact set S satisfies an (ε, R_1)-condition at a point P_0, with $0 < \varepsilon \leq \varepsilon_1$ and $0 < R_1 \leq R_0'''$, and the ω_{0p} have their significance above, then there exists a sequence $\{c_j(p)\}$ of orthogonal matrix functions with $c_j \in C^2(S^*, 3r_j/2)$ such that*

$$(10.5.9) \qquad \text{(i)} \quad |\nabla\, c_j(p)| \leq \frac{1}{20}\, \eta_1\, r_j^{-1};$$

$$(10.5.10) \qquad \text{(ii)} \quad |c_j(p) - c_{j-1}(p)| \leq C_6(2\varepsilon + \eta_1/4\varrho) + (1 + \varrho)\, \eta_1/8;$$

$$(10.5.11) \qquad \text{(iii)} \quad \textit{if } p \in S^*,\ j \geq 1, \textit{ and } \textstyle\sum^j(p) = \omega_{0p}[\sum_0^j(p)],\ \sum_0^j(p)$$

being the ν-plane in R_N through 0 spanned by $c_{j1}(p), \ldots, c_{j\nu}(p)$, then

$$D[S \cap B(p, 6r_j), \quad \textstyle\sum^j(p) \cap B(p, 6r_j)] \leq \eta_1\, r_j.$$

Here, we define

$$(10.5.12) \quad r_0 = R_1/8, \quad r_j = r_0\, \varrho^j, \quad S^* = S \cap \overline{B(P_0, 2R_1 - 6r_0)}.$$

Proof. We define $c_0(p)$ as a constant matrix such that $c_{01}, \ldots, c_{0\nu}$ spans $\omega_{0P_0}^{-1}(\Sigma)$. Since $\Sigma_0^0(p) = \omega_{0P_0}^{-1}(\Sigma)$, it follows from the (ε, R_1) condition and Lemma 10.5.5 (with $r_1 = d(p, P_0) \leq 10r_0$, $r_2 = 6r_0$, $r_3 = d(p, \Sigma) \leq 8\,\varepsilon\, r_0$) that

$$D[S \cap B(p, 6r_0), \Sigma \cap B(p, 6r_0)] +$$
$$+ D[\Sigma \cap B(p, 6r_0), \Sigma^0(p) \cap B(p, 6r_0)] \leq K_3 \cdot 16r_0 \cdot 6r_0 + K_4 \cdot 8\,\varepsilon\, r_0$$

which implies (iii) for $j = 0$, provided that

$$96K_3\, r_0 + 8K_4\, \varepsilon \leq \eta_1. \tag{10.5.13}$$

Obviously (i) holds.

Now suppose that $c_{j-1}(p)$, $j \geq 1$, has been defined to satisfy (i) and (iii) above. We define $c_j(p)$ as follows: Let $\{p_{ji}\}$ be an r_j set (Def. 10.5.3) for

$$S' = S \cap \overline{B(P_0, 2R_1)}.$$

For each i, select *a* geodesic ν-plane $\Sigma_i^j = \omega_{0\,p_{ij}}(\Sigma_{i0}^j)$ through p_{ji} such that

$$D[S \cap B(p_{ji}, 8r_j), \quad \Sigma_i^j \cap B(p_{ji}, 8r_j)] \leq 8\,\varepsilon\, r_j\,^{1}. \tag{10.5.14}$$

It follows that

$$\begin{aligned} \Sigma_i^j \cap B(p_{ji}, 8r_j) &\subset (S, 8\varepsilon\, r_j) \cap B(p_{ji}, 8r_j) \\ &\subset (S, 8\varepsilon\, r_j) \cap B(p_{ji}, 6r_{j-1}) \cap B(p_{ji}, 8r_j) \\ &\subset (\Sigma^{j-1}(p_{ji}), 8\varepsilon\, r_j + \eta_1\, r_{j-1}) \cap B(p_{ji}, 8r_j). \end{aligned} \tag{10.5.15}$$

Applying $\omega_{0p_{ij}}^{-1}$ and Lemma 10.5.1 to (10.5.15), we find that

$$d_0(\Sigma_{i0}^j, \Sigma_0^{j-1}(p_{ji})) \leq (1 + 8\,\eta_0\, r_j)\,(\varepsilon + \eta_1/8\varrho) \leq 2\varepsilon + \eta_1/4\varrho. \tag{10.5.16}$$

($\Sigma_{i0}^j = \omega_{0p_{ij}}^{-1}(\Sigma_i^j)$, etc.) provided that

$$8\,\eta_0\, r_0 \leq 1, \tag{10.5.17}$$

so that we must have

$$2\varepsilon + \eta_1/4\varrho \leq d_2. \tag{10.5.18}$$

If (10.5.18) holds, we may then define

$$c_j(p_{ji}) = \Lambda[c_{j-1}(p_{ji}), \Sigma_{i0}^j]. \tag{10.5.19}$$

From Lemma 10.5.3 and the fact that

$$c_{j-1}(p_{ji}) = \Lambda[c_{j-1}(p_{ji}), \Sigma_0^{j-1}(p_{ji})], \tag{10.5.20}$$

we conclude, using (10.5.16) also, that

$$|c_j(p_{ji}) - c_{j-1}(p_{ji})| \leq C_6(2\varepsilon + \eta_1/4\varrho). \tag{10.5.21}$$

---

[1] We shall omit the bars over the various sets.

Using (10.5.14) and Lemma 10.5.5 with $r_1 \leq 6r_j$, $r_2 = 2r_j$, and $r_3 = 8\varepsilon r_j$, we find that if $d(p_{ji}, p_{ji'}) \leq 6r_j$, then

$$(10.5.22)\quad \begin{aligned} \Sigma_i^j \cap B(p_{ji}, 2r_j) &\subset (S, 8\varepsilon r_j) \cap B(p_{ji}, 2r_j) \\ &\subset (S, 8\varepsilon r_j) \cap B(p_{ji'}, 8r_j) \cap B(p_{ji}, 2r_j) \\ &\subset (\Sigma_{i'}^j, 16\varepsilon r_j) \cap B(p_{ji}, 2r_j) \\ &\subset (\tilde{\Sigma}_{i'}^j, 16\varepsilon r_j + K_3 \cdot 16 r_j^2 + K_4 \cdot 8\varepsilon r_j) \end{aligned}$$

$\tilde{\Sigma}_{i'}^j$ being the geodesic ν-plane centered at p_{ji} and parallel to $\Sigma_{i'}^j$ (with respect to $\{\omega_{0p}\}$). Proceeding as in (10.5.16), we obtain

$$(10.5.23)\qquad (d_0\, \Sigma_{i0}^j, \Sigma_{i'0}^j) \leq (16 + 8K_1)\, \varepsilon + 16K_3 r_j$$

provided that (10.5.17) holds. From (10.5.9) for $j - 1$, we find that

$$(10.5.24)\quad |c_{j-1}(p_{ji}) - c_{j-1}(p_{ji'})| \leq \frac{3}{10}\varrho\,\eta_1 \text{ if } d(p_{ji}, p_{ji'}) \leq 6r_j.$$

Therefore we conclude from Lemma 10.5.3 and equations (10.5.19), (10.5.23), and (10.5.24) that

$$(10.5.25)\quad |c_j(p_{ji}) - c_j(p_{ji'})| \leq C_6\left[\frac{3}{10}\varrho\,\eta_1 + 16\varepsilon + 8K_4\varepsilon + 16K_3 r_j\right]$$

provided that

$$(10.5.26)\qquad \frac{3}{10}\varrho\,\eta_1 + 16\varepsilon + 8K_4\,\varepsilon + 16K_3 r_0 \leq d_2.$$

If (10.5.26) holds we conclude from Lemma 10.5.2(b) that there exists an orthogonal matrix function of class $C^2(S', 3r_j/2)$ which is an extension of c_j as so far defined with

$$(10.5.27)\quad \begin{aligned} |\nabla c_j(p)| &\leq C_5 C_6\left(\frac{3}{10}\varrho\,\eta_1 + 16\varepsilon + 8K_4\,\varepsilon + 16K_3 r_0\right) r_j^{-1} \\ &\leq \frac{1}{20}\eta_1 r_j^{-1} \end{aligned}$$

if we choose

$$(10.5.28)\quad C_5 C_6 (6\varrho\,\eta_1 + 320\varepsilon + 160K_4\,\varepsilon + 320K_3 r_0) \leq \eta_1 \qquad (C_5, C_6 \geq 1).$$

We must now prove (iii). Let $p \in S^*$. Then there is a p_{ji} such that $d(p, p_{ji}) \leq r_j$. Using Lemma 10.5.5 with $r_1 \leq r_j$, $r_2 = 6r_j$, and $r_3 = 8\varepsilon r_j$, we find that ($\tilde{\Sigma}_i^j$ being $\{\omega_{0p}\}$-parallel to Σ_i^j through p)

$$(10.5.29)\quad D[\tilde{\Sigma}_i^j \cap B(p, 6r_j), \Sigma_i^j \cap B(p, 6r_j)] \leq K_4 \cdot 8\varepsilon r_j + 42K_3 r_j^2.$$

Using Lemma 10.5.1(d) and (i), we conclude that

$$(10.5.30)\quad D[\tilde{\Sigma}_i^j \cap B(p, 6r_j), \Sigma^j(p) \cap B(p, 6r_j)] \leq \frac{3}{10}\eta_1 r_j (1 + 6\eta_0 r_j).$$

Using (10.5.29), (10.5.30), (10.5.17) and (10.5.28), we see that

$$(10.5.31)\quad D[\Sigma_i^j \cap B(p, 6r_j), \Sigma^j(p) \cap B(p, 6r_j)] \leq \frac{113}{160}\eta_1 r_j.$$

Then

$$
\begin{aligned}
& S \cap B(p, 6r_j) \subset S \cap B(p_{ji}, 7r_j) \cap B(p, 6r_j) \\
(10.5.32) \quad & \subset (\Sigma_i^j, 8\varepsilon r_j) \cap B(p, 6r_j) \subset (\Sigma^j(p), \eta_1 r_j) \cap B(p, 6r_j).
\end{aligned}
$$

In like manner, we see that

$$
\begin{aligned}
(10.5.33) \quad & \Sigma^j(p) \cap B(p, 6r_j) \subset (\Sigma_i^j, \tfrac{113}{160}\eta_1 r_j) \cap B(p, 6r_j) \cap B(p_{ji}, 7r_j) \\
& \subset (S, \eta_1 r_j) \cap B(p, 6r_j).
\end{aligned}
$$

Part (iii) follows from (10.5.32) and (10.5.33).

Since $d(p, p_{ji}) \leq \frac{5}{2} r_j$, we obtain, from (i) and (10.5.21),

$$
(10.5.34) \quad |c_j(p) - c_{j-1}(p)| \leq C_6(2\varepsilon + \eta_1/4\varrho) + (1 + \varrho)\eta_1/8
$$

from which (ii) follows.

All the conditions in equations (10.5.12), (10.5.17), (10.5.18), (10.5.26), and (10.5.28) can be satisfied by requiring

$$
\begin{aligned}
(10.5.35) \quad & \eta_0 R_1 \leq 1, \quad 18 C_5 C_6 \varrho \leq 1, \quad (960 + 480 K_4) C_5 C_6 \varepsilon \leq \eta_1 \\
& \eta_1 \leq d_2 \varrho, \quad 120 K_3 C_5 C_6 R_1 \leq \eta_1
\end{aligned}
$$

since $\varrho < 1/2$.

Proof of Theorem 10.5.1. We suppose that the locally compact set S satisfies an (ε, R_1)-condition at P_0 and that ϱ, η_2, and θ_1 are positive numbers satisfying

$$
(10.5.36) \quad 0 < \varrho \leq \varrho_0(\mathfrak{M}) < \tfrac{1}{2}, \quad 0 < \eta_2 \leq \tfrac{1}{24}, \quad 0 < \theta_1 \leq \pi/12.
$$

We suppose that τ and the family $\{\omega_{0p}\}$ have been chosen to correspond to P_0 as in Lemma 10.5.6 with $\omega_{0P_0}(\Sigma_0^0) = \Sigma$, Σ_0^0 being the ν-plane in R_N spanned by $e_1, \ldots, e_\nu$. We suppose that $0 < \eta_1 \leq d_3(\mathfrak{M})\varrho$, $0 < \varepsilon \leq \varepsilon_1(\eta_1, \varrho, \mathfrak{M})$, $0 < R_1 \leq R_0(\eta_1, \varrho, \mathfrak{M})$, and that $\{c_j(p)\}$ is a sequence of orthogonal matrix functions satisfying the conditions of Lemma 10.5.6, the r_j being defined in (10.5.12). We retain the notations of that lemma. For each j, we define

$$
(10.5.37) \quad \lambda_j = \eta_2 r_0 \sum_{k=j+1}^{\infty} \varrho^k, \quad S_j = S^* \cap B(P_0, 2R_1 - 6r_0 - \lambda_j), \; j \geq 0.
$$

For each j, we choose the r_j-set $\{p_{ji}\}$ for S^* of Lemma 10.5.6 and define the mappings $\sigma_j \in C^2(S^*, 3r_j/2)$ by

$$
\begin{aligned}
(10.5.38) \quad & \sigma_j(p) = \omega_{jp}[\eta_j(p)], \quad \eta_j(p) = \sum_i \lambda_{ji}(p) \sum_{k=\nu+1}^{N} [\omega_{jp}^{-1}(p_{ji})]^k e_k, \\
& \omega_{jp}(y) = \omega_{0p}[c_j(p)\, y], \quad j \geq 1,
\end{aligned}
$$

where, for each j, the $\lambda_{ji}(p)$ are defined by (10.5.4) in terms of the p_{ji}. We then define the mappings $\tau_j \in C^2(T_0^0)$ and the T_j by

$$
\begin{aligned}
& \tau_0 = \omega_{0P_0} \,|\, T_0^0, \quad \tau_j = \sigma_j \tau_{j-1}, \quad T_0^0 = \Sigma_0^0 \cap B(0, 2R_1 - 6r_0 - \lambda_0) \\
(10.5.39) \quad & T_j = \tau_j(T_0^0) = \sigma_j(T_{j-1}).
\end{aligned}
$$

We divide the remainder of the proof into several parts.

Remarks. To orient the reader, we notice that if $\mathfrak{M} = R_N$ and $S \subset \Sigma = \Sigma_0^0$, then we may take

$$\omega_{0P_0} = I, \quad \omega_{0z}(y) = y + z, \quad \omega_{0z}^{-1}(y) = y - z, \quad c_j(z) = I.$$

Then, since the $p_{ji} = z_{ji} \in \Sigma_0^0$ all the $z_{ji}^k = 0$ if $k > \nu$, so that

$$\eta_j(z) = \sum_i \lambda_{ji}(z) \sum_{k=\nu+1}^{N} (z_{ji}^k - z^k)\, e_k = -\sum_{k=\nu+1}^{N} z^k e_k$$

since $\sum \lambda_{ji}(z) = 1$. Thus

$$\sigma_j(z) = z - \sum_{k=\nu+1}^{N} z^k e_k = \sum_{k=1}^{\nu} z^k e_k$$

and is the projection of z on S.

Part 1. *If η_1 and hence ε and R_1 are small enough, we verify the following facts by induction on j*

(10.5.40)
(i) $d[\sigma_j(p), p] \leq \eta_2 r_j$ *for* $p \in T_{j-1}$, $j \geq 1$;
(ii) $T_j \subset (S_j, \eta_3 r_{j+1})$, $\eta_3 = \eta_2/4 \leq 1/96$, $j \geq 0$
(iii) *the angle at p between $\Sigma^j(p)$ and T_j is $\leq \theta_1$ if $p \in T_j$*
(iv) $\frac{1}{2} d(p, q) \leq d[\sigma_j(p), \sigma_j(q)] \leq 2\, d(p, q)$ *if* $p, q \in T_{j-1}, j \geq 1$.

Proof of part 1. For $j = 0$, $T_0 = \Sigma \cap \overline{B(P_0, 2R_1 - 6r_0 - \lambda_0)}$ and $S_0 = S \cap B(P_0, 2R_1 - 6r_0 - \lambda_0)$ so (ii) follows provided merely that

$$8\,\varepsilon\,\varrho^{-1} \leq \eta_3. \tag{10.5.41}$$

To prove (iii), we use the normal coordinate system $\omega_{00} = \omega_{0P_0}$. Clearly

$$\omega_{00}^{-1}(T_0) = T_0^0, \quad \omega_{00}^{-1}[\Sigma^0(p)] = \omega_{00}^{-1}\,\omega_{0p}(T_0^0). \tag{10.5.42}$$

Thus $\omega_{00}^{-1}[\Sigma^0(p)]$ is part of the locus of the equation

$$w = U(y; p, P_0), \quad y \in \Sigma_0^0. \tag{10.5.43}$$

From § 10.1, we conclude that

$$\left| \sum_{l=\nu+1}^{N} \sum_{m=1}^{\nu} \frac{\partial w^l}{\partial y^m} \mu^l \lambda^m \right| \leq Z_1\, d(p, P_0), \quad |\lambda| = |\mu| = 1 \tag{10.5.44}$$

so that (iii) follows provided merely that

$$Z_2\, d(p, P_0) \leq \theta_1. \tag{10.5.45}$$

Suppose now that (10.5.40) is true for $j - 1$. Suppose $p \in T_{j-1}$. From (ii) for $j - 1$, we conclude that there is a point $p' \in S_j \subset S^*$ within a distance $\leq \eta_3 r_j$ of p and hence one of the p_{ji}, call it q, is within a distance $(1 + \eta_3) r_j$ of p. From the (ε, R_1)-condition and the construction in Lemma 10.5.6, we see that

$$S^* \cap B(q, 8r_j) \subset (\Sigma^j(q), 8\,\varepsilon\, r_j) \cap B(q, 8r_j) \tag{10.5.46}$$

so that all the p_{ji} in $B(q, 8r_j) \in (\sum^j(q), 8\varepsilon r_j)$ and (since $p \in (S^*, \eta_3 r_j)$)

$$p \in (\textstyle\sum^j(q), 8\varepsilon r_j + \eta_3 r_j). \tag{10.5.47}$$

So we use the normal coordinate system ω_{jq} and define

$$\begin{aligned} &\omega_{jq}^{-1}(p) = z, \quad \lambda_{ji0}(z) = \lambda_{ji}(p), \quad c_{j0}(z) = c_j(p) \\ &t_j(z) = \omega_{jq}^{-1}\sigma_j[\omega_{jq}(z)], \quad t_{j0} = z - \sum_{k=\nu+1}^{N} z^k e_k = \sum_{k=1}^{\nu} z^k e_k. \end{aligned} \tag{10.5.48}$$

We shall show in part 2 below that

$$\begin{aligned} t_j(z) = t_{j0}(z) + t_j^*(z), \quad t_j^*(0) = 0, \quad |\nabla t_j^*(z)| \\ \leq \eta_4(\varepsilon, \varrho, \eta_1, R_0'''; \mathfrak{M}). \end{aligned} \tag{10.5.49}$$

From (10.5.47), it follows that the distance

$$d(z, \textstyle\sum_0^0) \leq (1 + Z_3 r_j)(8\varepsilon r_j + \eta_3 r_j) \leq 2\eta_3 r_j \tag{10.5.50}$$

if the last inequality holds. Thus $|t_{j0}(z) - z| = d(z, \sum_0^0)$, so that

$$\begin{aligned} &|t_j(z) - z| \leq (\eta_4 + 2\eta_3) r_j \\ &d[p, \sigma_j(p)] \leq (1 + Z_3 r_j)(\eta_4 + 2\eta_3) r_j \leq \eta_2 r_j \end{aligned} \tag{10.5.51}$$

if the last inequality holds. Since $d[\sigma_j(p), q] \leq (1 + \eta_2 + \eta_3 + 8\varepsilon) r_j \leq 8r_j$, we see from (10.5.46) and the construction that

$$\begin{aligned} d[\sigma_j(p), S^*] &\leq D[S^* \cap B(q, 8r_j), \textstyle\sum^j(q) \cap B(q, 8r_j) + \\ &+ d[\sigma_j(p), \textstyle\sum^j(q)] \leq 8\varepsilon r_j + (1 + Z_3 r_j)\, d[t_j^*(z), \sum_0^0] \\ &\leq 8\varepsilon r_j + (1 + Z_3 r_j)\eta_4 r_j \leq \eta_3 r_{j+1} \end{aligned} \tag{10.5.52}$$

(since $t_{j0}(z) \in \sum_0^0$) provided that

$$\varrho^{-1}[8\varepsilon + (1 + Z_3 r_0)\eta_4] \leq \eta_3. \tag{10.5.53}$$

Thus (i) and (ii) are verified.

To verify (iii), we note first that the angle at p between $\sum^{j-1}(p)$ and T_{j-1} is $\leq \theta_1$ so the angle at z between the tangent planes to $\omega_{jq}^{-1}[\sum^{j-1}(p)]$ and $\omega_{jq}^{-1}(T_{j-1})$ at z is $\leq 2\theta_1$. Now, evidently, $\omega_{jq}^{-1}[\sum^{j-1}(p)]$ is the locus of the equations (see (10.5.38) and § 10.1)

$$w = \omega_{jq}^{-1}\omega_{j-1,p}(y) = [c_j(q)]^{-1} U[c_{j-1}(p)\, y; p, q], \quad y \in \textstyle\sum_0^0. \tag{10.5.54}$$

From § 10.1, we see first that

$$U_{\Upsilon m}^k[c_{j-1}(p)\, y; p, q] = \delta_m^k + \varepsilon_m^k[c_{j-1}(p)\, y; p, q], \quad |\varepsilon| \leq Z_4 r_j.$$

Letting ∇w and ε denote the matrices $\partial w^i/\partial y^s$ and ε_m^k, respectively, we obtain

$$\begin{aligned} \nabla w &= [c_j(q)]^{-1}[I + \varepsilon]\, c_{j-1}(p) = [c_j(q)]^{-1} c_{j-1}(p) + \varepsilon_1 (|\varepsilon_1| \leq Z_5 r_j) \\ &= I + [(c_j(q)]^{-1}[c_j(p) - c_j(q)] + [c_j(q)]^{-1}[c_{j-1}(p) - c_j(p)] + \varepsilon_1. \end{aligned}$$

Since, on our locus $\omega_{jq}^{-1}[\sum^{j-1}(p)]$, y ranges only over $\sum_0^0$, it follows that the angle between that locus and $\sum_0^0$ is

$$\leq Z_6[r_j + (\eta_1/20) + C_6(2\varepsilon + \eta_1/4\varrho) + (1 + \varrho)\eta_1/8] \leq \theta_1 \tag{10.5.55}$$

if the last inequality holds. Thus the angle between $\omega_{jq}^{-1}(T_{j-1})$ and $\sum_0^0$ is $\leq 3\theta_1$.

Now let dy be a vector tangent at z to $\omega_{jq}^{-1}(T_{j-1})$ and let dw be the corresponding vector tangent to $\omega_{jq}^{-1}(T_j)$ at $t_j(z)$. Let us write $dy = dy_1 + dy_2$ where dy_1 is the projection of dy on $\sum_0^0$. Then

$$\begin{aligned} dw &= dy_1 + \nabla t_j^*(z)\, dy = dy_1 + dw^*, \quad |dw^*| \leq \eta_4 |dy| \\ 2^{-1/2}|dy| &\leq \cos 3\theta_1 \cdot |dy| \leq |dy_1| \leq |dy| \end{aligned} \tag{10.5.56}$$

since $3\theta_1 \leq \pi/4$ and the angle between $\sum_0^0$ and the tangent plane to $\omega_{jq}^{-1}(T_{j-1})$ at z is $\leq 3\theta_1$. An easy argument, using (10.5.56) shows that

$$2^{-1}|dy| \leq |dw| \leq 2|dy|,$$

so that (iv) holds, and the angle between $\sum_0^0$ and the tangent plane to $\omega_{jq}^{-1}(T_j)$ at $t_j(z)$ is $\leq \arcsin \eta_4 \sqrt{2}$. Now, $\omega_{jq}^{-1}[\sum^j(p')]$ is seen, as in (10.5.54), to be the locus of the equations

$$w = [c_j(q)]^{-1}\, U\,[c_j(p')\,y; p', q], \quad y \in \textstyle\sum_0^0, \quad p' = \sigma_j(p). \tag{10.5.57}$$

Thus the angle between $\sum_0^0$ and the tangent plane at $t_j(z)$ to $\omega_{jq}^{-1}[\sum^j(p')]$ is $\leq Z_7\eta_1$ so that the angle at p' between T_j and $\sum^j(p')$ is

$$\leq (1 + Z_3 r_j)\cdot Z_7\eta_1 + \arcsin \eta_4\sqrt{2} \leq \theta_1 \tag{10.5.58}$$

if the last inequality holds. This proves (iii).

Part 2. *Proof of* (10.5.49). For j and q fixed, we define

$$\begin{aligned} u(y,z) &= \omega_{jq}^{-1}\,\omega_{jp}(y) = c_{j0}^{-1}(0)\, U\,[c_{j0}(z)\,y; \omega_{jq}(z), \omega_{jq}(0)] \\ y &= v(w,z) \Leftrightarrow w = u(y,z), \\ v(w,z) &= \omega_{jp}^{-1}[\omega_{jq}(w)] = c_{j0}^{-1}(z)\, U\,[c_{j0}(0)\,w; \omega_{jq}(0), \omega_{jq}(z)] \end{aligned}$$

(see (10.5.38), (10.5.54), and § 10.1). Then we see that

$$\begin{aligned} \eta_{j0}(z) &= \sum_i \lambda_{ji0}(z) \sum_{k=\nu+1}^{N} v^k(z_{ji}, z)\, e_k \quad (= \eta_j(p)) \\ u(y,0) &= y, \quad u(0,z) = z, \quad v(z,z) = 0, \quad v(w,0) = w \\ t_j(z) &= \omega_{jq}^{-1}\,\omega_{jp}[\eta_{j0}(z)] = u[\eta_{j0}(z), z], \quad z_{ji} = \omega_{jq}^{-1}(p_{ji}). \end{aligned} \tag{10.5.59}$$

From (10.5.59) and the uniform bounds for the second derivatives of $U(Y; p, q)$ with respect to its arguments, and from the bounds in (10.5.46) and Lemma 10.5.6 and the extension lemma (Lemma 10.5.2), we find that

$$\begin{aligned} \eta_{j0}(0) &= 1\cdot \sum_{k=\nu+1}^{N} v^k(0,0)\, e_k = 0 \quad (z_{ji} = 0 \text{ for } p_{ji} = q) \\ \eta_{j0}(z) &= \sum_i \lambda_{ji0}(z) \sum_{k=\nu+1}^{N} [v^k(z_{ji}, z) - v^k(0,z)]\, e_k + \sum_{k=\nu+1}^{N} v^k(0,z)\, e_k \\ &= \sum_i \sum_{k=\nu+1}^{N} \lambda_{ji0}(z)\, [z_{ji}^k + \delta_{1ji}^k(z)]\, e_k + \sum_{k=\nu+1}^{N} [-z^k + \delta_2^k(z)]\, e_k \\ &= -\sum_{k=\nu+1}^{N} z^k e_k + \eta_{j0}^*(z), \quad \eta_{j0}^*(z) = \sum_{k=\nu+1}^{N} \{\delta_2^k(z) + \sum_i \lambda_{ji0}(z)\, [z_{ji}^k + \\ &\qquad + \delta_{1ji}^k(z)]\}\, e_k \end{aligned}$$

$$\delta_2^k(z) = -[v^k(z, z) - v^k(z, 0) - v^k(0, z) + v^k(0, 0)],$$
$$\delta_{1ji}^k(z) = [v^k(z_{ji}, z) - v^k(z_{ji}, 0) - v^k(0, z) + v^k(0, 0)]$$
$$|z_{ji}^k| \leq Z_8\, \varepsilon\, r_j; \quad |\nabla\, \eta_{j0}^*(z)| \leq Z_9\, \varepsilon + Z_{10}\, r_j$$
$$\nabla_z\, t_j(z) = I + \nabla_z\, \eta_{j0}(z) + \varepsilon_2(z), \quad t_j(0) = u(0, 0) = 0,$$
$$|\varepsilon_2| \leq Z_{11}\, r_j$$
$$t_j(z) = t_{j0}(z) + t_j^*(z), \quad |\nabla\, t_j^*(z)| \leq Z_{12}\, r_j + Z_{13}\, \varepsilon$$

which proves the result.

Part 3. *The maps τ_j, defined in* (10.5.39), *converge uniformly to a homeomorphism τ from T_0^0 onto a topological ν-disc $\bar{S} \subset S^*$.* From (10.5.39) and (10.5.40), it follows that

$$d[\tau_j(y), \tau_{j-1}(y)] = d[\sigma_j[\tau_{j-1}(y)], \quad \tau_{j-1}(y)] \leq \eta_2\, r_0\, \varrho^j$$

so that the τ_j converge uniformly on T_0^0 to a map τ and

$$d[\tau_j(y), \tau(y)] \leq \lambda_j \left(= \eta_2\, r_0 \sum_{k=j+1}^{\infty} \varrho^k\right). \tag{10.5.60}$$

Suppose, now, that $\tau(x) = \tau(y)$. Then, from (10.5.60), we obtain

$$d[\tau_j(x), \tau_j(y)] \leq 2\lambda_j = 2\eta_2\, r_0\, \varrho^{j+1}/(1 - \varrho).$$

But then it follows from (10.5.40) (iv) that

$$d[\tau_0(x), \tau_0(y)] \leq \eta_2\, r_0 (2\varrho)^{j+1}/(1 - \varrho)$$

for every j. Thus $x = y$, since $\varrho < 1/2$.

Part 4. *If ε and R_1 are small enough, $\bar{S} \supset K(P_0, 4\, r_0)$.* From Lemma 10.5.4, we conclude that $\omega_{00}^{-1}(S)$ satisfies an (ε', R_1') condition at 0. By assuming R_1 small enough, we may assume that $\omega_{00}^{-1}(S)$ satisfies an (ε', R_1)-condition at 0, where we assume

$$36\varepsilon' \leq 1, \quad \lambda_0 \leq r_0.$$

We shall drop the prime on ε and assume that all the sets are in R_N; we assume $\sum = \sum_0^0$ and $T_0 = T_0^0$ and use the simpler notation.

In order to prove the result in this part, we select a $y \in S \cap B(P_0, 4r_0)$ and define a continuous map φ with domain T_0 as follows:

$$\varphi(x) = \begin{cases} \tau(x) & \text{if } |\tau(x) - y| \leq 2r_0 \\ x & \text{if } |\tau(x) - y| \geq 3r_0 \\ r_0^{-1}\{[|\tau(x) - y| - 2r_0]\, x + [3r_0 - |\tau(x) - y|]\, \tau(x)\}, & \text{otherwise.} \end{cases}$$

Since $\varphi(x)$ is on the segment from x to $\tau(x)$, we see that

$$|\varphi(x) - \tau(x)| \leq r_0, \quad |\varphi(x) - x| \leq r_0,$$

on account of (10.5.60). Accordingly if $x \in \partial T_0 = \sum \cap\, \partial B(P_0, 2R_1 - 6r_0 - \lambda_0)$, then

$$|\tau(x) - y| = |[\tau(x) - x] + (x - y)| \geq |x - y| - |\tau(x) - x|$$
$$\geq 2R_1 - 7r_0 - 4r_0 - r_0 = 4r_0$$

so that $\varphi(x) = x$ on ∂T_0. Since $|\varphi(x) - \tau(x)| \leq r_0$, it follows that if $|\varphi(x) - y| \leq r_0$ then $|\tau(x) - y| \leq 2r_0$ so $\varphi(x) = \tau(x)$. Thus

$$\varphi(T_0) \cap B(y, r_0) \subset \bar{S}.$$

Now, suppose $y \notin \bar{S}$. Then, for some n,

$$\varphi(T_0) \cap B(y, 2^{-n} r_0) \quad \text{is empty.} \tag{10.5.61}$$

If this holds for $n = 0$, then $B(y, r_0) \cap \varphi(T_0)$ is empty. But since $\varphi(x)$ is between x and $\tau(x)$ and $\tau(x) \in S^* \subset (\Sigma, 8\varepsilon r_0)$, it follows that $\varphi(T_0) \subset (\Sigma, 8\varepsilon r_0) \subset (\Sigma', 16\varepsilon r_0)$, Σ' being $\| \Sigma$ through y. Since $16\varepsilon < 1/2$, it follows that if π denotes the projection on Σ', then $\varphi_0 = \pi\varphi$ is a continuous map from T_0 into Σ which is the identity on ∂T_0 and whose range contains no points in $B(y_0, r_0/2) \cap \Sigma \subset$ the disc $T_0 (y_0 = \pi y)$. But this is impossible. If (10.5.62) holds for some $n > 0$, then

$$\varphi(T_0) \cap \overline{B(y, 2t)} = \varphi(T_0 \cap [\overline{B(y, 2t)} - B(y, t)], \quad t = 2^{-n} r_0.$$

But, if $\Sigma(y, 2t)$ denotes a ν-plane through y whose existence is guaranteed by the (ε, R_1)-condition, then

$$\varphi(T_0) \cap \overline{B(y, 2t)} \subset S \cap \overline{B(y, 2t)} \subset (\Sigma(y, 2t), 2t\varepsilon) \cap (\bar{B}_{2t} - B_t)$$

and this set may be mapped continuously onto a subset of $(\Sigma(y, 2t), 2t\varepsilon) \cap \partial B_{2t}$ by a map ω_n which is the identity outside and on ∂B_{2t}. Thus (setting $B_r = B(y, r)$, etc.)

$$(\omega_n \varphi)(T_0) \cap \bar{B}_{4t} = \{\varphi(T_0) \cap (\bar{B}_{4t} - B_{2t})\} \cup V_{n-1}$$

$$V_{n-1} = \omega_n\{\varphi(T_0) \cap (\bar{B}_{2t} - B_t)\} \subset (\Sigma(y, 2t), 2t\varepsilon) \cap \partial B_{2t}.$$

But now, using the (ε, R_1)-condition, we find that

$$(\Sigma(y, 2t), 2t\varepsilon) \cap \partial B_{2t} \subset (S, 4t\varepsilon) \cap \partial B_{2t} = (S, 4t\varepsilon) \cap \bar{B}_{4t} \cap \partial B_{2t}$$
$$\subset (\Sigma(y, 4t), 8t\varepsilon) \cap \partial B_{2t},$$
$$\varphi(T_0) \cap \bar{B}_{4t} \subset (\Sigma(y, 4t), 4t\varepsilon) \cap \bar{B}_{4t}$$

for an appropriate $\Sigma(y, 4t)$. Since ε is small, we see that that the set $(\omega_n \varphi)(T_0) \cap B_{4t}$ can be mapped continuously into a subset of $(\Sigma(y, 4t), 4t\varepsilon) \cap \partial B_{4t}$ by a map ω_{n-1} which is the identity outside and on ∂B_{4t}. This procedure may be repeated and we conclude finally that

$$(\omega_1 \ldots \omega_n \varphi)(T_0) \cap \overline{B(y, r_0)} \subset (\Sigma(y, r_0), r_0\varepsilon) \cap \partial B(y, r_0)$$

for an appropriate $\Sigma(y, r_0)$. Now

$$(\Sigma(y, r_0), r_0\varepsilon) \cap \partial \overline{B(y, r_0)} \subset (S, 2r_0\varepsilon) \cap \overline{B(P_0, 8r_0)} \cap \partial B(y, r_0)$$
$$\subset (\Sigma, 10r_0\varepsilon) \cap \overline{B(P_0, 8r_0)} \cap \partial B(y, r_0) \subset (\Sigma', 18\varepsilon r_0) \cap \partial B(y, r_0),$$

Σ' being $\| \Sigma$ through y. Since $\omega_1 \ldots \omega_n$ is the identity outside and on $\partial B(y, r_0)$, it follows that $\pi\omega_1 \ldots \omega_n \varphi$ has the same property as the former mapping $\varphi_0 = \pi\varphi$. Since this is impossible, $\bar{S} \supset S \cap \overline{B(P_0, 4r_0)}$.

In order to prove further smoothness properties of these discs, it is necessary to consider their ν-dimensional Lebesgue areas.

Lemma 10.5.7. *Let us suppose that* S, $\bar{S}$, *the* T_j, *the* τ_j *and* τ, *the* c_j, *the* p_{ji}, *the* $\sum^j(p)$, *and the positive numbers* $\varepsilon, \varrho, \theta_1, R_0', \eta_1$, *and* η_2 *have their significance above and that* $\varepsilon, \varrho, \theta_1, R_0', \eta_1$ *and* η_2 *are sufficiently small. Then for each* $\eta_6 > 0$, *there is a* $\delta > 0$ *such that*

$$\liminf_{j\to\infty} L_\nu(\tau_j|\Gamma) \le (1+\eta_6)\, \Lambda^\nu[\tau(\Gamma)] \text{ if } \Lambda^\nu[\tau(\partial\Gamma)] = 0 \tag{10.5.62}$$

provided that all the numbers above are $<\delta$, Γ *being any open set for which* $\bar{\Gamma} \subset T_0$; L_ν *denotes the Lebesgue area.*

Proof. For a fixed j and each i, we define D_{ji} as the set of all $p \in (S^*, 3r_j/2)$ such that $d(p, p_{ji}) \le d(p, p_{ji'})$ for all i' and let $\sum_{ji} = \sum^j(p_{ji})$. Each D_{ji} is bounded by portions of $\partial(S^*, 3r_j/2)$ and loci of equations of the form $d(p, p_{ji}) = d(p, p_{ji'})$. These latter loci are approximately planes which are approximately normal (depending on η_1 and r_j) to the geodesic ν-plane $\sum_{ji}$; we call these loci the *projecting faces.*

Since $\{p_{ji}\}$ is an r_j-set for S^*, $B(p_{ji}, r_j/2) \subset D_{ji}$. On the other hand since any $p \in D_{ji} \cap \bar{S} \in U\, B(p_{jk}, r_j)$ with $d(p, p_{ji}) \le$ every $d(p, p_{jk})$, we obtain

$$B(p_{ji}, r_j/2) \subset D_{ji}, \quad D_{ji} \cap \bar{S} \subset B(p_{ji}, r_j). \tag{10.5.63}$$

From (10.5.63) and (10.5.40) (iii), etc., it follows that $T_j \cap D_{ji}$ is almost parallel (depending on θ_1, η_2, and r_j) to $\sum_{ji} \cap D_{ji}$. Since we have seen that the projecting faces are almost normal to $\sum_{ji}$, we conclude that

$$L_\nu(T_j \cap D_{ji}) = \Lambda^\nu(T_j \cap D_{ji}) \le (1+\eta_6)^{1/2} \Lambda^\nu(D_{ji} \cap \textstyle\sum_{ji}) \tag{10.5.64}$$

if θ_1, η_1, η_2, and R_1, and hence ε (if ϱ is chosen $< \varrho_0(\mathfrak{M})$) are small enough. Moreover if $B(p_{ji}, r_j) \cap \bar{S}$ is not empty and $B(p_{ji}, r_j) \cap \partial\bar{S}$ is empty, it follows from Lemma 10.4.7 and Lemma 10.5.6 that the projection of $S \cap \overline{B(p_{ji}, r_j)}$ = the projection of $\bar{S} \cap \overline{B(p_{ji}, r_j)}$ (as defined in Definition 10.4.2)

$$\supset \textstyle\sum_{ji} \cap B\left(p_{ji}, r_j\sqrt{1 - Z^2\eta_1^2/36}\right)$$

Z being the constant in Lemma 10.4.7. Thus, since $\sum_{ji} \cap B(p_{ji}, r_j) \supset \sum_{ji} \cap D_{ji}$ and the projecting faces are nearly normal to $\sum_{ji}$, we conclude that

$$\Lambda^\nu(\bar{S} \cap D_{ji}) \ge (1+\eta_6)^{-1/2} \Lambda^\nu(D_{ji} \cap \textstyle\sum_{ji}). \tag{10.5.65}$$

From (10.5.64) and (10.5.65) we deduce that

$$L_\nu(T_j \cap D_{ji}) \le (1+\eta_6)\, \Lambda^\nu(\bar{S} \cap D_{ji}). \tag{10.5.66}$$

Let $\Gamma_{ji}^j = \tau_j^{-1}(T_j \cap D_{ji})$, $\Gamma_{ji} = \tau^{-1}(\bar{S} \cap D_{ji})$. Since τ and τ_j are homeomorphisms, it follows that all the Γ_{ji}, for those i for which $\Gamma_{ji}^j \cap \bar{\Gamma}$

is not empty, lie in (Γ, s_j), where $s_j \to 0$. Thus, summing over such i and using (10.5.66), we obtain

$$L_\nu(\tau_j \mid \Gamma) \leq \sum_i L_\nu(\tau_j \mid \Gamma_{ji}^j) \leq (1+\eta_6) \sum_i \Lambda^\nu[\tau(\Gamma_{ji})] \leq$$
$$\leq (1+\eta_6)\, \Lambda^\nu\{\tau(\Gamma, s_j)\}$$

from which the lemma follows by letting $j \to \infty$.

We wish to conclude this section with the following theorem:

Theorem 10.5.2. *Suppose that ε, ϱ, θ_1, R_0', η_1, and η_2 are fixed sufficiently small numbers, that S, $\bar{S}$, τ, and the τ_j have their previous significance, and that* (10.5.62) *holds for some fixed (finite) $\eta_6 > 0$. Then*

$$L_\nu(\tau \mid \Gamma) \leq \Lambda^\nu[\tau(\Gamma)], \text{ if } \Gamma \text{ is open and } \Gamma \subset T_0^0.$$

We shall prove this theorem after stating several definitions and proving several lemmas. The methods of proof are suggested by those of some deep results of FEDERER ([1], [2]). We present simplified proofs of some additional results, which we don't require here, in § 10.8.

Definition 10.5.4. Given $a \in R_N$ and $\varepsilon > 0$. We let $C_N'(a, \varepsilon)$ denote the totality of N-cubes of the form $|z^i - a^i - \varepsilon\,\delta^i)| \leq \varepsilon$, where δ is a sequence of even integers. We let $C_{N-1}'(a, \varepsilon)$ be the totality of the $(N-1)$-cells on the boundaries of the N-cells of $C_N'(a, \varepsilon)$, and denote by $C_{N-2}'(a, \varepsilon)$ the totality of $(N-2)$-cells on the boundaries of the $(N-1)$-cells of $C_{N-1}'(a, \varepsilon)$, etc. Similarly we let $C_N''(a, \varepsilon)$ denote the totality of N-cells of the form $|z^i - a^i - \varepsilon\,\theta^i| \leq \varepsilon$ where θ is a sequence of odd integers, and let $C_{N-1}''(a, \varepsilon)$ denote the totality of $(N-1)$-cells on the boundaries of the cells of $C_N''(a, \varepsilon)$, etc. We sometimes allow $C_k'(a, \varepsilon)$ or $C_k''(a, \varepsilon)$ to denote the union of the point sets corresponding to the collections. If $a = 0$ and $\varepsilon = 1$, we denote $C_k'(0, 1)$ by C_k', etc.

Lemma 10.5.8. (a) *γ_{kq} is a cell of $C_k' \Leftrightarrow$ it is of the form*

$$|z^i - \delta^i| \leq 1, \quad i \in s_{kq} \quad \textit{and} \quad z^i = \theta^i \quad \textit{if} \quad i \notin s_{kq}$$

s_{kq} being some set of k distinct integers $\leq N$; here the δ^i are even and the θ^i odd. The corresponding statement holds for the cells of each C_k''.

(b) *For each k, $\nu + 1 \leq k \leq N$, and each q, the center of $\gamma_{kq} \in C_{N-\nu-1}''$. If $k = \nu + 1$, $\gamma_{kq} \cap C_{N-\nu-1}''$ is the center of γ_{kq}. If $k < \nu + 1$, then $\gamma_{kq} \cap C_{N-\nu-1}''$ is empty.*

(c) *If we define $r_k : C_k' - C_{N-\nu-1}'' \to C_{k-1}'$ by defining it on each $\gamma_{kq} - C_{N-\nu-1}''$ as the radial projection from the center of γ_{kq} onto its boundary, then $r_k(C_k' - C_{N-\nu-1}'') \subset C_{k-1}' - C_{N-\nu-1}''$, $k \geq \nu + 1$.*

Proof. (a) is easily proved by induction downward. To prove (b) and (c) we may assume that $\gamma_{kq} = \gamma_k'$ where

$$\gamma_k' : |z^i| \leq 1 \text{ for } i = 1, \ldots, k; \quad z^i = 1 \text{ for } k+1 \leq i \leq N.$$

Now if $\gamma_{N-\nu-1}'' \in C_{N-\nu-1}''$, $\gamma_{N-\nu-1}''$ lies along a plane $z^i = \delta^i$, δ^i even, for $\nu + 1$ values of i. If $\gamma_{N-\nu-1}'' \cap \gamma_k'$ is not to be empty, these i must all be

$\leq k$. For each such set of i, there are $2^{k-\nu-1}$ cells of $C''_{N-\nu-1}$ which intersect γ_k, in which the z^j must satisfy

$$|z^j - \theta^j| \leq 1, \quad \theta^j = \pm 1, \quad 1 \leq j \leq k, \quad j \neq \text{any } i.$$

Thus $\gamma'_k \cap C''_{N-\nu-1}$ consists of all the $(k-\nu-1)$-cells for which $z^{k+1} = \cdots = z^N = 1$, $z^i = 0$ for $\nu + 1$ values of $i \leq k$, and $|z^j| \leq 1$ for the remaining j. Parts (b) and (c) now follow easily.

Definition 10.5.5. We define

$$\varrho_\nu : R_N - C''_{N-\nu-1} \to R_N \text{ by } \varrho_\nu = r_{\nu+1} \circ \ldots \circ r_N$$

where the r_k are defined in Lemma 10.5.8 (c). We define B as the unit cube in R_N, i.e. $B = \{y : |y^i| < 1, i = 1, \ldots, N$, and define

$$u_\nu(y) = (|y^{i_1} - \delta^{i_1}|, \ldots, |y^{i_{\nu+1}} - \delta^{i_{\nu+1}}|)$$

where δ is that sequence of even integers such that $y - \delta \in B$ and $(i_1, \ldots, i_N)$ is a permutation of $(1, \ldots, N)$ for which

$$|y^{i_1} - \delta^{i_1}| \leq |y^{i_2} - \delta^{i_2}| \leq \cdots \leq |y^{i_N} - \delta^{i_N}|.$$

Given a and ε, we define the maps ω, τ_a, and r_a by

$$\omega(y) = \varepsilon y, \quad \tau_a(y) = y + a, \quad r_a = \tau_a \circ \omega \circ \varrho_\nu \circ \omega^{-1} \cdot \tau_{-a}.$$

Lemma 10.5.9. (a) *If $\eta \in R_1 - C''_{N-\nu-1}$, then $\varrho_\nu(\eta) \in C'_\nu$ and is on the boundary of the N-cell of the form $\eta - \delta \in B$ in which η lies.*

(b) $\int_B |u_\nu(v - \eta)|^{-\nu} \, d\Lambda^N(\eta) = C_7(\nu, N)\, \Lambda^N(B)$.

(c) $\eta \in R^N - C''_{N-\nu-1} \Leftrightarrow u_\nu(\eta) \neq 0$.

(d) *If $y = a + \varepsilon\delta + \varepsilon\eta$, where $\eta \in B$ and δ is (as usual) an N-sequence of even integers, then $r_a(y)$ is the point on $C'(a, \varepsilon)$ given by*

$$r_a(y) = a + \varepsilon\delta + \varepsilon\varrho_\nu(\eta), \quad \varrho_\nu(\eta) \in \partial B,$$

and

$$|D r_a(y)| \leq C_8(\nu) \cdot |u_\nu(\eta)|^{-1}$$

where D denotes the maximum directional derivative.

Proof. (a) This follows from Lemma 10.5.8 and its proof. To prove (b), we note that u is periodic of period 2 in each η^i. Thus we have

$$\int_B |u_\nu(v - \eta)|^\nu \, d\Lambda^N(\eta) = \int_B |u_\nu(\eta)|^{-\nu} \, d\Lambda^N(\eta).$$

The result follows by direct evaluation which is made easy by the symmetry. (c) follows from the definitions. To prove (*d*), we note that

$$r_a(y) = a + \varepsilon[\delta + \varrho_\nu(\eta)], \quad \eta \in B.$$

From the homogeneity, we see that

$$|D_y r_a(y)| = |D_\eta \varrho_\nu(\eta)|.$$

To calculate a bound for the latter, it is sufficient, because of the symmetry, to assume that $0 \leq \eta^1 \leq \eta^2 \leq \cdots \leq \eta^N \leq 1$. Then

$$\varrho^\nu(\eta) = \left(\frac{\eta^1}{\eta^{\nu+1}}, \ldots, \frac{\eta^\nu}{\eta^{\nu+1}}, 1, \ldots, 1\right), \quad \eta^{\nu+1} > 0;$$

we must have $\eta^{\nu+1} > 0$ if $\eta \notin C''_{N-\nu-1}$. In our case

$$u_\nu(\eta) = (\eta^1, \ldots, \eta^{\nu+1})$$

and the bound $C_8 |u_\nu(\eta)|^{-1}$ is easily obtained for $|D\varrho^\nu(\eta)|$.

Lemma 10.5.10. *Suppose G is a bounded open set in R_ν, τ is a homeomorphism from $\bar{G}$ into $\mathfrak{M}$, $\Lambda^\nu[\tau(\bar{G})] < \infty$, $\Lambda^\nu[\tau(\partial G)] = 0$, $\tau(\bar{G}) \subset B(P_0, R)$, $R \leq R_0$, and $\tau = \omega \circ z$ where ω is a normal coordinate system with domain $B(0, R)$ and $\omega(0) = P_0$. Suppose also that there exists a sequence $\{\tau_n\}$, each $\tau_n = \omega \circ z_n \in C^1(\bar{G})$, such that $\tau_n \to \tau$ uniformly on $\bar{G}$ and such that, for some $\eta > 0$*

$$\lim_{n\to\infty} L_\nu(\tau_n \mid \Gamma) \leq (1 + \eta)\, \Lambda^\nu[\tau(\Gamma)], \quad \textit{if } \bar{\Gamma} \subset G \textit{ and } \Lambda^\nu[\tau(\partial\Gamma)] = 0.$$

Let $\varphi : \mathfrak{M} \to \mathfrak{M}$ be a piecewise smooth mapping with domain $\supset \tau(\bar{G})$ such that $|D\varphi(p)| \leq \chi(p)$, where χ is continuous near $\tau(\bar{G})$. Then

$$L_\nu(\varphi \circ \tau) \leq (1 + \eta) \int_{\tau(\bar{G})} [\chi(p)]^\nu \, d\Lambda(e_p).$$

Proof. Let $\mathfrak{G}$ be those cells r in R_N such that $\Lambda^\nu[z(\bar{G}) \cap \partial r] = 0$. For each $\sigma > 0$, we may cover $z(\bar{G})$ with a finite number of closed cells $\bar{r}_i$ of $\mathfrak{G}$ of diameter $< \sigma$, the open cells $r_i^{(o)}$ being disjoint. For each i for which $\bar{r}_i \cap z(\partial G)$ is empty, let $\Gamma_i = z^{-1}(r_i^{(o)})$. Since z and ω are homeomorphisms, the diameters of the Γ_i and of the $\tau(\Gamma_i) \to 0$ as $\sigma \to 0$. Moreover, for each such i, $\Lambda^\nu[\tau(\partial\Gamma_i)] = \Lambda^\nu\{\omega(\partial r_i \cap z(\bar{G})\} = 0$. For such i and all n

$$\begin{aligned} L_\nu(\varphi \circ \tau_n \mid \Gamma_i) &= \Lambda^\nu[\varphi \circ \tau_n(\Gamma_i)] \leq [\chi(p_i)]^\nu \Lambda^\nu[\tau_n(\Gamma_i)] \\ &= [\chi(p_i)]^\nu L_\nu(\tau_n \mid \Gamma_i) \leq (1 + \eta)\,[\chi(p_i)]^\nu \Lambda^\nu[\tau(\Gamma_i)] + \varrho_n, \\ \varrho_n &\to 0 \end{aligned} \tag{10.5.67}$$

where $\chi(p_i)$ denotes the maximum value of $\chi(p)$ for $p \in \tau(\bar{\Gamma}_i)$. The result follows by adding the results in (10.5.67) for those i for which $r_i \cap z(\partial G)$ is empty, taking the lim inf and then letting $\sigma \to 0$.

Lemma 10.5.11. *Suppose G is a bounded open set in R_ν, τ is continuous on $\bar{G}$, $L_\nu(\tau) < \infty$, $\tau(\bar{G}) \subset B(P_0, R)$, $R \leq R_0$, and $\tau = \omega \circ z$ where ω is a normal coordinate system with domain $B(0, R)$ and $\omega(0) = P_0$. Suppose Z is the union of a finite number of $(\nu - 1)$-planes in R_N, and $Q = z^{-1}[Z \cap z(\bar{G})]$. Then*

$$L_\nu(\tau \mid G - Q) = L_\nu(\tau).$$

Proof. It is sufficient to prove this for Z a single $(\nu - 1)$-plane. Let ω_{00} be the normal coordinate system related to ω by a rotation of axes such that Z is part of the $(N + 1 - \nu)$ plane $z^i = z_0^i$, $i = \nu, \ldots, N$. By approximations from the interior, we may also assume that G is the union of a finite number of domains of class C^1 which have disjoint closures. Moreover, for each ε and each σ, $0 < \sigma < 1$, we can find an open set

$\Gamma \subset G - Q$ such that Γ consists of a finite number of domains of class C^1 with disjoint closures and

$$L_\nu(\tau \mid \Gamma) > L_\nu(\tau \mid G - Q) - \varepsilon/2, \quad r[\tau(x)] < \sigma/2 \quad \text{for} \quad x \in \bar{G} - \bar{\Gamma},$$

$r(p)$ being the geodesic distance from p to $W = \omega_{00}(Z)$. We may assume $\omega = \omega_{00}$; for each $q \in B(P_0, R_0/3)$, we let ω_{0q} be the normal coordinate system centered at q and related to ω_{00} as in § 10.1. To points p near W we may assign coordinates w as follows: Let $q = \pi_W(p)$, the (geodesic) projection of p on W; then $\omega_{00}^{-1}(q) = (w^1, \ldots, w^{\nu-1}, z_0^\nu, \ldots, z_0^N)$. Then p must lie on the geodesic $(N+1-\nu)$-plane W through q so that $p = \omega_{0q}(0, \ldots, 0, w^\nu, \ldots, w^N)$. We define ω^* by $p = \omega^*(w)$; clearly ω^* is of class C^2 near Z.

Now, we define a sequence $\{\zeta_j\}$, each $\zeta_j \in C^1(\bar{\Gamma})$ and $\in C_1^0(\bar{G})$ (i.e. Lipschitz) such that $\zeta_j \to \tau$ uniformly on $\bar{G}$ and

$$L_\nu(\zeta_j \mid \Gamma) \to L_\nu(\tau, \Gamma). \tag{10.5.68}$$

To do this, we first approximate uniformly to z on $\bar{\Gamma}$ by $\{\zeta_j^*\}$, each $\zeta_j^* \in C_1(\bar{\Gamma})$, so that (10.5.68) holds with $\zeta_j = \omega\, \zeta_j^*$ and then approximate uniformly to z on $\bar{G}$ by any sequence $\{\eta_j\}$, each $\eta_j \in C^1(\bar{G})$. Now, for each j, $\zeta_j^* - \eta_j \mid \bar{\Gamma}$ can be extended to satisfy a Lipschitz condition on $\bar{G}$. Let ξ_j^* be such an extension and define

$$\xi_j(x) = \begin{cases} \xi_j^*(x), & \text{if } |\xi_j^*(x)| \le 2\omega_j \\ 2\omega_j\, |\xi_j^*(x)|^{-1}\, \xi_j^*(x), & \text{if } |\xi_j^*(x)| \ge 2\omega_j \end{cases}$$

$$\omega_j = \max_{x \in \bar{\Gamma}} |\xi_j^*(x)|$$

$$\varrho_j(x) = \eta_j(x) + \xi_j(x)^1, \quad \zeta_j(x) = \omega \circ \varrho_j(x).$$

Clearly $\varrho_j(x) = \zeta_j^*(x)$ on $\bar{\Gamma}$ and each ϱ_j is LIPSCHITZ on $\bar{G}$. Since $\eta_j \,|\bar{\Gamma} \to z|\bar{\Gamma}$, it follows that $\xi_j^* \,|\bar{\Gamma} \to 0$ so $\omega_j \to 0$ and $\xi_j \to 0$, so that $\varrho_j \to z$ (all convergence uniform) so $\{\zeta_j\}$ satisfies the conditions.

Finally, let us define $\varphi_\sigma : \mathfrak{M} \to \mathfrak{M}$ by

$$\varphi_\sigma(p) = \begin{cases} p, & \text{if } r(p) \ge \Sigma = \sigma^{1/2}, \\ \pi_W(p), & \text{if } r(p) \le \sigma, \\ p', & \text{if } \sigma \le r(p) \le \Sigma, \end{cases}$$

p' being the point h of the way from $\pi_W(p)$ along the geodesic to p where $h = (\Sigma - \sigma)^{-1}[r(p) - \sigma]$. With the aid of the ω^* coordinate system, it is easy to see that φ_σ is piecewise smooth and that

$$|D\,\varphi_\sigma(p)| \le \chi_\sigma, \quad \chi_\sigma = \text{a const.}, \quad \lim_{\sigma \to 0} \chi_\sigma = 1$$

$$d[\varphi_\sigma(p), p] \le \sigma, \quad \text{both for } p \in \mathfrak{M}.$$

---

[1] In case $\zeta_j^* - \eta_j \mid \Gamma \equiv 0$ on Γ, set $\xi_j(x) = \xi_j^*(x) \equiv 0$ on $\bar{G}$.

We now define $\tau_j = \varphi_\sigma \circ \zeta_j$ and we see that if j is large enough for $|\zeta_j(p) - \tau(p)| \leq \sigma/2$, then τ_j is Lipschitz and $\tau_j(x) \in W$ for all $x \in \bar{G} - \Gamma$. Therefore, for such j,

$$L_\nu(\tau_j) = L_\nu(\tau_j \mid \Gamma) \leq \chi_\sigma^\nu L_\nu(\zeta_j \mid \Gamma) \tag{10.5.69}$$
$$|\tau_j(p) - \tau(p)| \leq 3\sigma/2.$$

The lemma follows from (10.5.69) and (10.5.68) and the arbitrariness of σ and ε.

Proof of the theorem. Since $\bar{S}$ minimizes $\Lambda^\nu(X)$ among all compact $X \in \mathfrak{C}(\partial\bar{S}, L)$, where $L = H_{\nu-1}(\partial\bar{S})$ (Theorems 10.3.4 and 10.4.4), it follows that $\psi(P) = 1$ almost everywhere on $\bar{S}$. From Theorem 10.4.1, it follows that ψ is upper semicontinuous. Let $\eta > 0$ and choose positive numbers ϱ, θ_1, R_0', η_1, η_2, and ε so small that the η_6 of Lemma 10.5.7 is $< \eta$. From Theorem 10.4.5 it follows that we may take $\varepsilon_0 > 0$ but so small that if $\psi(P) < 1 + \varepsilon_0$, then $\bar{S}$ satisfies an (ε, R_2)-condition at P with $R_2 \leq R_0'$, if P is interior to $\bar{S}$. Let W be the subset of $\bar{S} - \partial\bar{S}$ where $\psi(P) \geq 1 + \varepsilon_0/2$; then W is closed in $\bar{S} - \partial\bar{S}$, $\Lambda^\nu(W) = 0$, and $X = \tau^{-1}(W)$ is closed in $B(0, R) \equiv (T_0^0)^{(0)}$, $R = 2R_1 - 6r_0 - \lambda_0$. We assume that $P_0 \in \bar{S}$ and that $\bar{S} \subset B(P_0, 2R_1)$ and that ω_{00} is a normal coordinate system with domain $\overline{B(0, R_0)}$ and $\omega_{00}(0) = P_0$. We let $Z = \omega_{00}^{-1}(W) = z(X)$. Let G be any open subset of $(T_0^0)^{(0)}$. We cover each x_0 of $G - X$ by a neighborhood $\mathfrak{N}$ with $\bar{\mathfrak{N}} \subset G - X$ such that there exists a sequence $\{\tau_n\}$, each $\tau_n \in C^1(\bar{\mathfrak{N}})$ with $\tau_n = \omega_{00} \circ z_n$, which converges uniformly to $\tau \mid \bar{\mathfrak{N}}$, and for which

$$\lim_{n\to\infty} L_\nu(\tau_n \mid \Gamma) \leq (1 + \eta)\, \Lambda^\nu[\tau(\Gamma)], \quad \text{if} \tag{10.5.70}$$
$$\Gamma \subset \mathfrak{N}, \quad \Lambda^\nu[\tau(\partial\Gamma)] = 0.$$

If $x_0 \in X$, we take $\mathfrak{N} = G$.

Let F be any compact subset of G. We may cover $z(F)$ by a finite number of closed cells $\bar{r}_i$ with disjoint interiors such that $\Lambda^\nu[z(\bar{G}) \cap \partial r_i] = 0$ and $r_i \cap z(\partial G)$ is empty for each i, each also being of diameter $< \sigma$. Since τ and z are homeomorphisms, we may take σ so small that each of the sets $\bar{\Gamma}_i = z^{-1}(\bar{r}_i)$ is in some one neighborhood $\mathfrak{N}$, provided $\bar{\Gamma}_i \cap F$ is not empty. So we let $\bar{G}'$ be the union of these $\bar{\Gamma}_i$; clearly $\bar{G}' \supset F$. It is sufficient to show that $L_\nu(\tau, G') = \Lambda^\nu[\tau(G')]$ for such G'.

Let Y be the union of these ∂r_i. The Γ_i have the property (10.5.70) possessed by the $\mathfrak{N}$. Let δ be an arbitrary positive number and choose a in R_N with $|a^i| < h$, where $h > 0$ but so small that

$$\int_{V_0 \cap z(G')} [\chi_0(z)]^\nu \, d\Lambda^\nu(e_z) < \delta, \quad V_0 = (Y, 5Nh), \quad \chi_0 = \chi_{a,h}$$

$$\chi_0(z) = \{C_9 \,|u_\nu[h^{-1}(z - a)]|^{-1} - 1\}\, q_2(z) + 1, \quad C_9 = C_8 + 3.$$

$$q_2(z) = 1 \text{ on } (Y, 4Nh) \text{ and } q_2(z) = 0 \text{ on } R_N - (Y, 5Nh)$$
$$0 \leq q_2(z) \leq 1 \quad \text{on} \quad (Y, 5Nh) - (Y, 4Nh),$$

and q_2 has Lipschitz constant $(Nh)^{-1}$. To see that this choice is possible, we note that (using the notation of Lemma 10.5.9)

$$\begin{aligned} &\int_{\omega(B)} d\Lambda^N(a) \int_{V_0 \cap z(\bar{G}')} [\chi_0(z)]^\nu d\Lambda^\nu(e_z) \\ &= h^N \int_B d\Lambda^N(b) \int_{V_0 \cap z(\bar{G}')} C_9^\nu |u_\nu(h^{-1}z - b)|^{-\nu} d\Lambda^\nu(e_z) \\ &= h^N \cdot C_9^\nu \cdot C_7(\nu, N) \Lambda^N(B) \Lambda^\nu[V_0 \cap z(\bar{G}')] \\ &= \Lambda^N[\omega(B)] \cdot C_7 \cdot C_9^\nu \cdot \Lambda^\nu[V_0 \cap z(\bar{G}')] \end{aligned}$$

using Lemma 10.5.9. Since $\Lambda^{\nu+1}[z(\bar{G}')] = 0$, it follows also that $\Lambda^{\nu+1}[P^J \circ z(\bar{G}')] = 0$ for each $(\nu+1)$-sequence J. Thus the N-measure of the totality of $(N-\nu-1)$-planes of the form $z^J = a^J$ which intersect the set $z(\bar{G}')$ is zero. Hence we may also choose a so that $z(\bar{G}') \subset R_N - C''_{N-\nu-1}(a, h)$. Moreover by choosing σ small enough, we may ensure that each Γ_i which intersects X is such that $r_i \subset$ a neighborhood U of $z[X \cap \bar{G}']$ such that

$$\int_{U \cap \tau(\bar{G}')} |\chi_0(z)|^\nu d\Lambda^\nu(e_z) < \delta. \tag{10.5.71}$$

Having chosen ε and a as above, we define $q_1(z)$ as we defined $q_2(z)$ except with $4Nh$ and $5Nh$ replaced, respectively, by $3Nh$ and $4Nh$. Then we define

$$\varphi_0(y) = y + q_1(y)[r_a(y) - y], \quad \zeta = \varphi_0 \circ z, \quad \tau' = \omega_{00} \circ \zeta.$$

We note that

$$\begin{aligned} &\varphi_0(y) = y \quad \text{for} \quad y \in R_N - (Y, 4Nh) \\ &|\varphi_0(y) - y| < Nh, \quad y \in R_N \\ &|\zeta(x) - z(x)| < Nh \end{aligned} \tag{10.5.72}$$
$$|D\varphi_0(y)| \leq 1 + (Nh)^{-1} \cdot Nh + 1 + |D r_a(y)| \leq |\chi_0(y)|, \quad y \in R_N.$$

Now, let Ω be the collection of components Γ of $G' - \zeta^{-1}[C'_{\nu-1}(a, h)]$. Then we conclude from Lemma 10.5.11 that

$$L_\nu(\tau' \mid G') = \sum_{\Gamma \in \Omega} L_\nu(\tau' \mid \Gamma). \tag{10.5.73}$$

If Γ does not intersect $z^{-1}(Y)$ nor X then it is part of one of the $\Gamma_i \subset$ some $\mathfrak{R}$ with $\bar{\mathfrak{R}} \subset G - X$, and we conclude from Lemma 10.5.10 that

$$L_\nu(\tau' \mid \Gamma) \leq L(1+\eta) \int_{z(\Gamma)} |\chi_0(z)|^\nu d\Lambda^\nu(e_z), \tag{10.5.74}$$

since

$$\tau' = \varphi \circ \tau, \ \varphi = \omega_{00} \varphi_0 \omega_{00}^{-1}, \ |D\varphi(p)| \le (1 + 2\eta_0 R_1)^2 |D\varphi_0[\omega_{00}^{-1}(p)]|, \quad p \in V$$

$$|D\varphi(p)| \equiv 1, \quad p \notin V, \quad V = \omega_{00}(V_0), \quad L = (1 + 2\eta_0 R_1)^{3\nu}.$$

If Γ does intersect $z^{-1}(Y)$, let $x_0 \in \Gamma \cap z^{-1}(Y)$. Then $z(x_0) \in Y$, so that (by (10.5.72)) $\zeta(x_0) \in (Y, Nh)$ and hence $\in$ a component ν cube (of side $2h$) E of $C'_\nu(a, h) - C'_{\nu-1}(a, h)$. Let x_1 be any point of Γ and join x_0 to x_1 by a path x_t, $0 \le t \le 1$. As long as $\zeta(x_t) \in (Y, 3Nh)$, $\zeta(x_t) \in E$, and as long as $\zeta(x_t) \in E$, it $\in (Y, 3Nh)$. Consequently $\zeta(\Gamma) \subset E \subset (Y, 3Nh)$ and $z(\Gamma) \subset (Y, 4Nh) \subset V_0$. For such a Γ, we can conclude only that

$$L_\nu(\tau' \mid \Gamma) \le L(1 + \eta_6) \int_{z(\Gamma)} |\chi_0(z)|^\nu \, d\Lambda^\nu(e_z). \tag{10.5.75}$$

If Γ intersects X but not $z^{-1}(Y)$, then $z(\Gamma) \subset U$ and we can conclude that (10.5.75) holds. Substituting the results (10.5.74) and (10.5.75) in (10.5.73), we obtain

$$\begin{aligned} L_\nu(\tau' \mid G') &\le (1 + \eta) \Lambda^\nu[\tau(\bar{G}')] + (2 + \eta + \eta_6) \cdot L \cdot \int_{V_0 \cap z(\bar{G}')} |\chi_0(z)|^\nu \, d\Lambda^\nu(e_z) + \\ &\quad + (1 + \eta_6) \int_{U \cap z(\bar{G}')} |\chi_0(z)|^\nu \, d\Lambda^\nu(e_z) \\ &\le (1 + \eta_6) \Lambda^\nu[\tau(\bar{G}')] + (3 + \eta + 2\eta_6) \cdot \delta \cdot L. \end{aligned} \tag{10.5.76}$$

$$\max |\tau'(x) - \tau(x)| \le Nh.$$

Since, for a given σ, η, and δ, h may be arbitrarily small, we obtain (10.5.76) with τ' replaced by τ. Then the result follows for $\bar{G}'$ from the arbitrariness of η and δ.

10.6. The Reifenberg cone inequality

Suppose z is a polynomial N-vector which defines a mapping from $\overline{B(0, R)}$ onto $\bar{S}$, suppose that $z_0 = P_0 \in R_N$, and $R(P_0)$ is the minimum value of $|z(s) - z_0|$ for s on $\partial B(0, R)$. Let

$$U(s) = |z(s) - z_0|^2;$$

then U is a polynomial. From recent results by WHITNEY [3], it follows that the locus of $U_{,\alpha}(s) = 0$, $\alpha = 1, \ldots, \nu$, has only finitely many components. From a theorem of A. P. MORSE [1], it follows that U is constant on each of these components. Thus, except for a finite number of values of r, $0 < r < R(P_0)$, the locus $l(r, P_0, z) = S \cap \partial B(P_0, r)$ consists of a finite number of regular orientable analytic manifolds without boundary. The remainder of this section is devoted to the proof of the following theorem:

Theorem 10.6.1 (The REIFENBERG cone inequality). *There exist positive numbers ϱ_0, ε_0, and k, $k < 1$, with the following property: If z, r, and*

$l(r, P_0, z)$ *have their significance above, r is not exceptional, and there is a ν-plane $\sum$ through P_0 such that*

(10.6.1)
$$\Lambda^{\nu-1}[l(r, P_0, z)] < \nu(1 + \varepsilon_0)\,\gamma_\nu\,r^{\nu-1}, \quad \text{and}$$
$$D[l(r, P_0, z), \sum \cap \partial B(P_0, r)] < \varrho_0\, r$$

then there is a set Y^* with $b(Y^*, A) \supset H_{\nu-1}(A)$, $A = l(r, P_0, z)$, such that

(10.6.2) $\Lambda^\nu(Y^*) \leq (1 - k)\,\nu^{-1}\,r\,\Lambda^{\nu-1}[l(r, P_0, z)] + k\,\gamma_\nu\,r^\nu.$

Suppose we let Y_r be the cone with vertex P_0 and base $l(r, P_0, z)$. Then the inequality (10.6.2) becomes

(10.6.2') $\Lambda^\nu(Y^*) \leq (1 - k)\,\Lambda^\nu(Y_r) + k\,\gamma_\nu\,r^\nu.$

On account of the homogeneity, we may assume that $r = 1$.

We may let $\mathfrak{M}_1$ be the collection of regular analytic manifolds, without boundary, in $B(0, R)$ which is the counterimage of $l(1, P_0, z)$. We choose an (x, y) coordinate system with origin at P_0 and $x^1, \ldots, x^\nu$ axes in $\sum$. We may then represent Y_1 parametrically on $I \times \mathfrak{M}_1$, $I = [0, 1]$, by

(10.6.3) $x = r\tilde{\xi}(p), \; y = r\tilde{\eta}(p), \; 0 \leq r \leq 1, \; |\tilde{\xi}|^2 + |\tilde{\eta}|^2 = 1, \; p \in \mathfrak{M}_1.$

It is convenient to let Y denote the infinite cone obtained by extending the rays of Y_1 to infinity and to represent Y by

(10.6.4)
$$x = r\,\xi(p), \quad y = r\,\eta(p), \quad r \geq 0,$$
$$\xi(p) = |\tilde{\xi}(p)|^{-1}\,\tilde{\xi}(p), \quad \eta(p) = |\tilde{\xi}(p)|^{-1}\,\tilde{\eta}(p), \quad p \in \mathfrak{M}.$$

We may approximate to $\tilde{\xi}$ and $\tilde{\eta}$ (or ξ and η) uniformly *together with their first derivatives* by piecewise analytic functions $\tilde{\xi}_n$ and $\tilde{\eta}_n$, each pair of which is such that the cone Y_n is composed of a finite number of "simplicial cones", each of which is the union of all rays joining P_0, hereafter called 0, to the points of a $(\nu - 1)$-simplex in a $(\nu - 1)$-plane not passing through 0. We call such a locus Y_n a *polyhedral cone*. Our method of proof of the theorem will consist in constructing a set Y_n^* as desired in the theorem for each Y_n and then showing that the Y_n^* tend to Y^* in the point-set sense, in such a smooth way that $\Lambda^\nu(Y_n^*) \to \Lambda^\nu(Y^*)$. We may assume that each simplicial cone in each Y_n has a $1 - 1$ projection on the x-plane $\sum$.

On a polyhedral cone Y, we shall denote by $x(P)$ and $y(P)$ the x and y coordinates of the point P on Y. We have seen that each of the manifolds composing Y is orientable; we suppose that each such manifold is oriented. This induces an orientation on each simplicial cone. If P is interior to such a simplicial cone, we define $\gamma(P) = \pm 1$ according as the simplicial cone and its projection on $\sum$ are similarly or oppositely oriented. On any simplicial cone of Y, we may consider the y^k as linear functions of the x^j and we define $|J_y(P)|^2$ as the sum of the squares of the Jacobians of all orders $\leq \min(\nu, N-\nu)$ of the y^k with respect to the x^j. The area element on Y at P is then

$$\sqrt{1+|J_y(P)|^2}\, d\Sigma(P), \quad d\Sigma(P) = |dx(P)| \tag{10.6.5}$$

where $|dx(P)|$ is the ν-area element on Σ. We also define

$$\begin{gathered} |D_y(P)|^2 = \sum_{j=1}^{\nu} \sum_{k=1}^{N-\nu} \left(\frac{\partial y^k}{\partial x^j}\right)^2 \\ \varepsilon = \int_{\tilde{Y}_1} \sqrt{1+|J_y(P)|^2}\, d\Sigma(P) - \gamma_\nu = \Lambda^\nu(\tilde{Y}_1) - \gamma_\nu \end{gathered} \tag{10.6.6}$$

where $\tilde{Y}_1$ is the part of Y in the cylinder $|x(P)| \leq 1$[1]. Comparing this notation with that in (10.6.4), we see that

$$x(P) = r\,\xi(p), \quad y(P) = r\,\eta(P). \tag{10.6.7}$$

We shall prove the theorem for all ϱ and ε which are sufficiently small, ϱ being a bound for $|y(P)|$ on $\tilde{Y}_1$; we may also assume that

$$\varepsilon < \varrho^{6\nu}. \tag{10.6.8}$$

We first notice the following fact:

Lemma 10.6.1. *Suppose Y is a polyhedral cone of the type described and suppose we define*

$$\bar{y}(x) = \sum_{x(P)=x} \gamma(P)\, y(P) \tag{10.6.9}$$

for points x not on the image of the boundary of any simplicial cone of Y. Then $\bar{y}$ coincides with a single-valued function which is continuous on R_ν and linear on each of a finite number of simplicial cones. Moreover

$$\sum_{x(P)=x} \gamma(P) = \varkappa, \quad \text{a.e. on} \quad Y, \varkappa = \pm 1. \tag{10.6.10}$$

Proof. The projections on the x-space of the simplicial boundaries of all the simplicial cones of Y divide that space into infinite conical regions which can be further subdivided into a finite number of simplicial cones.

Let us consider two of these simplicial cones Δ_1 and Δ_2 which have an $(\nu-1)$-dimensional simplicial cone Δ in common. The points P of Y whose projections lie in Δ_1 form a finite number δ_{1j}, $j = 1, \ldots, J_1$, of simplicial cones, each of which is a part of one of the original ones of Y. The same is true of Δ_2; let δ_{2j}, $j = 1, \ldots, J_2$, be those correspoinding to Δ_2. From the nature of Y, it follows that each δ_{1j} abuts on a unique δ_{2j} or another δ_{1j} and the same is true of each δ_{2j}. Thus, we may assume (by reordering if necessary) that each δ_{1j} abuts on δ_{2j} for $j = 1, \ldots, J_3$ and

[1] The ε here has no connection with the ε's used in other sections.

then that δ_{1, J_3+2p-1} abuts on δ_{1, J_3+2p} for $p = 1, \ldots, (J_1 - J_3)/2$, and $(\delta_{2, J_3+2q-1}$ abuts on δ_{2, J_3+2q}, $q = 1, \ldots, (J_3 - J_2)/2$. Then

$$\bar{y}(x) = \begin{cases} \sum_{j=1}^{J_3} \gamma(P_j)\, y(P_j) \pm \sum_{p=1}^{A} [y(P_{J_3+2p}) - y(P_{J_3+2p-1})], & x \in \Delta_1, \\ \sum_{j=1}^{J_3} \gamma(P_j)\, y(P_j) \pm \sum_{q=1}^{B} [y(P_{J_3+2q}) - y(P_{J_3 2q-1})], & x \in \Delta_2, \end{cases}$$

(10.6.11) $$x(P_j) = x, \quad 2A = J_3 - J_1, \quad 2B = J_3 - J_2.$$

Clearly $\gamma(P_j)$ has the same value on δ_{1j} and δ_{2j} but has opposite sign on P_{J_3+2p} and P_{J_3+2p-1}. Consequently as $x \to$ a point x_0 on Δ, the second sums $\to 0$ and the first sums approach the same value. Since $\bar{y}(x)$ is obviously linear interior to Δ_1 and Δ_2, the continuity of $\bar{y}$ follows.

Since $\gamma(P_j)$ has the same value on a δ_{1j} and δ_{2j} and has the opposite sign on P_{J_3+2p} and P_{J_3+2p-1} and on P_{J_3+2q} and P_{J_3+2q-1}, we see that

(10.6.12) $$\begin{aligned} \sum_{x(P_j)=x} \gamma(P_j) &= \sum_{j=1}^{J_3} \gamma(P_j), \quad x \in \Delta_1 \\ \sum_{x(P_j)=x} \gamma(P_j) &= \sum_{j=1}^{J_3} \gamma(P_j), \quad x \in \Delta_2. \end{aligned}$$

It follows that the sum in (10.6.10) has the same value $\varkappa$ on Δ_1 and Δ_2 and hence for almost all x in R_ν. From the hypothesis (10.6.1) (which holds for Y_n on account of the uniform convergence of the derivatives) it follows from Lemma 10.4.7, etc., that $\varkappa = \pm 1$.

Remark. *Hereafter we assume that Y (and the Y_n) are oriented so that*

(10.6.10′) $$\sum_{x(P)=x} \gamma(P) = +1 \quad \text{a.e.}$$

For Y a polyhedral cone as above we define $u(0) = 0$ and

(10.6.13) $$u(x) = \int_{B(x, \varrho|x|)} \varphi^*_{\varrho|x|}(t - x)\, \bar{y}(t)\, dt$$

where φ is a non-negative mollifier $\in C_c^\infty[B(0, 1)] \subset C_c^\infty(R_\nu)$ and

(10.6.14) $$\varphi^*_\varrho(y) = \varrho^{-\nu} \varphi(\varrho^{-1} y)$$

(see Theorem 3.1.3). If we represent Y by

(10.6.15) $$x = X(s), \quad y = Y(s), \quad s \in \mathfrak{R} = R_1^+ \times \mathfrak{M}_1,$$

we see from (10.6.9) and the definition of γ that

(10.6.16) $$u(x) = \int_{\mathfrak{R}} \varphi^*_{\varrho|x|}[X(s) - x]\, Y(s)\, dX(s).$$

In the form (10.6.16), we see that as $Y_n \to Y$, $u_n \to u$ uniformly with all its derivatives on any bounded closed set in R_ν which does not contain the origin. Clearly u is homogeneous of the first degree in x and $\in C^\infty(R_\nu - \{0\})$.

For each Y_n and for Y, we construct our desired

$$Y_n^* = V_{n1} \cup V_{n2} \cup U_{n3} \cup V_{n4}, \quad Y^* = V_1 \cup V_2 \cup V_3 \cup V_4$$

where V_1 is the part of Y_1 outside the cylinder $|x| = 1/2$ and V_2 can be represented on $[1/4, 1/2] \times \mathfrak{M}_1$ by

$$(10.6.17) \qquad V_2: \begin{cases} x = r\,\xi(p), & 1/4 \le r \le 1/2, \quad p \in \mathfrak{M}_1 \\ y = Z(r, p) = (4r - 1)\, r\,\eta(p) + (2 - 4r)\, u[r\,\xi(p)]. \end{cases}$$

It is shown in Lemma 10.6.12 below that if $|\nabla u|$ is sufficiently small (this is shown in Lemma 10.6.8 below) then there is a nearby Cartesian coordinate system $('x, 'y)$, which depends smoothly on ∇u (evaluated on $|x| = 1$, say, since u is homogeneous of the first degree) such that if U denotes the cone

$$(10.6.18) \qquad U: y = u(x) \quad \text{or} \quad 'y = 'u('x)$$

then

$$\int_{B_3} \frac{\partial\, 'u^k}{\partial\, 'x^i}\, dx' = 0, \quad B_3 = B(0, 1/8), \quad 1 \le i \le \nu, \quad 1 \le k \le N - \nu.$$

(10.6.19)

Then V_3 is the part of U for which $|x'| \ge 1/8$ and $|x| \le 1/4$ and V_4 is defined by

$$(10.6.20) \qquad \begin{aligned} V_4: 'y = v('r, '\theta) &\equiv '\overline{u} + (8r')^2 ['u(1/8, '\theta) - '\overline{u}], \\ 0 \le 'r = |'x| &\le 1/8, \quad '\theta = |'x|^{-1}\, 'x. \end{aligned}$$

Corresponding formulas hold for V_{n1}, etc. The parametric representations of the V_{nj} converge uniformly on their respective domains together with their first derivatives to those of the V_j so that $\Lambda^\nu(Y_n^*) \to \Lambda^\nu(Y^*)$. It is shown in Lemma 10.6.11 below that

$$\Lambda^\nu(U_n \cap W) \le h_1(\varrho, \varepsilon) \cdot \varepsilon + \Lambda^\nu(Y_n \cap W), \quad \lim_{\varepsilon, \varrho \to 0} h_1(\varrho, \varepsilon) = 0,$$

for any bounded measurable W. Using this, it is shown in Lemma 10.6.10 below that $\Lambda^\nu(V_{n2}) \le h_2(\varrho, \varepsilon) \cdot \varepsilon + \Lambda^\nu$ (corresponding part of Y_n). Finally, in Lemma 10.6.14 below, it is shown that

$$(10.6.21) \quad \Lambda^\nu(V_{n4}) \le \frac{4\nu}{4\nu + 1} \Lambda^\nu \text{ (corresponding part of } U_n).$$

Thus, whereas replacing the part of Y_{n1} outside $|'x| = 1/8$ by $V_{n1} \cup V_{n2} \cup V_{n3}$ may increase the area by $h(\varrho, \varepsilon) \cdot \varepsilon$ where $h(\varrho, \varepsilon) \to 0$ as $\varrho, \varepsilon \to 0$, replacing the part of Y_n inside $|'x| = 1/8$ reduces the area by an amount $\lambda(\nu, N) \cdot \varepsilon$, $\lambda > 0$. This leads to the theorem.

However, $W_1 \equiv V_3 \cup V_4$ is evidently a ν-cell with boundary

$$(10.6.22) \qquad C_1: y = u(x), \quad |x| = 1/4,$$

and $W_0 = V_1 \cup V_2$ may be represented on $[0, 1] \times \mathfrak{M}_1$ by equations like

$$(10.6.23) \quad W_0: x = X(\sigma, p), \quad y = Y(\sigma, p), \quad \sigma \in [0, 1], \quad p \in \mathfrak{M}_1$$

in such a way that $\sigma = 0$ yields a homeomorphism of $\mathfrak{M}_1$ onto $l(r, P_0, z) \equiv C_0$ and $\sigma = 1$ yields a mapping onto C_1 which, however, need not be a homeomorphism. We wish to show that

$$b(Y^*, C_0) \supset H_{\nu-1}(C_0), \quad Y^* = W_0 \cup W_1. \tag{10.6.24}$$

To do this, we first apply Theorems 10.3.9 and 10.3.10 with

$$X = Y^*, \quad X_1 = W_0, \quad X_2 = W_1, \quad A_1 = C_0 \cup C_1, \quad A_2 = C_1,$$
$$A = C_0, \quad B = A_1.$$

Those theorems yield the result (since $b(W_1, C_1) = H_{\nu-1}(C_1)$ by Theorem 10.3.4)

$$\begin{aligned} i(C_0 \cup C_1, C_0)_* \, b(Y^*, C_0) &= b(W_0, C_0 \cup C_1) + \\ &+ i(C_0 \cup C_1, C_1)_* \, H_{\nu-1}(C_1). \end{aligned} \tag{10.6.25}$$

If we now apply Theorem 10.3.13 with

$$X' = I \times \mathfrak{M}, \quad A'_0 = 0 \times \mathfrak{M}_1, \quad A'_1 = 1 \times \mathfrak{M}_1, \quad A' = A'_0 \cup A'_1,$$
$$A_0 = C_0, \quad A_1 = C_1, \quad X = W_0,$$

the mapping $f: X' \to W_0$ being given by (10.6.23), we conclude that if $L_0 = H_{\nu-1}(C_0)$ there is a subgroup L_1 of $H_{\nu-1}(C_1)$ such that

$$K + i(C_0 \cup C_1, C_0)_* \, L_0 = K + i(C_0 \cup C_1, C_1)_* \, L_1,$$
$$K = b(W_0, C_0 \cup C_1).$$

That is, if $h \in H_{\nu-1}(C_0)$ and $k_1 \in K$, there exists a k_2 in K and an l in $H_{\nu-1}(C_1)$ such that

$$i(C_0 \cup C_1, C_0)_* \, h = (k_2 - k_1) + i(C_0 \cup C_1, C_1)_* \, l$$

which proves (10.6.24) by showing (cf. (10.6.25)) that

$$i(C_0 \cup C_1, C_0)_* \, H_{\nu-1}(C_0) \subset b(W_0, C_0 \cup C_1) + i(C_0 \cup C_1, C_1)_* \, H_{\nu-1}(C_1)$$

the right side being that of (10.6.25). Of course this is also true for each n.

We now prove the inequalities which we have mentioned for a polyhedral cone Y_1. We use the notation of (10.6.5)—(10.6.7).

Lemma 10.6.2. (a) $\sqrt{1 + a_1^2 + \cdots + a_n^2} - 1 \leq \sum_{i=1}^{n} \left[\sqrt{1 + a_i^2} - 1\right]$;

(b) $\sqrt{1 + [\lambda a + (1 - \lambda) b]^2} \leq \lambda \sqrt{1 + a^2} + (1 - \lambda) \sqrt{1 + b^2}$, $0 \leq \lambda \leq 1$,

(c) $\sqrt{1 + \lambda^2 a^2} - 1 \leq \lambda \left[\sqrt{1 + a^2} - 1\right]$, $0 \leq \lambda \leq 1$.

(d) $\sqrt{1 + (a_1 + \cdots + a_n)^2} \leq \sum_{i=1}^{n} \sqrt{1 + a_i^2}$.

Proof. (a) For $A \geq 0$ and $B \geq 0$, we define

$$f(A, B) = (1 + A)^{1/2} + (1 + B)^{1/2} - (1 + A + B)^{1/2} - 1$$

and show by differentiation that $f_A \geq 0$ so $f \geq 0$. The result for more terms follows by induction. (b) just follows from the fact that $\sqrt{1 + x^2}$

is a convex function. (c) follows by setting $b = 0$ in (b). To prove (d) set

$$f(a_1, \ldots, a_n) = \sum_i \sqrt{1 + a_i^2} - \sqrt{1 + (a_1 + \cdots + a_n)^2}.$$

It follows easily that each $f_{a_i} \leq 0$ and that $f(a_1, \ldots, a_n) \to 0$ as all $a_i \to +\infty$.

Lemma 10.6.3. *Suppose f is homogeneous of degree zero. Then*

$$\int_{B_1} \left\{ \int_{B(x, \varrho|x|)} \left[\int_0^1 \varrho^{-\nu} |x|^{-\nu} |f[(1-\lambda)x + \lambda t]| \, d\lambda \right] dt \right\} dx \leq C_{10}(\nu) \int_{B_1} |f(x)| \, dx.$$

Proof. Let $g(x, \tau, \lambda) = |f(x + \lambda \varrho |x| \tau)|$. Then the integral above is

$$\int_{B_1} \left\{ \int_{B_1} \left[\int_0^1 g(x, \tau, \lambda) \, d\lambda \right] d\tau \right\} dx$$

$$= \int_{B_1} \left\{ \int_0^1 \left[\int_{G(\tau,\lambda)} |f(\eta)| \, d\eta \right] d\lambda \right\} d\tau, \quad G(\tau, \lambda) \subset B(0, 1 + \lambda \varrho),$$

where, for τ and λ fixed, $G(\tau, \lambda)$ is the image of B_1 under the $1-1$ transformation

$$\eta = x + \lambda \varrho \tau |x|.$$

That this transformation is $1-1$ follows since

$$|x_2 - x_1| \cdot (1 - \lambda \varrho |\tau|) \leq |\eta_2 - \eta_1| \leq |x_2 - x_1| \cdot (1 + \lambda \varrho |\tau|).$$

The result follows easily.

Lemma 10.6.4. *Let us define (almost everywhere)*

$$\nu(x) = \sum_{x(P) = x} |\gamma(P)|,$$

define X_1 as the subset of B_1 where $\nu(x) = 1$ and X_2 as the subset of B_1 where $\nu(x) > 1$. Then

$$\int_{X_2} \nu(x) \, dx \leq 2\varepsilon, \quad \int_{B_1} [\nu(x) - 1] \, dx = \int_{X_2} [\nu(x) - 1] \, dx \leq \varepsilon.$$

Proof. Clearly

$$\gamma_\nu + \varepsilon = \Lambda^\nu(\tilde{Y}_1) \geq \int_{B_1} \nu(x) \, dx = \gamma_\nu + \int_{X_2} [\nu(x) - 1] \, dx.$$

The result follows from the fact that

$$\nu(x) \leq 2[\nu(x) - 1] \quad \text{for} \quad \nu(x) \geq 2.$$

Lemma 10.6.5. *If $|x| \leq 1$, then $|u(x)| \leq C_{11}(\nu, N) \varrho$ (if $\varrho \leq 1/2$).*

Proof. From the homogeneity, it follows that is is sufficient to prove this for $|x| = 1 - \varrho$, if $\varrho \leq 1/2$. From (10.6.9) and the definition of $\nu(x)$ we see that

$$|\bar{y}(t)| \leq \varrho \, \nu(t). \tag{10.6.26}$$

From (10.6.13), it then follows that

$$|u(x)| \leq \varrho + Z_1 \varrho^{-\nu} \int_{X_2} \varrho \cdot [\nu(t) - 1] \, dt \leq \varrho [1 + Z_1 \varrho^{-\nu} \varepsilon]$$

from which the result follows, since $\varepsilon < \varrho^{6\nu}$. Z_1 depends also on the mollifier φ.

Lemma 10.6.6. *Suppose μ is a measure over a set E, $\mu(E)$ is finite, and f is non-negative and integrable (μ) over E. Then*

$$\min\left\{[\mu(E)]^2, \left[\int_E f\,d\mu\right]^2\right\} \leq 3[\mu(E)]\cdot\int_E\left(\sqrt{1+f^2}-1\right)d\mu.$$

Proof. From the Schwarz inequality with

$$f = \left(f/\sqrt{\sqrt{1+f^2}+1}\right)\cdot\sqrt{\sqrt{1+f^2}+1}, \text{ we obtain}$$

$$\left[\int_E f\,d\mu\right]^2 \leq \left[\int_E\left(\sqrt{1+f^2}-1\right)d\mu\right]\cdot\left[\int_E\left(\sqrt{1+f^2}-1+2\right)d\mu\right]$$
$$\leq A^2+2AB, \quad A = \int_E\left(\sqrt{1+f^2}-1\right)d\mu, \quad B=\mu(E).$$

If $A^2+2AB \leq 3AB$, the lemma is true. If $A^2+2AB > 3AB$, then $A > B$ and $B^2 < 3AB$.

Lemma 10.6.7. $\int_{\bar{Y}_1}|u[x(P)]-y(P)|^2\,d\sum(P) \leq C_{12}(\nu,N)\cdot\varrho^2\varepsilon.$

Proof. Let us denote the integral by I. Then

$$(10.6.27)\qquad \begin{aligned} I &\leq I_1+\int_{X_2}(C_{10}+1)^2\varrho^2\nu(x)\,dx \leq I_1+Z_2(\nu,N)\,\varrho^2\,\varepsilon,\\ I_1 &= \int_{X_1}|u(x)-\bar{y}(x)|^2\,dx,\end{aligned}$$

using Lemmas 10.6.4 and 10.6.5 and (10.6.26). Now if $x\in X_1$

$$|B(x,\varrho|x|)|\cdot|u(x)-\bar{y}(x)|$$
$$\leq\begin{cases} Z_3\cdot\varrho|x|\cdot\int_{B(x,\varrho|x|)}\left\{\int_0^1|D\bar{y}[(1-\lambda)x+\lambda t]\,d\lambda\right\}dt\\ Z_3(\varphi,\nu,N)\cdot\varrho|x|\cdot|B(x,\varrho|x|)|,\end{cases}$$

on account of Lemma 10.6.5. Using Lemma 10.6.6 with E the set (t,λ) with $t\in B(x,\varrho|x|)$ and $0\leq\lambda\leq 1$ and $d\mu = dt\,d\lambda$, we obtain

$$(10.6.28)\qquad \begin{aligned} &|B(x,\varrho|x|)|\cdot|u(x)-\bar{y}(x)|^2\\ &\leq 3Z_3^2\varrho^2|x|^2\int_{B(x,\varrho|x|)}\left\{\int_0^1\left[\sqrt{1+|D\bar{y}[(1-\lambda)x+\lambda t]|^2}-1\right]d\lambda\right\}dt.\end{aligned}$$

Using Lemma 10.6.3, we therefore conclude that

$$(10.6.29)\qquad \int_{X_1}|u(x)-\bar{y}(x)|^2\,dx \leq Z_4\,\varrho^2\int_{B_1}\left(\sqrt{1+|D\bar{y}(x)|^2}-1\right)dx.$$

But, from (10.6.9) it follows that

$$(10.6.30)\qquad |D\bar{y}(x)| \leq \sum_{x(P)=x}|Dy(P)|.$$

Thus, from Lemma 10.6.2 (d), we conclude that

$$\int_{B_1} \left[\sqrt{1 + D\bar{y}^2} - 1\right] dx \leq \int_{\tilde{Y}_1} \sqrt{1 + J\, y^2}\, d\Sigma(P) - \gamma_\nu = \Lambda^\nu(\tilde{Y}_1) - \gamma_\nu = \varepsilon.$$

(10.6.31)

The result follows from (10.6.27)—(10.6.31).

Lemma 10.6.8. (i) $|D u(x)| \leq C_{13}(\nu, N)\, \varepsilon^{1/3}$, if $\varrho \leq 1/2$;

(ii) $|u(x)| \leq |x| \cdot |D u(x)|$.

Proof. (ii) follows from (i) and the homogeneity. From (10.6.13), it follows that

$$u(x) = \int_{B_1} \varphi(\lambda)\, \bar{y}(x + \lambda \varrho\, |x|)\, d\lambda \quad (\lambda = (\varrho\, |x|)^{-1}(t - x))$$

so that

(10.6.32) $$\frac{\partial u^k}{\partial x^\alpha} = \int_{B_1} \varphi(\lambda) \left[\frac{\partial \bar{y}^k}{\partial x^\alpha} + \varrho \sum_{\beta=1}^\nu \lambda^\beta \frac{\partial \bar{y}^k}{\partial x^\beta} |x|^{-1} x^\alpha\right] d\lambda$$

from which it follows, by setting $t = x + \varrho\, |x|\, \lambda$, that

(10.6.33) $$|D u(x)| \leq (1 + Z_1 \varrho) \int_{B(x, \varrho|x|)} \varphi^*_{\varrho|x|}(t - x)\, |D \bar{y}(t)|\, dt.$$

Since $D u$ and $D \bar{y}$ are homogeneous of degree 0, it is sufficient to prove (i) for $|x| = 1 - \varrho$. Then, since $B(x, \varrho\, |x|) \subset B_1$, we conclude from the Schwarz inequality as in the proof of Lemma 10.6.6 that

(10.6.34) $$\left[\int_{B(x, \varrho|x|)} \varphi^*_{\varrho|x|}(t - x)\, |D \bar{y}(t)|\, dt\right]^2$$
$$\leq \left[\int_{B(x, \varrho|x|)} \varphi^*_{\varrho|x|}(t - x)\left(\sqrt{1 + |D \bar{y}(t)|^2} - 1\right) dt\right] \times$$
$$\times \left[\int_{B(x, \varrho|x|)} \varphi^*_{\varrho|x|}(t - x)\left(\sqrt{1 + |D \bar{y}(t)|^2} - 1 + 2\right) dt\right]$$
$$\leq Z \varrho^{-\nu} \varepsilon (2 + Z \varrho^{-\nu} \varepsilon).$$

The result follows from (10.6.34), (10.6.8), and (10.6.31).

Lemma 10.6.9. $\int_B \sqrt{1 + |D u(x)|^2}\, \nu(x)\, dx \leq \int_B \sum_{x(P)=x} \sqrt{1 + |D y(P)|^2} \times$
$$\times dx + C \varrho \varepsilon, \quad B = B_a - B_b, \quad 0 < b < a \leq 1.$$

Proof. Since $D u$, ν, and $D y$ are homogeneous of degree 0, it is sufficient to prove this with B replaced by B_1.

From (10.6.33) and the definition of $\bar{y}$, we obtain

$$|D u(X)| \leq (1 + Z_1 \varrho) \int_{Y(X, \varrho|X|)} |D y(P)|\, d\mu(P; X)$$

(10.6.35) $$= (1 + Z_1 \varrho) \int_{B(X, \varrho|X|)} \varphi^*_{\varrho|X|}(t - X) \sum_{x(P)=t} |D y(P)|\, dt$$

$$(d\mu(P; X) = \varphi^*_{\varrho|X|}[x(P) - X]\, d\Sigma(P))$$

where $Y(X, \varrho\, |X|)$ consists of all P on Y such that $x(P) \in B(X, \varrho\, |X|)$.

Using the device of (10.6.34) with $B(X, \varrho|X|)$ replaced by $Y(X, \varrho|X|)$ and $|D\bar{y}(t)|$ replaced by $|Dy(P)|$, we find that

$$\int_{B_1} \left(\sqrt{1+|Du|^2}-1\right)dx \leq \frac{1}{2}\int_{B_1} |Du(X)|^2\,dX$$

$$\leq \frac{1}{2}(1+Z_1\varrho)^2 \int_{B_1} w(X)\,[w(X)+2\omega(X)]\,dX,$$

$$w(X) = \int_{B(X,\varrho|X|)} \varphi^*_{\varrho|X|}(t-X)\left\{\sum_{x(P)=t}\left[\sqrt{1+|Dy(P)|^2}-1\right]\right\}dt \tag{10.6.36}$$

$$\omega(X) = \int_{B(X,\varrho|X|)} \varphi^*_{\varrho|X|}(t-X)\,\nu(t)\,dt$$

$$= 1 + \int_{B(X,\varrho|X|)} \varphi^*_{\varrho|X|}(t-X)\,[\nu(t)-1]\,dt.$$

Clearly w and ω are homogeneous of degree 0 so we may estimate them assuming that $|X| = 1-\varrho$. Then, since $|Dy(P)| \leq |J_y(P)|$, we see first, using (10.6.36). Lemma 10.6.4, and (10.6.8), that

$$\begin{aligned} &w(X) \leq Z_2\varrho^{-\nu}\varepsilon, \quad \omega(X) \leq 1 + Z_2\varrho^{-\nu}\cdot 2\varepsilon \\ &w(X) + 2\omega(X) \leq 2(1+Z_3\varrho) \end{aligned} \tag{10.6.37}$$

so that

$$I \equiv \int_{B_1} \left(\sqrt{1+|Du|^2}-1\right)dx \leq (1+Z_4\varrho)\int_{B_1} w(X)\,dX. \tag{10.6.38}$$

If, in (10.6.36), we set $t = X + \varrho|X|\tau$, integrate, and change the order of integration, we obtain

$$\int_{B_1} w(X)\,dX = \int_{B_1} \varphi(\tau)\left\{\int_{B_1}\left\langle \sum_{x(P)=X+\varrho|X|\tau}\left[\sqrt{1+|Dy(P)|^2}-1\right]\right\rangle dX\right\}d\tau. \tag{10.6.39}$$

If we set, for each τ in B_1, $\eta = X + \varrho|X|\tau$, we see that the transformation and its inverse are Lipschitz and the Jacobian $|dX/d\eta| \leq 1 + Z_5\varrho$; moreover, for each τ the image of $B_1 \subset B_{1+\varrho}$. Accordingly we obtain (since Dy is homogeneous of degree 0)

$$I \leq (1+Z_6\varrho)\int_{B_1} \sum_{x(P)=\eta}\left[\sqrt{1+|Dy(P)|^2}-1\right]d\eta. \tag{10.6.40}$$

Now, using Lemma 10.6.8, we may write

$$\int_{B_1} \sqrt{1+|Du(x)|^2}\,\nu(x)\,dx \leq \int_{B_1} \sqrt{1+|Du(x)|^2}\,dx +$$

$$+ (1+Z_7\varrho)\int_{B_1} [\nu(x)-1]\,dx \leq \gamma_\nu + \int_{B_1}\left\{\sum_{x(P)=\eta}\sqrt{1+|Dy(P)|^2}\right\} \times$$

$$\times\, d\eta - \int_{B_1} \nu(x)\,dx + \int_{B_1} [\nu(x)-1]\,dx +$$

$$+ Z_6\varrho\int_{B_1}\left\{\sum_{x(P)=\eta}\left[\sqrt{1+|Dy(P)|^2}-1\right]\right\}d\eta + Z_7\varrho\int_{B_1}[\nu(x)-1]\,dx$$

from which the result follows ($|Dy(P)| \leq |Jy(P)|$, Lemma 10.6.4).

Lemma 10.6.10. *Suppose $Z = V_2$ and is defined by* (10.6.17). *Then*

$$\Lambda^\nu(Z) \le \Lambda^\nu(Y_2) + \varphi(\varrho, \varepsilon) \cdot \varepsilon, \quad \lim_{\varrho, \varepsilon \to 0} \varphi(\varrho, \varepsilon) = 0$$

where Y_2 is the part of Y for which $1/4 \le |x| \le 1/2$.

Proof. On each simplicial cone of Y, we may consider y and z as function of x. From (10.6.17) and the notation of (10.6.5), etc., we obtain

$$z(x) = (4r - 1)\, y(x) + (2 - 4r)\, u(x)$$

$$\frac{\partial z^k}{\partial x^i} = (4r - 1) \frac{\partial y^k}{\partial x^i} + (2 - 4r) \frac{\partial u^k}{\partial x^i} + d^k_{1i}, \; d^k_{1i} = 4(y^k - u^k)(x^i/r)$$

$$\frac{\partial(x^k, x^l)}{\partial(x^i, x^j)} = (4r - 1)^2 \frac{\partial(y^k, y^l)}{\partial(x^i, x^j)} + d^{kl}_{2ij},$$

$$\frac{\partial(z^{k_1}, z^{k_2}, z^{k_3})}{\partial(x^{i_1}, x^{i_2}, x^{i_3})} = (4r - 1)^3 \frac{\partial(y^{k_1}, y^{k_2}, y^{k_3})}{\partial(x^{i_1}, x^{i_2}, x^{i_3})} + d^k_{3I}, \ldots$$

where each d_p is a sum of Jacobians, of order $< p$, of the y^k with respect to the x^i multiplied by products of components of $D u$ and $y - u$. Regarding the Jacobians, etc., as components of a vector, we obtain

$$|J z(P)| \le (4r - 1)\, |J y(P)| + (2 - 4r)\, |D u[x(P)]| + |d(P)|.$$

Then, for each $\eta > 0$, we have

$$|J z(P)|^2 \le (1 + \eta)\, \{(4r - 1)\, |J y(P)| + (2 - 4r)\, |D u[x(P)]|\}^2 + \\ + (1 + \eta^{-1})\, |d(P)|^2.$$

Hence (refer to Lemma 10.6.2)

$$\int_{Y_2} \left[\sqrt{1 + |J z(P)|^2} - 1\right] d\Sigma \le \int_{Y_2} \left[\sqrt{1 + (1 + \eta^{-1})\, |d(P)|^2} - 1\right] d\Sigma +$$

$$+ \int_{Y_2} \left[\sqrt{1 + (1 + \eta) \{(4r - 1)\, |J y(P)| + (2 - 4r)\, |D u[x(P)]|\}^2} - 1\right] d\Sigma$$

$$\le (1 + \eta^{-1})^{1/2} \int_{Y_2} \left[\sqrt{1 + |d(P)|^2} - 1\right] d\Sigma +$$

$$+ (1 + \eta)^{1/2} \int_{Y_2} \left[\sqrt{1 + \{(4r - 1)\, |J y(P)| + (2 - 4r)\, |D u[x(P)]|\}^2} - 1\right] d\Sigma$$

$$\le (1 + \eta^{-1})^{1/2} \int_{Y_2} \left[\sqrt{1 + |d(P)|^2} - 1\right] d\Sigma +$$

$$+ [(1 + \eta)^{1/2} - 1] \left\{(4r - 1) \int_{Y_2} \left[\sqrt{1 + |J y(P)|^2} - 1\right] d\Sigma + \right.$$

$$\left. + (2 - 4r) \int_{Y_2} \sqrt{\left[1 + |D u[x(P)]|^2\right.} - 1\Big] d\Sigma \right\} +$$

$$+ (4r - 1) \int_{Y_2} \left[\sqrt{1 + |J y(P)|^2} - 1\right] d\Sigma +$$

$$+ (2 - 4r) \int_{Y_2} \left[\sqrt{1 + |D u[x(P)]|^2} - 1\right] d\Sigma.$$

$$\therefore \int_{Y_2} \sqrt{1 + |J z(P)|^2}\, d\Sigma \le (4r - 1) \int_{Y_2} \sqrt{1 + |J y(P)|^2}\, d\Sigma +$$

$$+ (2 - 4r) \int_{Y_2} \sqrt{1 + |D u[x(P)]|^2}\, d\Sigma + \varepsilon \cdot \varphi_1(\varrho, \varepsilon)$$

from which the result follows, using Lemma 10.6.9.

Lemma 10.6.11. *If U is the cone given by* (10.6.18), *then*

$$\Lambda^\nu(U \cap W) \le \Lambda^\nu(Y \cap W) + h(\varrho, \varepsilon) \cdot \varepsilon, \quad \lim_{\varrho, \varepsilon \to 0} h(\varrho, \varepsilon) = 0$$

for any bounded Borel set W.

Proof. It is sufficient to prove this for W the cylinder $|x| \le 1$. Since $|J u|^2 = |D u|^2$ plus the squares of all the Jacobians of the u^i with respect to the x^j of order two or higher, we see from Lemma 10.6.8 that

$$\Lambda^\nu(U_1) = \int_{B_1} \sqrt{1 + |J u(x)|^2}\, dx \le \int_{B_1} \sqrt{1 + |D u(x)|^2}\, dx + Z \varepsilon^{4/3}$$

$$\le \int_{B_1} \sum_{x(P)=x} \sqrt{1 + |D y(P)|^2}\, dx + Z \varepsilon^{4/3} + C \varrho \varepsilon \le \Lambda^\nu(Y_1) + Z \varepsilon^{4/3} + C \varrho \varepsilon$$

using Lemma 10.6.9.

Let us now consider the cone U given in the (x, y) system by (10.6.18) where $u \in C^\infty(R_\nu - \{0\}$ and u is homogeneous of the first degree. We wish to prove the following lemma:

Lemma 10.6.12. *If $|D u(x)|$ is sufficiently small, there is a nearby $('x, 'y)$ Cartesian coordinate system which depends smoothly on $\nabla u(x)$, evaluated on $|x| = 1$, such that*

$$\int_{B_1} \frac{\partial' u^i}{\partial' x^k}\, dx' = 0, \quad i = 1, \ldots p = N - \nu, \quad k = 1, \ldots, \nu. \tag{10.6.41}$$

Remark. It is clear that (10.6.41) will then hold with B_1 replaced by $B_a = B(0, a)$ for any $a > 0$.

Proof. Suppose we let

$$'x^k = c^k_l x^l + d^k_r y^r, \quad 'u^i = e^i_l x^l + f^i_r y^r \tag{10.6.42}$$

be a positive orthogonal transformation. Then U is given parametrically in the $('x, 'y)$-system by

$$'x^k = c^k_l x^l + d^k_r u^r(x), \quad 'u^i = e^i_l x^l + f^i_r u^r(x). \tag{10.6.43}$$

Then

$$\frac{\partial' u^i}{\partial' x^k} = \sum_l (-1)^{k+l} \frac{\partial' u^i}{\partial x} \cdot \frac{\partial('x^1, \ldots, 'x^{k-1}, 'x^{k+1}, \ldots, 'x^\nu)}{\partial(x^1, \ldots, x^{l-1}, x^{l+1}, \ldots, x^\nu)} \cdot \left(\frac{\partial x'}{\partial x}\right)^{-1}. \tag{10.6.44}$$

Thus, the equations (10.6.41) become

$$\begin{aligned} &\int_B \frac{\partial' u^i}{\partial' x^k}\, dx' \\ &= \int_{B^*} \sum (-1)^{k+l} \frac{\partial' u^i}{\partial' x^l} \cdot \frac{\partial('x^1, \ldots, 'x^{k-1}, 'x^{k+1}, \ldots, 'x^\nu)}{\partial(x^1, \ldots, x^{l-1}, x^{l+1}, \ldots, x^\nu)}\, dx = 0 \end{aligned} \tag{10.6.45}$$

where the derivatives $\partial' u^i/\partial x^l$, $\partial' x^k/\partial x^l$, etc., may be found from (10.6.43) and B^* is the region in the x-space correspoinding to the unit sphere B in the $'x$-space by the first equations in (10.6.43). To determine the matrices c, d, e, and f, we have the $N(N+1)/2$ equations which require the matrix

$$\begin{pmatrix} c & d \\ e & f \end{pmatrix} \quad (N = \nu + p) \tag{10.6.46}$$

to be orthogonal. Next we have the νp equations (10.6.45), which are of the form

$$\mathfrak{F}_k^i(c, d, e, f; u) = 0, \quad i = 1, \ldots, p, \quad k = 1, \ldots, \nu \tag{10.6.47}$$

where $\mathfrak{F}_k^i$ is a polynomial in the c, d, e, f with coefficients which are integrals over B^* of various jacobians of the u^i with respect to the x^j. This leaves

$$\frac{(\nu + p)^2 - \nu - p}{2} - \nu p = \frac{\nu^2 - \nu}{2} + \frac{p^2 - p}{2}$$

equations which we can use to prescribe the c_l^k with $l > k$ and the f_j^i with $j > i$.

By replacing u by $u + \lambda v$, it is easy to see that the Fréchet differentials

$$\begin{aligned} &\mathfrak{F}_k^i(c, d, e, f; u, v) \\ &= \lim_{\lambda \to 0} \lambda^{-1} [\mathfrak{F}_k^i(c, d, e, f; u + \lambda v) - \mathfrak{F}_k^i(c, \ldots, f; u)] \end{aligned} \tag{10.6.48}$$

are continuous for $\| u \|$ sufficiently small, where

$$\| u \| = \max_{|x|=1} |D u(x)|. \tag{10.6.49}$$

We have already noted the dependence of $\mathfrak{F}_k^i$ on (c, d, e, f). The integrand in (10.6.45) is

$$\det \begin{pmatrix} 1 + c_1^1 + d_j^1 u^j_{,1} & c_1^2 + d_j^2 u^j_{,1} \ldots e_1^i + (\delta_j^i + f_j^i) u^j_{,1} \ldots & c_1^\nu + d_j^\nu u^j_{,1} \\ c_2^1 + d_j^1 u^j_{,2} & 1 + c_2^2 + d_j^2 u^j_{,2} \ldots e^i + (\delta_j^i + f_j^i) u^j_{,2} \ldots & c_2^\nu + d_j^\nu u^j_{,2} \\ \cdot & \cdot \quad \cdots \quad \cdot \quad \cdots & \cdot \\ c_\nu^1 + d_j^1 u^j_{,\nu} & c_\nu^2 + d_j^2 u^j_{,\nu} \ldots e_\nu^i + (\delta_j^i + f_j^i) u^j_{,\nu} \ldots & 1 + c_\nu^\nu + d_j^\nu u^j_{,\nu} \end{pmatrix}; \tag{10.6.50}$$

we have replaced c_k^k by $1 + c_k^k$ and f_j^i by $\delta_j^i + f_j^i$.

If we now regard all the c, d, e, f as small and neglect products of two or more of them and two or more of the u-derivatives, the determinant in (10.6.50) reduces to that of the matrix

$$(e_k^i + u^i_{,k} + f_j^i u^j_{,k}). \tag{10.6.51}$$

If we also replace B^* by B and define

$$U_k^i = |B|^{-1} \int_B u^i_{,k}(x)\, dx \tag{10.6.52}$$

the equations for the c, d, e, f become

$$\begin{aligned} &e_k^i + U_k^i + f_j^i U_k^j = 0, \quad c_k^k = f_i^i = 0 \\ &c_k^l = -c_l^k \ \text{ if } \ k \neq l, \quad f_i^j = -f_j^i \ \text{ if } \ j \neq i \\ &e_k^i = -d_i^k, \quad i, j = 1, \ldots, p, \quad k, l = 1, \ldots, \nu. \end{aligned} \tag{10.6.53}$$

It is clear that we may prescribe the f_j^i with $j > i$ and the c_l^k with $l > k$ and the remaining c_l^k and f_j^i are determined. Then the e_k^i are determined from the U_j^k and f_j^i and then the d_i^k are determined. Since the equations (10.6.53) are obtained by linearizing the actual equations, and since the derivatives and FRÉCHET differentials are continuous, it follows that the equations (10.6.47) together with the orthogonality relations yield a unique solution in which the matrix (10.6.46) is sufficiently near the identity provided that the prescribed e_l^k and f_j^i are sufficiently near 0 as is $\|u\|$.

We now prove:

Lemma 10.6.13. *Suppose that $u \in C^1(R_\nu - \{0\})$ and is homogeneous of degree 1. Suppose also that*

$$(10.6.54) \qquad \int_B u^i_{,k}(x)\,dx = 0, \quad i = 1, \ldots, p, \quad k = 1, \ldots, \nu.$$

Let v be defined by

$$(10.6.55) \qquad \begin{aligned} v(r, \theta) &= \bar{u}(0) + r^2[u(1, \theta) - \bar{u}(0)], \\ \bar{u}(0) &= |\Sigma|^{-1} \int_\Sigma u(1, \theta)\, d\Sigma(\theta). \end{aligned}$$

Then

$$(10.6.56) \qquad \int_B |Dv|^2\,dx \le \frac{2\nu}{2\nu + 1} \int_B |Du|^2\,dx.$$

Proof. $v_r = 2r[u(1, \theta) - \bar{u}(p)]$

$$(10.6.57) \qquad \begin{aligned} r^{-1}\nabla_\theta v(r, \theta) &= r\nabla_\theta u(1, \theta) \\ |Dv|^2 &= 4r^2|u(1, \theta) - \bar{u}(0)|^2 + r^2|\nabla_\theta u(1, \theta)|^2. \end{aligned}$$

From homogeneity,

$$u(r, \theta) = r\,u(1, \theta), \quad u_r(r, \theta) = u(1, \theta), \quad r^{-1}\nabla_\theta u(r, \theta) = \nabla_\theta u(1, \theta).$$

Hence

$$\int_B |Dv|^2\,dx = \frac{1}{\nu + 2} \int_\Sigma \{4|u(1, \theta) - \bar{u}(0)|^2 + |\nabla_\theta u(1, \theta)|^2\}\,d\Sigma(\theta)$$

$$\int_B |Du|^2\,dx = \frac{1}{\nu} \int_\Sigma \{|u(1, \theta)|^2 + |\nabla_\theta u(1, \theta)|^2\}\,d\Sigma(\theta).$$

If we now introduce the spherical harmonics q_{nk} (see after Lemma 10.6.14 below) with

$$q_{01} = |\Sigma|^{-1/2}, \quad q_{1k} = C_{\nu k} \cdot |x|^{-1} x^k, \quad k = 1, \ldots, \nu;$$

our hypotheses and definitions imply that

$$u^i(1, \theta) = \bar{u}^i(0) + \sum_{\substack{n, k \\ n \ge 2}} a^i_{nk}\, q_{nk}(\theta).$$

It follows (see below) that

$$\int_B |Du|^2\,dx = \frac{1}{\nu}\Big\{|\Sigma|\cdot|\bar u(0)|^2 + \sum_{\substack{n,k\\ n\ge 2}} |a_{nk}|^2[1 + n(n+\nu-2)]\Big\}$$

$$\int_B |Dv|^2\,dx = \frac{1}{\nu+2}\Big\{\sum_{\substack{n,k\\ n\ge 2}} |a_{nk}|^2[4 + n(n+\nu-2)]\Big\}$$

$$= \frac{1}{\nu+2}\Big\{\sum_{\substack{n,k\\ n\ge 2}} |a_{nk}|^2[1 + n(n+\nu-2)]\Big[1 + \frac{3}{1+n(n+\nu-2)}\Big]\Big\}$$

$$\le \frac{\nu}{(\nu+2)}\Big[1 + \frac{3}{2\nu+1}\Big]\cdot\int_B |Du|^2\,dx = \frac{2\nu}{2\nu+1}\int_B |Du|^2\,dx.$$

Lemma 10.6.14. *There is a number $\eta(\nu, N) > 0$ such that if $u \in C^1(R - \{0\})$, $|Du| \le \eta$, u satisfies* (10.6.54), *and v is defined by* (10.6.55) *then*

$$\int_B \big(\sqrt{1+|Jv|^2} - 1\big)\,dx \le \frac{4\nu}{4\nu+1}\int_B \big(\sqrt{1+|Ju|^2} - 1\big)\,dx. \tag{10.6.58}$$

Proof. For if $|Du| \le \eta$, it follows from (10.6.57) that $|Dv| \le 2\eta$ on B. Thus

$$\int_B \big(\sqrt{1+|Jv|^2} - 1\big)\,dx \le \frac{1}{2}\int_B |Jv|^2\,dx \le \frac{1}{2}(1 + Z_1\eta^2)\int_B |Dv|^2\,dx$$

$$\le \frac{\nu(1+Z_1\eta^2)}{2\nu+1}\int_B |Du|^2\,dx \le \frac{2\nu(1+Z_1\eta^2)}{(2\nu+1)(1-Z_2\eta^2)}\int_B \big(\sqrt{1+|Du|^2} - 1\big)\,dx$$

since

$$\sqrt{1+\xi^2} - 1 \ge \frac{1}{2}\xi^2 - \frac{1}{8}\xi^4$$

for all ξ and $|Du| \le |Ju|$. The result (10.6.58) follows easily.

The spherical harmonics. We begin by noticing that if $n \ne p$ and H_n and H_p are homogeneous harmonic polynomials of respective degrees n and p, then

$$\begin{aligned} 0 &= \int_B H_n \Delta H_p\,dx = -\int_B H_{n,\alpha} H_{p,\alpha}\,dx + \int_{\partial B} H_n H_{pr}\,dS \\ &= \int_{\partial B} (H_n H_{pr} - H_p H_{nr})\,dS. \end{aligned} \tag{10.6.59}$$

From the homogeneity it follows that

$$H_{nr} = r H_{nr} = n H_n, \quad H_{pr} = r H_{pr} = r H_{pr} = p H_p \quad \text{if} \quad r = 1.$$

Thus we see that

$$\int_{\partial B} (p - n) H_n H_p\,dS = 0 \quad \text{or} \quad \int_{\partial B} H_n H_p\,dS = 0. \tag{10.6.60}$$

So we define the q_{nk} on $\partial B = \sum$, for each n, to be a complete n.o. set, each of which is the restriction to $\sum$ of a (homogeneous) harmonic polynomial of degree n. Then all the q_{nk} for all n form a complete n.o. set for $L_2(\sum)$.

Let us suppose that u is representable as a finite sum of the form

$$u(r,\theta) = \sum_{n,k} a_{nk}(r)\, q_{nk}(\theta), \quad a_{nk}(r) = r^n b_{nk}(r) \tag{10.6.61}$$

where each $b_{nk}(r)$ is an even polynomial in r. Since $r^n q_{nk}$ is a harmonic polynomial, u is a polynomial. Then

$$\int_B |Du|^2 dx = \int_B (u_r^2 + r^{-2}|\nabla_\theta u|^2)\, dx = \int_{\partial B} u\, u_r\, d\textstyle\sum - \displaystyle\int_B u \Delta u\, dx. \tag{10.6.62}$$

If u is given by (10.6.61), it follows that

$$\begin{aligned} u_r &= \sum_{n,k} a'_{nk}(r)\, q_{nk}(\theta) \\ \Delta u &= u_{rr} + (\nu - 1) r^{-1} u_r + r^{-2} \Delta_{2\theta} u \end{aligned} \tag{10.6.63}$$

where $\Delta_{2\theta}$ is the Beltrami operator on $\sum$. Applying (10.6.63) to $u = H_{nk} = r^n q_{nk}$, we obtain

$$\begin{aligned} 0 &= n(n+\nu-2)\, r^{n-2} q_{nk} + r^{n-2} \Delta_{2\theta} q_{nk}, \\ \Delta_{2\theta} q_{nk} &= -n(n+\nu-2)\, q_{nk}. \end{aligned} \tag{10.6.64}$$

Thus

$$\Delta u = \sum_{n,k} [a''_{nk} + (\nu-1) r^{-1} a'_{nk} - n(n+\nu-2) r^{-2} a_{nk}]\, q_{nk}. \tag{10.6.65}$$

Substituting (10.6.61), (10.6.63), and (10.6.65) in (10.6.62), we obtain

$$\begin{aligned} \int_B |Du|^2 dx &= \int_0^1 r^{\nu-1} \int_\Sigma (u_r^2 + r^{-2}|\nabla_\theta u|^2)\, d\textstyle\sum\, dr \\ &= \int_0^1 r^{\nu-1} \sum_{n,k} [a'^2_{nk} + n(n+\nu-2) r^{-2} a^2_{nk}]\, dr. \end{aligned}$$

From the arbitrariness of n, it follows that

$$\int_\Sigma |\nabla_\theta u(1,\theta)|^2 d\textstyle\sum = \displaystyle\sum_{n,k} n(n+\nu-2)\, a^2_{nk}(1).$$

10.7. The local differentiability

In this section we prove the following main theorem:

Theorem 10.7.1. *Suppose $X_0 = S \cup A$ is a minimizing set as in Theorem* 10.4.4. *There exists an $\varepsilon_0 > 0$ with the following property: If $P_0 \in S = X_0 - A$ and $\psi(P_0) < 1 + \varepsilon_0$, there exists for each μ, $0 < \mu < 1$, a neighborhood of P_0 on S which is a regular ν-cell of class C^3_μ. Thus the subset of points $P \in S$ where $\psi(P) = 1$ is open on S and $\psi(P_0) = 1$. If $\mathfrak{M}$ is of class C^n_μ for some $n \geq 4$ and some μ, $0 < \mu < 1$, the ν-disc may be taken to be of class C^n_μ. If $\mathfrak{M}$ is of class C^∞ or analytic, the ν-disc may be taken to be of class C^∞ or analytic, respectively.*

From Theorem 10.4.4, it follows that $\psi(P)$ (see § 10.4) $= 1$ almost everywhere on S. From Theorem 10.4.5, it follows that if $\psi(P_0) < 1 + \varepsilon_1$ (for $\varepsilon_1 = \varepsilon_1(\varepsilon^*, \nu, \mathfrak{M})$ sufficiently small), then S satisfies an (ε^*, R_1) condition at P_0. If ε^* and R_1 are sufficiently small, it follows that a neighborhood of P_0 on S is a topological ν-disc $\bar{S}$. From Lemma 10.4.4, it follows that if $\eta_0 > 0$ and ε_1 and R_0' are small enough, we may take $\bar{S}$ so small that $\psi(P) \leq 1 + \eta_0$ for P on $\bar{S}$. Let

$$\tau : \overline{B(0, R)} \text{ onto } \bar{S}, \quad \tau = \omega_{00} \circ z, \tag{10.7.1}$$

where ω_{00} is a normal coordinate system centered at P_0. Then we conclude from Theorem 10.5.2 that

$$L_\nu(\tau) \leq \Lambda^\nu(\bar{S}), \quad L_\nu(\tau \mid \Gamma) \leq \Lambda^\nu[\tau(\Gamma)], \quad \text{if } \Gamma \text{ is open and } \Gamma \subset B(0, R). \tag{10.7.2}$$

Lemma 10.7.1. *The equality holds in* (10.7.2).

Proof. For suppose for some Γ that the inequality holds. Then $\exists$ a sequence $\{\tau_n\}$, where $\tau_n \in C^1(\bar{\Gamma})$ and $\tau_n \to \tau$ uniformly on $\bar{\Gamma}$ such that

$$\lim_{n\to\infty} L_\nu(\tau_n \mid \Gamma) = d < \Lambda^\nu[\tau(\bar{\Gamma})];$$

we may assume that each τ_n is locally regular. Then it follows that $L_\nu[\tau_n \mid \Gamma] = \Lambda^\nu[\tau_n(\Gamma)]$. We may now set up the functions $\psi(r, P, \tau_n)$, $\varphi(r, P, \tau_n)$ and their limits just as in § 10.4. The developments of that section may be repeated and lead to a set X_0' for which $\Lambda^\nu(S') = d$, where $S' = X_0' - \tau(\partial\Gamma)$. We now modify the τ_n as follows: We select a sequence of analytic domains $\{G_n\} \ni \bar{G}_n \subset G_{n+1}$ for each n and $\cup G_n - \Gamma$. By going along the inner normals to ∂G_n a short distance ϱ_n, we form another analytic domain $G_n' \subset G_n$. We define $\tau_n^*(x) = \tau_n(x)$ for $x \in G_n'$, $\tau_n^*(x) = \tau(x)$ for $x \in \bar{\Gamma} - G_n$, and

$$\tau_n^*[(1-t)\,x' + t\,x] = \omega_{00}[(1-t)\,z_n(x') + t\,z(x)], \quad 0 \leq t \leq 1,$$

$$\tau_n = \omega_{00}\, z_n,$$

where x' and x are extremities of a normal segment to ∂G_n and ω_{00} is a normal coordinate system centered at P_0. This alteration does not affect the functions $\psi(r, P, z_n)$ and $\varphi(r, P, z_n)$ if we merely restrict $r < R(P) - \sigma_n$ where $\sigma_n \to 0$. But the sets $\tau_n^*(\bar{\Gamma})$ and $\tau(\bar{\Gamma})$ now all have the same boundary. Thus X_0' has the same boundary as $\tau(\bar{\Gamma})$. But, since $d < \Lambda^\nu[\tau(\Gamma)]$, this contradicts the minimizing character of X_0 and hence $\tau(\Gamma)$. The proof for the whole surface is the same.

Proof of the theorem. Suppose that $P_0 \in S$, $\psi(P_0) < 1 + \varepsilon_2$, and $\bar{S}$ is a ν-disc containing P_0 for which (10.7.2) holds. Since ψ is upper semi-continuous, we may find an $r_3 > 0$ (see Lemma 10.4.4) such that

$$\psi^+(r, P) \equiv \Lambda^\nu[\bar{S} \cap B(P, r)] \leq (1 + 2\varepsilon_2)\, h(r), \quad P \in B(P_0, r_3), \quad 0 < r < r_3. \tag{10.7.3}$$

From Theorem 10.4.5, it follows that there is an r_4, $0 < r_4 \leq r_3$ such that with each $P \in B(P_0, r_4)$ and each r, $0 < r < r_4$, there is a geodesic ν-plane $\sum(P, r)$ through P such that

$$D[\bar{S} \cap B(P, r), \textstyle\sum(P, r) \cap B(P, r)] \leq \varrho_0 \, r/3, \tag{10.7.4}$$

ϱ_0 being the constant in the cone inequality.

Let us suppose that $\bar{S}$ is represented as in (10.7.1). Then we may approximate uniformly to z on $\overline{B(0, R)}$ by polynomials z_n so that

$$\lim_{n\to\infty} L_\nu(\tau_n) = L_\nu(\tau) = \Lambda^\nu(\bar{S}), \quad \tau_n = \omega_0 \circ z_n.$$

Choose P interior to $\bar{S}$, let ω_P be a normal coordinate system centered at P, and let us define

$$\bar{S}_n = \tau_n[B(0, R)], \quad l(r, P, \bar{S}_n) = \bar{S}_n \cap \partial B(P, r),$$
$$l_0(r, P, \bar{S}_n) = \omega_P^{-1}[l(r, P; \bar{S}_n].$$

From the cone inequality it follows that there are positive numbers ϱ_0 (above) and ε_3 with the following property: For each r for which there is a ν-plane $\sum$ in R_N through 0 such that

$$D[l_0(r, P, \bar{S}_n), \textstyle\sum \cap \partial B(0, r)] \leq \varrho_0 \, r, \quad \text{and}$$
$$\Lambda^{\nu-1}[l_0(r, P, S_n)] \leq (1 + \varepsilon_3) \, \nu \, \gamma_\nu \, r^{\nu-1},$$

there exists a set Y_0^* with $b(Y^*, A_n) \supset H_{\nu-1}(A_n)$, where

$$A_n = l(r, P, \bar{S}_n), \quad Y^* = \omega_P(Y_0^*),$$

such that

$$\Lambda^\nu(Y_0^*) \leq (1 - k) \, \nu^{-1} \, r \, \Lambda^{\nu-1}[l_0(r, P, \bar{S}_n)] + k \, \gamma_\nu \, r^\nu.$$

From the uniform convergence and (10.7.4), it follows that if we choose r_1 and r_2, with $0 < r_1 \leq r_2 \leq r_4$, then (if $1 + \eta_0 \, r \leq 3/2$)

$$D[l(r, P, \bar{S}_n), \textstyle\sum(P, r) \cap \partial B(P, r)] \leq 2\varrho_0 \, r/3 \quad r_1 \leq r \leq r_2$$
$$D[l_0(r, P, \bar{S}_n), \textstyle\sum \cap \partial B(0, r)] \leq \varrho_0 \, r \quad n \geq N.$$

If we let d_n be the inf. of $\Lambda^\nu(X)$ for all sets X for which $b(X, A_n) \supset H_{\nu-1}(A_n)$, it then follows, as in the proof of the lemma above that $d_n \to \Lambda^\nu(\bar{S})$. Then, we conclude from the cone inequality, the inequalities in § 10.1, and the argument in the proof of Lemma 10.4.2 that

$$\psi(r, P, S_n) \leq \begin{cases} \eta_n + t^{-1} \, r(1 + h\, r) \, \varphi_r(r, P, \bar{S}_n) + k \, \gamma_\nu \, r^\nu (1 + \eta_0 \, r)^\nu, & \text{if (a)} \\ \eta_n + \nu^{-1} \, r(1 + h\, r) \, \varphi_r(r, P, S_n), & \text{in all cases} \end{cases}$$

$$\begin{cases} r_1 \leq r \leq r_2, \ t = (1 - k)^{-1} \nu, \ n \geq N, \ P \in B(P_0, r_4), \ \lim_{n\to\infty} \eta_n = 0 \\ \text{(a)} \quad \varphi_r(r, P, \bar{S}_n) < (1 + \varepsilon_3) \, \nu \, \gamma_\nu \, r^{\nu-1} (1 + \eta_0 \, r)^{1-\nu}, \\ 1 + h \, r \geq (1 + \eta_0 \, r)^{2\nu-1}. \end{cases} \tag{10.7.5}$$

For n and P fixed, define $k_n(r) = \max[0, \psi(r, P, S_n) - \eta_n]$.

Since φ_r and χ (below) are ≥ 0 and $\varphi_r \leq \psi_r$ (Lemma 10.4.1, 2, etc.) and φ is A.C., we notice first that

$$k_n(r) = q_n(r) + s_n(r), \quad s_n'(r) = 0 \quad \text{a.e.} \tag{10.7.6}$$

where q_n is A.C. and s_n is non-negative and non-decreasing. Then, from (10.7.5), we conclude that (a.e.)

$$k_n(r) \leq \begin{cases} t^{-1} r(1 + h r)\, k_n'(r) + k\chi, & \text{if } k_n'(r) < \omega(r) \\ \nu^{-1} r(1 + h r)\, k_n'(r), & \text{in all cases} \end{cases} \tag{10.7.7}$$

$$\chi(r) = \gamma_\nu r^\nu (1 + \eta_0 r)^\nu, \quad \omega(r) = (1 + \varepsilon_3)\, \nu\, \gamma_\nu r^{\nu-1} (1 + \eta_0 r)^{1-\nu}.$$

Since $(1 + \eta_0 r)^{2\nu-1} \leq 1 + h r$, it follows that

$$t^{-1} r(1 + h r)\, \omega(r) + k\chi \leq \nu^{-1} r(1 + h r)\, \omega(r). \tag{10.7.8}$$

We now define $p_n(r)$ as the unique solution in C^1 of the equation (obtained by requiring the equality in (10.7.7))

$$p_n' = \begin{cases} t\, r^{-1}(1 + h r)^{-1}(p_n - k\chi), & \text{if } p_n < t^{-1} r(1 + h r)\,\omega + k\chi \\ \nu\, r^{-1}(1 + h r)^{-1} p_n, & \text{if } p_n > \nu^{-1} r(1 + h r)\,\omega \\ \omega(r), & \text{if } t^{-1} r(1 + h r)\,\omega + k\chi \leq p_n \leq \nu^{-1} r(1 + h r)\,\omega \end{cases} \tag{10.7.9.}$$

$$p_n(r_2) = k_n(r_2).$$

By considering all possible cases, we conclude that

$$k_n'(r) \geq p_n'(r) \text{ whenever } k_n(r) \geq p_n(r), \quad r_1 \leq r \leq r_2, \quad n \geq N. \tag{10.7.10}$$

So, if we define $u_n(r) = k_n(r) - p_n(r)$, we have

$$u_n(r_2) = 0 \text{ and } u_n'(r) \geq 0 \text{ whenever } u_n(r) \geq 0, \quad r_1 \leq r \leq r_2, \tag{10.7.11}$$

$(n \geq N)$. It follows that $u_n(r) \leq 0$, $r_1 \leq r \leq r_2$, $n \geq N$; for if, for some n and $\bar r$ with $r_1 \leq \bar r < r_2$, we would have $u_n(\bar r) > 0$, then $u_n(r)$ would be > 0 for r in an interval $[\bar r, \bar r + \delta]$ (on account of (10.7.6)) in which $u_n'(r) \geq 0$ a.e. From (10.7.6), it would follow that $\bar r + \delta = r_2$ which would be impossible since $u_n(r_2) = 0$.

If the first line of (10.7.9) holds for $r_1 \leq r \leq r_2$, then

$$\begin{aligned} p_n - \chi &= t^{-1} r(1 + h r)(p_n' - \chi_n') + (1 - k)\,\gamma_\nu r^{\nu+1} z(r), \\ z(r) &= (1 + \eta_0 r)^{\nu-1}(h + \eta_0 + 2h\eta_0 r). \end{aligned} \tag{10.7.12}$$

Integrating (10.7.12), we obtain

$$\begin{aligned} p_n(r) - \chi(r) \leq &\left(\frac{1 + h r_2}{1 + h r}\right)^t \cdot \left(\frac{r}{r_2}\right)^t [p_n(r_2) - \chi(r_2)] + \\ &+ Z_1 \frac{r^t}{(1 + h r)^t} r_2^{\nu+1-t} \\ &\text{if } t < \nu + 1. \end{aligned} \tag{10.7.13}$$

Now, since $1 + h r \geq (1 + \eta_0 r)^{2\nu-1}$,

$$t^{-1} r(1 + h r)\,\omega - (1 - k)\,\chi \geq (1 - k)\,\varepsilon_3\,\chi(r). \tag{10.7.14}$$

Now, the right side of (10.7.13) will be $<$ that of (10.7.14) iff

$$\begin{aligned}(10.7.15)\quad &[h_t(r_2)]^{-1}[p_n(r_2)-\chi(r_2)]+Z_1 r_2^{\nu+1-t}<(1-k)\,\varepsilon_3 f(r)\\ &f(r)=[h_t(r)]^{-1}\chi(r),\quad h_t(r)=r^t(1+h\,r)^{-t}.\end{aligned}$$

Now

$$(10.7.16)\qquad \frac{f'(r)}{f(r)}=\nu\,r^{-1}\left(1+\frac{\eta_0\,r}{1+\eta_0\,r}\right)-t\,r^{-1}(1+h\,r)^{-1}$$

$$=-t\,r^{-1}(1+h\,r)^{-1}\left[1-(1-k)(1+h\,r)\left(\frac{1+2\eta_0\,r}{1+\eta_0\,r}\right)\right]<0 \text{ if } r\le Z_2.$$

So $f(r)$ takes on its minimum for $r=r_2$. Then (10.7.15) holds for all $r\le r_2$ if

$$(10.7.17)\quad p_n(r_2)<[1+(1-k)\,\varepsilon_3]\,\chi(r_2)-Z_1 r_2^{\nu+1}(1+h\,r_2)^{-t}.$$

Since (Lemma 10.4.2), $[h(r)]^{-1}k_n(r)$ is non-decreasing for each n and $k_n(r)\to\psi(r,P)$ and $[h(r)]^{-1}\psi(r,P)<1+2\,\varepsilon_2$ by (10.7.3), we see that, if ε_2 is small enough, we may choose r_2 so small and N' so large that (10.7.17) holds with $p_n(r_2)=k_n(r_2)$ and $n>N'$. Accordingly $k_n(r)\le p_n(r)$, where p_n is bounded as in (10.7.13). Passing to the limit, we see that (since r_1 was arbitrary)

$$(10.7.18)\qquad \psi^+(r,P)<\chi(r)+(1-k\,\varepsilon_3)\,[h_t(r_2)]^{-1}[h_t(r)],\quad 0<r\le r_2.$$

From (10.7.18) and (10.7.5), we conclude that there are numbers Z_1 and β, depending only on ν and $\mathfrak{M}$ such that

$$\begin{aligned}(10.7.19)\qquad &h(r)<\psi^+(r,P)\le(1+Z_1 r^\beta)\,h(r)\ \text{for all}\\ &B(P,r)\subset B(P_0,32r_0),\quad 32r_0\le\min(r_2,r_4)\end{aligned}$$

and r_0 is sufficiently small to satisfy the further conditions in (10.7.21) and (10.7.22) below. From Theorem 10.4.5, it follows that S satisfies an (ε_0^*,r_0) condition at P_0 with

$$\begin{aligned}(10.7.20)\qquad &\varepsilon_0^*=\text{larger of } 32\,c_1^{-1}r_0 \text{ and } Z_2 r_0^\alpha,\\ &Z_2=(2^{-1}\cdot c_2^{-1}\cdot Z_1\cdot 32^\beta)^{1/h},\quad \alpha=\beta/h>0\end{aligned}$$

provided that r_0 satisfies the following conditions ($r_2=32r_0$)

$$(10.7.21)\quad 32r_0\le R_0'',\ \ 32r_0\le c_1\,\varepsilon_0^*,\ \ \varepsilon_2=2^{-1}\cdot Z_1\cdot 32^\beta\cdot r_0^\beta\le c_2\cdot(\varepsilon_0^*)^h.$$

We see that $\varepsilon_0^*=Z_2 r_0^\alpha$ provided that

$$(10.7.22)\qquad r_0^{1-\alpha}\le 32^{-1}c_1 Z_2.$$

Now, suppose $0<\lambda<1$. For each k, we conclude from (10.7.19) that

$$\begin{aligned}(10.7.23)\qquad &h(r)\le\psi^+(r,P)\le(1+Z_1\cdot 32^\beta\lambda^{k\beta}r_0^\beta)\,h(r)\ \text{for all}\\ &B(P,r)\subset B(P_k,32\lambda^k r_0)\ \text{where}\ P_k\in B(P_0,32r_0-32\lambda^k r_0).\end{aligned}$$

Thus, from Theorem 10.4.5 and equations (10.7.21) and (10.7.22) it follows that $\bar S$ satisfies an $(\varepsilon_k^*,\lambda^k r_0)$ condition at any $P_k\in B(P_0,32r_0-$ $-32\lambda^k r_0)$, where

$$(10.7.24)\qquad \varepsilon_k^*=Z_2\,\lambda^{k\alpha}r_0^\alpha.$$

Using the definition, we see that for each k, the following is true: For each P in any $B(P_k, 2\lambda^k r_0)$ and each $R \leq \lambda^k r_0$. There exists a geodesic ν-plane $\sum(P, R)$ centered at P, such that

$$D[\bar{S} \cap B(P, R), \quad \textstyle\sum(P, R) \cap B(P, R)] < \varepsilon_k^* \cdot R. \tag{10.7.25}$$

For a given k, $\cup B(P_k, 2\lambda^k r_0) = B(P_0, 32 r_0 - 30\lambda^k r_0)$ and the intersection of these for all $k > 0$ is just $B(P_0, 2r_0)$. If $P \in B(P_0, 2r_0)$, the conclusions above hold for all k. So, if we assume that

$$\lambda^{k+1} r_0 < R \leq \lambda^k r_0$$

we see that

$$\varepsilon_k^* < Z_3 R^\alpha, \quad Z_3 = \lambda^{-\alpha} Z_2.$$

Since this is true for each k, we conclude that

$$\begin{aligned} &D[\bar{S} \cap B(P, R), \textstyle\sum(P, R) \cap B(P, R)] \leq Z_3 R^{1+\alpha}, \\ &P \in B(P_0, 2r_0), \quad 0 \leq R \leq r_0. \end{aligned} \tag{10.7.26}$$

From the proof of Lemma 10.5.4, it follows that (10.7.26) holds with $\bar{S}$ and $\sum(P, r)$ replaced by $\omega_{00}^{-1}(\bar{S})$ and $\omega_{00}^{-1}[\sum(P, r)]$, respectively. Accordingly, for the remainder of the proof, we shall assume that $\bar{S}$ stands for $\omega_{00}^{-1}(\bar{S})$ and that $\sum(P, r)$, etc., are actual ν-planes in R_N.

Next, we observe, using (10.7.26), that

$$\begin{aligned} \textstyle\sum(P, r/2) \cap B(P, r/2) &\subset [\bar{S} \cap B(r/2), Z_3(r/2)^{1+\alpha}] \cap B(P, r/2) \\ &\subset [\bar{S} \cap B(P, r), Z_3(r/2)^{1+\alpha}] \cap B(P, r/2) \\ &\subset [\textstyle\sum(P, r) \cap B(P, r), Z_3 r^\alpha + Z_3(r/2)^\alpha] \cap B(P, r/2). \end{aligned} \tag{10.7.27}$$

Accordingly

$$d[\textstyle\sum(P, r), \sum(P, r/2)] \leq Z_4 r^\alpha, \quad Z_4 = 2Z_3(1 + 2^{-1-\alpha}). \tag{10.7.28}$$

Replacing r by $2^{-n} r$ for $n = 1, 2, \ldots$, in (10.7.21), we conclude that there is a ν-plane $\sum(P, 0)$ through P such that

$$d[\textstyle\sum(P, r), \sum(P, 0)] \leq Z_5 r^\alpha, \quad Z_5 = Z_4(1 + 2^{-\alpha} + 2^{-2\alpha} + \cdots), \tag{10.7.29}$$

so that

$$\begin{aligned} &D[\bar{S} \cap B(P, r), \textstyle\sum(P, 0) \cap B(P, r)] \leq Z_6 r^{1+\alpha}, \quad Z_6 = Z_3 + Z_5 \\ &0 < r \leq r_2, \quad P \in B(P_0, r_2) \end{aligned} \tag{10.7.30}$$

provided that r_2 is small enough.

Next, if P and $P' \in B(P_0, r_2)$ and $|PP'| = r/2$, we see, by proceeding as in (10.7.20) and using (10.7.23)

$$\begin{aligned} \textstyle\sum(P', 0) \cap B(P', r/2) &\subset [\bar{S} \cap B(P', r/2), Z_6(r/2)^{1+\alpha}] \cap B(P', r/2) \\ &\subset [\textstyle\sum(P, 0) \cap B(P, r), Z_6 r^{1+\alpha} + Z_6(r/2)^{1+\alpha}] \cap B(P', r/2) \\ &\subset [\textstyle\sum'(P, 0) \cap B(P, r)], 2Z_6 r^{1+\alpha} + 2Z_6(r/2)^{1+\alpha}) \cap B(P', r/2) \end{aligned}$$

since $P' \in [\sum(P, 0) \cap B(P, r), Z_6 r^{1+\alpha} + Z_6(r/2)^{1+\alpha}]$; here $\sum'(P, 0)$ is the ν-plane through $P' \parallel \sum(P, 0)$. Thus

$$d[\textstyle\sum(P, 0), \sum(P', 0)] \leq Z_7(r/2)^\alpha. \tag{10.7.31}$$

From (10.7.23), it follows that if r_2 is small enough and Π_P is the $(N-\nu)$-plane through $P \perp \sum(P, 0)$ and Π' is another $(N-\nu)$-plane through P with $d_0(\Pi', \Pi_P) \leq 1/2$, say, then P is the only intersection in $B(P, r_2)$ of Π' and $\bar{S}$. By taking Π' as the plane through $P \| \Pi_{P_0}$, we see that $\bar{S} \cap B(P_0, r_2/2)$ has a $1-1$ projection on $\sum(P_0, 0)$. It follows easily that $\sum(P, 0)$ is the actual tangent ν-plane to $\bar{S}$ at P. If we choose a (ξ, η) coordinate system with origin at P_0 and ξ axes in $\sum(P_0, 0)$, we may represent the part of $\bar{S}$ near P_0 in nonparametric form $\eta = \eta(\xi)$. From (10.7.24) it follows that $\eta \in C_h^1$. The higher differentiability follows from the results of §§ 6.4 and 6.6 since the EULER equations form an elliptic system.

10.8. Additional results of Federer concerning Lebesgue ν-area

In this section we present an account of those results of FEDERER which the writer originally thought were relevant for our problem but which he later found not to be essential. Since the results are important and not easily accessible, we include them there together with the necessary background material.

Our presentation leans heavily on the notion of the order of a point with respect to a mapping which was introduced in § 9.2. The use of algebraic topology has been reduced to a minimum. We begin by extending the definition of the order function and introducing FEDERER's essential multiplicity function.

Definition 10.8.1. We suppose that $\mathfrak{M}$ is an oriented ν-dimensional manifold of class C^1 which is diffeomorphic to an analytic manifold, G is a domain on $\mathfrak{M}$ the closure of which, $\bar{G}$, $\subset \mathfrak{M}$, and we suppose that $\tau: z = z(x)$ is a continuous mapping from G into R_ν, and $z_0 \in R_\nu$. If $z_0 \in \tau(\partial G)$, we define $O[z_0, \tau(\partial G)] = 0$; otherwise, we define

$$O[z_0, \tau(\partial G)] = \lim_{n\to\infty} O[z_0, \tau(\partial G_n)] \tag{10.8.1}$$

where $\{G_n\}$ is any sequence of domains of class C^1 such that $\bar{G}_n \subset G_{n+1}$ for each n and $\cup G_n = G$; it is clear that the limit is independent of the sequence. We define

$$M(z_0, \tau) = \sup \sum_i |O[z_0, \tau(\partial G_i)]|$$

for all finite or countable sequences $\{G_i\}$ of disjoint subdomains of G.

Lemma 10.8.1. *Suppose G, τ, and each τ_n satisfy the conditions oj Definition 10.8.1 and τ_n converges uniformly to τ on $\bar{G}$. Then*

(a) *$O[z, \tau(\partial G)]$ is constant on each component of $R_\nu - \tau(\partial G)$.*

(b) $O[z, \tau(\partial G)] = \lim_{n\to\infty} O[z, \tau_n(\partial G)], \quad z \notin \tau(\partial G)$

(c) $M(z, \tau) \leq \liminf_{n\to\infty} M(z, \tau_n)$

(d) $L_\nu(z, G) \geq \int_{R_\nu} M(z, \tau)\, dz$

(e) *If* $\tau_t: G \to R_\nu$, $0 \leq t \leq 1$, *is a homotopy,* $z \in R_\nu$, *and* $\{x \mid \tau_t(x) = z\} = \{x \mid \tau_0(x) = z\}$, *then* $M(z, \tau_1) = M(z, \tau_0)$.

Proof. (a) and (b) follow from Lemmas 9.2.3 and 9.2.3′. To prove (c), we let $D_1, \ldots, D_m$ be any finite sequence of disjoint sub-domains of G. Then, since $O[z, \tau(\partial D)] = 0$ if $z \in \tau(\partial D)$,

$$\sum_{i=1}^{m} |O[z, \tau(\partial D_i)]| \leq \liminf_{n\to\infty} \sum |O[z, \tau_n(\partial D_i)] \leq \liminf_{n\to\infty} M(z, \tau_n).$$

The result follows. To prove (d), we choose $\{z_n\}$ of class C^1 on a domain $\supset \bar{G}$ (on $\mathfrak{M}$) so that $z_n \to z$ uniformly and $L_\nu(z_n, G) \to L_\nu(z, G)$. For each n, the inequality holds (use (9.2.19)) and it holds in the limit by lower-semicontinuity.

To prove (e), let $D_1, \ldots, D_n$ be disjoint subdomains of G such that $z \notin \tau_0(\partial D_i)$ for any i. Then $\tau_t^{-1}(z) \cap \cup \partial D_i$ is empty and

$$\sum_i |O[z, \tau_t(\partial D_i)]|$$

is continuous in t. Thus $M(z, \tau_1) \leq M(z, \tau_0)$. The argument is symmetric.

Lemma 10.8.2. *Let* $\sigma: y = \zeta(x)$ *be a continuous map from* $\partial B(0,1)$ *into* $\partial B(0, 1)$, $B(0, 1) \subset R_k$, $k \geq 2$, *let* y_0 *and* $y_1 \in \partial B(0, 1)$ *with* $y_1 \neq y_0$, *let* $\gamma = \sigma^{-1}(y_0)$, *and let* Γ *be a domain on* $\partial B(0, 1)$ *such that* $\gamma \subset \Gamma$ *but* $y_1 \notin \sigma(\bar{\Gamma})$. *Suppose also that* $\tau: z = \eta(y)$ *is a diffeomorphism from* $\partial B(0, 1) - \{y_1\}$ *onto* R_{k-1}. *Finally, suppose that* $O[0, \sigma\{\partial B(0, 1)\}] = 0$. *Then*

$$O[\tau y_0, \tau \sigma(\partial \Gamma)] = 0. \tag{10.8.2}$$

Proof. Suppose ζ_n converges uniformly to ζ on $\partial B(0,1)$, each ζ_n being a simplicial map (see below). If n is large enough the hypotheses hold for (σ_n, ζ_n). It is clear, then, that we may as well assume that ζ is already simplicial. This means that the domain and range spheres are each subdivided into small curvilinear simplices Δ_i and G_j, each of which is the radial projection on $\partial B(0, 1)$ of the actual simplex having the given vertices. The simplices are supposed to fit together in the usual way as do the simplices of a polyhedron; the subdivision is then completely determined by its vertices. Finally the map ζ maps each simplex Δ_i of its domain berycentrically onto some simplex $\bar{G}_j$ on the range sphere (several different Δ_i may be mapped onto the same $\bar{G}_j$); that is the points of the flat simplices corresponding by radial projection correspond in that way.

If we introduce local coordinates $\theta_i^1, \ldots, \theta_i^{k-1}$ on each domain simplex, the formula for the order becomes (since $|\zeta| = 1$)

$$O[0, \sigma\{\partial B(0, 1)\}] = \Gamma_k^{-1} \sum_i \int_{\Delta_i} \zeta^\alpha (-1)^{\alpha-1} (\partial \zeta'_\alpha / \partial \theta_i)\, d\theta_i.$$

The quantity $(-1)^{\alpha-1}(\partial\zeta'_\alpha/\partial\theta_i)$ is proportional to the normal to the $(k-1)$-surface $\zeta = \zeta(\theta_i)$. Since this lies on the $(k-1)$-sphere $\partial B(0, 1)$, the normal either points towards or away from the origin; the direction does not change within any simplex Δ_i. Thus

$$O[0, \sigma\{\partial B(0, 1)\}] = \Gamma_k^{-1} \sum_i \int_{\Delta_i} \varepsilon_i (d\textstyle\sum/dS)\, dS, \quad \varepsilon_i = \pm 1$$

where dS and $d\sum$ are the $(k-1)$-area elements on the domain and range spheres, respectively. Clearly, we may write

$$O[0, \sigma\{\partial B(0, 1)\}] = \Gamma_k^{-1} \sum_j A^{k-1}(G_j) \cdot \omega_j$$

$$\omega_j = \sum_{i \in s(j)} \varepsilon_i, \quad s(j) = \{i \mid \sigma(\Delta_i) = G_j\};$$

for each y interior to G_j, ω_j is the number of x such that $\zeta(x) = y$, each point being counted with its multiplicity. We shall now show that each $\omega_j = O[0, \sigma\{\partial B(0, 1)\}]$.

To do this, let G and G' be two adjacent range simplices having the $(k-2)$-simplex γ in common. Let $\Delta^1, \ldots$ be those Δ_i carried into G and $\Delta'^1, \ldots$ be those carried into G'. From the nature of the mapping, we see that we may order these simplices so that Δ^p and Δ'^p are adjacent for $p = 1, \ldots, P_1$, say; the remaining Δ^p can be arranged in adjacent pairs, the same being true of the remaining Δ'^p. It is clear that $\varepsilon^p = \varepsilon'^p$ for each $p \le P_1$ and that the ε^p and ε'^p for each of the remaining adjacent pair have opposite signs. Thus

$$\omega = \sum_{p=1}^{P_1} \varepsilon^p = \sum_{p=1}^{P_1} \varepsilon'^p = \omega'.$$

Hence all the ω_j are equal and hence equal to $O[0, \sigma(S)]$, $S = \partial B(0, 1)$. In our case, the order vanishes.

Now, if we consider the piecewise analytic map $\tau\sigma$, we see again that if $y_0 \notin \tau(\partial G_j)$ for any j, the number of times that $\tau\sigma(x) = y_0$ with positive multiplicity equals the number of times this happens with negative multiplicity. Approximating to Γ by smooth regions and using Lemma 9.2.3 and its proof, we obtain the result (10.8.2). If $y_0 \in$ some $\tau(\partial G_j)$, the same result holds by continuity.

We now prove the following special case of HOPF's *extension theorem* (ALEXANDROFF-HOPF, pp. 498—508):

Theorem 10.8.1. *Suppose that G and τ satisfy the conditions of Definition 10.8.1 with $\nu \ge 2$, and suppose that*

$$y_0 \notin \tau(\partial G) \text{ and } O[y_0, \tau(\partial G)] = 0.$$

Then $\exists$ a map $\tau^*: \bar{G} \to R_\nu$, such that $z^*(x) = z(x)$, $x \in \partial G$, and $y_0 \notin \tau^*(\bar{G})$.

Proof. We shall prove this and the following statement A by induction on $\nu \ge 2$:

A-Suppose σ: $y = \zeta(x)$ *is a map from* ∂G *into* $S = S^{\nu-1} = \partial B(0, 1)$ ($B(0, 1) \subset R_\nu, \nu \geq 2$) *such that* $O[0, \sigma(\partial G)] = 0$, $\bar{G}$ *being diffeomorphic to* $\overline{B(0, 1)}$. *Then* $\exists$ *a homotopy* σ_t: $y = h(x, t)$, *where*

$$h(x, 0) = \zeta(x), \quad h(x, t) \in S, \quad h(x, 1) = \bar{y}, \quad x \in \partial G, \quad 0 \leq t \leq 1,$$

$\bar{y}$ *being a fixed (independent of* x*) point of* S.

We first show that, for each given ν, statement A implies the theorem. Clearly, we may assume that the manifold $\mathfrak{M}$ of Definition 10.8.1 is analytic and in some R_N. Let us choose analytic domains G_t where $\bar{G}_0 \subset G$ and ∂G_t is obtained by proceeding a distance $\varrho\, t$ along the normals to ∂G_0, ϱ being small, $0 \leq t \leq 1$, $\bar{G}_1 \subset G_0$. We may choose z_1^* analytic and near z on $\bar{G}_1$ and then define

$$z_1(x) = z(x) \text{ on } \bar{G} - G_0, \quad z_1(x) = z_1^*(x) \text{ on } \bar{G}_1,$$
$$z_1(x) = (1 - t)\, z_1^*(x) + t\, z(x) \text{ on } G_t, \quad 0 \leq t \leq 1;$$

$\bar{G} - G_0$ is supposed to be so close to ∂G and z_1^* so close to z that $y_0 \notin \tau(\bar{G} - G_0) \cup \tau(\partial G_1)$. If $\partial z_1/\partial x \equiv 0$ on G_1, then $\tau_1(\bar{G}_1)$ is of measure 0 and we may find a y_1 arbitrarily close to y_0 which is not in $\tau_1(\bar{G}_1)$ and τ^* may be formed by following τ_1 by a radial projection of $\overline{B(y_1, \varrho_1)} - \{y_1\}$ onto $\partial B(y_1, \varrho_1)$ where $y_0 \in B(y_1, \varrho_1)$ and $\overline{B(y_1, \varrho_1)} \cap \tau(\partial G)$ is empty. Otherwise we can find a y_1 arbitrarily close to y_0 such that $\tau_1^{-1}(y_1)$ is a finite set of points $\subset G_1$. We may then choose a domain Γ of class $C^1 \ni \tau_1^{-1}(y_1) \subset \Gamma \subset \bar{\Gamma} \subset G_1$, and $\bar{\Gamma}$ is diffeomorphic with $\overline{B(0, 1)}$ (begin by passing an arc of class C^2 through the points, going slightly beyond the first and last points, then begin with a thin tube about that arc). On $\bar{\Gamma}$, we introduce polar coordinates r, $0 \leq r \leq 1$, and p on $\partial B(0, 1)$ by mapping on $\overline{B(0, 1)}$. Choosing ϱ_1 as above, we define $z^*(x) = z_1(x)$ outside $\bar{\Gamma}$ and

$$z^*(r, p) = y_1 + \varrho_1\, h(p, 1 - r).$$

We now prove A for $\nu = 2$. Clearly we may assume that $G = B(0, 1)$. Then, if we introduce angular coordinates θ and φ on the domain and range circles, respectively, we may write

$$\sigma: \varphi = \varphi(\theta), \quad 0 \leq \theta \leq 2\pi.$$

If $\varphi \in C^1$, then the proof of Lemma 10.8.2 shows that

$$(10.8.3) \quad O[0, \sigma(S)] = \frac{1}{2\pi} \int_0^{2\pi} \varphi'(\theta)\, d\theta = [\varphi(2\pi) - \varphi(0)]/2\pi = 0$$
$$(S = \partial B(0, 1)).$$

If $\varphi(\theta)$ is merely continuous, we see that (10.8.3) holds by approximating. We may then define

$$\sigma_t: \varphi = (1 - t)\, \varphi(\theta) + t\, \bar{\varphi}, \quad \bar{\varphi} = \frac{1}{2\pi} \int_0^{2\pi} \varphi(\theta)\, d\theta = \varphi(\bar{\theta}).$$

Now, let us assume that statement A and our theorem are true for $2 \leq \nu \leq k$ and let us consider statement A for $\nu = k + 1$. Again, we may assume that $\partial G = S = \partial B(0, 1)$. We first show that there is a map σ_1: $y = \zeta_1(x)$ from S to S which is homotopic to τ and which is such that some point $y_0 \in S$ is not in $R(\tau_1)$. Then it is clear that σ_1 is homotopic to the map σ_2: $y = \bar{y}$ where $\bar{y}$ is on S and diametrically opposite to y_0. If σ already omits some point y_0, we may take $\sigma_1 = \sigma$. Otherwise, let σ^* (: $y = \zeta^*(x)$) be an analytic map from S to S such that $|\zeta^*(x) - \zeta(x)| < 1/3$ for $x \in S$. The homotopy

$$\zeta^*(x, t) = \frac{(1-t)\,\zeta(x) + t\,\zeta^*(x)}{|(1-t)\,\zeta(x) + t\,\zeta^*(x)|}, \quad x \in S, \quad 0 \leq t \leq 1,$$

shows that ζ^* is homotopic to ζ. If $\partial \zeta^*/\partial x \equiv 0$, the range of σ^* has measure 0 and we may take $\zeta_1 = \zeta^*$. Otherwise, $\sigma^*(Z)$ has measure 0, Z being the set where $\partial \zeta^*/\partial x = 0$, and we may choose y_0 and y_1 in S with $y_0 \neq y_1$ so that $\sigma^{*-1}(y_0)$ is a finite set. Then we may choose a smooth domain $\Gamma \supset \sigma^{*-1}(y_0)$ such that $\sigma^*(\bar{\Gamma})$ does not contain y_1. If we now map $S - \{y_1\}$ diffeomorphically onto R_k by a map ω: $z = \eta(y)$, it follows from Lemma 10.8.2 that $O[\omega y_0, \omega\,\sigma(\partial\Gamma)] = 0$ so that there exists a map τ_1: $\bar{\Gamma} \to R_k$ such that τ_1 and $\omega\,\sigma$ coincide on $\partial\Gamma$ and $\omega(y_0) \notin \tau_1(\bar{\Gamma})$. If we define

$$\sigma_1(x) = \omega^{-1}\tau_1(x), \quad x \in \bar{\Gamma}, \quad \sigma_1(x) = \sigma^*(x), \quad x \in S - \bar{\Gamma},$$

we see that $y_0 \notin \sigma_1(S)$. If we define

$$\left.\begin{aligned} &\sigma(x, t) = \sigma^*(x), \quad x \in S - \Gamma \\ &\sigma(x, t) = \omega^{-1}[(1-t)\,\omega\,\sigma^*(x) + t\,\tau_1(x)], \quad x \in \bar{\Gamma} \end{aligned}\right\} 0 \leq t \leq 1,$$

we see that σ_1 is homotopic to σ^* and hence to σ.

Lemma 10.8.3. *Suppose G satisfies the conditions of Definition* 10.8.1, $U \subset R_\nu$, $\nu \geq 2$, $U = \{y \mid |y^\alpha| < 1, \ \alpha = 1, \ldots, \nu\}$ *and*: $\bar{G} \to \bar{U}$ *is a mapping* $\ni 0 \notin \tau(\partial G)$.

(a) *Suppose* $O[0, \tau(\partial G)] = 0$. *Then* $\exists$ *a map* τ_1: $\bar{G} \to \partial U$ *such that* $\tau_1(x) = \tau(x)$ *for* $x \in \partial G \cap \tau^{-1}(\partial U)$.

(b) *Suppose* $O[0, \tau(\partial G)] = \varkappa\, m$, $\varkappa = \pm 1$, $0 < m < +\infty$. *Then* $\exists$ *disjoint compact sets* $\bar{Q}_1, \ldots, \bar{Q}_m$ *in G and a mapping* τ_1: $\bar{G} \to \bar{U} \ni \tau_1(x) = \tau(x)$ *on* $(\partial G \cap \tau^{-1}(\partial U))$, $(\tau_1 \mid \bar{Q}_i)$ *is a diffeomorphism from* $\bar{Q}_i$ *onto* $\bar{U}$ *with Jacobian having the sign of* $\varkappa$, $i = 1, \ldots, m$, *and*

$$\tau_1[\bar{G} - \bigcup_i Q_i^{(0)}] \subset \partial U.$$

Proof. (a) The mapping τ_1 may be obtained from that of Theorem 10.8.1 by following it with a radial projection from 0 onto U; clearly if $\tau(x) \in \partial U$, $\tau_1(x) = \tau(x)$.

(b) We choose the analytic domains G_t as in the first part of the proof of Theorem 10.8.1 and approximate closely on $\bar{G}_1$ to τ by an analytic τ'; as before we construct $\tau^* = \tau$ on $\bar{G} - G_1$, $\tau^* = \tau'$ on $\bar{G}_0$ and $\tau^*(x) = (1-t)\,\tau(x) + t\,\tau'(x)$ on ∂G_t. By translating τ' slightly, if necessary, we may ensure that $0 \notin \tau^*(Z)$, Z being the subset of $\bar{G}_0$ where $\partial\tau^*/\partial x = 0$; since $m \neq 0$, $\partial\tau^*/\partial x \neq 0$. Then, $\tau^{*-1}(0) = \{x_1', \ldots, x_n'\}$ and $\partial\tau^*/\partial x \neq 0$ at each x_i'. There must be at least m of these, call them $x_1, \ldots, x_m$, at each of which $\partial\tau^*/\partial x$ has the sign of $\varkappa$. If we choose ϱ, $0 < \varrho < 1$, small enough and set $U_\varrho = \{y \mid |y^\alpha| \leq \varrho,\ \alpha = 1, \ldots, \nu\}$, then $\tau^{*-1}(U_\varrho) \supset \bigcup_i \bar{Q}_i$ where $\tau^*(\bar{Q}_i) = \bar{U}_\varrho$ and $x_i \in Q_i$, $i = 1, \ldots, m$. If we let $\Gamma = G - \bigcup \bar{Q}_i$, we see that $O[0, \tau^*(\partial\Gamma)] = 0$ so we conclude from Theorem 10.8.1 that there is a map τ_2: $\bar{\Gamma} \to R_\nu - \{0\}$ which coincides with τ^* on $\partial\Gamma$; by following this with a map which is a radial projection from 0 onto ∂U_ϱ if $y \in U_\varrho$ and is the identity outside U_ϱ, we may ensure that τ_2: $\bar{\Gamma} \to R_\nu - U_\varrho$. Let $\tau_3 = \tau_2$ on $\bar{\Gamma}$ and $\tau_3 = \tau^*$ on each $\bar{Q}_i$. We then obtain τ_1 by following τ_3 by σ where σ is the radial expansion from U_ϱ onto U for $y \in U_\varrho$ and $\sigma(y)$ is the radial projection of y on ∂U if $y \in \bar{U} - U_\varrho$.

Definition 10.8.2. For each sequence $I = (i_1, \ldots, i_\nu)$ in which the i_α are integers and $1 \leq i_1 < i_2 < \cdots < i_\nu \leq N$, we define the projection P^I from $R_N\,(N \geq \nu)$ onto R_ν by

$$w^\alpha(z) = z^{i_\alpha}, \quad \alpha = 1, \ldots, \nu, \text{ where } w = P^I(z). \tag{10.8.4}$$

We now recall the notations introduced in Definition 10.5.4 and Lemma 10.5.8. In addition to the statements proved in Lemma 10.5.8, we prove those in the following lemma:

Lemma 10.8.4. *Suppose $\bar{G}$ satisfies the usual conditions*

(a) *If* ω: $\bar{G} \to C_k' - C_{N-\nu-1}''$ *and, for some* x_0 *and* I, $P^I \circ \omega(x_0) = 0$, *then* $P^I \circ r_k[\omega(x_0), t] = 0$ *for* $0 \leq t \leq 1$, *where*

$$r_k(\omega, t) = (1-t)\,\omega + t\,r_k(\omega), \tag{10.8.5}$$

r_k being defined in Lemma 10.5.8.

(b) *If* ω: $\bar{G} \to C_k' - C_{N-\nu-1}''$ *and, for some* x_0 *and* I, $P^I \circ \omega(x_0) \neq 0$, *then* $P^I \circ r_k[\omega(x_0), t] \neq 0$, $0 \leq t \leq 1$.

Proof. (a) We assume that $\omega(x_0) \in$ some γ_{kq}, where

$$\begin{aligned} &\gamma_{kq}: |\omega^j - \delta^j| \leq 1, \quad j = j_1, \ldots, j_k, \quad \omega^j = \theta^j, \quad j \neq \text{any } j_p, \\ &\delta^j \text{ even}, \ \theta^j \text{ odd}, \ 1 \leq j_1 < j_2 < \cdots < j_k \leq N. \end{aligned} \tag{10.8.6}$$

Since $\omega^{i_\alpha}(x_0) = 0$, the i_α are among the j_p. We may, without loss of generality, assume that the $\delta^j = 0$ and the $\theta^j = 1$. By renumbering the axes, we may assume that $\gamma_{kq} = \gamma_k$, as defined in the proof of Lemma 10.5.8(a), and assume that $i_\alpha = \alpha$. But, since $\omega(x_0) \notin C_{N-\nu-1}''$, we must have $\omega^i(x_0) \neq 0$, $i = \nu + 1, \ldots, k$. Clearly the first ν coordinates of $r_k[\omega(x_0)]$ will still be 0 and the others will not.

To prove (b), we again assume that $\omega(x_0) \in$ some γ_{kq} as defined in (10.8.6). If the i_α are not included in the j_p, then $\omega^{i_\alpha}(x_0) = \theta^j$ and $r_k[\omega(x_0), t]$ has i_α-coordinate $= \theta^j \neq 0$ for $0 \le t \le 1$. If the i_α are included in the j_p but all the δ^{j_p} are not 0, the result still holds. But, if all the $\delta_j = 0$, we may again assume that $i_\alpha = \alpha$ for $1 \le \alpha \le \nu$ and $j_p = p$, $1 \le p \le k$. Since $P^I \circ \omega(x_0) \neq 0$, not all the $\omega^{i_\alpha}(x_0)$ are zero and this still holds after applying $r_k(\omega, t)$.

Lemma 10.8.5. *Suppose G satisfies the conditions of Definition* 10.8.1. *Suppose $a \in R_N$, $\varepsilon > 0$, $z: \bar{G} \to R_N - C''_{N-\nu-1}(a, \varepsilon)$ is a mapping $\ni z(\partial G) \subset R_N - C''_{N-\nu}(a, \varepsilon)$, and Δ denotes the totality of ν-termed sequences δ of even integers.*

Then $\exists$ a Lipschitz map $u: \bar{G} \to R_N$ such that

$$|u(x) - z(x)| \le 3\varepsilon\sqrt{N}, \quad x \in \bar{G}, \tag{10.8.7}$$

$$L_\nu(u, G) \le \sum_I \sum_{\delta \in \Delta} 2^\nu \varepsilon^\nu M[P^I \circ z, P^I(a) + \varepsilon\delta]. \tag{10.8.8}$$

Proof. If the right side is $+\infty$, there is nothing to prove. In view of an obvious reduction by homothetic transformations, we may also assume that $a = 0$ and $\varepsilon = 1$.

By virtue of Lemma 10.8.4 and our hypothesis, it follows that the maps $\omega_j: \bar{G} \to C'_{N-j} - C''_{N-\nu-1}$ given by

$$\omega_j = r_{N-j+1} \circ \cdots \circ r_N \circ z, \quad j = 1, \ldots, N - \nu$$

are all defined and continuous. If we define $w = \omega_{N-\nu}$, it follows from Lemma 10.8.4 and from Lemma 10.8.1 (e) that

$$M(P^I \circ w, \delta) = M(P^I \circ z, \delta), \quad w(\bar{G}) \subset C'_\nu \tag{10.8.9}$$

for all I and δ.

Let $\mathfrak{F}$ be the family of all components Γ' of $w^{-1}(C'_\nu - C'_{\nu-1})$. Since the intersection of $C''_{N-\nu}$ with one of the open ν-cells K of $C'_\nu - C'_{\nu-1}$ is just the center of that cell and since w is continuous on a compact set, it follows from the Heine-Borel theorem that the sub-family $\mathfrak{F}_1$ of those Γ' which contain points of the compact set $w^{-1}[C''_{N-\nu} \cap (C'_\nu - C'_{\nu-1})]$ is finite. Each component Γ' of $\mathfrak{F}_1$ is open on $\bar{G}$ but it can happen that $\Gamma' \cap \partial G$ is not empty, in which case $w(\Gamma' \cap \partial G) \subset$ some K; otherwise $w(\partial\Gamma') \subset \partial K$. Let us define $\Gamma = \Gamma' - \partial G$; then Γ is also a domain ($= \Gamma'$ if $\Gamma' \cap \partial G$ is empty). Since $z(\partial G) \cap C''_{N-\nu}$ is empty, it follows from Lemma 10.8.4 that the center c_K of K is not on $w(\partial G)$.

So let us choose a Γ, let K be the open ν-cell of $C'_\nu - C'_{\nu-1}$ containing $w(\Gamma)$, let I be the unique ν-sequence $\ni P^I(K)$ is a ν-cell U_δ ($= \{\omega \mid |\omega^\alpha - \delta^\alpha| < 1,\ \delta^\alpha$ even, $\alpha = 1, \ldots, \nu\}$), $\delta = P^I(c_K)$ and let ω_Γ be defined by $\omega_\Gamma(x) = P^I \circ \omega(x) - P^I c_K$; clearly $\omega_\Gamma: \bar{\Gamma} \to \bar{U}$, $U = U_0$. Then we note first of all (see 10.8.9) that

$$\sum_{\Gamma \subset \mathfrak{S}(I,\delta)} |O[0, \omega_\Gamma(\partial\Gamma)]| \le M(P^I \circ z, \delta) \tag{10.8.10}$$

where $S(I, \delta)$ is the set of all $\Gamma \ni w(\Gamma) \subset K$ for some $K \ni P^I K = U_\delta$. From Lemma 10.8.3, it follows that we may choose a map $v_\Gamma : \bar{\Gamma} \to \bar{K} \ni$ (a) if $O[0, \omega_\Gamma(\partial\Gamma)] = 0$, then $v_\Gamma(\bar{\Gamma}) \subset \partial K$, and (b) if $|O[0, \omega_\Gamma(\partial\Gamma)]| = m$, $0 < m < +\infty$, there are domains $Q_1, \ldots, Q_m$ with the $\bar{Q}_i$ disjoint and in Γ such that $v_\Gamma(\bar{\Gamma}) \subset \bar{K}$, $v_\Gamma \mid \bar{Q}_i$ is a diffeomorphism of $\bar{Q}_i$ onto $\bar{K}$, and $v_\Gamma(\bar{\Gamma} - \bigcup Q_i) \subset \partial K$. In both cases, $v_\Gamma(x) = w(x)$ for $x \in \partial\Gamma \cap w^{-1}(\partial K)$. Let us define $v'(x) = v_\Gamma(x)$ for x on each $\bar{\Gamma}$ of $\mathfrak{F}_1$ and $v'(x) = w(x)$ elsewhere on $\bar{G}$. Then, since if $\Gamma \in \mathfrak{F} - \mathfrak{F}_1$, it is clear that $w(\bar{\Gamma})$ does not contain the center c_K of any cell K in $C'_\nu - C'_{\nu-1}$, it follows that $v'(\bar{G})$ contains no such c_K. So, let us define $v(x) = r_\nu[v'(x)]$, r_ν being defined on each $\bar{K} - c_K$ as the radial projection from c_K onto ∂K. Then

$$|v(x) - z(x)| \leq 2\sqrt{N}, \quad x \in \bar{G}, \tag{10.8.11}$$

since if $z(x) \in$ some $K_N \in C'_N$, $v(x)$ is on ∂K_N. Moreover there are a finite number R_i, $i = 1, \ldots, P$, of domains such that the $\bar{R}_i$ are disjoint and $\subset G$, and $v \mid \bar{R}_i$ is a diffeomorphism from $\bar{R}_i$ onto some cell $\bar{K}$ of C'_ν. Finally

$$\begin{gathered} v(x) \in C'_{\nu-1} \quad \text{for} \quad x \in \bar{G} - \bigcup_i R_i \\ \sum_I \sum_\delta M(P^I \circ v, \delta) \leq \sum_I \sum_\delta M(P^I \circ z, \delta). \end{gathered} \tag{10.8.12}$$

Unfortunately v is not Lipschitz (probably). To remedy this, we begin by extending v to a slightly larger domain $D \supset \bar{G}$. For each sufficiently small ϱ, we form the "mollified" functions v_ϱ by defining a function $K_\varrho(x, \xi)$ on $\mathfrak{M}$ for x and ξ near D so that

$$K_\varrho(x, \xi) \geq 0, \quad \int_{\mathfrak{M}} K_\varrho(x, \xi)\, d\xi = 1, \quad K_\varrho(x, \xi) = 0 \quad \text{if} \quad |x - \xi| \geq \varrho,$$

where $K_\varrho \in C^0$ in (x, ξ) and C^2 in x, and $|x - \xi|$ denotes the geodesic distance along $\mathfrak{M}$. We then define

$$v_\varrho^i(x) = \int_{\mathfrak{M}} K_\varrho(x, \xi)\, v^i(\xi)\, d\xi, \quad x \in \bar{G}.$$

Then, for $\varrho > 0$, each $v^i \in C^2$, v_ϱ converges uniformly to v on $\bar{G}$, and ∇v_ϱ converges uniformly to ∇v on each compact subset of each R_i. Moreover, for each $\varkappa > 0$, we can choose a sufficiently small $\varrho > 0$ such that v_ϱ furnishes a diffeomorphism from some $\bar{R}'_i \subset R_i$ onto the corresponding $K_{i\varkappa}$, the cell of side $2 - 2\varkappa$ concentric with $\bar{K}_i$ $(= v(\bar{R}_i))$, and carries all points outside the R'_i into the set $\overline{(C'_{\nu-1}, \varkappa)}$. We then define $u = \tau \circ r_{\nu+1} \circ \ldots \circ r_N \circ v_\varrho$ where τ is defined on each $\bar{K} \in C'_\nu$ as the expansion from $\bar{K}_\varkappa$ onto $\bar{K}$ for y on $\bar{K}_\varkappa$ and define $\tau(y)$ as the radial projection of y from c_K into ∂K, if $y \in \bar{K} - \bar{K}_\varkappa$. Then u has all the preceding properties of v and is also Lipschitz. That u is the desired vector follows from its form, (10.8.12), and Theorems 9.1.3, 9.1.4, and 9.2.1.

Lemma 10.8.6. *Suppose $\bar{G}$ satisfies the conditions of Definition* 10.8.1, $z:\bar{G}\to R_N$, Δ *has its preceding significance,* $\varepsilon>0$, *and* R *is the set of* $a\in R_N \ni$

$$\sum_I\sum_{\delta\in\Delta} 2^\nu \varepsilon^\nu M[P^I o\, z, P^I a+\varepsilon\delta] \le \sum_I \int_{R_\nu} M(P^I o\, z, y)\, dy. \tag{10.8.13}$$

Then $\Lambda^N(R)>0$.

Proof. Letting S and T be the cubes in R_N and R_ν, respectively, with center at 0 and side 2ε, we obtain

$$\begin{aligned}
&\int_S \sum_{I,\delta} 2^\nu \varepsilon^\nu M[P^I o\, z, P^I(a)+\varepsilon\delta]\, d\Lambda^N(a)\\
&= \sum_I 2^\nu\varepsilon^\nu \int_S \sum_\delta M[P^I o\, z, P^I(a)+\varepsilon\delta]\, d\Lambda^N(a)\\
&= \sum_I 2^N \varepsilon^N \int_T \sum_\delta M[P^I o\, z, y+\varepsilon\delta]\, d\Lambda^\nu(y)\\
&= 2^N\varepsilon^N \sum_I \int_{R_\nu} M[P^I o\, z, y]\, dy\\
&= \int_S \Big\{\sum_I \int_{R_\nu} M[P^I o\, z, y]\, dy\Big\}\, d\Lambda^N(a).
\end{aligned}$$

The result follows.

Theorem 10.8.2. *Suppose $\bar{G}$ satisfies the usual conditions,* $z:\bar{G}\to R_N$, $N\ge\nu$, *and*

$$\Lambda^{\nu+1}[z(\bar{G})]=0 \quad \textit{and} \quad \Lambda^\nu[z(\partial G)]=0. \tag{10.8.14}$$

Then

$$L_\nu(z,G)\le \sum_I L_\nu(P^I o\, z, G). \tag{10.8.15}$$

If $N=\nu$, *then*

$$L_\nu(z,G)=\int_{R_\nu} M(z,y)\, dy. \tag{10.8.16}$$

Proof. Let $\varepsilon>0$. From (10.8.14) we conclude that $\Lambda^{\nu+1}[P^J o\, z(G)]=0$ for each $(\nu+1)$-sequence J, so that the N-measure of the set of all $N-\nu-1$ dimensional planes of the form $x^J=a^J$ which intersect $z(\bar{G})$ is zero. Consequently the set of all $a\in R_N\ni z(\bar{G})\cap C''_{N-\nu-1}(a,\varepsilon)$ is not empty has measure zero. In like manner, the set of a in $R_N\ni z(\partial G)\cap C''_{N-\nu}(a,\varepsilon)$ is not empty has N measure 0. From Lemma 10.8.6, it then follows that there is an a in R_N such that (10.8.13) holds and $z(\bar{G})\cap C''_{N-\nu-1}(a,\varepsilon)$ and $z(\partial G)\cap C''_{N-\nu}(a,\varepsilon)$ are both empty. From Lemma 10.8.5 and (10.8.13), it follows that $\exists$ a Lipschitz map $u:\bar{G}\to R_N$ such that (10.8.7) holds and

$$L_\nu(u,G)\le \sum_I \int_{R_\nu} M[P^I o\, z, y]\, dy.$$

Since this u exists for each $\varepsilon > 0$, it follows that

$$L_\nu(z, G) \le \sum_I \int_{R_\nu} M(P^I \circ z, y)\, dy. \tag{10.8.17}$$

In case $N = \nu$, (10.8.16) follows from (10.8.17) (I must be $(1, 2, \ldots, \nu)$) and Lemma 10.8.1 (d). Then (10.8.15) follows from (10.8.16) and (10.8.17).

Definition 10.8.3. By an *admissible measure* Φ we mean one which is the difference of two non-negative measures, for each of which Borel sets are measurable; a set S is measurable $\Leftrightarrow$ for each $\varepsilon > 0$, $\exists$ a compact set F and an open set $O \ni F \subset S \subset O$ and $\Phi^{\pm}(O) - \Phi^{\pm}(F) < \varepsilon$, Φ^+ and Φ^- being defined as follows:

$$\|\Phi\|(e) = \sup \sum_i |\Phi(e_i)|$$

for all measurable partitions $e = \bigcup e_i$. We then define

$$\Phi^+(e) = [\|\Phi\|(e) + \Phi(e)]/2, \quad \Phi^-(e) = [\|\Phi\|(e) - \Phi(e)]/2.$$

Definition 10.8.4. A continuous mapping $f\colon X \to Y$ is said to be monotone $\Leftrightarrow f^{-1}(y)$ is a continuum or a point for each $y \in f(X)$. A continuous mapping $f\colon X \to Y$ is said to be *light* $\Leftrightarrow f^{-1}(y)$ is completely disconnected for each $y \in f(X)$. Suppose $z : X \to Z$ is continuous, X being a *connected metric* space. Let Y be the collection of continua of constancy of z and, if γ_1 and γ_2 are two such, define $d_z(\gamma_1, \gamma_2) = \inf$ diam $[z(\gamma)]$ for every continuum γ which intersects both γ_1 and γ_2.

Lemma 10.8.7. *With this definition, Y is a metric space, the mapping $m : X$ onto Y, defined by the condition that $m(x)$ be the continuum of constancy containing x is monotone, and the mapping $\zeta : Y \to Z$ defined by $\zeta(\eta) = z[m^{-1}(\eta)]$ is light. If $\zeta : Y \to Z$ is a light mapping, Y and Z being compact and metric, then $\exists$ a function $\varphi(\varrho) \to 0$ as $\varrho \to 0 \ni$ if γ is a continuum in Z of diameter $\le \varrho$, then any component of $\zeta^{-1}(\gamma)$ has diameter $\le \varphi(\varrho)$.*

Proof. The first statements are evident. To prove the last, let $\{\gamma_n\}$ be a sequence of continua in Z whose diameters $\to 0$ and, for each n, let η_n be a component of $\zeta^{-1}(\gamma_n)$ and suppose diam $\eta_n \ge \sigma_0 > 0$. By choosing a subsequence, still called $\{n\}$, we may assume that $\gamma_n \to \gamma$ and $\eta_n \to \eta$. Clearly η is a continuum of diameter $\ge \sigma_0$ and, by continuity, $\zeta(\eta) = \gamma$. But since diam $\gamma_n \to 0$ we see that γ is a point. But this contradicts the lightness of ζ.

Definition 10.8.5. We suppose that $\bar{G}$ satisfies the conditions of Definition 10.8.1 and that $z : \bar{G} \to R_N$, $L_\nu(z, G) < \infty$, $\Lambda^{\nu+1}[z(\bar{G})] = 0$, $\Lambda^\nu[z(\partial G)] = 0$, and $z = \zeta \circ m$ where $m : \bar{G}$ onto X is monotone, and $\zeta : X \to R_N$ is light. Let $\mathfrak{G}$ be the totality of cells r in R_N such that $\Lambda^\nu[z(\bar{G}) \cap \partial r] = 0$; evidently all cells r, none of whose faces lie along a certain countable number of hyperplanes belong to $\mathfrak{G}$. Let us define Ω_0 as the totality of components of $\zeta^{-1}(r)$ and $\mathfrak{U}_0$ as the totality of components of $z^{-1}(r) - \partial G$ for all $r \in \mathfrak{G}$. We define Ω_1^I as the totality of components of $(P^I \circ \zeta)^{-1}(r)$ and $\mathfrak{U}_1$ as the totality of components of $(P^I \circ z)^{-1}(r)$ for all $r \subset R_\nu$.

Lemma 10.8.8. *For each I, there is a unique measure Φ_0^I, defined on X, such that, for each open set $\omega \subset X$,*

$$\Phi_0^I(\omega) = \int_{R_\nu} O[y, P^I \circ z(\partial U)]\, dy, \quad U = m^{-1}(\omega) - \partial G. \tag{10.8.18}$$

We assume that $\bar{G}$, X, z, m, and ζ satisfy all the conditions above.

Proof. We begin by defining $\Phi_0^I(\omega)$ by (10.8.18) for $\omega \in \Omega_0$ so that $U \in \mathfrak{U}_0$. Now if $\omega, \omega_1, \ldots, \omega_n$ all $\in \Omega_0$, $U = m^{-1}(\omega) - \partial G$, $U_i = m^{-1}(\omega_i) - \partial G$, the ω_i and U_i are disjoint, and $\bar{\omega} = \bigcup_i \bar{\omega}_i$, $\bar{U} = \bigcup \cdot \bar{U}_i$, then we see by choosing a smaller smooth domain in each U_i that, if $y \notin P^I \circ z(\partial U) \cup P^I \circ z(\partial U_i)$,

$$\begin{aligned} O[y, P^I \circ z(\partial U)] &= \sum_{i=1}^n O[y, P^I \circ z(\partial U_i)], \\ \Phi_0^I(\omega) &= \sum_{i=1}^n \Phi_0^I(\omega_i). \end{aligned} \tag{10.8.19}$$

From this, it follows that, if we define

$$\|\Phi_0^I\|(\omega) = \sup \sum_i |\Phi_0^I(\omega_i)|$$

for generalized partitions of ω like those above, $\|\Phi_0^I\|$ satisfies (10.8.19) for the same types of partitions. Next we may define

$$\Phi_0^{I+}(\omega) = [\|\Phi_0^I\|(\omega) + \Phi_0^I(\omega)]/2, \quad \Phi_0^{I-}(\omega) = [\|\Phi_0^I\|(\omega) - \Phi_0^I(\omega)]$$

and then define (ω_i disjoint and $\in \Omega_0$)

$$\Phi_0^{I\pm}(\omega) = \sup \sum_{i=1}^n \Phi_0^{I\pm}(\omega_i), \quad \bigcup_i \bar{\omega}_i \subset \omega, \quad \omega \text{ open},$$

$$\Phi_0^{I\pm}(\sigma) = \inf \sum_{i=1}^n \Phi_0^{I\pm}(\omega_i), \quad \sigma \subset \bigcup_i \bar{\omega}_i, \quad \sigma \text{ closed}.$$

Since ζ is light and there are arbitrarily small cells $r \in \mathfrak{G}$, there are $\omega \in \Omega_0$ of arbitrarily small diameter. Thus if σ is closed (and hence compact) and ω is open, it follows easily that

$$\Phi_0^{I\pm}(\omega - \sigma) = \Phi_0^{I\pm}(\omega) - \Phi_0^{I\pm}(\sigma).$$

We then define the exterior measure $\Phi_{0e}^{I\pm}(S)$ as the inf of $\Phi_0^{I\pm}(\omega)$ for open $\omega \supset S$ and the interior measures $\Phi_{0i}^{I\pm}(S)$ as the sup of $\Phi_0^{I\pm}(\sigma)$ for closed $\sigma \subset S$ and the measures can be developed classically.

Lemma 10.8.9. *Suppose G, z, m, ζ, and X satisfy all the conditions above. For each I, let Z^I be the set of $\xi \in X \ni \xi \in$ a continuum γ of positive diameter such that $P^I \circ \zeta(\gamma)$ is a point. Then Z^I is of Borel type F_σ and*

$$\Lambda^\nu[(P^I \circ \zeta)(Z^I)] = 0, \quad \|\Phi_0^I\|[(\zeta^I)^{-1}\{\zeta^I(Z^I)\}] = 0. \tag{10.8.20}$$

Proof. If we let Z_k^I be the set of $\xi \in X \ni \xi \in$ a continuum γ of diameter $\geq 1/k$ such that $P^I \circ \zeta(\gamma)$ is a point, then Z_k is seen to be closed. For a given I and a given $y \in (P^I \circ \zeta)(Z^I)$, we see that $\zeta(Z^I) \cap (P^I)^{-1}(y) \supset$ a continuum of positive diameter, so that $\Lambda^1[\zeta(X) \cap (P^I)^{-1}(y)] > 0$. Hence $\Lambda^\nu[(P^I \circ \zeta)(Z^I)] = 0$, for otherwise it would follow from Theorem 10.2.3 that $\Lambda^{\nu+1}[\zeta(X)] = \Lambda^{\nu+1}[z(\bar{G})] > 0$. Let Z_k^I be defined as above, suppose ω is open on X and $\omega \supset (\zeta^I)^{-1}\{\zeta^I(Z_k^I)\}$ and suppose $U = m^{-1}(\omega) - \partial G$. Then from the definitions, we find that

$$\|\Phi_0^I\|\{(\zeta^I)^{-1}[\zeta^I(Z_k^I)]\} \leq \|\Phi_0^I\|(\omega) \leq \int_R M[y,(P^I \circ z)\,|\,U]\,dy$$

$$\leq \int_{(P^I \circ z)(U)} M[y, P^I \circ z]\,dy.$$

From Theorem 10.8.2, it follows that $M(y, P^I \circ z) \in L_1(R_\nu)$ so the last result follows by expressing $(\zeta^I)^{-1}\{\zeta^I(Z_k^I)\}$ as the intersection of a decreasing sequence $\{\omega_n\}$ of open sets.

Since our goal in this section is really Theorem 10.8.3 below, we content ourselves in what follows immediately with situations in which $\bar{G} = G$ *is a compact manifold of class* C^1 *without boundary.*

Lemma 10.8.10. *Suppose* $\bar{G} = G$ *is a compact manifold of class* C^1 *without boundary,* $z: G \to R_\nu$ *is continuous with* $L_\nu(z, G) < \infty$, *and otherwise suppose,* z, X, ζ, *and* m *satisfy their usual conditions. Suppose* $z_n \to z$ *uniformly on* G *with* $z_n \in C^1(G)$ *and* $L_\nu(z_n, G) \leq K$ *for all* n *and suppose* φ *is continuous on* X. *Then*

$$\lim_{n\to\infty} \int_G \varphi[m(x)] \frac{\partial z_n}{\partial x}\,dx = \int_X \varphi(\xi)\,d\Phi_0(e_\xi), \tag{10.8.21}$$

Φ_0 *being the measure* Φ_0^I, $I = (1, 2, \ldots, \nu)$, *of Lemma* 10.8.8.

Proof. Since the limit depends only on the limit vector z, it is sufficient to prove this for some subsequence. We may therefore assume that the measures Ψ_n defined by

$$\Psi_n(e) = \int_{z_n^{-1}(e)} \left|\frac{\partial z_n}{\partial x}\right| dx$$

converge weakly to some measure Ψ. Let $\mathfrak{G}$ denote the cells of continuity of Ψ. If $r \in \mathfrak{G}$ and $\varepsilon > 0$, we may choose r' and r'' with $\bar{r}' \subset r \subset \bar{r} \subset r''$ such that

$$\int_{r''-r'} M(y, z)\,dy < \varepsilon, \quad \Psi_n(\bar{r}'' - r') < \varepsilon \quad \text{for all } n. \tag{10.8.22}$$

Let ω be an open subset of $\zeta^{-1}(r)$ which consists of components of $\zeta^{-1}(r)$, and let $U = m^{-1}(\omega) = z^{-1}(r)$. Then $z(\partial U) \subset \partial r$ and $z(U) \subset r$.

Also, since $z_n(\bar{U}) \subset r''$ and $z_n(\partial U) \subset r'' - \bar{r}'$ for n large

$$\int_U \frac{\partial z_n}{\partial x} dx = \int_{r'} O[y, z_n(\partial U)]\, dy + \int_{r''-r'} O[y, z_n(\partial U)]\, dy, \quad n > N,$$

$$\Phi_0(\omega) = \int_{r'} O[y, z(\partial U)]\, dy + \int_{r-r'} O[y, z(\partial U)]\, dy.$$

Since $O[y, z_n(\partial U)] = O[y, z(\partial U)]$ for $y \in r'$, it follows from (10.8.22) that

$$\left| \int_U \frac{\partial z_n}{\partial x} dx - \Phi_0(\omega) \right| < 2\varepsilon \quad \text{for } n > N.$$

Since ζ is light it follows from Lemma 10.8.7 that we may write $X = \cup \bar{\omega}_i$, this being a finite union of $\bar{\omega}_i$ where the $\bar{\omega}_i$ have arbitrarily small diameter and the ω_i are disjoint and of the type discussed above. The lemma follows.

Theorem 10.8.3 (FEDERER's convergence property [2]). *Suppose $\bar{G} = G$ is a compact manifold of class C^1 without boundary, suppose $z: G \to R_N$ is continuous with $L_\nu(z, G) < \infty$ and, suppose z, X, m, and ζ satisfy their usual conditions. Suppose that $z_n \to z$ uniformly on G with $L_\nu(z_n, G) \leq K$ for all n, each $z_n \in C^1(G)$. Suppose that the φ_I are continuous on X. Then*

$$\lim_{n\to\infty} \int_G \sum_I \varphi_I[m(x)] \frac{\partial z_n^I}{\partial x} dx = \int_X \sum_I \varphi_I(\xi)\, d\Phi_0^I(e_\xi), \tag{10.8.23}$$

where the Φ_0^I are the measures of Lemma 10.8.8.

Proof. Let us factor $P^I \circ \zeta = \eta^I \circ h_I$ and let $h_I(X) = X_I$. Then $h_I \circ m$ is monotone from G to X_I and η^I is light from X_I to R_ν and $P^I \circ z = \eta^I \circ (h^I \circ m)$. Let Φ_0^{*I} be the measure of Lemma 10.8.10 on X_I. Then, since $L_\nu(P^I \circ z, G) \leq K$ for each I, it follows from that lemma that, if the φ_I^* are continuous on X_I,

$$\lim_{n\to\infty} \int_G \sum_I \varphi_I^*[(h^I \circ m)(x)] \frac{\partial z_n}{\partial x} dx = \sum_I \int_{X_I} \varphi_I^*(\xi)\, d\Phi_0^{*I}(e_\xi). \tag{10.8.24}$$

Moreover, *for a subsequence* still called z_n, we have *some* measures Φ^I on X such that, if the φ_I are continuous on X,

$$\lim_{n\to\infty} \int_G \sum_I \varphi_I[m(x)] \frac{\partial z_n^I}{\partial x} dx = \int_X \sum_I \varphi_I(\xi)\, d\Phi^I(e_\xi). \tag{10.8.25}$$

Next we note that if ω is a component in X_I of $\eta_I^{-1}(r)$ where $r \in \mathfrak{G}^I$ and to $P^I \mathfrak{G}$ (so to speak), then $h_I^{-1}(\omega)$ consists of a finite number of disjoint open sets ω_i in X, and a comparison of the definitions shows that $\Phi_0^{*I}(\omega) = \sum \Phi_0^I(\omega_i)$. It follows that

$$\Phi_0^{*I}(e) = \Phi_0^I[h_I^{-1}(e)], \quad e \text{ a BOREL set} \subset X_I. \tag{10.8.26}$$

In addition, by choosing the $\varphi_I(\xi) = \varphi_I^*[h_I(\xi)]$ in (10.8.25), we conclude that (10.8.26) holds with Φ_0^I replaced by Φ^I. But now, since $\|\Phi_0^I\|(Z^I) = 0$ and h_I is topological off of Z^I, it follows rather easily that $\Phi^I = \Phi_0^I$. Since the limit in (10.8.23) is independent of the subsequence $\{z_n\}$, the theorem follows.

In the paper [2], FEDERER proves the following two theorems which we shall merely state, since we haven't found a way to simplify their proofs and do not need them in their full generality.

Theorem 10.8.4. *Suppose $\bar{G} = G$, z, X, M, and ζ satisfy their usual conditions and that the Φ_0^I are the measures of Lemma 10.8.8. Then $\exists$ a simple ν-vector $v^I(\xi)$ defined for $\|\Phi\|$-almost all ξ on X for which $|v(\xi)|$ is an integer and*

$$\Phi_0^I(A) = \int_A v^I(\xi)\, dS(e_\xi), \quad S(e) = \Lambda^\nu[\zeta(e)],$$

where A and e are Borel subsets of X.

Theorem 10.8.5. *Suppose $\bar{G} = G$, z, X, m, ζ, and Φ_0^I satisfy the conditions above. Then, if A is any open subset of X,*

$$\|\Phi(A)\| = L_\nu[z \mid m^{-1}(A)].$$

That is, Lebesgue area is additive.

Bibliography

AGMON, S.: The L_p approach to the Dirichlet problem. Ann. Scuola Norm. Sup. Pisa **13,** 405—448 (1959).

AGMON, S., DOUGLIS, A., and L. NIRENBERG: [1] Estimates near the boundary for the solutions of elliptic differential equations satisfying general boundary values I, Comm. Pure Appl. Math. **12,** 623—727 (1959).

— [2] Estimates near the boundary for the solutions of elliptic differential equations satisfying general boundary values II, Comm. Pure Appl. Math. **17,** 35—92 (1964).

AKHIEZER, N. T.: The calculus of variations (trans. from Russian by ALINE H. FRINK), New York: Blaisdell Publishing Co. 1962.

ALEXANDROFF, P., and H. HOPF: Topologie. Berlin: Springer 1935.

ARONSZAJN, N., KRZYWICKI, A., and J. SZARSKI: A unique continuation theorem for exterior differential forms on Riemannian manifolds. Arkiv för Math. **4,** 417—453 (1962).

ARONSZAJN, N., and K. T. SMITH: [1] Functional spaces and functional completion. Ann. Inst. Fourier (Grenoble) **6,** 125—185 (1956).

— [2] Theory of Bessel potentials I, Studies in eigenvalue problems. Technical Report No. 22, University of Kansas, 1959.

BERNSTEIN, S.: [1] Sur la nature analytique des solutions des équations aux deriveés partielles du second ordre, Math. Ann. **59**, 20—76 (1904).

— [2] Démonstration du théorème de M. HILBERT sur la nature analytique des solutions des équations du type elliptique sans l'emploi des series normales. Math. Zeit **28,** 330—348 (1928).

BESICOVITCH, A. S.: Parametric surfaces. Bull. Amer. Math. Soc. **56,** 288—296 (1950).

BICADZE, A. V.: On the uniqueness of the solution of the Dirichlet problem for elliptic partial differential equations. Uspehi Mat. Nauk (N. S.) **3,** 211—212 (1948).

BLISS, G. A.: Lectures on the calculus of variations. University of Chicago Press, 1946.

BOCHER, M.: Introduction to Higher Algebra. New York,The Macmillan Co. 1931.

BOCHNER, S.: [1] Analytic mapping of compact Riemann spaces into Euclidean space. Duke Math. J. **3,** 339—354 (1937).

— [2] Harmonic surfaces in Riemann metric, Trans. Amer. Math. Soc. **47,** 146—154 (1940).

BOERNER, H.: Über die Legendresche Bedingung und die Feldtheorien in die Variationsrechnung der mehrfachen Integrale. Math. Z. **46,** 720—742 (1940) (and two earlier papers).

BOLZA, O.: Lectures on the calculus of variations. New York: Chelsea Publishing Co.

BROWDER, F. E.: [1] On the spectral theory of elliptic differential operators, I. Math. Ann. **142**, 22—130 (1961).
— [2] On the spectral theory of elliptic differential operators, II. Math. Ann. **145**, 81—226 (1962) (and other papers).
— [3] Non-linear elliptic boundary value problems. Bull. Amer. Math. Soc. **69**, 862—874 (1963).
CALDERON, A. P.: Lebesgue spaces of differentiable functions and distributions. Proc. Symp. Pure Math. IV, Amer. Math. Soc., Providence, R. I., 33—49 (1961).
—, and A. ZYGMUND: [1] On the existence of certain singular integrals. Acta Math. **88**, 85—139 (1952).
— [2] Singular integral operators and differential equations. Amer. J. Math. **79**, 901—921 (1957).
CALKIN, J. W.: Functions of several variables and absolute continuity I. Duke Math. J. **6**, 170—185 (1940).
CARATHEODORY, C.: [1]Über die Variationsrechnung bei mehrfachen Integralen, Acta Szeged **4**, 193—216, (1929).
— [2] Variationsrechnung und partielle Differentialgleichungen erster Ordnung. Leipzig: Teubner **1935**.
CESARI, L.: [1] La nozione di integrale sopra una superficie in forma parametrica. Ann. Scuola Norm. Sup. Pisa **13**, 73—117 (1947).
— [2] Condizioni sufficienti per la semicontinuità degli integrali sopra una superficie in forma parametrica. Ann. Scuola Norm. Sup. Pisa **14**, 47—79 (1948).
— [3] Condizioni necessarie per la semicontinuità degli integrali sopra una superficie in forma parametrica. Ann. Mat. Pura Appl. (4) **29**, 199—224 (1949).
— [4] An existence theorem of the calculus of variations for integrals on parametric surfaces. Amer. J. Math. **74**, 265—292 (1952).
CONNER, P. E.: [1] The Green's and Neumanns' problems for differential forms on Riemannian manifolds. Proc. Nat. Acad. Sci., U. S. A., **40**, 1151—1155 (1954).
— [2] The Neumann's problem for differential forms on Riemannian manifolds. Mem. Amer. Math. Soc. no. **20** (1956), 56 pp.
CORDES, H.: [1] Über die erste Randwertaufgabe bei quasilinearen Differentialgleichungen zweiter Ordnung in mehr als zwei Variablen. Math. Ann. **131**, 278—312 (1956).
— [2] Vereinfachter Beweis der Existenz einer a priori Hölder-konstanten. Math. Ann. **138**, 155—178 (1959).
— [3] Zero order a priori estimates for solutions of elliptic differential equations. Proc. Sympos. Pure Math. **4**, 157—166, Amer. Math. Soc., Providence, R.I., 1962.
COURANT, R.: [1] Plateau's problem and Dirichlet's principle. Ann. of Math. **38**, 679—724 (1937).
— [2] The existence of minimal surfaces of given topological structure under prescribed boundary conditions. Acta Math. **72**, 51—98 (1940).
— [3] Dirichlet's principle, conformal mapping, and minimal surfaces. Interscience, New York, 1950.
DANSKIN, J. M.: On the existence of minimizing surfaces in parametric double integral problems in the calculus of variations, Riv. Mat. Univ. Parma **3**, 43—63 (1952).
DEDECKER, P.: Calcul des variations et topologie algébrique, Thèse. Université de Liege, Faculté des Sciences, 1957.
DE DONDER, T.: Théorie invariantive du calcul des variations. Brussels: Hayez 1935.
DE GIORGI, E.: [1] Sulla differenziabilità e l'analiticità delle estremali degli integrali multipli regolari, Mem. Accad. Sci. Torino **3**, 25—43 (1957).

DE GIORGI, E.: [2] Frontiere orientate die misura minima. Seminario Mat. Scuola Norm. Pisa, 1960—61.

DEMERS, M. R., and H. FEDERER: On Lebesgue area, II, Trans. Amer. Math. Soc. **90**, 499—522 (1959).

DENY, J.: Les potentiels d'énergie fini. Acta Math. **82**, 107—183 (1950).

DE RHAM, G., and K. KODAIRA: Harmonic integrals (mimeographed notes). Institute for Advanced Study, 1950.

DOUGLAS, J.: Solution of the problem of Plateau. Trans. Amer. Math. Soc. **33**, 263—321 (1931).

DOUGLIS, A., and L. NIRENBERG: Interior estimates for elliptic systems of partial differential equations. Comm. Pure Appl. Math. **8**, 503—538 (1955).

DUFF, G. F. D., and D. C. SPENCER: Harmonic tensors on Riemannian manifolds with boundary, Ann. of Math. **56**, 128—156 (1952).

EELLS, J., and J. H. SAMPSON: Harmonic mappings of Riemannian manifolds. Amer. J. Math. **86**, 109—160 (1964).

EILENBERG, S., and N. E. STEENROD: Foundations of Algebraic Topology. Princeton; N. J.: Princeton University Press, 1952.

EVANS, G. C.: [1] Fundamental points of potential theory. Rice Inst. Pamphlets No. **7**. Rice Institute, Houston, Texas, 252—359 (1920).

— [2] Potentials of positive mass, I. Trans. Amer. Math. Soc. **37**, 226—253 (1935).

FEDERER, H.: [1] On Lebesgue area. Ann. of Math. **61**, 289—353 (1955).

— [2] Currents and area. Trans. Amer. Math. Soc. **98**, 204—233 (1961).

—, and W. H. FLEMING: Normal and integral currents, Ann. of Math. **72**, 458—520 (1960).

FLEMING, W. H.: [1] Irreducible generalized surfaces. Riv. Mat. Univ. Parma **8**, 251—281 (1957).

— [2] On the oriented Plateau problem. Rend. Circ. Mat. Palermo (2) **11**, 69—90 (1962).

FLEMING, W. H., and L. C. YOUNG: [1] A generalized notion of boundary, Trans. Amer. Math. Soc. **76**, 457—484 (1954).

— [2] Representation of generalized surfaces as mixtures. Rend. Circ. Mat. Palermo (2) **5**, 117—144 (1956).

— [3] Generalized surfaces with prescribed elementary boundary, Rend. Circ. Mat. Palermo (2) **5**, 320—340 (1956).

FRIEDMAN, A.: [1] On the regularity of the solutions of non-linear elliptic and parabolic systems of partial differential equations. Jour. Math. Mech. **7**, 43—60 (1958).

— [2] Generalized functions and partial differential equations. Englewood Cliffs, N. J. Prentice Hall Publishing Co.: 1963.

FRIEDRICHS, K. O.: [1] On the identity of weak and strong extensions of differential operators, Trans. Amer. Math. Soc. **55**, 132—151 (1944).

— [2] On the differentiability of the solutions of linear elliptic equations. Comm. Pure Appl. Math. **6**, 299—326 (1953).

— [3] On differential forms on Riemannian manifolds. Comm. Pure Appl. Math. **8**, 551—558 (1955).

FUBINI, G.: Il principio di minimo e i teoremi di esistenza per i problemi di contorno relativi alle equazioni alle derivate parziali di ordine pari. Rend. Circ. Mat. Palermo **23**, 58—84 (1907).

FUNK, P.: Variationsrechnung und ihre Anwendung in Physik und Technik. Berlin · Göttingen · Heidelberg: Springer 1962.

GAFFNEY, M. P.: [1] The harmonic operator for exterior differential forms. Proc. Nat. Acad. Sci., U.S.A. **37**, 48—50 (1951).

GAFFNEY, M. P.: [2] The heat equation method of Milgram and Rosenbloom for open Riemannian manifolds. Ann. of Math. **60**, 458—466 (1954).

— [3] Hilbert space methods in the theory of harmonic integrals. Trans. Amer. Math. Soc. **78**, 426—444 (1955).

GARABEDIAN, P. R., and D. C. SPENCER: Complex boundary value problems. Trans. Amer. Math. Soc. **73**, 223—242 (1952).

GÅRDING, L.: [1] Dirichlet's problem for linear elliptic partial differential equations. Math. Scand. **1**, 55—72 (1953).

— [2] Some trends and problems in linear differential equations. Proc. Int. Congr. Math., 1958.

GEL'MAN, I. V.: The minimum problem for a non-linear functional, Leningrad Gos. Ped. Inst. Uč. Zap. **166**, 255—263 (1958).

GEVREY, M.: Démonstration du théorème de Picard-Bernstein par la méthode des contours successifs; prolongement analytique, Bull, des Sciences Math., (2) **50**, 113—128 (1926).

GILBARG, D.: Boundary value problems for non-linear elliptic equations in n variables. Proc. Sympos. Non-linear Problems, Madison, Wis., 1962.

GOURSAT, E.: Leçons sur l'intégration des équations aux dérivées partielles du premier ordre. Paris: Librairie scientifique J. Hermann, 1921.

GRAUERT, L.: On Levi's problem and the imbedding of real analytic manifolds, Ann. of Math. **68**, 460—472 (1958).

GRAVES, L. M.: The Weierstrass condition for multiple integral variation problems. Duke Math. J. **5**, 656—660 (1939).

HAAR, A.: Über das Plateausche Problem. Math. Ann. **97**, 127—158 (1927).

HADAMARD, J.: Leçons sur la propagation des ondes. Paris, p. 253 (1903).

HESTENES, M. R.: Applications of the theory of quadratic forms in Hilbert space to the calculus of variations. Pacific J. Math. **1**, 525—581 (1951).

—, and E. J. MCSHANE: Theorem on quadratic forms and its application in the calculus of variations. Trans. Amer. Math. Soc. **47**, 501—512 (1940).

HILBERT, D.: Über das Dirichletsche Prinzip. Jber. Deutsch. Math. Verein. **8**, 184—188 (1900).

HIRSCHFELD, O.: Über eine Transformation von Variations- und Randwertproblemen, S.-B. Berlin, Math. Ges. 110—122, 33. Jahrgang.

HODGE, W. V. D.: [1] A Dirichlet problem for harmonic functionals with applications to analytic varieties, Proc. London, Math. Soc. (2), **36**, 257—303 (1934).

— [2] The theory and applications of harmonic integrals, second edition. Cambridge University Press, 1952.

HÖLDER, E.: Beweise einiger Ergebnisse aus der Theorie der 2 Variation mehrfacher Extremalintegrale, Math. Ann. **148**, 214—225 (1962).

HOLMGREN, E.: Über Systeme von linearen partiellen Differentialgleichungen. Öfversigt af kongl. Vetenskaps-Akademiens Förhandlingar **58**, 91—103 (1901).

HOPF, E.: [1] Elementare Bemerkungen über die Lösungen partieller Differentialgleichungen zweiter Ordnung von elliptischen Typen, S.-B. Preuss. Akad. Wiss. **19**, 147—152 (1927).

— [2] Zum analytischen Charakter der Lösungen regulärer zwei dimensionaler Variationsprobleme. Math. Zeit. **30**, 404—413 (1929).

— [3] Über den Funktionalen, insbesondere den analytischen Charakter der Lösungen elliptischer Differentialgleichungen zweiter Ordnung, Math. Zeit. **34**, 194—233 (1932).

HÖRMANDER, L.: [1] Linear differential operators. Berlin · Göttingen · Heidelberg: Springer 1963.

HÖRMANDER, L.: [2] L^2 estimates and existence theorems for the $\bar{\partial}$ operator. Acta Math. **113**, 89—152 (1965).

JENKINS, H.: On quasi-linear elliptic equations which arise from variational problems: J. Math. Mech. **10**, 705—727 (1961).

—, and J. SERRIN: Variational problems of minimal surfaces type I. Arch. Rational Mech. Anal. **12**, 185—212 (1963).

JOHN, F.: [1] On linear partial differential equations with analytic coefficients. Comm. Pure Appl. Math. **2**, 209—253 (1949).

— [2] Derivatives of continuous weak solutions of linear elliptic equations. Comm. Pure Appl. Math. **6**, 327—335 (1953).

— [3] Plane waves and spherical means applied to partial differential equations. New York: Interscience Publishers 1955.

KIPPS, T. C.: The parametric problem for double integrals in the calculus of variations, Univ. of California, Berkeley, Dept. of Math., Tech. Rep. No. 23, ONR Contract Nonr-222 (37).

KODAIRA, K.: Harmonic fields in Riemannian manifolds. Ann. of Math. **50**, 587—665 (1949).

KOHN, J. J.: [1] Harmonic integrals on strongly pseudoconvex manifolds I. Annals of Math. **78**, 112—148 (1963).

— [2] Harmonic integrals on strongly pseudo-convex manifolds II. Annals of Math. **79**, 450—472 (1964).

—, and L. NIRENBERG: Non-coercive boundary value problems. Comm. Pure Appl. Math. **18**, 443—491 (1965).

—, and D. C. SPENCER: Complex Neumann problems. Ann. of Math. **66**, 89—140 (1957).

LADYZENSKAYA, O. A., and N. URAL-'TSEVA: [1] Quasi-linear elliptic equations and variational problems with many independent variables, Russian Math. Surveys, London Math. Soc. **16**, 17—91 (1961).

— [2] On the smoothness of weak solutions of quasi-linear equations in several variables and of variational problems. Comm. Pure Appl. Math. **14**, 481—495 (1961).

— [3] Linear and quasi-linear differential equations of elliptic type, Publishing House "Science", Moscow, 1964 (Russian).

LAX, P.: On Cauchy's problem for hyperbolic equations and the differentiability of the solutions of elliptic equations, Comm. Pure Appl. Math. **8**, 615—633 (1955).

LEBESGUE, H.: [1] Intégrale, longueur, aire (thesis), Ann. Mat. Pura Appl. **7**, 231—359 (1902).

—, [2] Sur le problème de Dirichlet, Ren. Circ. Mat. Palermo **24**, 371— 402 (1907).

LE PAGE, J. T.: Sur les champs géodesiques des intégrales multiples. Bull. Acad. Roy. Belg. Cl. Sci. Vs **27**, 27—46 (1941).

LERAY, J., and J. L. LIONS: Unpublished.

LEVI, B.: Sul principio di Dirichlet, Rend. Circ. Mat. Palermo **22**, 293—359 (1906).

LEVI, E. E.: Sulle equazioni lineari totalement ellipticne alle derivate parziali. Rend. Palermo **24**, 275—317 (1907).

LEWY, H.: [1] Neuer Beweis des Analytischen Charakters der Lösungen elliptischer Differentialgleichungen. Math. Ann. **101**, 609—619 (1929).

— [2] On minimal surfaces with partially free boundary. Comm. Pure Appl. Math. **4**, 1—13 (1951).

LICHTENSTEIN, L.: [1] Über den analytischen Charakter der Lösungen Zweidimensionaler Variationsproblem. Bull. Acad. Sci. Cracovie, Cl. Sci. Mat. Nat. A, 915—941 (1912).

— [2] Zur Theorie der konformen Abbildung. Konforme Abbildung nichtanalytischer singularitätenfreier Flächenstücke auf ebene Gebiete, Bull. Acad. Sci. Cracovie, Cl. Sci. Mat. Nat. A, 192—217 (1916).

LICHTENSTEIN, L.: [3] Neuere Entwicklung der Potentialtheorie, Encyklopädie der mathematischen Wissenschaften. Vol. 2:3:1, pp. 177—377, particularly p. 209 (1918).

LIONS, J. L.: [1] Théorèmes de traces et d'interpolation I, Ann. Sc. Norm. Pisa **13**, 389—403 (1959).

— [2] Équations differentielles opérationelles et problèmes aux limites. Berlin · Göttingen · Heidelberg: Springer 1961.

—, and MAGENES: Problèmes aux limites non homogènes VII. Ann. Mat. Pura Appl. (4) **63**, 201—224 (1963).

LOPATINSKII, Y. B.: On a method of reducing boundary problems for a system of differential equations of elliptic type to regular equations. Ukrain. Mat. Zurnal **5**, 123—151 (1953).

MCSHANE, E. J.: [1] On the necessary condition of Weierstrass in the multiple integral problem of the calculus of variations I; Ann. of Math. **32**, 578—590 (1931).

— [2] On the necessary condition of Weierstrass in the multiple integral problem of the calculus of variations, II. Ann. of Math. **32**, 723—733 (1931).

— [3] On the semi-continuity of d uble integrals in the calculus of variations. Ann. of Math. **33**, 460—484 (1932).

— [4] Integrals over surfaces in parameןric form, Ann. of Math. **34**, 815—838 (1933).

— [5] Parametrizations of saddle surfaces with applications to the problem of Plateau. Trans. Amer. Math. Soc. **35**, 718—733 (1934).

MALGRANGE, B.: Plongement des variétés analytiques réelles, Bull. Soc. Math. France **85**, 101—113 (1957).

MANDELBROJT, S.: Séries de Fourier et classes quasi-analytique de fonctions: Collection Borel, Paris, 1935.

MEYERS, N.: Quasi-convexity and lower semicontinuity of multiple variational integrals of any order. Trans. Amer. Math. Soc. **119**, 125—149 (1965).

MILGRAM, A., and P. ROSENBLOOM: [1] Harmonic forms and heat conduction I: Proc. Nat. Acad. Sci. U.S.A. **37**, 180—184 (1951).

— [2] Harmonic forms and heat conduction, II. Proc. Nat. Acad. Sci. U.S.A. **37**, 435—438 (1951).

MINTY, G. J.: On the solvability of non-linear functional equations of "montonic" type, Pac. J. Math. **14**, 249—255 (1964).

MIRANDA, C.: [1] Sui sistemi di tipo ellittico di equazioni lineari a derivate parziali del primo ordine, in n variabili independente. Atti Accad. Naz. Lincei. Mem. Cl. Sci. Fis. Mat. Nat. Ser I (8) **3**, 85—121 (1952).

— [2] Equazioni alle derivate parziali di tipo ellittico. Ergeb. der Math. N.F.H. 2 Berlin. Göttingen. Heidelberg. Springer (1955).

MORREY, C. B., JR.: [1] A class of representations of manifolds, I. Amer. J. Math. **55**, 683—707 (1933).

— [2] A class of representations of manifolds, II. Amer. J. Math. **56**, 275—293 (1934).

— [3] An analytic characterization of surfaces of finite Lebesgue area, I. Amer. J. Math. **57**, 692—702 (1935).

— [4] On the solutions of quasi-linear elliptic partial differential equations. Trans. Amer. Math. Soc. **43**, 126—166 (1938).

— [5] Functions of several variables and absolute continuity II. Duke Math. J. **6**, 187—215 (1940).

— [6] Existence and differentiability theorems for the solutions of variational problems for multiple integrals. Bull. Amer. Math. Soc. **46**, 439—458 (1940)

MORREY, C. B., JR.: [7] Multiple integral problems in the calculus of variations and related topics. Univ. of California Publ. in Math., new ser. **1,** 1—130 (1943).
— [8] The problem of Plateau on a Riemannian manifold, Ann. of Math. **49,** 807—851 (1948).
— [9] Quasi-convexity and the lower semicontinuity of multiple integrals. Pacific J. Math. **2,** 25—53 (1952).
— [10] Second order elliptic systems of differential equations. Ann. of Math. Studies No. 33, Princeton Univ. Press, 101—159, 1954.
— [11] A variational method in the theory of harmonic integrals II, Amer. J. Math. **78,** 137—170 (1956).
— [12] On the analyticity of the solutions of analytic non-linear elliptic systems of partial differential equations, I. Amer. Jour. Math. **80,** 198—218, II. 219—234 (1958).
— [13] The analytic embedding of abstract real analytic manifolds, Ann. of Math. **68,** 159—201 (1958).
— [14] Second order elliptic equations in several variables and Hölder continuity. Math. Z. **72,** 146—164 (1959).
— [15] Multiple integral problems in the calculus of variations and related topics. Ann. Scuola Norm. Pisa (III) **14,** 1—61 (1960).
— [16] Existence and differentiability theorems for variational problems for multiple integrals, Partial Differential Equations and Continuum Mechanics. Univ. of Wisconsin Press, Madison, 241—270, 1961.
— [17] The parametric variational problem for double integrals, Comm. Pure Appl. Math. **14,** 569—575 (1961).
— [18] Some recent developments in the theory of partial differential equations, Bull. Amer. Math. Soc. **68,** 279—297 (1962).
— [19] The $\bar{\partial}$-Neumann problem on strongly pseudo-convex manifolds. Outlines of the Joint Soviet-American Symposium on Partial Differential Equations, Novosibirsk, 171—178 (1963).
— [20] The $\bar{\partial}$-Neumann problem on strongly pseudo-convex manifolds. Differential Analysis. The Tata Institute, Bombay, 1964.
— [21] The higher dimensional PLATEAU problem on a Riemannian manifold. Proc. Nat. Acad. Sci., U.S.A. **54,** 1029—1035 (1965).
—, and EELLS, JAMES, JR.: A variational method in the theory of harmonic integrals. Ann. of Math. **63,** 91—128 (1956).
—, and L. NIRENBERG: On the analyticity of the solutions of linear elliptic systems of partial differential equations. Comm. Pure Appl. Math. **10,** 271—290 (1957).
MORSE, A. P.: [1] The behavior of a function on its cricical set, Ann. of Math. **40,** 62—70 (1939).
— [2] A theory of covering and differentiation. Trans. Amer. Math. Soc. **55,** 205—235 (1944).
MORSE, M.: [1] The calculus of variations in the large. Amer. Math. Soc. Colloq. Pub., vol. **18,** New York (1934).
— [2] Functional topology and abstract variational theory. Ann. of Math. **38,** 386—449 (1937).
— [3] Functional topology and abstract variational theory, Mémor. Sci. Math. no. **92** (1939).
— [4] The first variation in minimal surface theory, Duke Math. J. **6,** 263—289 (1940).
—, and C. B. TOMPKINS: [1] Existence of minimal surfaces of general critical type. Ann. of Math. **40,** 443—472 (1939).
— [2] Unstable minimal surfaces of higher topological types, Proc. Nat. Acad. Sci. U.S.A. **26,** 713—716 (1940).

MORSE, and C. B. TOMPKINS: [3] Minimal surfaces not of minimum type by a new mode of approximation. Ann. of Math. **42**, 62—72 (1941).

— [4] Unstable minimal surfaces of higher topological structure. Duke Math. J. **8**, 350—375 (1941).

MOSER, J.: A new proof of De Giorgi's theorem concerning the regularity problem for elliptic differential equations, Comm. Pure Appl. Math. **13**, 457—468 (1960).

NASH, J.: [1] C^1 isometric imbeddings. Ann. of Math. **60**, 383—396 (1954).

— [2] The imbedding problem for Riemannian manifolds. Ann. of Math. **63**, 20—63 (1956).

— [3] Continuity of the solutions of parabolic and elliptic equations. Amer. J. Math. **80**, 931—954 (1958).

NEWLANDER, A., and L. NIRENBERG: Complex analytic coordinates on almost complex manifolds. Ann. of Math. **65**, 391—404 (1957).

NIKODYM, O.: Sur une classe de fonctions considerées dans l'étude du problème de Dirichlet, Fund. Math. **21**, 129—150 (1933).

NIRENBERG, L.: [1] On nonlinear elliptic partial differential equations and Hölder continuity, Comm. Pure Appl. Math. **6**, 103—156 (1953).

— [2] Remarks on strongly elliptic partial differential equations. Comm. Pure Appl. Math. **8**, 648—674 (1955).

— [3] On elliptic partial differential equations. Ann. Scuola Norm. Pisa (III) **13**, 1—48 (1959).

PALAIS, R., and S. SMALE: A generalized Morse theory. Bull. Amer. Math. Soc. **70**, 165—172 (1964).

PARS, L.: An introduction to the calculus of variations. London: Heinemann 1962.

PAUC, C. Y.: La méthode metrique en calcul des variations. Paris: Hermann 1941. Especially p. 54.

PETROWSKY, I.: Sur l'analyticité des solutions des systèmes d'équations différentielles. Rec. Math. N. S. Mat. Sbornik **5** (47), 3—70 (1939).

RADEMACHER, H.: Über partielle und totale Differenzierbarkeit von Funktionen mehrerer Variablen, und über die Transformation der Doppelintegrale. Math. Ann. **79**, 340—359 (1918).

RADO, T.: [1] Das Hilbertsche Theorem über den analytischen Charakter der Lösungen der partiellen Differentialgleichungen zweiter Ordnung. Math. Zeit. **25**, 514—589 (1926).

— [2] Über zweidimensionale reguläre Variationsprobleme. Math. Ann. **101**, 620—632 (1929).

— [3] On the problem of least area and the problem of Plateau, Math. Z. **32**, 763—796 (1930).

— [4] Length and area, Amer. Math. Soc. Colloq. Publ. Vol. 30, Providence, R. I., 1948.

RADO, T., and P. V. REICHELDERFER: Continuous transformations in analysis. Berlin · Göttingen · Heidelberg: Springer 1955.

REIFENBERG, E. R.: [1] Solution of the Plateau problem for m-dimensional surfaces of varying topological type. Acta Math. **104**, 1—92 (1960).

— [2] An epiperimetric inequality related to the analyticity of minimal surfaces. Ann. of Math. **80**, 1—14 (1964).

— [3] On the analyticity of minimal surfaces, Ann. of Math. **80**, 15—21 (1964).

RELLICH, R.: Ein Satz über mittlere Konvergenz, Nachr. Akad. Wiss. Göttingen, Math.-Phys. Kl., 30—35 (1930).

ROSENBLOOM, L. C.: Linear partial differential equations. Surveys in Applied Math. **5**, New York, 1958.

SAKS, S.: Theory of the integral, 2nd revised ed., Eng. translation by L. C. Young. Hafner Pub. Co., New York.

SCHAUDER, J.: Über lineare elliptische Differentialgleichungen zweiter Ordnung. Math. Z. **38**, 257—282 (1934).

SCHECHTER, M.: On L^p estimates and regularity, Am.r. J. Math. **85**, 1—13 (1963).

SEELEY, R. T.: [1] Singular integrals on compact manifolds. Amer. J. Math. **81**, 658—690 (1959).

— [2] Regularisation of singular integral operators on compact manifolds. Amer. J. Math. **83**, 265—275 (1961).

— [3] Unpublished.

SERRIN, J.: [1] On a fundamental theorem of the calculus of variations, etc., Acta Math. **102**, 1—32 (1959).

— [2] On the definition and properties of certain variational integrals, Trans. Amer. Math. Soc. **101**, 139—167 (1961).

SHIFFMAN, M.: Differentiability and anlyticity of solutions of double integral variational problems. Ann. of Math. **48**, 274—284 (1947).

SHUTRICK, H. B.: Complex extensions: Quart. J. Math. Oxford, ser. (2) **9**, 189—201 (1958).

SIGALOV, A. G.: [1] Regular double integrals of the calculus of variations in non-parametric form. Doklady Akad. Nauk SSSR **73**, 891—894 (1950) (Russian).

— [2] Two dimensional problems of the calculus of variations. Uspehi Mat. Nauk (N. S.) **6**, 16—101 (1951). Amer. Math. Soc. Translation No. 83.

— [3] Two-dimensional problems of the calculus of variations in non-parametric form, transformed into parametric form. Mat. Sb. N. S. 38 (**80**), 183—202 (1956). Amer. Math. Soc. Translations, Ser. 2, **10**, 319—340 (1958).

SILVERMAN, E.: Definitions of Lebesgue area for surfaces in metric spaces, Rivista Mat. Univ. Parma **2**, 47—76 (1951).

SMALE, S.: Morse theory and a non-linear generalization of the Dirichlet problem, mimeographed notes. Columbia Univ., New York.

SOBOLEV, S. L.: [1] On a theorem of functional analysis, Mat. Sb. (N. S.) **4**, 471—497 (1938).

— [2] Applications of functional analysis in mathematical physics. Transl. Math. Mon. Vol 7, Amer. Math. Soc., Procvidene, R. I., 1963.

STAMPACCHIA, G.: [1] I problemi al contorno per le equazioni differenziali di tipo ellittico, Atti VI Congr. Un. Mat. Ital., Naples, 21—44 (1959); Edition Cremonense, Rome, 1960.

— [2] Problemi al contorno ellittici, con dati discontinui, dotati die soluzioni hölderiane. Ann. Mat. Pura Appl. (IV) **51**, 1—38 (1960).

— [3] On some regular multiple integral problems in the calculus of variations. Comm. Pure Appl. Math. **16**, 383—421 (1963).

TONELLI, L.: [1] Sui massimi e minimi assoluti del calcolo delle variazione, Rend. Circ. Mat. Palermo **32**, 297—337 (1911).

— [2] Sul caso regolare nel calcolo delle variazioni. Rend. Circ. Mat. Palermo **35**, 49—73 (1913).

— [3] Sur une méthode directe du calcul des variations. Rend. Circ. Mat. Palermo **39**, 233—264 (1915).

— [4] La semicontinuità nel calcolo delle variazioni. Rend. Circ. Mat. Palermo **44**, 167—249 (1920).

— [5] Fondamenti del calcolo delle variazioni, Vols. 1—3, Zanichelli, Bologna.

— [6] Sulla quadratura delle superficie. Atti Reale Accad. Lincei(6)**3**,633—638(1926).

— [7] Sur la semi-continuité des intégrales double du calcul des variations. Acta Math. **53**, 325—346 (1929).

TONELLI, L.: [8] L'estremo assoluto degli integrali doppi. Ann. Scuola Norm. Sup. Pisa (2) **3**, 89—130 (1933).

VAN HOVE, L.: Sur l'extension de la condition de Legendre du calcul des variations aux intégrales multiples a plusieurs fonctions inconnues. Nederl. Akad. Wetensch. **50**, 18—23 (1947).

VIŠIK, I. M.: Simultaneous quasi-linear elliptic equations with lower order terms Doklady Akad. Nauk **144**, 13—16 (1962).

—, and S. L. SOBOLEV: Nouvelle formulation générale des problèmes aux limites... Doklady Akad. Nauk **111**, 521—523 (1956).

VOLEVICH, L. R.: On the theory of boundary value problems for general elliptic systems. Eng. translation. Soviet Math. **4**, 97—100 (1963).

WEYL, H.: [1] Geodesic fields, Ann. of Math. **36**, 607—629 (1935).

— [2] The method of orthogonal projection in potential theory, Duke Math. J. **7**, 411—444 (1940).

WHITNEY, H.: [1] Differentiable manifolds. Ann. of Math. **37**, 645—680 (1936).

— [2] Geometric integration theory, Princeton University Press, Princeton, N. J., 1957.

— [3] Elementary structure of real algebraic varieties. Ann. of Math. **66**, 545—556 (1957).

—, and F. BRUHAT: Quelques propriétés fondamentales des ensembles analytiques réelles. Comment. Math. Helv. **33**, 132—160 (1959).

YOUNG, L. C.: [1] Some applications of the Dirichlet integral to the theory of surfaces, Trans. Amer. Math. Soc. **64**, 317—335 (1948).

— [2] A variational algorithm. Riv. Mat. Univ. Parma **5**, 255—268 (1954).

— [3] Generalized surfaces of finite topological type. Memoirs Amer. Math. Soc. No. 17 (1955).

— [4] Some new methods in two dimensional variational problems with special reference to minimal surfaces, Comm. Pure Appl. Math. **9**, 625—632 (1956).

— [5] Contours on generalized and extremal varieties. J. Math. Mech. **11**, 615—646 (1962).

ZYGMUND, A.: Trigonometrical series. 2nd ed. Chelsea Pub. Co., New York, 1952.

Index

Admissible boundary coordinate
  system, 300
Admissible coordinate system, 300
Admissible measure, 489
Algebraic boundary $b(X, A)$, 411
Area integral, 1

Basis (for complex 1 forms), 331
Basis, orthogonal, 332
Boundary $b\,M$ of M, 300
Boundary operator, 319
Boundary values, 76
Bounded slope condition, 98

CALDERON's extension theorem, 74
CALDERON-ZYGMUND inequalities, 56, 58
CAUCHY inequality, 35
Class $\mathfrak{C}(A, L)$, 411
Closed forms $\varphi\,(d\,\varphi = 0)$, 299
Coerciveness inequality, 253
$\bar{\partial}$ cohomology, 322
Complementing condition, 212
Complex-analytic manifold, 317
Complex DIRICHLET integral $d(\varphi, \psi)$,
  320
Complex improper integral, 174
Contractible, 414
Convex function, 21
Convex set, 21
Cover (in the sense of Vitali), 409
$d(\Gamma)$, $d^*(\Gamma)$, 374, 380

DINI condition, 54
Direct methods, 16
DIRICHLET data, 252
DIRICHLET growth condition, 32
DIRICHLET growth lemma, 79
DIRICHLET integral, 1
DIRICHLET integral for forms, 291, 320
DIRICHLET's principle, 5
DIRICHLET problem, 44
Distance between FRÉCHET varieties,
  351
  between mappings $[D(z_1, z_2)]$, 350
  $d_0(\Sigma, \Sigma')$, 439
  geodesic, 378
  point set, 406
Distribution derivatives, 20
Domain B_{hR}, 174
  B_{0hR}, 181
  G_{hr}, 181
  Lipschitz (of class C_1^0), 25, 77
  of class C^n, C_μ^n, C^∞, analytic, 4
  of class D', 384
  strongly Lipschitz, 72
  with regular boundary 72,

Elementary function, 43
Elliptic systems (of differential
  equations), 210
EULER's equations, 2, 7, 8
Exterior *co*-differential, 290
Exterior derivative, 290
Exterior differential *r*-forms, 288
Extremal, 32

FEDERER's convergence property, 492
FEDERER's multiplicity function, 480
Field of extremals, 14
First differential, 31
First variation, 7.8
Form, closed, 299
  complex, of type (p, q), 318
  even, 288
  exterior differential, 288
  in H_p^1, $\mathfrak{L}_p$, 288
  $\mathfrak{L}_2$-flat, 299
  normal part of a, 301 302
  odd, 288

Form, potential of a, 295, 314
tangential part of a, 301, 302
FRÉCHET variety, 351
on a manifold, 379
oriented, 353
FRIEDRICHS mollifier, 20
Function, absolutely continuous in the sense of TONELLI, 19, 67
admissible, 8
essentially homogeneous of degree p, 47
mollified, 20
monotone in the sense of LEBESGUE, 16, 17
of class C^n_μ, 4
of class $H^1_p(G)$, $H^m_n(G)$, 20
quasi-convex, 112
strongly in $H^1_s(\partial R)$, 114
strongly quasi-convex, 114

GAFFNEY-GÅRDING inequality, 293
GÅRDING's inequality, 253
Generalized derivatives, 62
Generalized surface, 354
Geodesic cones $C(P, S)$, $C(\Pi, S)$, 409, 410
Geodesic k-plane, 406
centered at P, 406
Gradient, ∇z, 1
GREEN's function, 44

Harmonic field, 294
Harmonic form, 316
Harmonic function, 40
HAUSDORFF outer measure, 350, 408
HERMITEan metric, 317
HILBERT's invariant integral, 14
HILBERT transform, 55
HÖLDER condition, 4
HOPF's extension theorem, 482, 483

Inner product, localized (of forms), 302
of complex forms, 319
of complex forms, localized, 319
of forms in $\mathfrak{L}_2$, 288
of forms in H^1_2, 289
$D(\varphi, \psi)$, 320
$((\varphi, \psi))$, $((\varphi, \psi))^-_z$, 333
Integral in parametric form, 2, 349
Integrals over oriented FRÉCHET varieties, 353
Integrand, normal, 92
of a parametric problem, 349
Integrand, of type I, 96
of type II, 97

JACOBI condition, 15
JACOBI's equation, 13
JENSEN's inequality, 21

KODAIRA decomposition, 286, 298, 312, 322
$l(\Gamma)$, $l^*(\Gamma)$, 374, 380

LAPLACE's equation, 43
LEBESGUE area of a FRÉCHET surface, 352
LEGENDRE condition, 2, 10
LEGENDRE-HADAMARD condition, 11
Linear functionals on $H^m_p(G)$, 70
Linear sets $\mathfrak{C}, \mathfrak{D}$, 312
LIPSCHITZ condition, 4
Lipschitz convergence, 113
LIPSCHITZ domain (of class C^0_1), 25
Localized inner product, 301
of complex forms, 319

Manifold with boundary (of class C^n_μ, etc.,), 300
Mapping, absolutely continuous, 379, 382
light, 489
monotone, 489
of class $\underline{H^1_p}$, 382
of class $\overline{H^1_p}$, 379
$r_k : C'_k - C''_{N-\nu-1} \to C'_{k-1}$, 453
$\varrho_\nu : R_N - C''_{N-\nu-1} \to R_N$, 454
$u_\nu(y)$, 454
Maximum principle, 40, 61
Mean value properties of harmonic functions, 40, 41
Measurable mappings, 378
Minimizing sequences, 16
Minimum conditions (on a general elliptic system), 215
Minus potential (of a form), 310
Multinomial coefficient C_α, 63
n-th gradient $\nabla^n z$, 4
$\bar{\partial}$-NEUMANN problem, 320

Norm $|||u|||_{sR}$, 337
$'|||u|||_{sR}$, 338
$|||\varphi|||_s$ for complex forms, 342
Normal coordinate system, 402
centered at a point q on $\mathfrak{M}$, 402
Normal part (of a form), 301, 302

Operator $\mathfrak{d}$, for complex forms, 319
$\bar{\mathfrak{d}}$, for complex forms, 319
L, for complex forms, 321
N, for complex forms, 321
$\square$, for complex forms, 320
Order $O[z, \tau(\partial G)]$, 356, 357, 359, 480
Oriented FRÉCHET variety, 353
Orthogonal basis, 332

Parallel geodesic planes, 443
Parametric representation, 351
Plus potential (of a form), 310
Polyhedral cone, 460
POINCARÉ's inequality, 69
Point set distance $D(S_1, S_2)$, 406
POISSON's equation, 47
POISSON's integral formula, 45, 46
Potential, 47
Potential of a form, 259, 314
Principal part (of a differential operator), 210
Problem of PLATEAU in R_N, 375, 376
Projection on a geodesic p-plane, 435
Projection P^I, 485
Properly elliptic system, 211

Quasi-linear equations, 8
Quasi-potential, 86

Reflection principle, 54
Regular mapping, 64
Regular parametric problem, 32
Regular variational problem, 8, 11
REIFENBERG cone inequality, 459, 460
REIFENBERG's (ε, R_1) condition, 433
RELLICH's theorem, 75
RIEMANNian manifold of class C^k, C^k_μ, 288
Root condition, 211

Second variation, 10
Set (S, ϱ), 406
Sets $C'_k(a, \varepsilon)$, $C''_k(a, \varepsilon)$, 453
Simplicial cone, 460
Singular integrals, 50
Slope functions, 14
SOBOLEV lemmas, 78, 80
SOBOLEV's embedding theorem, 80
SOBOLEV spaces, 19, 20
Space $\mathfrak{A}$, 318
$\mathfrak{A}_0$, $\mathfrak{A}^{pq}$, 319
$\mathfrak{B}_k(G; L; \gamma; \delta)$, 194
$\mathfrak{D}$, of complex forms, 320
$\mathfrak{D}^{pq}$, 21
$\mathfrak{H}$ of complex harmonic fields $(\bar{\delta}\varphi = \vartheta\varphi = 0)$, 321
$\mathfrak{H}^{pq}$, 321
$H^m_p(G)$, 20, 288
$H^m_{p0}(G)$, 68
$\mathfrak{H}^s$, 343
$\mathfrak{L}_p$, 288
$\mathfrak{L}^{pq}$, L, 319
$L_p(E, \mu)$ (of mappings), 378
$\mathfrak{P}^+_2$ and $\mathfrak{P}^-_2$, 309
Spherical harmonics, 73, 474
Stokes' theorem, 290
Strongly elliptic system, 252
Strongly LIPSCHITZ domain, 72
Strongly pseudo-convex boundary, 318
Strong relative minimum, 12
Supplementary conditions, 211

Tangential coordinate system, 321
Tangential part (of a form), 301, 302
Topological type (of a FRÉCHET variety), 351
Transversality conditions, 2

Uniformly elliptic system, 211

Weak convergence in $H^m_p(G)$, 70
Weak convergence of maps in H^1_p, 385
Weakly differentiable functions, 97
Weak solutions, 33
WEIERSTRASS condition, 2, 12
WEIERSTRASS E- function, 12
WEYL's lemma, 41, 42

Printing: Krips bv, Meppel, The Netherlands
Binding: Stürtz, Würzburg, Germany

CIM

M. Aigner Combinatorial Theory ISBN 978-3-540-61787-7
A. L. Besse Einstein Manifolds ISBN 978-3-540-74120-6
N. P. Bhatia, G. P. Szegő Stability Theory of Dynamical Systems ISBN 978-3-540-42748-3
J. W. S. Cassels An Introduction to the Geometry of Numbers ISBN 978-3-540-61788-4
R. Courant, F. John Introduction to Calculus and Analysis I ISBN 978-3-540-65058-4
R. Courant, F. John Introduction to Calculus and Analysis II/1 ISBN 978-3-540-66569-4
R. Courant, F. John Introduction to Calculus and Analysis II/2 ISBN 978-3-540-66570-0
P. Dembowski Finite Geometries ISBN 978-3-540-61786-0
A. Dold Lectures on Algebraic Topology ISBN 978-3-540-58660-9
J. L. Doob Classical Potential Theory and Its Probabilistic Counterpart ISBN 978-3-540-41206-9
R. S. Ellis Entropy, Large Deviations, and Statistical Mechanics ISBN 978-3-540-29059-9
H. Federer Geometric Measure Theory ISBN 978-3-540-60656-7
S. Flügge Practical Quantum Mechanics ISBN 978-3-540-65035-5
L. D. Faddeev, L. A. Takhtajan Hamiltonian Methods in the Theory of Solitons ISBN 978-3-540-69843-2
I. I. Gikhman, A. V. Skorokhod The Theory of Stochastic Processes I ISBN 978-3-540-20284-4
I. I. Gikhman, A. V. Skorokhod The Theory of Stochastic Processes II ISBN 978-3-540-20285-1
I. I. Gikhman, A. V. Skorokhod The Theory of Stochastic Processes III ISBN 978-3-540-49940-4
D. Gilbarg, N. S. Trudinger Elliptic Partial Differential Equations of Second Order ISBN 978-3-540-41160-4
H. Grauert, R. Remmert Theory of Stein Spaces ISBN 978-3-540-00373-1
H. Hasse Number Theory ISBN 978-3-540-42749-0
F. Hirzebruch Topological Methods in Algebraic Geometry ISBN 978-3-540-58663-0
L. Hörmander The Analysis of Linear Partial Differential Operators I – Distribution Theory and Fourier Analysis ISBN 978-3-540-00662-6
L. Hörmander The Analysis of Linear Partial Differential Operators II – Differential Operators with Constant Coefficients ISBN 978-3-540-22516-4
L. Hörmander The Analysis of Linear Partial Differential Operators III – Pseudo-Differential Operators ISBN 978-3-540-49937-4
L. Hörmander The Analysis of Linear Partial Differential Operators IV – Fourier Integral Operators ISBN 978-3-642-00117-8
K. Itô, H. P. McKean, Jr. Diffusion Processes and Their Sample Paths ISBN 978-3-540-60629-1
T. Kato Perturbation Theory for Linear Operators ISBN 978-3-540-58661-6
S. Kobayashi Transformation Groups in Differential Geometry ISBN 978-3-540-58659-3
K. Kodaira Complex Manifolds and Deformation of Complex Structures ISBN 978-3-540-22614-7
Th. M. Liggett Interacting Particle Systems ISBN 978-3-540-22617-8
J. Lindenstrauss, L. Tzafriri Classical Banach Spaces I and II ISBN 978-3-540-60628-4
R. C. Lyndon, P. E Schupp Combinatorial Group Theory ISBN 978-3-540-41158-1
S. Mac Lane Homology ISBN 978-3-540-58662-3
C. B. Morrey Jr. Multiple Integrals in the Calculus of Variations ISBN 978-3-540-69915-6
D. Mumford Algebraic Geometry I – Complex Projective Varieties ISBN 978-3-540-58657-9
O. T. O'Meara Introduction to Quadratic Forms ISBN 978-3-540-66564-9
G. Pólya, G. Szegő Problems and Theorems in Analysis I – Series. Integral Calculus. Theory of Functions ISBN 978-3-540-63640-3
G. Pólya, G. Szegő Problems and Theorems in Analysis II – Theory of Functions. Zeros. Polynomials. Determinants. Number Theory. Geometry ISBN 978-3-540-63686-1
W. Rudin Function Theory in the Unit Ball of $\mathbb{C}^n$ ISBN 978-3-540-68272-1
S. Sakai C*-Algebras and W*-Algebras ISBN 978-3-540-63633-5
C. L. Siegel, J. K. Moser Lectures on Celestial Mechanics ISBN 978-3-540-58656-2
T. A. Springer Jordan Algebras and Algebraic Groups ISBN 978-3-540-63632-8
D. W. Stroock, S. R. S. Varadhan Multidimensional Diffusion Processes ISBN 978-3-540-28998-2
R. R. Switzer Algebraic Topology: Homology and Homotopy ISBN 978-3-540-42750-6
A. Weil Basic Number Theory ISBN 978-3-540-58655-5
A. Weil Elliptic Functions According to Eisenstein and Kronecker ISBN 978-3-540-65036-2
K. Yosida Functional Analysis ISBN 978-3-540-58654-8
O. Zariski Algebraic Surfaces ISBN 978-3-540-58658-6